REACTIVE DISTILLATION DESIGN AND CONTROL

REACTIVE DISTILLATION DESIGN AND CONTROL

WILLIAM L. LUYBEN
Lehigh University

CHENG-CHING YU
National Taiwan University

AIChE 100

WILEY

A JOHN WILEY & SONS, INC., PUBLICATION

Published by John Wiley & Sons, Inc., Hoboken, New Jersey
Published simultaneously in Canada

Library of Congress Cataloging-in-Publication Data:

Luyben, William L.
Reactive distillation design and control/William L. Luyben, Cheng-Ching Yu.
p. cm.
Includes index.
ISBN 978-0-470-22612-4 (cloth)
1. Distillation apparatus--Design and construction. 2. Chemical process control. 3. Distillation. 4. Reactivity (Chemistry) I. Yu, Cheng-Ching, 1956– II. Title.
TP159.D5L895 2008
660′.28425--dc22 2008007189

10 9 8 7 6 5 4 3 2 1

This book is dedicated to Albert, Jessica and Patricia (CCY) and to all my former graduate students who have carried on the Lehigh tradition of engineering excellence (WLL).

CONTENTS

PREFACE xvii

1 INTRODUCTION 1

1.1 History 2
1.2 Basics of Reactive Distillation 3
1.3 Neat Operation Versus Excess Reactant 7
1.4 Limitations 8
1.4.1 Temperature Mismatch 8
1.4.2 Unfavorable Volatilities 9
1.4.3 Slow Reaction Rates 9
1.4.4 Other Restrictions 9
1.5 Scope 9
1.6 Computational Methods 10
1.6.1 Matlab Programs for Steady-State Design 10
1.6.2 Aspen Simulations 10
1.7 Reference Materials 11

PART I STEADY-STATE DESIGN OF IDEAL QUATERNARY SYSTEM 15

2 PARAMETER EFFECTS 17

2.1 Effect of Holdup on Reactive Trays 20
2.2 Effect of Number of Reactive Trays 22
2.3 Effect of Pressure 24
2.4 Effect of Chemical Equilibrium Constant 27
2.5 Effect of Relative Volatilities 29
2.5.1 Constant Relative Volatilities 30
2.5.2 Temperature-Dependent Relative Volatilities 30
2.6 Effect of Number of Stripping and Rectifying Trays 32
2.7 Effect of Reactant Feed Location 33
2.7.1 Reactant A Feed Location (N_{FA}) 33
2.7.2 Reactant B Feed Location (N_{FB}) 35
2.8 Conclusion 36

3 ECONOMIC COMPARISON OF REACTIVE DISTILLATION WITH A CONVENTIONAL PROCESS 37

3.1 Conventional Multiunit Process 38
3.1.1 Assumptions and Specifications 38
3.1.2 Steady-State Design Procedure 40
3.1.3 Sizing and Economic Equations 42
3.2 Reactive Distillation Design 43
3.2.1 Assumptions and Specifications 44
3.2.2 Steady-State Design Procedure 45
3.3 Results for Different Chemical Equilibrium Constants 47
3.3.1 Conventional Process 47
3.3.2 Reactive Distillation Process 54
3.3.3 Comparisons 61
3.4 Results for Temperature-Dependent Relative Volatilities 61
3.4.1 Relative Volatilities 62
3.4.2 Optimum Steady-State Designs 64
3.4.3 Real Chemical Systems 69
3.5 Conclusion 70

4 NEAT OPERATION VERSUS USING EXCESS REACTANT 71

4.1 Introduction 72
4.2 Neat Reactive Column 72
4.3 Two-Column System with Excess B 75
4.3.1 20% Excess B Case 76
4.3.2 10% Excess B Case 78

4.4 Two-Column System with 20% Excess of A 81
4.5 Economic Comparison 85
4.6 Conclusion 86

PART II STEADY-STATE DESIGN OF OTHER IDEAL SYSTEMS 87

5 TERNARY REACTIVE DISTILLATION SYSTEMS 89

5.1 Ternary System Without Inerts 90
5.1.1 Column Configuration 90
5.1.2 Chemistry and Phase Equilibrium Parameters 90
5.1.3 Design Parameters and Procedure 92
5.1.4 Effect of Pressure 94
5.1.5 Holdup on Reactive Trays 94
5.1.6 Number of Reactive Trays 94
5.1.7 Number of Stripping Trays 94
5.2 Ternary System With Inerts 99
5.2.1 Column Configuration 99
5.2.2 Chemistry and Phase Equilibrium Parameters 99
5.2.3 Design Parameters and Procedure 100
5.2.4 Effect of Pressure 102
5.2.5 Control Tray Composition 103
5.2.6 Reactive Tray Holdup 105
5.2.7 Effect of Reflux 107
5.2.8 Chemical Equilibrium Constant 109
5.2.9 Feed Composition 109
5.2.10 Number of Reactive Trays 113
5.2.11 Number of Rectifying and Stripping Trays 113
5.3 Conclusion 116

6 TERNARY DECOMPOSITION REACTION 119

6.1 Ternary Decomposition Reaction: Intermediate-Boiling Reactant 120
6.1.1 Column Configuration 120
6.1.2 Chemistry and Phase Equilibrium Parameters 120
6.1.3 Design Parameters and Procedure 121
6.1.4 Holdup on Reactive Trays 123
6.1.5 Number of Reactive Trays 124
6.1.6 Number of Rectifying and Stripping Trays 126
6.1.7 Location of Feed Tray 126
6.2 Ternary Decomposition Reaction: Heavy Reactant with Two-Column Configurations 127
6.2.1 Column Configurations 127
6.2.2 Chemistry and Phase Equilibrium Parameters 128
6.2.3 Design Parameters and Procedure 128

6.2.4 Reactive Holdup 129
6.2.5 Number of Reactive Trays 131
6.2.6 Number of Rectifying Trays 132

6.3 Ternary Decomposition Reaction: Heavy Reactant with One-Column Configurations 134
6.3.1 Feasibility Analysis 134
6.3.2 Column Configuration 139
6.3.3 Design Parameters and Procedure 139
6.3.4 Reactive Tray Holdup 139
6.3.5 Number of Reactive Trays 139
6.3.6 Number of Rectifying Trays 140
6.3.7 Location of Feed Tray 143
6.3.8 Comparison Between These Two Flowsheets 143

6.4 Conclusion 143

PART III STEADY-STATE DESIGN OF REAL CHEMICAL SYSTEMS 145

7 STEADY-STATE DESIGN FOR ACETIC ACID ESTERIFICATION 147

7.1 Reaction Kinetics and Phase Equilibria 147
7.1.1 Reaction Kinetics 147
7.1.2 Phase Equilibria 149

7.2 Process Flowsheets 153
7.2.1 Type I Flowsheet: MeAc 153
7.2.2 Type II Flowsheet: EtAc and IPAc 156
7.2.3 Type III Flowsheet: BuAc and AmAc 157

7.3 Steady-State Design 158
7.3.1 Design Procedure 158
7.3.2 Optimized Design 160

7.4 Process Characteristics 168
7.4.1 Type I: MeAc 168
7.4.2 Type II: EtAc and IPAc 168
7.4.3 Type III: BuAc and AmAc 170

7.5 Discussion 175

7.6 Conclusion 177

8 DESIGN OF TAME REACTIVE DISTILLATION SYSTEMS 179

8.1 Chemical Kinetics and Phase Equilibrium 180
8.1.1 Chemical Kinetics 180
8.1.2 Phase Equilibrium Using Aspen Plus 181
8.1.3 Conceptual Design 186

8.2 Component Balances 194
8.3 Prereactor and Reactive Column 195
8.3.1 Base Case Design of Reactive Column 195
8.3.2 Effect of Design Parameters on Reactive Column 199
8.4 Pressure-Swing Methanol Separation Section 208
8.5 Extractive Distillation Methanol Separation Section 209
8.6 Economic Comparison 210
8.7 Conclusion 212

9 DESIGN OF MTBE AND ETBE REACTIVE DISTILLATION COLUMNS 213

9.1 MTBE Process 213
9.1.1 Phase Equilibrium 214
9.1.2 Reaction Kinetics 214
9.1.3 Aspen Plus Simulation Issues 214
9.1.4 Setting up the Aspen Plus Simulation 215
9.1.5 Effect of Design Parameters 221
9.1.6 Chemical Equilibrium Model 229
9.2 ETBE Process 231
9.2.1 Kinetic Model 231
9.2.2 Process Studied 232
9.2.3 User Subroutine for ETBE 232
9.2.4 Chemical Equilibrium Model 234
9.2.5 Effects of Design Parameters 236
9.3 Conclusion 237

PART IV CONTROL OF IDEAL SYSTEMS 239

10 CONTROL OF QUATERNARY REACTIVE DISTILLATION COLUMNS 241

10.1 Introduction 242
10.2 Steady-State Design 243
10.3 Control Structures 245
10.4 Selection of Control Tray Location 246
10.5 Closed-Loop Performance 247
10.5.1 CS7-R Structure 247
10.5.2 CS7-RR Structure 248
10.6 Using More Reactive Trays 249
10.6.1 Steady-State Design 249
10.6.2 SVD Analysis 250
10.6.3 Dynamic Performance of CS7-RR 253

10.7 Increasing Holdup on Reactive Trays 254
10.8 Rangeability 256
10.9 Conclusion 259

11 CONTROL OF EXCESS REACTANT SYSTEMS **261**

11.1 Control Degrees of Freedom 261
11.2 Single Reactive Column Control Structures 263
11.2.1 Two-Temperature Control Structure 265
11.2.2 Internal Composition Control Structure 272
11.3 Control of Two-Column System 278
11.3.1 Two-Temperature Control 279
11.3.2 Temperature/Composition Cascade Control 285
11.4 Conclusion 292

12 CONTROL OF TERNARY REACTIVE DISTILLATION COLUMNS **293**

12.1 Ternary System Without Inerts 293
12.1.1 Column Configuration 293
12.1.2 Control Structure CS1 296
12.1.3 Control Structure CS2 300
12.1.4 Control Structure CS3 303
12.2 Ternary System With Inerts 310
12.2.1 Column Configuration 310
12.2.2 Control Structure CS1 310
12.2.3 Control Structure CS2 314
12.2.4 Control Structure CS3 320
12.2.5 Conclusion for Ternary A + B ⇔ C System 322
12.3 Ternary A ⇔ B + C System: Intermediate-Boiling Reactant 324
12.3.1 Column Configuration 324
12.3.2 Control Structure CS1 326
12.3.3 Control Structure CS2 329
12.3.4 Control Structure CS3 334
12.4 Ternary A ⇔ B + C System: Heavy Reactant With Two-Column Configuration 334
12.4.1 Column Configuration 334
12.4.2 Control Structure CS1 334
12.4.3 Control Structure CS2 335
12.5 Ternary A ⇔ B + C System: Heavy Reactant With One-Column Configuration 342
12.5.1 Column Configuration 342
12.5.2 Control Structure CS1 342
12.5.3 Control Structure CS2 344
12.5.4 Control Structure CS3 345
12.5.5 Conclusion for Ternary A ⇔ B + C System 352

PART V CONTROL OF REAL SYSTEMS 353

13 CONTROL OF REACTIVE DISTILLATIONS FOR ACETIC ACID ESTERIFICATION 355

13.1 Process Characteristics 355
13.1.1 Process Studies 355
13.1.2 Quantitative Analysis 356
13.2 Control Structure Design 362
13.2.1 Selection of Temperature Control Trays 363
13.2.2 Control Structure and Controller Design 366
13.2.3 Performance 368
13.2.4 Alternative Temperature Control Structures 376
13.3 Extension to Composition Control 380
13.4 Conclusion 388

14 PLANTWIDE CONTROL OF TAME REACTIVE DISTILLATION SYSTEM 389

14.1 Process Studied 389
14.1.1 Prereactor 390
14.1.2 Reactive Column C1 391
14.1.3 Extractive Column C2 391
14.1.4 Methanol Recovery Column C3 397
14.2 Control Structure 397
14.2.1 Prereactor 397
14.2.2 Reactive Distillation Column C1 399
14.2.3 Extractive Distillation Column C2 399
14.2.4 Methanol Recovery Column C3 401
14.3 Results 403
14.4 Conclusion 406

15 CONTROL OF MTBE AND ETBE REACTIVE DISTILLATION COLUMNS 407

15.1 MTBE Control 407
15.1.1 Steady State 407
15.1.2 Control Structure with C4 Feedflow Controlled 408
15.1.3 Control Structure with Methanol Feedflow Controlled 416
15.2 ETBE Control 418
15.2.1 Control Structure with Flow Control of C4 Feed 419
15.2.2 Control Structure with Flow Control of Ethanol Feed 424

PART VI HYDRID AND NONCONVENTIONAL SYSTEMS 429

16 DESIGN AND CONTROL OF COLUMN/SIDE REACTOR SYSTEMS 431

16.1 Introduction 431
16.2 Design for Quaternary Ideal System 433
16.2.1 Assumptions and Specifications 434
16.2.2 Reactor and Column Equations 435
16.2.3 Design Optimization Procedure 436
16.2.4 Results and Discussion 437
16.2.5 Reactive Column with Optimum Feed Tray Locations 445
16.3 Control of Quaternary Ideal System 446
16.3.1 Dynamic Tubular Reactor Model 446
16.3.2 Control Structures 447
16.4 Design of Column/Side Reactor Process for Ethyl Acetate System 458
16.4.1 Process Description 458
16.4.2 Conceptual Design 459
16.5 Control of Column/Side Reactor Process for Ethyl Acetate System 474
16.5.1 Determining Manipulated Variables 475
16.5.2 Selection of Temperature Control Trays 479
16.5.3 Controller Design 481
16.5.4 Performance 481
16.5.5 Extension to Composition Control 485
16.5.6 Comparison with Reactive Distillation Temperature Control 485
16.6 Conclusion 485

17 EFFECTS OF BOILING POINT RANKINGS ON THE DESIGN OF REACTIVE DISTILLATION 487

17.1 Process and Classification 487
17.1.1 Process 487
17.1.2 Classification 490
17.2 Relaxation and Convergence 492
17.3 Process Configurations 495
17.3.1 Type I: One Group 496
17.3.2 Type II: Two Groups 501
17.3.3 Type III: Alternating 507
17.4 Results and Discussion 511
17.4.1 Summary 511
17.4.2 Excess Reactant Design 514
17.5 Conclusion 518

18 EFFECTS OF FEED TRAY LOCATIONS ON DESIGN AND CONTROL OF REACTIVE DISTILLATION 519

18.1 Process Characteristics 519
18.1.1 Modeling 521
18.1.2 Steady-State Design 522
18.1.3 Base Case 522
18.1.4 Feed Locations Versus Reactants Distribution 523
18.1.5 Optimal Feed Locations 527
18.2 Effects of Relative Volatilities 529
18.2.1 Changing Relative Volatilities of Reactants 529
18.2.2 Changing Relative Volatilities of Products 530
18.2.3 Summary 532
18.3 Effects of Reaction Kinetics 533
18.3.1 Reducing Activation Energies 533
18.3.2 Effects of Preexponential Factor 536
18.4 Operation and Control 538
18.4.1 Optimal Feed Location for Production Rate Variation 538
18.4.2 Control Structure 539
18.4.3 Closed-Loop Performance 541
18.5 Conclusion 544

APPENDIX CATALOG OF TYPES OF REAL REACTIVE DISTILLATION SYSTEMS 545

REFERENCES 563

INDEX 573

PREFACE

Most chemical processes involve two important operations (reaction and separation) that are typically carried out in different sections of the plant and use different equipment. The reaction section of the process can use several types of reactors [continuous stirred-tank reactor (CSTR), tubular, or batch] and operate under a wide variety of conditions (catalyzed, adiabatic, cooled or heated, single phase, multiple phases, etc.). The separation section can have several types of operations (distillation, extraction, crystallization, adsorption, etc.), with distillation being by far the most commonly used method. Recycle streams between the two sections of these conventional multiunit flowsheets are often incorporated in the process for a variety of reasons: to improve conversion and yield, to minimize the production of undesirable byproducts, to improve energy efficiency, and to improve dynamic controllability.

Instead of conducting the reaction and separation in separate units and vessels, it is sometime possible to combine these operations in a single vessel. This is called reactive distillation or catalytic distillation, which is the subject of this book.

Economic and environmental considerations have encouraged industry to focus on technologies based on process "intensification." This is an area of growing interest that is defined as any chemical engineering development that leads to smaller inventories of chemical materials and higher energy efficiency. Reactive distillation is an excellent example of process intensification. It can provide an economically and environmentally attractive alternative to conventional multiunit flowsheets in some systems.

One important inherent advantage of reactive distillation is the feature of simultaneous production and removal of products. For reversible chemical reactions, the removal of the product components drives the reaction toward the product side. Thus, the chemical equilibrium constraint on conversion can be overcome and high conversions can be achieved, even in cases with small chemical equilibrium constants. Of course, the relative volatilities among the reactants and the products must be such that the products can be fairly easily removed from the region in the column where the reaction is occurring and reactants are not lost from this region.

Figure P.1 The Eastman methyl acetate reactive distillation column. (Courtesy of Jeff Siirola.)

An important limitation of reactive distillation is the need for a match between the temperature favorable for reaction and the temperature favorable for separation. Because both operations occur in a single vessel operating at a single pressure, the temperatures in a reactive distillation column are set by vapor–liquid equilibrium and tray compositions. If these temperatures are low and produce low specific reaction rates for the reaction kinetics involved, very large holdups (or large amounts of catalyst) will be required. If these temperatures are high and correspond to very small chemical equilibrium constants (as can occur with exothermic reversible reactions), it may be difficult to achieve the desired conversion. High temperatures may also promote undesirable side reactions. In either the low- or high-temperature case, reactive distillation may not be economical. As a result, the design of reactive distillation columns is much more sensitive to pressure than a conventional distillation column.

A small number of industrial applications of reactive distillation have been around for many decades. One of the earliest was a DuPont process in which dimethyl terephthalate was reacted with ethylene glycol in a distillation column to produce methanol and ethylene terephthalate. The methanol was removed from the top of the column. The ethylene terephthalate, which was used for polyester production, was removed from the bottom.

However, there were few applications of reactive distillation until about two decades ago. The publication of a very influential paper by engineers from Eastman Chemical[1] produced a surge of interest in reactive distillation in both industry and academia. The Eastman reactive distillation column (see Fig. P.1.) produces methyl acetate out the top and water out the bottom, with methanol fed into the lower part of the column and acetic acid fed in the upper part. Jeff Siirola reports that this single reactive column replaced a conventional multiunit process that consumed 5 times more energy and whose capital investment was 5 times that of the reactive column.[2] The methyl acetate reactive distillation column provides an outstanding example of innovative chemical engineering.

Several hundred papers and patents have appeared in the area of reactive distillation, which are too numerous to discuss. A number of books have dealt with the subject such as (1) *Distillation, Principles and Practice* by Stichlmair and Fair,[3] (2) *Conceptual Design of Distillation Systems* by Doherty and Malone,[4] and (3) *Reactive Distillation—Status and Future Directions* by Sundmacher and Kienle.[5] These books deal primarily with the steady-state design of reactive distillation columns. Conceptual approximate design approaches are emphasized, but there is little treatment of rigorous design approaches using commercial simulators. The issues of dynamics and control structure development are not covered. Few quantitative economic comparisons of conventional multiunit processes with reactive distillation are provided.

The purpose of this book is to present a comprehensive treatment of both steady-state design and dynamic control of reactive distillation systems using rigorous nonlinear models. Both generic ideal chemical systems and actual chemical systems are studied. Economic comparisons between conventional multiunit processes and reactive distillation are presented. Reactive distillation columns in isolation and in plantwide systems are considered. There are many parameters that affect the design of a reactive distillation column. Some of these effects are counterintuitive because they are different than in conventional distillation. This is one of the reasons reactive distillation is such a fascinating subject.

We hope this book will be useful for both students and practitioners. We have attempted to deal with many of the design and control challenges in reactive distillation systems in a quantitative way.

WILLIAM L. LUYBEN

CHENG-CHING YU

[1]V. H. Agreda, L. R. Partin, and W. H. Heise, High-purity methyl acetate via reactive distillation, *Chem. Eng. Prog.* **86**(2), 40–46 (1990).

[2]J. J. Siirola, Industrial applications of chemical process synthesis, *Adv. Chem. Eng.* **23**, 1062 (1996).

[3]J. G. Stichlmair and J. R. Fair, *Distillation, Principles and Practice*, Wiley–VCH, New York, 1998, p. 252.

[4]M. F. Doherty and M. F. Malone, *Conceptual Design of Distillation Systems*, McGraw–Hill, New York, 2001, Chapter 10.

[5]K. Sundmacher and A. Kienle, Editors, *Reactive Distillation—Status and Future Directions*, Wiley–VCH, New York, 2003.

ACKNOWLEDGMENTS

CCY would like to thank the students of National Taiwan University and National Taiwan University Sci. and Tech. who make the exploration of reactive distillation fun and full of surprise. In particular, S. B. Hung and Y. T. Tang turn Aspen Plus and Aspen Dynamics into accessible and friendly to average users. Literature surveys provided by J. K. Cheng amazed us on the scope of application of reactive distillation. The feasibility study of J. S. Chen leads us to a new territory. The project of "Green Chemical Process Technology" provides the support for the long-term research on process intensification with emphasis on reactive separation which turns out to be fruitful. Collaboration with Professors H. P. Huang and M. J. Lee and Y. C. Liu of ITRI are delightful and the collected effort makes the research useful. Consultations of Professors Doherty and Malone over the years are also appreciated.

WLL would like to acknowledge the contributions of Muhammad Al-Arfaj and Devrim Kaymak in their PhD dissertation studies of reactive distillation systems. At the time Muhammad began his studies, there were fewer than a half dozen papers that dealt with the control of reactive distillation. His was indeed pioneering work. Devrim picked up where Muhammad left off, significantly extending and broadening the exploration of reactive distillation systems. The contributions of these two young men form major sections of the chapters in this book.

CHAPTER 1

INTRODUCTION

The development of the chemical industry over the last two centuries has provided modern civilization with a whole host of products that improve the well-being of the human race. The result has been a better quality of life, longer life expectancy, more leisure time, rapid transportation to anywhere in the world (and outer space), healthier food, more comfortable homes, better clothing, and so forth.

A major factor in this development has been inexpensive energy and inexpensive raw materials. Coal was the major energy source in the 19th century. Petroleum and natural gas were the major sources in the 20th century. Crude oil offers definite advantages over coal in terms of ease of production and transportation from its origins to the points of consumption. Natural gas also has an inherent advantage over coal because of the hydrogen to carbon ratio. Natural gas is mostly methane (CH_4) with an H/C ratio of 4, but coal's H/C ratio is approximately 1. This means that coal produces much more carbon dioxide when these fuels are burned. Therefore, as an energy source, coal contributes more to greenhouse gases and global warming problems. In addition, coal contains sulfur compounds that require expensive stack-gas cleanup facilities.

Only a fool (or weatherman or economist) would dare to predict what lies ahead. However, the era of inexpensive energy is definitely over because of the rapid growth in demand in developing countries and the increasing difficulty and expense of finding and producing new supplies. It is clear that our modern society must undergo dramatic and perhaps painful changes in lifestyle that will sharply reduce per capita energy consumption in order to achieve a sustainable supply of energy.

The end of the era of cheap energy has had a major impact in the chemical industry. Significant modifications of the processes to produce chemicals have been made to reduce energy consumption. New and innovative processing methods have been developed

Reactive Distillation Design and Control. By William L. Luyben and Cheng-Ching Yu

and commercialized. Extensive use of heat integration has cut energy consumption in some processes by factors of 2 or 3.

Reactive distillation is an excellent example of process innovation. In a conventional chemical plant, there are reaction sections and separation sections. These have their own vessels and equipment, but they are often linked together by material and energy recycles. In reactive distillation, separation and reaction occur in the same vessel. This can result in significant reductions in both energy and equipment in systems that have appropriate chemistry and appropriate vapor–liquid phase equilibrium.

In this chapter we introduce the subject of reactive distillation by covering some of the basic aspects of this interesting and challenging process.

1.1 HISTORY

As mentioned in the Preface, a small number of industrial applications of reactive distillation have been around for many decades. One of the earliest was a DuPont process in which dimethyl terephthalate was reacted with ethylene glycol in a distillation column to produce methanol and ethylene terephthalate. The reactants were fed into the middle of the column where the reversible reaction occurred. The more volatile, low-boiling methanol product was removed from the top of the column, and the high-boiling ethylene terephthalate product was removed from the bottom. The removal of the products from the reaction zone drove the reversible reaction toward the product side. This is one of the fundamental advantages of reactive distillation. Low chemical equilibrium constants can be overcome and high conversions achieved by the removal of products from the location where the reaction is occurring.

There were few early applications of reactive distillation. About two decades ago, engineers at Eastman Chemical published a very influential paper.[1] This seminal paper produced a surge of interest in reactive distillation in both industry and academia.

The Eastman reactive distillation column (see Preface, Fig. P.1) has reactant feedstreams of methanol and acetic acid. Methanol is more volatile than acetic acid and is fed into the lower part of column. The heavier acetic acid is fed into the upper part of the column. As the lighter methanol works it way up the column, it comes in contact with the heavier acetic acid that is coming down the column. The two react to form methyl acetate and water. Methyl acetate is the most volatile component in the system, so it goes into the vapor stream flowing up the column. This keeps the concentration of methyl acetate low in the liquid phase where the reversible reaction is occurring. Thus, the reaction is driven toward the product side and high conversion is achieved despite a modest equilibrium constant.

Jeff Siirola reports that this single reactive column replaced a conventional multiunit process that consumed 5 times more energy and whose capital investment was 5 times that of the reactive column.[2] The methyl acetate reactive distillation column has become the prize example of the application of reactive distillation. It provides an outstanding example of innovative chemical engineering.

Over the last two decades there have been a number of other industrial applications of reactive distillation. The most important from the standpoint of the number of installations

[1] V. H. Agreda, L. R. Partin, and W. H. Heise, High-purity methyl acetate via reactive distillation, *Chem. Eng. Prog.* **86**(2), 40–46 (1990).

[2] J. J. Siirola, Industrial applications of chemical process synthesis, *Adv. Chem. Eng.* **23**, 1062 (1996).

and production capacity is methyl tertiary butyl ether (MTBE), which is used in gasoline blending. A mixed C4 hydrocarbon stream from a refinery light-ends debutanizer column contains isobutene and other C4 components (isobutane, *n*-butane, and *n*-butene), which are not involved in the reaction. This mixed C4 stream is fed into a reactive distillation column along with methanol. The isobutene reacts with the methanol to form MTBE. The heavy MTBE is removed at the bottom, and the chemically inert C4s go out the top. The use of MTBE in gasoline is being phased out because of environmental problems. Other similar esters are being substituted [ethyl tertiary butyl ether (ETBE) and *tert*-amyl methyl ether (TAME)], which are also produced using reactive distillation. All of these applications will be discussed in detail in subsequent chapters.

1.2 BASICS OF REACTIVE DISTILLATION

Reactive distillation is attractive in those systems where certain chemical and phase equilibrium conditions exist. We will discuss some of its limitations in Section 1.4. Because there are many types of reactions, there are many types of reactive distillation columns. In this section we describe the ideal classical situation, which will serve to outline the basics of reactive distillation.

Consider the system in which the chemical reaction involves two reactants (A and B) producing two products (C and D). The reaction takes place in the liquid phase and is reversible.

$$\mathrm{A} + \mathrm{B} \Leftrightarrow \mathrm{C} + \mathrm{D}$$

For reactive distillation to work, we should be able to remove the products from the reactants by distillation. This implies that the products should be lighter and/or heavier than the reactants. In terms of the relative volatilities of the four components, an ideal case is when one product is the lightest and the other product is the heaviest, with the reactants being the intermediate boiling components.

$$\alpha_{\mathrm{C}} > \alpha_{\mathrm{A}} > \alpha_{\mathrm{B}} > \alpha_{\mathrm{D}}$$

Figure 1.1 presents the flowsheet of this ideal reactive distillation column. In this situation the lighter reactant A is fed into the lower section of the column but not at the very bottom. The heavier reactant B is fed into the upper section of the column but not at the very top. The middle of the column is the reactive section and contains N_{RX} trays. Figure 1.2 shows a single reactive tray on which the net reaction rate of the reversible reaction depends on the forward and backward specific reaction rates (k_F and k_B) and the liquid holdup (or amount of catalyst) on the tray (M_n). The vapor flowrates through the reaction section change from tray to tray because of the heat of the reaction.

As component A flows up the column, it reacts with descending B. Very light product C is quickly removed in the vapor phase from the reaction zone and flows up the column. Likewise, very heavy product D is quickly removed in the liquid phase and flows down the column.

The section of the column above where the fresh feed of B is introduced (the rectifying section with N_R trays) separates light product C from all of the heavier components, so a distillate is produced that is fairly pure product C. The section of the column below where the fresh feed of A is introduced (the stripping section with N_S trays) separates

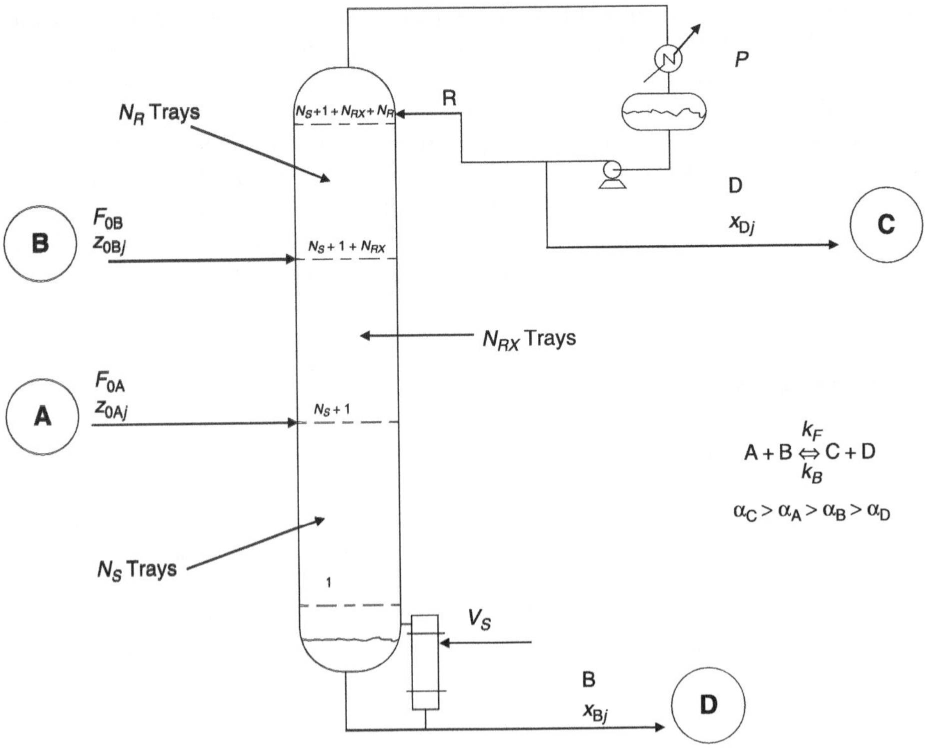

Figure 1.1 Ideal reactive distillation column.

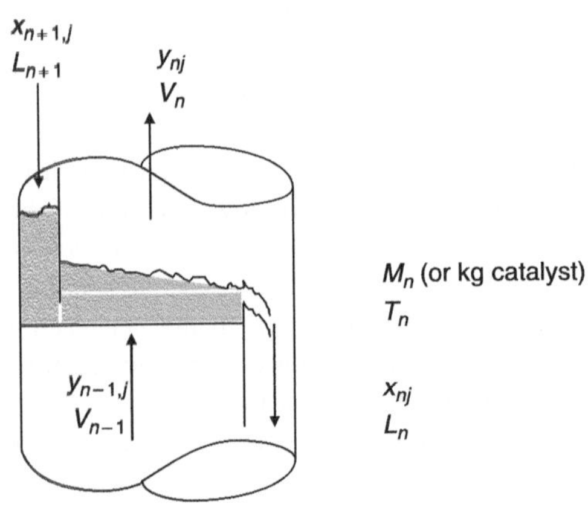

$$R_n = M_n\,(k_{F(T_n)} x_{nA} x_{nB} - k_{B(T_n)} x_{nC} x_{nD})$$

$$V_n = V_{n-1} + \frac{-\lambda R_n}{\Delta H_v}$$

Figure 1.2 Reactive tray.

heavy product D from all of the lighter components, so a bottom is produced that is fairly pure product D. The reflux flowrate and the reboiler heat input can be manipulated to maintain these product purities. Figure 1.3 gives typical composition profiles for this ideal case. The specific numerical case has 30 total trays, consisting of 10 stripping trays, 10 reactive trays, and 10 rectifying trays. Trays are numbered from the bottom. Note that the concentrations of the reactants peak at their respective feed trays (tray 11 for A, tray 20 for B). The purities of the two products are both 95 mol%, with B the major impurity in the bottoms and A the major impurity in the distillate.

One of the most important design parameters for reactive distillation is column pressure. Pressure effects are much more pronounced in reactive distillation than in conventional distillation. In normal distillation, the column pressure is selected so that the separation is made easier (higher relative volatilities). In most systems this corresponds to low pressure. However, low pressure implies a low reflux-drum temperature and low-temperature coolant. The typical column pressure is set to give a reflux-drum temperature high enough (49 °C, 120 °F) to be able to use inexpensive cooling water in the condenser and not require the use of much more expensive refrigeration.

In reactive distillation, the temperatures in the column affect both the phase equilibrium and chemical kinetics (Fig. 1.4). A low temperature that gives high relative volatilities may

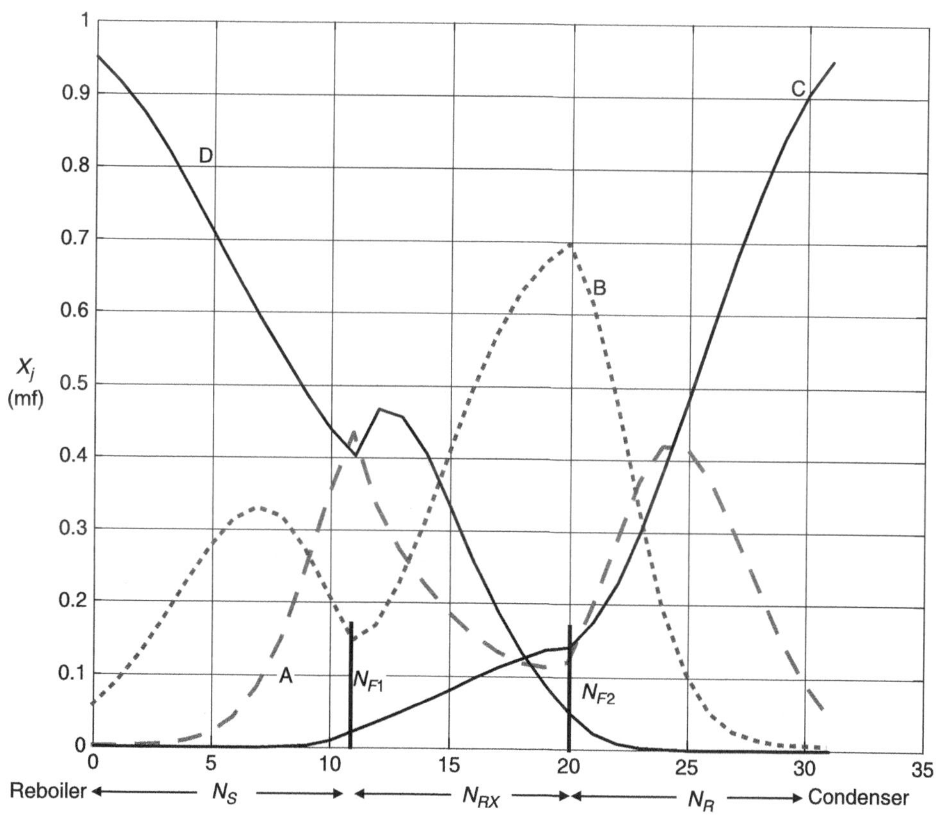

Figure 1.3 Base case composition profiles (95% purities).

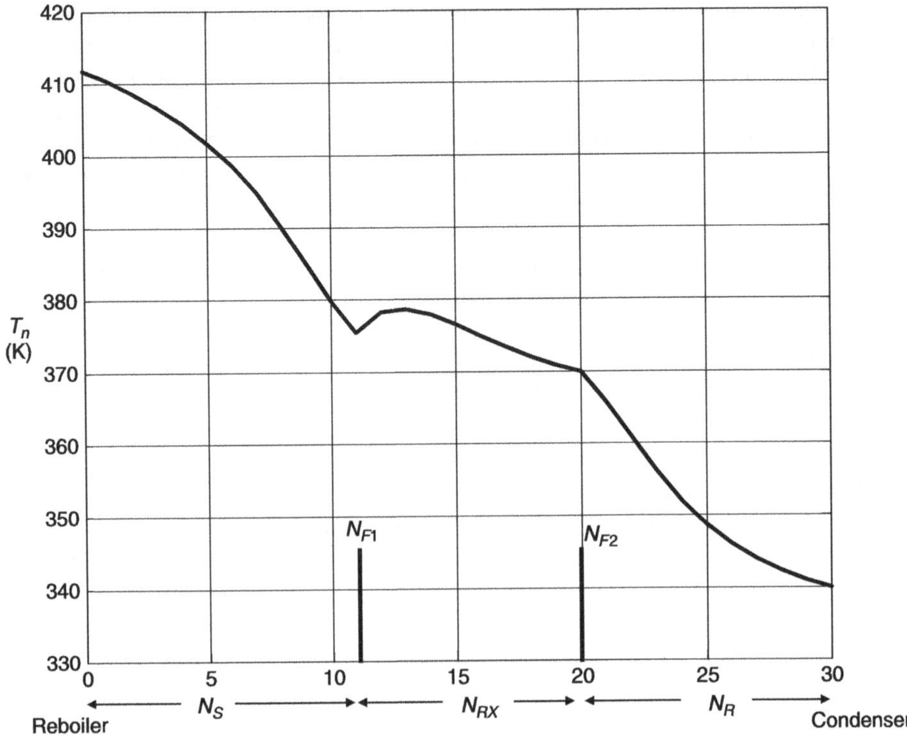

Figure 1.4 Temperature profile.

give small specific reaction rates that would require very large liquid holdups (or amounts of catalyst) to achieve the required conversion. In contrast, a high temperature may give a very small chemical equilibrium constant (for exothermic reversible reactions), which makes it more difficult to drive the reaction to produce products. High temperatures may also promote undesirable side reactions. Thus, selecting the optimum pressure in a reactive distillation column is very important. This will be illustrated in subsequent chapters.

Reactive distillation is also different from conventional distillation in that there are both product compositions and reaction conversion specifications. The many design degrees of freedom in a reactive distillation column must be adjusted to achieve these specifications while optimizing some objective function such as total annual cost (TAC). These design degrees of freedom include pressure, reactive tray holdup, number of reactive trays, location of reactant feedstreams, number of stripping trays, number of rectifying trays, reflux ratio, and reboiler heat input.

Another design aspect of reactive distillation that is different from conventional is tray holdup. Holdup has no effect on the steady-state design of a conventional column. It certainly affects dynamics but not steady-state design. Column diameter is determined from maximum vapor-loading correlations after vapor rates have been determined that achieve the desired separation. Typical design specifications are the concentration of the heavy key component in the distillate and the concentration of the light key component in the bottoms. However, holdup is very important in reactive distillation because reaction rates directly depend on holdup (or the amount of catalyst) on each tray. This means that the holdup must be known before the column can be designed and before the column diameter is known. As a result, the design procedure for reactive distillation is iterative. A tray holdup is assumed and the column is designed to achieve the desired conversion and product purities. The diameter of the column is calculated from maximum vapor-loading correlations. Then the required height of liquid on the reactive trays to give the assumed tray holdup is calculated. Liquid heights greater than 10–15 cm (4–6 in.) are undesirable because of hydraulic pressure-drop limitations. Thus, if the calculated liquid height is too large, a new and smaller tray holdup is assumed and the design calculations repeated. An alternative, which may be more expensive in terms of capital cost, is to make the column diameter larger than that required by vapor loading.

1.3 NEAT OPERATION VERSUS EXCESS REACTANT

The reactive distillation column described in the previous section was designed to operate "neat" (precisely the correct amounts of reactants are fed to the column to satisfy the stoichiometry of the chemistry and there are only small amounts of unreacted reactants that leave in the streams leaving the column). Only a single column is required, so both capital investment and energy cost are minimized. However, it can be difficult to control a reactive column that operates in this neat mode. The problem is the need to feed in exactly enough of both reactants, down to the last molecule, to make sure that there is no excess of either reactant. If the balance is not absolutely perfect, the reactant that is in excess will gradually build up in the column, and it will not be possible to maintain product purities. This build-up may take hours or days, but eventually the column control structure will not be able to hold the products at their specified compositions.

One might think that this problem can be very easily overcome by simply ratioing the flowrates of the two fresh reactant feeds. This strategy works in computer simulations, but it does not work in a real plant environment. The reasons why ratio schemes are not effective are inaccuracies in flow measurements, which are always present, and/or changes in the compositions of the feedstreams. Either cause will result in an imbalance of the stoichiometry. Therefore, it is necessary to have some way to determine the amount of at least one of the reactants inside the column so that feedback control can be used to adjust a fresh feed flowrate. Sometimes temperatures or liquid levels can be used. Sometimes a direct composition measurement on a tray in the column is required. This issue is the heart of the reactive distillation control problem and will be quantitatively studied in detail in subsequent chapters.

An alternative flowsheet, which is more costly but easier to control, uses two distillation columns. As illustrated in Figure 1.5, the first is a reactive distillation column into which an excess of one of the reactants is fed (component B), along with the second fresh feed of component A. The total B fed to the reactive column is 10–20% in excess of the stoichiometric amount needed to react with the moles of A being fed. The amount of this excess is determined by the variability in the compositions of the two fresh feeds and by the flow measurement inaccuracies. Reactant A is the "limiting reactant" in this column and its conversion is high. The conversion of reactant B is not high in the reactive column. Because not all of the B is consumed by the reaction, the excess comes out of the bottom of the column with product component D. This binary mixture is fed to the second distillation column, the recovery column, which produces component D out the bottom and component B out the

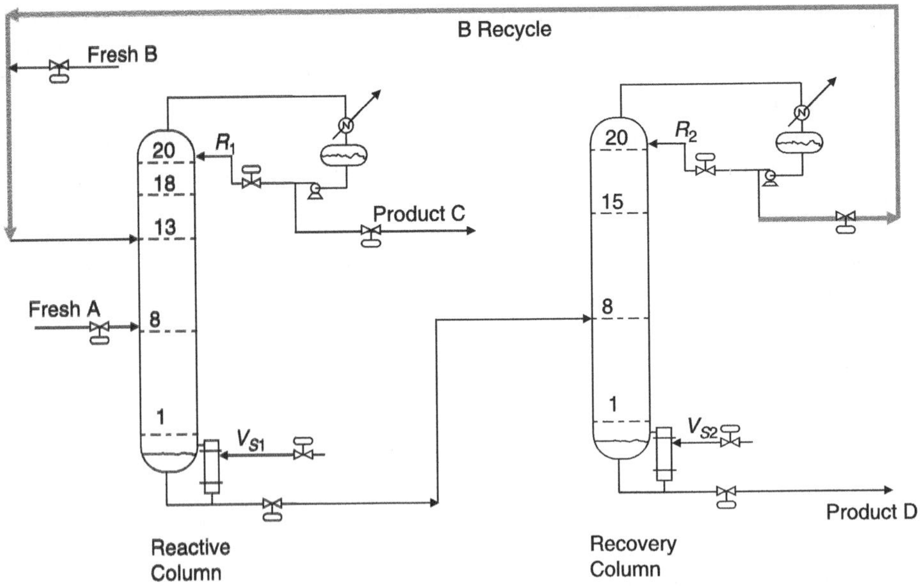

Figure 1.5 Two-column process with excess reactant.

top. The distillate is recycled back to the reactive column, the fresh feed of B is added to the recycle stream, and the total is fed to the reactive column.

The control of this system is easy because the inventory of B in the system can be inferred from the liquid level in the reflux drum of the recovery column. If too much B is being fed into the system, it will accumulate in the reflux drum because the total B fed to the reactive column is fixed. Thus, a simple level controller on the recovery column reflux drum adjusting the flowrate of the fresh feed of B into the system can achieve the required balancing of the stoichiometry. Note that the overall conversion of B is high, considering the entire process, despite having a low "per pass" conversion in the reactive column.

These two alternative flowsheets (neat vs. excess) are quantitatively compared in Chapter 4 in terms of steady-state design and in Chapter 11 in terms of dynamic controllability.

1.4 LIMITATIONS

Although reactive distillation sounds like a great idea, its area of application is fairly restricted. Both the chemistry and the vapor–liquid equilibrium phase equilibrium must be suitable.

1.4.1 Temperature Mismatch

The principle limitation is that there must be no mismatch in the temperatures that are favorable for reaction and the temperatures that are favorable for separation. Because reaction and separation both occur in a single vessel at essentially a single pressure, the temperatures throughout the column are fixed by tray compositions. Both the reactions and vapor–liquid equilibrium see the same temperatures.

Contrast this with what can be done in a conventional multiunit flowsheet. The reactors can be operated at their optimum pressures and temperatures that are selected to be the most favorable for their given chemical kinetics. The distillation columns can be operated at their optimum pressures and temperatures that are selected to be the most favorable for their vapor–liquid equilibrium properties.

1.4.2 Unfavorable Volatilities

The second major limitation for the application of reactive distillation is that the relative volatilities of the components must be such that the reactants can be contained in the column and the products can be easily removed from the top and/or from the bottom.

For example, suppose we wished to produce acetic acid and methanol from methyl acetate and water (the reverse of the methyl acetate process). Now the reactant methyl acetate is the lightest, and it would be very difficult to keep it in the reactive zone and not have much of it escape into the distillate with the methanol that is being produced. This process would not be suitable for reactive distillation.

1.4.3 Slow Reaction Rates

Another limitation for reactive distillation is the need for reasonably large specific reaction rates. If the reactions are very slow, the required tray holdups and number of reactive trays would be too large to be economically provided in a distillation column.

1.4.4 Other Restrictions

Reactive distillation is limited to liquid-phase reactions because there is very little holdup in the vapor phase. The heats of reaction must be modest to prevent large changes in vapor and liquid rates through the reactive zone. A highly exothermic reaction could completely dry up the trays.

1.5 SCOPE

Any book reflects the experiences and prejudices of the authors. We both come from a background of design and control with an emphasis on practical engineering solutions to real industrial problems. Thus, this book contains no elegant mathematics or complex methods of analysis.

Our emphasis is on rigorous simulations, not approximate methods. Rigorous models are used for steady-state design and dynamic analysis of a variety of different types of reactive distillation columns. Several types of ideal systems are studied as well as several real chemical systems.

Steady-state designs of reactive distillation columns are developed that are economically "optimum" in terms of total annual cost, which includes both energy and capital costs. The economics of reactive distillation columns are quantitatively compared with conventional multiunit processes over a range of parameter values (chemical equilibrium constants, specific reaction rates, and relative volatilities). Then effective control structures are developed for these types of reactive distillation columns.

1.6 COMPUTATIONAL METHODS

The rigorous steady-state and dynamic models used in this book are solved using Matlab programs or Aspen Technology simulation software (Aspen Plus and Aspen Dynamics).

1.6.1 Matlab Programs for Steady-State Design

For the ideal chemical cases, a dynamic model is simulated in Matlab. This model consists of ordinary differential equations for tray compositions and algebraic equations for vapor–liquid equilibrium, reaction kinetics, tray hydraulics, and tray energy balances. The dynamic model is used for steady-state design calculations by running the simulation out in time until a steady state is achieved. This dynamic relaxation method is quite effective in providing steady-state solutions, and convergence is seldom an issue.

Specifying the conversion usually sets the product purities. The unreacted reactants will be impurities in the product streams. For the base case $A + B \Leftrightarrow C + D$ system, the distillate will contain most of the unreacted A, and the bottoms will contain most of the unreacted B. For example, suppose 100 mol of both A and B are fed. If the conversion is 95%, there will be 5 mol of A and B that will leave the column in the products. Most of the lighter reactant A will leave in the distillate with product C. Most of the heavier reactant B will leave in the bottoms. There will be some B in the distillate and some A in the bottoms. However, there will be essentially no D in the distillate and no C in the bottoms.

If the distillate and bottoms impurity levels are equal, there will be 5 mol of impurities in each product stream in this example. Then, the total distillate will be 100 mol. The amount of C in the distillate will be 95 mol, so the composition of the distillate is 95 mol% C. Likewise, the total bottoms will be 100 mol. The amount of D in the bottoms will be 95 mol, so the composition of the bottoms is 95 mol% D.

With all feed conditions and the column configuration specified (number of trays in each section, tray holdup in the reactive section, feed tray locations, pressure, and desired conversion), there is only one remaining degree of freedom. The reflux flowrate is selected. It is manipulated by a distillate composition controller to drive the distillate composition to 95 mol% C. The vapor boilup is manipulated to control the liquid level in the base. Note that the distillate and bottoms flowrates are known and fixed as the dynamic model is converged to the steady state that gives a distillate composition of 95 mol% C. The composition of the bottoms will be forced by the overall component balance to be 95 mol% D.

Similar approaches are used for other chemical systems with different stoichiometry. In most cases the columns converged to steady-state conditions in about 15–20 h of process time, which takes about 5–10 min on current personal computers.

1.6.2 Aspen Simulations

Aspen Plus is used for the steady-state designs of the real chemical systems. Convergence problems can occur because of the difficulty of trying to solve the large set of very nonlinear simultaneous algebraic equations. Another problem is that the current version of

Aspen Plus does not permit the use of activities in the reaction rate expressions. "User subroutines" are used to incorporate this feature when necessary.

Aspen Dynamics is used to study dynamics and control of the real systems. The type of reactions that can be used are limited (they must be kinetic and of power law form). These restrictions make the use of Aspen products somewhat less convenient than we would like.

1.7 REFERENCE MATERIALS

There are many reactive distillation systems and many recent publications and patents. Doherty and Malone give 61 chemical systems (see their table 10.5) and cite 134 references in their chapter on reactive distillation.[3] An updated literature survey shows that there were 1105 publications and 814 US patents between 1971 and 2007.

Figure 1.6 is an updated version of a figure by Malone and Doherty.[4] The numbers in the figure are search results from the Engineering Index and the U.S. Patent Office using the following keywords: reactive distillation, catalytic distillation, catalytic reactive distillation, reactive rectification, reactive separation, reactive packing, reaction column, and reacting distillation.

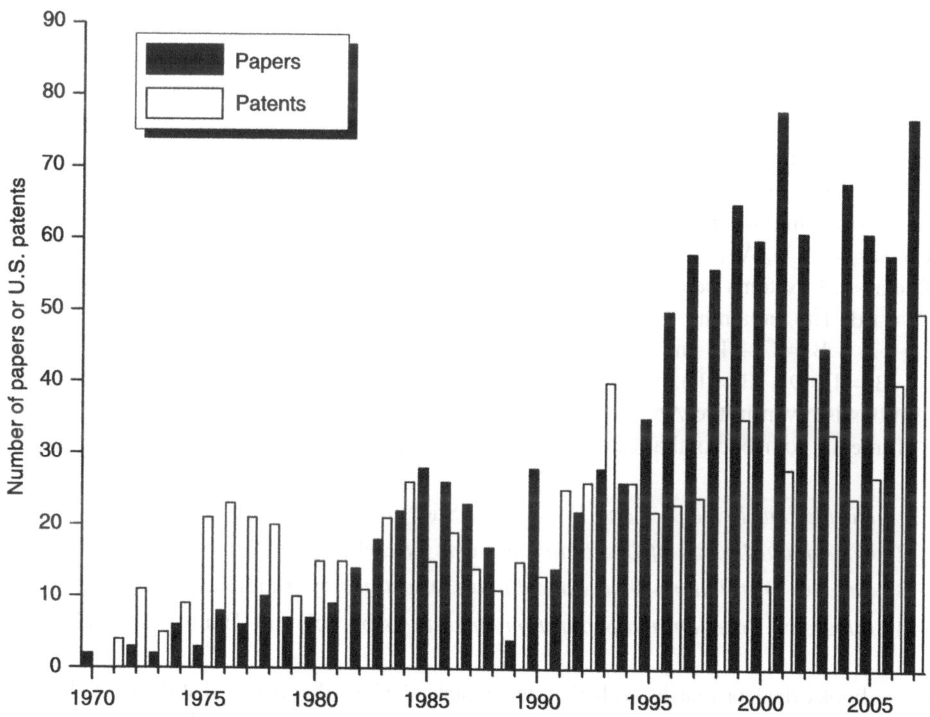

Figure 1.6 Publications or U.S. patents of reactive distillation for the period 1971–2007 as listed in the Engineering Index and U.S. Patent Office (through December 31, 2007).

[3] M. F. Doherty and M. F. Malone, *Conceptual Design of Distillation Systems*, McGraw-Hill, New York, 2001, Chapter 10.

[4] M. F. Doherty and M. F. Malone, Reactive distillation, *Ind. Eng. Chem. Res.* **39**, 3953 (2000).

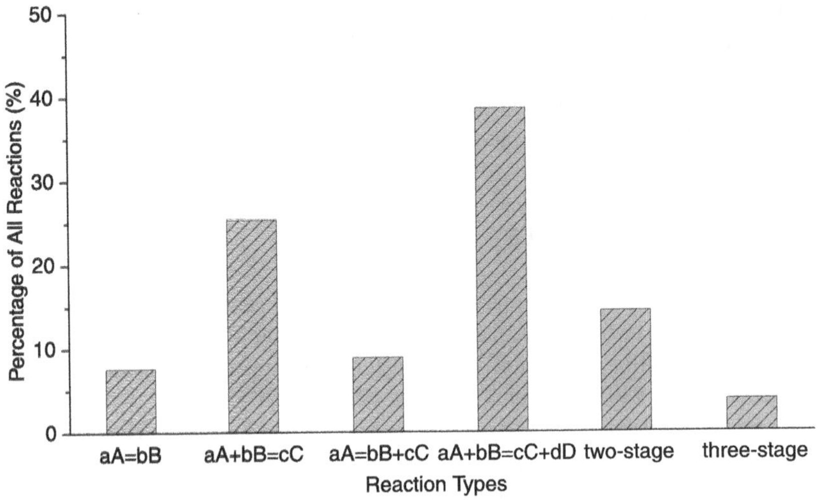

Figure 1.7 Percentage of different reaction types using reactive distillation in the literature.

A literature search using Compendex showed some interesting chronological features. The search was limited to only journal articles in English. From 1969 to 1994 there were only 35 citations in reactive distillation design and a mere six in reactive distillation control. From 1995 to 2007 there were 435 citations in reactive distillation design and 106 in reactive distillation control. This clearly indicates the recent level of interest, particularly in control.

For reactive distillation, a literature survey shows a total of 236 reaction systems. If these are classified into reaction types, 91 systems belong to the aA + bB = cC + dD class (e.g., A + B = C + D, A + 2B = C + 2D, etc.), 60 are of the form of the general aA + bB = cC class, 21 systems belong to the aA = bB + cC class, and 18 of them are of the form aA = bB. The remaining 33 reaction systems fall into the category of a two-stage reaction (e.g., A + B = C + D and B + C = D + E) or a three-stage reaction (e.g., A + B = C, C + B = D, D + B = E). These are illustrated in Figure 1.7. A complete listing of these reactions is given in the Appendix.

There are four books that deal with reactive distillation, among other subjects:

1. *Distillation, Principles and Practice* by Stichlmair and Fair[5]
2. *Conceptual Design of Distillation Systems* by Doherty and Malone[3]
3. *Reactive Distillation—Status and Future Directions* by Sundmacher and Kienle[6]
4. *Integrated Reaction and Separation Operations* by Schmidt-Traub and Gorak[7]

These books deal primarily with the steady-state design of reactive distillation columns. Conceptual approximate design approaches are emphasized. There is little treatment of rigorous design approaches using commercial simulators. The issues of dynamics and control

[5] J. G. Stichlmair and J. R. Fair, *Distillation, Principles and Practice*, Wiley–VCH, New York, 1998.

[6] K. Sundmacher and A. Kienle, Editors, *Reactive Distillation—Status and Future Directions*, Wiley–VCH, New York, 2003.

[7] H. Schmidt-Traub and A. Gorak, *Integrated Reaction and Separation Operations*, Springer, New York, 2006.

structure development are not covered. Few quantitative economic comparisons of conventional multiunit processes with reactive distillation are provided. Schmidt-Traub and Gorak discuss the control of a batch reactive distillation column and give experimental results. Some aspects of the control of reactive distillation systems are discussed in *Distillation Design and Control Using Aspen Simulation* by Luyben.[8]

[8]W. L. Luyben, *Distillation Design and Control Using Aspen Simulation*, Wiley, New York, 2006, Chapter 9.

PART I

STEADY-STATE DESIGN OF IDEAL QUATERNARY SYSTEM

In the next three chapters we will explore various aspects of the ideal quaternary chemical system introduced in Chapter 1. This system has four components: two reactants and two products. The effects of a number of kinetic, vapor–liquid equilibrium, and design parameters on steady-state design are explored in Chapter 2. Detailed economic comparisons of reactive distillation with conventional multiunit processes over a range of chemical equilibrium constants and relative volatilities are covered in Chapter 3. An economic comparison of neat versus excess-reactant reactive distillation designs is discussed in Chapter 4.

CHAPTER 2

PARAMETER EFFECTS

The effects of several important design and chemical parameters on the steady-state design of the ideal chemical system with four components are considered in this chapter. The impact of some parameters is similar to that experienced in conventional distillation. However, in some cases the effects are counterintuitive and unique to reactive distillation. The approach is to see the effect of changing one parameter at a time, while holding all other parameters at their base case values. The base case values of kinetic and vapor–liquid equilibrium parameters are given in Table 2.1. Table 2.2 gives design parameters and steady-state values of process variables for the base case.

The chemistry involves a reversible, liquid-phase, exothermic reaction.

$$A + B \Leftarrow C + D \tag{2.1}$$

The forward and backward specific reaction rates have Arrhenius temperature dependence.

$$k_F = \alpha_F e^{-E_F/RT} \tag{2.2}$$

$$k_B = \alpha_B e^{-E_B/RT} \tag{2.3}$$

The net reaction rate on a reactive tray depends on the liquid concentrations in mole fractions and liquid holdup in kilomoles on that tray.

$$\Re_n = M_{RX}(k_F x_{nA} x_{nB} - k_B x_{nC} x_{nD}) \tag{2.4}$$

In the base case, the forward specific reaction rate is 0.008 kmol s^{-1} kmol^{-1} at 366 K with an activation energy of 30 kcal/mol. The chemical equilibrium constant K_{EQ} is 2 at 366 K, so the backward specific reaction rate at 366 K is 0.004 kmol s^{-1} kmol^{-1}. The backward reaction rate has an activation energy of 40 kcal/mol, which corresponds to a heat of

Reactive Distillation Design and Control. By William L. Luyben and Cheng-Ching Yu

TABLE 2.1 Kinetic and Vapor–Liquid Equilibrium Parameters for Base Case

Activation energy (kcal/mol)		
Forward	30	
Backward	40	
Specific reaction rate at 366 K (kmol s^{-1} $kmol^{-1}$)		
Forward	0.008	
Backward	0.004	
Heat of reaction (kcal/mol)	−10	
Heat of vaporization (kcal/mol)	6.944	
Molecular weights (g mol^{-1})	50	
Vapor pressure constants	A_j	B_j
A	12.34	3862
B	11.45	3862
C	13.04	3862
D	10.96	3862

reaction of −10 kcal/mol. Because the activation of the backward reaction is larger than the forward, the chemical equilibrium constant decreases with increasing temperature.

In the base case, the relative volatilities are constant and equal between adjacent components.

$$\alpha_C = 8 \qquad \alpha_A = 4 \qquad \alpha_B = 2 \qquad \alpha_D = 1 \tag{2.5}$$

Pure component vapor pressures are calculated from a two-parameter Antoine equation.

$$\ln P_j^S = A_j - B_j/T \tag{2.6}$$

where the vapor pressure is in bars and the temperature is Kelvin. Table 2.1 gives the values of the vapor pressure constants A_j and B_j for each of the four components. Note that the B_j

TABLE 2.2 Steady-State Conditions and Design Parameters for Base Case

Fresh feed flowrate of A (mol/s)	12.6	
Fresh feed flowrate of B (mol/s)	12.6	
Distillate flowrate (mol/s)	12.6	
Bottoms flowrate (mol/s)	12.6	
Vapor boilup (mol/s)	28.91	
Reflux flowrate (mol/s)	33.55	
Overhead vapor flowrate (mol/s)	46.15	
Stripping trays	5	
Reactive trays	9	
Rectifying trays	5	
Liquid holdup on reactive trays (mol)	1000	
Pressure (bar)	8	
Product Composition (Mole Fraction)	Distillate	Bottoms
A	0.0326	0.0174
B	0.0174	0.0326
C	0.9500	0
D	0	0.9500

constants are the same for all components when the relative volatilities do not change with the temperature.

Table 2.2 provides design parameters and steady-state conditions for the base case. Figure 2.1 demonstrates the reactive column and the terminology. The two pure fresh feed flowrates are 12.6 mol/s for both A and B. The operating pressure is 8 bar, and the conversion is 95%. There are 19 total trays: 9 reactive, 5 stripping, and 5 rectifying. The fresh feed of A (F_{0A}) is fed onto the first reactive tray immediately above the stripping section. The fresh feed of B (F_{0B}) is fed on the top reactive tray immediately below the rectifying section.

The holdup on the reactive trays is 1000 mol. The vapor boilup required to achieve the desired 95 mol% purities of the two products at base case conditions is 28.91 mol/s, and the corresponding reflux flowrate is 33.55 mol/s. These purities correspond to a 95% conversion.

The vapor leaving the top of the column is larger than the vapor boilup, despite the fact that equimolal overflow has been assumed. This is due to the exothermic heat of reaction generating more vapor on the reactive trays.

$$V_n = V_{n-1} - \frac{\lambda}{\Delta H_V} \mathcal{R}_n \tag{2.7}$$

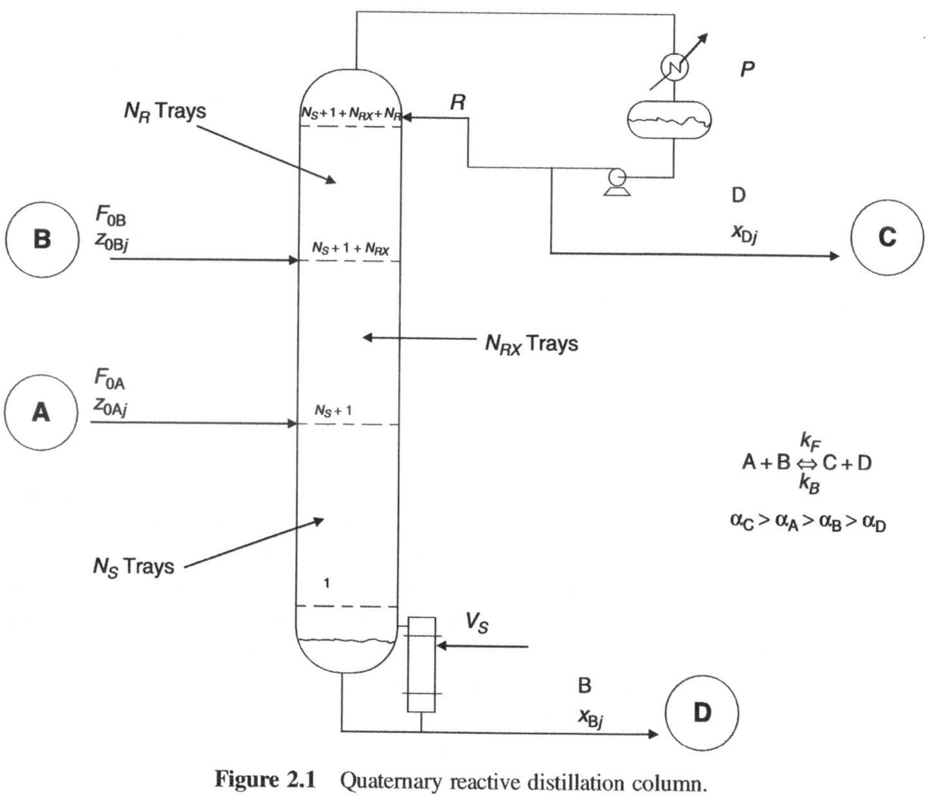

Figure 2.1 Quaternary reactive distillation column.

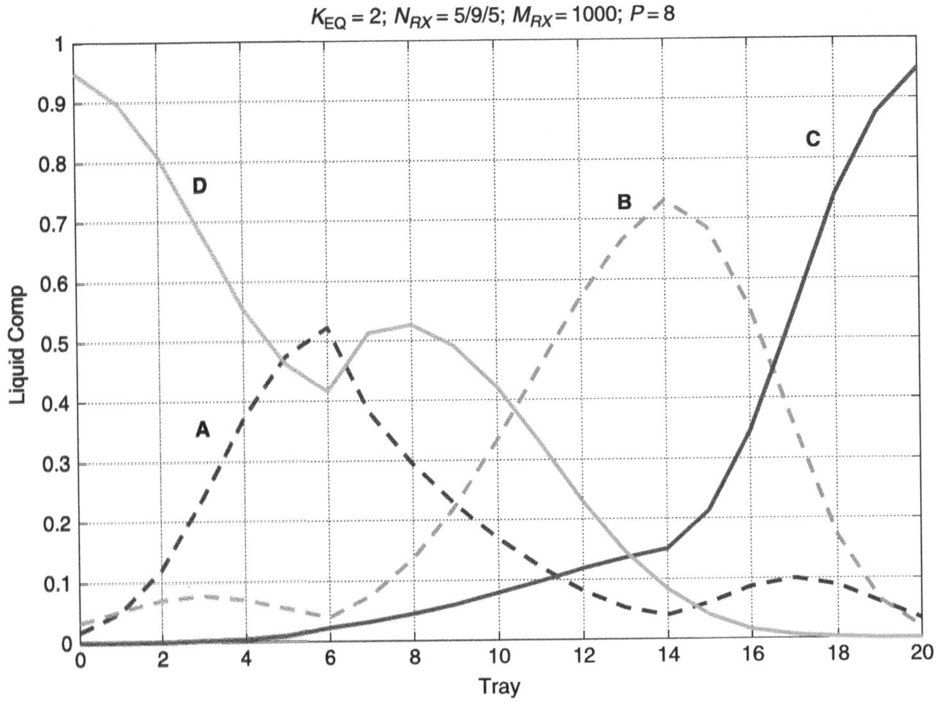

Figure 2.2 Composition profiles for base case.

In the base case, the overhead vapor is 46.15 mol/s and the vapor boilup is 28.91 mol/s. This increase of 17.23 mol/s reflects the reaction of 12.6 (0.95) mol/s of reactants with a −10 kcal/mol heat of reaction and a 6.944 kcal/mol heat of vaporization. Of course, there is a corresponding decrease in the liquid flowrates down the column in the reactive zone.

Figure 2.2 presents the liquid composition profiles in the column under base case conditions. The compositions of reactants A and B peak at their respective feed trays. The products contain impurities of both reactants. There is more A impurity in the distillate than B impurity. The reverse is true in the bottoms.

2.1 EFFECT OF HOLDUP ON REACTIVE TRAYS

We now investigate the impact of changes in various parameters from those used in the base case. The first parameter studied is the holdup of liquid on the reactive trays. As we would expect, the larger the holdup, the easier it is to achieve the desired conversion.

Figure 2.3 shows how vapor boilup (V_S), reflux (R), and product impurities change as reactive tray liquid holdup (M_{RX}) is increased. Below a holdup of about 260 mol, the desired 95% conversion cannot be achieved in this reactive column with the given tray configuration and operating pressure. The reflux and vapor boilup become very large as this limit is approached. As holdup is increased, vapor boilup and reflux steadily decrease. There is little change for holdups greater than 1000 mol. Larger holdups would require either a higher liquid height on the reactive trays or a larger diameter column. The diameter

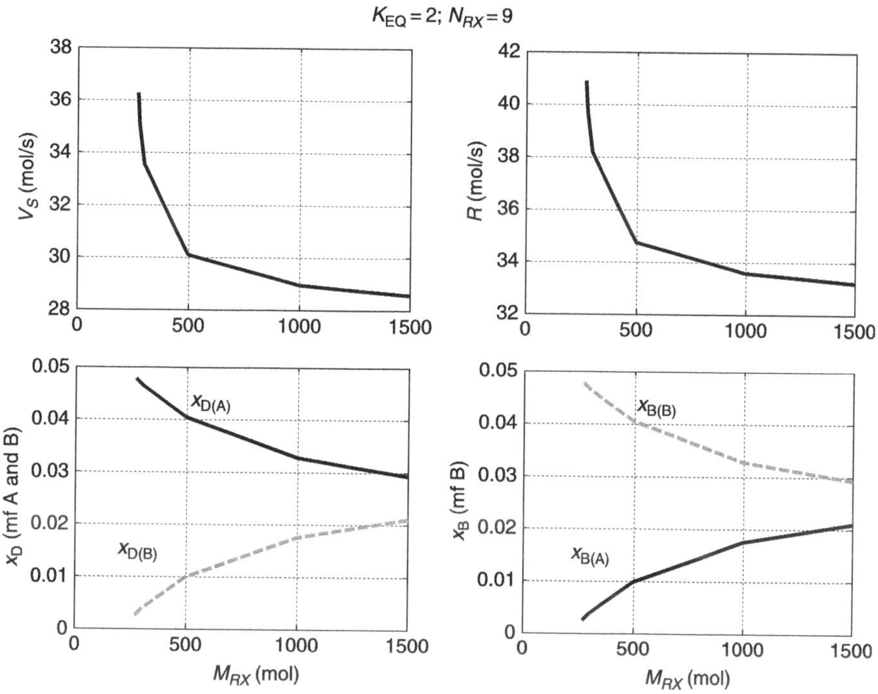

Figure 2.3 Effect of reactive tray holdup.

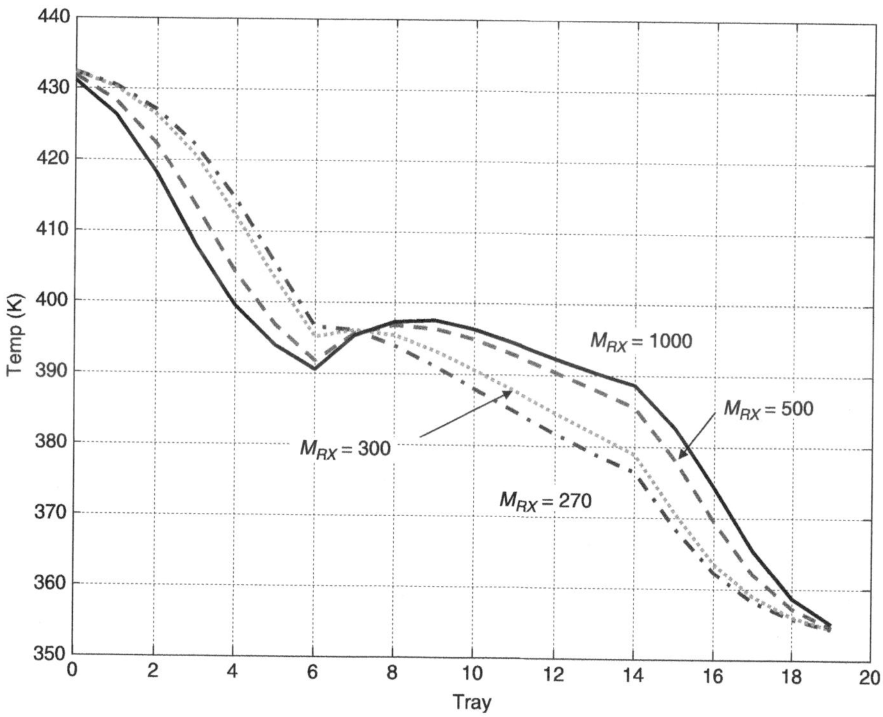

Figure 2.4 Effect of reactive tray holdup on temperature profiles.

of this column, established by vapor-loading limitations, is 0.8 m. The 1000 mol of liquid holdup correspond to a reasonable liquid height of 12 cm, using a density of 800 g/L and a molecular weight of 50 g/mol.

The vapor boilup reflects the energy cost for operating the column, so larger tray holdups reduce energy consumption.

Note that the impurities in the product change with tray holdup. The larger vapor and liquid rates needed with the small holdups give more separation, so there is less of heavier reactant B in the distillate and less of lighter reactant A in the bottoms. The fractionation decreases as liquid and vapor rates decrease, so the products contain more of the reactant component that is not adjacent to the product. Remember that the purities of both products remain at 95 mol%.

Figure 2.4 shows how temperature profiles are affected by tray holdup. The profiles tend to become steeper as holdup decreases because of the increase in fractionation.

2.2 EFFECT OF NUMBER OF REACTIVE TRAYS

The previous section showed no surprises regarding the effect of tray holdup. In this section we look at changing the number of reactive trays. Intuition would lead us to think that the more trays the better. This is certainly the case in conventional distillation. However, as we will see, this is not the case with a steady-state reactive distillation column for this type of reaction (two reactants, two products).

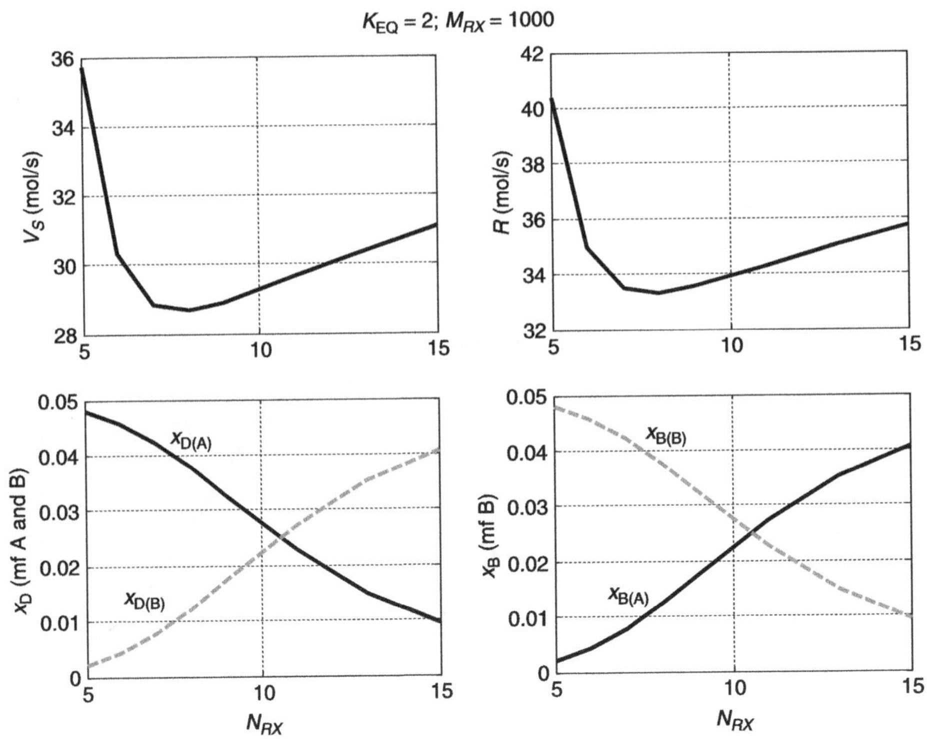

Figure 2.5 Effect of number of reactive trays.

Figure 2.5 demonstrates the effect of changing the number of reactive trays N_{RX} with all other parameters held constant at base case values. The important graph is the upper left one that shows how vapor boilup (or energy) changes as the number of reactive trays is varied. It is quite unexpected that there is a minimum in this curve, which says there is an optimum value for the number of reactive trays in terms of energy consumption. This is certainly different than in conventional distillation in which adding more trays always reduces energy consumption. Figure 2.6 gives temperature profiles with several numbers of reactive trays.

The explanation of this counterintuitive result can be deduced from the composition profiles shown in Figure 2.7. With few reactive trays, a lot of vapor boilup is required to keep the reactants from leaving the column. The top left graph shows that the concentration of A is quite high in the rectifying section of the column with an N_{RX} value of 7. Likewise, the top right graph shows that the concentration of B is quite high in the stripping section of the column for this case. This occurs because the consumption of the reactants is not as large as they move up or down through the reactive zone because there are few trays. Therefore, a lot of vapor boilup and reflux is needed in order to keep reactant B from dropping out of the bottom and to keep reactant A from going out of the top.

The situation is just the reverse with a large number of reactive trays. The concentrations of the reactants at the ends of the column where they are fed become large because much of the other reactant has been consumed on the many trays in the reactive zone before it reaches the opposite end. There is little B arriving near the bottom of the reactive zone, so the concentration of A is large near where it is fed. Likewise, there is little A arriving near the top of the reactive zone, so the concentration of B is large near where it is fed.

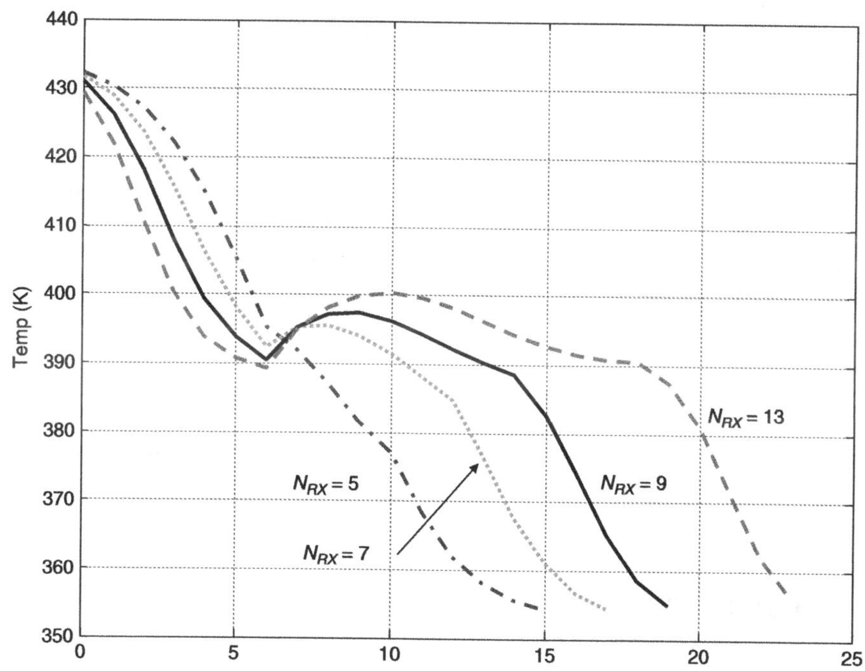

Figure 2.6 Effect of number of reactive trays on temperature profile.

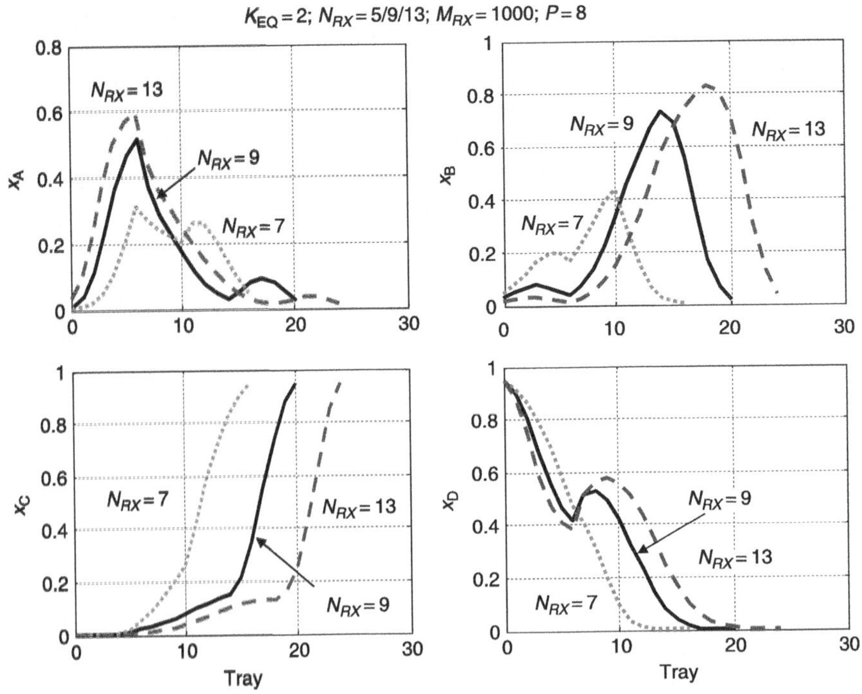

Figure 2.7 Effect of number of reactive trays on composition profiles.

These high concentrations require more vapor boilup and reflux to keep reactant B from escaping out of the top and reactant A from dropping out of the bottom of the column.

This phenomenon is one of the most intriguing aspects of reactive distillation design. Adding more reactive trays does not always produce a more economical design, even if the increase in capital cost is ignored.

There are also control implications. As we will see in later chapters, the dynamic controllability of a reactive distillation column is improved by adding more reactive trays. Thus, as is true in many chemical processes, there is a conflict between steady-state design and dynamic controllability. The column with 9 reactive trays is the steady-state economic optimum. However, as we will demonstrate in Chapter 10, a column with 13 reactive trays provides better dynamic performance in terms of the ability to maintain conversion and product purities in the face of disturbances in throughput and feed compositions.

The reactive distillation process provides an excellent example of the ever-present interaction between design and control. Both steady-state and dynamic aspects of a chemical process must be considered at all stages of the development and commercialization of a chemical process: laboratory, pilot plant, and plant.

2.3 EFFECT OF PRESSURE

In conventional distillation, column pressure is usually selected to be as low as possible while still being able to use cooling water in the condenser. This is because relative volatilities increase with decreasing temperature in many chemical systems. The other situation

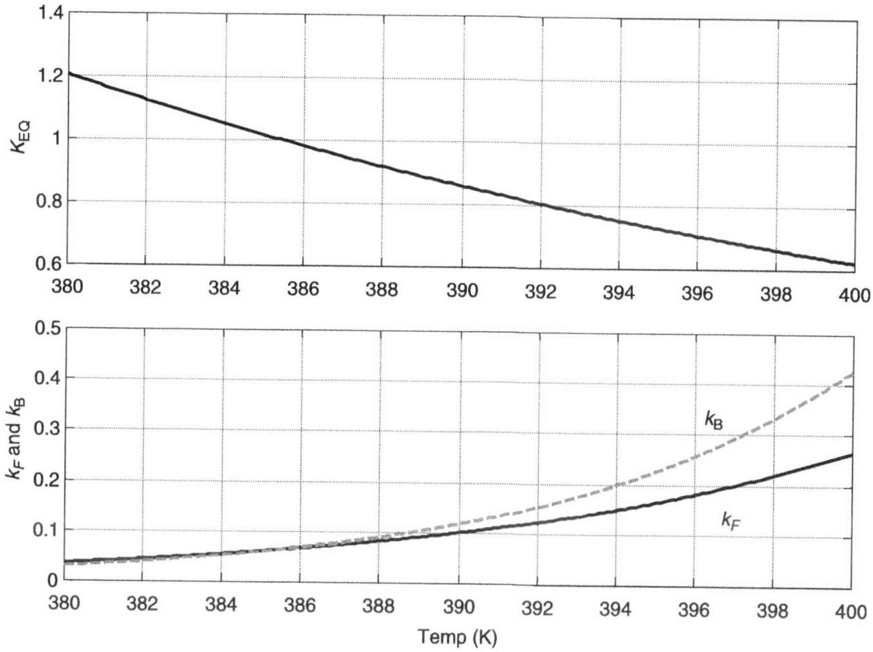

Figure 2.8 Temperature dependence of kinetic parameters.

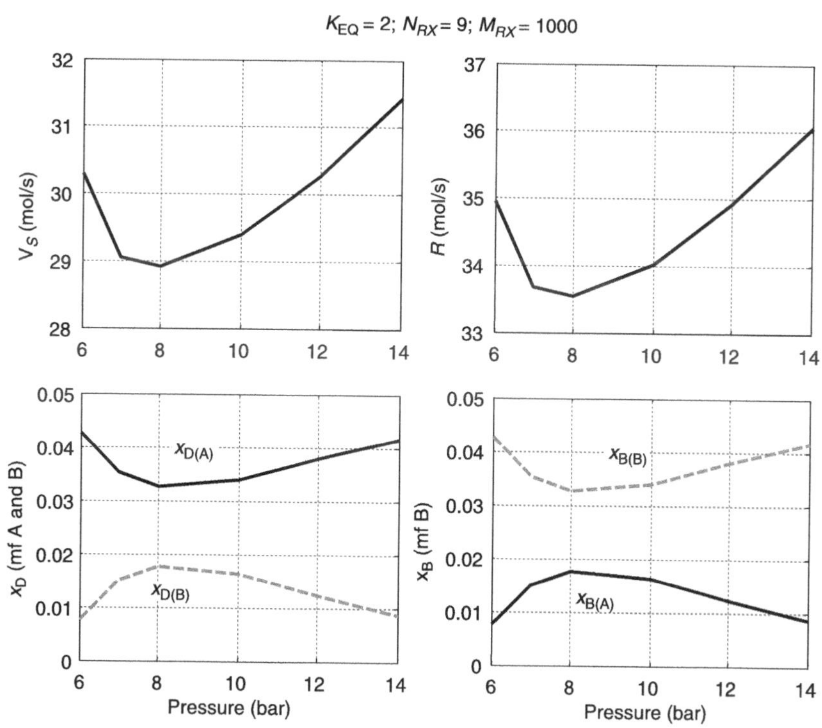

Figure 2.9 Effect of pressure.

that affects the selection of pressure is a high-temperature limitation. In this case the pressure is chosen to keep the temperature in the base of the column, where the temperature is the highest, below this maximum. Pressure affects only the vapor–liquid equilibrium in conventional distillation.

However, in reactive distillation, pressure affects both chemical kinetics and vapor–liquid equilibrium. Therefore, the optimum pressure may not correspond to the minimum attainable while still using cooling water in the condenser. For example, the optimum pressure for the base case is 8 bar, as we will demonstrate. The corresponding reflux-drum temperature is 353 K (80 °C, 176 °F), which is well above the temperature that could be achieved using 305 K (32 °C, 90 °F) cooling water.

Because relative volatilities are constant in the base case, there is no effect of pressure on the ease or difficulty of separation. The system with relative volatilities that do change with temperature will be explored in Section 2.5. In this situation a mismatch can occur between temperatures that are favorable for reaction and temperatures that are favorable for separation.

In the constant relative volatility case, pressure only affects the specific reaction rates and the chemical equilibrium constant. Increasing pressure increases temperatures. Figure 2.8 shows the effects of temperature on the chemical kinetics. Because the reaction is exothermic, the chemical equilibrium constant decreases with increasing temperature. The backward specific reaction rate increases more quickly with increasing temperature than the forward specific reaction rate because of its larger activation energy.

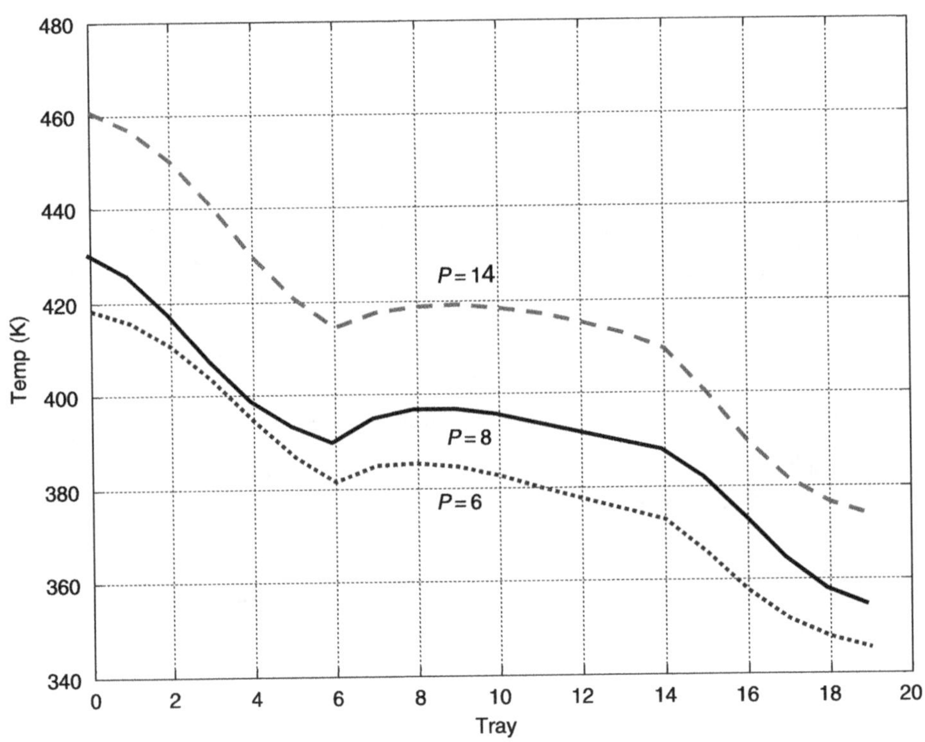

Figure 2.10 Effect of pressure on temperature profiles.

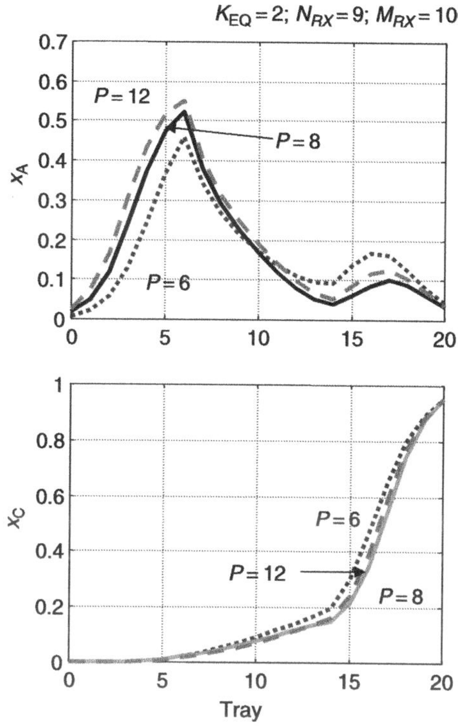

Figure 2.11 Effect of pressure on composition profiles.

These curves indicate that low pressure and the resulting low temperatures would result in low reaction rate, which would require large holdups on reactive trays. Alternatively, for a fixed number of trays and a fixed holdup, the concentrations of the reactants in the reactive zone would have to be large at low temperatures. This would require a large vapor boilup and reflux flowrate to keep the reactants from escaping out of the top or bottom.

In contrast, high pressure and temperatures would result in a lower chemical equilibrium constant, which would also require large vapor boilup and reflux flowrates.

Figure 2.9 shows the effect of the operating pressure. There is an optimum pressure at which the energy consumption is minimized, as shown in the top left graph. The impurities of reactants in the two products reflect the changes in fractionation as the vapor and liquid rates change with pressure. Less fractionation permits more of the nonadjacent component to go into the product; that is, there is more B in the distillate product (mostly C) and more A in the bottoms product (mostly D) at the smaller vapor and liquid rates.

Figure 2.10 displays temperature profiles. The higher the pressure is, the higher the temperatures throughout the column. Figure 2.11 provides the composition profiles at three different pressures. The solid lines are for the optimum case with a pressure of 8 bar.

2.4 EFFECT OF CHEMICAL EQUILIBRIUM CONSTANT

In this section the value of the chemical equilibrium constant at 366 K $[(K_{EQ})_{366}]$ is changed to explore its effect on the steady-state design of the reactive column. The

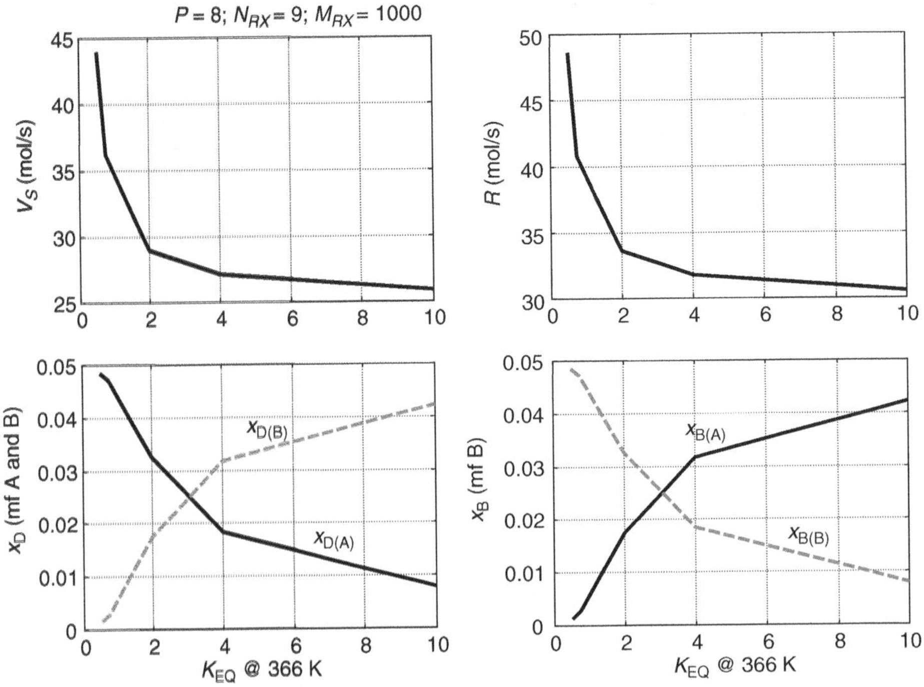

Figure 2.12 Effect of K_{EQ}.

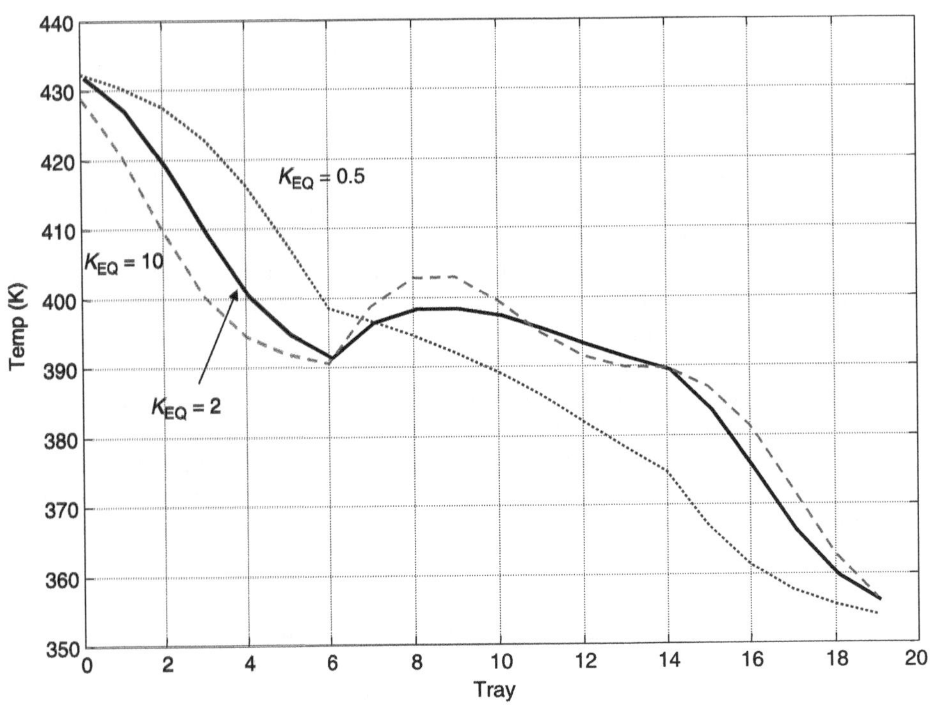

Figure 2.13 Effect of K_{EQ} on temperature profiles.

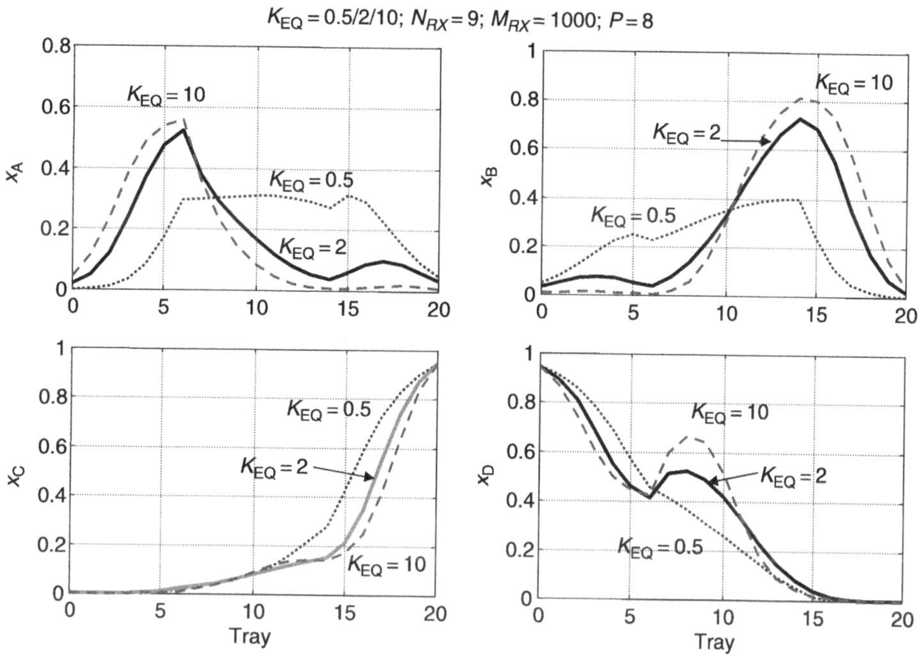

Figure 2.14 Effect of K_{EQ} on composition profiles.

kinetic parameters of the forward reaction are held constant. The preexponential factor of the backward reaction is adjusted to give the desired specific reaction rate at 366 K.

$$(k_B)_{366} = \frac{(k_F)_{366}}{(K_{EQ})_{366}} \tag{2.8}$$

All other parameters are kept the same as the base case: column configuration, pressure, tray holdup, and conversion.

Figure 2.12 shows the effect of varying the chemical equilibrium constant. As $(K_{EQ})_{366}$ increases, the required vapor boilup and reflux to maintain conversion and product purities decrease because driving the reaction to the product side becomes easier. The compositions of the impurities in the two products shift as the fractionation decreases (lower vapor and liquid rates in the column) to having more of the nonadjacent reactant components in the product. Figure 2.13 gives the temperature profiles for three values of K_{EQ}.

Figure 2.14 contains the composition profiles at the same three values of K_{EQ}. The low values of the chemical equilibrium constant require large concentrations of reactants in the reactive zone to achieve conversion.

2.5 EFFECT OF RELATIVE VOLATILITIES

Two types of changes in relative volatilities are considered in this section. The first is a simple change from the value of 2 for the relative volatilities of adjacent components. As expected, energy requirements increase as the separation becomes more difficult. The second type of change is when the volatilities are temperature dependent. In this case,

TABLE 2.3 Coefficients and Vapor Boilup Required for Constant α_{CA} and α_{BD} Cases

$\alpha_{CA} = \alpha_{BD}$	Vapor Pressure Coeff.	Component A	B	C	D	Vapor Boilup (mol/s)
1.75	A_j	12.34	11.65	12.91	11.09	33.05
	B_j	3862	3862	3862	3862	
2	A_j	12.34	11.65	13.04	10.96	28.91
	B_j	3862	3862	3862	3862	
2.5	A_j	12.34	11.65	13.26	10.73	25.15
	B_j	3862	3862	3862	3862	
3	A_j	12.34	11.65	13.44	10.55	23.14
	B_j	3862	3862	3862	3862	

the pressure in the column may have to be lowered from the optimum that applies when the relative volatilities are constant.

2.5.1 Constant Relative Volatilities

In the base case the volatilities are $\alpha_C = 8$, $\alpha_A = 4$, $\alpha_B = 2$, and $\alpha_D = 1$. The ratios of the volatilities of adjacent components are all equal to 2. We keep the relative volatilities of the reactants A and B unchanged from the base case, so the vapor pressure coefficients A_j and B_j for both reactants stay the same. However, the vapor pressure coefficients for the two products are changed over a range of values. Table 2.3 gives the values of the A_j and B_j coefficients used in the various cases. Note that all of the B_j coefficients are the same because constant relative volatilities are assumed. Figure 2.15 demonstrates how the vapor pressures depend on the temperature for the constant relative volatility case (Fig. 2.15*a*) and for the case when the relative volatilities between A and C and between B and D are changed to 1.25 (Fig. 2.15*b*).

For a fixed number of trays, it becomes impossible to maintain conversion if the relative volatilities are too small. The limit is about 1.75 when the base case numbers of trays are used in the stripping, reacting, and rectifying sections. As expected, energy requirements (last column, Table 2.3) decrease as relative volatilities increase.

2.5.2 Temperature-Dependent Relative Volatilities

Relative volatilities are fairly constant in hydrocarbon systems with similar chemical structures. However, in many systems the temperature dependences of vapor pressures of individual components are not the same. This typically results in a decrease in relative volatility as temperature increases. If higher temperatures favor chemical kinetics and low temperatures give high relative volatilities, a mismatch may occur that can make reactive distillation unattractive. We will quantify these economic issues in Chapter 3.

The reflux-drum temperature in a conventional column would typically be 320 K because this permits the use of cooling water in the condenser. This would minimize pressure and maximize relative volatilities. The temperature in the reflux drum of the base case reactive distillation column operating at 8 bar is about 394 K (see Fig. 2.10). This is well above the 320 K minimum and has been selected because it favors chemical kinetics. No issues of vapor–liquid equilibrium were involved in this selection because the relative volatilities were assumed to be constant and not affected by temperature.

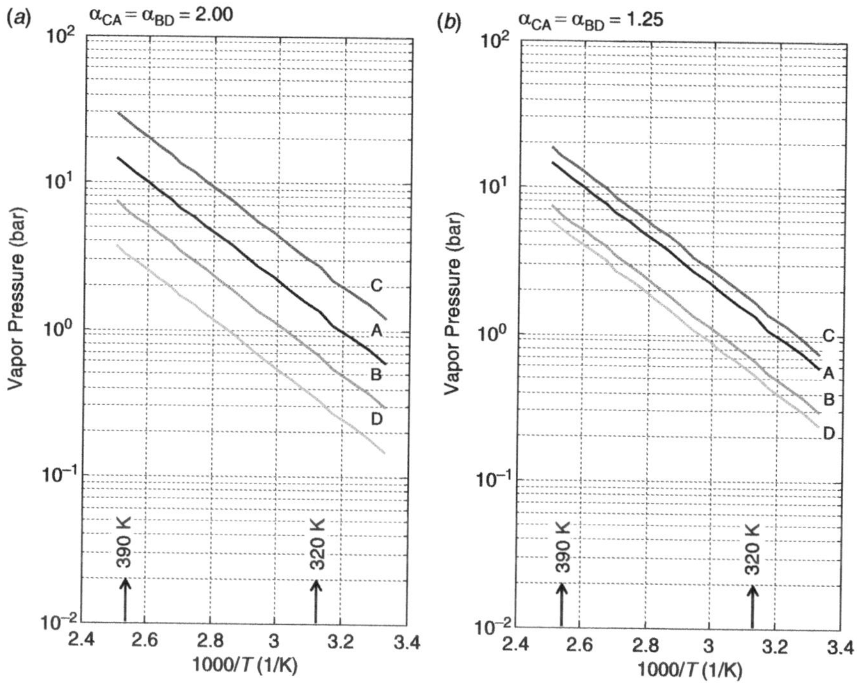

Figure 2.15 Vapor pressures.

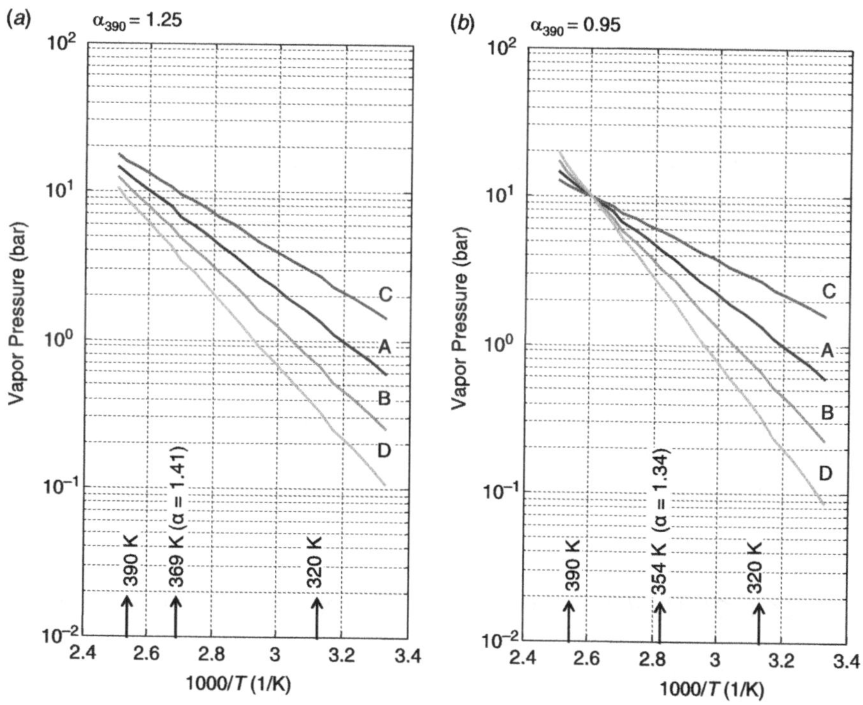

Figure 2.16 Temperature-dependent vapor pressures.

TABLE 2.4 Vapor Pressure Constants for Temperature Dependent α_{ij} Case

α_{390}	Vapor Pressure Coef.	Component				Vapor	Boilup
		A	B	C	D	8 bar (mol/s)	7 bar (mol/s)
1.75	A_j	12.34	12.38	12.29	12.44	33.73	38.07
	B_j	3862	4100.1	3623.9	4338.1		
1.85	A_j	12.34	12.08	12.60	11.82	32.91	33.12
	B_j	3862	4001.0	3723.0	4140.0		
2	A_j	12.34	11.65	13.04	10.96	28.91	29.05
	B_j	3862	3862	3862	3862		

Now we demonstrate what happens when the relative volatilities of the components are temperature dependent. This is achieved by assuming that the relative volatility values of adjacent components are 2 when the temperature is 320 K but become smaller as the temperature increases. The value of the relative volatility of adjacent components at 390 K (α_{390}) is selected as a way to define this temperature dependence. The value of α_{390} is reduced from 2 (which corresponds to the constant relative volatility case) to lower values.

The vapor pressure coefficients of component A are kept constant for all α_{390} cases. Therefore, the slope of the vapor pressure line for component A is the same for all cases, as shown in Figure 2.16 for two different values of α_{390}. The lines are not parallel and get closer as the temperature increases. As the right graph shows, the vapor pressure lines actually cross each other at a certain temperature for the extreme $\alpha_{390} = 0.95$ case. Of course, the column could never operate in this temperature range because there would be no separation. In this situation, the column pressure would have to be lowered to keep temperatures low enough for reasonable separation. However, chemical reaction rates would be smaller, so more reactive trays may be required to achieve the specified conversion. In an extreme case, reactive distillation may not be feasible.

The B_j vapor pressure coefficients of the other three components are calculated in order to reduce the relative volatilities to the specified values at 390 K. Table 2.4 gives the values of all of the vapor pressure constants for the cases considered. The last two columns in Table 2.4 give the energy requirements for the various cases at two different pressures. Increasing temperature dependence corresponds to decreasing values of α_{390}. Vapor boilup requirements increase as α_{390} decreases. The base case column with its fixed number of trays cannot achieve the desired 95% conversion for α_{390} values less than $\sim$1.75. Decreasing the 8-bar base case optimum pressure to 7 bar increases vapor boilup because the lower pressure decreases reaction rates.

We will return to this problem in Chapter 3 and show how the design of the reactive column must be altered to handle cases with temperature dependence of the relative volatilities. As expected, the optimum column pressure decreases and the number of reactive trays increase as the values of α_{390} decrease. Reactive distillation becomes economically unattractive for α_{390} values of less than $\sim$1.2.

2.6 EFFECT OF NUMBER OF STRIPPING AND RECTIFYING TRAYS

The number of stripping trays is assumed to be equal to the number of rectifying trays because the relative volatilities between the key components are the same in each

TABLE 2.5 Effect of N_S and N_R on Vapor Boilup

$N_S = N_R$	V_S (mol/sec)
2	40.74
3	30.44
4	29.19
5	28.91
6	28.85
7	29.86
10	28.92
15	29.13

section. We would expect that adding more trays would make the fractionation easier and require less vapor boilup.

The base case number is $N_S = N_R = 5$. Table 2.5 shows how vapor boilup changes with the number of fractionating trays. As expected from our experience with conventional distillation, with fewer than three trays, vapor boilup begins to increase sharply.

There was a very slight increase when a very large numbers of trays were used, which was unexpected. This was because there was less heavier B near the top of the column and less lighter A near the bottom, which makes the separation somewhat more difficult. As we will see in Chapter 5, which studies a different chemical system, this unexpected effect can become more pronounced.

2.7 EFFECT OF REACTANT FEED LOCATION

The final design parameter to be studied in this chapter is the locations of the two fresh feedstreams. Up to this point we have assumed that the lighter reactant fresh feed F_{0A} is introduced on the bottom tray of the reactive zone ($N_S + 1$) and the heavier reactant fresh feed F_{0B} is introduced on the top tray of the reactive zone ($N_S + 1 + N_{RX}$). This configuration seems like a logical choice. However, the fresh feeds could be introduced on trays inside the reactive zone. The question is how this affects the design, primarily in terms of energy consumption because this is our major economic performance criterion.

2.7.1 Reactant A Feed Location (N_{FA})

The base case location of the fresh feed of reactant A is on tray 6 because there are five stripping trays. Figures 2.17–2.19 show the effect of moving the feed tray up into the reactive zone. The upper left graph in Figure 2.17 demonstrates that there is an initial decrease in the vapor boilup required to achieve the specified 95% conversion and impurities. Thus, the optimum feed tray location in this example is *not* at the very bottom of the reactive zone.

The lower graphs in Figure 2.17 show that the concentration of impurity A in the distillate increases as the feed tray is moved up in the column and the concentration of B increases in the bottoms. This seems reasonable because the A is being introduced closer to the top and further away from the bottom. Figure 2.18 gives the temperature profiles for different values of N_{FA}. The temperatures in the stripping section increase and the temperatures in the rectifying section decrease as the feed tray is moved up the column. Because

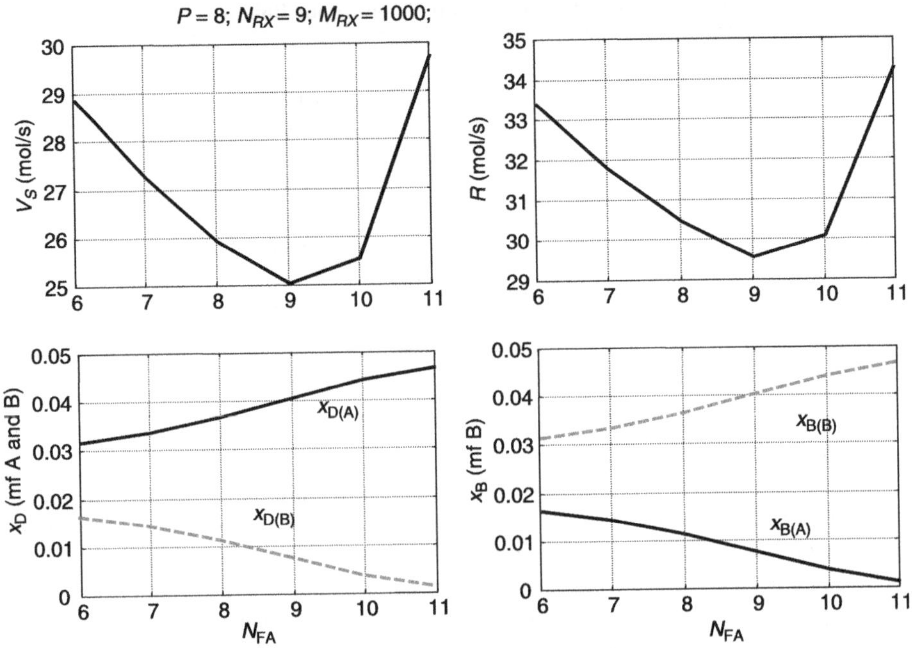

Figure 2.17 Effect of location of fresh feed of A.

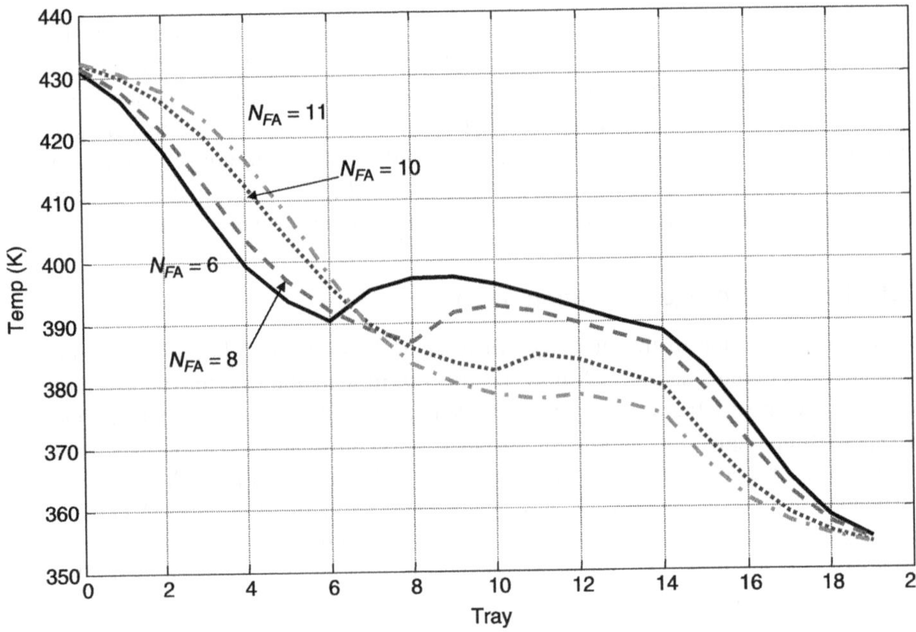

Figure 2.18 Effect of location of fresh feed of A on temperature profile.

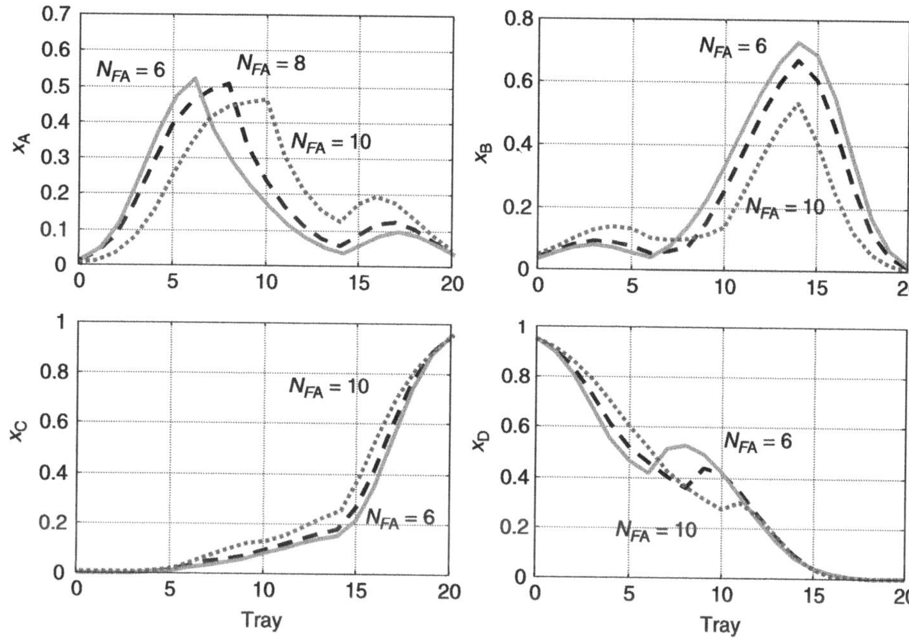

Figure 2.19 Effect of location of fresh feed of A on composition profiles.

A is lighter than B, the lower concentrations of A in the stripping section result in higher temperatures. The reverse is true in the rectifying section. The composition profiles throughout the column are provided in Figure 2.19.

2.7.2 Reactant B Feed Location (N_{FB})

The base case location of the fresh feed of reactant B is on tray 14 because there are five stripping trays and nine reactive trays. Several cases were run with lower feed tray locations. There was little effect on the vapor boilup as indicated in Table 2.6. The location of the heavy feed does not seem to have as large an effect as that of the light feed unless the feed is introduced quite far down in the reactive zone.

TABLE 2.6 Effect of N_{FB} on Vapor Boilup

N_{FB}	V_S (mol/s)
14	28.91
13	28.68
12	28.39
11	28.08
10	28.00
9	29.21

2.8 CONCLUSION

The impacts of a number of important kinetic, phase equilibrium, and design parameters have been explored in the chapter. Most of the parameters have effects that are logical and follow our experience with conventional distillation.

However, there are two parameters that give counterintuitive results. The most important of these is the number of reactive trays. More trays does not necessarily improve the steady-state economics of a reactive distillation column. However, the effect of more reactive trays on the dynamic performance of a reactive column is distinctly different, as we will demonstrate in Chapter 10.

CHAPTER 3

ECONOMIC COMPARISON OF REACTIVE DISTILLATION WITH A CONVENTIONAL PROCESS

This chapter presents detailed economic comparisons of two alternative flowsheets. In the first, a single reactive distillation is operated in neat mode. In the second, a conventional multiunit process with independent reaction and separation sections is designed. Both flowsheets are optimized in terms of their TACs, which reflect both energy and capital costs.

The two designs are performed over a range of values for two parameters: the chemical equilibrium constant (K_{EQ}) and temperature-dependent relative volatilities. The chemical reaction is the same ideal quaternary exothermic, reversible, liquid-phase reaction considered in Chapter 2: $A + B \Leftrightarrow C + D$. In the reactive distillation system, the design optimization variables include pressure, number of reactive trays, and number of total trays. In the conventional system, the design optimization variables include reactor temperature, reactor size, and reactor composition. Systematic design procedures for both flowsheets are developed as three-dimensional optimization problems, using several heuristic rules and engineering assumptions. The effects of the three selected design parameters on the optimum design variables and TACs are studied for both process designs. Both flowsheets have identical feeds and produce identical products. However, they have different energy requirements and different capital investments for the different equipment required.

Results show that reactive distillation is significantly less expensive (by a factor of up to 3) than the conventional process for all values of the chemical equilibrium constant. However, when the relative volatilities show strong temperature dependence, reactive distillation becomes more expensive because of the mismatch between the temperatures that are favorable for reaction and the temperatures that are favorable for vapor–liquid equilibrium.

The reactive distillation process was described in Chapter 2, and the multiunit conventional process is described in this chapter.

Reactive Distillation Design and Control. By William L. Luyben and Cheng-Ching Yu

3.1 CONVENTIONAL MULTIUNIT PROCESS

Figure 3.1 provides a detailed flowsheet of the process and the notation used. The reaction occurs in a CSTR with molar holdup V_R. There are two fresh feedstreams F_{0A} and F_{0B} that contain pure reactants A and B, respectively, and a recycle stream D_2 returns from a downstream distillation column. The reactor effluent contains a multicomponent mixture because complete one-pass conversion is not achieved. Two columns are needed to separate the two products from the intermediate-boiling reactants. The reactor effluent F with composition z_j is fed into the first distillation column to separate product C from unreacted reactants A and B and heavy product D. Product C goes out in the distillate of the first column with the desired 95 mol% purity, and the other components go out in the bottoms, which is fed to the second column. This column produces a bottoms stream of D with the desired 95 mol% purity and a distillate of unreacted reactants A and B that is recycled back to the reactor with specific amounts of product impurities $x_{D2,C}$ and $x_{D2,D}$.

The temperature in the reactor T_R can be set completely independent of the conditions in the two distillation columns, so the best reactor temperature can be used that is most favorable for the reaction. The pressures and temperatures in the two distillation columns can be set at their best values that are favorable for vapor–liquid phase equilibrium conditions.

There are 12 design degrees of freedom for this multiunit process. Subtracting the number of specifications and assumed heuristic relationships from the degrees of freedom gives the number of design optimization variables.

3.1.1 Assumptions and Specifications

An approximate heuristic design procedure is used for the columns in the conventional process. The optimum number of trays is assumed to be equal to 2 times the minimum, and the optimum reflux ratio is assumed to be 1.2 times the minimum. The assumptions and specifications use up 9 of the 12 design degrees of freedom, leaving 3 that can be used to optimize the process. Three design optimization variables were selected:

1. Molar holdup in the reactor V_R
2. Composition (mole fraction) of reactant B in reactor z_B
3. Reactor temperature T_R

The chemical equilibrium constant at 366 K [$(K_{EQ})_{366}$] and the relative volatilities (constant or temperature dependent) are specified for each case. Equimolal overflow is assumed in the distillation columns, which means that neither energy balances nor total balances are needed on the trays for steady-state calculations. Other assumptions are isothermal operation of the reactor, theoretical trays, saturated liquid feed and reflux, total condensers, and partial reboilers in the columns. Additional assumptions and specifications are the following:

1. In all cases the net production rates are set by fixing the fresh feed flowrate of pure component A at F_{0A} to 12.6 mol/s and B at F_{0B} to 12.6 mol/s.
2. The amount of reactant A lost in product stream D_1 is constant at $A_{loss} = 0.63$ mol/s ($x_{D1,A} = 0.05$ mol fraction). There is no B and D going overhead ($x_{D1,B} = x_{D1,D} = 0$).

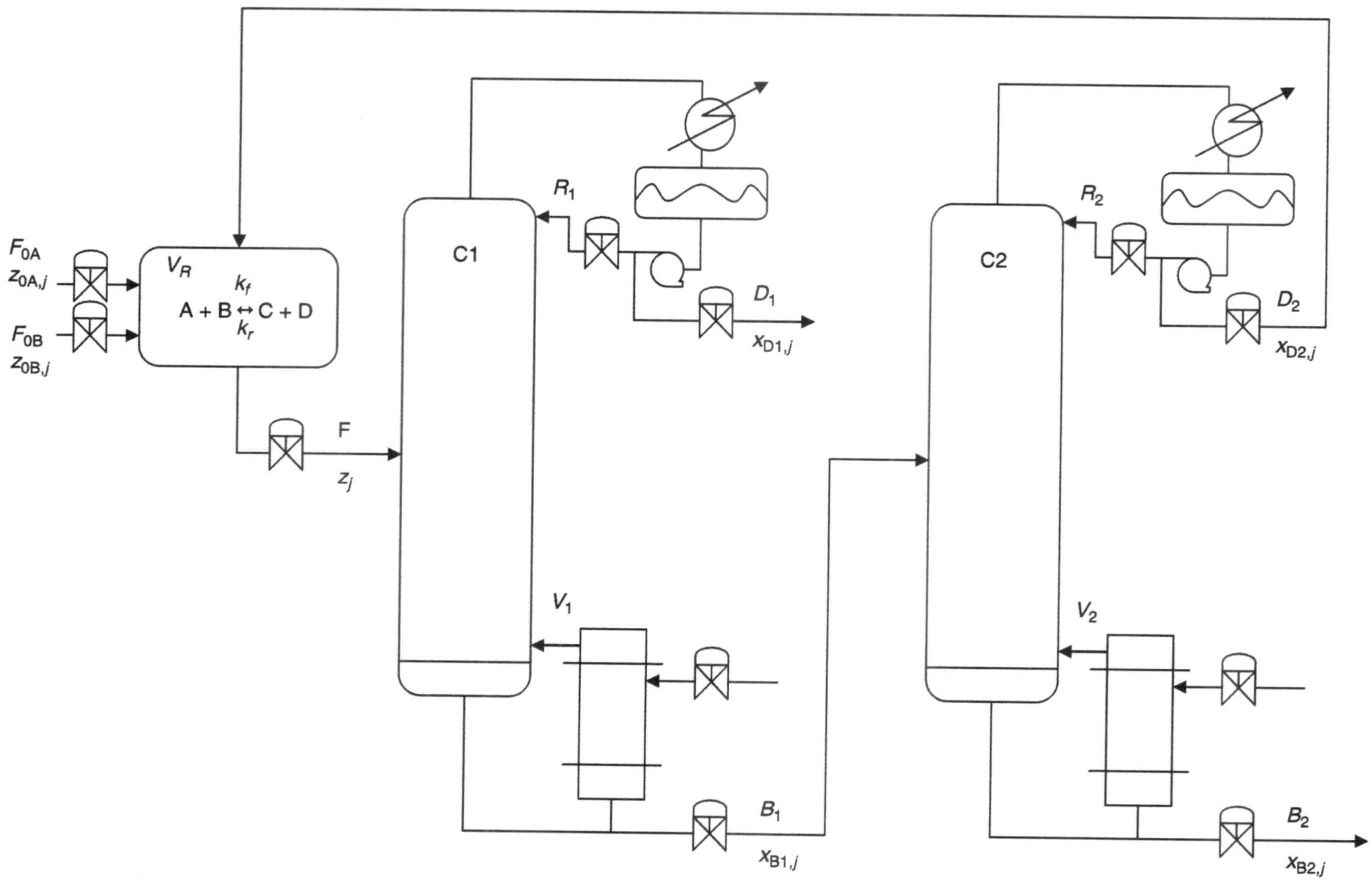

Figure 3.1 Conventional process.

3. The bottoms product stream B_2 contains some amount of component B as an impurity, $B_{\mathrm{loss}} = 0.63$ mol/s ($x_{B2,\mathrm{B}} = 0.05$ mole fraction). There is no A or C going out the bottoms ($x_{B2,\mathrm{A}} = x_{B2,\mathrm{C}} = 0.$).
4. The impurity of product C in the recycle stream D_2 is $x_{D2,\mathrm{C}} = 0.05$.
5. The impurity of component D in D_2 is $x_{D2,\mathrm{D}} = 0.05$.
6. As a result of the previous two assumptions and the fact that C and D have identical stoichiometric coefficients, the concentrations of components C and D in the reactor (z_C and z_D) are equal for all cases ($z_\mathrm{C} = z_\mathrm{D}$) but vary in value from case to case.
7. The column pressures are set using the vapor pressures P^S of pure components and liquid compositions in the reflux drum $x_{D,j}$ at 320 K (so that cooling water can be used in the condenser).
8. In each column, the reflux ratio RR is 1.2 times the minimum reflux ratio $\mathrm{RR}_{\min}$ calculated via the Underwood equations.
9. In each column, the number of stages N_T is twice the minimum number of stages $N_{T\min}$ calculated via the Fenske equation.
10. Kirkbride's method is used to find the optimal feed tray location N_F.

Several different specifications in items 2–5 will be used to investigate the effects of product quality, conversion, and recycle impurities on the economically optimal steady-state design. However, these do not affect the general structure of the design procedure.

A grid-search optimization strategy is used in this work to find the optimum values of the three design optimization variables. Other methods could be used, such as gradient nonlinear programming techniques. However, the grid method is more robust because a numerical programming method can easily drive the process into an infeasible region in which the specified purities and production rates cannot be achieved.

3.1.2 Steady-State Design Procedure

The production rates of components C and D (R_C and R_D) in the reactor are given by

$$R_\mathrm{C} = R_\mathrm{D} = V_R(k_F z_\mathrm{A} z_\mathrm{B} - k_B z_\mathrm{C} z_\mathrm{D}) = (12.6)(0.95) \tag{3.1}$$

Using the assumption of equal compositions of z_C and z_D in the reactor, the concentration z_A can be calculated by rearranging Eq. (3.1).

$$z_\mathrm{A} = \frac{4\left(\frac{k_F}{k_B}\right) z_\mathrm{B} + 2(1 - z_\mathrm{B}) \pm \sqrt{\left(4\left(\frac{k_F}{k_B}\right) z_\mathrm{B} + 2(1 - z_\mathrm{B})\right)^2 - 4\left((1 - z_\mathrm{B})^2 + 4\frac{R_\mathrm{C}}{V_R k_B}\right)}}{2} \tag{3.2}$$

The following steps in the design procedure are employed, utilizing the material balances and specifying necessary variables.

1. Fix the value of T_R at a small value (to be varied).
2. Fix the value of reactor holdup V_R at a small value (to be varied).
3. Specify the value of reactor composition z_B (to be varied).
4. Calculate z_A from Eq. (3.2).

5. Calculate the product concentrations z_C and z_D in the reactor:

$$z_C = z_D = \frac{1 - z_A - z_B}{2} \tag{3.3}$$

6. Calculate the distillate product flowrate and compositions for the first column.

$$D_1 = R_C + A_{loss} \tag{3.4}$$
$$x_{D1,A} = A_{loss}/D_1 \tag{3.5}$$
$$x_{D1,C} = R_C/D_1 \tag{3.6}$$

7. Calculate the bottoms flowrate and compositions for the second column.

$$B_2 = R_D + B_{loss} \tag{3.7}$$
$$x_{B2,B} = B_{loss}/B_2 \tag{3.8}$$
$$x_{B2,D} = R_D/B_2 \tag{3.9}$$

8. A steady-state total molar balance can be used because the reactions are equimolar. This balance, the two steady-state component balances, and a mole fraction summation around the reactor are

$$F_{0A} + F_{0B} + D_2 = F \tag{3.10}$$
$$F_{0A} + D_2 x_{D2,A} = F z_A + R_C \tag{3.11}$$
$$F_{0B} + D_2 x_{D2,B} = F z_B + R_D \tag{3.12}$$
$$x_{D2,A} + x_{D2,B} + x_{D2,C} + x_{D2,D} = 1 \tag{3.13}$$

These four equations can be solved for the four unknowns: recycle stream D_2, reactor effluent flowrate F, and the reactant compositions in second distillate streams $x_{D2,A}$ and $x_{D2,B}$. Combining Eqs. (3.10)–(3.13) yields

$$D_2 = \frac{(F_{0A} + F_{0B})(z_A + z_B - 1) + R_C + R_D}{(1 - x_{D2,C} - x_{D2,D}) - z_A - z_B} \tag{3.14}$$

$$F = F_{0A} + F_{0B} + D_2 \tag{3.15}$$

$$x_{D2,A} = \frac{F z_A + R_C - F_{0A}}{D_2} \tag{3.16}$$

$$x_{D2,B} = (1 - x_{D2,C} - x_{D2,D}) - x_{D2,A} \tag{3.17}$$

9. The total molar balance and component balances around the first column enable the calculation of the bottoms flowrate and compositions for that column.

$$B_1 = F - D_1 \tag{3.18}$$

$$x_{B1,j} = \frac{F z_j - D_1 x_{D1,j}}{B_1} \tag{3.19}$$

10. Now the feed, bottoms, and distillate flowrates and compositions are known for both columns. The Fenske equation is used to calculate the minimum number of trays. The actual number of trays is set to 2 times the minimum. The Underwood equations are used to calculate the minimum reflux ratio. For estimating the column diameter, heat exchanger areas, and energy requirements from the vapor rate in the column, the actual reflux ratio is set to 1.2 times the minimum.
11. Based on these data, the TAC is calculated by combining the energy cost with the annual capital cost, using a payback period. All the terms and equations related to sizing of the process equipment and the economic parameters are given in Section 3.1.3.
12. The value of z_B is varied over a wide range, and steps 4–11 are repeated for each value of z_B, generating its corresponding TAC.
13. Then the value of the reactor holdup is varied over a range, and steps 3–12 are repeated.
14. Finally, the value of the reactor temperature is changed over a wide range, and steps 2–13 are repeated for each temperature. The minimum for the TAC is selected as the optimum economically steady-state design for the given $(K_{EQ})_{366}$ value and the given relative volatility relationship.

3.1.3 Sizing and Economic Equations

The economic basis for the calculations involves an objective function that sums the annual capital and energy costs of the process assuming a payback period (β_{pay}) for capital. The TAC is defined as

$$\text{TAC} = \text{energy cost} + \frac{\text{capital investment}}{\beta_{pay}} \tag{3.20}$$

Table 3.1 summarizes the economic parameters and the basis of the equipment sizing calculations. The capital costs of individual equipment and the energy cost are estimated using the following equations:

$$\text{reactor cost} = 52920 D_R^{1.066} L_R^{0.802} \tag{3.21}$$

$$\text{column cost} = 17640 D_C^{1.066} L_C^{0.802} \tag{3.22}$$

$$\text{tray cost} = 229 D_C^{1.55} N_T \tag{3.23}$$

TABLE 3.1 Sizing and Economic Basis

Reboiler	
Heat transfer coefficient U_R	0.568 kJ s^{-1} K^{-1} m^{-2}
Temperature difference ΔT_R	34.8 K
Condenser	
Heat transfer coefficient U_C	0.852 kJ s^{-1} K^{-1} m^{-2}
Temperature difference ΔT_C	13.9 K
Energy cost	4.7 \$ $(10^6$ kJ$)^{-1}$
Payback period β_{pay}	3.0 years

$$\text{heat exchangers cost} = 7296A_R^{0.65} + 7296A_C^{0.65} \tag{3.24}$$

$$\text{energy cost} = 0.6206\Delta H_v V_S \tag{3.25}$$

where L is the length (m), A is the area (m^2), V_S is the vapor rate (mol/s), and ΔH_v is the heat of vaporization (kJ/mol).

The terms involved in the TAC equations are calculated from the following set of assumptions and guidelines:

1. The diameter of the reactor is calculated from

$$D_R = (0.00003977 V_R)^{0.3333} \tag{3.26}$$

2. The reactor length is assumed to be twice the diameter.

$$L_R = 2D_R \tag{3.27}$$

3. Assuming an F factor of 1 in English engineering units, the diameter of the column is calculated from

$$D_C = 1.735 \times 10^{-2}\left(\frac{M_w T}{P}\right)^{0.25} V_{NT}^{0.5} \tag{3.28}$$

4. The column height is calculated assuming 0.61 m (2 ft) tray spacing and allowing 20% more height for base level volume.

$$L_C = 0.73152 N_T \tag{3.29}$$

5. The heat transfer areas of the reboiler and the condenser are calculated using the steady-state vapor rates and the heat of vaporization.

$$A_R = 0.0042\frac{V_S \Delta H_v}{U_R \Delta T_R} \tag{3.30}$$

$$A_C = 0.0042\frac{V_{NT} \Delta H_v}{U_C \Delta T_C} \tag{3.31}$$

In the reactive column, the vapor rate in the top tray V_{NT} is higher than the vapor rate in the reboiler V_S because of the liquid vaporized through the reactive section. Thus, the heat transfer areas of the reboiler and the condenser are calculated using two different vapor rates.

6. Both processes are assumed to be equally reliable and to operate for 365 days per year.

3.2 REACTIVE DISTILLATION DESIGN

Figure 3.2 gives the reactive distillation column and the terminology used. The column is fed with two pure reactant fresh feedstreams: F_{0A} and F_{0B}. The column has three zones. There are N_S trays and a partial reboiler in the stripping section. Above this section, there is a reactive zone with N_{RX} reactive trays. The third section is the rectifying section with N_R trays and a total condenser.

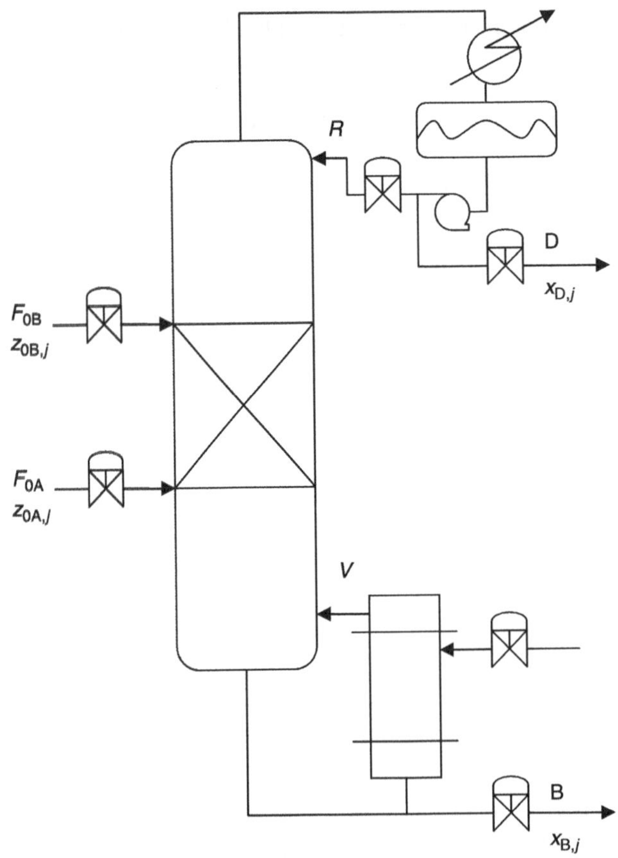

Figure 3.2 Reactive distillation column.

Light reactant A is fed to the bottom tray of the reactive zone, and heavy reactant B is introduced at the top of the reactive section. Light product C leaves in the distillate, and heavy product D is removed in the bottoms. Because the light reactant goes up through the column after being fed on the bottom tray of the reactive zone, there will be little component A in the bottoms. Likewise, the heavy reactant goes down through the column after being fed on the top tray of the reactive zone, and there will be little component B in the distillate. Thus, the key components are light product C and light reactant A for the rectifying section. For the stripping section, the key components are heavy product D and heavy reactant B.

3.2.1 Assumptions and Specifications

The optimization of a reactive distillation column has a large number of design variables. The following specifications and assumptions are made to reduce the number of design optimization variables in the economic steady-state design:

1. The design objective is to obtain 95% conversion for fixed fresh feed flowrates of 12.6 mol/s.

2. The product purity of the distillate stream is $x_{D,C} = 0.95$.
3. The product impurity of the bottoms stream is $x_{B,D} = 0.95$.
4. The two feed points are at the two ends of the reactive zone.
5. The holdups on the reactive trays are 1000 mol.
6. The numbers of stripping and rectifying trays are assumed to be equal ($N_S = N_R$) because the relative volatilities between the components being separated in the two sections are the same. In the rectifying section the separation is between C and A, where the relative volatility is $\alpha_C/\alpha_A = 2$; in the stripping section the separation is between B and D, where the relative volatility is $\alpha_B/\alpha_D = 2$.

Based on these specifications and simplifying assumptions, there are three design optimization variables:

1. Column pressure P
2. Number of reactive trays N_{RX}
3. Number of the separating (stripping/rectifying) trays N_S (or N_R)

Items 2 and 3 in the specifications list will be changed during the study to investigate the effects of product quality and conversion on the economically optimal steady-state design. However, this does not affect the general structure of the design procedure.

3.2.2 Steady-State Design Procedure

Simultaneous solution of the very large set of nonlinear and algebraic equations that describe a reactive distillation column is difficult, especially with the high degree of nonlinearity attributable to reaction kinetics. The relaxation method is efficient and robust in solving this large set of equations. This method is used to calculate steady-state mole fractions and temperature profiles throughout the column. In general, relaxation methods use the equilibrium stage model equations in unsteady-state form and integrate them numerically until the steady-state solution is found (all time derivatives = 0). Liquid holdups on trays are assumed constant, that is, instantaneous hydraulics. The net reaction rate for component j on tray n in the reactive zone is given by

$$R_{n,j} = \nu_j M_n (k_{Fn} x_{n,A} x_{n,B} - k_{Bn} x_{n,C} x_{n,D}) \tag{3.32}$$

where ν_j is the stoichiometric coefficient of component j.

The steady-state vapor and liquid rates are constant through the stripping and rectifying sections because equimolal overflow is assumed. However, these rates change through the reactive zone because of the exothermic reaction. The heat of reaction vaporizes some liquid on each tray in that section; therefore, the vapor rate increases up through the reactive trays and the liquid rate decreases down through the reactive trays.

$$V_n = V_{n-1} - \frac{\lambda}{\Delta H_V} R_{n,C} \tag{3.33}$$

$$L_n = L_{n+1} + \frac{\lambda}{\Delta H_V} R_{n,C} \tag{3.34}$$

The dynamic component balances for the column are

reflux drum:

$$\frac{d(x_{D,j}M_D)}{dt} = V_{NT}y_{NT,j} - D(1 + RR)x_{D,j} \tag{3.35}$$

rectifying and stripping trays:

$$\frac{d(x_{n,j}M_i)}{dt} = L_{n+1}x_{n+1,j} + V_{n-1}y_{n-1,j} - L_n x_{n,j} - V_n y_{n,j} \tag{3.36}$$

reactive trays:

$$\frac{d(x_{n,j}M_n)}{dt} = L_{n+1}x_{n+1,j} + V_{n-1}y_{n-1,j} - L_n x_{n,j} - V_n y_{n,j} + R_{n,j} \tag{3.37}$$

feed trays:

$$\frac{d(x_{n,j}M_n)}{dt} = L_{n+1}x_{n+1,j} + V_{n-1}y_{n-1,j} - L_n x_{n,j} - V_n y_{n,j} + R_{n,j} + F_n z_{n,j} \tag{3.38}$$

column base:

$$\frac{d(x_{B,j}M_B)}{dt} = L_1 x_{1,j} - Bx_{B,j} - V_S y_{B,j} \tag{3.39}$$

With the equimolal overflow assumption mentioned above, all of the vapor rates (V_n) throughout the stripping section are equal to V_S, and all of the liquid rates (L_n) are equal to L_S. Analogously, all V_n beginning from the top feed tray throughout the rectifying section and total condenser are V_R, and all L_n are equal to L_R.

The vapor–liquid equilibrium is assumed ideal. Column pressure P is optimized for each case. With pressure P and tray liquid compositions $x_{n,j}$ known at each point in time on each tray, the temperature T_n and the vapor compositions $y_{n,j}$ can be calculated. This is a bubblepoint calculation and can be solved by a Newton–Raphson iterative convergence method.

$$P = \sum_{j=1}^{NC} x_{n,j} P^S_{j(T_n)} \tag{3.40}$$

$$y_{n,j} = \frac{P^S_j}{P} x_{n,j} \tag{3.41}$$

With a fixed value of the kinetic parameter $(K_{EQ})_{366}$ and given relative volatility relationships, the design procedure is the following:

1. Fix the column pressure P at a small value (to be varied).
2. Fix the number of stripping trays N_S and rectifying N_R trays at a small value (to be varied).
3. Specify the number of the reactive trays N_{RX} (to be varied).
4. The flowrates of the two fresh feeds are fixed at 12.6 mol/s.
5. The flowrates of the distillate and bottoms are fixed at 12.6 mol/s.
6. The vapor boilup V_S is manipulated by a proportional controller to control the level in the column base. The reflux-drum level is not controlled.

7. The reflux flowrate is manipulated by a proportional-integral controller to drive the composition of product (mol% C) in the distillate $x_{D,C}$ to its desired value. This also sets the purity of the product (mol% D) in the bottoms $x_{B,D}$ and the conversion χ in the reactive zone to their desired values.
8. The vapor compositions and temperatures on each tray are computed using bubble-point calculations.
9. The reaction rates are computed with Eq. (3.32).
10. The vapor and liquid flowrates are calculated using Eqs. (3.33) and (3.34).
11. The time derivatives of component material balances are evaluated using Eqs. (3.35)–(3.39).
12. All of the differential equations are integrated using the simple Euler algorithm.
13. Steps 6–12 are repeated until the system achieves a convergence criterion, which is that the largest time derivative of any component on any tray is less than

$$\max\left|\frac{d_{x_{n,j}}}{dt}\right| \leq 10^{-6}$$

14. With the vapor boilup now known, the column is sized and the TAC is calculated by combining the energy cost with the annual capital cost.
15. The number of reactive trays is varied over a wide range, and steps 4–14 are repeated for each value of N_{RX}, generating its corresponding TAC.
16. Then the number of the stripping and rectifying trays is varied over a range, and steps 3–15 are repeated.
17. Finally, the value of the column pressure is changed over a wide range, and steps 2–16 are repeated for each single pressure. The minimum TAC is selected as the optimum economical steady-state design for the given value of $(K_{EQ})_{366}$ and the given relative volatility relationship.

Based on the values of the optimization variables chosen, the column can leave the feasible region. This can be easily observed from the continuously increasing vapor boilup, and the simulation can be restarted with new optimization variables.

3.3 RESULTS FOR DIFFERENT CHEMICAL EQUILIBRIUM CONSTANTS

A wide range of $(K_{EQ})_{366}$ values is explored in this section. Optimum economic steady-state designs of both the reactive distillation process and the conventional multiunit process are developed and compared in terms of TAC.

3.3.1 Conventional Process

Base Case $(K_{EQ})_{366}$. Figure 3.3 gives a three-dimensional plot of the TAC versus reactor V_R volume and reactor z_B composition for the base case $(K_{EQ})_{366} = 2$ with constant $T_R = 367$ K. The reactor holdup and the reactor composition of component B were varied by increments of 2.5 kmol and 0.025 mol fraction, respectively, yielding a total of 360 different designs.

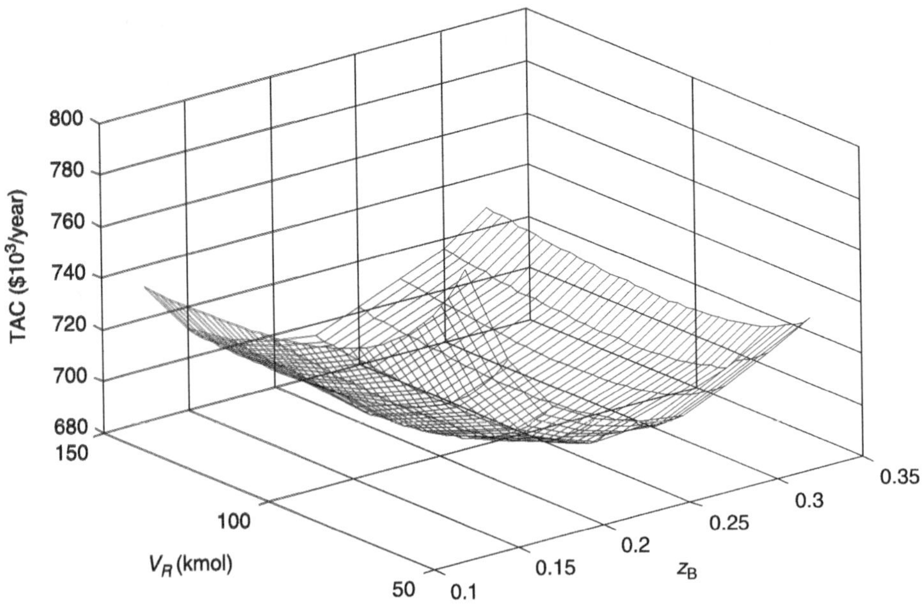

Figure 3.3 Conventional process TAC at 367 K reactor temperature.

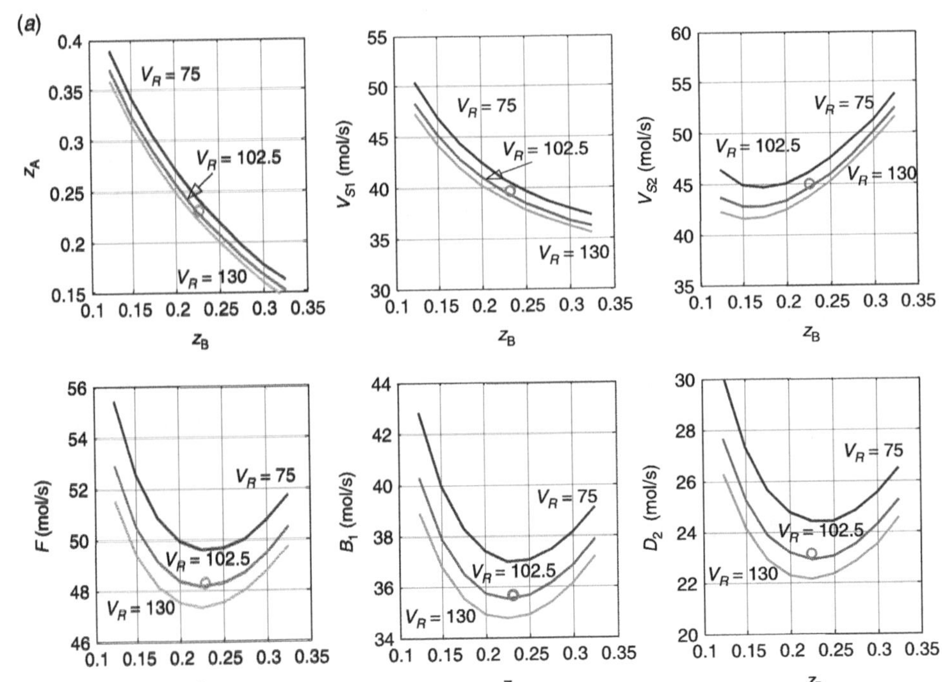

Figure 3.4 Effect of z_B and V_R at 367 K reactor temperature.

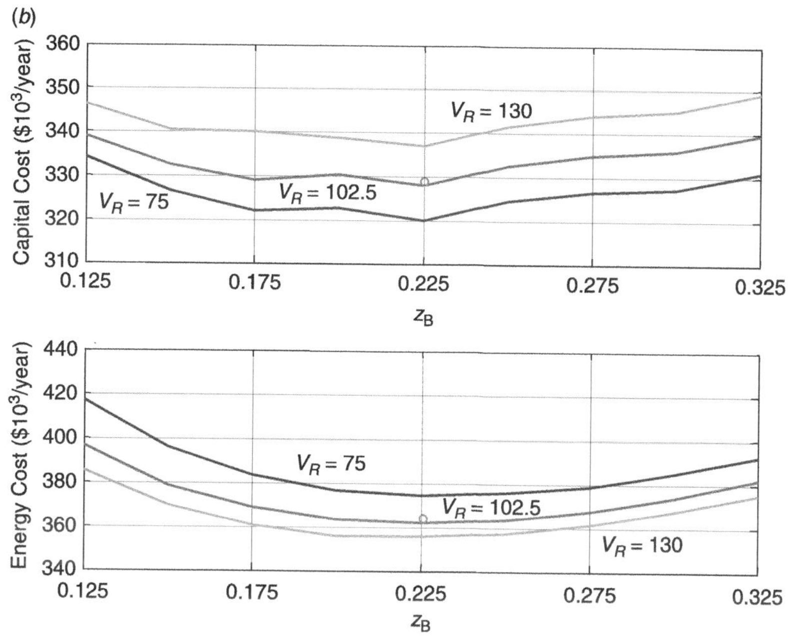

Figure 3.4 (*Continued*).

More details are provided in Figure 3.4*a*, which illustrates how several design parameters are affected by the variation of the optimization variables V_R and z_B. Figure 3.4*b* shows how annual capital cost and annual energy cost vary with these two variables. Both figures are plotted versus z_B for three different values of V_R with fixed $T_R = 367$ K.

Figure 3.4*a* demonstrates that the reactor effluent has a higher concentration of reactant components as the reactor size gets smaller. Thus, there is a need for more vapor boilup in the second column (V_{S2}) and more recycle (D_2) to achieve the desired product specifications with smaller reactor holdups.

Increasing z_B decreases z_A for a given reactor holdup. The vapor boilup in the first column V_{S1} decreases because there is less A to be separated from C. As the value of z_B increases, the vapor boilup in the second column V_{S2} starts to increase because there is more B to take overhead. The process can be operated at any point on one of these curves with the same production rate for the given V_R, but the reactor effluent F, the flowrate of the bottoms stream B_1, and the flowrate of the recycle stream D_2 will vary, as shown in the lower graphs of Figure 3.4*a*. There is a value of z_B that produces a minimum reactor effluent flowrate, and this occurs in the region where the compositions of both reactants are similar because the reaction rate depends on the product of the two reactant concentrations $z_A z_B$. The optimum design point is shown on each graph as an open circle.

Figure 3.4*b* demonstrates that a larger reactor size results in higher capital cost but lower energy cost because of lower vapor boilup requirements for all values of z_B. For all reactor holdups there is a minimum in both the capital cost and the energy cost curves at the specific value of z_B at which reactor concentrations of z_A and z_B are similar. Thus, there should be a minimum in the total cost around this value of z_B because of the tradeoff between increasing capital cost and decreasing energy cost.

Figure 3.5 provides results for the base case using a range of reactor temperatures as the third optimization variable. All of the results use the economical optimal steady-state values of V_R and z_B at each of the given temperatures. The specific reaction rates in the reactor increase with higher temperatures. Figure 3.5*a* shows that increasing temperature decreases the reactor holdup V_R for a specified rate of production of C and D. This means that higher temperature results in lower reactor cost as demonstrated in Figure 3.5*b*. In contrast, the column cost, heat exchanger cost, and energy cost all depend on the vapor boilups in the system. It can be seen from Figure 3.5*a* that there are minimums in the vapor boilup curves at certain reactor temperatures. Because the optimum V_R decreases with increasing reactor temperature, the per-pass conversion in the reactor decreases. Thus, reactor effluent F has more reactant components A and B at higher temperatures. Another reason why the reactant compositions in the reactor effluent increase is that the K_{EQ} decreases with increasing temperature for exothermic reactions. More reactants fed to the columns result in an increase in reflux and vapor boilup to achieve the desired purities of product streams. Therefore, these separation costs increase when the reactor temperature gets high. Because of the decreasing reactor cost, the total capital cost decreases with the increase in temperature until a certain value. However, Figure 3.5*b* shows that after this temperature value, increasing column cost and heat exchanger cost are more dominant and result in increasing total capital cost.

There is a minimum in the TAC curve at the optimum reactor temperature because of the classical tradeoff between reactor cost and separation cost. The optimum reactor

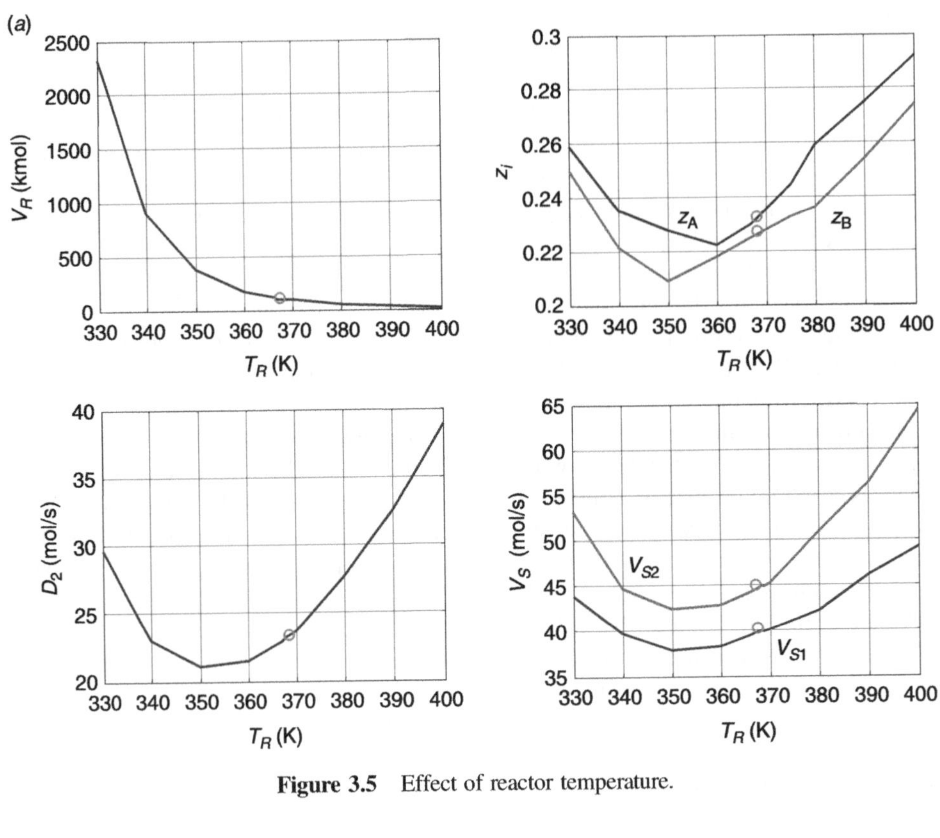

Figure 3.5 Effect of reactor temperature.

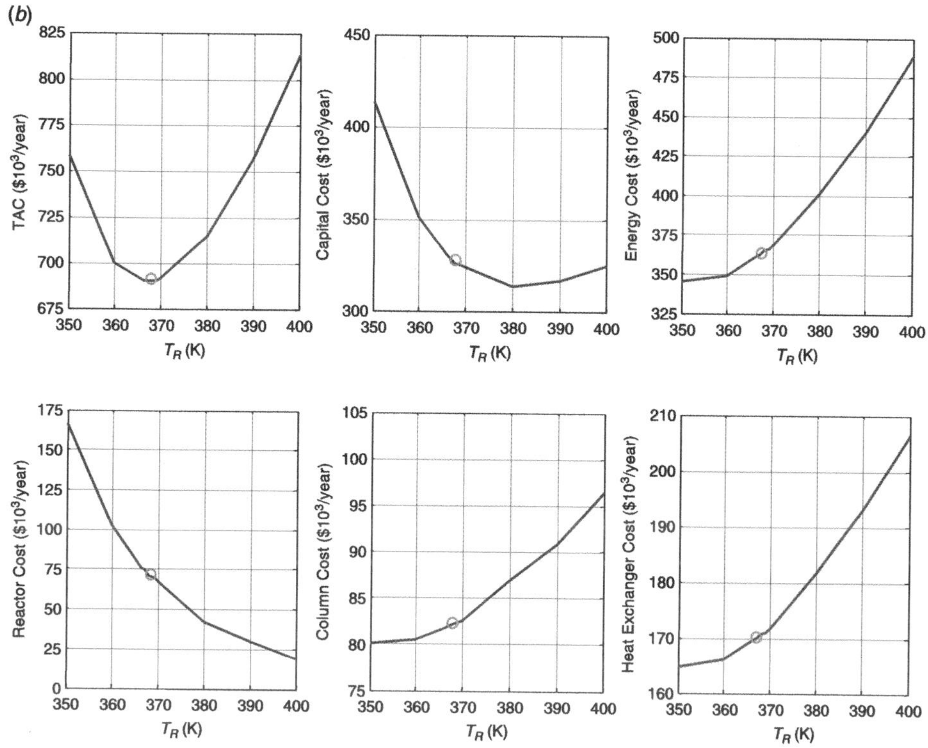

Figure 3.5 *(Continued)*.

temperature for the base case with a $(K_{EQ})_{366}$ value of 2 is 367 K. The optimum values of other optimization variables at the optimum reactor temperature are $V_R = 102.5$ kmol and $z_B = 0.225$. Other important design parameters are the recycle flowrate ($D_2 = 22.89$ mol/s) and the composition of component A in the reactor ($z_A = 0.2296$).

Effect of Changing $(K_{EQ})_{366}$. Figure 3.6 provides the results for different $(K_{EQ})_{366}$ cases. The upper left graph shows that larger values of $(K_{EQ})_{366}$ decrease the TAC of the system. The optimum temperature increases as the value of $(K_{EQ})_{366}$ increases. Reactor holdups decrease for all kinetic cases as the temperatures increase. The lower graphs show that higher values of $(K_{EQ})_{366}$ result in lower values of the recycle stream and vapor boilup. There are minimums in the curves of the recycle stream and the vapor boilup of second column at certain reactor temperatures. For each kinetic case, the shapes of the curves are similar, but they have minimums at different temperatures. For both low and high temperatures, the increase of these flowrates is caused by the higher reactant compositions in the reactor effluent, as shown in Figure 3.4 for base case $(K_{EQ}) = 2$. As mentioned previously, higher temperatures give smaller V_R and lower K_{EQ}, which give higher reactant concentrations. However, low temperatures give small specific reaction rates, and more reactants leave the reactor. Therefore, more vapor boilup is required in the columns to attain product streams with specified purities.

Optimization results for the conventional process for all cases are summarized in Table 3.2. These are the optimum designs in terms of the reactor temperature, reactor holdup, and composition of component B in the reactor. Optimal operating conditions

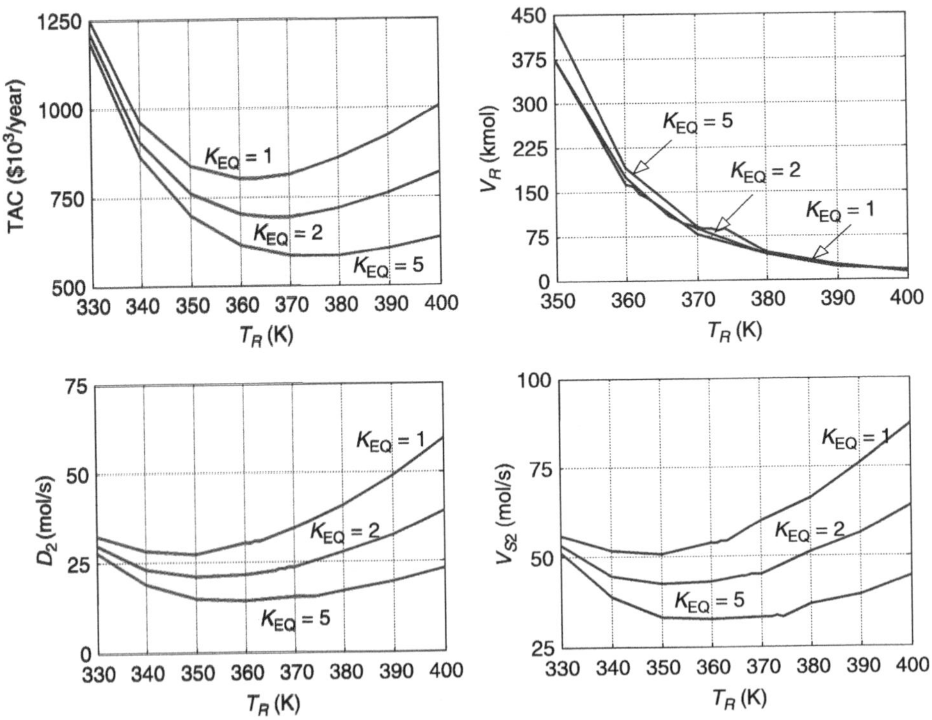

Figure 3.6 Effect of $(K_{EQ})_{366}$.

for the same cases are given in Table 3.3. The column pressures are set using the vapor pressures P^S of pure components at 320 K and liquid compositions in the reflux drum $x_{D,j}$. In a conventional multiunit process, the column temperatures can be set independently so as to optimize column efficiency. The reactor temperature can then be independently adjusted to its optimum. This is not the case in reactive distillation because both reaction and separation are occurring in the same vessel, operating at a single pressure.

TABLE 3.2 Optimization Results of Conventional Process

	$(K_{EQ})_{366}$					
	0.5	1.0	2.0	5.0	10.0	50.0
Design variables						
T_R (K)	356.0	362.0	367.0	373.0	379.0	395.0
V_R (kmol)	222.5	145.0	102.5	82.5	60.0	25.0
z_B	0.275	0.250	0.225	0.175	0.150	0.110
Capital cost ($10³)						
Reactor	358.6	274.7	221.3	193.3	158.6	91.9
Heat exchanger	630.1	565.7	509.6	444.2	407.2	341.3
Column	293.3	268.1	245.8	217.9	202.4	175.0
Tray	9.5	8.3	7.3	6.1	5.5	4.4
Energy cost ($10³/year)	502.9	425.4	361.9	292.8	256.3	195.7
TAC ($10³/year)	933.4	797.7	689.9	580.0	514.2	399.9

TABLE 3.3 Optimal Operation Conditions of Conventional Design

	$(K_{EQ})_{366}$					
	0.5	1.0	2.0	5.0	10.0	50.0
Reactor						
F (mol/s)	66.02	55.81	48.09	40.39	36.65	31.20
V_R (kmol)	222.50	145.00	102.50	82.50	60.00	25.00
T_R (K)	356.00	362.00	367.00	373.00	379.00	395.00
k_F (s^{-1})	0.0025	0.0051	0.0090	0.0174	0.0329	0.1653
k_B (s^{-1})	0.0034	0.0044	0.0046	0.0045	0.0053	0.0091
z_A	0.3005	0.2662	0.2296	0.1947	0.1655	0.1035
z_B	0.2750	0.2500	0.2250	0.1750	0.1500	0.1100
z_C	0.2122	0.2419	0.2727	0.3151	0.3422	0.3933
z_D	0.2122	0.2419	0.2727	0.3151	0.3422	0.3933
Column 1						
B (mol/s)	53.42	43.21	35.49	27.79	24.05	18.60
D (mol/s)	12.60	12.60	12.60	12.60	12.60	12.60
V_S (mol/s)	50.56	44.57	39.56	34.67	31.66	25.62
R (mol/s	37.96	31.97	26.96	22.07	19.06	13.02
P (bar)	2.57	2.57	2.57	2.57	2.57	2.57
$x_{D,A}$	0.0500	0.0500	0.0500	0.0500	0.0500	0.0500
$x_{D,B}$	0.0000	0.0000	0.0000	0.0000	0.0000	0.0000
$x_{D,C}$	0.9500	0.9500	0.9500	0.9500	0.9500	0.9500
$x_{D,D}$	0.0000	0.0000	0.0000	0.0000	0.0000	0.0000
$x_{B,A}$	0.3596	0.3292	0.2933	0.2603	0.2261	0.1397
$x_{B,B}$	0.3399	0.3229	0.3049	0.2543	0.2286	0.1845
$x_{B,C}$	0.0382	0.0354	0.0322	0.0273	0.0238	0.0161
$x_{B,D}$	0.2623	0.3124	0.3695	0.458	0.5215	0.6596
N_T	13	13	13	13	13	13
N_F	6	6	7	7	8	8
D_C (m)	1.09	1.03	0.97	0.90	0.86	0.78
A_R (m^2)	74.36	65.56	58.18	50.99	46.57	37.68
A_C (m^2)	124.11	109.42	97.11	85.10	77.72	62.90
Column 2						
B (mol/s)	12.60	12.60	12.60	12.60	12.60	12.60
D (mol/s)	40.82	30.61	22.89	15.19	11.45	6.00
V_S (mol/s)	66.14	54.15	44.42	33.28	27.82	19.80
R (mol/s)	25.33	23.54	21.54	18.09	16.37	13.80
P (bar)	1.00	0.99	0.98	1.00	1.00	0.97
$x_{D,A}$	0.4706	0.4648	0.4548	0.4762	0.4749	0.4331
$x_{D,B}$	0.4294	0.4352	0.4452	0.4238	0.4251	0.4669
$x_{D,C}$	0.0500	0.0500	0.0500	0.0500	0.0500	0.0500
$x_{D,D}$	0.0500	0.0500	0.0500	0.0500	0.0500	0.0500
$x_{B,A}$	0.0000	0.0000	0.0000	0.0000	0.0000	0.0000
$x_{B,B}$	0.0500	0.0500	0.0500	0.0500	0.0500	0.0500
$x_{B,C}$	0.0000	0.0000	0.0000	0.0000	0.0000	0.0000
$x_{B,D}$	0.9500	0.9500	0.9500	0.9500	0.9500	0.9500
N_T	13	13	13	13	13	13
N_F	8	8	7	7	6	6
D_C (m)	1.58	1.43	1.30	1.12	1.03	0.87
A_R (m^2)	97.29	79.65	65.34	48.96	40.92	29.12
A_C (m^2)	162.38	132.93	109.06	81.71	68.30	48.61

Effects of Conversion and Recycle Impurity. The effect of the design value of conversion is explored by increasing the conversion from 95% to 99%. Increasing conversion means that product purities increase (less unreacted A and B are lost in the product streams), and the optimum values of all of the design parameters change. Figure 3.7 provides results for two values of conversion χ over a range of $(K_{EQ})_{366}$ values. The vapor boilup required to achieve 99% conversion increases, as does the flowrate of the recycle stream D_2. Reactor holdups are almost the same for both conversions. These results illustrate that, as expected, the TAC of the system increases when a higher conversion is required.

Figure 3.8 shows the effect of impurities in the D_2 recycle stream on the TAC for the base case, where $(K_{EQ})_{366} = 2$. The impurities of both components C and D are simultaneously varied from 0.001 to 0.125. The TAC decreases with higher amounts of impurities. The reactor holdup also decreases, and the recycle flowrate and vapor boilup of the second column increase. Although there is an initial large decrease in TAC as the impurities are increased from 0.001, the change of the TAC is not very significant after an impurities value of ~0.05. Therefore, $x_{D2,C} = x_{D2,D} = 0.05$ is used as the base impurity level for other kinetic cases.

3.3.2 Reactive Distillation Process

Figure 3.9 demonstrates the effects of the number of separation stages ($N_S = N_R$) on the economical steady-state design of a reactive distillation column with three different operating pressures. The results are given for the base case $(K_{EQ})_{366} = 2$ with constant reactive trays $N_{RX} = 9$. The graph in the middle shows that the column cost increases as the number of

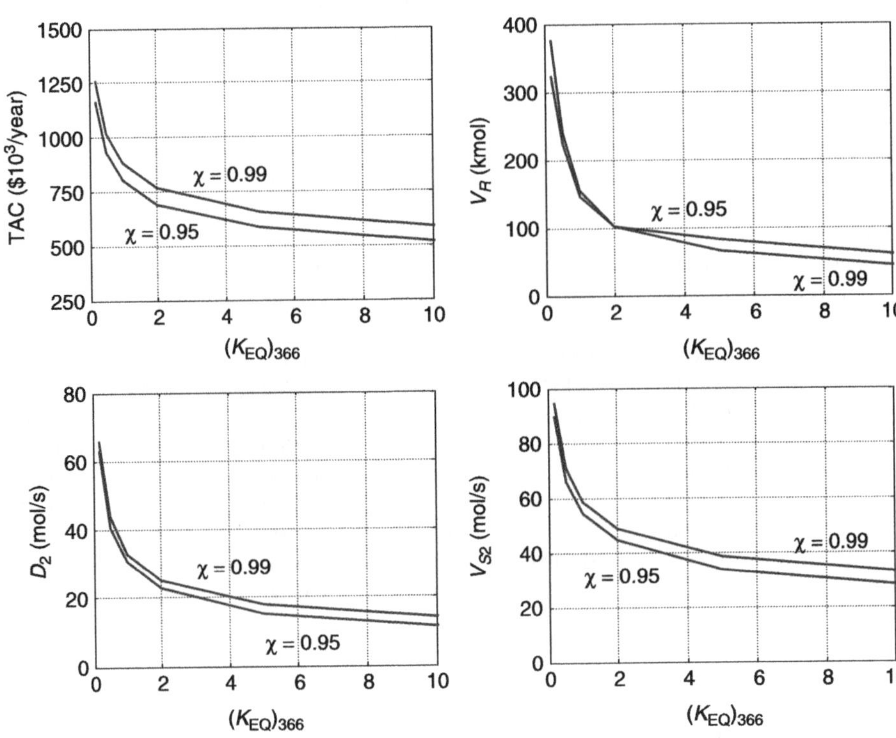

Figure 3.7 Effect of conversion and $(K_{EQ})_{366}$.

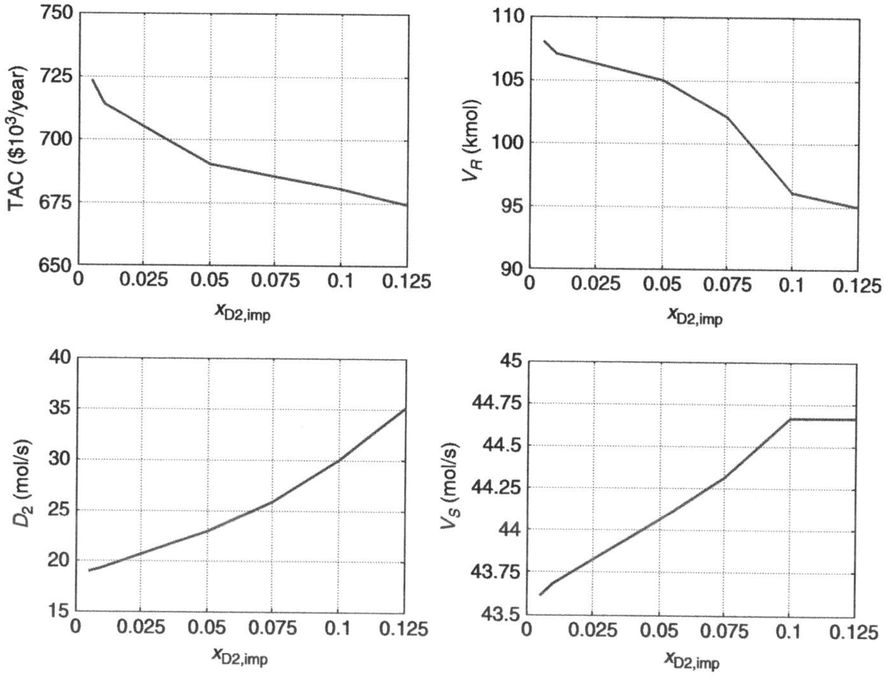

Figure 3.8 Effect of recycle impurities.

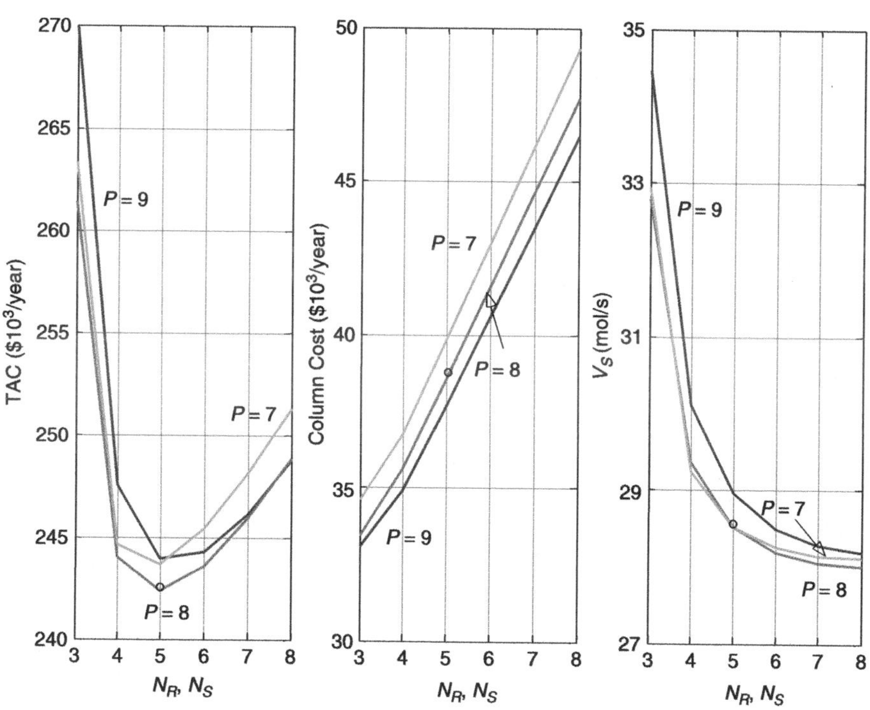

Figure 3.9 Effect of separation trays and pressure on reactive distillation.

separation trays increases. It also shows that decreasing operating pressure results in an increase of column cost for any number of separation trays because decreasing pressure results in lower vapor density, which increases column diameter. The right graph indicates that the higher the number of separation stages, the lower vapor boilup required. More stripping and rectifying stages provide the required separation while using less energy.

Because the temperature of the reactive zone increases with higher column pressure, it reduces the chemical equilibrium constant, which results in an increase of reactant compositions leaving the reactive zone. Therefore, the right graph in Figure 3.9 also shows that operating at higher pressure increases the vapor boilup requirement to meet desired product specifications. The left graph reveals that there is a minimum in the TAC curve at a certain number of separation trays because of the tradeoff between increasing column cost and decreasing costs related to the vapor boilup (energy cost and heat exchanger costs). The overlapping TAC curves for different pressures show that there is an optimum pressure. This occurs because of the competing effects of temperature on the reaction rates and the chemical equilibrium constant. For this case, five stripping and five rectifying trays are optimum for all pressures.

Figure 3.10 studies the effects of the number of reactive trays N_{RX} for the base case with a constant number of separation stages $N_R = N_S = 5$. The graph in the middle shows that increasing the number of reactive trays results in increasing column cost, and higher pressures lead to lower column cost for any values of N_{RX}. The right graph shows the effects of the number of reactive stages on the vapor boilup. As the number of reactive trays increases, the amount of vapor boilup required initially decreases. However, increasing the number of reactive trays above a certain number results in an increase in the required vapor boilup. This interesting and counterintuitive phenomenon was discussed and explained in

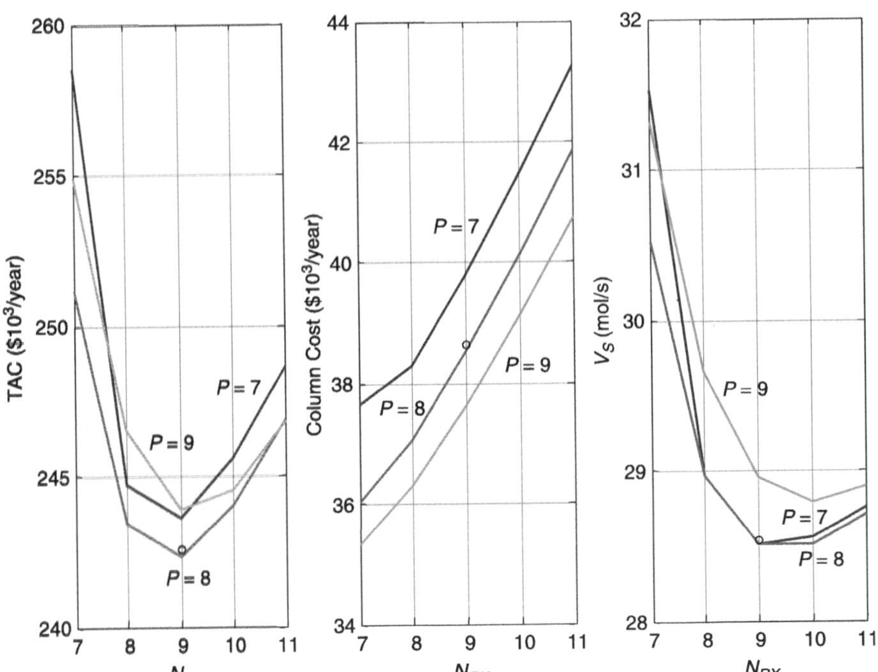

Figure 3.10 Effect of reactive trays and pressure on reactive distillation.

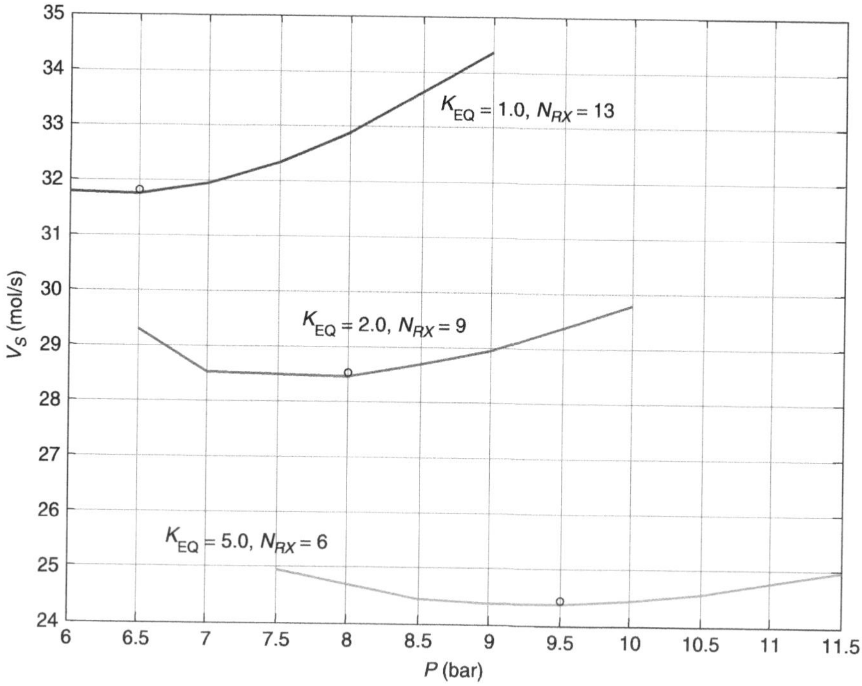

Figure 3.11 Effect of pressure and K_{EQ}.

Chapter 2. Thus, there is an optimum number of reactive trays for each pressure that minimizes vapor boilup, energy cost, and heat exchanger cost. The left graph of Figure 3.10 shows that there is a minimum in the TAC curve at a certain number of reactive trays because of the tradeoff between increasing column cost and decreasing vapor boilup. For this case, nine reactive trays is the optimum number for all pressures. This figure also indicates that the optimum pressure is 8 bar.

Figure 3.11 shows the effect of operating pressure on vapor boilup for three different kinetic cases [$(K_{EQ})_{366} = 1$, 2, and 5] with the number of reactive trays that is optimum for each case. There is a different optimum pressure for each case that minimizes the vapor boilup. The higher $(K_{EQ})_{366}$ cases require less vapor boilup and operate at higher column pressure. The higher equilibrium constant pushes the reaction to the right. The K_{EQ} decreases with increasing temperature for an exothermic reaction. Therefore, the column can operate at higher pressures when $(K_{EQ})_{366}$ is larger. It is possible to conclude from this figure that the sensitivity to pressure increases as the value of $(K_{EQ})_{366}$ decreases. Therefore, operating at the optimum pressure is more important in systems with small values of $(K_{EQ})_{366}$ than in systems with high values. The optimum pressures for the three cases $(K_{EQ})_{366} = 1$, 2 and 5 are 6.5, 8.0, and 9.5 bar, respectively.

Figure 3.12 gives economic optimum steady-state design results for three different $(K_{EQ})_{366}$ cases. For each case the optimum pressure that minimizes vapor boilup is used and the number of reactive trays is changed over a range to find the minimum TAC. The upper graph shows that there is a minimum in the TAC curve at some optimum number of reactive stages. The optimum number of reactive stages decreases with increasing values of $(K_{EQ})_{366}$. Column costs and TAC decrease as the value of $(K_{EQ})_{366}$ becomes larger.

Results for five different kinetic cases are provided in Table 3.4. The table includes the optimum design parameters and costs in terms of optimization variables: number of

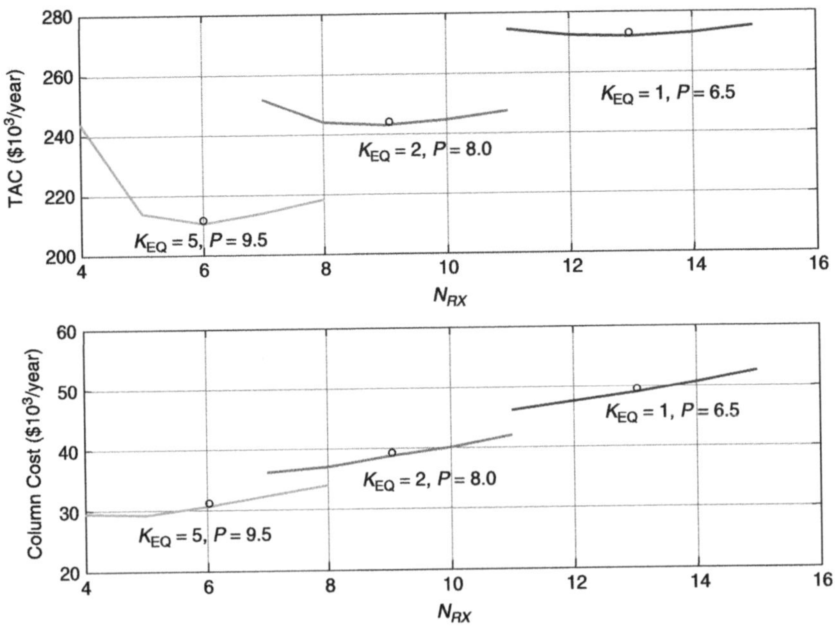

Figure 3.12 Effect of N_{RX} and K_{EQ} on costs.

TABLE 3.4 Optimization Results of Reactive Distillation Design

	$(K_{EQ})_{366}$					
	0.5	1.0	2.0	5.0	10.0	50.0
Design variables						
$N_R = N_S$	6	5	5	5	6	5
N_{RX}	18	13	9	6	4	3
P (bar)	5.5	6.5	8.0	9.5	11.0	16.0
Design temp. (K)						
Base	415.8	423.1	432.8	441.0	449.0	467.6
Bottom reactive	382.0	386.7	394.0	400.7	407.7	422.5
Top reactive	365.1	383.3	393.9	402.6	409.7	428.4
Reflux drum	341.5	346.6	353.2	358.9	363.8	377.3
Design parameters						
N_T	30	23	19	16	16	13
V_S (mol/s)	34.86	31.74	28.91	24.36	21.37	16.24
R (mol/s)	39.50	36.38	33.55	29.00	26.01	20.88
D_C (m)	0.935	0.873	0.805	0.738	0.688	0.590
A_R (m^2)	51.28	46.68	41.93	35.83	31.44	23.89
A_C (m^2)	127.90	120.23	112.30	102.12	94.79	82.21
Capital cost ($\$10^3$)						
Heat exchanger	265.1	252.8	239.7	222.3	209.2	185.6
Column	195.5	146.8	115.5	91.8	85.1	61.1
Tray	6.2	4.3	3.1	2.3	2.1	1.3
Energy cost ($\$10^3$/year)	150.2	136.8	122.8	105.0	92.1	70.0
TAC ($\$10^3$/year)	305.9	271.4	242.3	210.5	190.9	152.7

separation trays, number of reactive trays, and column pressure. The temperatures at the top and bottom of different sections are also given in this table.

Figure 3.13 displays the composition profile of the optimum design for the base case $(K_{EQ})_{366} = 2$. The highest composition of reactant A is at the bottom of the reactive zone where it is fed. Reactant B has its highest composition at the top of the reactive zone, which is also its feed tray. The composition of A decreases up through the reactive zone, but the composition of product C increases. The reverse occurs for reactant B and product D through the reactive zone. The rest of the column operates as a separation unit. Thus, the composition of heavy product D increases down through the stripping section, and the composition of light product C increases up through the rectifying section.

Figure 3.14 shows the temperature profiles of the optimum designs for three different kinetic cases: $(K_{EQ})_{366} = 1$, 2, and 5. The temperature profiles of each case are similar with higher temperatures for higher values of $(K_{EQ})_{366}$ because of the higher optimum pressures. There are fairly significant temperature breaks around tray 4 for all cases, and these tray temperatures can be used in control schemes to infer bottoms purity. A similar break occurs at different tray numbers near the top of the column for each case. These could be used to infer distillate purity. For all kinetic cases, the temperatures show little change in the reactive zone.

Figure 3.15 illustrates the effect of conversion on the TAC for different pressures. The conversion is increased to 99%, and the optimal results at each pressure are given for base kinetic case $(K_{EQ})_{366} = 2$. To achieve the required purity, the optimum number of separation stages is $N_S = N_R = 7$ in this instance, which is 2 more than in the 95% conversion system. The optimum number of reactive stages is 13, which is 4 more than the base case.

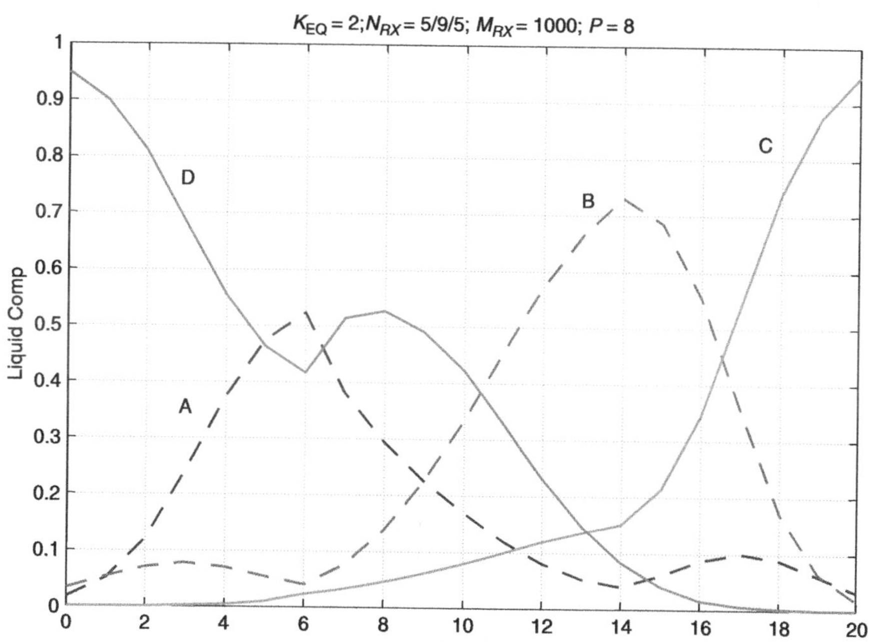

Figure 3.13 Base case composition profiles.

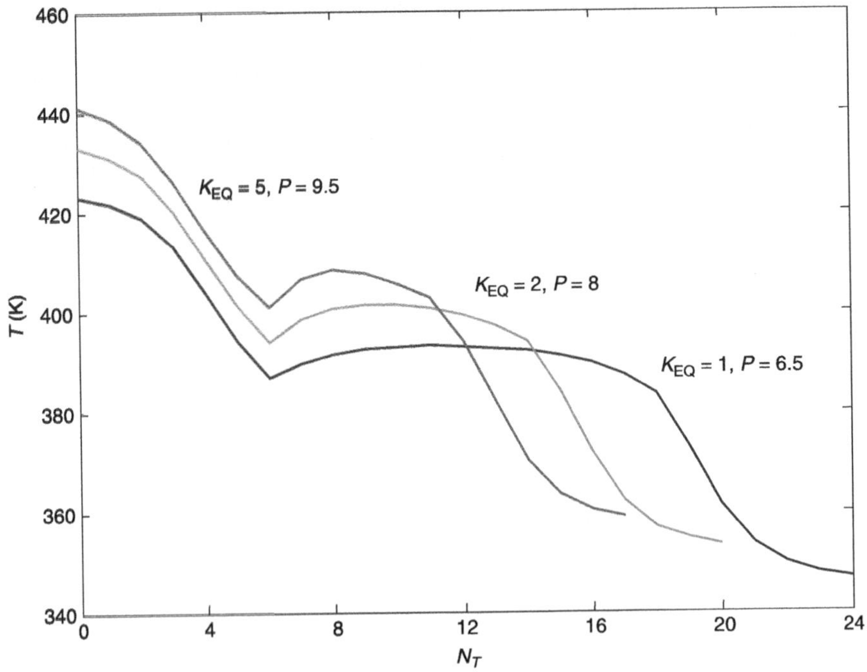

Figure 3.14 Temperature profiles.

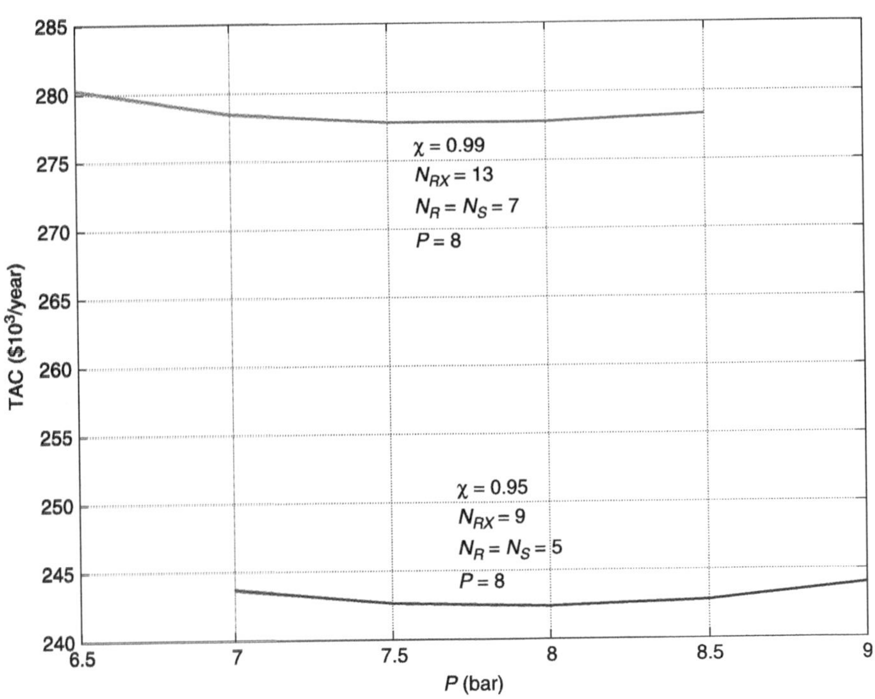

Figure 3.15 Effect of conversion on reactive distillation.

TABLE 3.5 Economic Comparison of Two Different Process Flowsheets

	$(K_{EQ})_{366}$					
	0.5	1.0	2.0	5.0	10.0	50.0
Capital cost ($\$10^3$)						
Conventional design	1291.5	1116.8	984.0	861.6	773.6	612.6
Reactive column design	466.8	403.9	358.4	316.4	296.4	248.1
Energy cost ($\$10^3$/year)						
Conventional design	502.9	425.4	361.9	292.8	256.3	195.7
Reactive column design	150.2	136.8	122.8	105.0	92.1	70.0
TAC ($\$10^3$/year)						
Conventional design	933.4	797.7	689.9	580.0	514.2	399.9
Reactive column design	305.9	271.4	242.3	210.5	190.9	152.7

The optimum pressure is the same (8 bar) as the 95% conversion case. These results indicate that the TAC of the system increases when a higher conversion is required.

3.3.3 Comparisons

Direct comparisons of the conventional multiunit process with the reactive column process at their economic optimum steady-state designs are given in Table 3.5 for five different kinetic cases. The results indicate that the TACs of both design configurations decrease as the value of $(K_{EQ})_{366}$ increases. The results also show that the reactive distillation column configuration has lower capital cost and energy cost than the conventional configuration for all kinetic cases. These costs result in lower TAC for the reactive distillation columns compared to the reactor/column/recycle systems.

3.4 RESULTS FOR TEMPERATURE-DEPENDENT RELATIVE VOLATILITIES

In the previous section, the optimum economic steady-state designs of reactive distillation columns were quantitatively compared with conventional multiunit systems for a wide range of chemical equilibrium constants. Relative volatilities ($\alpha = 2$) were assumed constant. Reactive distillation was shown to be much less expensive than the conventional process. In this section we explore how temperature-dependent relative volatilities affect the designs of these two systems.

A fundamental difference between the two flowsheets is the ability in the conventional process to adjust reactor temperature and distillation column temperatures completely independently, which is not possible in the reactive distillation process. In the conventional system, the reactor temperature can be set at an optimum value and distillation temperatures can be independently set at their optimum values by adjusting column pressures. In reactive distillation, these temperatures are not independent because only one pressure can be set in the vessel. Therefore, the design of a reactive distillation requires a tradeoff between temperatures conducive for reaction (kinetics and equilibrium constants) and temperatures favorable for vapor–liquid separation. The temperature dependency of the relative volatilities will illustrate this important difference between the two processes.

Results presented below show that, when relative volatilities are temperature dependent and decrease significantly as temperatures approach those required for reasonable reaction rates, the reactive distillation process becomes more expensive than the conventional flowsheet. A single value of the chemical equilibrium constant is used in this section: $(K_{EQ})_{366} = 2$.

3.4.1 Relative Volatilities

For an ideal mixture, Raoult's law relates the partial pressure of a component in the vapor phase to its liquid-phase composition and vapor pressure. The vapor pressure P^S of a component is a function of temperature and can be calculated from a two-parameter Antoine equation over a limited temperature range:

$$\ln P_j^S = A_j - \frac{B_j}{T} \tag{3.42}$$

where A_j and B_j are constants for component j. A component having a higher vapor pressure at a given temperature than another component is more volatile. Relative volatility α_{ij} for an ideal mixture is equal to the ratio of the vapor pressure of component i to the vapor pressure of component j.

$$\alpha_{ij} = \frac{P_i^S}{P_j^S} \tag{3.43}$$

If the vapor pressures of an ideal mixture have the same temperature dependence (B_j constants are the same for all components), the relative volatilities are constant.

To study the effect of the temperature dependency of relative volatilities on the steady-state design of both flowsheets, the relative volatilities between adjacent components are assumed to be 2 at a temperature of 320 K but become smaller at higher temperatures. The 320 K temperature is chosen because it corresponds to the normal reflux-drum temperature that can be achieved in a distillation column while using cooling water in the condenser. The distillation columns of the conventional system can operate at this temperature and the same pressures as used in the base case conditions.

However, the relative volatilities of all components are now assumed to be temperature dependent. This is achieved by reducing the relative volatility between all adjacent components at higher temperatures. The selected reference temperature is 390 K because the average reactive zone temperature for the base case is about this value. Parameter α_{390} quantifies this temperature dependency. It is the volatility of the adjacent component at 390 K. Remember that at 320 K, the volatility of all adjacent components is 2.

Figure 3.16 provides vapor pressure lines for two cases. The left graph is the base case where all of the relative volatilities are constant at 2 and are not functions of temperature. The lines are parallel because the B_j coefficients are the same for all components. The relative volatilities are independent of temperature, so they are constant throughout the distillation column, despite having temperatures that change from tray to tray in the column and having different column pressures.

The right graph in Figure 3.16 gives vapor pressure curves for the case in which $\alpha_{390} = 0.95$. The lines are not parallel but get closer as the temperature increases. The vapor pressure lines actually cross each other at a certain temperature for this $\alpha_{390} = 0.95$ case. This means that the relative volatilities between adjacent components will have switched

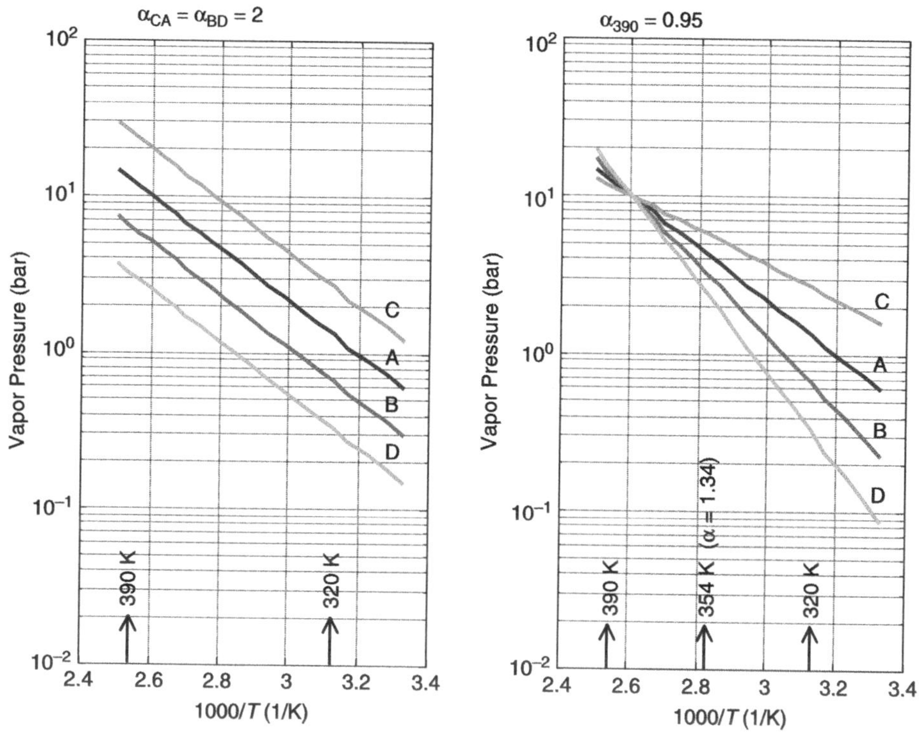

Figure 3.16 Relative volatilities.

at some temperature that is slightly lower than 390 K. Operating the column in this temperature range would be infeasible because the reactants would tend to go out the ends of the column and the products would tend to stay in the reactive zone. Table 3.6 gives the values of all of the vapor pressure constants for the six cases considered.

In the conventional flowsheet, the relative volatilities will be slightly lower as we move down the column because of the increase in tray temperatures. In the reactive distillation

TABLE 3.6 Vapor Pressure Constants for Temperature-Dependent α_{ij} Case

α_{390}	Const.	A	B	C	D
0.95	A_{VP}	12.34	15.80	8.89	19.26
	B_{VP}	3862.00	5189.23	2534.77	6516.46
1.00	A_{VP}	12.34	15.52	9.18	18.68
	B_{VP}	3862.00	5097.78	2626.22	6333.56
1.10	A_{VP}	12.34	14.99	9.71	17.62
	B_{VP}	3862.00	4927.86	2796.14	5993.72
1.25	A_{VP}	12.34	14.27	10.42	16.20
	B_{VP}	3862.00	4699.95	3024.05	5537.90
1.50	A_{VP}	12.34	13.26	11.44	14.17
	B_{VP}	3862.00	4374.90	3349.10	4887.80
2.00	A_{VP}	12.34	11.65	13.04	10.96
	B_{VP}	3862.00	3862.00	3862.00	3862.00

column, it will not be possible to operate at a pressure that gives favorable reaction conditions (~390 K) because these temperatures would make it impossible to maintain the required separation.

3.4.2 Optimum Steady-State Designs

The design objective is to obtain 95% conversion for fixed fresh feed flowrates ($F_{0\mathrm{A}}$ and $F_{0\mathrm{B}}$) of 12.6 mol/s and product purities of both components C and D of 95 mol%. The assumptions, specifications, and steady-state design procedures used for both process flowsheets are the same as used earlier in this chapter. There are three optimization variables for the conventional multiunit process: molar holdup in the reactor V_R, composition of reactant B in the reactor z_B, and reactor temperature T_R.

For reactive distillation in the constant relative volatility case, there are three optimization variables: column pressure P, number of reactive trays N_{RX}, and number of separating (stripping/rectifying) trays ($N_S = N_R$). For the constant relative volatility case, the numbers of stripping and rectifying trays are assumed equal because the relative volatilities between the components being separated in the two sections are the same. However, this simplifying assumption cannot be applied for the temperature-dependent relative volatility case because the temperatures in the stripping and rectifying sections are different. Therefore, one more optimization variable is needed in this case: the optimum values of both the number of stripping N_S and the number of rectifying trays N_R must be determined.

Conventional Process. In each of the distillation columns of the conventional flowsheet, geometric average relative volatilities are calculated from the reflux-drum and base temperatures. The operating pressure is fixed by specifying the reflux-drum temperature at 320 K so that cooling water can be used in the condenser. The vapor pressures of the pure components and liquid compositions in the reflux drum and column base are known, so the relative volatilities can be calculated at both locations and averaged.

$$(\alpha)_{\mathrm{geometric}} = \sqrt{\alpha_{\mathrm{reflux}}\alpha_{\mathrm{base}}} \tag{3.44}$$

Table 3.7 provides the optimum design results for the conventional process over a range of temperature-dependent relative volatilities. Because the column relative volatilities are only slightly lower than those of the constant relative volatility case, we assume that the three design optimization variables are the same as in the constant relative volatility case. The slightly lower relative volatilities produce small increases in the number of trays, the reflux ratios, and the vapor boilups in both columns. There is a small increase in the recycle flowrate (D_2).

Reflux-drum temperatures are constant for both columns for different values of α_{390}. However, base temperatures actually decrease slightly as the reference relative volatilities decrease from case to case. This occurs because of the way we modified the vapor pressures. The vapor pressure constants for component A were held constant. The B_j coefficients of the other three components were modified to make the relative volatilities decrease with temperature. This produces a higher vapor pressure for component D at higher temperatures than it did in the constant relative volatility situation.

Figure 3.17 shows how several variables change in the two distillation columns as the α_{390} parameter is changed. The top two graphs give the temperatures in the base (T_B)

TABLE 3.7 Optimization Results for Conventional Process (Temperature-Dependent α_{ij})

Design Parameters	α_{390}					
	0.95	1.00	1.10	1.25	1.50	2.00
Column 1						
N_{T1}	16	16	16	15	15	13
V_{S1} (mol/s)	47.40	47.13	46.55	45.58	43.17	39.56
R_1 (mol/s)	34.80	34.53	33.95	32.98	30.57	26.96
D_{C1} (m)	1.06	1.05	1.05	1.04	1.01	0.97
A_{R1} (m^2)	69.72	69.30	68.47	67.04	63.49	58.18
A_{C1} (m^2)	116.36	115.71	114.28	111.89	105.98	97.11
Column 2						
N_{T2}	16	16	16	15	15	13
V_{S2} (mol/s)	51.55	51.17	50.52	49.57	47.98	44.42
R_2 (mol/s)	28.66	28.28	27.63	26.69	25.10	21.54
D_{C2} (m)	1.38	1.38	1.37	1.36	1.33	1.30
A_{R2} (m^2)	75.82	75.27	74.30	72.92	70.58	65.34
A_{C2} (m^2)	126.55	125.62	124.01	121.70	117.80	109.06
Capital cost ($\$10^3$)						
Reactor	221.3	221.3	221.3	221.3	221.3	221.3
Heat exchanger	567.1	564.7	560.0	552.8	537.5	509.6
Column	313.8	312.7	310.6	291.9	285.4	245.8
Tray	10.1	10.0	9.9	9.1	8.9	7.3
Energy cost ($\$10^3$/year)	426.4	423.6	418.3	410.0	392.8	361.9
TAC ($\$10^3$/year)	797.2	793.2	785.6	768.4	743.8	689.9

[a]For all α_{390} cases, $V_R = 102.5$ kmol, $P_1 = 2.57$ bar, and $P_2 = 1.03$ bar.

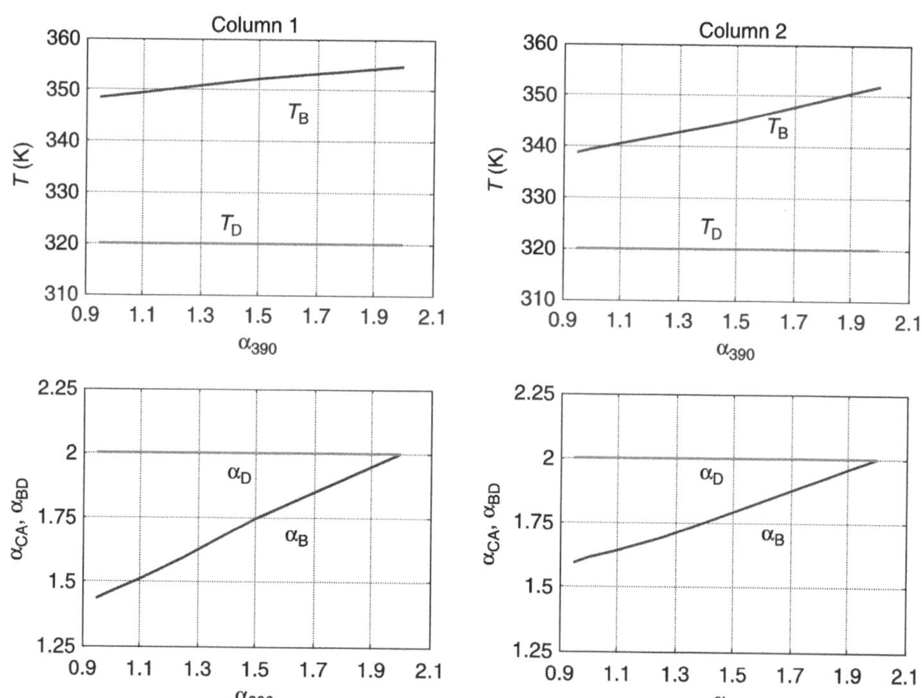

Figure 3.17 Conventional process.

and in the reflux drum (T_D) of the two columns. Reflux-drum temperatures are constant at 320 K, so the relative volatilities at the top of the columns are equal to 2 (bottom two graphs, Fig. 3.17). However, in the base of the column the higher temperatures result in somewhat smaller relative volatilities as the α_{390} parameter decreases.

Reactive Distillation. Figure 3.18 and Table 3.8 give optimum design results for the reactive distillation process for a range of temperature-dependent relative volatilities. As the α_{390} parameter decreases, the optimum pressure decreases. This occurs because lower pressure helps the vapor–liquid equilibrium because it lowers temperatures and hence increases relative volatilities. However, a lower temperature is unfavorable for reaction because the reaction rates are too small. The result is a rapid increase in the required number of reactive trays.

Note that the optimum number of stripping trays is larger than the optimum number of rectifying trays. This is caused by the higher temperatures in the lower part of the column, which means lower relative volatilities.

Although cases are considered in which the reference α_{390} approaches and even drops below 1, the actual relative volatilities in the reactive column do not get too close to 1. If they did, the required separation would become impossible. Figure 3.19 illustrates this point. The average relative volatility through the reactive zone is 2 for the base case ($\alpha_{390} = 2$), because it is constant. However, the average relative volatilities in the reactive zone become smaller as the value of α_{390} decreases. The decrease of the column temperature prevents the relative volatilities from becoming too small. As shown in the top right

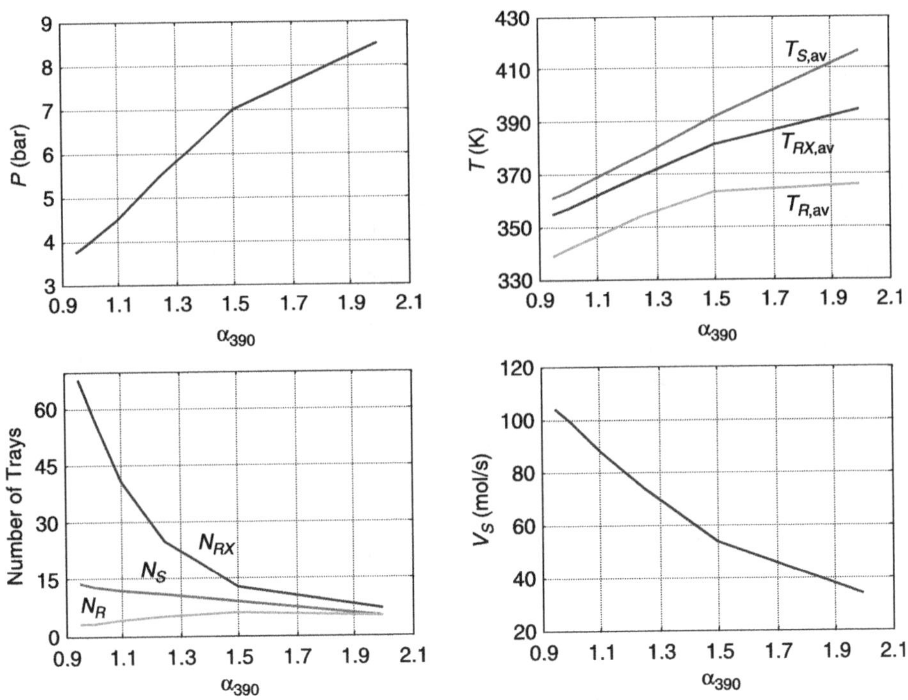

Figure 3.18 Reactive distillation process.

TABLE 3.8 Optimization Results of Reactive Distillation Design (Temperature-Dependent α_{ij})

	α_{390}					
	0.95	1.00	1.10	1.25	1.50	2.00
Design variables						
N_S	14	13	12	11	9	5
N_{RX}	68	58	41	25	13	9
N_R	3	3	4	5	6	5
P (bar)	3.75	4.00	4.50	5.50	7.00	8.0
Design temp. (K)						
Base	363.0	365.8	371.4	381.4	398.8	432.8
Reactive zone (avg.)	354.4	356.6	361.4	369.3	380.7	394.2
Reflux drum	335.9	338.1	341.8	347.9	353.7	353.2
Design parameters						
N_T	85	74	57	41	28	19
V_S (mol/s)	99.15	94.03	83.58	68.50	48.82	28.51
R (mol/s)	103.78	98.66	88.22	73.14	53.46	33.14
D_C (m)	1.53	1.48	1.37	1.21	1.00	0.805
A_R (m^2)	145.83	138.30	122.93	100.75	71.81	41.93
A_C (m^2)	285.71	273.14	247.49	210.48	162.18	112.3
Capital cost (\$10^3)						
Heat exchanger	474.1	459.5	428.9	382.4	316.7	239.7
Column	763.0	656.4	490.9	329.3	198.8	115.5
Tray	37.7	31.0	21.2	12.5	6.4	3.1
Energy cost (\$10^3/year)	427.2	405.2	360.2	295.2	210.4	122.8
TAC (\$10^3/year)	852.2	787.5	673.8	536.6	384.4	242.3

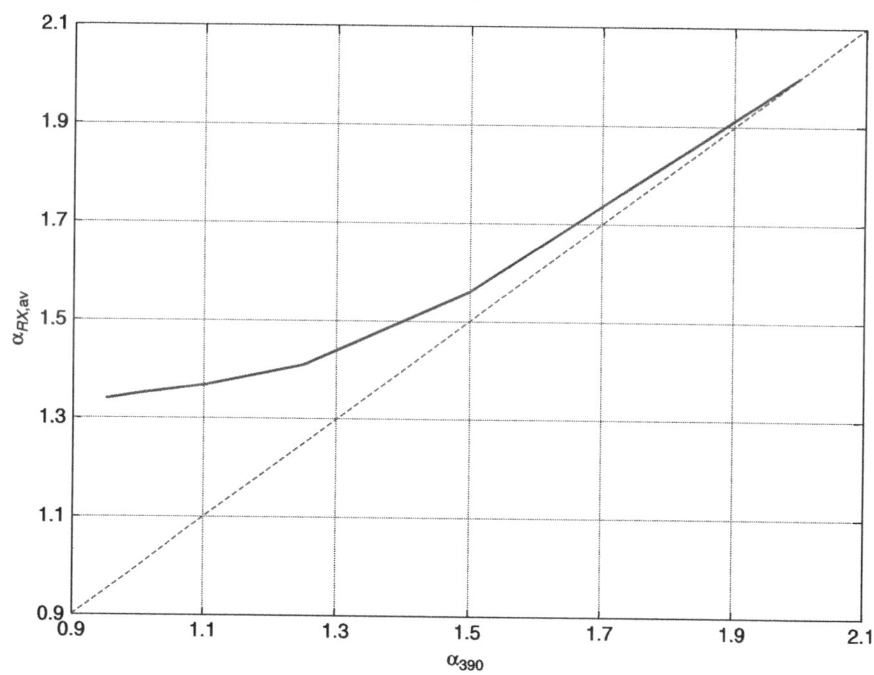

Figure 3.19 Reactive distillation process.

graph in Figure 3.18, the average temperature in the reactive zone $T_{RX,av}$ for the $\alpha_{390} = 0.95$ case is 354 K. This gives relative volatilities of 1.34. Remember that, at any temperature, all of the relative volatilities are the same among the adjacent components.

Results for different α_{390} cases are displayed in Table 3.8. These are the optimum designs in terms of the four design optimization variables: column pressure and the number of stripping, rectifying, and reactive trays. Reducing the relative volatility increases both capital and energy costs.

Comparison. The top graph in Figure 3.20 gives a direct comparison of the TACs of both processes for the temperature-dependent cases. There is a small increase in TAC for the conventional multiunit process as the relative volatilities decrease, but there is a very rapid increase for the reactive distillation process.

The lower graph in Figure 3.20 shows how the reactor temperature in the conventional process and the average temperature in the reactive section of the reactive distillation column change as temperature-dependence changes. At base case conditions with $\alpha_{390} = 2$ (no temperature dependence), the optimum reactor temperature in the conventional process is 367 K, and we assumed this did not change for other values of α_{390}. At base case conditions with $\alpha_{390} = 2$ (no temperature dependence), the optimum average reaction zone temperature in the reactive distillation column is 394 K. This is higher than the conventional reactor temperature. Because the reactive distillation column is removing products from the reaction zone, a smaller chemical equilibrium constant can be tolerated (higher temperature). The conventional process, in which no products are removed from the reaction zone, is favored by a lower reactor temperature because it gives a higher chemical equilibrium constant.

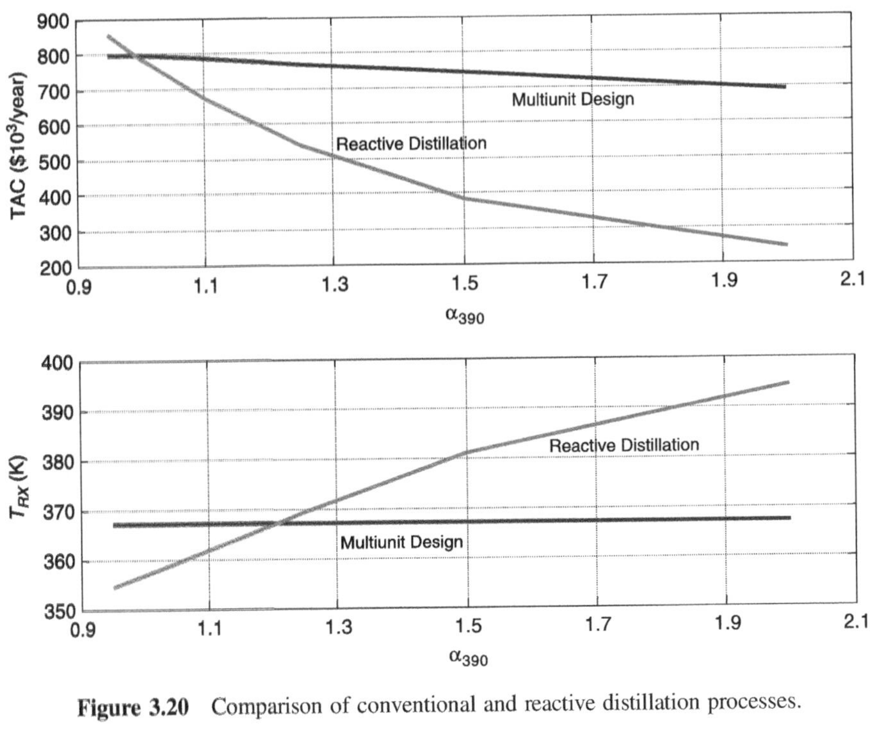

Figure 3.20 Comparison of conventional and reactive distillation processes.

However, as the value of α_{390} decreases, the optimum average reaction zone temperature in the reactive distillation decreases because the separation is becoming more difficult as temperature increases.

These results clearly illustrate the fundamental difference between a conventional process and a reactive distillation process. In a conventional process, the temperature for reaction and the temperatures for separation can be independently set. This is not true for reactive distillation. Therefore, reactive distillation is not economical for systems in which the temperatures for reaction and for separation are not similar.

3.4.3 Real Chemical Systems

Some confirmation of the results presented above for the ideal quaternary chemical system can be seen in several articles that compare conventional processes with reactive distillation for real chemical systems. Two articles were found that make such a comparison and provide sufficient detail about process conditions.

ETBE Process. An article by Sneesby et al.[1] provides a description of the conventional ETBE process. There are two reactors in series, the first operating at 90 °C and the second at 50–60 °C. The lower temperature in the second reactor gives a higher equilibrium constant because the reaction of ethanol and isobutene to produce ETBE is exothermic.

The temperature in the reactive zone of the ETBE reactive distillation process is about 70 °C. Thus, the conventional and reactive distillation processes have similar temperatures. Therefore, we would expect the reactive distillation process to be more economical, which is indeed the case.

Toluene Disproportionation Process. A article by Stitt[2] compares a conventional process to produce benzene from toluene with a reactive distillation process. Several steady-state economic indicators are used to show that the reactive distillation process "does not prove to be a fruitful development opportunity ... due to economic considerations."

The reactor temperature in the conventional process is 400 °C (vapor-phase reaction). The separation section consists of several distillation columns operating at normal temperature levels for benzene/toluene separation.

The reactive distillation column has a temperature of about 280 °C in the reactive zone. To achieve this temperature, the column must operate at 30 bar.

The difference between the conventional reactor temperature and the reactive distillation temperature indicates that the optimum temperature for reaction is different than the optimum temperature for separation. Therefore, we would expect that the reactive distillation column would not be superior to the conventional process, which is precisely what Stitt found.

[1]M. G. Sneesby, M. O. Tadé, R. Datta, and T. N. Smith, ETBE synthesis via reactive distillation. 1. Steady-state simulation and design aspects, *Ind. Eng. Chem. Res.* **36**, 1855–1869 (1997).

[2]E. H. Stitt, Reactive distillation for toluene disproportionation: A technical and economical evaluation, *Chem. Eng. Sci.* **57**, 1537–1543 (2002).

3.5 CONCLUSION

The economics of reactive distillation have been quantitatively compared with those of conventional multiunit processes with separate reaction and separation sections. With favorable chemistry and relative volatilities, reactive distillation is less expensive than a conventional process. However, if a mismatch occurs in the temperatures conducive for good reaction kinetics and the temperatures conducive for good vapor–liquid equilibrium, reactive distillation is not an attractive alternative.

CHAPTER 4

NEAT OPERATION VERSUS USING EXCESS REACTANT

The reactive distillation columns considered in previous chapters were all operated in neat mode. The two reactants are fed in exactly the correct amounts to satisfy the stoichiometry of the reaction. The control system must be able to detect any imbalance, which will inevitably result in a gradual buildup of one of the reactants and a loss of conversion and product purities.

An alternative to operating neat is to operate the reactive column with an excess of one of the reactants. This eliminates the need to perfectly balance the reactant feeds in the reactive column itself, which makes the control of the reactive column easier. However, this mode of operation may have the disadvantage of requiring the recovery and recycle of the reactant that is in excess. The flowsheet typically consists of a two-column system: a reactive column and a recovery column.

We would expect that a single reactive column that is operated neat will have lower capital investment and energy costs than a two-column system. The purpose of this chapter is to give a quantitative comparison of these two alternative processes.

The neat one-column system may require a composition analyzer to detect the inventory of one of the reactants so that the fresh feed can be adjusted. Because composition analyzers are expensive and sometimes unreliable, the economic advantages of the one-column system need to be assessed.

The process considered is the ideal quaternary system with the reversible reaction $A + B \Leftrightarrow C + D$. The relative volatilities are favorable for reactive distillation, that is, the reactants are intermediate boilers between light product C and heavy product D. Relative volatilities are constant at 2 between adjacent components.

We study two-column systems in which an excess of either A or B is fed to the reactive column. If there is an excess of B, the distillate from the reactive column is product C. The bottoms stream is a mixture of components B and D, which are separated in the second column. The bottom of the recovery column B_2 is product D. The distillate of the second column D_2, which is mostly component B but also contains some component D, is recycled back to the first column and mixed with fresh feedstream F_{0B}.

If there is an excess of A, the bottom of the reactive column is product D. The distillate stream is a mixture of components A and C, which are separated in the second column. The distillate of the recovery column is product C. The bottom of the second column, which is mostly component A but also contains some component C, is recycled back to the first column and mixed with fresh feedstream F_{0A}.

4.1 INTRODUCTION

Many industrial reactive distillation systems do not use stoichiometric amounts of reactants. An excess (10–20% above the stoichiometric amount) of one of the reactants is fed to the reactive column. There may be kinetic reasons for using an excess in some systems. These include suppressing undesirable side reactions, reducing catalyst requirements, and increasing conversion. However, even in the absence of kinetic reasons, the use of an excess of one of the reactants makes the control problem easier because the fresh feed flowrates of the components do not have to be precisely balanced in the reactive column. Achieving this exact balance may require the use of expensive and high maintenance on-line composition analyzers in some systems. In addition, the variability of product quality may be larger in the neat operation process because there are fewer manipulated variables available and there is only one column to contain disturbances.

One example of an industrial system with excess reactant is the ETBE system. A 10–20% excess of ethanol is fed to the column. If the excess ethanol can be included in the ETBE bottoms product from the column and blended into gasoline with the ETBE, there may be no economic penalty and no need for recovering the excess ethanol. However, in other systems, the excess reactant must be removed from the product and recycled. This involves an additional separation step, so capital investment and energy costs are increased.

In the two-column process, the recovery column acts conceptually as an on-line analyzer: a higher recycle flowrate means that more of the excess reactant is leaving the reactive column, so the fresh feed flowrate of that reactant must be decreased. An effective control structure for the two-column system is to flow control the sum of the recycle stream and the appropriate fresh feedstream. When the recycle flowrate increases, the fresh feed flowrate is decreased to keep the total constant. Thus, the scheme changes the fresh feed flowrate to accommodate changes in the component inventory of the reactant. From a steady-state economic perspective, the two alternative processes (one column and two columns) have different capital investments and different operating costs. From a dynamic perspective, the two processes show different dynamic behavior and require different control structures. The economic design differences are quantitatively explored in this chapter. The control of these types of systems is discussed in Chapter 11.

4.2 NEAT REACTIVE COLUMN

The process considered is the ideal quaternary system with the reversible exothermic reaction occurring on the reactive trays $A + B \Leftrightarrow C + D$ with constant and favorable relative volatilities.

$$\alpha_C > \alpha_A > \alpha_B > \alpha_D$$

Reactants A and B are intermediate boiling between the products. Therefore, fresh feedstream F_{0A} containing reactant A is fed at the bottom of the reactive zone, and fresh feedstream F_{0B} containing reactant B is fed at the top of the reactive zone. The reactive section contains N_{RX} trays. The rectifying section above the reactive section contains N_R trays and the stripping section below the reactive section contains N_S trays. The optimum economic design found in Chapter 3 for the single reactive column operating in neat mode is used (see Table 4.4).

Figure 4.1 shows the one-column reactive distillation column flowsheet with stream information and equipment sizes. Specification products are produced at both ends of the column. Conversion is 95%, and product purities are 95 mol%. Fresh feeds are 12.6 mol/s. Note that the production rate of both products is 12.6 mol/s with equal amounts of the two reactants lost in the two products.

There are five stripping trays, five rectifying trays, and nine reactive trays. The column operating pressure is 8 bar, and the holdup on the reactive trays is 1000 mol. The vapor boilup is 28.91 mol/s, and the column diameter is 0.805 m. Figures 4.2 and 4.3 give composition and temperature profiles in the column, respectively.

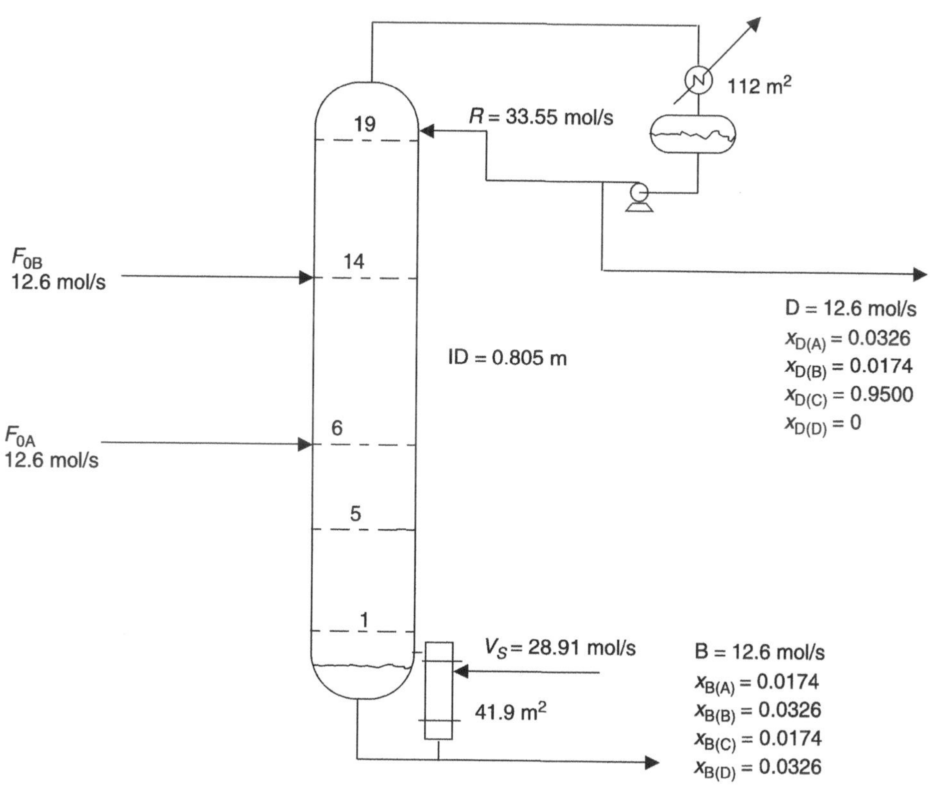

Figure 4.1 Reactive column for neat operation.

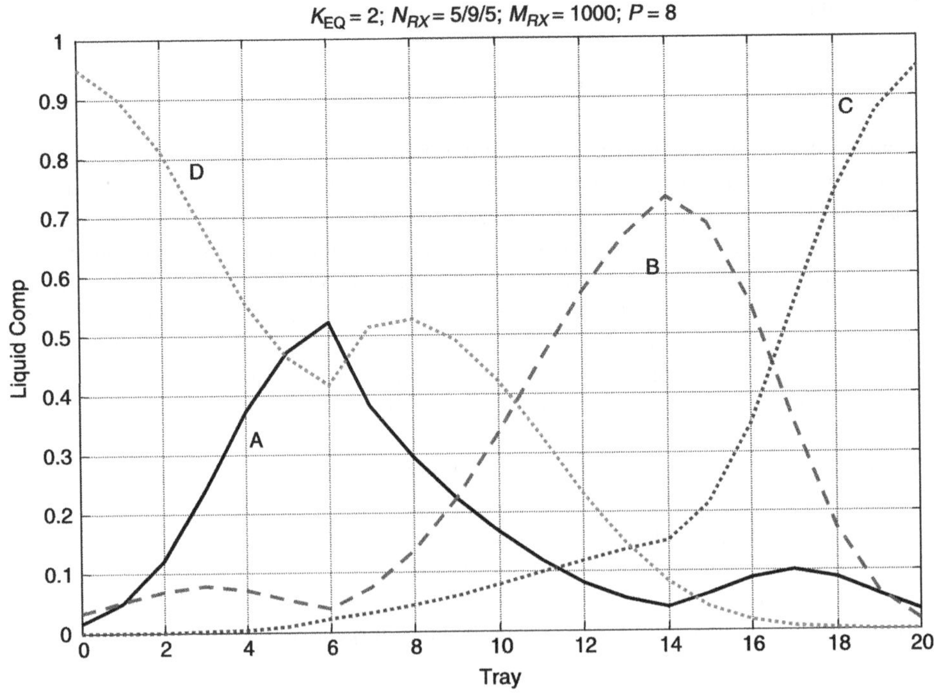

Figure 4.2 Composition profiles for neat operation.

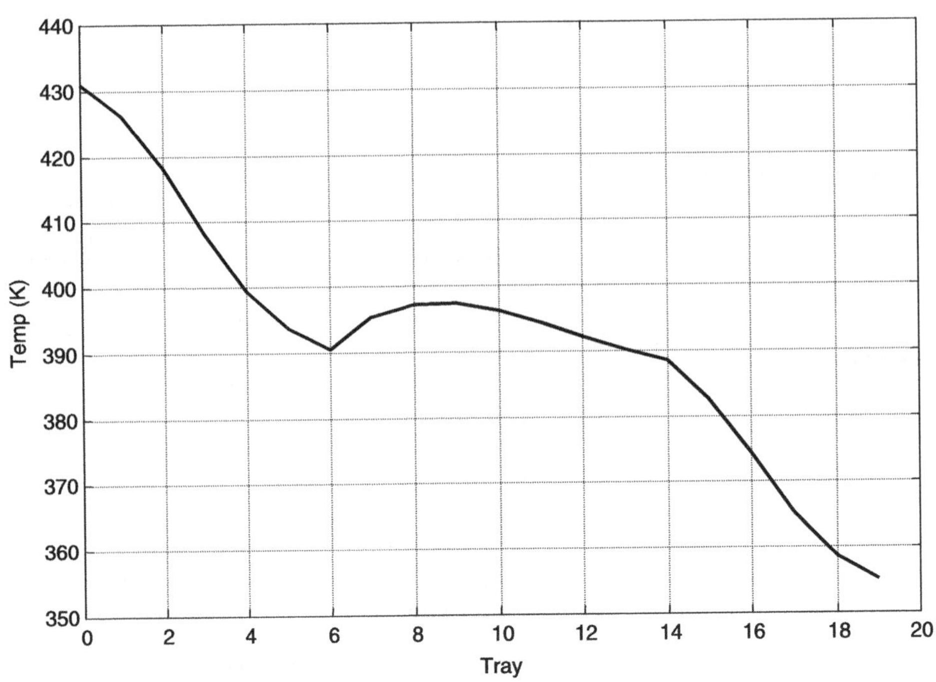

Figure 4.3 Temperature profile for neat operation.

As determined in a later section of this chapter, the total capital investment of this one-column process is \$358,300 and the annual energy cost is \$122,800 per year, giving a total annual cost (TAC) of \$242,300 per year.

4.3 TWO-COLUMN SYSTEM WITH EXCESS B

Figure 4.4 contains a typical flowsheet in which two columns are required. Fresh reactant A is fed to the first reactive column in stream F_{0A}. An excess of reactant B is also fed to this column. This total stream $F_{B,tot}$ consists of fresh feed F_{0B} and the recycle stream D_2 from recovery column, which contains mostly reactant component B with a little impurity of product component D. The amount of impurity depends on the design and operation of the recovery column and is a design optimization variable.

In this section B is selected as the excess component. If for some reason reactant A were chosen, the flowsheet would be different. This case is considered in Section 4.4. The selection of which reactant to keep in excess depends on kinetics, relative volatilities, heats of vaporization, and the costs of the two reactants. In either flowsheet, the excess reactant and one of the products are vaporized only once, so there is no obvious inherent energy difference between the two flowsheets.

The sizing relationships and economic parameters presented in Chapter 3 are used to determine the optimum design of the two-column system. The presence of the second column adds additional design optimization variables. To simplify the optimization, we assume that the reactive column has the same numbers of trays in each section as in the one-column process and operates at the same pressure. These appear to be reasonable assumptions because the TAC is fairly insensitive to the number of trays chosen in each section as long as the number of trays is kept fairly large (well above the minimum).

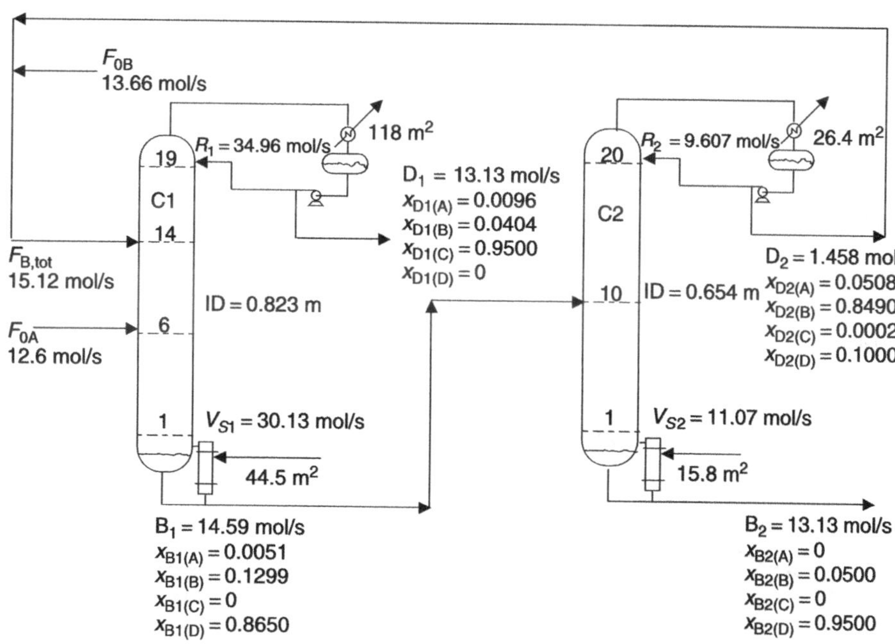

Figure 4.4 Reactive and recovery columns for 20% excess of reactant B.

The pressure in the recovery column is fixed at 1 bar, which permits the use of cooling water in the condenser (about 330 K in the reflux drum with mostly B). The number of trays in this column is varied to find the minimum TAC. Energy consumption is fairly insensitive to feed tray location in the recovery column, so the feed is introduced into the middle of the column.

The composition of impurity D in distillate D_2 from the recovery column is heuristically fixed at 10 mol% because case studies show that the TAC of the system is fairly insensitive to this parameter. We do not want to recycle too much product D back to the reactive column because this will drive the reaction in the wrong direction. In contrast, requiring a high purity of B in the distillate increases energy consumption. The fresh feed of A is constant at 12.6 mol/s in all of the flowsheets.

4.3.1 20% Excess B Case

The feed composition to the second column is mostly D, with an amount of B that depends on the excess reactant fed to the first column. Results with a 20% excess of B are given in Figure 4.4. The composition of the bottom from the first column is 12.99 mol% B. The total feed of B to the first column is 15.12 mol/s, and the fresh feed of B is 13.66 mol/s. The recycle flowrate D_2 is 1.458 mol/s.

Composition profiles in the reactive column are provided in Figure 4.5 for the 20% excess case. Note that the concentrations of A throughout the column are much lower than in the neat column (see Fig. 4.2). Figure 4.6 shows the composition profiles in the recovery column, and Figure 4.7 gives the temperature profiles in the two columns. Temperatures are lower in the recovery column because of its lower pressure (1 vs. 8 bar).

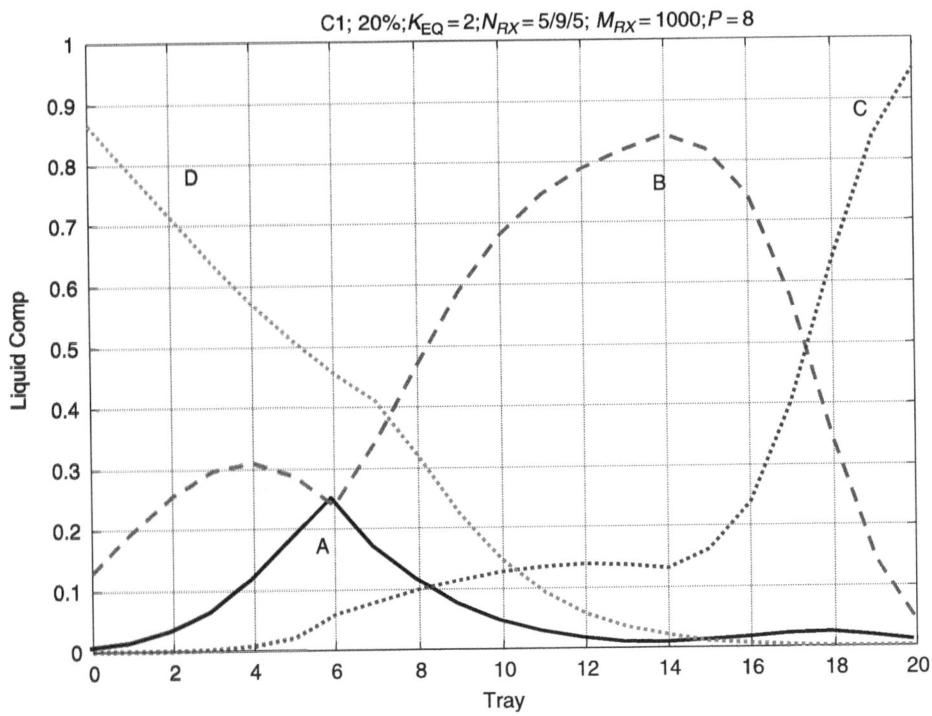

Figure 4.5 Reactive column for 20% excess of reactant B.

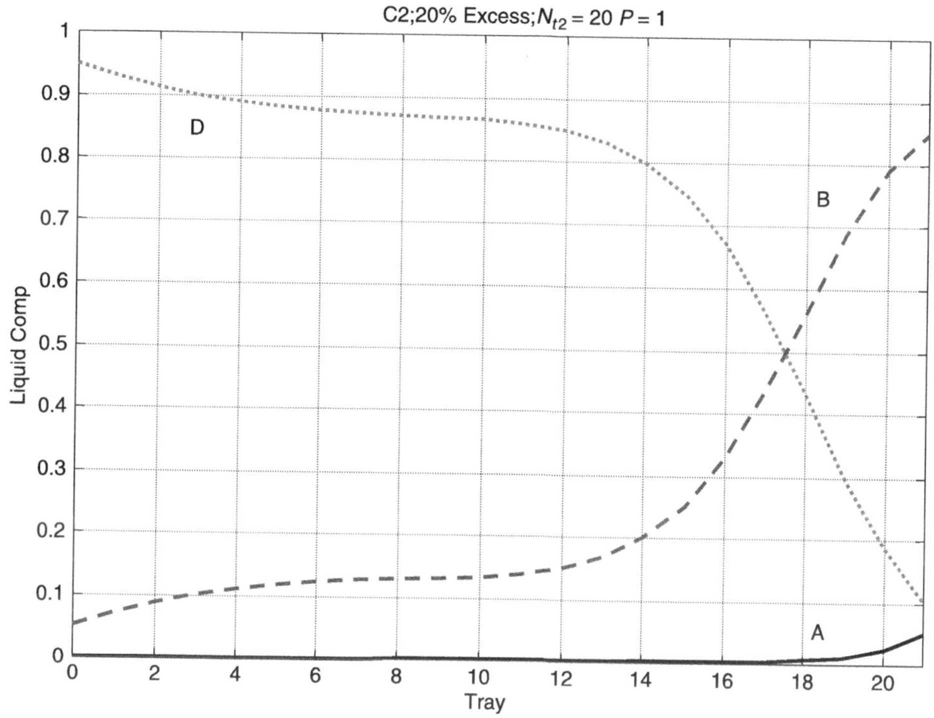

Figure 4.6 Recovery column for 20% excess of reactant B.

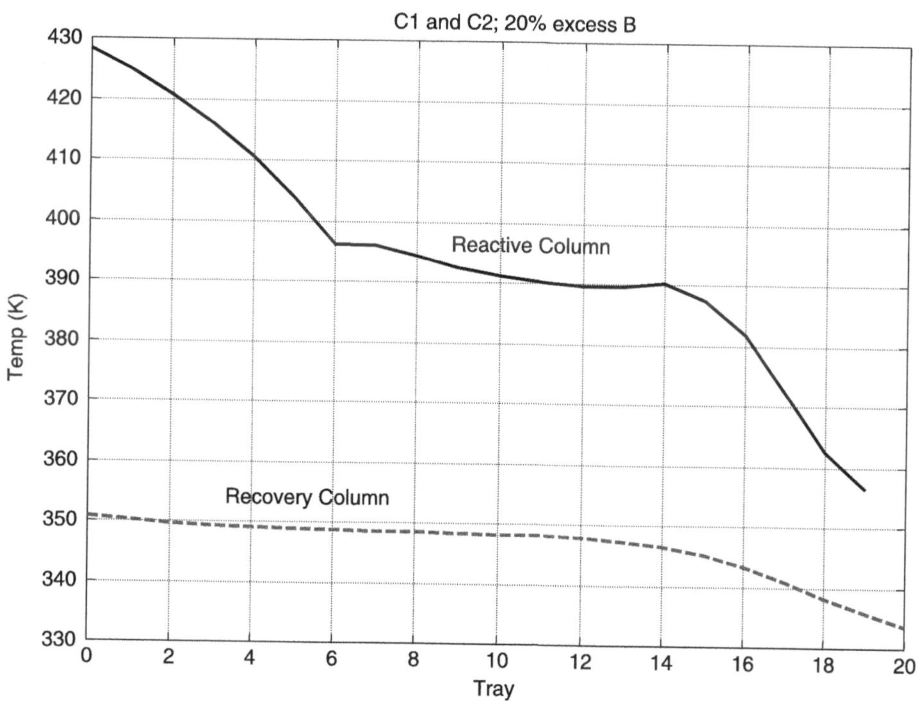

Figure 4.7 Reactive and recovery columns for 20% excess of reactant B.

TABLE 4.1 Effect of Number of Trays in Recovery Column for 20% Excess of B

N_{T2}	10	20	30	40
D_{C2} (m)	0.883	0.654	0.643	0.642
L_{C2} (m)	7.32	14.6	21.9	29.3
A_{C2} (m^2)	49.7	27.2	26.4	26.2
A_{R2} (m^2)	29.8	16.3	15.8	15.7
Capital cost (10^3 \$)				
Shell	76.2	96.5	131.2	164.9
Condenser and reboiler	158.6	107.4	105.0	104.7
Energy (10^3 \$/year)	87.0	47.7	46.1	45.9
Total annual (10^3 \$/year)	165.9	116.5	126.0	137.3

The fresh feed of A is constant at 12.6 mol/s in all flowsheets. However, the production rates of the two products and the fresh feed of B change somewhat with the amount of excess used. The reason for this is that less A is lost as more excess B is used. Figure 4.1 shows that

$$(12.6)(0.0326) + (12.6)(0.0174) = 0.63 \text{ mol/s}$$

of A are lost in the one-column neat operation in the two product streams. In the 20% excess case, Figure 4.4 shows that

$$(13.13)(0.0096) + (13.13)(0) = 0.126 \text{ mol/s}$$

of A are lost, essentially all of which is in distillate D_1. Thus, using more excess of B increases the conversion of A. At the same time, the losses of B increase because the purities of the two products remain at 95 mol%. The flowrates of the two product streams increase slightly from 12.6 to 13.13 mol/s.

Table 4.1 provides equipment and economic results for recovery column C2 with 20% excess for different numbers of trays N_{T2} in the recovery column. The minimum cost configuration has 20 trays.

4.3.2 10% Excess B Case

Results with a 10% excess of B are given in Figure 4.8. The composition of the bottom from the first column is 6.60 mol% B. The total feed of B to the first column is 13.86 mol/s, and the fresh feed of B is 13.56 mol/s. The recycle flowrate D_2 is 0.3056 mol/s. Composition profiles in the reactive column are displayed in Figure 4.9 for the 10% excess case. Figure 4.10 shows the composition profiles in the recovery column, and Figure 4.11 gives the temperature profiles in the two columns.

In the two-column case with 10% excess, Figure 4.8 shows that the loss of A in the two product streams is

$$(13.08)(0.0128) + (13.08)(0.0006) = 0.175 \text{ mol/s}$$

This should be compared with losses of 0.63 mol/s in the single neat column case and 0.126 mol/s in the 20% excess case. The conversion of A to products C and D increases as more excess B is used.

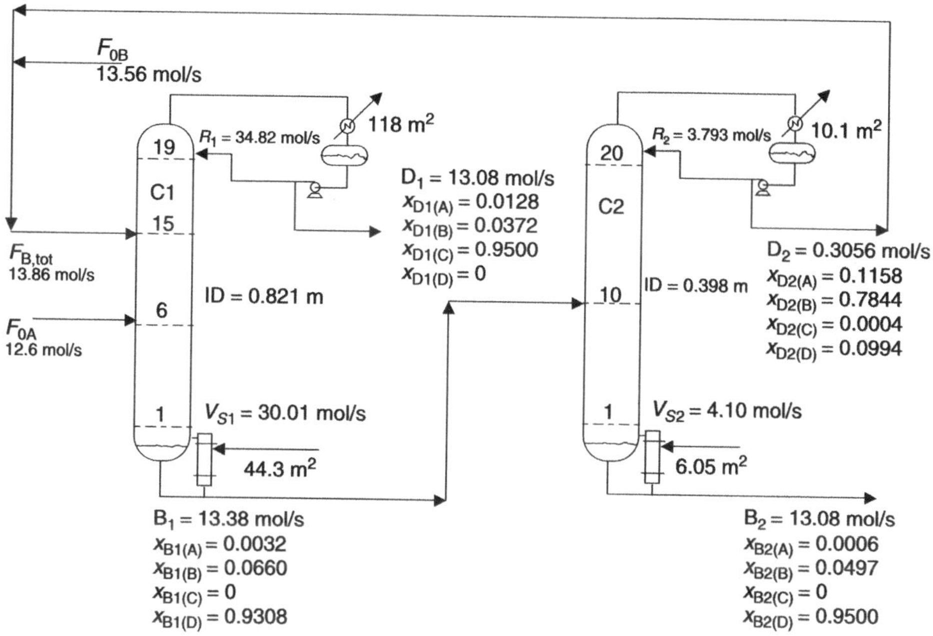

Figure 4.8 Reactive and recovery columns for 10% excess of reactant B.

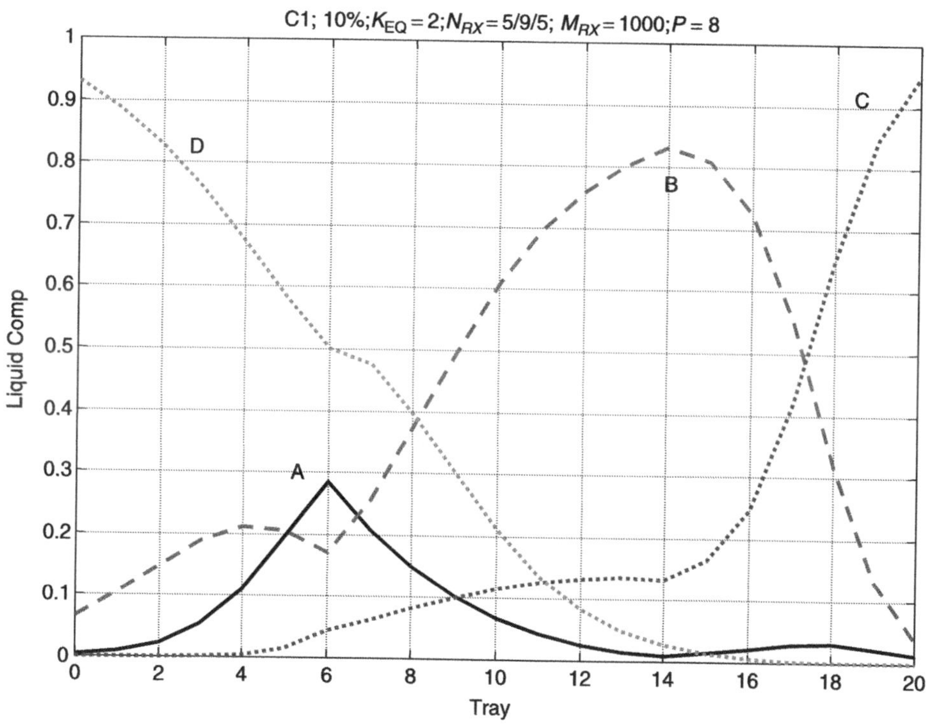

Figure 4.9 Reactive column for 10% excess of reactant B.

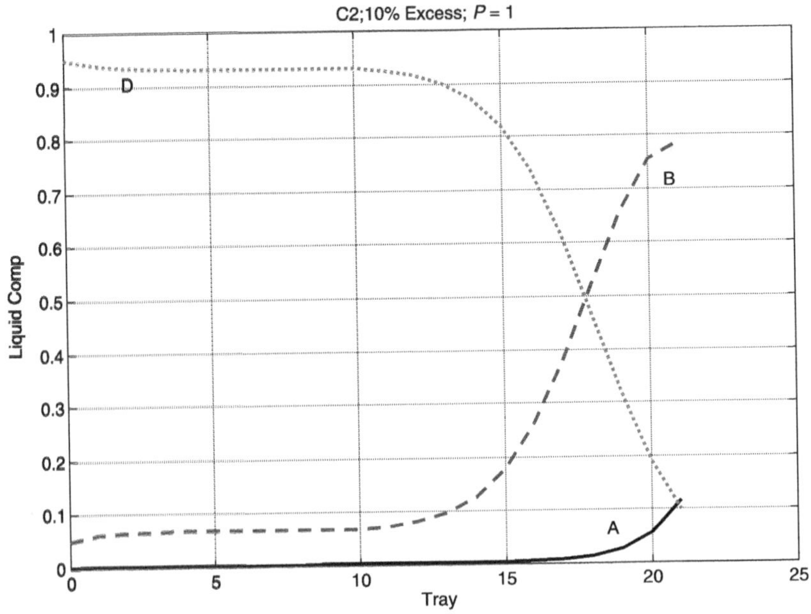

Figure 4.10 Recovery column for 10% excess of reactant B.

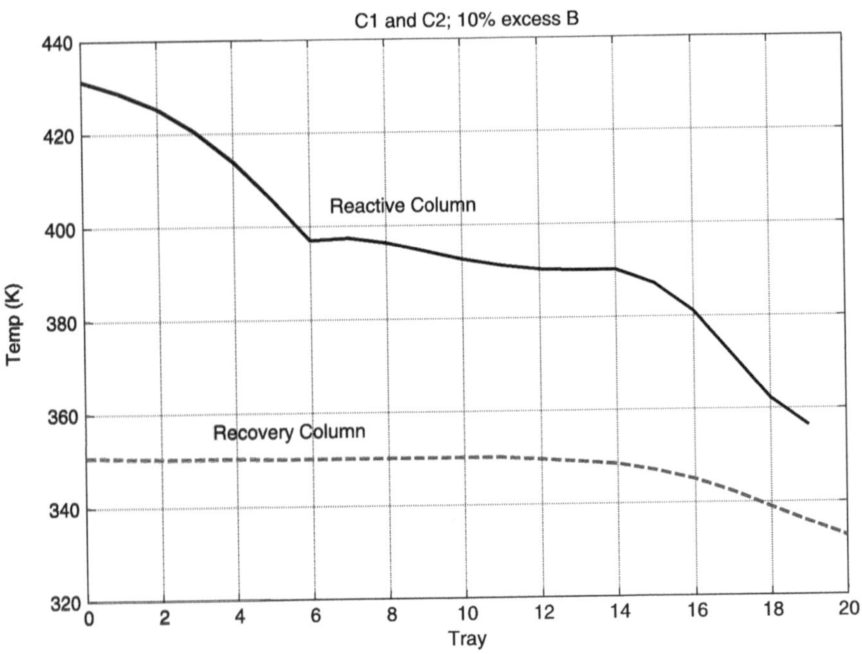

Figure 4.11 Reactive and recovery columns for 10% excess of reactant B.

TABLE 4.2 Effect of Number of Trays in Recovery Column for 10% Excess of B

N_{T2}	15	20	30
D_{C2} (m)	0.430	0.398	0.388
L_{C2} (m)	11.0	14.6	21.9
A_{C2} (m^2)	11.8	10.1	9.58
A_{R2} (m^2)	6.94	6.05	5.74
Capital cost (K$)			
Shell	49.0	56.9	76.5
Condenser and reboiler	62.0	56.3	54.4
Energy (10^3 $/year)	20.3	17.7	16.8
Total annual (10^3 $/year)	57.6	57.5	60.9

TABLE 4.3 Comparison of One- and Two-Column Flowsheets

	Neat Operation	10% Excess B	20% Excess B	20% Excess A
Reactive column C1				
N_{T1}	19	19	19	19
D_{C1} (m)	0.805	0.821	0.822	0.827
L_{C1} (m)	13.9	13.9	13.9	13.9
A_{C1} (m^2)	112.3	117.8	118.4	119.7
A_{R2} (m^2)	41.9	44.3	44.5	46.1
Capital costs in C1 (10^3 $)				
Shell	115.5	118.9	118.3	118.9
Condenser and reboiler	239.7	247.7	248.4	251.6
Energy (10^3 $/year)	122.8	129.3	129.7	134.6
Total annual (10^3 $/year)	242.3	252.3	253.2	259.2
Recovery column C2				
Energy cost (10^3 $/year)		17.7	47.7	70.9
TAC (10^3 $/year)		57.6	116.5	158.8
Total cost C1 and C2				
Energy (10^3 $/year)	122.8	147.0	177.6	205.5
Capital (10^3 $)	358.3	483.3	576.2	635.2
Annual (10^3 $/year)	242.3	308.1	369.6	417.2

Table 4.2 lists equipment and economic results for the 10% excess case for different numbers of trays N_{T2} in the recovery column. The minimum cost configuration has 20 trays. Table 4.3 compares the 10 and 20% excess B cases.

4.4 TWO-COLUMN SYSTEM WITH 20% EXCESS OF A

Figure 4.12 gives the flowsheet with a 20% excess of A fed to the reactive column. The number of trays, pressure, and holdup are kept the same in the reactive column as in the one-column case. The fresh feed of B is fixed at 12.6 mol/s, and the total flow $F_{A,tot}$ to the bottom of the reactive zone is fixed at 15.12 mol/s. This stream is the sum of the fresh feed F_{0A}, which is 12.76 mol/s, and the recycle B_2 from the base of the recovery column. The composition of C in this recycle stream is set at 10 mol%.

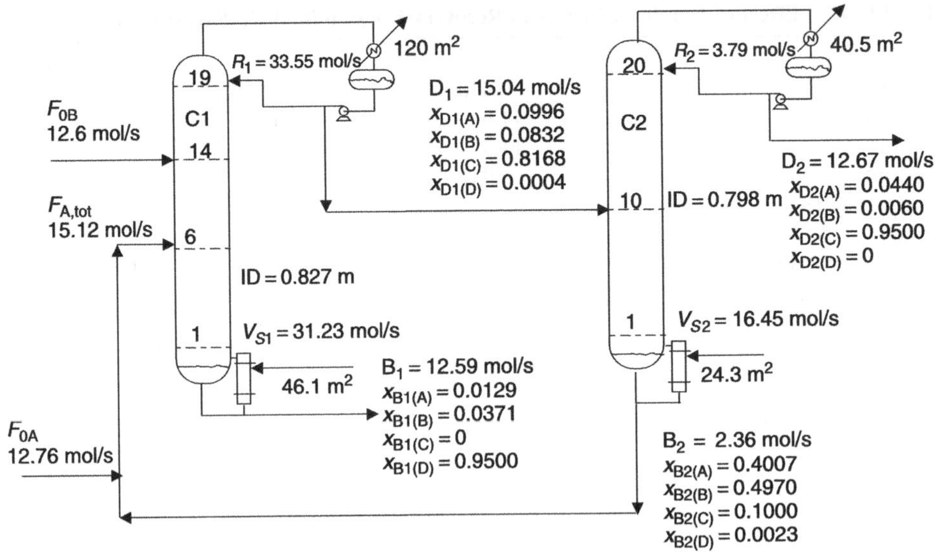

Figure 4.12 Reactive and recovery columns for 20% excess of reactant A.

Composition profiles in the reactive column are shown in Figure 4.13. The concentrations of A in the upper part of the column are higher than in the neat case (Fig. 4.2), and the concentrations of A in the lower part of the column are much higher than in the 20% excess B case (Fig. 4.13).

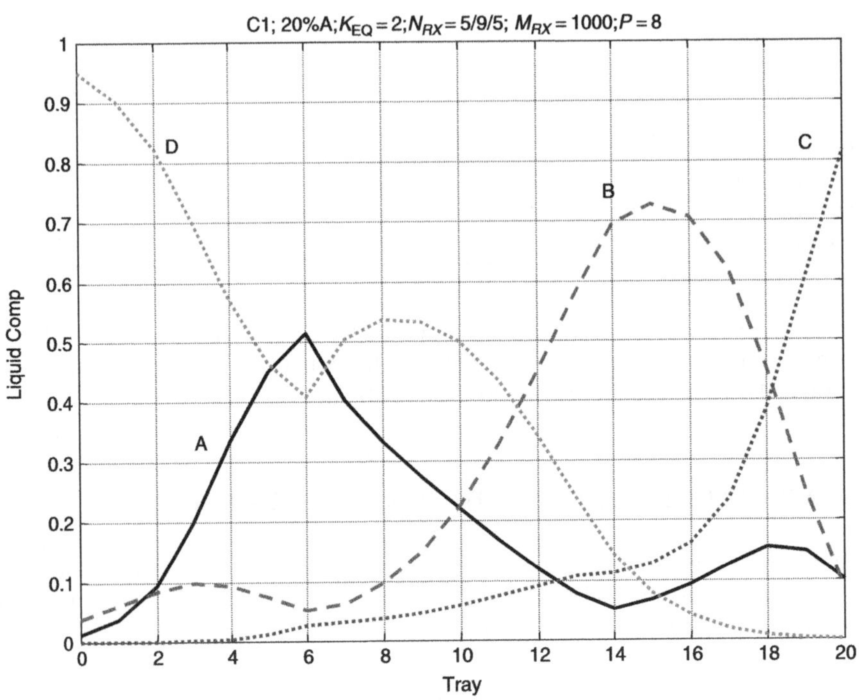

Figure 4.13 Reactive column for 20% excess of reactant A.

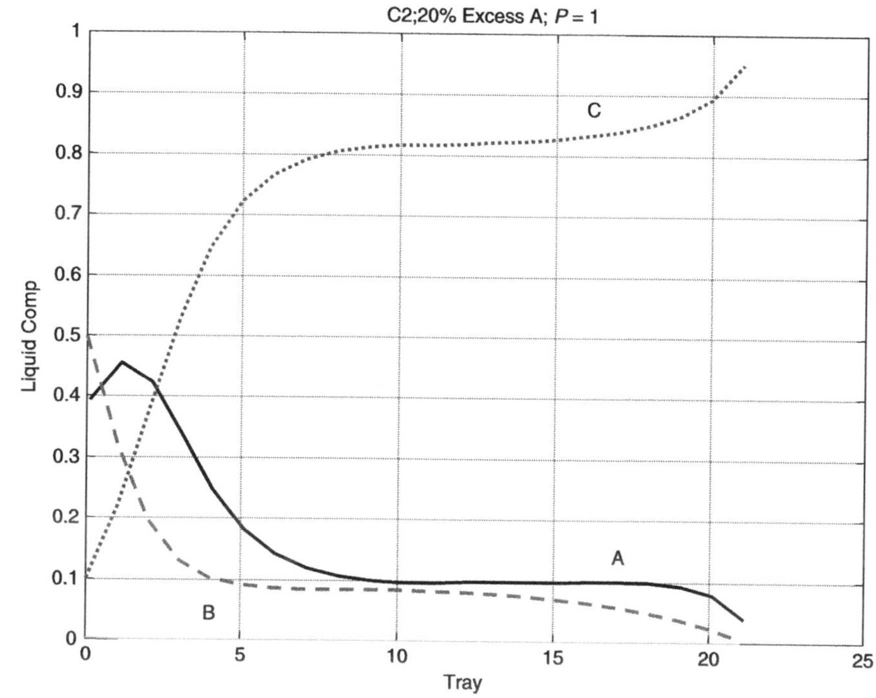

Figure 4.14 Recovery column for 20% excess of reactant A.

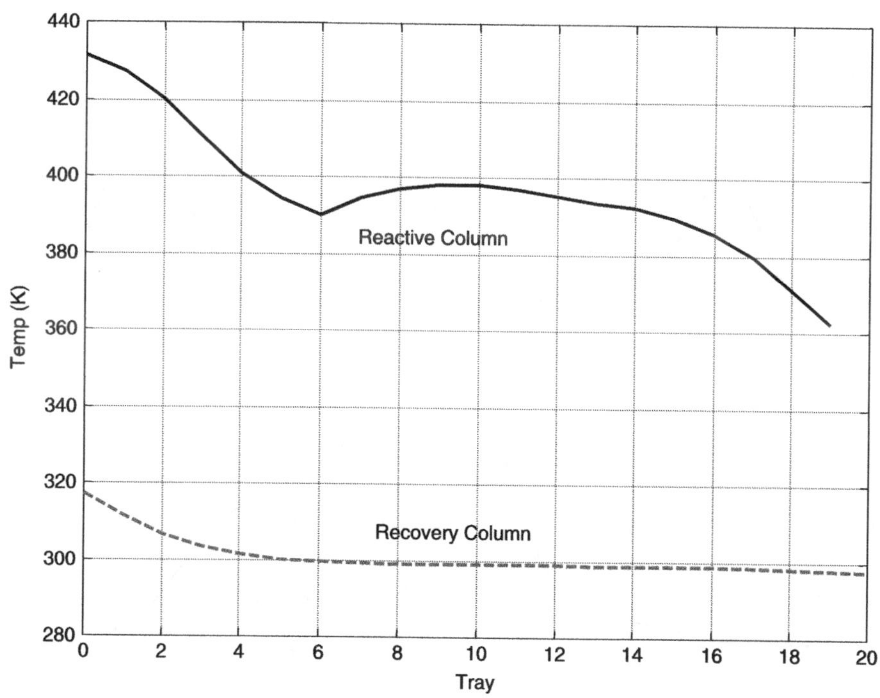

Figure 4.15 Reactive and recovery columns for 20% excess of reactant A.

TABLE 4.4 Recovery Column Design for 20% Excess of A or B

	Excess A	Excess B
N_{T2}	20	20
D_{C2} (m)	0.789	0.654
L_{C2} (m)	14.6	14.6
A_{C2} (m^2)	40.3	27.2
A_{R2} (m^2)	24.3	16.3
Capital cost (10^3 \$)		
Shell	119.2	96.5
Condenser and reboiler	138.9	107.4
Energy (10^3 \$/year)	70.9	47.7
Total annual (10^3 \$/year)	158.0	116.5

Figure 4.14 provides composition profiles in the recovery column, and Figure 4.15 shows the temperature profiles in the two columns. Table 4.4 compares the designs of the recovery column for 20% excess of A and 20% excess of B. The former is significantly more expensive in both energy and capital cost. This is due to the higher vapor boilup in the recovery column (16.45 vs. 11.07 mol/s) that results from the larger recycle flowrate (2.36 vs. 1.458 mol/s). The impurity of B in the bottom from the reactive column is 12.99 mol% in the 20% excess B case. In the 20% excess A case, the distillate of the reactive column contains significant impurities of both A and B (9.96 and 8.32 mol%, respectively). It appears that it is easier to keep A from going out the

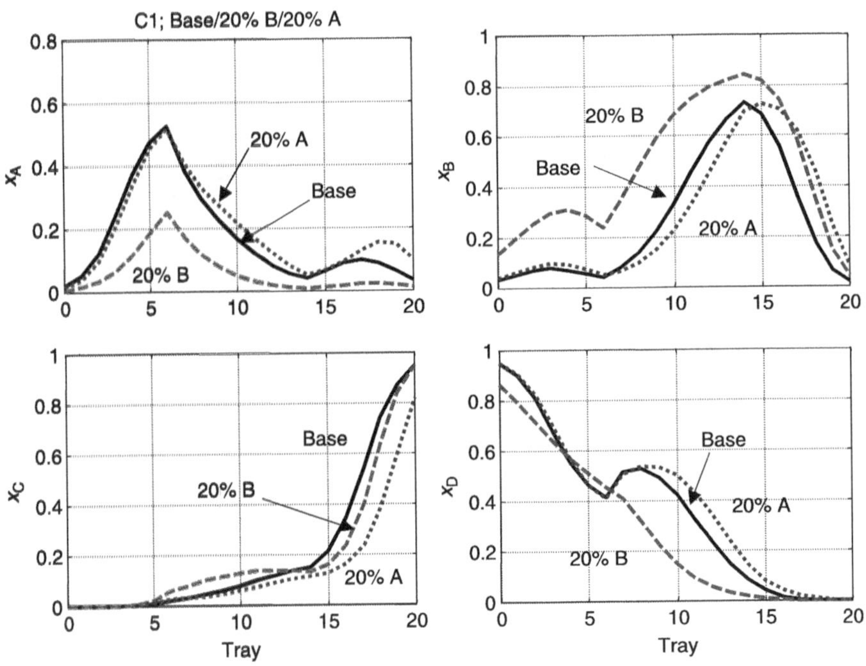

Figure 4.16 Reactive column for base and 20% excess of A and B.

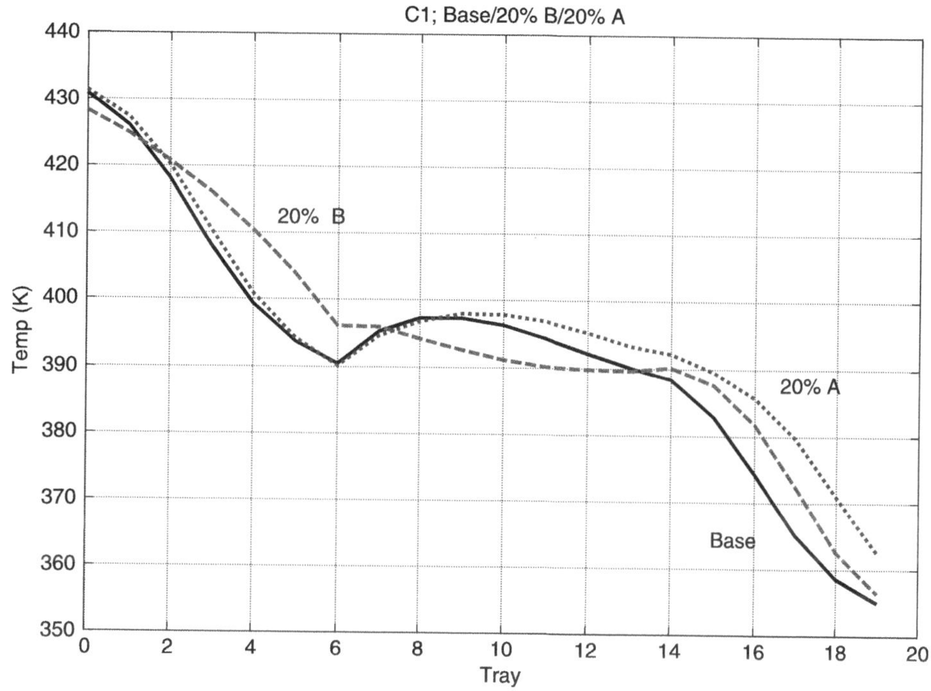

Figure 4.17 Reactive column for base and 20% excess of A and B.

bottom than to keep B from going out the top when an excess amount of the one of the two reactants is used.

Figure 4.16 compares the reactive column composition profiles in the two excess reactant cases with those for base case neat operation. The concentrations of A are lower than the base case when an excess of B is used and larger when an excess of A is used. The opposite is true for the concentrations of B. The concentration of C at the top of the column is lower when an excess of A used. The concentration of D at the bottom of the column is lower when an excess of B used.

Figure 4.17 is a comparison of the temperature profiles in the reactive column for the three cases. Temperatures are lower at the bottom of the column when an excess of B is used because of the lighter B in the base. At the top, the temperature is slightly higher because there is more A impurity in the C product than in the base case.

Temperatures are higher at the top of the column when an excess of A is used because the distillate contains more of the heavier A impurity in the C going overhead.

4.5 ECONOMIC COMPARISON

The TAC of the single neat reactive column is \$242,300 per year with an energy cost of \$122,800 per year, as shown in Table 4.3. The two-column design with a 10% excess of B has a TAC of \$308,100 per year with an energy cost of \$147,000 per year. The two-column design with a 20% excess of B has a TAC of $369,600 per year with an energy cost of $177,600 per year.

These results demonstrate the significant economic penalty for using the two-column reactive/recovery system. If a 20% excess of B is required to handle flow measurement and feed composition disturbances, the TAC increases by over 50% and energy cost increases by 45%. The economics are even worse when an excess of A is used.

The control of these two alternative flowsheets is studied in Chapter 11. The question to be answered is will the improvement in control justify the increase in costs when using a two-column system?

4.6 CONCLUSION

The use of excess reactant has been demonstrated to increase capital and energy costs when compared to neat operation. The dynamic control of these two flowsheets will be compared in Chapter 11.

PART II

STEADY-STATE DESIGN OF OTHER IDEAL SYSTEMS

The first three chapters have explored in a fair amount of detail the four-component quaternary system with the reaction A + B ⇔ C + D. This system has two reactants and two products. In the next two chapters we will study various aspects of two other types of ideal chemical systems. In Chapter 4 we investigated the impact of a number of kinetic and design parameters on the ideal ternary system with the reaction A + B ⇔ C with and without inerts in the system. In Chapter 5 we study systems with the reaction A ⇔ B + C in which there is only one reactant but two products. We will illustrate that the chemistry has an important effect on how the many kinetic and design parameters impact the reactive distillation column.

PART II

CHAPTER 5

TERNARY REACTIVE DISTILLATION SYSTEMS

The chemical system considered in previous chapters featured the classical quaternary two-reactant, two-product A + B ⇔ C + D reversible reaction. Some interesting phenomena were discussed. In particular, the effect of the number of reactive trays on energy consumption was demonstrated to be counterintuitive, that is, there is an optimum number of reactive trays that minimizes energy consumption.

In this chapter we explore a similar but somewhat different chemical system. The reaction is A + B ⇔ C, which is reversible, liquid phase, and exothermic. There is only one product instead of two. This seemingly small change in the chemistry alters the effects of several important design parameters of a reactive distillation column. Because there are three components, we label the system ternary.

This is a very important class of reactions. The majority of commercial reactive distillation columns deal with this type of reaction. The large volume gasoline additive chemicals produced by reactive distillation, such as MTBE, ETBE, and TAME, are this type. These three examples involve an alcohol and an iso-olefin (either C4 or C5). The reactants are methanol and isobutene for MTBE, ethanol and isobutene for ETBE, and methanol and isoamylenes for TAME.

Although there are only three components involved in the reaction, in many of the A + B ⇔ C systems there are more than three components in the column because the feedstreams contain other components. These components are "inert" from the standpoint of the reaction, but they are not inert from the standpoint of their effect on the vapor–liquid equilibrium in the column. These inert components are present in the olefin feedstreams that contain the reactive iso C4 and C5 olefins in the examples cited. The reason for their presence is the great difficulty in separating the desired iso-olefin from the other components. For example, in the MTBE and ETBE cases, isobutene is produced in a catalytic cracker in a refinery along with a number of other C4 components (isobutane, *n*-butane, and *n*-butene).

Reactive Distillation Design and Control. By William L. Luyben and Cheng-Ching Yu

All of these C4 components have very similar boiling points. Their separation using distillation would be very energy intensive. Therefore, the mixture of all of these C4 components is fed to the reactive distillation column, and the chemically inert components are removed as a product stream from the column.

We first discuss the ternary system without inerts to gain some insights into how changing from a two-product system to a one-product system impacts reactive distillation design. Unlike the quaternary column with distillate and bottoms products, the ternary column without inerts has only one product stream leaving the column. The effects of several design variables are shown to be quite different in the ternary system than those we observed in the quaternary system.

Then we explore the ternary system with inerts present in one of the feedstreams. The reactive distillation column now has two streams leaving the column. One contains product C and the other contains the inerts.

5.1 TERNARY SYSTEM WITHOUT INERTS

5.1.1 Column Configuration

In the two-product reaction system, the column has both bottoms and distillate products coming from the two ends of the column. The two reactant feedstreams are fed into the middle section of the column. With a one-product reaction system without inerts, the column has only a bottoms or a distillate product. If product component C is heavier than reactant components A and B, there is a bottoms stream but no distillate. The column operates at total reflux with all of the overhead vapor condensed and returned to the column as reflux. There is no need to have a rectifying section because there is no distillate and no need to maintain any composition at the top of the column.

Thus, the column configuration in the one-product ternary system without inerts is quite different than the two-product configuration. Figure 5.1 gives the flowsheet. There are only stripping and reactive sections.

5.1.2 Chemistry and Phase Equilibrium Parameters

Table 5.1 gives kinetic and vapor–liquid phase equilibrium parameters used in the numerical case considered in this chapter. The forward reaction rate parameters are the same as those used in the two-product case. The activation energy of the backward reaction is the same, but the preexponential factor of the backward reaction is changed because the backward reaction is only first order in the concentration of product C.

$$\mathcal{R}_n = M_{RX}(k_F x_{n\mathrm{A}} x_{n\mathrm{B}} - k_B x_{n\mathrm{C}}) \tag{5.1}$$

The backward reaction depends on only 1 mol fraction, so a smaller value of k_B is needed to achieve the same conversion (the product of 2 mol fractions is smaller than 1 mol fraction). Therefore, equilibrium constant K_{EQ} is increased to 20 from the previous value of 2.

The vapor pressures of reactants A and B are the same as in the two-product system. The vapor pressure of heavy product C is assumed to be the same as D in the quaternary system. Relative volatilities are constant.

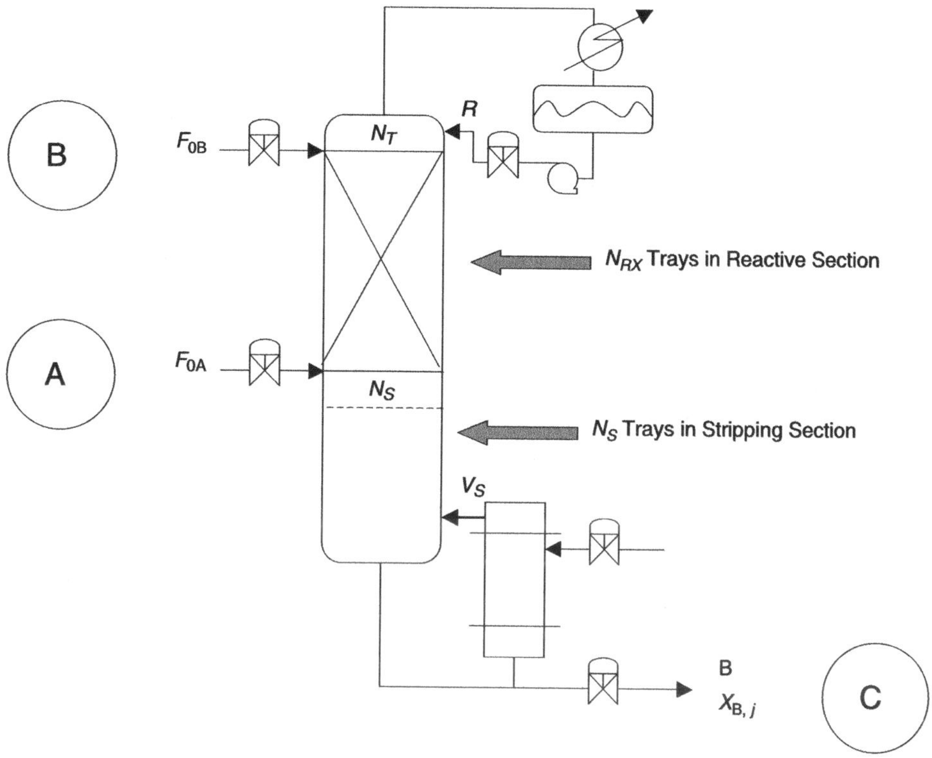

Figure 5.1 Ternary reactive distillation column without inerts.

TABLE 5.1 Kinetic and Vapor–Liquid Equilibrium Parameters for Base Case

Activation energy (kcal/mol)		
Forward	30	
Backward	40	
Specific reaction rate at 366 K (kmol s^{-1} $kmol^{-1}$)		
Forward	0.008	
Backward	0.0004	
Chemical equilibrium constant at 366 K	20	
Heat of reaction (kcal/mol)	−10	
Heat of vaporization (kcal/mol)	6.944	
Molecular weights (g/mol)		
A and B	50	
C	100	
Liquid density (kg/m^3)	700	
Vapor Pressure Constants	A_j	B_j
A	12.34	3862
B	11.45	3862
C	10.96	3862

5.1.3 Design Parameters and Procedure

In the quaternary case, the flowrates of the two fresh feeds, the distillate, and the bottoms were all set to 12.6 mol/s. Then the relaxation model adjusted the reflux flowrate to drive the purity of the bottoms product to the desired value, $x_{B,D} = 0.95$ mol fraction D for the 95% conversion case. The vapor boilup was adjusted to control the base level. This approach gave the same impurity levels in both product streams and the same conversions of the two reactants fed to the system.

The ternary system without inerts has a different column structure and requires a different approach for designing the column. There are still two feedstreams and a bottoms stream, but there is no distillate. In addition, the impurity in the bottoms product will be mostly the heavier of the two reactants, component B. This means that the flowrates of the two fresh feedstreams will *not* be equal. Moreover, the reaction is not equimolar. Two moles of reactants produce 1 mol of product. Thus, there is a decrease in the molar liquid flowrates in the reaction section that is attributable to the nonequimolar reaction.

The procedure used for this system is to fix the production rate of product C at 12.6 mol/s and the purity of the bottoms product at 0.98 mol fraction C. This means that the bottoms flowrate is 12.6/0.98 = 12.857 mol/s. The production rate requires that 12.6 mol/s of both A and B be consumed. Thus, at least this amount must be fed to the column. In addition, however, there is a loss of reactants in the bottoms to account for the 2 mol% impurity. It is mostly B, but there is also a small amount of A. The concentrations of A and B change with the various designs. Therefore, the flowrates of the fresh feeds are slightly different in each design. At each point in time during the dynamic simulation, the fresh feed flowrates are calculated from the fixed value of the bottoms flowrate B and the present value of the bottoms composition $x_{B,j}$, which changes with time until a steady-state solution is achieved.

$$\begin{aligned} F_{0A} &= 12.6 + Bx_{B,A} \\ F_{0B} &= 12.6 + Bx_{B,B} \end{aligned} \tag{5.2}$$

Because B is heavier than A, the fresh feed flowrate of B is somewhat larger than that of A. The reflux flowrate is changed to drive the bottoms composition to 98 mol% C. The vapor boilup controls the level in the base. There is no distillate. The reflux drum level is not controlled.

The base case considered has five stripping trays and nine reactive trays. The column operates at 8 bar and has 1000 mol of holdup on the reactive trays. Under these conditions, the bottoms composition is 0.25 mol% A and 1.75 mol% B. The resulting fresh feeds are $F_{0A} = 12.63$ mol/s and $F_{0B} = 12.82$ mol/s. The vapor boilup required to achieve this purity of the product is 62.03 mol/s, and the reflux flowrate is 80.17 mol/s, which is the overhead vapor rate. Table 5.2 gives conditions for the base case. Note that the reflux composition is mostly the lightest component A, but some of the other two components are also present. Figure 5.2 presents the composition profiles.

Another change made in the equations accounts for the reduction in moles as 1 mol of product is produced by the consumption of 2 mol of reactant. The liquid rates on the reactive trays are reduced by the rate of reaction and by the vaporization caused by the heat of reaction.

$$L_n = L_{n+1} - \mathfrak{R}_n + \frac{\lambda}{\Delta H_V}\mathfrak{R}_n \tag{5.3}$$

Remember the heat of reaction λ is negative.

TABLE 5.2 Steady-State Conditions and Design Parameters for Base Case

Fresh feed flowrate of A (mol/s)	12.63	
Fresh feed flowrate of B (mol/s)	12.82	
Distillate flowrate (mol/s)	0	
Bottoms flowrate (mol/s)	12.857	
Vapor boilup (mol/s)	62.03	
Reflux flowrate (mol/s)	80.17	
Overhead vapor flowrate (mol/s)	80.17	
Stripping trays	5	
Reactive trays	9	
Rectifying trays	0	
Liquid holdup on reactive trays (mol)	1000	
Pressure (bar)	8	
Vapor Pressure Constants	A_j	B_j
A	12.34	3862
B	11.65	3862
C	10.96	3862
Product Composition (Mole Fraction)	Reflux	Bottoms
A	0.8725	0.0025
B	0.0928	0.0175
C	0.0347	0.9800

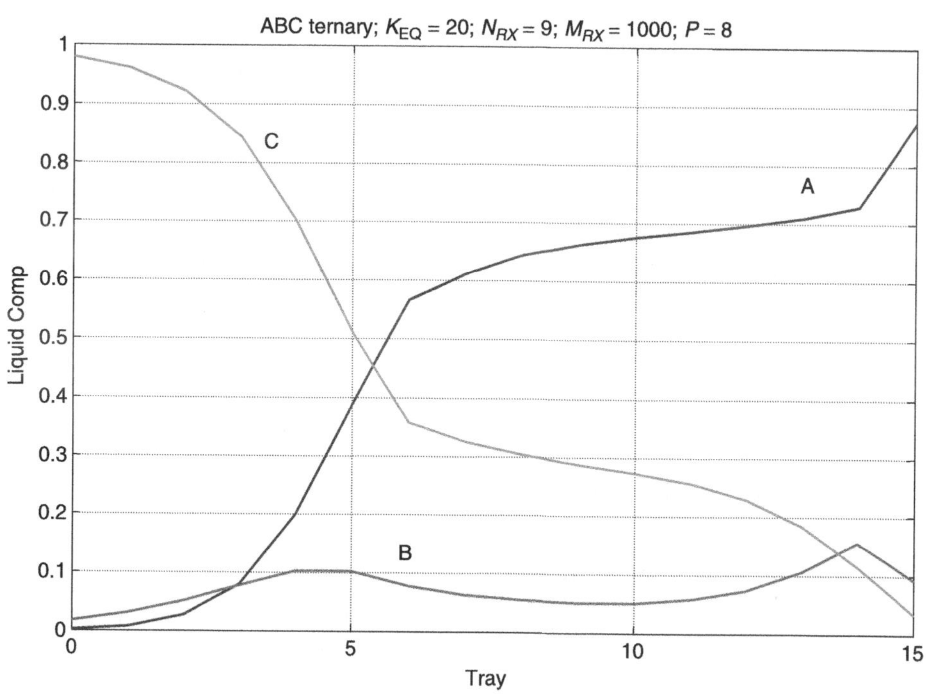

Figure 5.2 Ternary system composition profiles.

5.1.4 Effect of Pressure

Now that we have a design procedure and a base case, we are ready to see how various parameters affect the ternary reactive column without inerts. Figure 5.3 shows the effect of changing the column operating pressure. The top left graph in Figure 5.3*a* shows that the optimum pressure is 7 bar in this ternary system. Remember that the optimum pressure is 8 bar in the quaternary system. The lower pressure is a result of not having the very light product that tends to reduce temperatures in the quaternary system. The reaction zone temperatures in the ternary system at 7 bar are similar to the reaction zone temperatures in the quaternary system at 8 bar.

The lower graphs in Figure 5.3*a* give the compositions of reflux x_{Dj} and bottoms product x_{Bj}. They change very little with pressure. Temperature profiles at different pressures are given in Figure 5.3*b*.

5.1.5 Holdup on Reactive Trays

In the quaternary system, increasing the reactive tray holdup decreases energy consumption. The same is true in the ternary system, as demonstrated in Figure 5.4. Thus, adding more reactive tray holdup improves the steady-state designs of both systems. There are no counterintuitive effects of reactive tray holdup.

5.1.6 Number of Reactive Trays

In the quaternary system in Chapter 2, we demonstrated that there is an optimum number of reactive trays at which vapor boilup is minimized. Having too few or too many reactive trays increases the steady-state energy consumption. This effect is counterintuitive. Does the same occur in the ternary system?

Figure 5.5 shows that it does not. Adding more reactive trays improves the steady-state design of the ternary system because vapor boilup decreases as reactive trays are added. Figure 5.5 also shows that the compositions in the top and bottom of the column do not change much. Figure 5.6 illustrates that the same is true for the temperature profile. Figure 5.7 gives the composition profiles throughout the column for three values of N_{RX}.

Note the sharp liquid composition changes that occur between the top of the column on tray N_T and the reflux drum (stage $N_T + 1$). This is caused by the total reflux operation. The vapor composition on tray N_T is higher for lighter component A than the liquid composition. The reverse is true for heavier component B. Therefore, the liquid leaving the total condenser is richer in A than the top tray and leaner in B. Thus, the liquid composition profiles rise for A and drop for B at the top of the column.

These results demonstrate a fundamental difference between a two-product reactive column (quaternary system) and a one-product reactive column (ternary system). This is a good example of the complexity and challenges associated with reactive distillation.

5.1.7 Number of Stripping Trays

In the quaternary system in Chapter 2, we demonstrated that increasing the number of fractionating trays in the stripping and rectifying sections tends to decrease vapor boilup. In the ternary system, the column has only stripping trays. The impact of changing the number from the base case of $N_S = 5$ is shown in Figure 5.8.

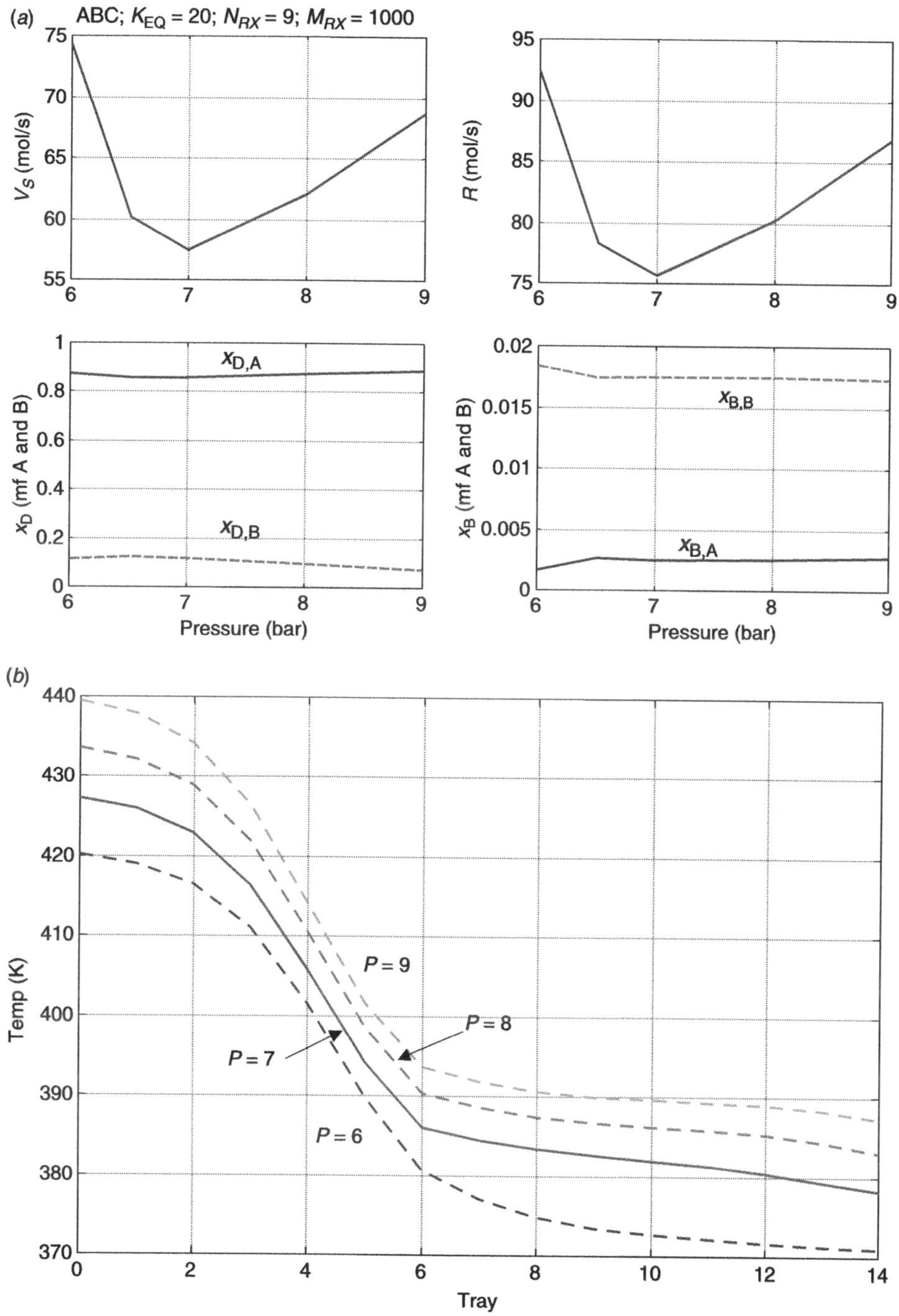

Figure 5.3 (*a*) Effect of pressure and (*b*) effect of pressure on temperature profiles.

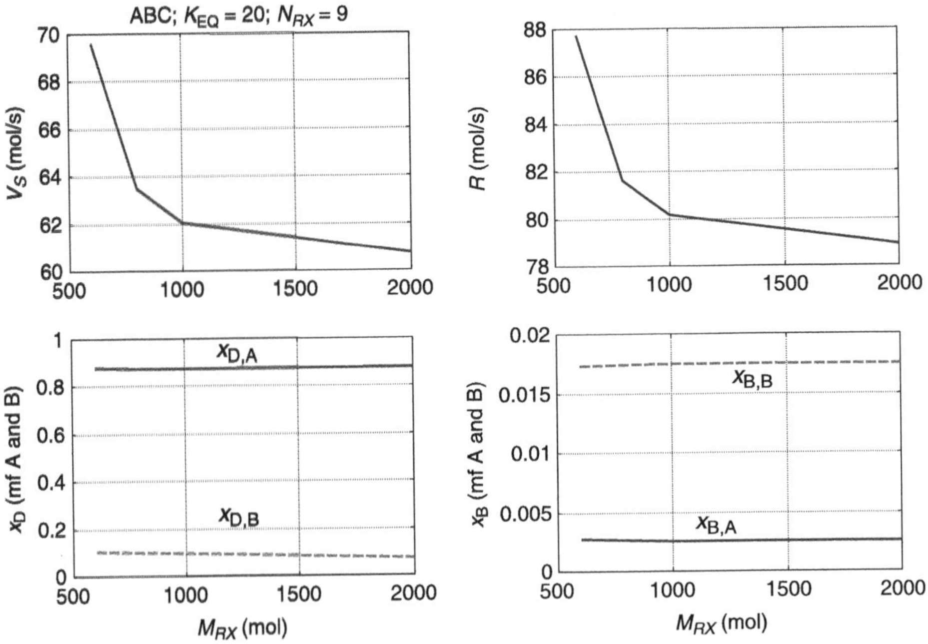

Figure 5.4 Effect of holdup on reactive trays.

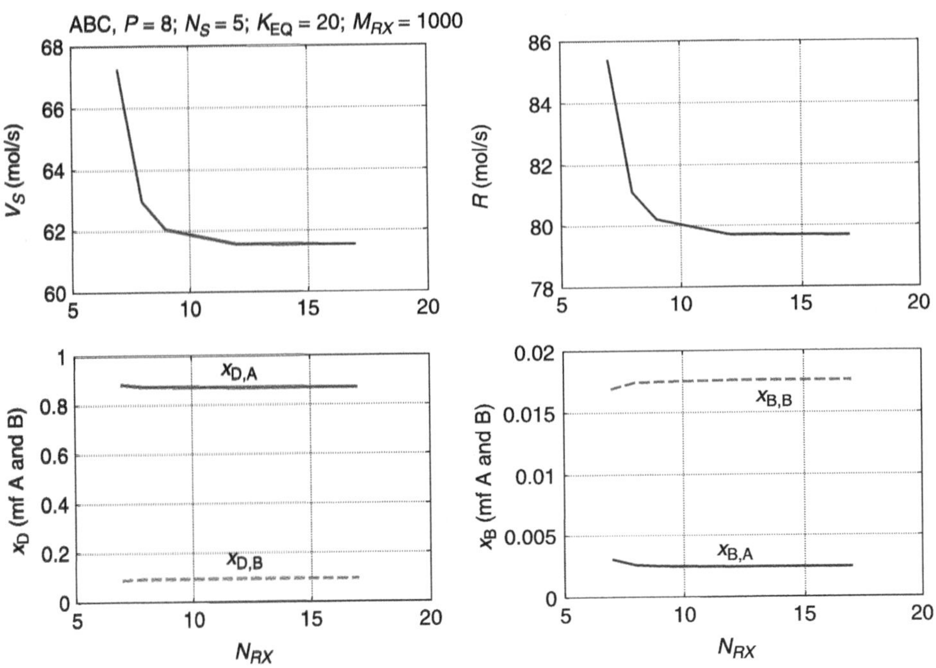

Figure 5.5 Effect of number of reactive trays.

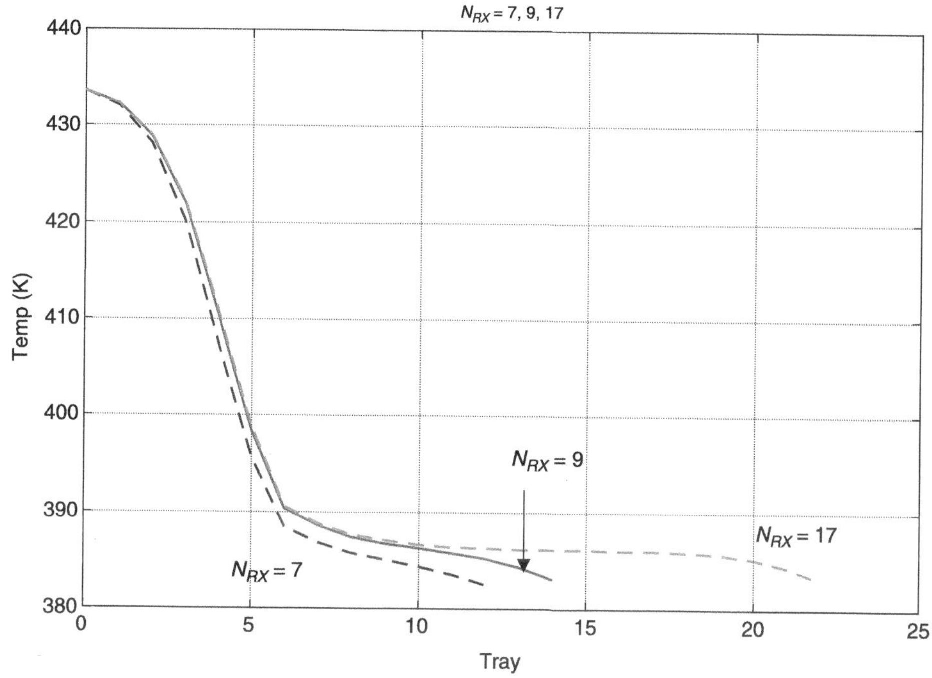

Figure 5.6 Effect of number of reactive trays on temperature profiles.

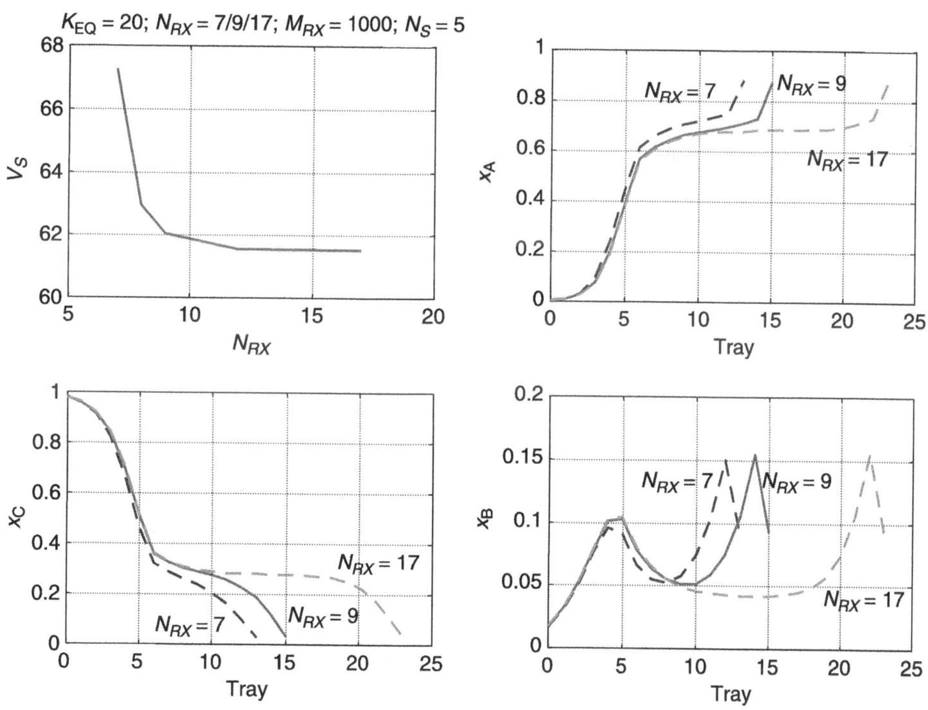

Figure 5.7 Effect of number of reactive trays on composition profiles.

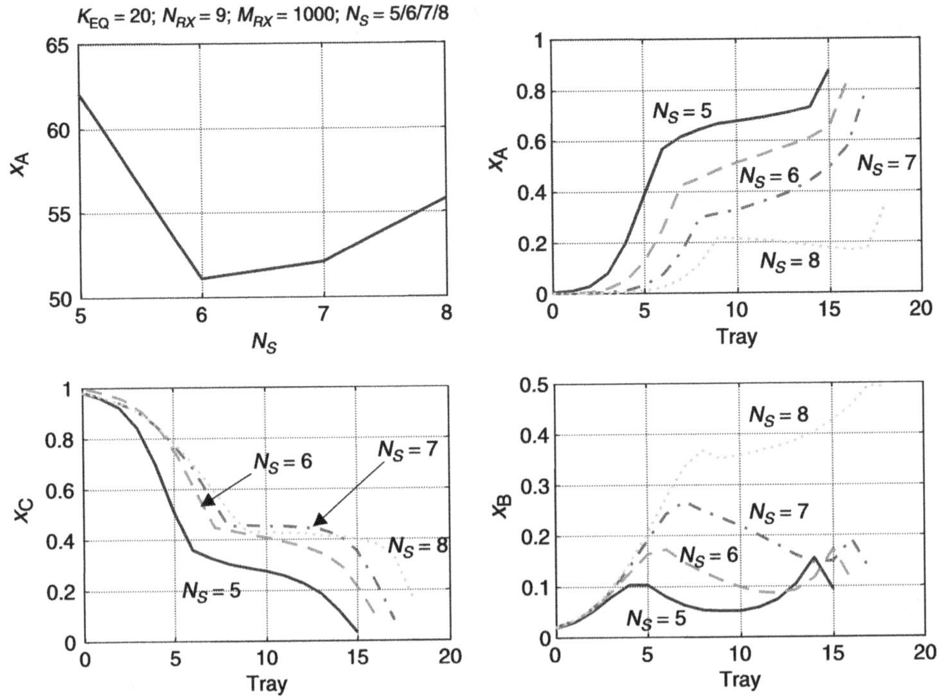

Figure 5.8 Effect of number of stripping trays.

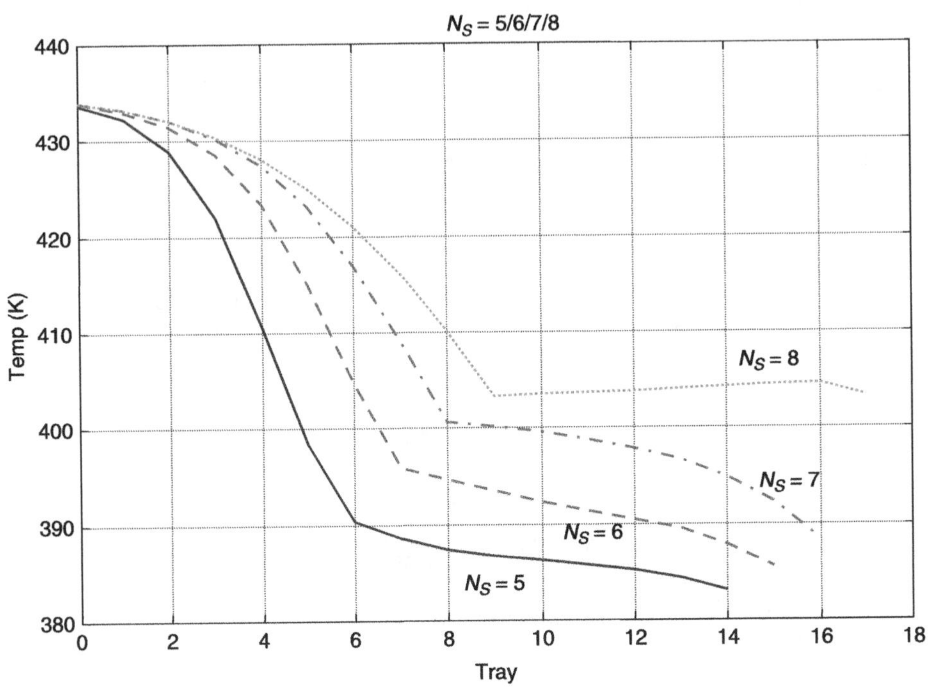

Figure 5.9 Effect of number of stripping trays on temperature profiles.

These results are strikingly different than those obtained in the two-product system. There is an optimum number of stripping trays. Increasing N_S from 5 to 6 causes a decrease in energy consumption, as shown in the upper left graph in Figure 5.8. However, further increases result in higher vapor boilup requirements.

The other three graphs in Figure 5.8 show what happens to the three composition profiles as N_S is increased, and these profiles help to explain this unexpected effect. With a large number of stripping trays, for example, $N_S = 8$, the composition of light component A is very small throughout the stripping section and the composition of heavier component B is high. Because B/C separation is more difficult than A/C separation, higher vapor boilups are required to maintain the same bottoms purity of 98 mol% C.

Figure 5.9 demonstrates how the temperature profile changes with different values of N_S. The smaller amounts of light component A produce higher temperatures in the column.

5.2 TERNARY SYSTEM WITH INERTS

The previous section considered the case in which the fresh feedstreams of both reactants A and B are pure. In most of the real commercial reactive distillation systems, lighter reactant A is fed with other components that are inert in terms of the reaction but have volatilities that are quite similar to component A. We will assume that fresh feedstream F_{0A} is a mixture of reactant A and an inert component I, which is not involved in the reaction. The volatility of I is assumed to be identical to that of A, so both of these components are lighter than the other reactant B and product C.

5.2.1 Column Configuration

In the ternary reaction system without inerts, the column has only a bottoms product in which heavy product C is removed and has only stripping and reactive zones. In the ternary reaction system with inerts, the column has both distillate and bottoms streams. Figure 5.10 gives the flowsheet of the reactive column.

Heavy product C comes out the bottom. Because inert component I has the same volatility as low-boiling A, it is removed from the column in a distillate stream. The reactive distillation column has all three zones: a stripping zone to keep light components A, I, and B from dropping out the bottom; a reactive zone in which the reaction occurs; and a rectifying zone to keep heavier component B from escaping out the top.

It is important to note that any A that leaves the top of the reactive zone will go overhead with inert I. The reactive zone cannot keep A from escaping. It can only keep reactant B from being lost in the distillate.

5.2.2 Chemistry and Phase Equilibrium Parameters

The chemical kinetics are slightly modified from those used in the case without inerts. Because the presence of the inert decreases the concentrations of the reactants, the reaction rates will be smaller. Therefore, $(K_{EQ})_{366}$ is increased to 50 in the base case. The effect of changing this parameter is demonstrated in the next section.

The volatility of I is identical to A, so the vapor pressure constants used for I are the same as those used for A (see Table 5.1).

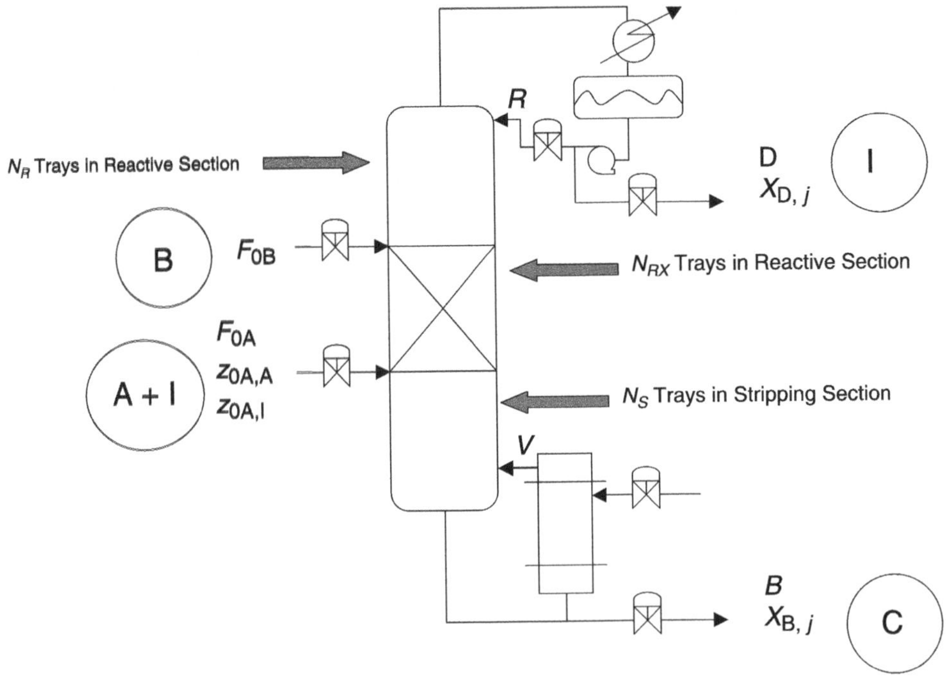

Figure 5.10 Ternary reactive distillation column with inerts.

5.2.3 Design Parameters and Procedure

In the *quaternary* system without inerts, the flowrates of both fresh feeds and both product streams are fixed, and the reflux is changed to drive the purity of one of the products to the desired specification. In the *ternary* system without inerts and without a distillate stream, the method for converging the column is very different. The flowrate and the composition of product C of the bottoms are fixed. There is no distillate. The flowrates of the two fresh feeds are calculated to provide the amounts needed for the reaction plus that lost in the bottoms. The reflux flowrate is changed to drive the bottoms composition to 98 mol% C. The vapor boilup is changed to control the level in the base. Reflux-drum level is not controlled.

The presence of inerts in the ternary systems requires a different method for converging the flowsheet. The impurity levels in both the bottoms and distillate streams are not known and change as design parameters change. The bottoms, which are 98 mol% C, can have impurities of any of the other three components: A, I, or B. The distillate, which is mostly I, can have impurities of either A or B.

The convergence method used in this ternary case with inerts is outlined in the following list and is quite different that those used in the other systems.

1. Fix the flowrate of fresh feed $F_{0B} = 12.6$ mol/s, which is pure B. This fixes the throughput in the process but does not fix the production rate of C because varying amounts of B are lost in both the distillate and bottoms products.
2. Fix the flowrate of reflux $R = 70$ mol/s. The effect of this design parameter will be explored.

3. Manipulate the bottoms flowrate to control the base level.
4. Manipulate the distillate flowrate to control the reflux-drum level.
5. Manipulate the vapor boilup to drive the composition of the bottoms to 98 mol% C.
6. Manipulate the flowrate of fresh feed F_{0A} to control the composition of A on the bottom tray in the reactive zone at 30 mol% A.

The last item requires some discussion. One of the important aspects of the design and/or control of a reactive distillation column operating in the neat mode is the need to balance the stoichiometry. The correct amounts of each reactant must be fed into the system. This usually requires some way of detecting any imbalance by measuring something inside the process that is an indication of a gradual buildup or depletion of one of the reactants. One way to achieve this is to measure an internal composition inside the column. Because the composition of A is highest at its feed tray, this location is selected.

The base case conditions are supplied in Table 5.3. Note that the number of reactive trays in the base case has been increased to 15 from the 9 considered in the system without inerts. Likewise, the holdup on the reactive trays has been increased to 2000 mol. The effects of these design parameters are explored in the following paragraphs. The composition $z_{0A(j)}$ of the F_{0A} fresh feed is 50 mol% A and 50 mol% I. This results in a much larger flowrate of this stream. The distillate is 97.2 mol% inerts. The main impurity is B at 2.04 mol%,

TABLE 5.3 Steady-State Conditions and Design Parameters for Base Case With Inerts

Fresh feed flowrate of A F_{0A} (mol/s)	24.51	
Composition of F_{0A} (mol% A/mol% I)	50/50	
Fresh feed flowrate of B F_{0B} (mol/s)	12.60	
Distillate flowrate D (mol/s)	12.58	
Bottoms flowrate B (mol/s)	12.39	
Vapor boilup V_S (mol/s)	65.10	
Reflux flowrate R (mol/s)	70	
Stripping trays N_S	5	
Reactive trays N_{RX}	15	
Rectifying trays N_R	5	
Internal control tray N_C	6	
Composition on control tray (mole fraction A)	0.30	
$(K_{EQ})_{366}$	50	
Liquid holdup on reactive trays M_{RX} (mol)	2000	
Pressure P (bar)	8	
Vapor Pressure Constants	A_j	B_j
A	12.34	3862
B	11.65	3862
C	10.96	3862
I	12.34	3862
Product Composition (Mole Fraction)	Distillate x_{Dj}	Bottoms x_{Bj}
A	0.0017	0.0017
B	0.0204	0.0163
C	0	0.9800
I	0.9719	0.0020

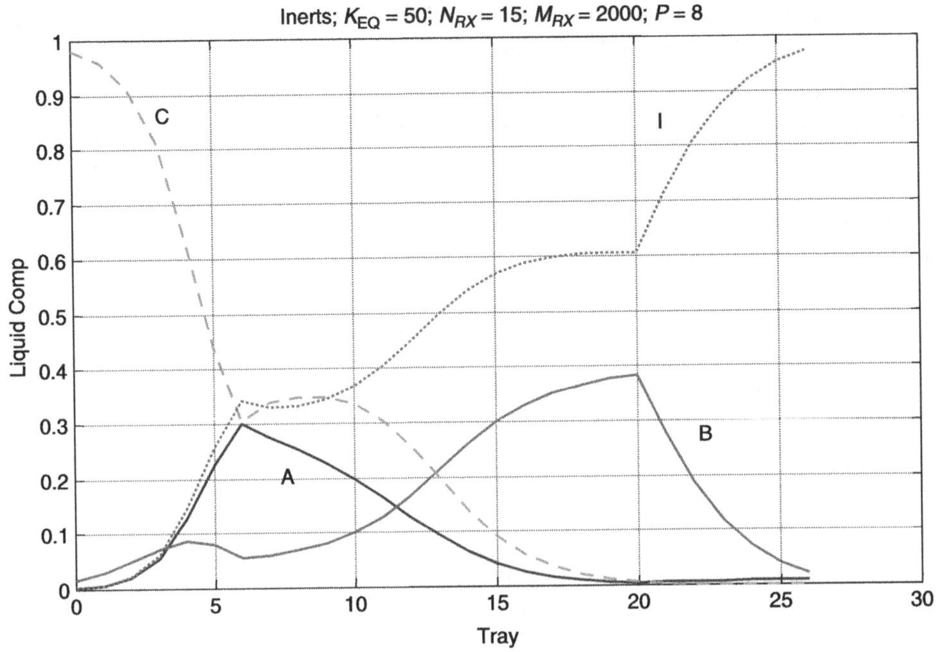

Figure 5.11 Composition profile with inerts.

but there is a small amount of A. The bottoms are 98 mol% C with the main impurity B being 1.63 mol% and small amounts of both A and I.

Figure 5.11 gives the base case liquid composition profiles. The concentration of A is the highest at tray 6 where it is fed. The concentration of inert I builds up throughout the reactive and rectifying stages, particularly in the rectifying stages where B is prevented from being lost in the overhead. Note that there is a slight increase in the concentration of A in the rectifying section. Any A that leaves the reactive zone cannot be prevented from going overhead with the inert because the two have the same volatility.

5.2.4 Effect of Pressure

Now that we have a design procedure and a base case, we are ready to see how various parameters affect the ternary reactive column with inerts. Figure 5.12 shows the effect of changing the column operating pressure. The pressure that minimizes vapor boilup is about 8.5 bar.

However, in this system there are other considerations that are equally, if not more, important than energy consumption. The top right graph in Figure 5.12 shows that pressures higher or lower than 8.5 bar require more fresh feed F_{0A} to keep the composition of A on tray 6 at 30 mol%. This increase in the amount of A (and I) fed into the column goes out in the distillate. The lower left graph in Figure 5.12 shows that the impurity of A in the distillate increases at higher or lower pressures. Remember that the amount of reactant B entering the column is constant because F_{0B} is fixed. The amount of C produced is essentially constant. This means that the additional reactant A fed into the system represents a loss of an expensive reactant, which usually has more economic impact than energy.

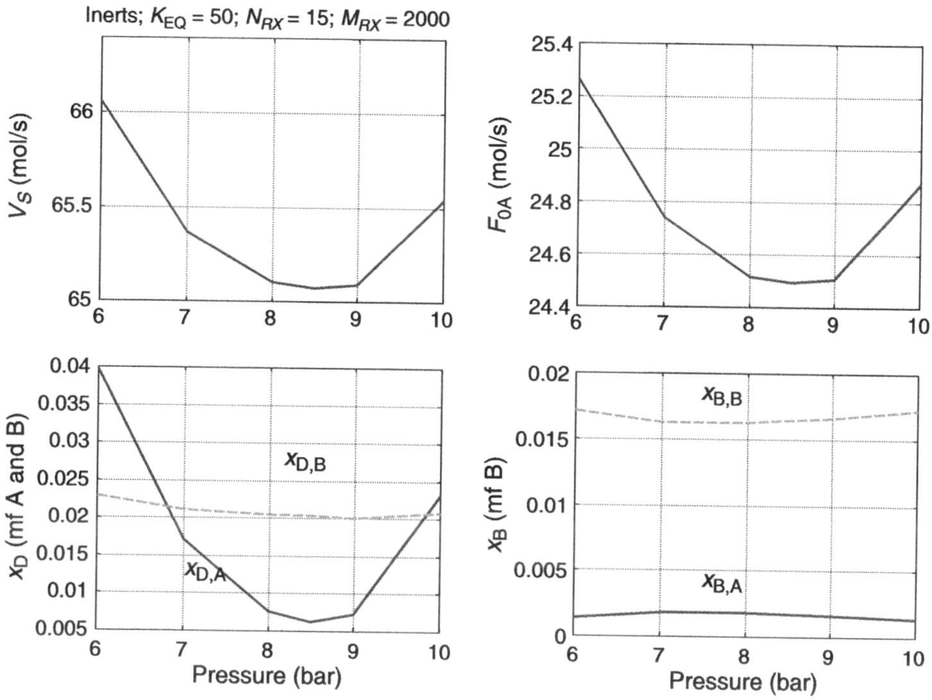

Figure 5.12 Effect of pressure with inerts.

One of the most important aspects of the design and operation of the ternary system with inerts is to minimize the losses of reactants that can leave with the inerts in the distillate. We will see how reactant losses change as design parameters are varied in the following sections.

Figure 5.13 presents the composition profiles at three pressures. The compositions of A in the reactive zone at both high and low pressures are increased because of the higher fresh feed rate F_{0A}.

5.2.5 Control Tray Composition

As stated in the design procedure, the composition of reactant A on the first tray in the reactive zone (tray 6 in the base case) is controlled by changing fresh feed F_{0A}. The composition on this tray is 30 mol% A in the base case. In this section we explore how changing the specified value of $x_{6,A}$ affects the design.

Figure 5.14 shows that there is a range of values of $x_{6,A}$ between about 20 and 30 mol% A over which there is very little change in vapor boilup (energy) and loss of reactants (yield). However, for smaller values of $x_{6,A}$, vapor boilup increases and fresh feed F_{0A} decreases. There is not enough A being fed to consume the fixed amount of B that is entering the system. Unreacted B goes out the top of the column, as shown in the rapid increase in the composition of B in the distillate (lower left graph, Fig. 5.14). The reduction in the production of C produces a reduction in the bottoms flowrate (lower right graph, Fig. 5.14).

At the other end of the range where the values of $x_{6,A}$ are $>$30 mol%, vapor boilup again increases. However, fresh feed F_{0A} increases instead of decreasing. This occurs because the

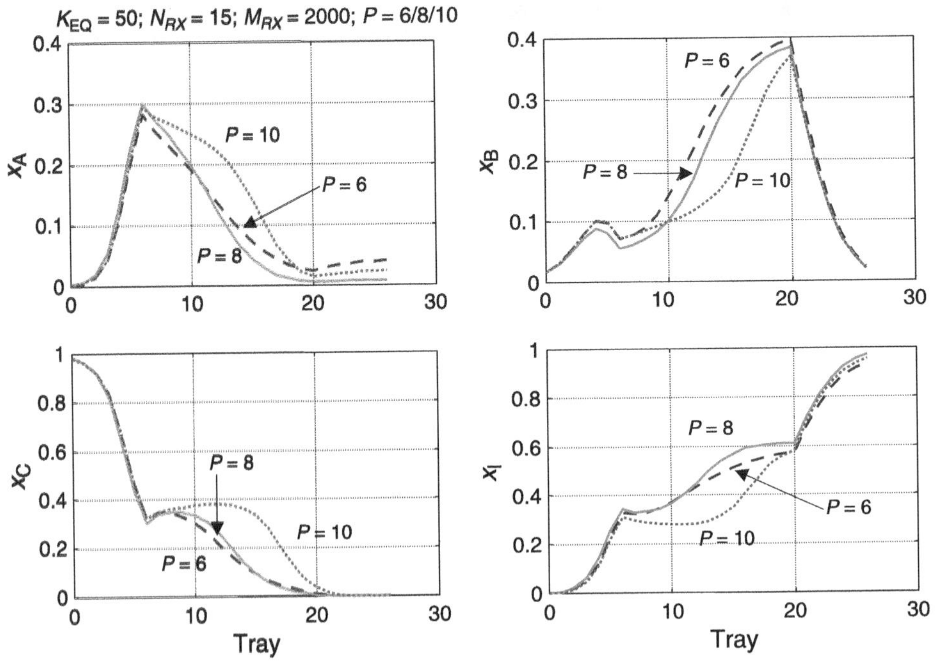

Figure 5.13 Effect of pressure on composition profiles.

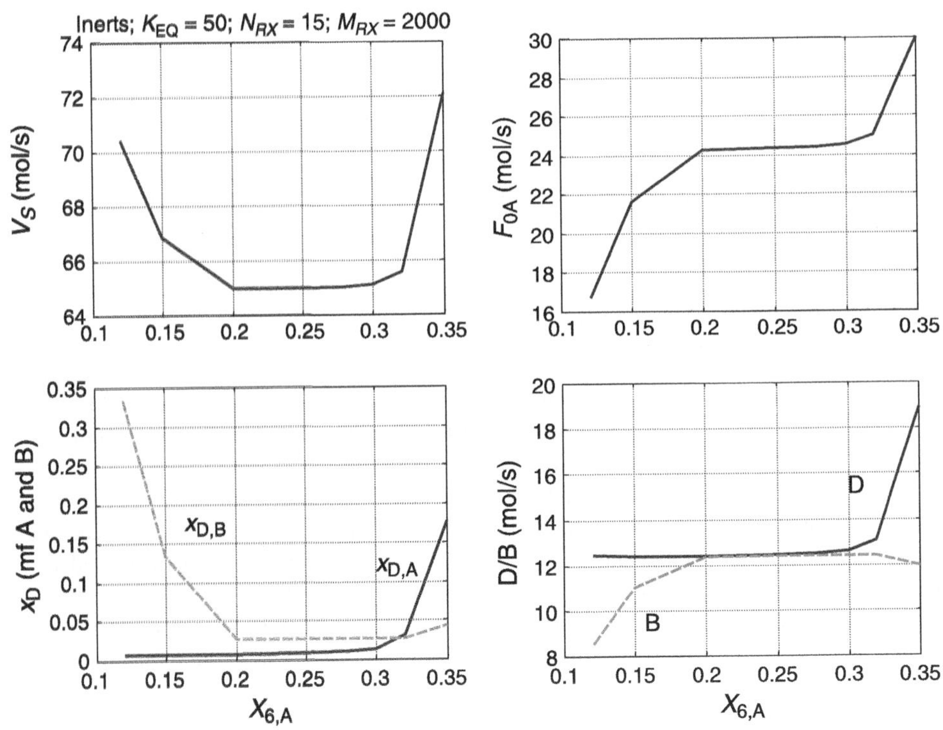

Figure 5.14 Effect of control tray composition.

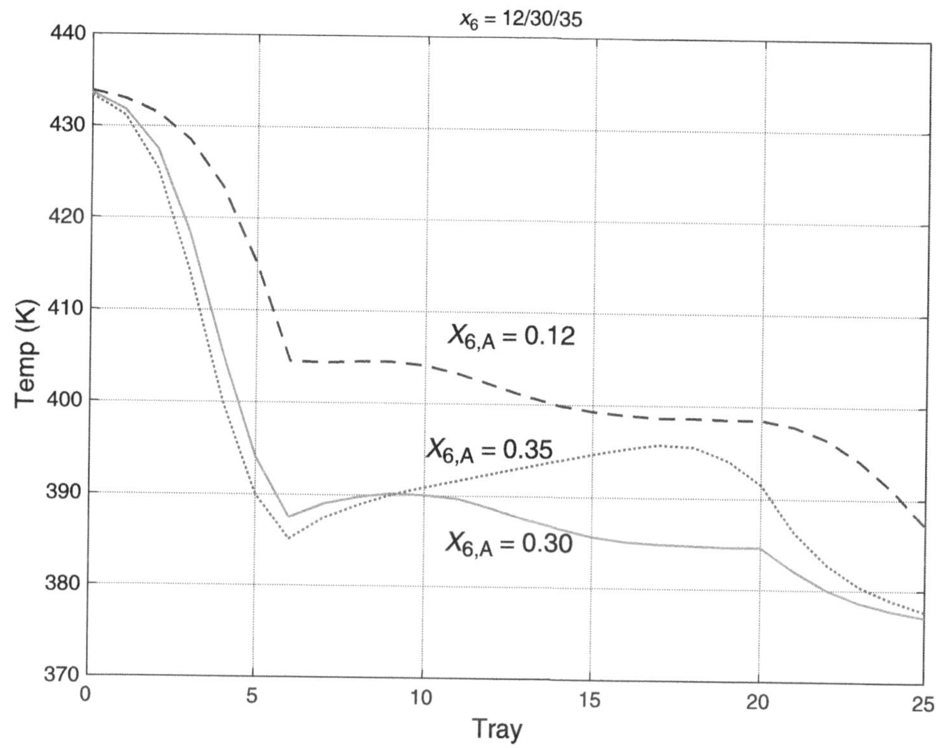

Figure 5.15 Effect of control tray composition on temperature profile.

high composition of A on tray 6 pushes too much A out the top of the column. The lower graphs in Figure 5.14 show that the composition of A in the distillate and the flowrate of the distillate both increase. There is a large loss of reactant A.

The temperature profiles for three values of $x_{6,A}$ are given in Figure 5.15. The low $x_{6,A}$ value of 12 mol% A corresponds to having less of lighter A and I components and more of heavier B throughout the column, so the temperatures are higher. The high $x_{6,A}$ value of 35 mol% A corresponds to having more of lighter A and I components in the stripping section of the column, so temperatures are lower there. However, the concentration of product C in the reactive zone increases because the reaction is driven to the product side by the high concentration of A. The result is that temperatures are somewhat higher in the reactive zone than they are in the base case $x_{6,A} = 30$ mol%. Figure 5.16 shows the composition profiles for all four components with the three values of $x_{6,A}$.

These results indicate that the design is fairly insensitive to the value of the A composition controlled on the internal tray.

5.2.6 Reactive Tray Holdup

In the quaternary system and in the ternary system without inerts, increasing reactive tray holdup improved reactive column performance in terms of reducing energy consumption. Figure 5.17 demonstrates that the same is true for the ternary system with inerts. However, in addition to the energy benefits, there is also an improvement in yield. Less

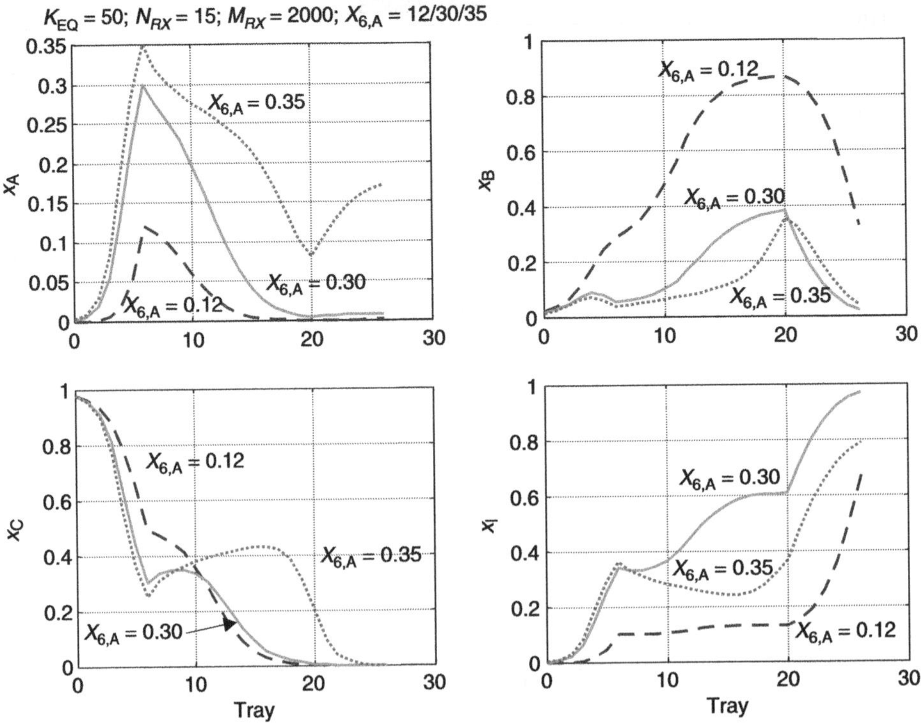

Figure 5.16 Effect of control tray composition on composition profiles.

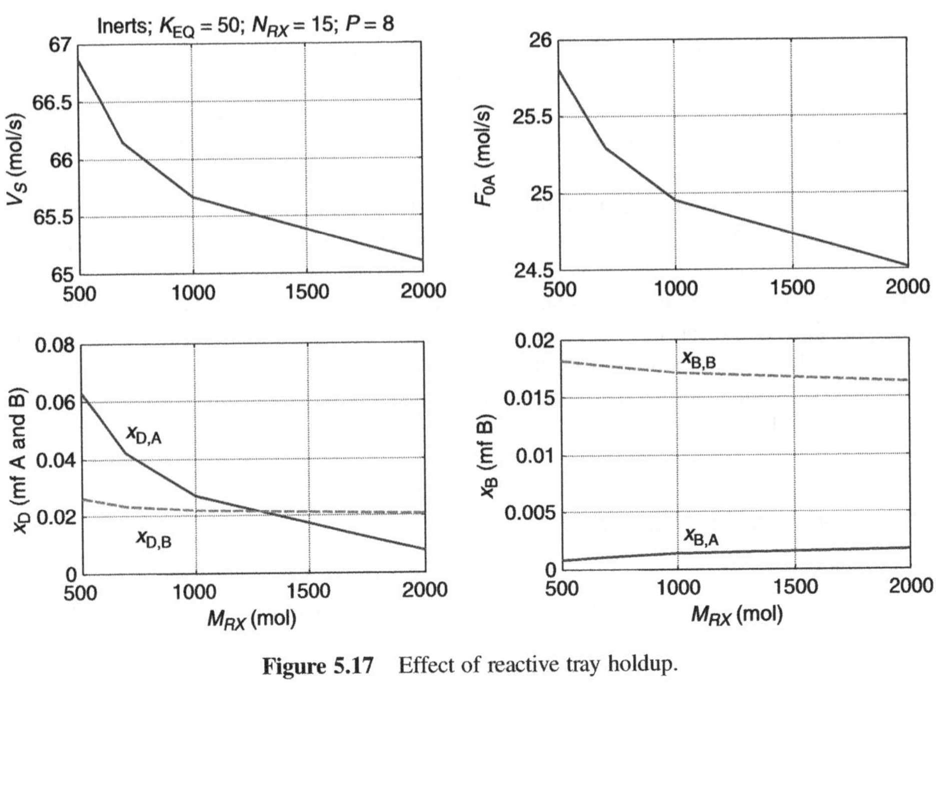

Figure 5.17 Effect of reactive tray holdup.

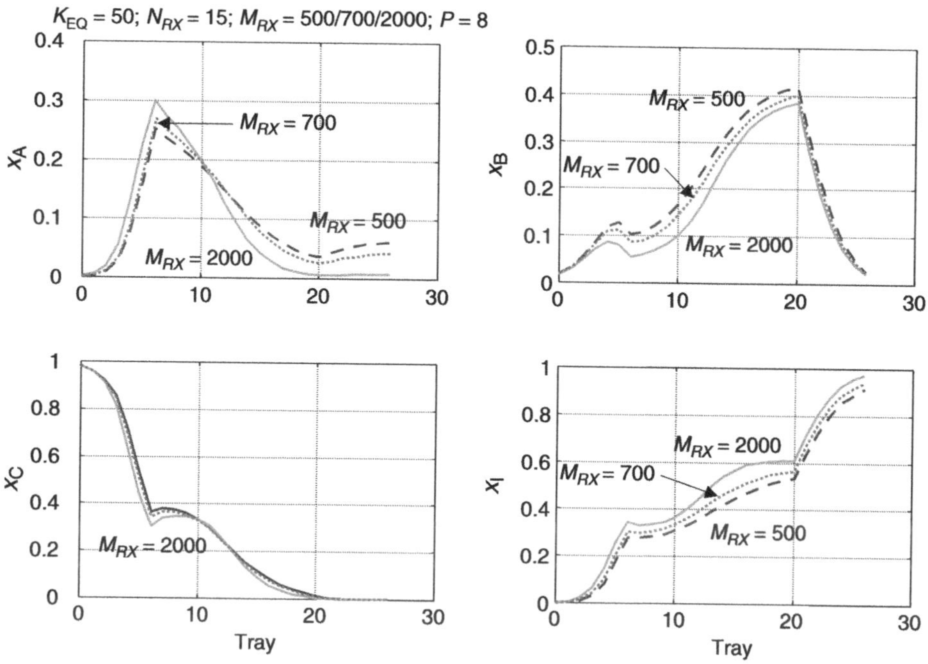

Figure 5.18 Effect of reactive tray holdup on composition profiles.

A is lost because less A is fed (upper right graph). and the concentration of A in the distillate is lower (lower left graph).

Figure 5.18 gives the composition profiles for three different reactive tray holdups. Increasing holdup reduces the concentrations of both reactants in the reactive zone, so less A is lost out the top.

5.2.7 Effect of Reflux

In the base case, the flowrates of both the fresh feed of B (F_{0B}) and reflux (R) are specified. The production rate is basically set by F_{0B}. In the base case the reflux is 70 mol/s. Increasing the reflux flowrate for a fixed flowrate of F_{0B} should require more vapor boilup, but it should improve the separation. If the improvement in separation results in a significant reduction in the loss of reactants, the optimum design might not correspond to the point of minimum vapor boilup.

The results given in Figure 5.19 show that vapor boilup is minimized at a reflux flowrate of 66 mol/s. However, at this reflux flowrate there is a very large loss of reactant A in the distillate. Both F_{0A} and $x_{D,A}$ increase rapidly for reflux flowrates lower than 69 mol/s. At higher reflux flowrates, F_{0A} and $x_{D,A}$ again increase. Note that using somewhat more reflux than that giving minimum vapor boilup involves a small penalty in energy and yield. However, using a reflux that is only a very small amount less than that giving minimum vapor boilup produces a very severe penalty in both energy and yield. Therefore, a reasonable value for the reflux flowrate is probably 70 mol/s because this provides a little cushion between the operating point and "going over the cliff" into a region of very high yield losses.

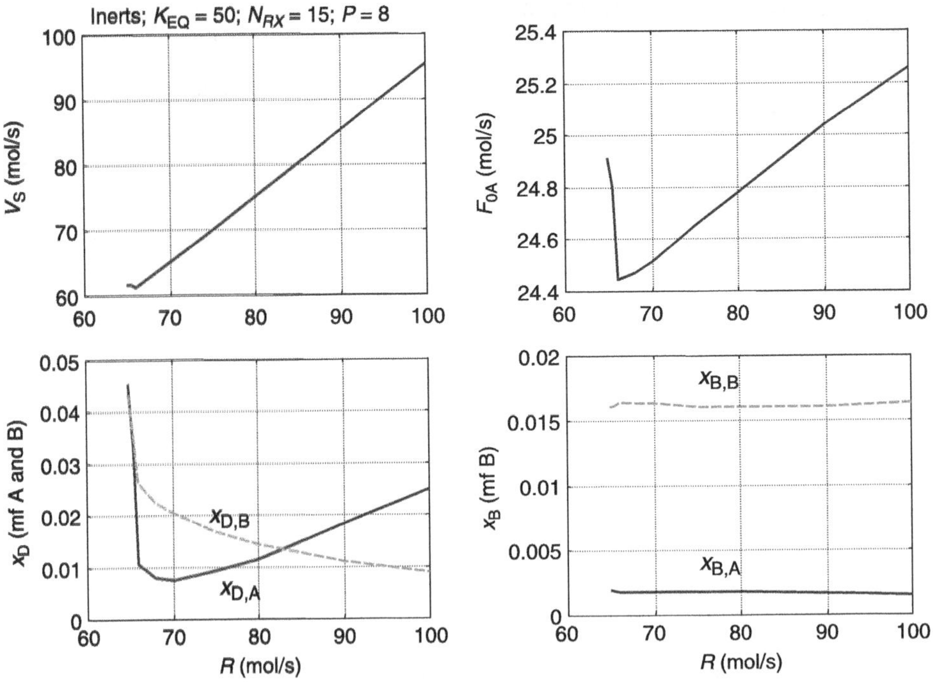

Figure 5.19 Effect of reflux.

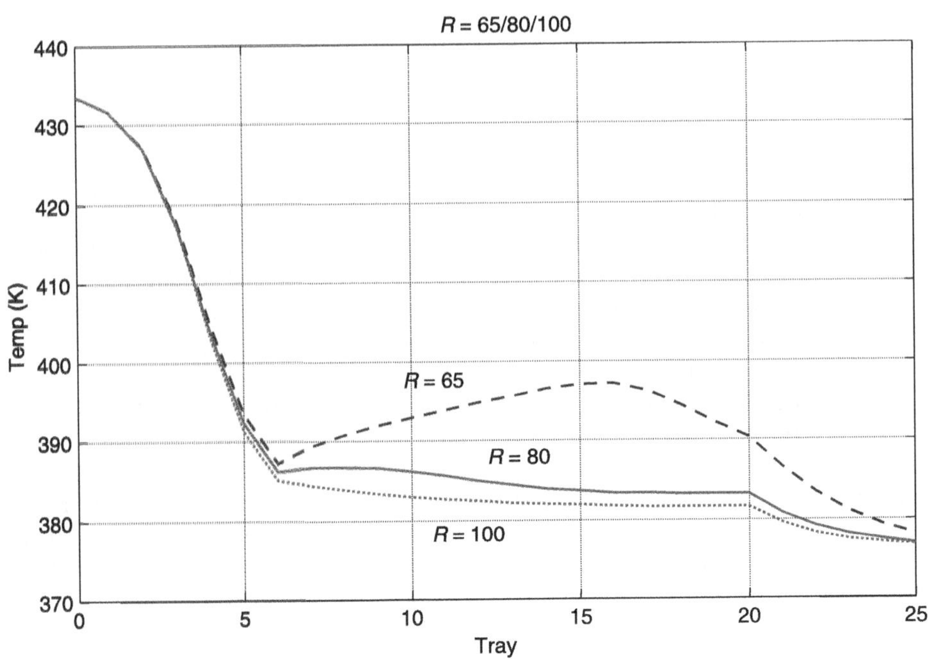

Figure 5.20 Effect of reflux on temperature profiles.

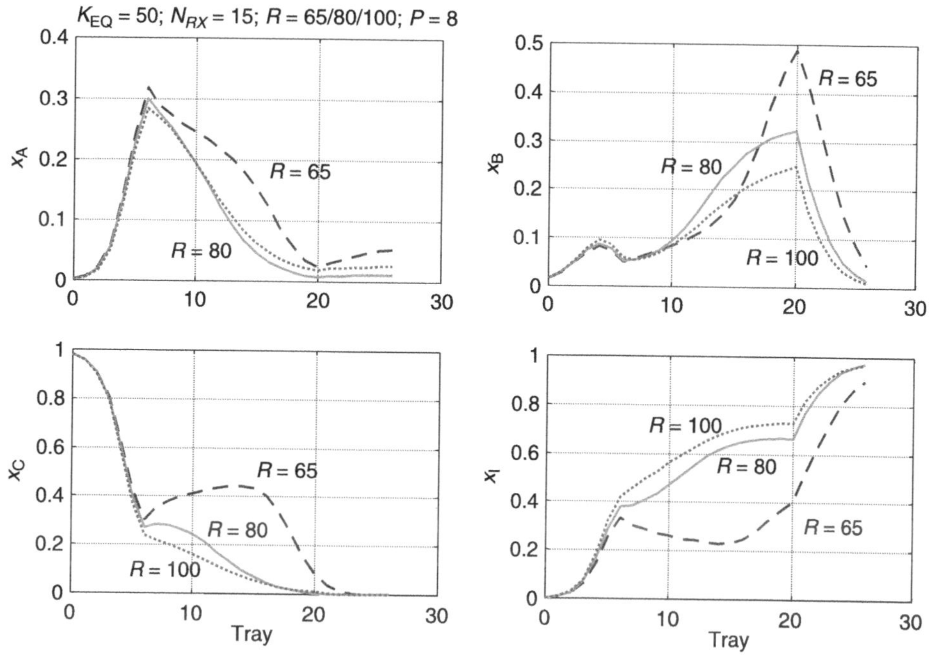

Figure 5.21 Effect of reflux on composition profiles.

Figures 5.20 and 5.21 present temperature and composition profiles for three value of reflux. Low reflux flowrates produce lower concentrations of light inert component I and higher concentrations of heavier components B and C in the column, so temperatures are higher. High reflux flowrates have the opposite effect, so temperatures are lower.

5.2.8 Chemical Equilibrium Constant

The chemical equilibrium constant in the base case is 50. Reducing the value of K_{EQ} makes it more difficult to achieve the desired high conversion. We would expect that vapor boilup would have to increase. The upper left graph in Figure 5.22 shows that this is indeed what happens.

In addition to increasing energy consumption, reducing K_{EQ} increases the loss of reactant. The upper right graph in Figure 5.22 shows that more A is fed, and the lower left graph shows that the impurities of both reactants increase in the distillate. Figures 5.23 and 5.24 provide the temperature and composition profiles, respectively, for three values of K_{EQ}. As the chemical equilibrium constant decreases, the concentrations of heavy component C on the trays in the column decrease and the concentrations of the light inert increase. The result is lower temperatures in the column.

5.2.9 Feed Composition

The concentration of inert in the fresh feed of reactant A is assumed to be $z_{0A(A)} = 0.50$ in all of the cases considered in the previous sections. Now we wish to see how this

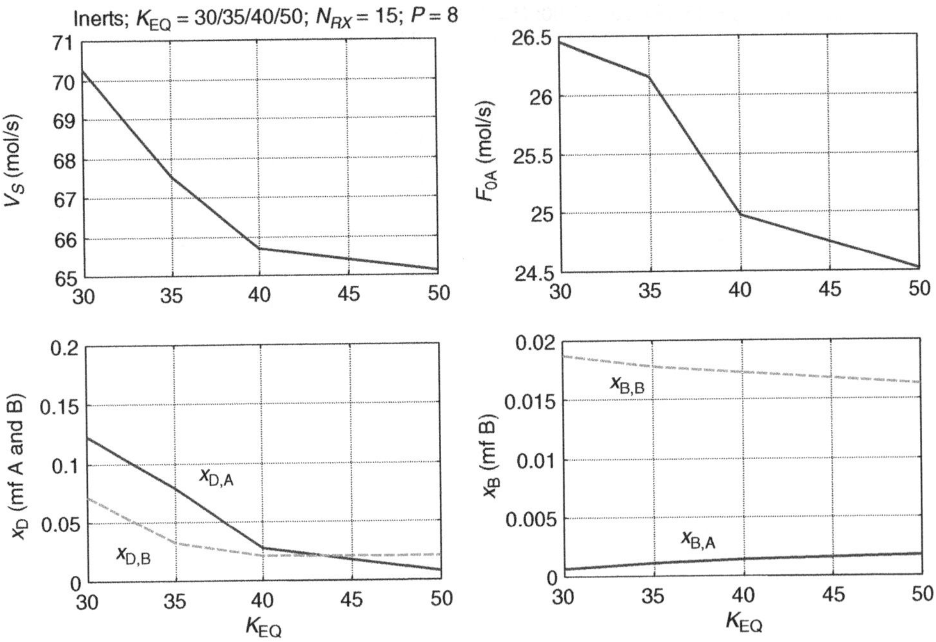

Figure 5.22 Effect of K_{EQ}.

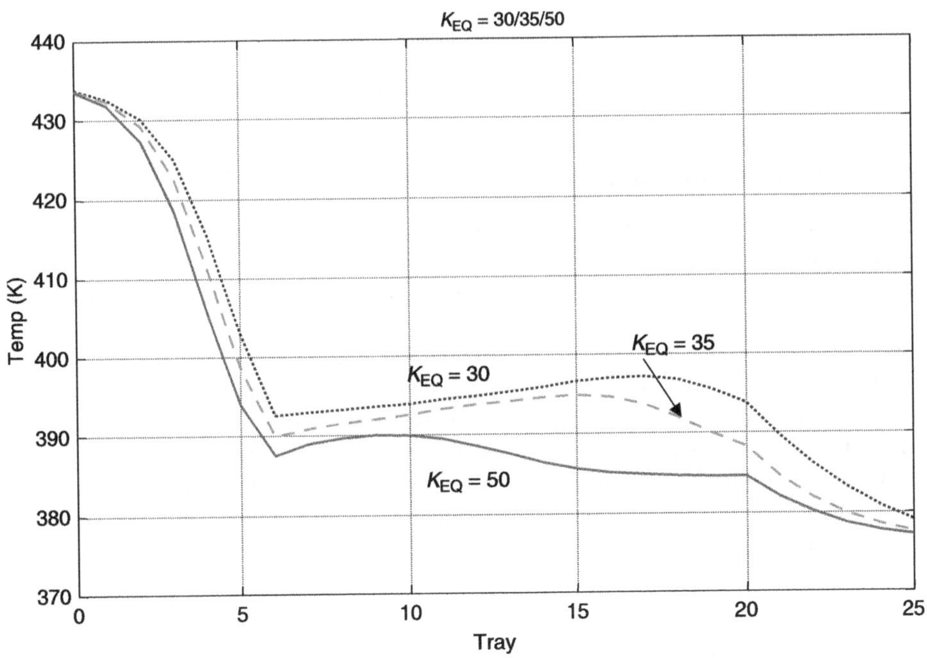

Figure 5.23 Effect of K_{EQ} on temperature profile.

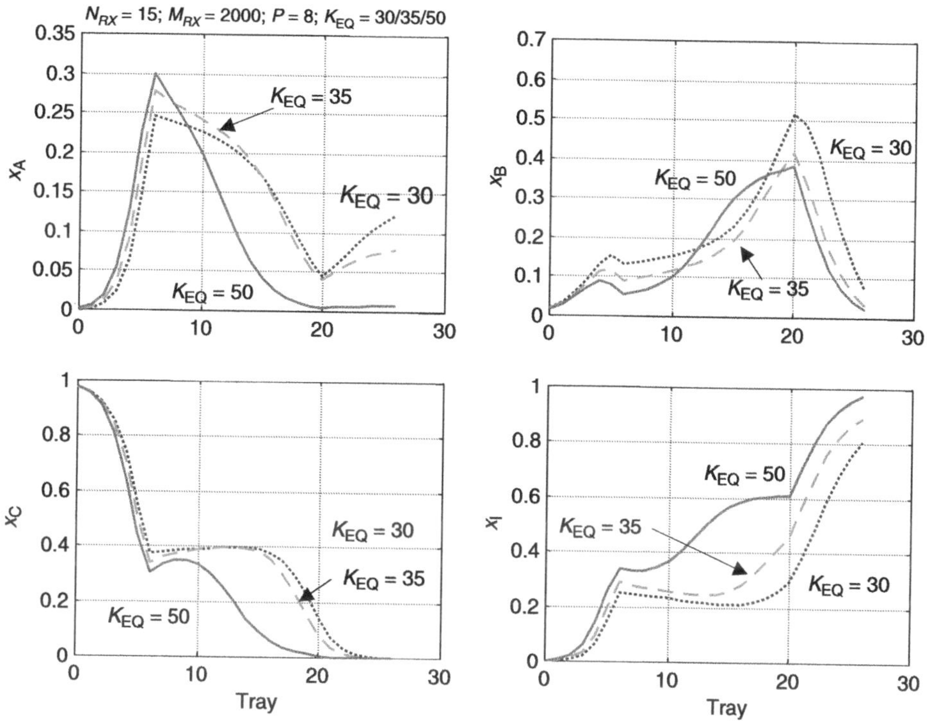

Figure 5.24 Effect of K_{EQ} on composition profiles.

composition affects the design of the reactive distillation column. Of course, the limiting case with $z_{0A(A)} = 1$ and $z_{0A(I)} = 0$ was already studied in Section 5.1 (no inerts). We would expect that it should be easier to achieve the desired conversion as the amount of inerts in the feed decreases.

Figure 5.25 shows that this is true. The upper graphs show that both vapor boilup and fresh feed F_{0A} decrease as less inert enters in the feed. Of course, the distillate flowrate also decreases because there is less inert to remove.

The specified value of the concentration of A on tray 6 has been changed as $z_{0A(A)}$ is changed. With no inerts in the system ($z_{0A(A)} = 1$), the composition on tray 6 is 10 mol% A. In Section 5.2.5 we found that a tray 6 composition of about 30 mol% A is reasonable for the case where $z_{0A(A)}$ is 0.5. Therefore, values of 20 and 25 mol% A are used for the specified composition on tray 6 for the cases with $z_{0A(A)} = 0.70$ and 0.60, respectively.

Figures 5.26 and 5.27 give the temperature and composition profiles for three values of $z_{0A(A)}$. The most striking feature in these plots is the higher temperature when $z_{0A(A)} = 0.7$, which is caused by the much higher concentrations of component B and the lower concentrations of I throughout the column (right graphs, Fig. 5.27). The B composition profile should be compared with that shown in Figure 5.2, which has no inerts. The same high concentration of B can be observed.

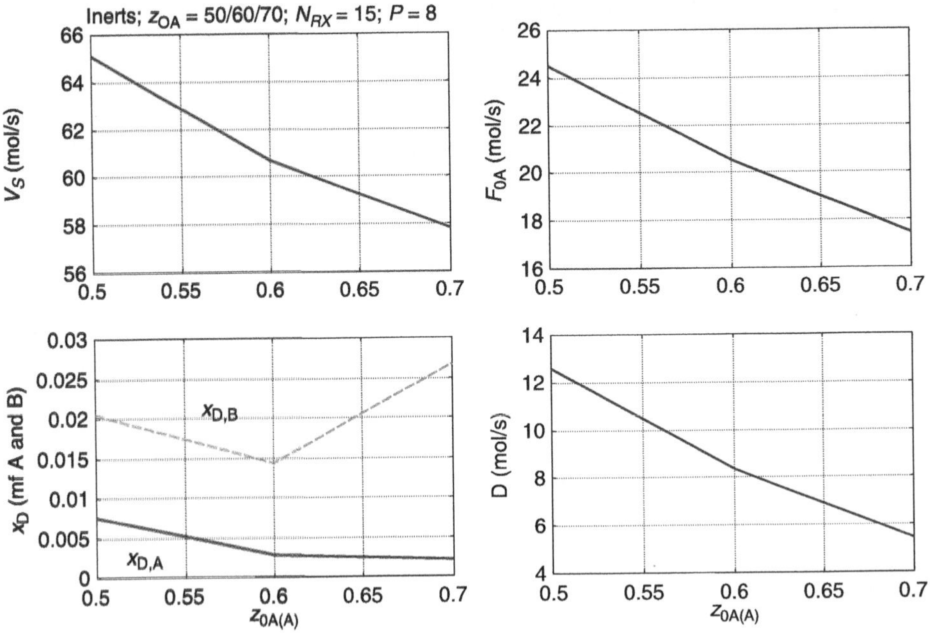

Figure 5.25 Effect of feed composition z_{0A}.

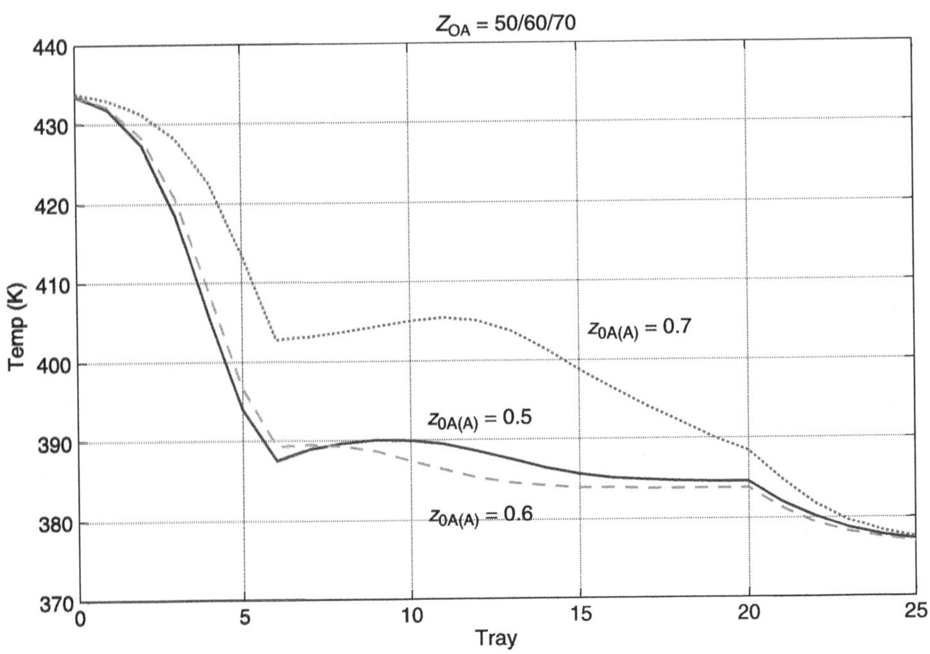

Figure 5.26 Effect of $z_{0A(A)}$ on temperature profile.

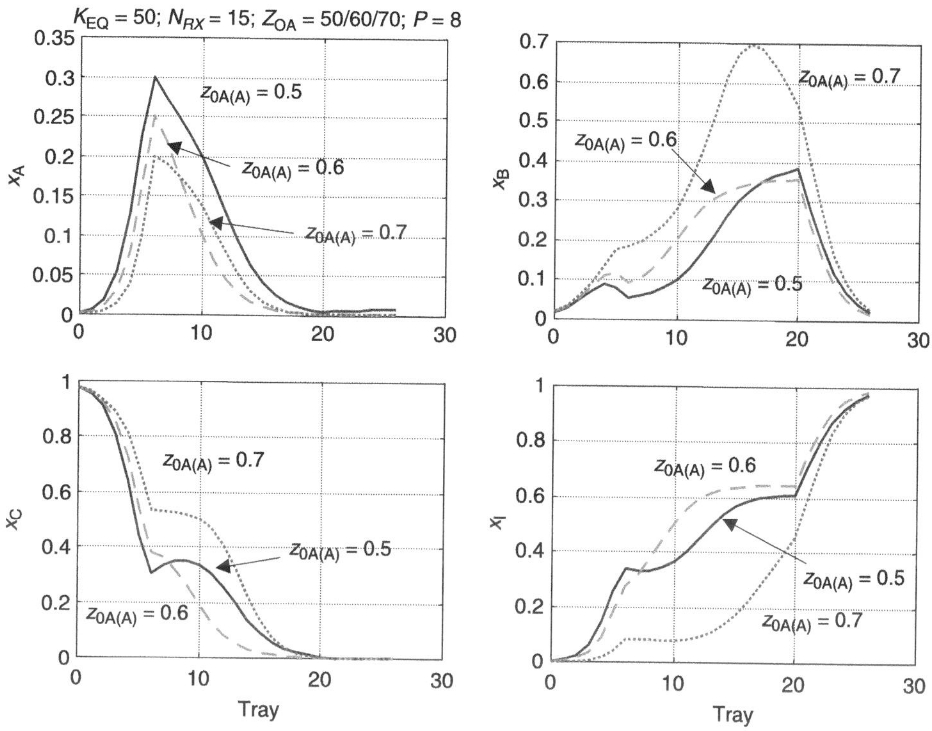

Figure 5.27 Effect of $z_{0A(A)}$ on composition profiles.

5.2.10 Number of Reactive Trays

Increasing the number of reactive trays N_{RX} improved energy consumption in the ternary system without inerts. When inerts are present, the same result is observed, as shown in Figure 5.28. Vapor boilup decreases as more reactive trays are added.

In addition, the loss of reactant decreases as N_{RX} increases. Both the amount of reactant A fed and its concentration in the distillate decrease as N_{RX} increases. Thus, there is a double benefit for having more reactive trays in the system with inerts. As we will demonstrate in Part IV of this book, there are additional dynamic benefits for having numerous reactive trays.

Temperature profiles are quite similar for all values of N_{RX}, as demonstrated in Figure 5.29.

5.2.11 Number of Rectifying and Stripping Trays

The final parameter explored in this chapter is the number of trays used in the two separation sections. In Chapter 2 we found that increasing the number of stripping and rectifying trays decreases energy consumption in the quaternary system. In Section 5.1.7 in this chapter we found that there is an optimum number of stripping trays in the ternary system without inerts. What are the effects for the ternary system with inerts?

Figure 5.30 provides the answer to this question. The base case value is $N_R = N_S = 5$. When this number is increased, vapor boilup decreases but only slightly. However, if

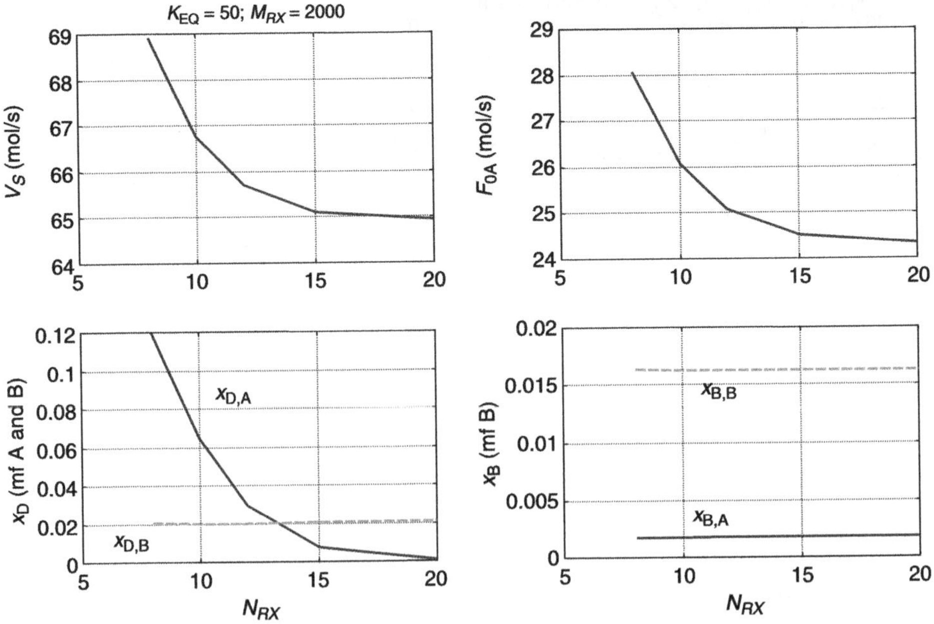

Figure 5.28 Effect of number of reactive trays.

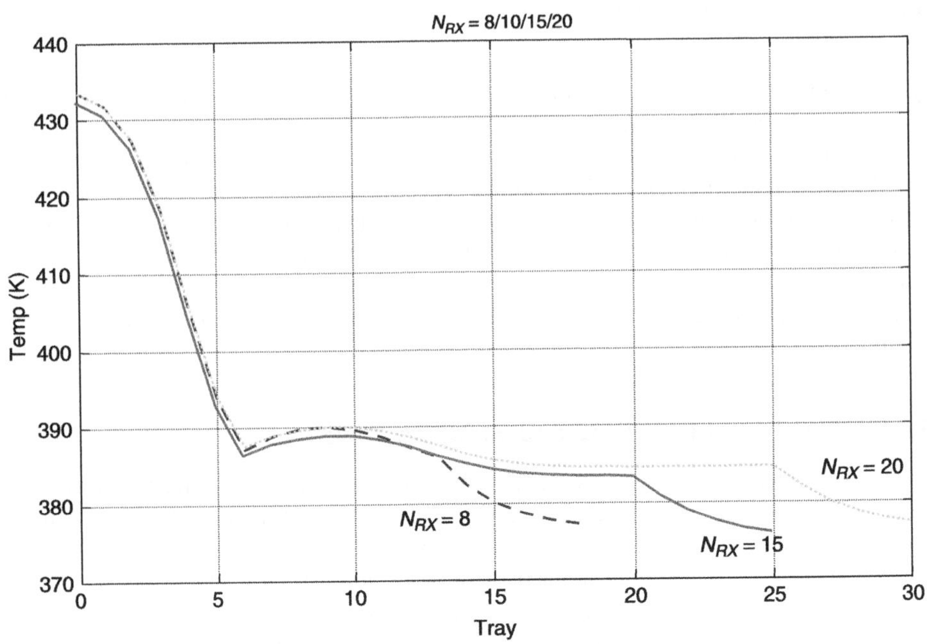

Figure 5.29 Effect of number of reactive trays on temperature profile.

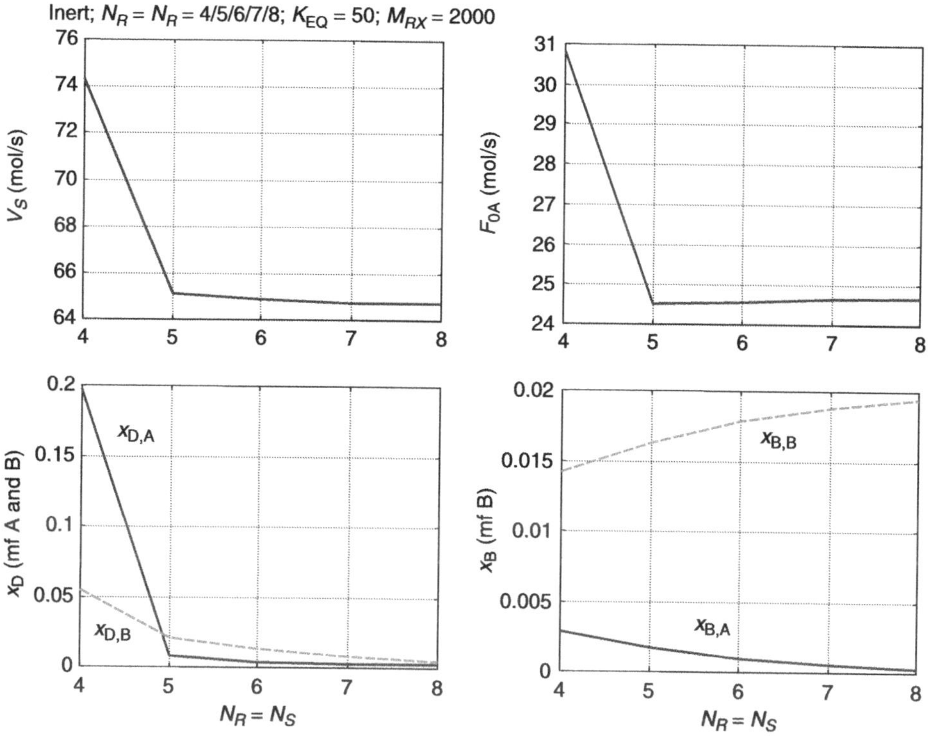

Figure 5.30 Effect of number of rectifying and stripping trays.

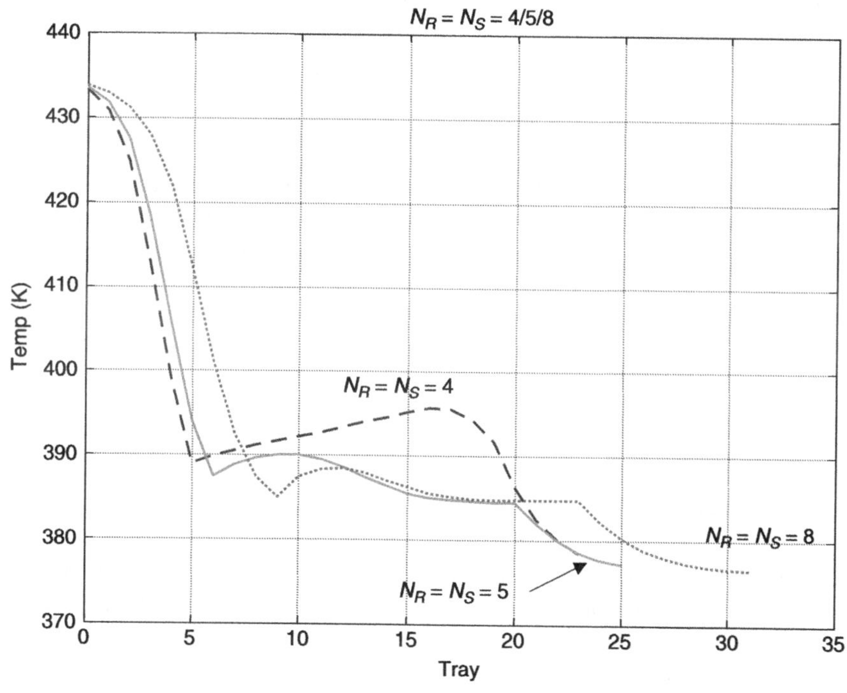

Figure 5.31 Effect of N_R and N_S on temperature profile.

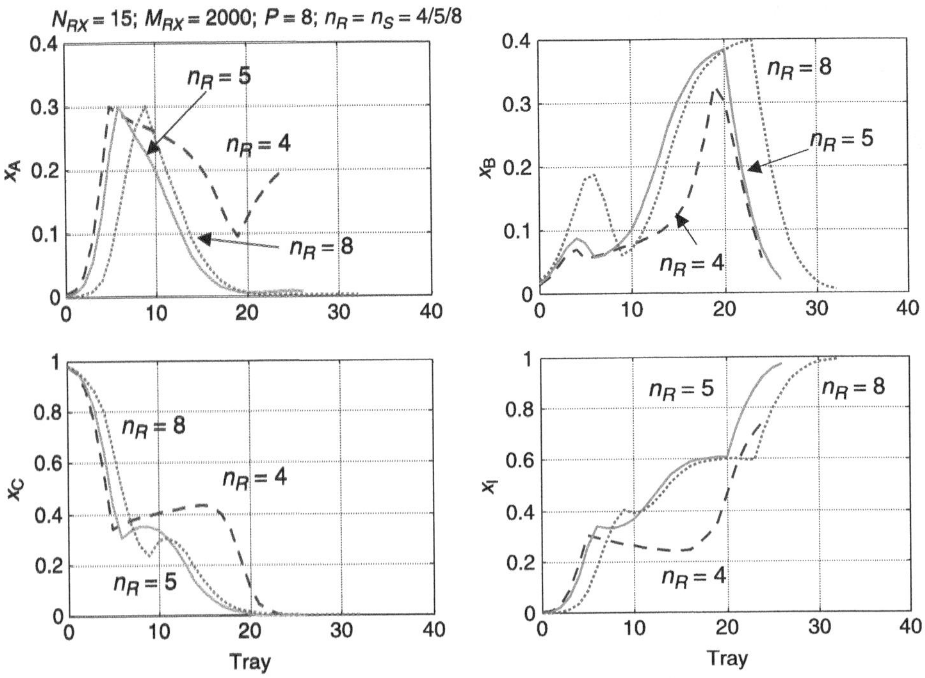

Figure 5.32 Effect of N_R and N_S on composition profiles.

this number is decreased from 5 to 4, there is a sharp jump in vapor boilup. There are also large increases in the fresh feed of A and of the impurity of A in the distillate. Thus, both energy and yield considerations indicate that there is a very hard lower constraint on the number of separation stages.

Figures 5.31 and 5.32 give temperature and composition profiles for three values of separation stages. The $N_R = N_S = 4$ case shows a composition profile with high concentrations of heavy C component in the reactive zone, which raises temperatures. The concentrations of A are also large, but the concentrations of I are small. Because A and I have the same volatility, their opposite changes in compositions tend to cancel each other.

Note the increase in the composition of A in the rectifying section for the $N_R = N_S = 4$ case. A large amount of A is escaping from the top of the reactive zone, but the rectifying trays are only able to keep heavier components B and C from leaving in the distillate.

5.3 CONCLUSION

Two different ternary systems have been explored in this chapter. Both have two reactants and one product. However, one has an additional component coming in with one of the feeds that is not involved in the reaction. It is inert as far as the chemistry is concerned. However, it dilutes the concentrations of the reactants and the products, so it does affect overall reaction rates.

Of more importance, the presence of the inert component has a major impact on both the structure of the column (two outlet streams are required instead of one) and the vapor–liquid phase equilibrium.

The effects of a number of kinetic and design parameters have been explored and found to change significantly from system to system. The dynamics and control of these ideal systems are studied in Part IV.

CHAPTER 6

TERNARY DECOMPOSITION REACTION

As opposed to the synthesis reaction A + B ⇔ C, we focus in this chapter on the decomposition reaction A ⇔ B + C. The reaction is reversible, liquid phase, and endothermic. Two different relative volatility rankings are studied. The components are labeled as light key (LK), intermediate key (IK), and heavy key (HK) according to their boiling point ranking. The first case is when the reactant is the intermediate key: IK ⇔ LK + HK. Two examples are *tert*-butyl alcohol, which decomposes into isobutene and water,[1] and cumene hydroperoxide, which decomposes into phenol and water.[2] The second case is when the reactant is the heavy key: HK ⇔ LK + IK. Notable examples include MTBE decomposition,[3] phenyl-ethanol decomposition,[4] and glycerol decomposition.[5] The relative volatility ranking has an important effect on the process configurations and design. These two cases will be explored separately to see how design parameters affect the reactive distillation design.

[1]Z. W. Qi and K. Sundmacher, Multiple product solutions of *tert*-butyl alcohol dehydration in reactive distillation, *Ind. Eng. Chem. Res.* **45**, 1613–1621 (2006).
[2]L. Doron and S. Jose, Apparatus for producing phenol using reactive distillation, US Patent 6,411,252 (2002).
[3]K. J. Huang and S. J. Wang, Design and control of a methyl tertiary butyl ether (MTBE) decomposition reactive distillation column, *Ind. Eng. Chem. Res.* **46**, 2508–2519 (2007).
[4]J. P. Lange and V. Otten, Dehydration of phenyl-ethanol to styrene: Zeolite catalysis under reactive distillation, *J. Catal.* **238**, 6–12 (2006).
[5]C. W. Chin, M. A. Dasari, G. J. Suppes, and W. R. Sutterlin, Dehydration of glycerol to acetol via catalytic reactive distillation, *AIChE J.* **52**, 3543–3548 (2006).

Reactive Distillation Design and Control. By William L. Luyben and Cheng-Ching Yu

6.1 TERNARY DECOMPOSITION REACTION: INTERMEDIATE-BOILING REACTANT

6.1.1 Column Configuration

This is the most favorable relative volatility ranking in reactive distillation: IK $\Leftrightarrow$ LK + HK. The reactant (IK) is kept in the reactive zone, which is placed in the middle of the column, and the light and heavy products (LK and HK) can be withdrawn from the top and bottom of the column. Thus, the reactant concentration remains high in the reactive zone (midsection of the column) and products are removed continuously via separation in the rectifying and stripping sections. Figure 6.1 gives the flowsheet in which the reactive zone is placed in the midsection of the column.

6.1.2 Chemistry and Phase Equilibrium Parameters

Table 6.1 provides the kinetic and vapor–liquid phase equilibrium parameters used in the numerical case considered in this chapter. The reaction on the nth tray can be expressed as

$$\Re_n = M_{RX}(k_F x_{n\mathrm{A}} - k_B x_{n\mathrm{B}} x_{n\mathrm{C}}) \tag{6.1}$$

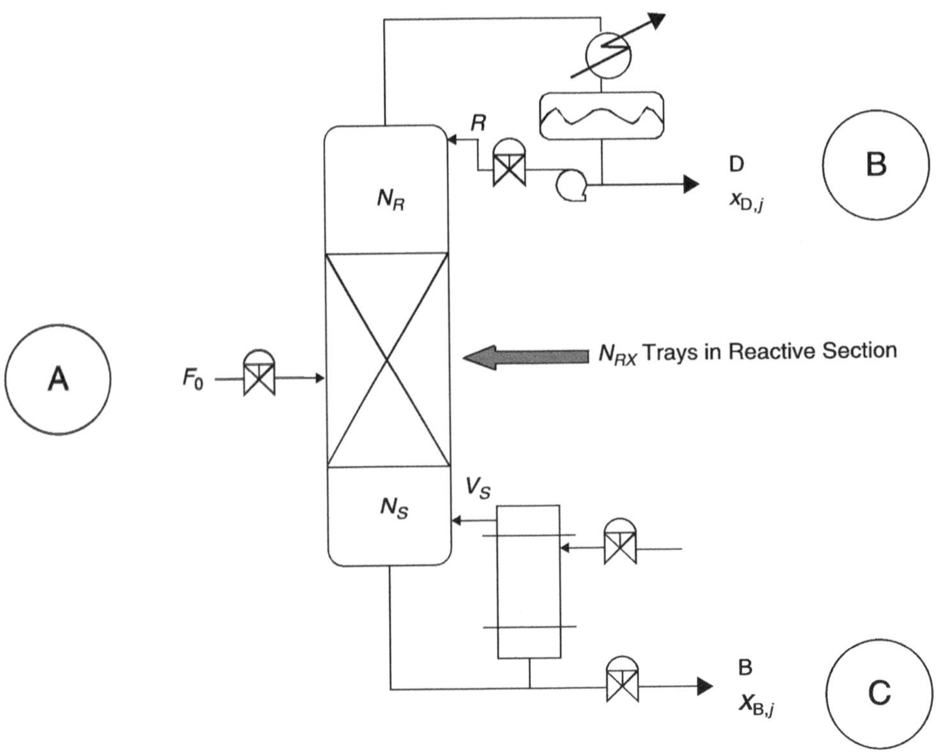

Figure 6.1 Ternary reactive distillation for decomposition reaction A $\Leftrightarrow$ B + C (IK $\Leftrightarrow$ LK + HK), with $\alpha_B/\alpha_A/\alpha_C = 4/2/1$.

TABLE 6.1 Kinetic and Vapor–Liquid Equilibrium Parameters

Activation energy (kcal/mol)		
Forward	40	
Backward	30	
Specific reaction rate at 366 K (kmol s^{-1} $kmol^{-1}$)		
Forward	0.008	
Chemical equilibrium constant at 366 K	0.2	
Heat of reaction (kcal/mol)	+10	
Heat of vaporization (kcal/mol)	6.944	
Molecular weights (A/B/C) (g/mol)	100/50/50	
Vapor Pressure Constants	A_j	B_j
A	11.65	3862
B	12.34	3862
C	10.96	3862

The forward reaction is first order in the mole fraction of A, and the backward reaction is proportional to the mole fractions of B and C. The equilibrium constant K_{EQ} is set to 0.2 as a result of an imbalance in reaction stoichiometry. Note also that this is an endothermic reaction as one would expect from a decomposition reaction (cracking). For phase equilibrium, constant relative volatility is assumed and the tray temperature can be computed from the tray liquid composition using the Antoine vapor pressure equations.

6.1.3 Design Parameters and Procedure

From a process configuration perspective, the decomposition reaction A ⇔ B + C with the reactant A being the intermediate boiler (e.g., relative volatility B/A/C = 4/2/1) is actually quite similar to the quaternary reaction system A + B ⇔ C + D with intermediate-boiling reactants A and B (e.g., C/A/B/D = 8/4/2/1). In both cases, the reactive zone is placed in the middle of the column with a rectifying section and a stripping section at the ends of the column. High purity heavy product C can be obtained from the bottoms, and light product B is further purified and withdrawn from the top. The vapor boilup and reflux ratio are candidate manipulated variables to control the product purities in both ends of the column. As compared to the quaternary system, there is an important difference: this is a single feed system. Therefore, the feed location can be assumed to be in the middle of the reactive zone.

The vapor rate (V_n) on a reactive tray is affected by the heat of reaction (λ). Knowing the heat of vaporization (ΔH_V), the vapor leaving reactive tray n can be expressed as

$$V_n = V_{n-1} - \frac{\lambda}{\Delta H_V} \mathfrak{R}_n \tag{6.2}$$

The reaction is endothermic, so λ is positive. Thus, the vapor rates decrease as we move up the column.

One mole of reactant produces 2 mol of products, and this tends to increase in the molar liquid flowrates in the reaction section. In addition, the liquid rate (L_n) is increased by the effects of the heat of reaction:

$$L_n = L_{n+1} + \frac{\lambda}{\Delta H_V} \mathfrak{R}_n + \mathfrak{R}_n \tag{6.3}$$

The decomposition reaction is endothermic because a large molecule decomposes into two smaller molecules, which requires extra energy to break the chemical bond. Because higher pressure leads to higher equilibrium conversion as well as higher forward/backward reaction rates, a high column pressure is always preferable as far as design is concerned. In this work, we limit the column temperature to 120 °C; consequently, the column pressure is set.

Before exploring the process parameters, the design procedure is outlined. The design variables are the numbers of reactive (N_{RX}), rectifying (N_R), and stripping (N_S) trays and the feed tray location (N_F). All combinations are exhausted to find the final design. The objective function for design is the minimum TAC, and the specifications are 98 mol% purities of the top and bottoms products. The design problem can be formulated as

$$\begin{aligned} &\underset{\mathbf{X}}{\text{minimize}}\ \text{TAC} \\ &\text{subject to:} \qquad x_{\text{product,C}} \geq 0.98 \\ &\qquad\qquad\qquad\ \ x_{\text{product,D}} \geq 0.98 \end{aligned} \tag{6.4}$$

where **X** is the vector of design variables and TAC is the total annual cost, defined as

$$\text{TAC} = \text{operating cost} + (\text{capital cost/payback period}) \tag{6.5}$$

Here, a payback period of 3 is used. The formula for the TAC computation is taken from Kaymak and Luyben[6]. The optimization is carried out in a sequential manner and the design steps are the following:

1. Place the reactive zone in the midsection of the column and pick a number for the reactive trays (N_{RX}).
2. Guess the number of trays in the rectifying section (N_R) and the stripping section (N_S).
3. Guess the feed tray location (N_F).
4. Estimate the holdup on the reactive trays (M_{RX}). Perform a dynamic simulation using relaxation until the product specifications are met, followed by column sizing to find the column diameter and the corresponding reactive holdup (M_{RX}). Iterate the simulation runs until the reactive holdup converges. A weir height of 10 cm is assumed.
5. Return to step 3 and modify the feed locations (N_F) until the TAC is minimized.
6. Return to step 2 and change N_R and N_S until the TAC is minimized.
7. Return to step 1 and vary N_{RX} until the TAC is minimized.

The optimized design is provided in Table 6.2. Figure 6.2 gives the composition profiles for the design. The effects of the design parameters are explored next.

[6]D. B. Kaymak and W. L. Luyben, Quantitative comparison of reactive distillation with conventional multiunit reactor/column/recycle systems for different chemical equilibrium constants, *Ind. Eng. Chem. Res.* **43**, 2493–2507 (2004).

TABLE 6.2 Steady-State Conditions and Design Parameters for TAC Optimum Case

Fresh feed flowrate of A (mol/s)	12.60	
Distillate flowrate (mol/s)	12.36	
Bottoms flowrate (mol/s)	12.35	
Vapor boilup (mol/s)	52.49	
Reflux flowrate (mol/s)	22.68	
Stripping trays	2	
Reactive trays	25	
Rectifying trays	7	
Liquid holdup on reactive trays (mol)	909.3	
Pressure (bar)	3	
Product Composition (Mole Fraction)	Distillate	Bottoms
A	0.0200	0.0200
B	0.9800	0.0000
C	0.0000	0.9800

6.1.4 Holdup on Reactive Trays

The effects of five different design variables on the vapor boilup rate are revealed in Figure 6.3. Note that the changes are made around the nominal steady state. The top left graph shows that increasing the holdup on the reactive trays improves the steady-state designs for the ternary decomposition reaction A $\Leftrightarrow$ B + C with relative volatility B/A/C = 4/2/1. Figure 6.4 shows how the temperature and compositions change with reactive tray holdup. The reactant concentration (x_A) increases as the holdup decreases, while the temperature remains fairly constant. In addition, note that the vapor rate decreases

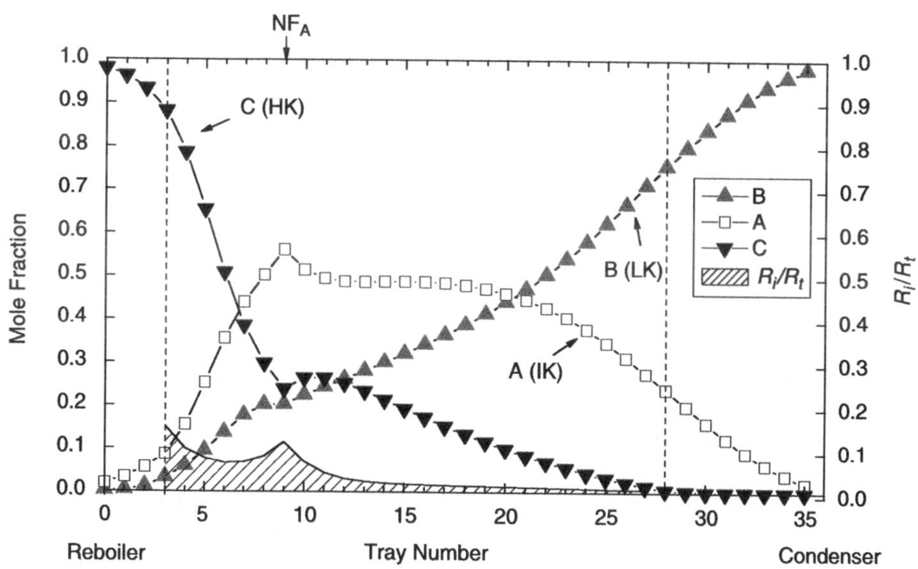

Figure 6.2 Composition profiles for A $\Leftrightarrow$ B + C with reactive zone (between two dashed lines), feed tray (indicated in NF_A), and fraction of total conversion (R_i/R_t; shaded area).

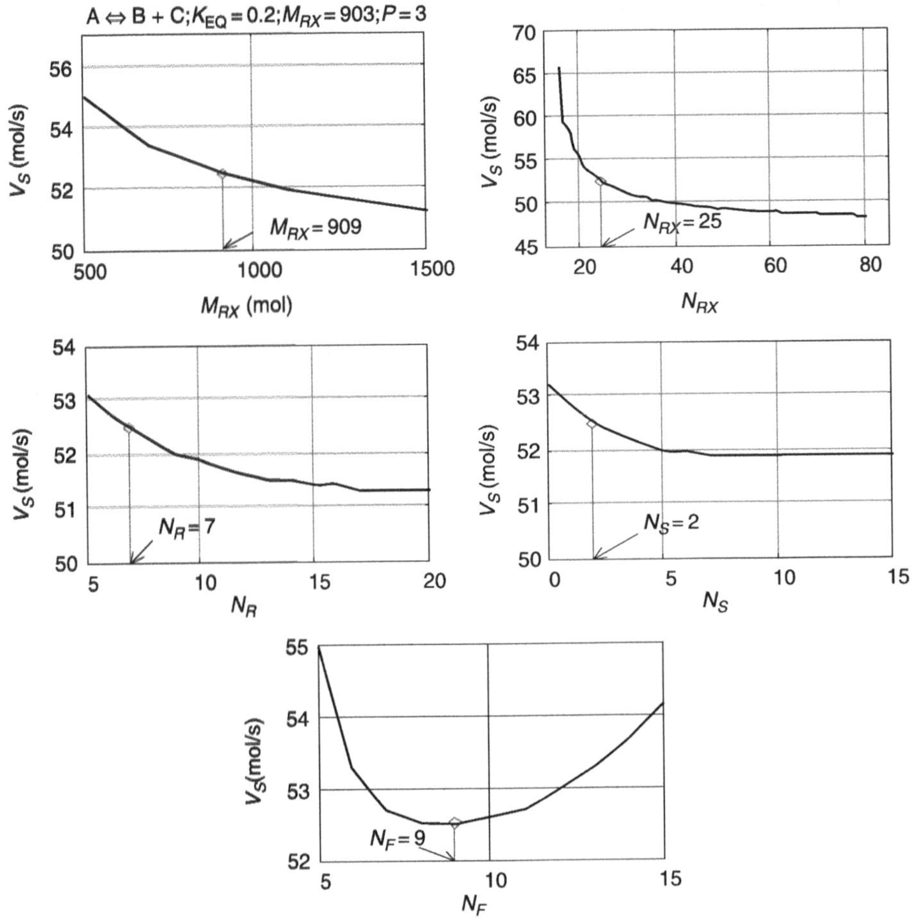

Figure 6.3 Effect of design variables on vapor boilup rate V_S with nominal steady-state values (vertical line).

as reactive holdup increases. This is the same result found in the quaternary system. There are no counterintuitive effects of reactive tray holdup.

6.1.5 Number of Reactive Trays

As we discussed in Chapters 2 and 5, there is an optimal number of reactive trays at which vapor boilup is minimized in the quaternary system. This effect does not occur in the ternary "synthesis" reaction system (A + B ⇔ C). Does it occur in the ternary "decomposition" reaction (A ⇔ B + C)?

The upper right graph in Figure 6.3 shows that adding more reactive trays (N_{RX}) improves the steady-state design of the ternary decomposition reaction. The vapor boilup rate decreases as reactive trays are added. Figure 6.5 also shows that the reactant concentration remains high throughout the reactive section as we increase the reactive trays. Because the tray numbers are all different, the column height is normalized by the total

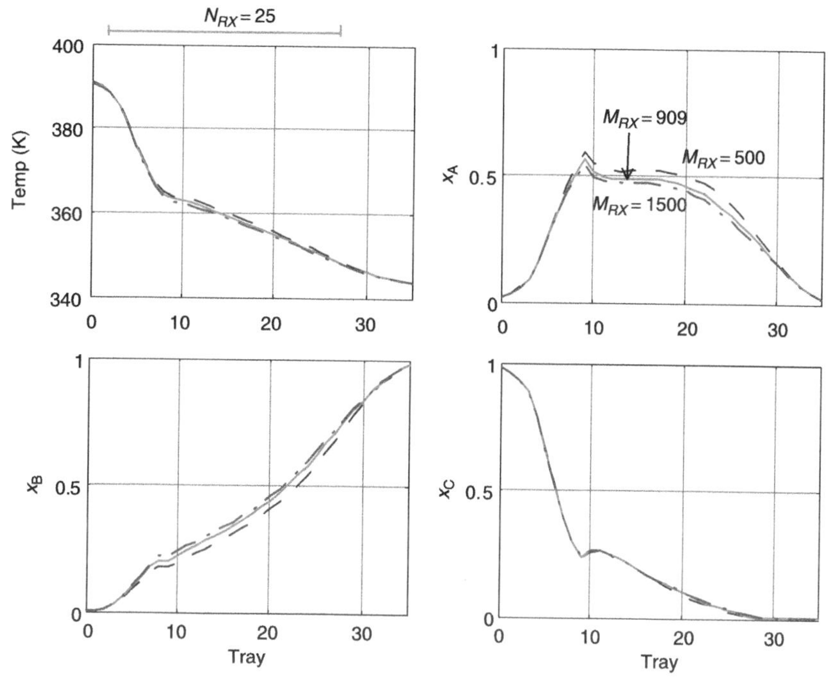

Figure 6.4 Effect of reactive holdup on temperature and composition profiles.

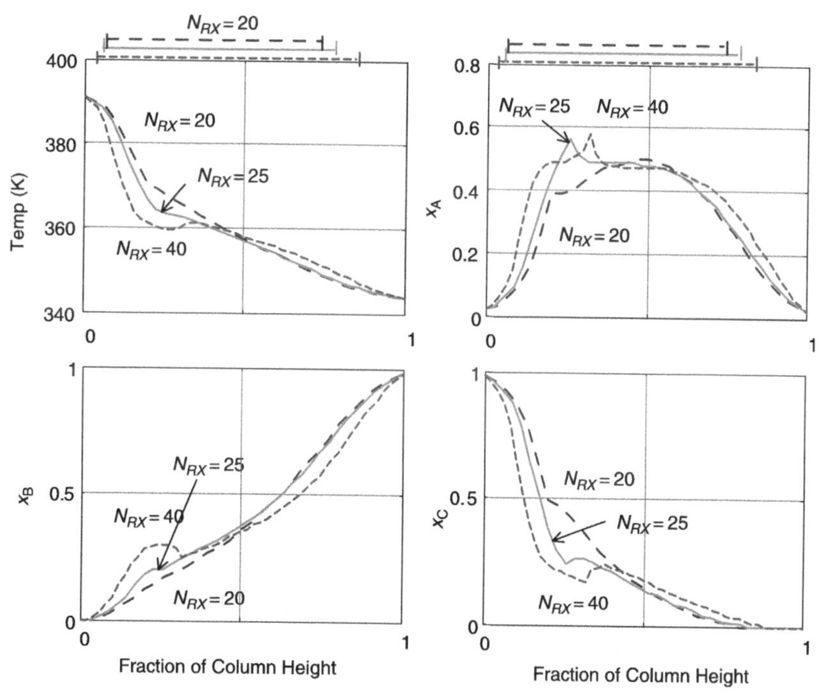

Figure 6.5 Effect of number of reactive trays (N_{RX}) on temperature and composition profiles where fraction of tray height is used instead of tray number; 0, reboiler; 1, column top. Width of reactive zone shown.

number of trays, and the corresponding reactive zone is indicated on the top of the graphs. As a result of a much wider reactive section with relatively uniform reactant concentration as N_{RX} increases, more light product (B) is observed toward the column base and more heavy product (C) is also found toward the column overhead.

6.1.6 Number of Rectifying and Stripping Trays

In the quaternary system, increasing the number of fractioning trays decreases vapor boilup, but in the ternary synthesis system, there is an optimal number of stripping trays. For the ternary decomposition with intermediate key components, the effects of rectifying and stripping sections on vapor boilup are shown in the middle two graphs in Figure 6.3. For the case of changing rectifying trays, Figure 6.6 demonstrates that adding rectifying trays lowers the profile of heavy component C (i.e., less C toward the top). Other than that, the reactant composition remains fairly constant in the reactive zone for all cases. Moreover, if we keep adding rectifying trays, little to no effect on the vapor boilup is observed as shown in Figure 6.3. Similar behavior can be seen for changing the stripping trays. That implies that as long as we have enough separation trays, adding more trays has little effect on reactive distillation performance.

6.1.7 Location of Feed Tray

Feed tray location is an important design parameter in reactive distillation, especially in the ternary decomposition system. As shown in the bottom graph in Figure 6.3, there

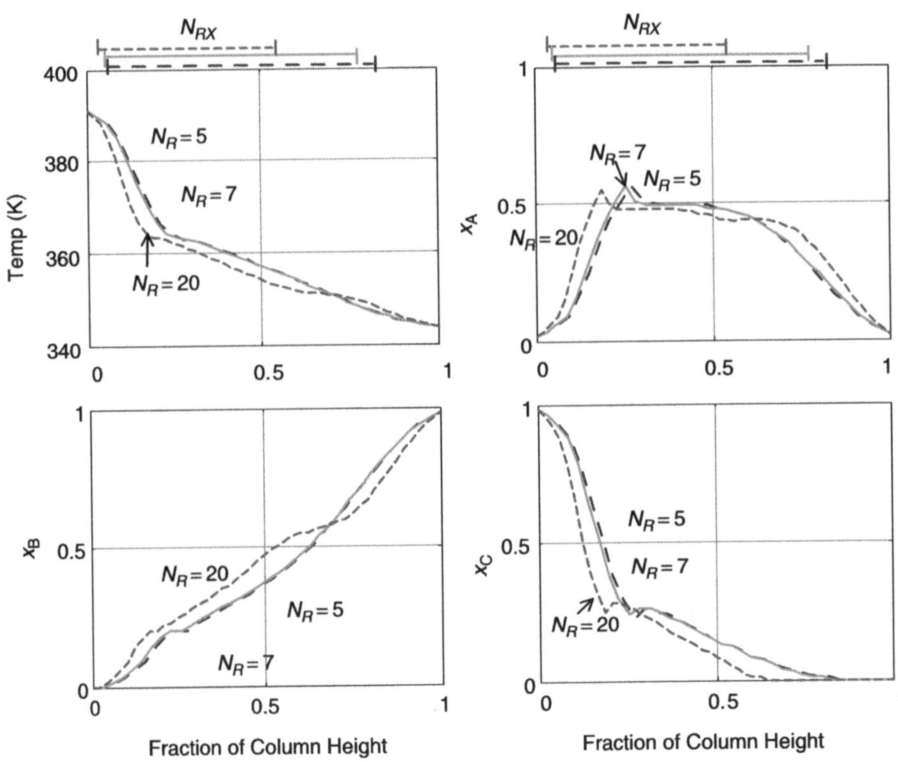

Figure 6.6 Effect of number of rectifying trays (N_R) on temperature and composition profiles.

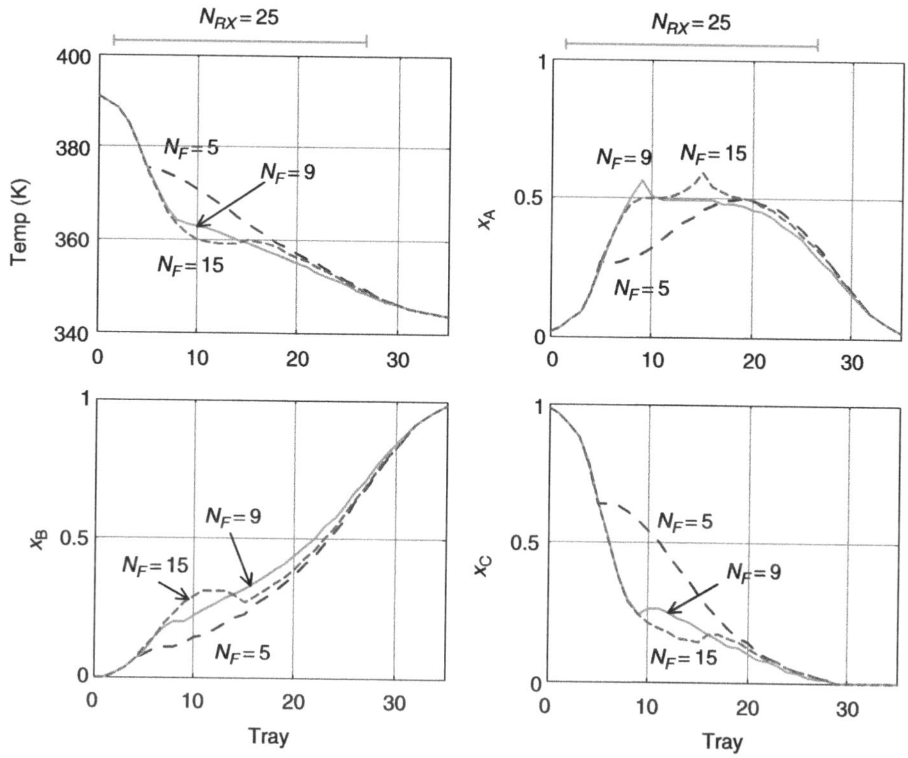

Figure 6.7 Effect of feed location (N_F) on temperature and composition profiles.

is an optimal feed tray location in this design, which is located in the lower section of the reactive zone. A feed tray located too high or too low increases vapor boilup. The reason for this is that a balance between temperature and composition is needed to facilitate chemical reaction. Figure 6.7 gives the temperature and composition profiles using various feed tray location. They show that the case of $N_F = 9$ gives a higher reactant concentration as well as a higher column temperature around tray 10 as compared to $N_F = 15$.

6.2 TERNARY DECOMPOSITION REACTION: HEAVY REACTANT WITH TWO-COLUMN CONFIGURATIONS

6.2.1 Column Configurations

Reactions with the heavy reactant decomposing into lighter products, HK $\Leftrightarrow$ LK + IK, is the more general case in ternary decomposition reactions. This type of relative volatility may have adverse effects on column design because the heavy reactant tends to stay in the lower part of the column. This makes it difficult to remove light products from the column base.

A straightforward approach is to simply use two columns. The heavy reactant is kept in the lower section of the reactive column, and both of the lighter products are

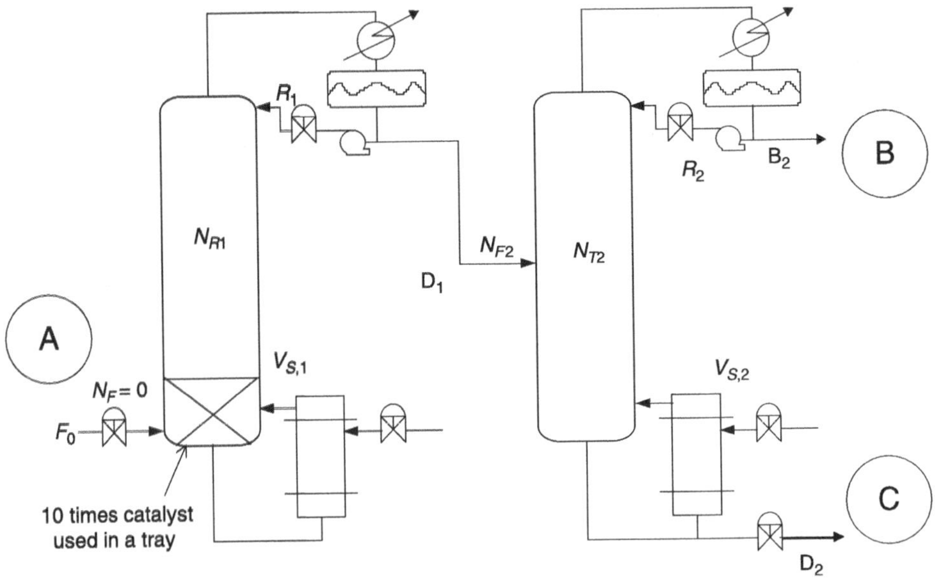

Figure 6.8 Two-column flowsheet for decomposition reaction A ⇔ B + C (HK ⇔ LK + IK), with $\alpha_B/\alpha_C/\alpha_A = 4/2/1$.

withdrawn together from the top. A second column separates the two products. Figure 6.8 gives the process flowsheet. Reactant A is fed to the column base of the first column, and products B and C are both taken out from overhead and separated in a second distillation column. The first column is a reactive distillation column with one feed and one product stream, and the reactive zone is placed in the lower section with a reactive column base where a large amount of catalyst can be used (10 times the catalyst used in a tray). In process simulation, this implies the reactive holdup in the column base ($M_{RX,\text{BASE}}$) is 10 times of that of a tray.

6.2.2 Chemistry and Phase Equilibrium Parameters

Table 6.3 lists the kinetics and vapor–liquid phase equilibrium parameters used in this case. The parameters are almost the same as the previous case, except for the volatility ranking; reactant A becomes the heaviest. Another difference is the chemical equilibrium constant, which is changed from 0.2 to 0.28.

6.2.3 Design Parameters and Procedure

The TAC is used to obtain the optimal design. For the design specifications in the second column, we set the light product to 98.5 mol% B in the distillate and the heavy product (intermediate key) to 98 mol% C in the bottoms. In the reactive distillation column, the conversion of A is set to 98% and the composition of the column overhead is 49.75 mol% B and 49.75 mol% C. Because this is a sequential configuration with the two columns not coupled together, design steps can be carried out separately.

TABLE 6.3 Kinetic and Vapor–Liquid Equilibrium Parameters

Activation energy (kcal/mol)		
Forward	40	
Backward	30	
Specific reaction rate at 366 K (kmol s^{-1} $kmol^{-1}$)		
Forward	0.008	
Chemical equilibrium constant at 366 K	0.2866	
Heat of reaction (kcal/mol)	+10	
Heat of vaporization (kcal/mol)	6.944	
Molecular weights (A/B/C) (g/mol)	100/50/50	
Vapor Pressure Constants	A_j	B_j
A	10.96	3862
B	12.34	3862
C	11.65	3862

For the reactive distillation column, the steps are the following:

1. Place the reactive zone in the base of the column with an initial estimate of reactive holdup and fix the number of reactive trays (N_{RX}).
2. Guess the number of trays in the rectifying section (N_R) and fix the feed tray location (N_F) location at the column base (tray 0).
3. Perform dynamic simulation using relaxation until the conversion is met, followed by column sizing to find the column diameter and corresponding reactive holdup (M_{RX}). Iterate simulation runs until the reactive holdup converges, using a weir height of 10 cm on the reactive trays.
4. Return to step 2 and change N_R until the TAC is minimized.
5. Return to step 1 and vary N_{RX} until the TAC is minimized.

For the distillation column, the steps are the following:

1. Guess a total number of trays and feed tray location.
2. Perform dynamic simulation until specifications are met.
3. Return to step 2 and change N_T and N_F until the TAC is minimized.

Table 6.4 is the optimized design, and the composition profiles are shown in Figure 6.9. The reactive distillation consists of 17 rectifying trays and 0 reactive trays. The only reactive section is the column base. Heavy reactant A shows a decreasing trend toward the column top, and both products B (LK) and C (IK) have the same composition in the distillate. The reflux ratio is <1 and the boilup rate is twice that of the reflux flow rate.

6.2.4 Reactive Holdup

Increasing the reactive holdup in the column base decreases the vapor boilup but by a very small fraction, as shown in the upper left graph in Figure 6.10. However, adding

TABLE 6.4 Steady-State Conditions and Design Parameters for TAC Optimum Case

Column 1 (Reactive Distillation Column)		
Fresh feed flowrate of A F_{0A} (mol/s)	12.60	
Distillate flowrate D_1 (mol/s)	25.08	
Vapor boilup $V_{S,1}$ (mol/s)	46.50	
Reflux flowrate R_1 (mol/s)	21.40	
Reactive trays $N_{RX,1}$	0	
Rectifying trays $N_{R,1}$	17	
$(K_{EQ})_{366}$	0.2866	
Liquid holdup in column base (10 M_{RX}) (mol)	8910	
Pressure P_1 (bar)	6	
Product Composition (Mole Fraction)	Distillate x_{Dj}	
A	0.0050	
B	0.4975	
C	0.4975	
Column 2 (Distillation Column)		
Distillate flowrate D_2 (mol/s)	12.54	
Bottoms flowrate B_2 (mol/s)	12.54	
Vapor boilup $V_{S,2}$ (mol/s)	38.80	
Reflux flowrate R_2 (mol/s)	26.30	
Stripping trays $N_{S,2}$	15	
Rectifying trays $N_{R,2}$	15	
Liquid holdup in column tray M_2 (mol)	1653.0	
Pressure P_2 (bar)	1	
Product Composition (Mole Fraction)	Distillate x_{Dj}	Bottoms x_{Bj}
A	0.0000	0.0100
B	0.9850	0.0100
C	0.0150	0.9800

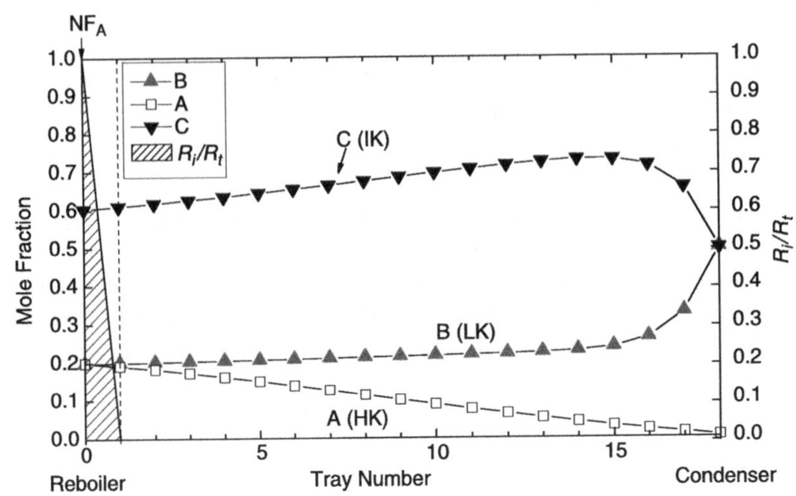

Figure 6.9 Composition profiles for reactive distillation of two-column flowsheet with reactive zone placed in column base, feed tray NF_A, and fraction of total conversion (R_i/R_t).

reactive trays and increasing their reactive holdup has the opposite effect, which will be explained next.

6.2.5 Number of Reactive Trays

The optimum design has no reactive trays, just a reactive base. However, if five reactive trays are used (upper right graph, Fig. 6.10), increasing the number of reactive trays has <1% influence on the vapor boilup. This is quite different from the system considered previously with an intermediate-boiling reactant. For a case with five reactive trays, Figure 6.11 shows that the reverse reaction dominates from tray 3 upward. Adding more trays does not help with the conversion. As opposed to the case of IK $\Leftrightarrow$ LK + HK, both products in the HK $\Leftrightarrow$ LK + IK case move in the same direction toward the top of the reactive distillation. High product concentrations promote the reverse reaction and lead to a net negative reaction in the reactive trays as shown in Figure 6.11. Additional reactive trays would keep the compositions at their chemical equilibrium values, which would not change from tray to tray. The temperature profile in Figure 6.12 demonstrates that there is very little change in temperature over

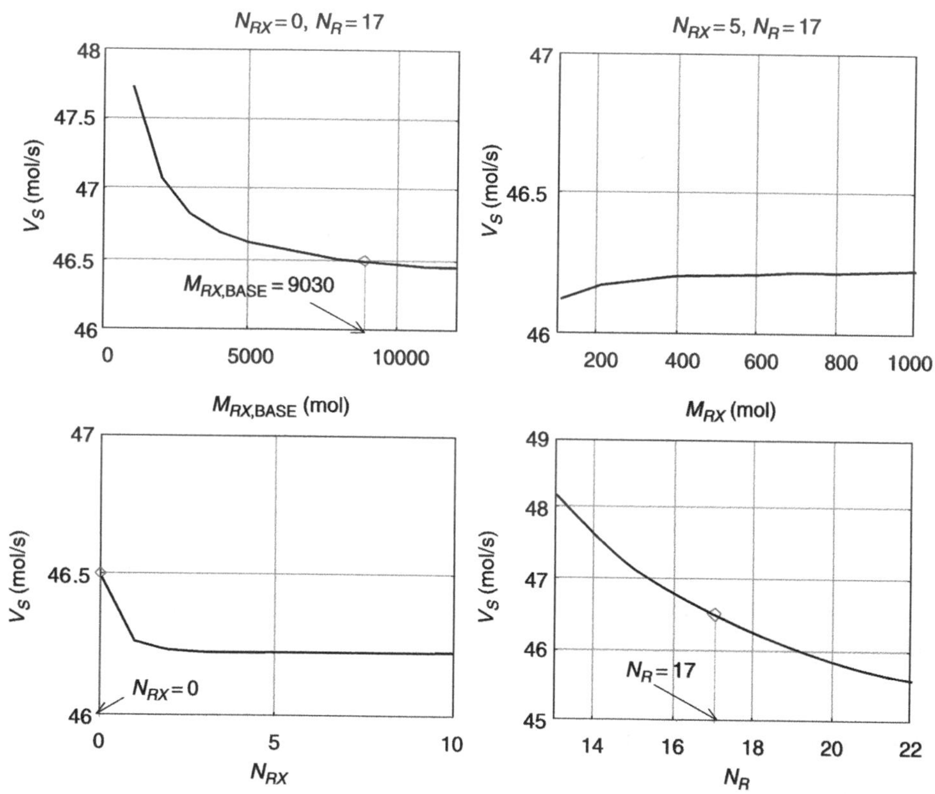

Figure 6.10 Effect of design variables on vapor boilup rate V_S in reactive distillation column with nominal steady-state values (vertical line).

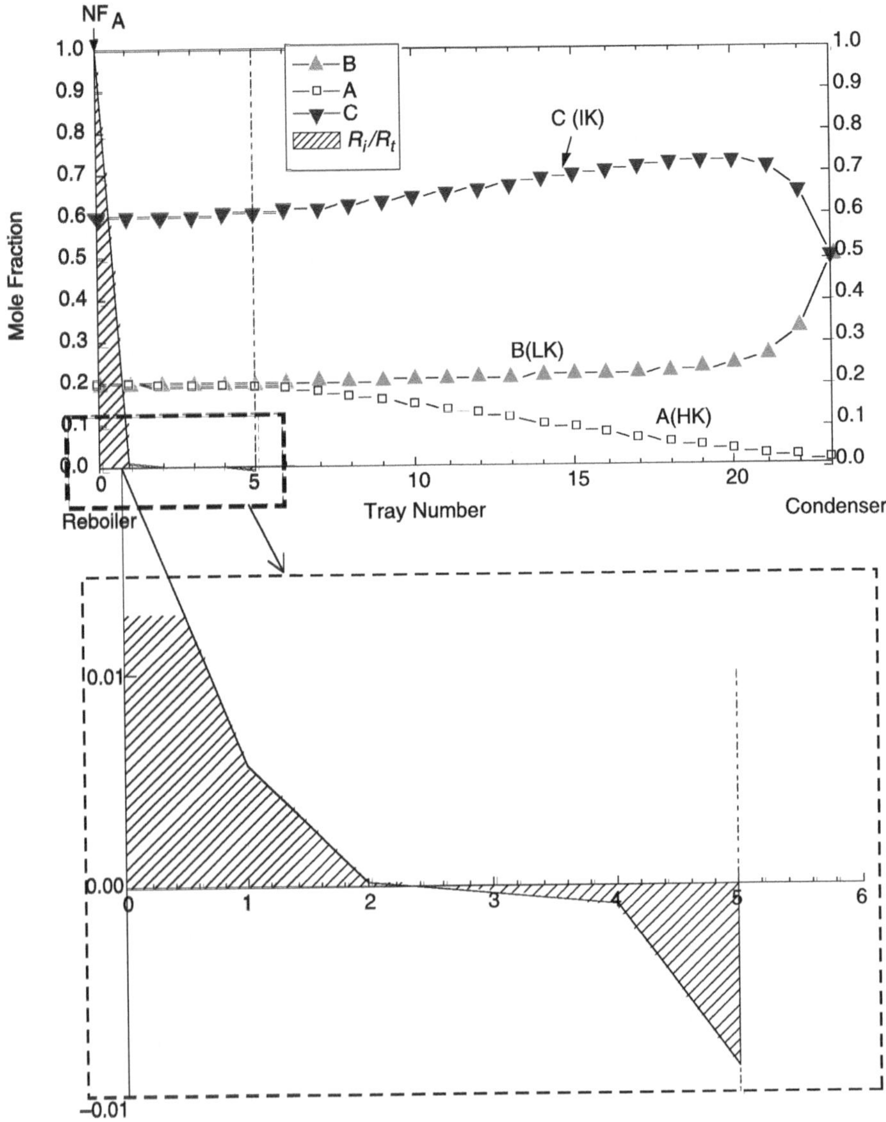

Figure 6.11 Composition profiles for reactive distillation with $N_{RX} = 5$ and showing negative net reaction on trays 3–5 as indicated by fraction of total conversion (R_i/R_t).

a wide section of the column, especially in the lower section. This implies that these additional trays are not necessary for this process (Fig. 6.10).

6.2.6 Number of Rectifying Trays

Adding more rectifying trays would decrease the vapor boilup (lower right graph, Fig. 6.10). Rectifying trays help keep the reactant in the reactive section as shown in Figure 6.13, but the difference is relatively small.

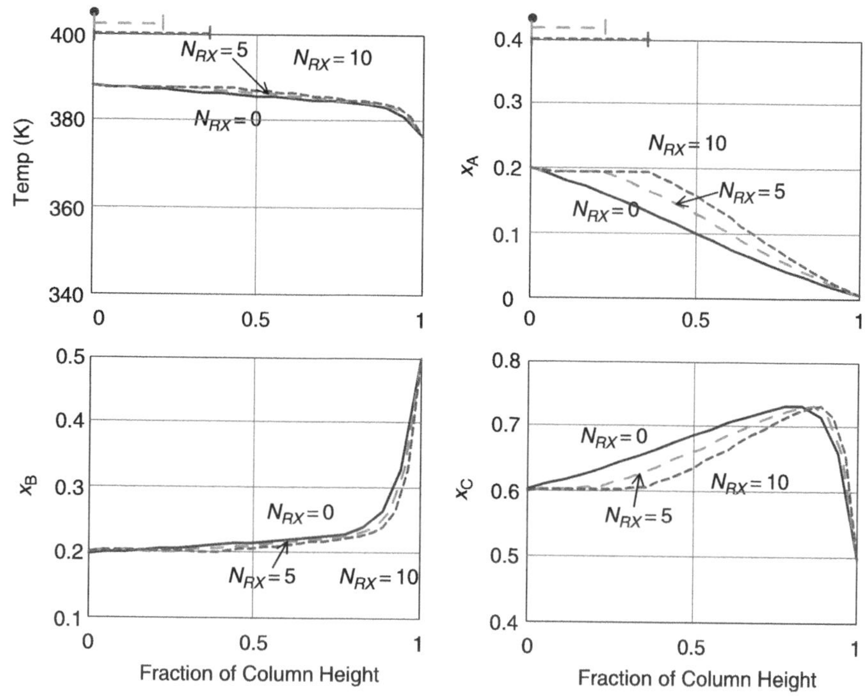

Figure 6.12 Effect of number of reactive trays (N_{RX}) on temperature and composition profiles and $N_{RX} = 0$, implying only reactive column base.

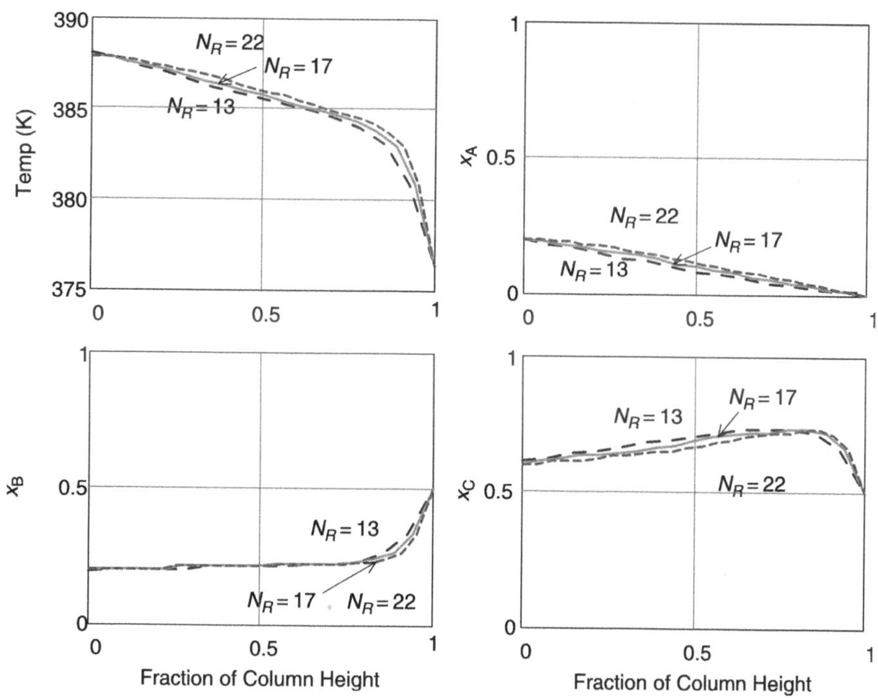

Figure 6.13 Effect of number of rectifying trays (N_R) on temperature and composition profiles.

6.3 TERNARY DECOMPOSITION REACTION: HEAVY REACTANT WITH ONE-COLUMN CONFIGURATIONS

We studied the effects of design parameters in the two-column configuration. It is always possible to produce high purity products using this two-column flowsheet (shown in Fig. 6.8). Now we consider a more simple process configuration that has only a single column, which is shown in Figure 6.14. If we consume most of the heavy reactant (HK, A) toward the bottoms of the column, the heavier product (IK, C) can be obtained from the column base, and the lighter product (LK, B) can be withdrawn from the overhead. Then, the question becomes, can the same purity level be obtained for both products using a single reactive distillation column? Feasibility analysis is useful in exploring this question.

6.3.1 Feasibility Analysis

Feasibility analysis helps us to determine whether a single reactive distillation process will produce on-spec products. It is carried out at the very beginning of the design and can provide guidance for conceptual design. One useful approach is the difference point method[7] in which tray-by-tray calculations are carried out to determine the feasibility.

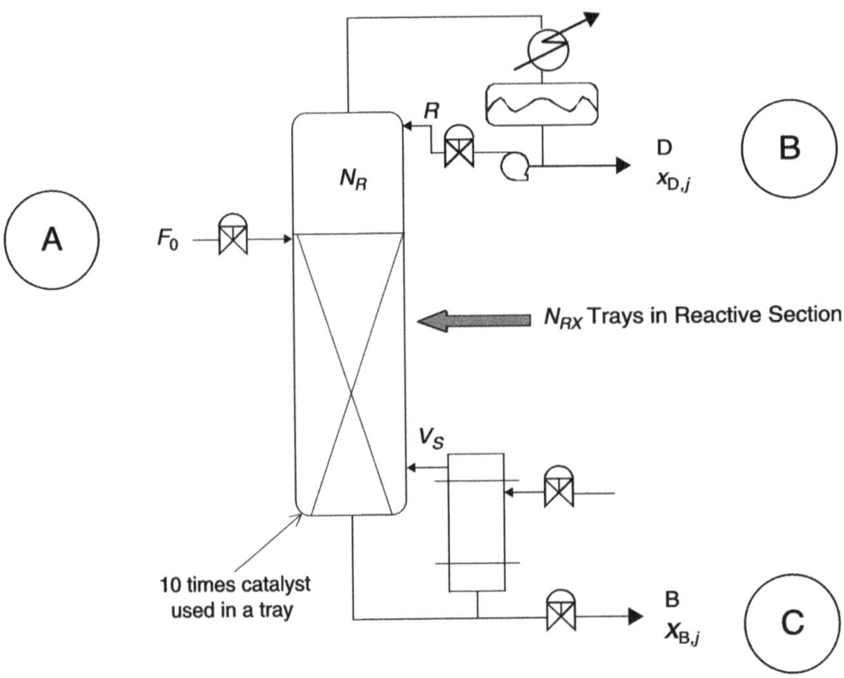

Figure 6.14 One-column flowsheet for decomposition reaction A ⇔ B + C (HK ⇔ LK + IK), with $\alpha_B/\alpha_C/\alpha_A = 4/2/1$.

[7]J. W. Lee, S. Hauan, K. M. Lien, and A. W. Westerberg, Difference points in extractive and reactive cascades. II. Generating design alternatives by the lever rule for reactive systems, *Chem. Eng. Sci.* **55**, 3161–3174 (2000).

An advantage of this method is that the interaction between reaction and separation can be visualized. The entire procedure can be carried out using relatively simple calculations. Moreover, the theory is based on basic material balances.

A brief introduction is given here, and readers can refer to the original literature for greater detail. Let us use the case A ⇔ B + C (HK ⇔ LK + IK) to illustrate the analysis. In the ternary composition space shown in Figure 6.15, H denotes the heavy key component, I is the intermediate key component, and L stands for the light key. We count the number of trays from bottoms up, that is, the reboiler is tray 0, followed by tray 1, and so on (Fig. 6.15*a*). Consider the material balance around the envelope shown in Figure 6.15*a*, where there are n trays with reaction in this section and all flowrates are molar flows. There are two streams leaving the system (vapor flowrate V_n and bottoms flowrate B) and one liquid flow entering the system (L_{n+1}). Here, ξ_n is the total extent of reaction on all of the trays from tray 1 to n, including the column base.

The variables y_n, x_B, and x_{n+1} are corresponding composition vectors of V_n, B, and L_{n+1}, respectively; υ is the vector chemical stoichiometric, that is, (B, C, A) = (L, I, H) = (1, 1, − 1); and υ_T is the sum of chemical stoichiometric, that is, $\upsilon_T = 1$. Thus, the overall molar balance and component material balances can be expressed as

$$L_{n+1} - V_n - B + \xi_n \upsilon_T = 0 \tag{6.6}$$

$$L_{n+1} \vec{x}_{n+1} - V_n \vec{y}_n - B \vec{x}_B + \xi_n \vec{\upsilon} = 0 \tag{6.7}$$

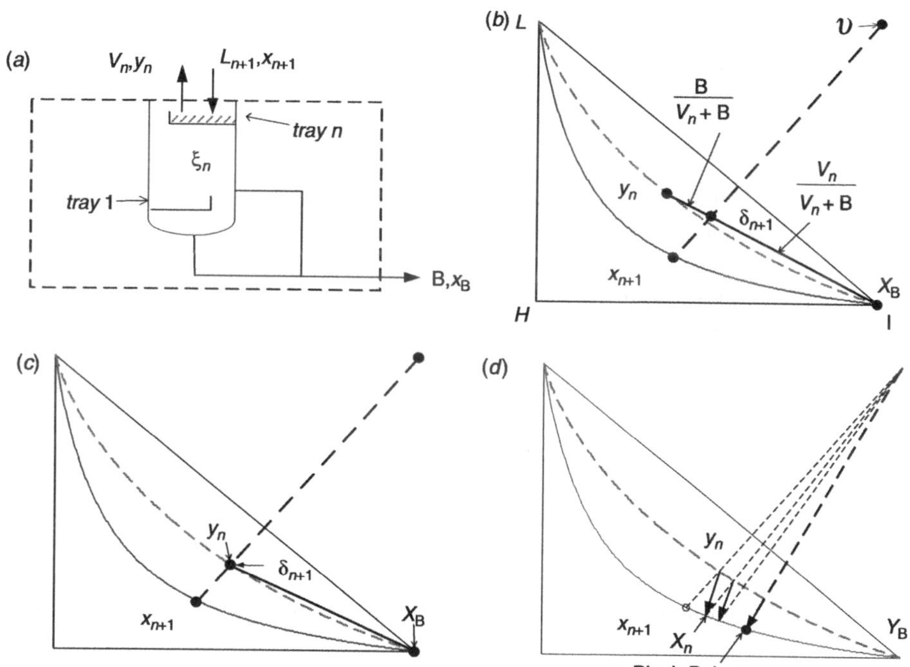

Figure 6.15 Feasibility analysis: (*a*) material balances on the envelope, (*b*) locate δ_{n+1} from Eq. (6.10) and y_n from Eq. (6.9), (*c*) under total boilup, and (*d*) stepping down column under total boilup.

Because $\upsilon_T = 1$, it will be dropped for subsequent development. Moving negative terms in Eqs. (6.6) and (6.7) to the right-hand side (RHS), we have

$$V_n + B = L_{n+1} + \xi_n \tag{6.8}$$

$$V_n \vec{y}_n + B\vec{x}_B = L_{n+1}\vec{x}_{n+1} + \xi_n \vec{\upsilon} \tag{6.9}$$

Dividing the left-hand side (LHS) of Eq. (6.9) with the LHS of Eq. (6.8) and the RHS of Eq. (6.9) with the RHS of Eq. (6.8), δ_{n+1} can be defined.

$$\delta_{n+1} = \frac{V_n \vec{y}_n + B\vec{x}_B}{V_n + B} = \frac{L_{n+1}\vec{x}_{n+1} + \xi_n \vec{\upsilon}}{L_{n+1} + \xi_n} \tag{6.10}$$

Rearranging the first two terms in Eq. (6.10), we have

$$\delta_{n+1} = \frac{V_n}{V_n + B}\vec{y}_n + \frac{B}{V_n + B}\vec{x}_B \tag{6.11}$$

Similarly, manipulating the first and third terms in Eq. (6.10), one obtains

$$\delta_{n+1} = \frac{L_{n+1}}{L_{n+1} + \xi_n}\vec{x}_{n+1} + \frac{\xi_n}{L_{n+1} + \xi_n}\vec{\upsilon} \tag{6.12}$$

Given the product specification (x_B), flowrates (V_n, B, L_{n+1}), and x_{n+1}, now ξ_n can be computed directly from the overall material balance [Eq. (6.6)]. Then, y_n can be computed from Eq. (6.7). Using the ternary mixing rule, the variables x_{n+1}, υ, and δ_{n+1} lie on a straight line in the ternary space. Applying the lever arm rule, δ_{n+1} can be located according to Eq. (6.12) as shown in the dashed straight line in Figure 6.15*b*. Next, y_n can be found graphically using the lever arm rule according to Eq. (6.11). If the column is operated under total boilup ($V_n/B \to \infty$), δ_{n+1} coincides with y_n. Thus, the points υ, y_n, and x_{n+1} form the straight line shown in Figure 6.15*c*. This is useful because we can perform the material balance at the limiting condition. Once y_n becomes available, we can find the corresponding x_n from the tie line (vapor–liquid equilibrium). Hence, one can step down the column all the way to x_B, if it is feasible. Figure 6.15*d* shows how one performs the tray-by-tray calculation toward the column base. However, when the direction of the reaction direction is the same as the separation direction (tie line), we have a pinch point and the desired product specification (x_B) will not be reached (Fig. 6.15*d*).

It is interesting to note that the pinch point in Figure 6.15*d* is exactly the "reactive azeotrope" that was discovered by Doherty and Malone and discussed in their book.[8] A mathematical approach is taken and the framework for the analysis is based on the "transformed composition" that is invariant from the standpoint of reaction. Let us

[8]M. F. Doherty and M. F. Malone, *Conceptual Design of Distillation Systems*, McGraw-Hill, New York, 2001.

use the HK $\Leftrightarrow$ LK + IK system to illustrate the approach. The transformation can be expressed as

$$X_I = \frac{x_I + x_H}{1 + x_H} \tag{6.13}$$

$$Y_I = \frac{y_I + y_H}{1 + y_H} \tag{6.14}$$

where x_I and x_H are liquid mole fractions and y_I and y_H are vapor mole fractions of the IK and HK, respectively. Because mole fractions add up to unity ($x_L + x_I + x_H = 1$), Eq. (6.13) can be rewritten as

$$X_I = \frac{x_I + x_H}{1 + x_H} = \frac{x_I + (1 - x_I - x_L)}{1 + (1 - x_I - x_L)} = \frac{1 - x_L}{(1 - x_L) + (1 - x_I)} \tag{6.15}$$

It is clear that X_I falls between zero and one ($1 \geq X_I \geq 0$). When $x_I = 1$, we have unity for the transformed variable $X_I = 1$ and, similarly, $x_L = 1$ gives $X_I = 0$. Consider the ternary space in Figure 6.16*a* where the solid line represents the chemical equilibrium line and the dashed line is the corresponding vapor line (in vapor liquid equilibrium with liquid composition in the chemical equilibrium line). The transformation of

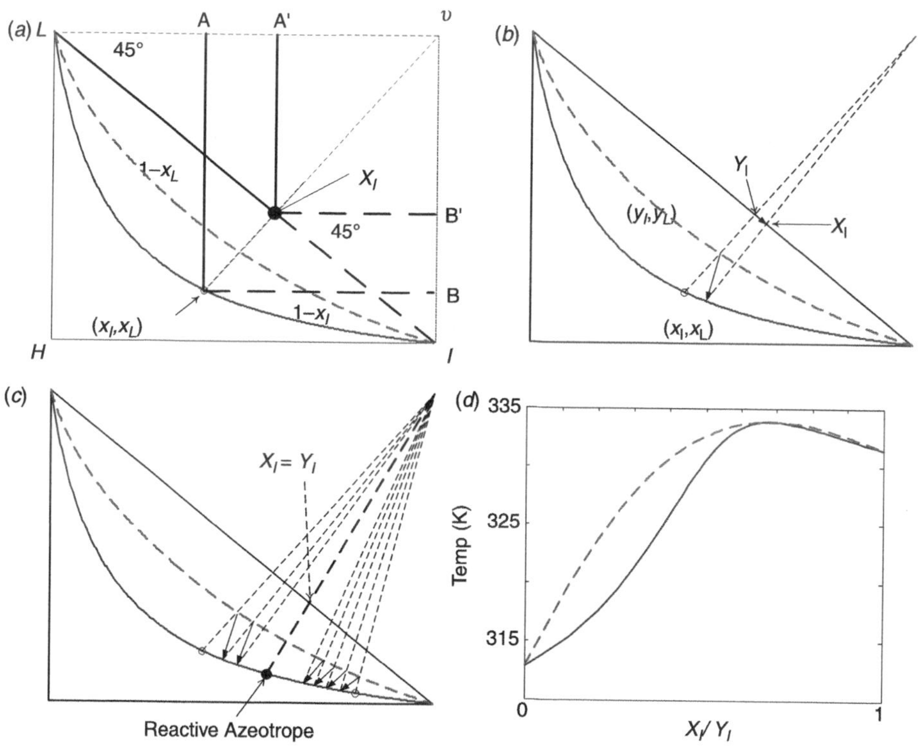

Figure 6.16 Transformed composition analysis: (*a*) projecting x_I and x_L onto *L-I* edge as X_1, (*b*) projecting corresponding vapor composition onto *L-I* edge as Y_1, (*c*) $X_1 = Y_1$ indicating reactive azeotrope, and (*d*) reactive azeotrope in *T-X-Y* diagram.

Eqs. (6.13) and (6.15) projects a composition in two-dimensional space (x_I, x_L) onto the L-I edge (X_I) along the direction of υ with the value of X_I. Let us use plane geometry to illustrate this. Assume X_I is the projection of (x_I, x_L). Figure 6.16*a* shows that the two rectangles formed by A-υ-B-(x_I, x_L) and A′-υ-B′-X_I are similar, and the two triangles formed by A′-X_I-L and B′-I-X_I are equilateral right triangles. The following condition is satisfied:

$$\frac{X_I}{1 - X_I} = \frac{r \cdot (1 - x_L)}{r \cdot (1 - x_I)} = \frac{1 - x_L}{1 - x_I} \tag{6.16}$$

where r is the ratio of the two lines formed by A-(x_I, x_L) and A′-X_I. Rearranging Eq. (6.16) gives exactly the same equation that we used in the transformation [Eq. (6.15)]. In other words, the transformation of Eqs. (6.13) and (6.14) is indeed the projection of (x_I, x_L) on the L-I edge, and Figure 6.16*a* shows that projection X_I is simply the intersection of the straight line formed by υ and (x_I, x_L).

Figure 6.16*b* provides the projections of liquid (X_I) and corresponding equilibrium vapor (Y_I) compositions for the transformed variables. The singular point with $X_i = Y_i$ at chemical equilibrium is the reactive azeotrope, which is shown in Figure 6.16*c*. The feature of this point is that the direction of the reaction is the same as the direction of the material balance line from a tray-by-tray calculation at an infinite reflux ratio. A more familiar expression for the reactive azeotrope is the maximum (or minimum) value on a T-X-Y diagram, except that X and Y are the transformed composition, as shown in Figure 6.16*d*. Therefore, given a chemical reaction system and corresponding vapor–liquid equilibrium, we can compute the reactive azeotrope numerically with simple transformation.

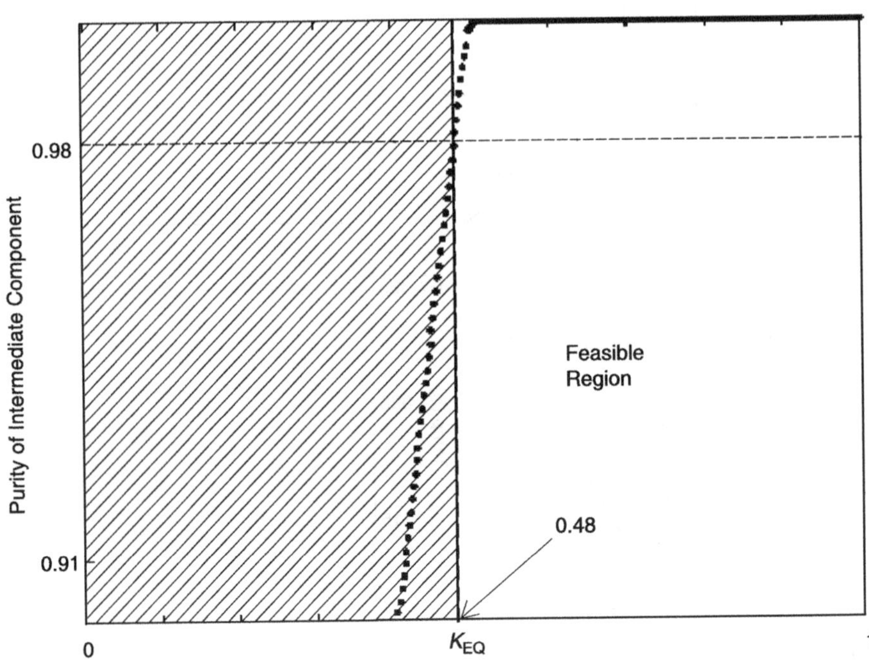

Figure 6.17 Bifurcation diagram for chemical equilibrium constant (K_{EQ}) and corresponding feasible region when product spec set to 0.98.

Certainly, the reactive azeotrope is the achievable bottoms product purity. Figure 6.17 shows the computed reactive azeotrope, assuming chemical equilibrium. Assuming the temperature is 393 K, the chemical equilibrium has to be >0.48 to reach 98% specification. That implies that a one-column configuration will be feasible for $K_{EQ} \geq 0.48$.

6.3.2 Column Configuration

Figure 6.14 gives a possible process flowsheet. This is a reactive distillation with a rectifying section. To keep the composition near chemical equilibrium, a reactive stripping section is always necessary. The base of the column has 10 times the amount of catalyst used on the reactive trays. Heavy key reactant is fed to the top of the reactive sections and the light product can be easily obtained from the top of the column. The purity of the intermediate key product increases stage-by-stage in the reactive stripping sections all the way to the bottoms.

6.3.3 Design Parameters and Procedure

By setting the product specifications at 98%, light key product in the top, and intermediate key product in the bottoms, the reflux ratio and the boilup rate are manipulated to control product purities. Distillate and bottoms flows are manipulated to control the inventory of reflux-drum and column base.

The design procedure based on the TAC is the following:

1. Place the reactive zone in the lower section of the column and fix the number of reactive trays (N_{RX}). Note that the column base is packed with 10 times the amount of catalyst used on a reactive tray.
2. Guess the number of trays in the rectifying section (N_R).
3. Fix the reactant feed tray (N_F).
4. Perform a dynamic simulation using relaxation via feedback control to meet the product specifications.
5. Return to step 3 and change N_F until the TAC is minimized.
6. Return to step 2 and change N_R until the TAC is minimized.
7. Return to step 1 and vary N_{RX} until the TAC is minimized.

The optimal design is provided in Table 6.5. Figure 6.18 gives the composition profiles.

6.3.4 Reactive Tray Holdup

Increasing the reactive tray holdup decreases the vapor boilup because the products are going toward the opposite direction as the result of consuming all of reactant A, which is shown in Figure 6.19. There is no counterintuitive effect.

6.3.5 Number of Reactive Trays

Increasing the number of reactive trays dramatically reduces the vapor boilup, which is illustrated in the upper right graph in Figure 6.19. This is the result of approaching the reactive azeotrope toward the bottoms of the column, which can only be achieved gradually by

TABLE 6.5 Steady-State Conditions and Design Parameters for TAC Optimum Case

Fresh feed flowrate of A F_{0A} (mol/s)	12.60	
Distillate flowrate D (mol/s)	12.60	
Bottom flowrate B (mol/s)	12.45	
Vapor boilup V_S (mol/s)	73.90	
Reflux flowrate R (mol/s)	44.10	
Reactive trays N_{RX}	79	
Rectifying trays N_R	11	
Feed tray location N_F	79	
$(K_{EQ})_{366}$	0.2866	
Liquid holdup on reactive tray M_{RX} (mol)	1067.6	
Pressure P (bar)	6	
Product Composition (Mole Fraction)	Distillate x_{Dj}	Bottoms x_{Bj}
A	0.0000	0.0080
B	0.9800	0.0120
C	0.0200	0.9800

widening the reactive stripping section. Figure 6.20 gives the effects of the number of reactive trays on the temperature and composition profiles.

6.3.6 Number of Rectifying Trays

Increasing the number of rectifying trays decreases the vapor boilup (lower left graph, Fig. 6.19), but not by much. The almost invariant temperature profiles from simulation confirm this insensitivity. There is no counterintuitive effect.

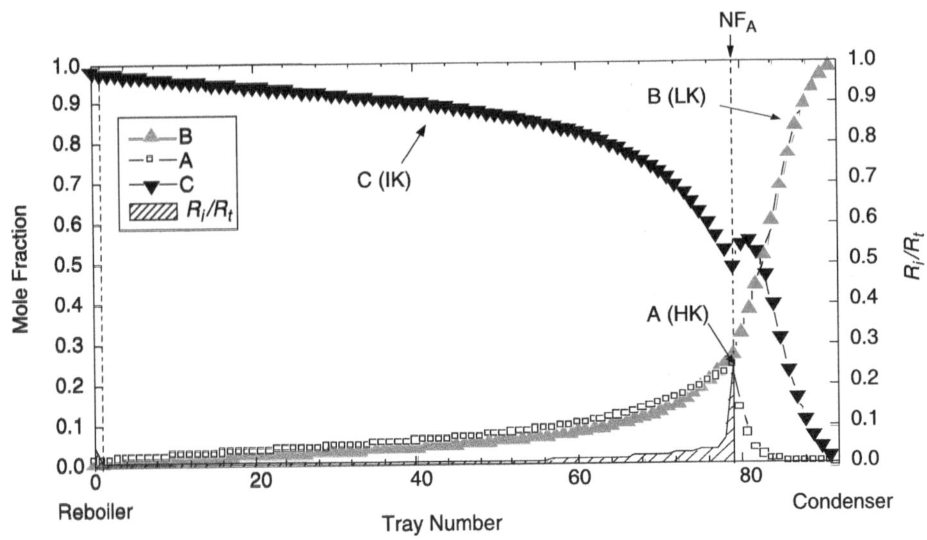

Figure 6.18 Composition profiles of one-column flowsheet with reactive zone (between dashed lines), feed tray NF_A, and fraction of total conversion (R_i/R_t; shaded area).

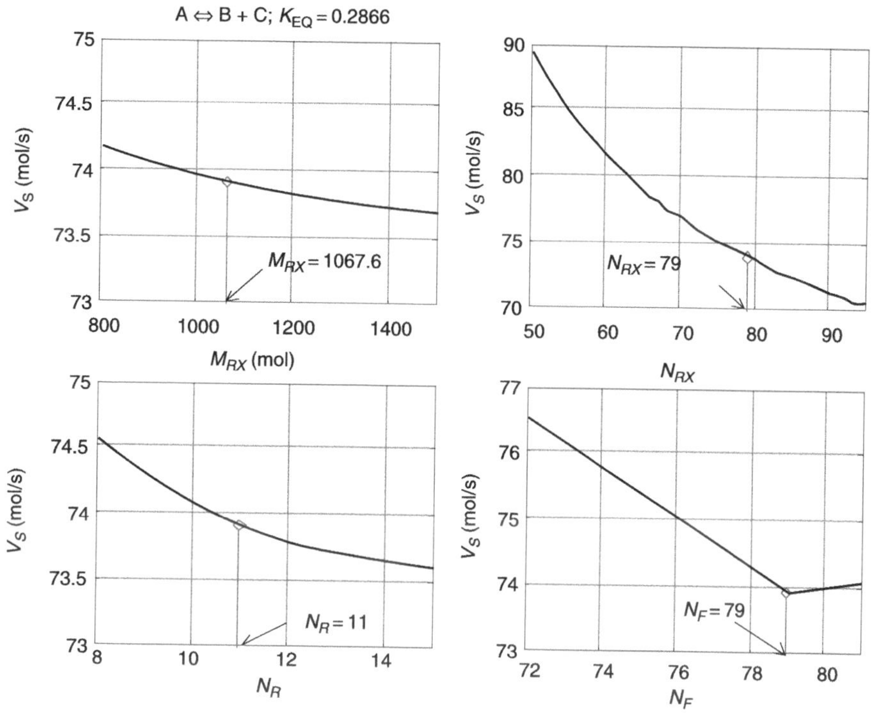

Figure 6.19 Effect of design variables on vapor boilup rate V_S in reactive distillation column with nominal steady-state values (vertical line).

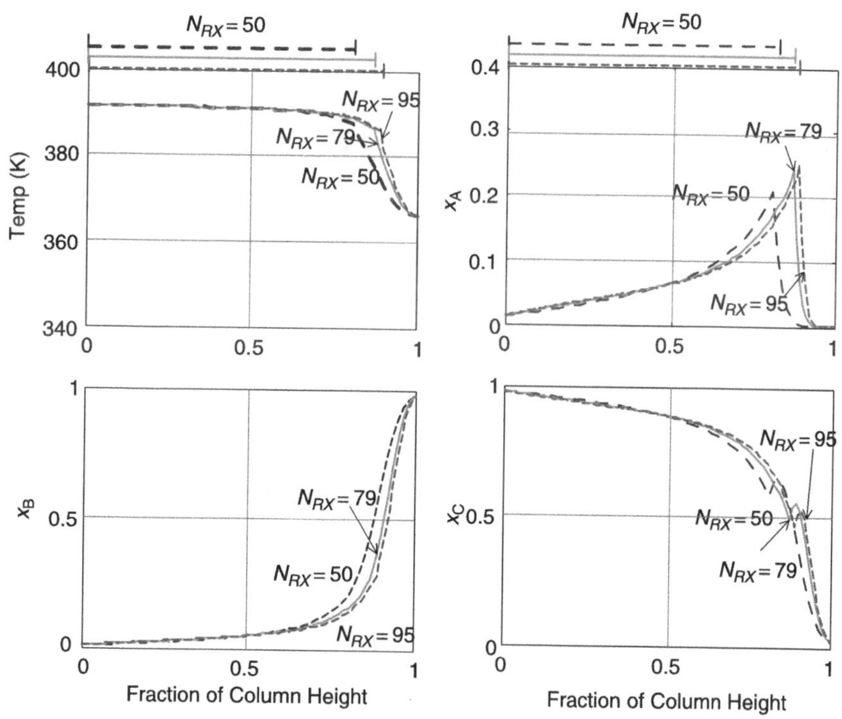

Figure 6.20 Effect of number of reactive trays (N_{RX}) on temperature and composition profiles.

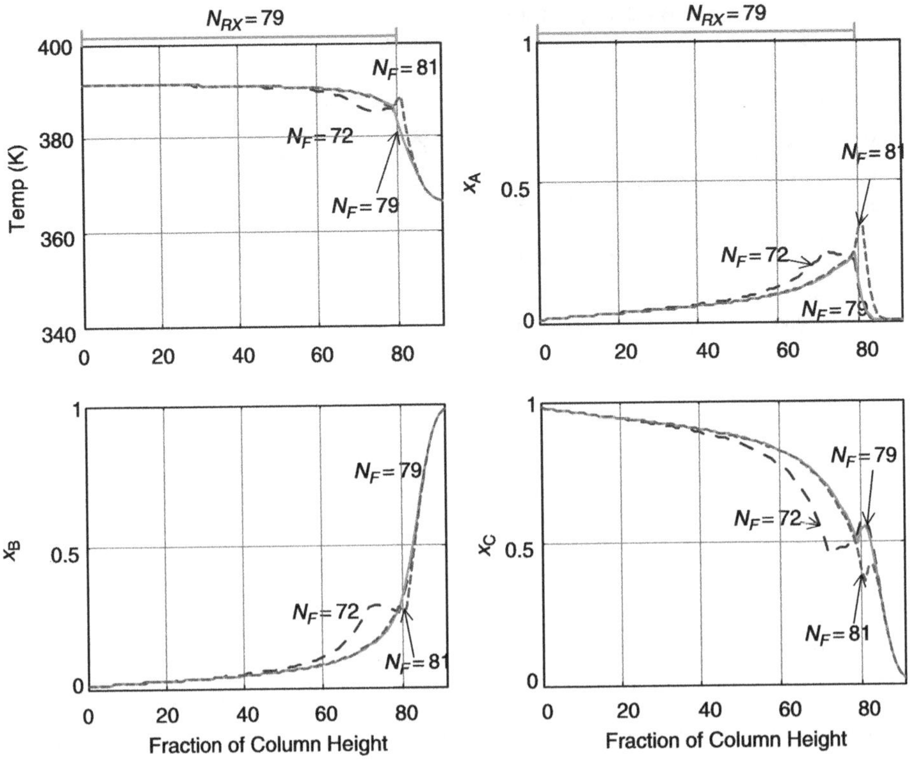

Figure 6.21 Effect of feed location (N_F) on temperature and composition profiles.

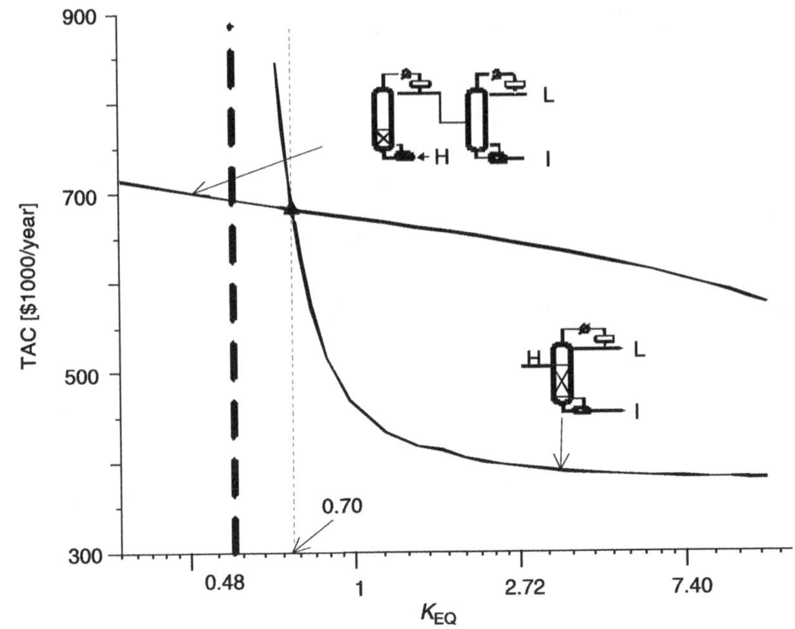

Figure 6.22 TAC comparison for the two-column and one-column flowsheets.

6.3.7 Location of Feed Tray

The lower right graph in Figure 6.19 shows that there is an optimal feed tray location. Figure 6.21 shows the effect of feed tray locations on temperature and composition profiles. This optimum location is the top of the reactive section. Above this point, the effect of the rectifying section to keep the reactant in the reactive sections is lessened, and below this point there is an effect similar to reducing the number of reactive trays. Either position will increase the vapor boilup.

6.3.8 Comparison Between These Two Flowsheets

After investigating two possible flowsheets, we will be interested in choosing the appropriate one for a given process. It turns out that the chemical equilibrium constant plays a critical role for selecting the better one, as described in Figure 6.22. The TAC of the one-column flowsheet decreases dramatically into the feasible region. When the chemical equilibrium constant is >0.70, the one-column configuration prevails. On the contrary, if the chemical equilibrium constant is small, the two-column configuration is more economical. When the equilibrium constant is reduced further, past the limit of feasibility, the two-column configuration is the only choice.

6.4 CONCLUSION

Two different relative volatility rankings for a ternary decomposition reaction were explored in this chapter. The behavior of systems with an intermediate-boiling reactant is similar to the quaternary system with intermediate-boiling reactants. Systems with heavy reactant are quite different. Two possible flowsheets (two-column and one-column configurations) were discussed. The two-column flowsheet is workable for all positive chemical equilibrium constants. The one-column configuration is feasible for systems with higher chemical equilibrium constants. Economical comparisons of these two flowsheets were also given.

6.3.7 Location of Feed Tray

[illegible]

PART III

STEADY-STATE DESIGN OF REAL CHEMICAL SYSTEMS

CHAPTER 7

STEADY-STATE DESIGN FOR ACETIC ACID ESTERIFICATION

The esterifications of acetic acid with five different alcohols, ranging from C_1 to C_5, are studied in this chapter. The purpose is to understand how process configurations vary as chemical systems change. First, qualitative relationships between process flowsheets and phase equilibria are established, and the process flowsheets are classified into types I–III for these five systems. Second, a systematic design procedure is devised to optimize the design based on the TAC, and dominant design variables are identified for different flowsheets. Third and finally, the economic potentials of these three flowsheets are explored and explanations are given. The results clearly indicate that it is possible to systemize the design of reactive distillation by qualitatively generating flowsheets from phase equilibria and by quantitatively completing the process flow diagram from a sequential design procedure.

7.1 REACTION KINETICS AND PHASE EQUILIBRIA

7.1.1 Reaction Kinetics

The esterification of acetic acid with different alcohols can be expressed in the following general form:

$$\text{acid} + \text{alcohol} \underset{k_B}{\overset{k_F}{\Longleftrightarrow}} \text{acetate} + \text{water} \tag{7.1}$$

The alcohols studied here include methanol (MeOH), ethanol (EtOH), isopropanol (IPOH), *n*-butanol (BuOH), and *n*-pentanol (amyl alcohol, AmOH). The corresponding products are methyl acetate (MeAc), ethyl acetate (EtAc), isopropyl acetate (IPAc),

Reactive Distillation Design and Control. By William L. Luyben and Cheng-Ching Yu

n-butyl acetate (BuAc), and amyl acetate (AmAc), respectively. We are interested in the solid-catalyzed reaction kinetics for the obvious reason that we can place the reactive zone in different sections of the reactive distillation columns. It is clear that this provides flexibility in process design as opposed to using homogeneous catalysis (e.g., sulfuric acid). The solid catalysts that are used are acidic ion-exchange resins such as Amberlyst 15 (Rohm and Haas) and Purolite CT179 (Purolite). The solid catalyst can be either immobilized via structured packings such as Katapak-S (Sulzer Chemtech)[1,2] or simply placed inside the tray with certain types of replacement mechanisms. Table 7.1 lists the reaction kinetics used in this study. The reaction rates are expressed in the pseudohomogeneous model or Langmuir–Hinshelwood and generally with the component represented in terms of activity, except for ethyl acetate and amyl acetate systems.

Moreover, they are all catalyst weight (m_{cat}) based kinetics. Despite the fact that the kinetic experiments were performed by different groups with somewhat different types of catalysis, we observe certain degrees of consistency in the kinetic data (Table 7.1). The equilibrium

TABLE 7.1 Kinetic Equations for Five Esterification Systems

System	Kinetic Model (Catalyst)	k_F ($T = 363$ K)	K_{EQ} ($T = 363$ K)
(i) MeAc	Pseudohomogeneous model (Amberlyst 15) $r = m_{cat}(k_F a_{HAc} a_{MeOH} - k_B a_{MeAc} a_{H_2O})$ $k_F = 2.961 \times 10^4 \exp\left(\frac{-49190}{RT}\right)$ $k_B = 1.348 \times 10^6 \exp\left(\frac{-69230}{RT}\right)$	2.49×10^{-3} (kmol kg_{cat}^{-1} s^{-1})	16.76
(ii) EtAc	Pseudohomogeneous model (Purolite CT179) $r = m_{cat}(k_F x_{HAc}^{1.5} x_{EtOH} - k_B x_{EtAc} x_{H_2O})$ $k_F = 4.24 \times 10^3 \exp\left(\frac{-48300}{RT}\right)$ $k_B = 4.55 \times 10^5 \exp\left(\frac{-66200}{RT}\right)$	4.78×10^{-4} (kmol kg_{cat}^{-1} s^{-1})	3.50
(iii) IPAc	Langmuir–Hinshelwood model (Amberlyst 15) $r = m_{cat} \frac{k_F(a_{HAc} a_{IPOH} - a_{IPAc} a_{H_2O}/K_{EQ})}{(1 + K_{HAc} a_{HAc} + K_{IPOH} a_{IPOH} + K_{IPAc} a_{IPAc} + K_{H_2O} a_{H_2O})^2}$ $k_F = 7.667 \times 10^{-5} \exp\left(23.81 - \frac{68620.43}{RT}\right)$ $K_{EQ} = 8.7$, $K_{HAc} = 0.1976$, $K_{IPOH} = 0.2396$, $K_{IPAc} = 0.147$, $K_{H_2O} = 0.5079$ Assumption: mol $H^+/kg_{cat} = 4.6 \times 10^{-3}$	2.26×10^{-4} (kmol kg_{cat}^{-1} s^{-1})	8.7
(iv) BuAc	Pseudohomogeneous model (Amberlyst 15) $r = m_{cat}(k_F a_{HAc} a_{BuOH} - k_B a_{BuAc} a_{H_2O})$ $k_F = 3.3856 \times 10^6 \exp\left(\frac{-70660}{RT}\right)$ $k_B = 1.0135 \times 10^6 \exp\left(\frac{-74241.7}{RT}\right)$	2.32×10^{-4} (kmol kg_{cat}^{-1} s^{-1})	10.9
(v) AmAc	Pseudohomogeneous model (Amberlyst 15) $r = m_{cat}(k_F C_{HAc} C_{AmOH} - k_B C_{AmAc} C_{H_2O})$ $k_F = 31.1667 \exp\left(\frac{-51740}{RT}\right)$ $k_B = 2.2533 \exp\left(\frac{-45280}{RT}\right)$	1.13×10^{-6} (m^6 $kmol^{-1}$ kg_{cat}^{-1} s^{-1})	1.6

[1] R. Taylor and R. Krishna, Modelling reactive distillation, *Chem. Eng. Sci.* **55**, 5183–5229 (2000).
[2] S. Steinigeweg and J. Gmehling, *n*-Butyl acetate synthesis via reactive distillation: Thermodynamic aspects, reaction kinetics, pilot-plant experiments, and simulation studies, *Ind. Eng. Chem. Res.* **41**, 5483–5490 (2002).

constant (K_{EQ}, at 363 K) ranges from 1.6 to 16.8, the forward rate constant (k_F, at 363 K) changes from (1.73 to 25) $\times 10^{-4}$, the activation energy of the forward reaction varies from 44,000 to 70,000 kJ/kmol, and the heats of reaction are almost negligible except for the methyl acetate and ethyl acetate systems. Table 7.1 also shows that the methyl acetate system has the most favorable reaction kinetics, which is an order of magnitude larger than the rest of the systems. In applying reaction kinetics to a reactive distillation, it is assumed that the solid catalyst occupies 50% of the tray holdup volume and a catalyst density of 770 kg/m^3 is used to convert the catalyst occupied space into catalyst weight (m_{cat}).

7.1.2 Phase Equilibria

To account for nonideal vapor–liquid equilibrium and possible vapor–liquid–liquid equilibrium (VLLE) for these quaternary systems, the NRTL model or UNIQUAC model is used for activity coefficients. Table 7.2 provides the model parameters for these five quaternary systems where the EtAc, IPAc, and AmAc systems are described by the NRTL model and MeAc and BuAc systems are represented using the UNIQUAC model. Because of the near atmospheric pressure, the only vapor phase nonideality considered is the dimerization of acetic acid as described by the Hayden–O'Connell second virial coefficient model. The Aspen Plus built-in association parameters are used to compute fugacity coefficients.

It should be emphasized that the quality of the model parameters (Table 7.2) is essential to generate a correct process flowsheet. Two important steps to validate model parameters are good prediction of azeotropes and a reasonable description of the LL envelopes for VLLE systems. Correct description of the existence of azeotropes and the ranking of

TABLE 7.2 Activity Coefficient Models Parameters for Five Esterification Systems

(i) MeAc System UNIQUAC[3]

Comp. i Comp. j	HAc (1) MeOH (2)	HAc (1) MeAc (3)	HAc (1) H_2O (4)	EtOH (2) MeAc (3)	MeOH (2) H_2O (4)	MeAc (3) H_2O (4)
a_{ij}	−0.97039	0.43637	0.051007	0.71011	−3.1453	−0.010143
a_{ji}	2.0346	−1.1162	0.29355	−0.72476	2.0585	−0.96295
b_{ij} (K)	−390.26	62.186	−422.38	−62.972	575.68	−593.7
b_{ji} (K)	−65.245	−81.848	98.12	−326.2	−219.04	265.83
c_{ij} (K^{-1})	0.0030613	−0.00027235	0.00024019	−0.001167	0.0060713	0.0021609
c_{ji} (K^{-1})	−0.003157	0.0013309	0.000076741	0.0023547	−0.0070149	−0.00020133

(ii) EtAc System NRTL[4]

Comp. i Comp. j	HAc (1) EtOH (2)	HAc (1) EtAc (3)	HAc (1) H_2O (4)	EtOH (2) EtAc (3)	EtOH (2) H_2O (4)	EtAc (3) H_2O (4)
a_{ij}	0	0	−1.9763	1.817306	0.806535	−2.34561
a_{ji}	0	0	3.3293	−4.41293	0.514285	3.853826
b_{ij} (K)	−252.482	−235.279	609.8886	−421.289	−266.533	1290.464
b_{ji} (K)	225.4756	515.8212	−723.888	1614.287	444.8857	−4.42868
c_{ij}	0.3000	0.3000	0.3000	0.1000	0.4000	0.3643

(Continued)

[3]T. Pöpken, L. Götze, and J. Gmehling, Reaction kinetics and chemical equilibrium of homogeneously and heterogeneously catalyzed acetic acid esterification with methanol and methyl acetate hydrolysis, *Ind. Eng. Chem. Res.* **39**, 2601–2611 (2000).

[4]Y. T. Tang, H. P. Huang, and I. L. Chien, Design of a complete ethyl acetate reactive distillation system, *J. Chem. Eng. Jpn.* **36**, 1352–1363 (2003).

TABLE 7.2 (*Continued*)

(iii) IPAc System NRTL[5]

Comp. i Comp. j	HAc (1) IPOH (2)	HAc (1) IPAc (3)	HAc (1) H_2O (4)	IPOH (2) IPAc (3)	IPOH (2) H_2O (4)	IPAc (3) H_2O (4)
b_{ij} (K)	−141.6447911	70.96532461	−110.5806744	191.0869653	20.05742325	415.478762
b_{ji} (K)	40.96255662	77.9005536	424.0603422	157.1039255	833.0422748	1373.462808
c_{ij}	0.3048	0.3014	0.2997	0.3000	0.3255	0.3000

(iv) BuAc System UNIQUAC[6]

Comp. i Comp. j	HAc (1) BuOH (2)	HAc (1) BuAc (3)	HAc (1) H_2O (4)	EtOH (2) BuAc (3)	BuOH (2) H_2O (4)	BuAc (3) H_2O (4)
b_{ij} (K)	66.315	150.193	172.92	−41.53679	−34.227	−345.098
b_{ji} (K)	−74.62672	−358.447	−265.6904	−12.4	−292.4746	−232.247

(v) AmAc System NRTL[7]

Comp. i Comp. j	HAc (1) AmOH (2)	HAc (1) AmAc (3)	HAc (1) H_2O (4)	AmOH (2) AmAc (3)	AmOH (2) H_2O (4)	AmAc (3) H_2O (4)
b_{ij} (K)	−316.8	−37.943	−110.57	−144.8	100.1	254.47
b_{ji} (K)	178.3	214.55	424.018	320.6521	1447.5	2221.5
c_{ij}	0.1695	0.2000	0.2987	0.3009	0.2980	0.2000

NRTL model:

$$\ln \gamma_i = \frac{\sum_{j=1}^{nc} \tau_{ji} G_{ji} x_j}{\sum_{k=1}^{nc} G_{ki} x_k} + \sum_{j=1}^{nc} \frac{x_j G_{ij}}{\sum_{k=1}^{nc} G_{kj} x_k} \left[\tau_{ij} - \frac{\sum_{k=1}^{nc} x_k \tau_{kj} G_{kj}}{\sum_{k=1}^{nc} G_{kj} x_k} \right]$$

$$G_{ij} = \exp(-\alpha_{ij}\tau_{ij}),\ \tau_{ij} = a_{ij} + b_{ij}/T,\ \alpha_{ij} = c_{ij},\ G_{ii} = 1, \text{ and } \tau_{ii} = 0$$

UNIQUAC model:

$$\ln \gamma_i = \ln \frac{\Phi_i}{x_i} + \frac{z}{2} q_i \ln \frac{\theta_i}{\Phi_i} + l_i - \frac{\Phi_i}{x_i} \sum_{j=1}^{nc} x_j l_j + q'_i \left[1 - \ln \left(\sum_{j=1}^{nc} \theta'_j \tau_{ji} \right) - \sum_{j=1}^{nc} \frac{\theta'_j \tau_{ij}}{\sum_{k=1}^{nc} \theta'_k \tau_{kj}} \right]$$

$$\Phi_i = \frac{r_i x_i}{\sum_{k=1}^{nc} r_k x_k},\ \theta_i = \frac{q_i x_i}{\sum_{k=1}^{nc} q_k x_k},\ \theta'_i = \frac{q'_i x_i}{\sum_{k=1}^{nc} q'_k x_k}$$

$$\tau_{ij} = a_{ij} + b_{ij}/T + c_{ij}T,\ l_i = \frac{z}{2}(r_i - q_i) - (r_i - 1), \text{ and } z = 10$$

azeotropic temperatures will lead to the generation of a correct residue curve map (RCM) and consequently to the placement of separators in the right sections. A reasonable prediction of an LL envelope can facilitate possible use of a decanter, which is often encountered in esterification reactive distillation systems. This is also the reason why the NRTL and UNIQUAC models are preferred for activity coefficients. Table 7.3 reveals that the

[5]S. B. Gadewar, M. F. Malone, and M. F. Doherty, Feasible region for a countercurrent cascade of vapor–liquid CSTRs, *AIChE J.* **48**, 800–814 (2002).

[6]G. Venimadhavan, M. F. Malone, and M. F. Doherty, A novel distillate policy for batch reactive distillation with application to the production of butyl acetate, *Ind. Eng. Chem. Res.* **38**, 714–722 (1999).

[7]S. F. Chiang, C. L. Kuo, C. C. Yu, and D. S. H. Wong, Design alternatives for amyl acetate process: Coupled reactor/column and reactive distillation, *Ind. Eng. Chem. Res.* **41**, 3233–3246 (2002).

TABLE 7.3 Predicted and Experimental Azeotropic Data for Five Esterification Systems

(i) MeAc	(ii) EtAc	(iii) IPAc	(iv) BuAc	(v) AmAc
MeOH/MeAc 53.65 °C (0.3407, 0.6593)	EtOH/EtAc/H_2O 70.09 °C (0.1069, 0.6073, 0.2858)	**IPOH/IPAc/H_2O** **74.22 °C** **(0.2377, 0.4092, 0.3531)**	**BuOH/BuAc/H_2O** **90.68 °C** **(0.0865, 0.206, 0.7075)**	**AmOH/AmAc/H_2O** **94.71 °C** **(0.0488, 0.1292, 0.822)**
54 °C *(0.359, 0.641)*	*70.23 °C* *(0.1126, 0.5789, 0.3085)*	***75.5 °C*** ***(0.1377, 0.4938, 0.3885)***	***89.4 °C*** ***(0.111, 0.135, 0.754)***	***94.9 °C*** ***(0.046, 0.107, 0.847)***
MeAc/H_2O 56.43 °C (0.8904, 0.1096)	**EtAc/H_2O** **70.37 °C** **(0.6869, 0.3131)**	**IPAc/H_2O** **76.57 °C** **(0.5981, 0.4019)**	**BuAc/H_2O** **90.96 °C** **(0.2823, 0.7177)**	**AmAc/H_2O** **94.90 °C** **(0.1696, 0.8304)**
56.4 °C *(0.8804, 0.1196)*	***70.38 °C*** ***(0.6885, 0.3115)***	***76.6 °C*** ***(0.5982, 0.4018)***	***90.2 °C*** ***(0.278, 0.722)***	***95.2 °C*** ***(0.166, 0.834)***
	EtOH/EtAc 71.81 °C (0.4572, 0.5428)	IPOH/IPAc 78.54 °C (0.5984, 0.4016)	**BuOH/H_2O** **92.62 °C** **(0.2451, 0.7549)**	**AmOH/H_2O** **95.80 °C** **(0.1512, 0.8488)**
	71.81 °C *(0.462, 0.538)*	*80.1 °C* *(0.6508, 0.3492)*	***92.7 °C*** ***(0.248, 0.752)***	***95.8 °C*** ***(0.146, 0.854)***
	EtOH/H_2O 78.18 °C (0.9016, 0.0984)	IPOH/H_2O 80.06 °C (0.6691, 0.3309)	BuOH/BuAc 116.85 °C (0.7847, 0.2153)	HAc/AmOH/AmAc 139.89 °C (0.2225, 0.6108, 0.1667)
	78.174 °C *(0.9037, 0.0963)*	*82.5 °C* *(0.6875, 0.3125)*	*116.2 °C* *(0.73, 0.27)*	
			HAc/BuOH/BuAc 121.58 °C (0.4181, 0.2396, 0.3423)	HAc/AmOH 140.07 °C (0.2585, 0.7415)
			HAc/BuOH 123.21 °C (0.5359, 0.4641)	

Heteroazeotropes are in bold, and experimental data of Horsley[8] are in italic.

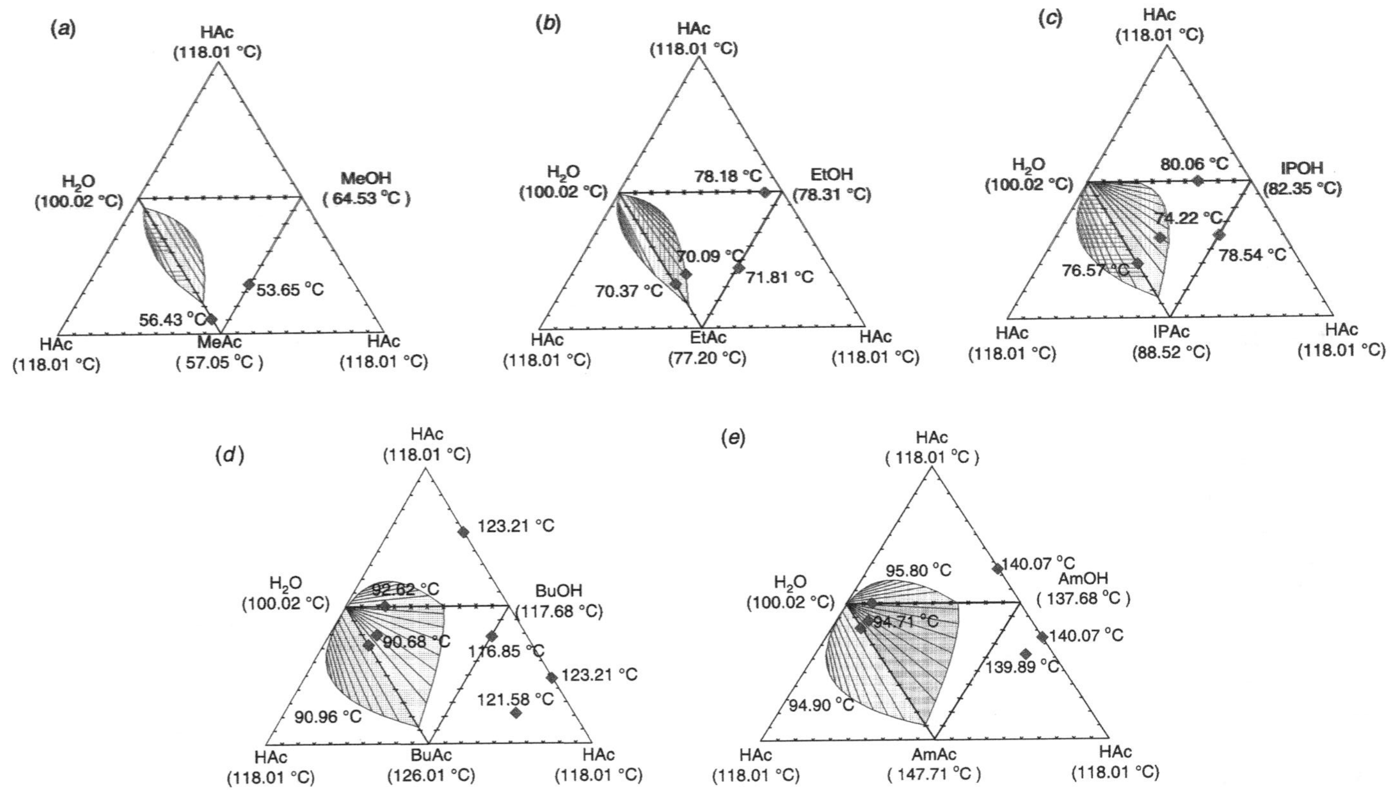

Figure 7.1 Azeotropes and two-liquid zone for five quaternary systems for acetic acid esterification: MeAc, EtAc, IPAc, BuAc, and AmAc.

model parameters give a reasonably good description of existing experimental azeotropic data[8] for all five systems. Moreover, the models also predict ternary azeotropes for four quaternary systems except the MeAc system (Table 7.3).

The phase equilibria results reveal that the MeAc quaternary system has two minimum boiling binary azeotropes (Fig. 7.1*a*), and this is the only system without any ternary azeotrope. For the EtAc and IPAc quaternary systems, the models predict three minimum boiling azeotropes and one minimum boiling ternary azeotrope (Figs. 7.1*b* and 7.1*c*). The BuAc quaternary system has three minimum boiling and one maximum boiling binary azeotropes and two ternary azeotropes (Fig. 7.1*d*), and the AmAc system gives three minimum boiling binary azeotropes and two ternary azeotropes (Fig. 7.1*e*). It is important to note that, for the EtAc, IPAc, BuAc, AmAc systems, the minimum temperature in the composition space is the *ternary minimum boiling azeotrope*. Significant LL envelopes are also observed for these five quaternary systems, especially for the acid-free ternary systems (Fig. 7.1). The MeAc, EtAc, and IPAc systems show type I LL envelopes and the BuAc and AmAc systems give type II envelopes as shown in Figure 7.1. The tie lines slope into the pure water node, and it becomes more evident as the carbon number in the alcohol increases. This means that relatively pure water can be recovered from the LL separation for four out of the five systems (except the MeAc system) and the separation (for achieving high purity water) becomes easier as the carbon number in the alcohol increases. For the EtAc system, the ternary minimum boiling azeotrope lies close to the edge of the LL envelope. For the IPAc, BuAc, and AmAc systems, the ternary minimum boiling azeotrope lies inside the LL envelope and it moves further inside the LL zone as the carbon number increases. As will be shown later, this has important implications in flowsheet development.

7.2 PROCESS FLOWSHEETS

In this section we attempt to generate a flowsheet for high purity acetates using the combination of a stripper, a rectifier, a reactive section, and possibly a decanter. The objective is to devise hybrid reactive distillation systems to produce commercial grade acetates. Figure 7.2 provides possible flowsheets for esterification systems.

7.2.1 Type I Flowsheet: MeAc

In reactive distillation the *heavy* reactant is typically fed from the *top* of the reactive section and the *light* reactant goes into the *bottom* part of the reactive zone. For the methyl acetate system, the heavy reactant is the acid (HAc) and the light reactant is the alcohol (MeOH). If the reactive zone consumes all of the acid, we are dealing with the separation of H_2O/MeOH/MeAc, an almost ternary system, in the stripping section. The RCM of Figure 7.3*a* indicates that the separation will lead to the pure H_2O node toward the bottom of the column. In contrast, if we react away most of the alcohol toward the top section of the reactive zone, in the rectifying section we are separating HAc/MeAc with a small amount of H_2O. Again, the RCM in Figure 7.3*a* shows that it is possible to obtain relatively high purity MeAc, despite being a saddle in the RCM notation. This is

[8] L. H. Horsley, *Azeotropic Data—III*, Advances in Chemistry Series No. 116, American Chemical Society, Washington, D.C., 1973.

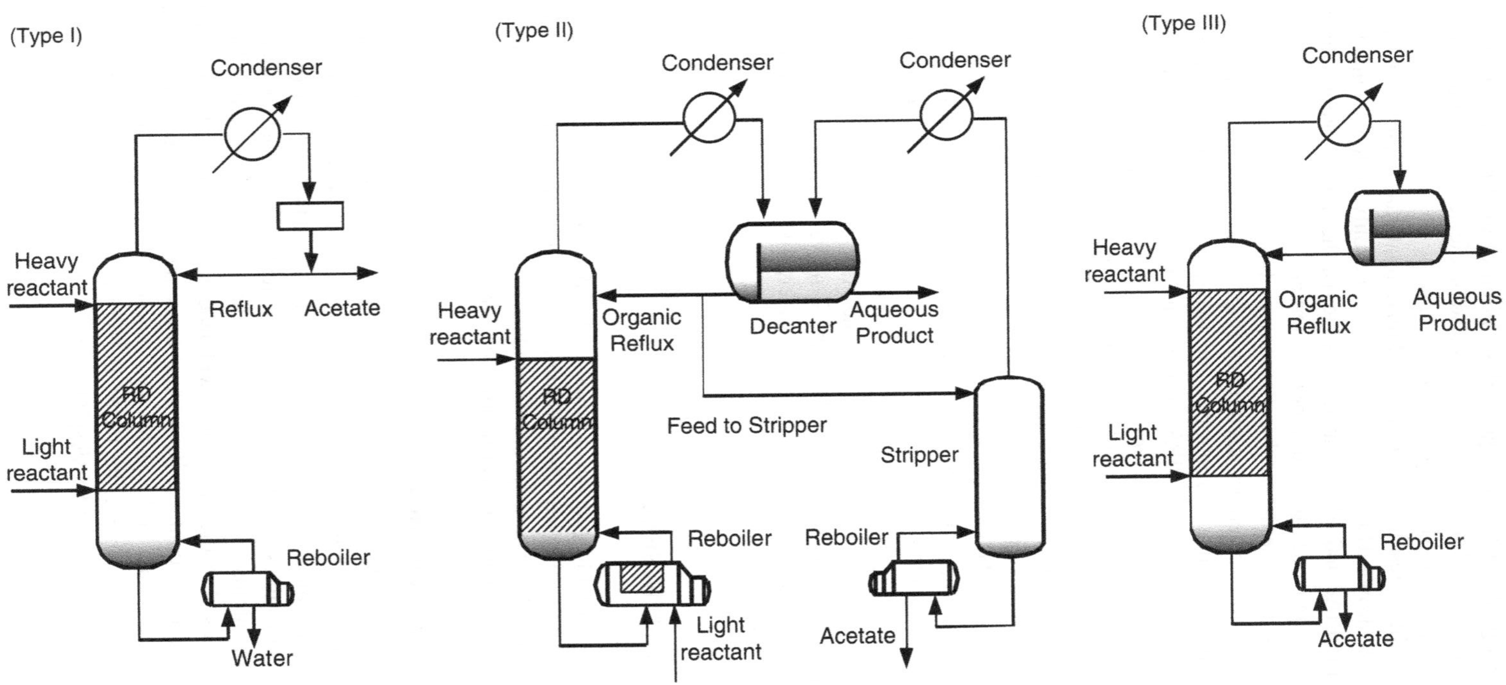

Figure 7.2 Three possible flowsheets for the five esterification systems.

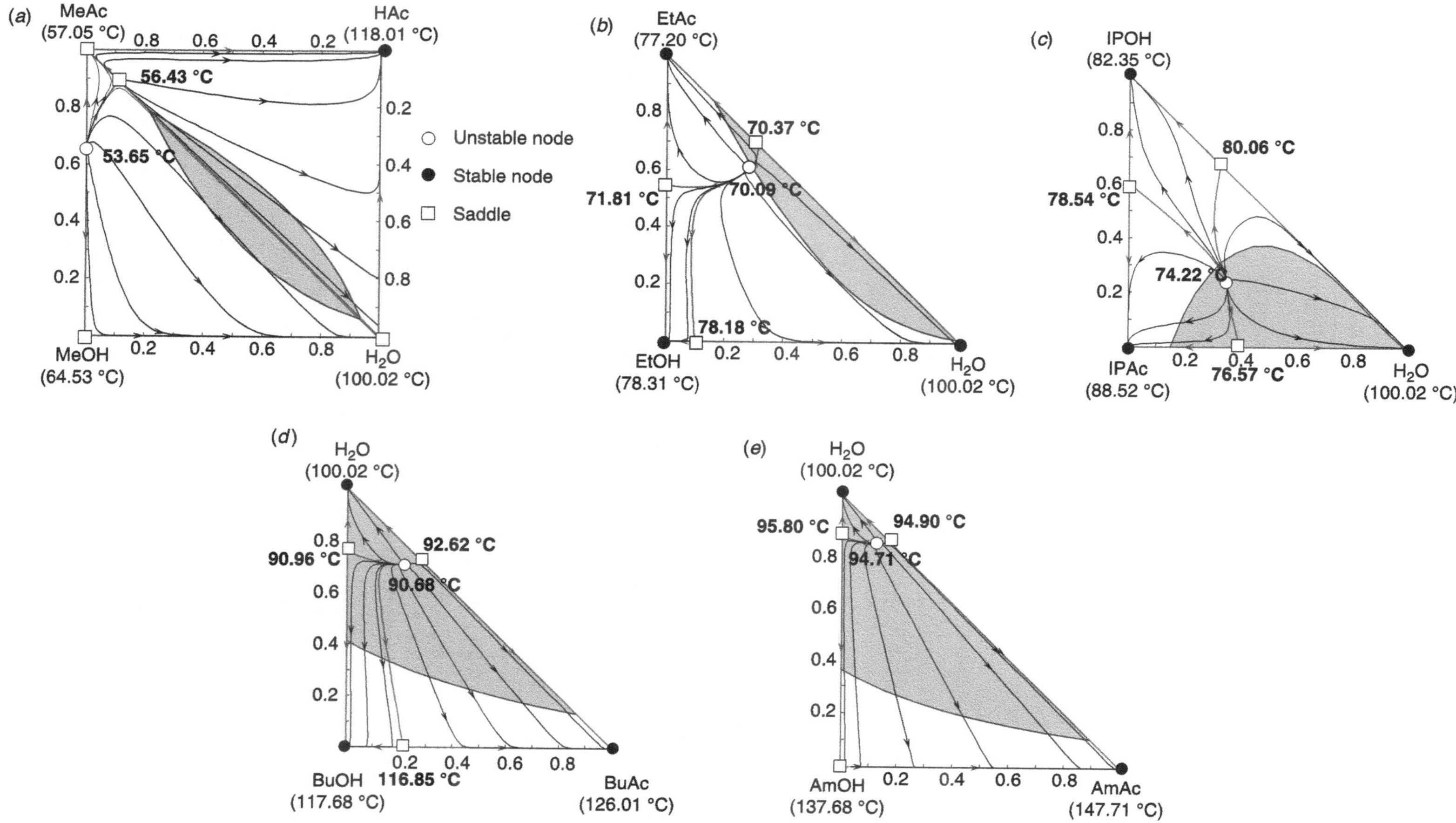

Figure 7.3 The RCM and LL envelope for five esterification systems shown in ternary spaces.

TABLE 7.4 Ranking of Azeotropic Temperatures and Pure Component Normal Boiling Point Temperatures

(i) MeAc	(ii) EtAc	(iii) IPAc	(iv) BuAc	(v) AmAc
MeOH/MeAc 53.65 °C	EtOH/EtAc/H_2O 70.09 °C	IPOH/IPAc/H_2O 74.22 °C	BuOH/BuAc/H_2O 90.68 °C	AmOH/AmAc/H_2O 94.71 °C
MeAc/H_2O 56.43 °C	EtAc/H_2O 70.37 °C	IPAc/H_2O 76.57 °C	BuAc/H_2O 90.94 °C	AmAc/H_2O 94.90 °C
MeAc 57.03 °C	EtOH/EtAc 71.81 °C	IPOH/IPAc 78.54 °C	BuOH/H_2O 92.62 °C	AmOH/H_2O 95.80 °C
MeOH 64.53 °C	EtAc 77.20 °C	IPOH/H_2O 80.06 °C	H_2O 100.02 °C	H_2O 100.02 °C
H_2O 100.02 °C	EtOH/H_2O 78.18 °C	IPOH 82.35 °C	BuOH/BuAc 116.85 °C	HAc 118.01 °C
HAc 118.01 °C	EtOH 78.31 °C	IPAc 88.52 °C	BuOH 117.68 °C	AmOH 137.68 °C
	H_2O 100.02 °C	H_2O 100.02 °C	HAc 118.01 °C	HAc/AmOH/AmAc 139.89 °C
	HAc 118.01 °C	HAc 118.01 °C	HAc/BuOH/BuAc 121.58 °C	HAc/AmOH 140.07 °C
			HAc/BuOH 123.21 °C	AmAc 147.71 °C
			BuAc 126.01 °C	

exactly the same methyl acetate process configuration that was first proposed by Agreda et al.[9] and studied by many others.[10,11]

This process flowsheet can be understood from boiling point temperature ranking, including azeotropes (Table 7.4). The heavy product of the reaction, H_2O, is withdrawn from the bottoms and the light product, MeAc, is taken out of the top. The azeotropes have little role in the separation sections except for the top product being a saddle (ultimately the RCM will end at the MeAc/H_2O azeotrope, 56.43 °C). This will cause some problems in control and operation. Nevertheless, at the design stage, it is possible to achieve reasonably high purity products at both ends with the type I flowsheet (Fig. 7.2*a*).

7.2.2 Type II Flowsheet: EtAc and IPAc

The boiling point temperature ranking suggest that we may use the type I flowsheet for EtAc production with EtAc and H_2O withdrawn from the top and the bottom of the column. Indeed, this configuration was simulated by several researchers,[12,13] but the purity level

[9]V. H. Agreda, L. R. Partin, and W. H. Heise, High-purity methyl acetate via reactive distillation, *Chem. Eng. Prog.* **86**(2), 40–46 (1990).

[10]M. F. Doherty and M. F. Malone, *Conceptual Design of Distillation Systems*, McGraw-Hill, New York, 2001.

[11]M. A. Al-Arfaj and W. L. Luyben, Comparative control study of ideal and methyl acetate reactive distillation, *Chem. Eng. Sci.* **57**, 5039–5050 (2002).

[12]H. Komatsu and C. D. Holland, A new method of convergence for solving reacting distillation problems, *J. Chem. Eng. Jpn.* **10**, 292–297 (1977).

[13]N. Vora and P. Daoutidis, Dynamic and control of an ethyl acetate reactive distillation column, *Ind. Eng. Chem. Res.* **40**, 833–849 (2001).

(~80%) is far from the product specification and further purification is needed. Analogous to the MeAc system, the separation in the stripping section involves the H_2O/EtOH/EtAc ternary system. It becomes obvious that the EtAc system has much more complicated phase behavior and, in particular, three distillation boundaries are found (Fig. 7.3*b*). Moreover, we also have the EtOH/H_2O (78.18 °C) azeotrope that may prevent us from obtaining high purity H_2O if the separation is performed along the EtOH–H_2O edge. Actually, Lee[14] shows that only one double-feed column (type I flowsheet) cannot produce pure EtAc, and further purification is needed. However, we also observe a ternary azeotrope, EtOH/EtAc/H_2O, which is the lowest temperature (70.09 °C) in the ranking (Table 7.4 and Fig. 7.1*b*). Furthermore, the azeotrope lies on the edge of the LL envelope with the tie lines sloping toward relatively high purity H_2O (Fig. 7.1*b*). Therefore, when most of the acid is consumed in the reactive section, we end up close to the ternary azeotrope toward the top of the column (see Fig. 7.3*b*). A decanter could be used to remove H_2O from the aqueous phase with the organic phase further processed to obtain high purity EtAc in a stripper. This is the basic process configuration presented by Tang et al.,[4] which is denoted here as the type II process (Fig. 7.2). Note that the products of both the first column (reactive zone + rectifier) and the second column (stripper) are all nodes (stable or unstable), and they possess much better operability compared to the MeAc system.

In contrast to MeAc and EtAc synthesis, the information about isopropyl acetate synthesis with acetic acid in a reactive distillation process is rarely found in the literature except for the vapor–liquid equilibrium and kinetics data.[10,15] The RCM of the IPAc system in Figure 7.3*c* clearly suggests that the type II process is an ideal candidate for IPAc synthesis using reactive distillation. The IPAc system is even more favorable with the type II flowsheet than the EtAc system because it has a larger LL envelope with a minimum boiling ternary azeotrope (74.22 °C) located further inside the LL zone with tie lines mostly pointing to the high purity H_2O corner. Once the H_2O is removed from the aqueous phase of the decanter, part of the organic distillate is fed to a stripper to remove high purity IPAc from the bottoms while recycling the overhead vapor back to the decanter (Figs. 7.2 and 7.3*c*). Note that the top products of both the reactive distillation (first column in Fig. 7.2) and the stripper (second column, Fig. 7.2) point toward the ternary azeotrope (an unstable node), and the bottoms product of the stripper is a stable node (IPAc corner, Fig. 7.3*c*).

7.2.3 Type III Flowsheet: BuAc and AmAc

The boiling point temperature rankings in Table 7.4 reveal that BuAc differs from IPAc in that the acetate is the highest boiler. This implies that we can remove the acetate from the column base. Moreover, the minimum boiling azeotrope between BuOH and H_2O is also a heterogeneous one (Fig. 7.1*d*), and this leads to a type II LL envelope in the BuAc/BuOH/H_2O ternary composition space. Figure 7.1*d* also reveals that the two-liquid zone in the ternary system constitutes more than 50% of the composition space. However, the BuAc and IPAc systems also share a common characteristic: a minimum boiling ternary azeotrope exists in the acetate/alcohol/water ternary system, and it is located inside the

[14]J. W. Lee, Feasibility studies on quaternary reactive distillation systems, *Ind. Eng. Chem. Res.* **41**, 4632–4642 (2002).

[15]L. S. Lee and M. Z. Kuo, Phase and reaction equilibria of the acetic acid–isopropanol–isopropyl acetate–water system at 760 mmHg, *Fluid Phase Equilib.* **123**, 147–165 (1996).

LL envelope (Fig. 7.1*d*). This immediately suggests a decanter should be placed in the column top to remove high purity pure water from the aqueous phase. Unlike the type II flowsheet, the heavy boiling acetate suggests that we should totally recycle the organic phase back to the column and withdraw the acetate from the bottoms. The scenario results in the type III flowsheet for the BuAc system that is shown in Figure 7.2. It is a reactive distillation column with the reactive zone located in the middle and with a decanter placed in the column top to perform LL separation. The removal of water should be relatively easy with the significant LL envelope and the favorable RCM as a result of the ternary azeotrope (Fig. 7.3*d*). This is exactly the flowsheet proposed by several different research groups.[16,17]

The amyl acetate system also shows a similar VLLE characteristic with an even larger LL envelope for the acetate/alcohol/water system, as shown in Figure 7.1*e*. The larger LL zone suggests that the LL separation could be a little easier for the AmAc system (compared to the BuAc system). This naturally leads to the type III flowsheet by removing high purity pure water from the top and heavy acetate from the bottoms. It is also observed that amyl alcohol is the second highest boiling pure component, which is different from the BuAc system where the acid is the heavy reactant.

It is interesting to see the gradual transition in the VLLE system (e.g., Figs. 7.1 and 7.3) as the carbon number in the alcohols increases, despite the discrete changes between chemical species. It is even more interesting to observe the jumps between process flowsheets (Fig. 7.2) as the VLLE characteristics (e.g., ranking of boiling points, Table 7.4) pass through some critical points. Qualitatively, it seems that we can determine the type of flowsheet by simply examining the VLLE behavior. These observations will be examined carefully in the next section.

Finally, it is also observed that, despite seemingly different process configurations, these three flowsheets (Fig. 7.2) share a common characteristic. They all consist of three major elements: a reactive section, a rectifier, and a stripper with somewhat different arrangements.

7.3 STEADY-STATE DESIGN

The TAC derived by Douglas[18] is used to evaluate different designs. It is defined as

$$\text{TAC} = \text{operating cost} + \frac{\text{capital cost}}{\text{payback period}} \tag{7.2}$$

where the operating cost includes the costs of steam, cooling water, and catalyst, and the capital cost covers the cost of the column, trays, and heat exchangers. A payback period of 3 is used and a catalyst life of 3 months is assumed.

7.3.1 Design Procedure

As pointed out earlier, despite having different process configurations, these flowsheets all consist of a rectifier, a stripper, and a reactive section. Obvious design parameters are the

[16] J. Gangadwala, S. Mankar, S. Mahajani, A. Kienle, and E. Stein, Esterification of acetic acid with butanol in the presence of ion-exchange resins as catalysts, *Ind. Eng. Chem. Res.* **42**, 2146–2155 (2003).

[17] S. G. Huang, C. L. Kuo, S. B. Hung, Y. W. Chen, and C. C. Yu, Temperature control of heterogeneous reactive distillation: Butyl propionate and butyl acetate esterification, *AIChE J.* **50**, 2203–2216 (2004).

[18] J. M. Douglas, *Conceptual Design of Chemical Process*, McGraw-Hill, New York, 1988.

number of rectifying trays (N_R), stripping trays (N_S), and reactive trays (N_{RX}). In addition to the tray numbers, another set of important design parameters is the feed tray locations (NF_{acid} and $NF_{alcohol}$). In theory, equimolar feed flowrates ($F_{acid} = F_{alcohol}$) should be economically optimal, but for the type II flowsheet the feed ratio (FR $= F_{acid}/F_{alcohol}$) is also an important design variable. Thus, potential design variables for these three flowsheets are N_R, N_S, N_{RX}, NF_{acid}, $NF_{alcohol}$, and FR. Once the column configuration is specified, we are left with 2 degrees of freedom for the types I and III processes and 3 degrees of freedom for the type II flowsheet. In the type I process, two product specifications are given to define the two-feed, two-product column; typically, reflux flow and heat input are used to meet the specifications. For the type II process, one-product purity, the organic reflux ratio, and no aqueous reflux are used to define the flowsheet. The organic reflux ratio in a sense defines the distribution of heat input ($Q_R/Q_{R,S}$) between the reactive distillation column and the stripper. Typically, this degree of freedom is used to drive the acid purity level to an acceptable level. Note that there is no product withdrawal from the bottoms of the reactive distillation column, so the boilup rate is the sum of the total feed and reflux flow and no degrees of freedom are left once the reflux flow is given. The third degree of freedom should be the aqueous reflux ratio. Because the water content in the aqueous phase is primarily determined by the shape of the LL zone, the aqueous reflux ratio has little effect on the product purity, so we assume no reflux for the aqueous phase. For the type III flowsheet, one-product purity and no aqueous reflux are used to define the flowsheet. This means that the top product is withdrawn from the aqueous phase and the organic phase is totally refluxed. For a type III process, generally we adjust the heat duty to achieve the desired product purity in the bottoms.

Once the dominant design variables are identified, a systematic design procedure can be devised for all three flowsheets. All of the simulations were carried out using Aspen Plus with the RadFrac module provided with Fortran subroutines for the reaction rates. For a system with a given production rate and product specifications, the design steps are the following:

1. Set the reactants' feed ratio to 1 initially (i.e., FR $= F_{acid}/F_{alcohol} = 1$).
2. Fix a number of reactive trays (N_{RX}).
3. Place the heavy reactant feed (NF_{heavy}) on the top of the reactive zone and introduce the light reactant feed (NF_{light}) on the lowest tray of the reactive zone.
4. Guess the tray numbers in the rectifying section (N_R) and the stripping section (N_S).
5. Change the reflux flow (R) and heat input (Q_R; type I flowsheet) or organic reflux flow (R) and stripper heat input ($Q_{R,S}$; type II flowsheet) or heat input (Q_R; type III flowsheet) until the product specification is met.
6. Go back to step 4 and change N_R and N_S until the TAC is minimized.
7. Go back to step 3 and find the feed locations (NF_{heavy} and NF_{light}) until the TAC is minimized.
8. Go back to step 2 and vary N_{RX} until the TAC is minimized.
9. Go back to step 1 and change the FR until the TAC is minimized (for type II flowsheet only).

It seems strange to devise a sequential design procedure when it may have been done simultaneously. Extensive experience on reactive distillation simulation indicates that the algorithm for the reactive distillation is much less robust compared to conventional distillation. It is more practical to carry out the simulation sequentially. In this work we assume the catalyst occupies 50% of the holdup volume in a reactive tray, the column diameter is sized using the short-cut method of Douglas,[18] and a weir height of 10.16 cm is assumed for the reactive tray. Thus, the catalyst weight on a reactive tray is fixed once the column diameter is determined.

7.3.2 Optimized Design

Type I: MeAc. For the MeAc system, we assume an equimolar feed ratio (FR = 1). The dominant variables for optimization become N_R, N_S, N_{RX}, NF_{heavy}, and NF_{light}. The top graph of Figure 7.4 shows that N_R and N_S have little impact on the TAC, despite showing a minimum as N_S varies. Note that this profile is obtained by fixing the feeds to the top and bottom of the reactive section (e.g., Fig. 7.2). Flat profiles are also observed as we change the number of reactive trays. The most dominant optimization variables are the *feed tray locations* as shown in Figure 7.4. The results indicate that the light reactant (MeOH) should be fed to tray 13 (counting the tray number from bottom up) and the heavy reactant (HAc) should be introduced on tray 36. Note that these two trays lie *inside* the reactive zone, not at the top or bottom. The optimum flowsheet is provided in Figure 7.5*a*.

Type II: EtAc and IPAc. The type II flowsheet differs from type I in that the reactive zone extends to the column base of the first column (called the reactive distillation column); therefore, a much larger holdup is expected in the bottom of the reactive distillation column (Fig. 7.2). The column base holdup is taken to be 10 times that of a reactive tray holdup. We also assume the feed ratio of the reactant can be changed, which leads to the following optimization variables: N_R, N_S, N_{RX}, NF_{heavy}, NF_{light}, and FR.

For EtAc, the results (Fig. 7.6) indicate that the *number of trays in the rectifying section* in the reactive distillation column (N_R) is one of the dominant optimization variables but the N_S shows little impact on the TAC once N_R is fixed. Variation of N_{RX} results in a rather small change in the TAC. Note that the top two graphs in Figure 7.6 are obtained by fixing the feeds at the top and bottom of the reactive section. For the feed tray locations, the optimal feed tray for the heavy reactant (HAc) is the column base, despite the small difference in the TACs. This is understandable because we have the largest catalyst holdup in the reboiler and this leads the same feed location for both reactants. It is surprising that another dominant variable for optimization is the *feed ratio*. Results show that a small excess of alcohol is preferred because this facilitates the consumption of acid to meet stringent acid specification. The reboiler duty in the reactive distillation column (Fig. 7.6) reveals this effect. Numerical simulation results indicate that the minimum number of reactive trays needed to achieve the desired purity (mole fraction of acetate = 0.99, acid purity <100 ppm) is 7 (i.e., $N_{RX,\min} = 7$). Note also that, for the same specification, the operating window for the feed ratio ranges from 0.962 to 0.98. This implies some feedback control is necessary to maintain the feed ratio.

For the IPAc system, we have the same optimization variables for the type II flowsheet. Again, simulation results reveal that N_S has little impact on the total annual cost and, as expected, both reactant feeds should be introduced to the reboiler. In addition,

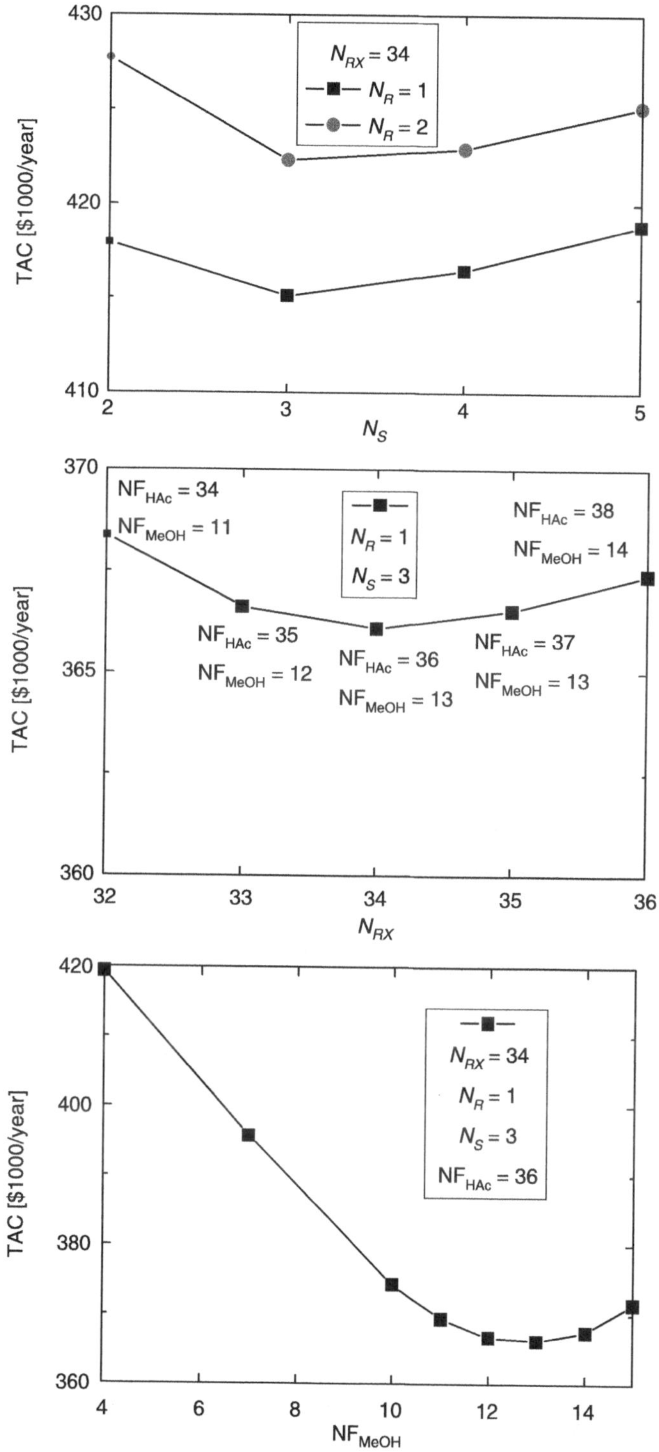

Figure 7.4 Effects of design variables on TAC for the MeAc system.

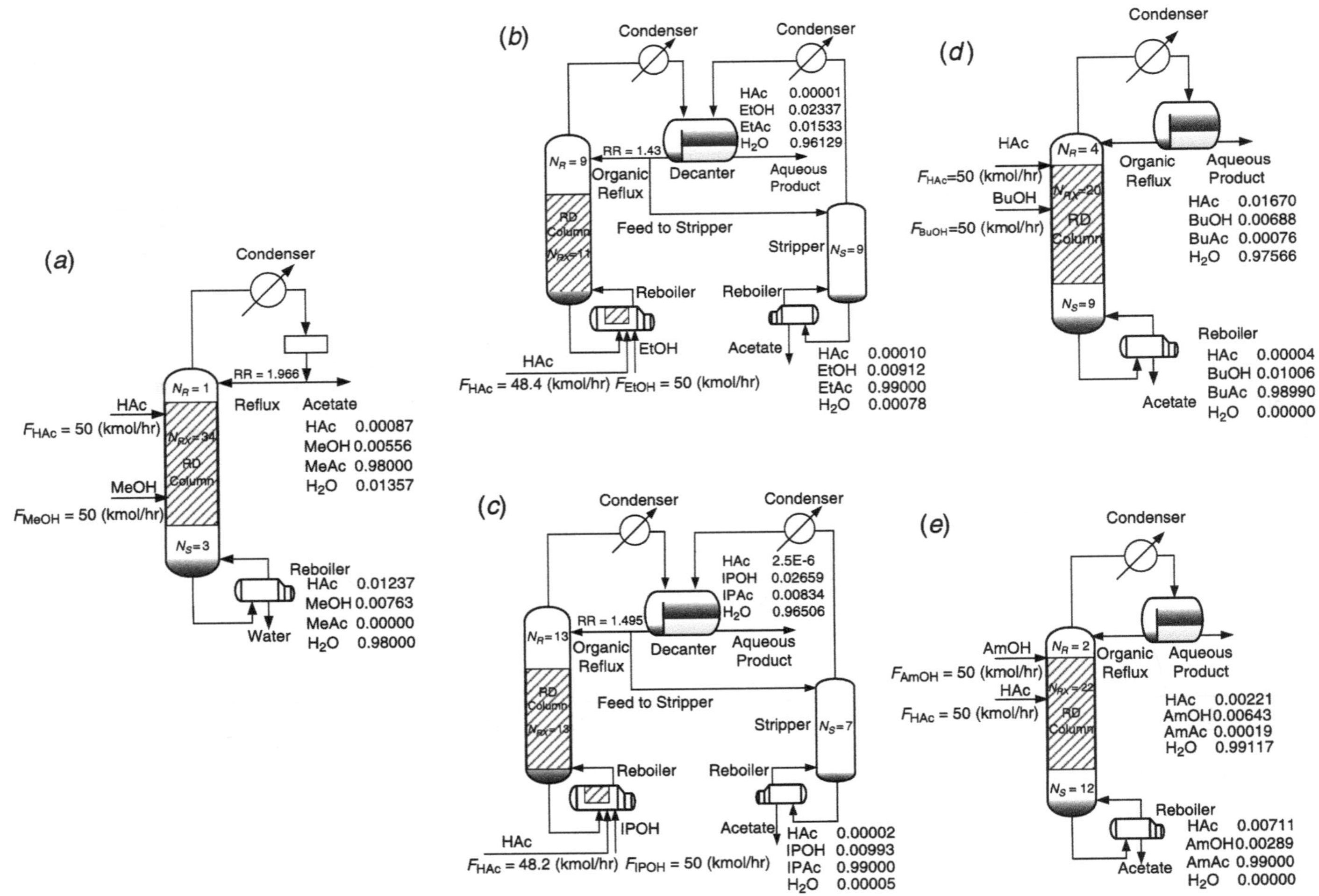

Figure 7.5 Optimized process flowsheets for the five esterification systems.

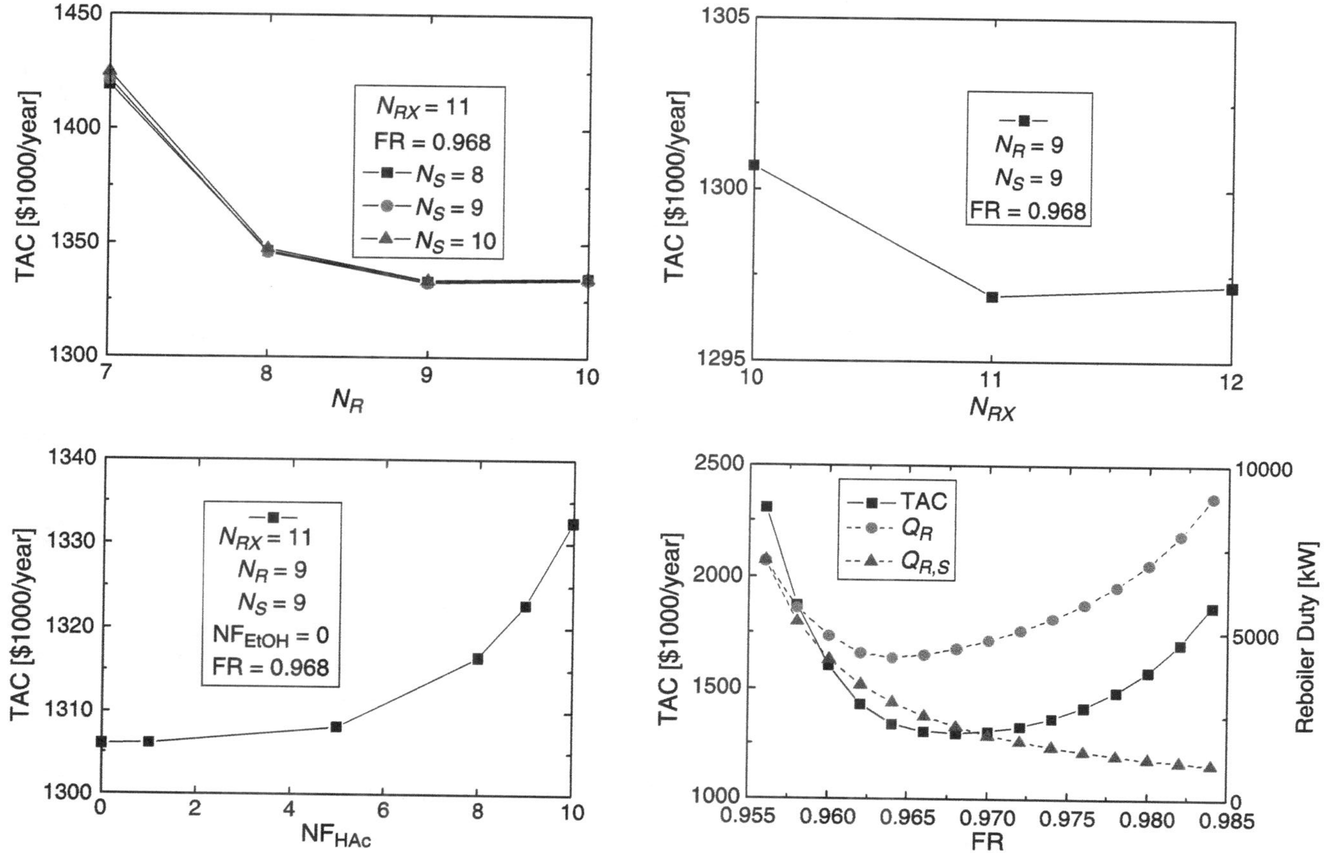

Figure 7.6 Effects of design variables on total annual cost for the EtAc system.

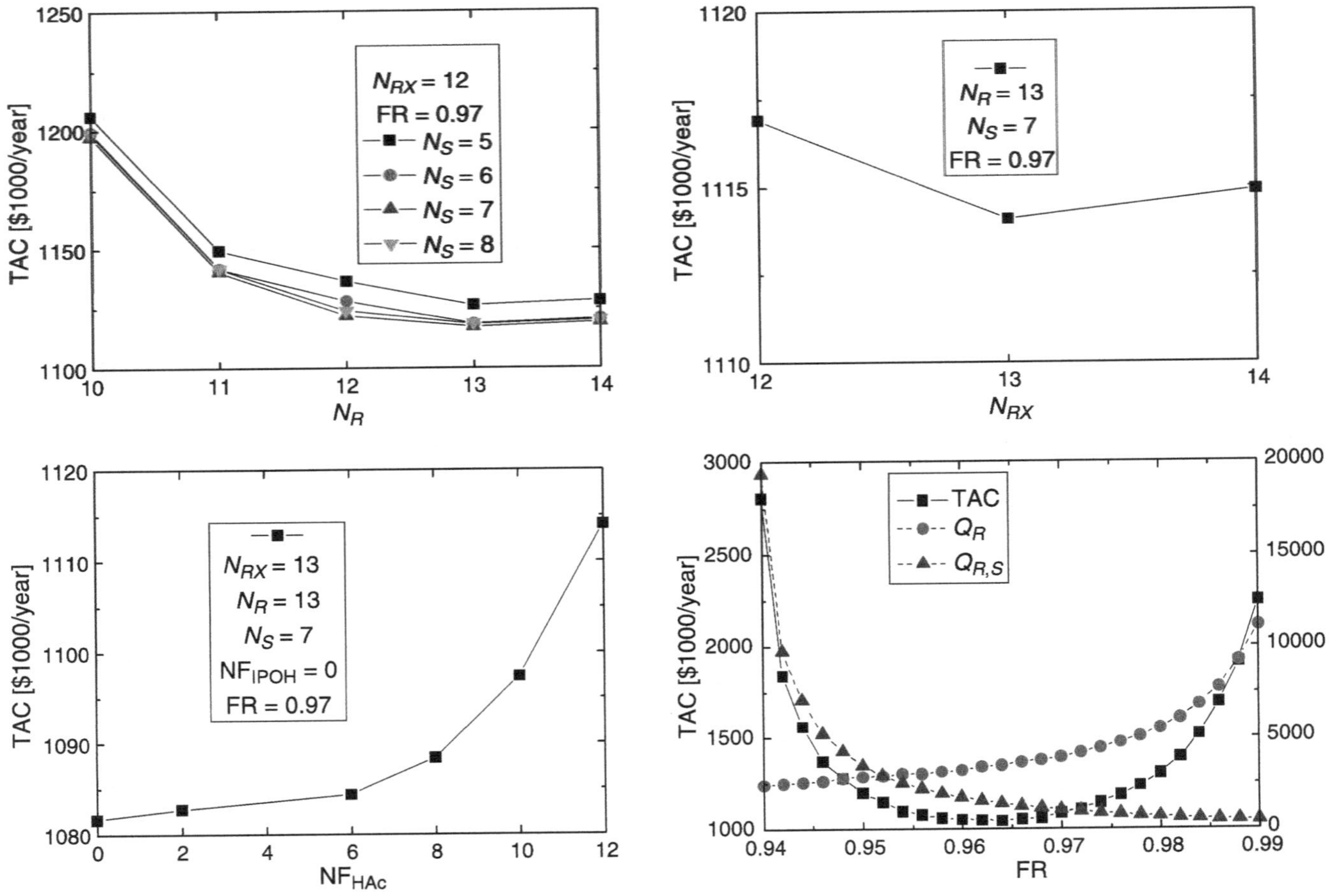

Figure 7.7 Effects of design variables on total annual cost for the IPAc system.

the *number of trays in the rectifying section* (N_R) is one of the dominant variables for optimization. Changes in N_{RX} result in a rather small variation in the TAC. Note that the top two graphs in Figure 7.7 are obtained by fixing the feeds at the top and bottom of the reactive section. The other important variable is the feed ratio as shown in Figure 7.7. The results indicate a slight excess of alcohol is favorable because it facilitates LL separation. For the IPAc process, the minimum number of reactive trays needed to achieve the desired purity (mole fraction of acetate $= 0.99$, acid impurity < 100 ppm) is 9 (i.e., $N_{RX,\min} = 9$). The operating window for the feed ratio ranges from 0.946 to 0.99 for the desired specification. Again, some feedback mechanism is needed to adjust the feed ratio.

Type III: BuAc and AmAc. The type III flowsheet has been studied by several researchers.[7] In theory, we have the following optimization variables: N_R, N_S, N_{RX}, NF_{heavy}, and NF_{light}. For the BuAc system, the top graph of Figure 7.8 shows that variations in N_R and N_S result in rather small changes in the TAC. Note that the TACs are obtained by fixing the feeds at the top and bottom of the reactive section. The bottom graph in

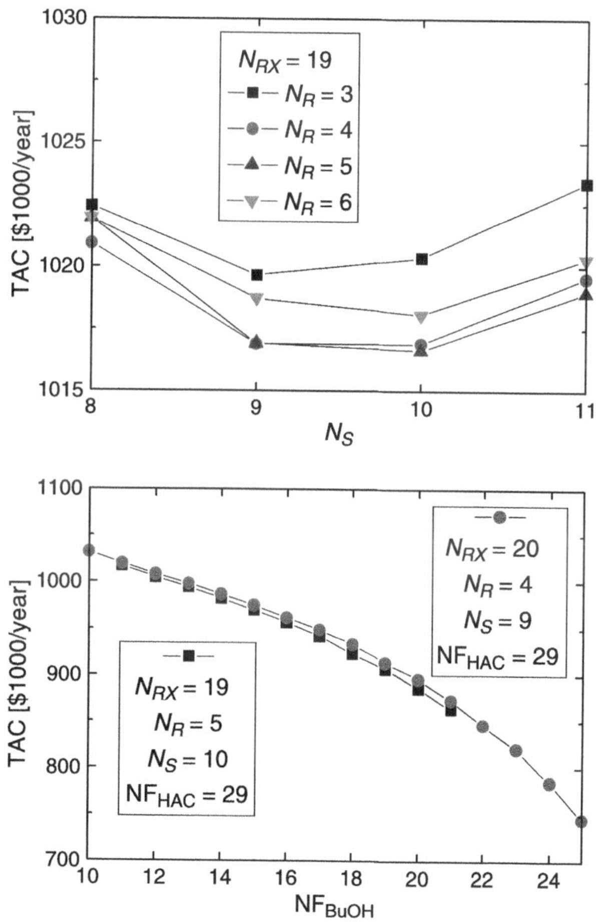

Figure 7.8 Effects of design variables on TAC for the BuAc system.

TABLE 7.5 Steady-State Operating Condition and TAC for Reactive Distillation (RD) Designs of Five Esterification Systems

System Column Configuration	(i) MeAc RD	(ii) EtAc RD	Stripper	(iii) IPAc RD	Stripper	(iv) BuAc RD	(v) AmAc RD
Total no. of trays including reboiler	39	20	10	26	8	34	37
No. of trays							
Stripping section (N_S)	3		9		7	9	12
Reactive section (N_{RX})	34	11		13		20	22
Rectifying section (N_R)	1	9		13		4	2
Reactive trays	4–37	0–10		0–12		10–29	13–34
Acetic acid feed tray	36	0		0		25	30
Alcohol feed tray	13	0		0		29	34
Feed flowrate (kmol/h)							
Acid	50.00	48.40		48.20		50.00	50.00
Alcohol	50.00	50.00		50.00		50.00	50.00
Product flowrate (kmol/h)							
Top	50.35	50.30		49.94		50.38	49.98
Bottom	49.65		48.10		48.26	49.62	50.02
X_D or $X_{D,\mathrm{aq}}$ (mf)							
Acid	0.00087	0.00001		2.5E-6		0.01670	0.00221
Alcohol	0.00556	0.02337		0.02665		0.00688	0.00643
Acetate	0.98000	0.01533		0.00835		0.00076	0.00019
Water	0.01357	0.96129		0.96500		0.97566	0.99117
X_B (mf)							
Acid	0.01237		0.00010		0.00002	0.00004	0.00711
Alcohol	0.00763		0.00912		0.00993	0.01006	0.00289
Acetate	$<10^{-8}$		0.99000		0.99000	0.98990	0.99000
Water	0.98000		0.00078		0.00005	$<10^{-8}$	$<10^{-8}$

Condenser duty (kW)	−1280.22	−4265.71	−1860.54	−3428.58	−1129.89	−2857.92	−1483.15
Subcooling duty (kW)		−833.82		−506.51		−461.35	−227.42
Reboiler duty (kW)	1035.71	4523.98	2195.68	3473.31	1370.90	3085.41	1532.48
Column diameter (m)	1.03	1.95	1.45	1.89	1.23	1.88	1.34
Weir height (m)	0.1016	0.1016	0.0508	0.1016	0.0508	0.0508	0.0508
Decanter temperature (°C)		40		50		50	50
Heat transfer area (m^2)							
Condenser	80.03	157.62	68.99	107.37	36.78	57.06	27.63
Subcooling		170.24		96.06		57.89	21.82
Reboiler	38.63	168.72	92.80	129.53	57.94	115.07	57.15
Damköhler number (*Da*)	28.88	29.61		13.08		16.86	4.35
Total capital cost ($1000)	730.78	2051.44		1731.74		1277.26	854.62
Column	323.30	511.57		539.56		547.92	408.93
Column trays	50.60	88.10		97.02		111.55	71.85
Heat exchangers	356.88	1451.77		1095.16		617.79	373.84
Total operating cost ($1000/year)	130.55	621.00		474.31		327.53	197.66
Catalyst cost	30.94	64.97		67.06		29.80	16.61
Energy cost	91.55	548.11		399.29		289.26	173.49
Wastewater treatment cost	8.061	7.925		7.965		8.470	7.562
TAC ($1000/year) (50 kmol/h)	374.14	1304.82		1051.56		753.28	482.54
TAC ($1000/year) (52,825 ton/year)	659.14	2005.82		1388.46		850.20	482.54

Figure 7.8 clearly shows that the *feed locations* are the most important optimization variables. Significant reductions in TAC are achieved by simply varying the feed locations. This can be understood because we need to arrange the feeds such that optimal reactant and temperature profiles can be achieved in the reactive section.

The sensitivity of the design parameters of the AmAc system is quite similar to that of the BuAc system. First, variations in N_R and N_S result in relatively small changes in the TAC. Second, the feed locations are the dominant optimization variables compared to other variables such as N_R, N_S, and N_{RX}.

The optimization results show an interesting fact that different flowsheets give rise to different dominant optimization variables: NF_{heavy} and NF_{light} for the type I flowsheet, N_R and FR for the type II flowsheet, and NF_{heavy} and NF_{light} for the type III flowsheet. Table 7.5 summarizes the optimal designs for these five systems, and all of the flowsheets are given in Figure 7.5.

7.4 PROCESS CHARACTERISTICS

7.4.1 Type I: MeAc

The normal boiling point temperature ranking of the MeAc process is shown in Table 7.4 for the pure components: HAc (118 °C) > H_2O (100 °C) > MeOH (64.5 °C) > MeAc (57.5 °C). The composition profiles in the upper graph in Figure 7.9 indicate that we are primarily separating MeAc/HAc in the rectifying section, instead of adjacent boilers (MeAc/MeOH). This clearly shows the advantage of reactive distillation reacting away the heaviest acid toward the lower part of the reactive zone and consuming the lighter alcohol toward the upper part of the reactive zone. This makes the separation easy.

The upper graph in Figure 7.9*a* also shows the fraction of total conversion in the reactive zone (shaded area). It reveals that 95% of the total conversion occurs in ~10 reactive trays from the total of 34 reactive trays. The rest of the reactive trays (especially toward the upper section of the reactive zone) seem to serve more for the purpose of separation. By consuming most of the MeOH, the separation is facilitated in an almost MeOH-free environment. The temperature profile in the MeAc reactive distillation column shows a nonmonotonic temperature distribution (lower graph, Fig. 7.9). The reasons come from *reaction* oriented thinking (as opposed to the *separation* oriented thinking) in the feed arrangement and deliberately driving the heavy reactant toward the top and light reactant to the bottom.

The three-dimensional composition profile in Figure 7.10 reveals a potential problem for the operation of the MeAc column: the top product MeAc is a saddle and a further increase in the MeAc/H_2O azeotrope in the MeOH-free composition space. Figure 7.11 shows the nonmonotonic behavior in the acetate composition as we change the reflux flowrate while keeping the bottoms water purity at 98%. Controlling a point near a saddle is not an easy task, and reports on this difficulty can be found in the literature.[11]

7.4.2 Type II: EtAc and IPAc

The type II flowsheet differs from type I in that the two columns (the reactive distillation column and the stripper) are separated by a decanter. Thus, this flowsheet should interact less from steady-state and dynamic points of view. The composition profile in Figure 7.12 indicates that the reactive section generates the necessary acetates, followed by a rectifying

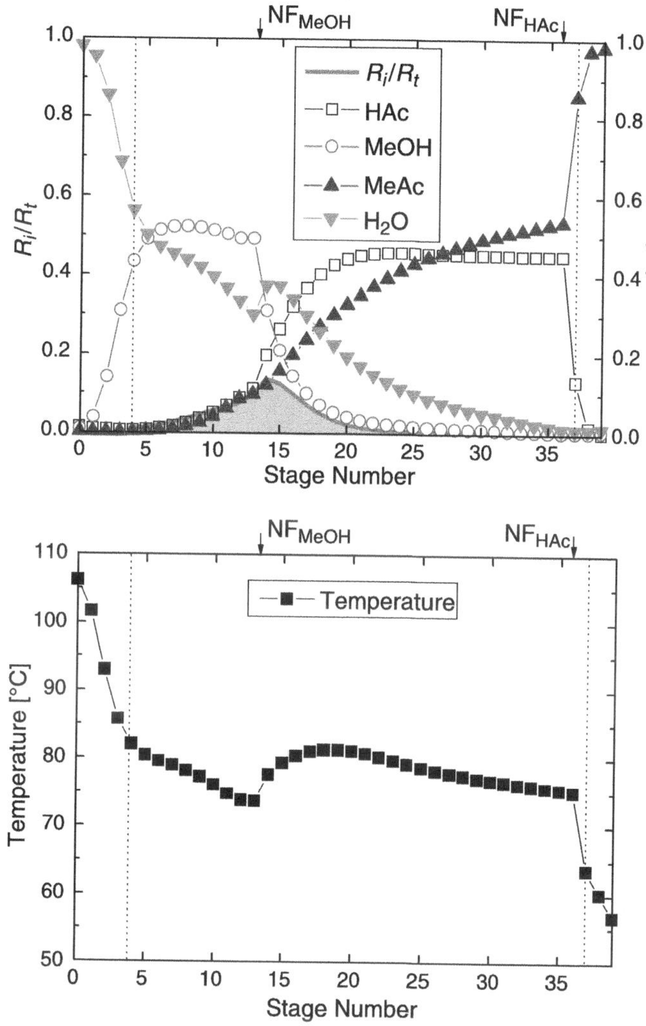

Figure 7.9 Composition and temperature profiles for the MeAc system. Fraction of total conversion (R_i/R_t) indicated in shaded area.

section in which the composition of the heavy acid (HAc) is kept to a parts per million level. The overhead vapor is condensed and the condensate falls within the LL envelope, which is further separated by the decanter, and high purity H_2O is removed from the aqueous as shown in Figure 7.10. Part of the organic phase is refluxed back to the reactive distillation column while the other part is fed to the stripper to purify EtAc in the left corner of the composition space (Figs. 7.10 and 7.12) and the top of the stripper is recycled back to the decanter. A straightforward separation is performed in the stripper, which is demonstrated in the composition profile (lower graph, Fig. 7.12). Very high purity EtAc can be obtained, and the purity level of the acid in the product stream depends on the amount of acid allowed in the overhead of the reactive distillation column.

Figure 7.12 also reveals that 80% of the conversion occurs in the column base of the reactive distillation column as a result of a large catalyst holdup, a high reactant

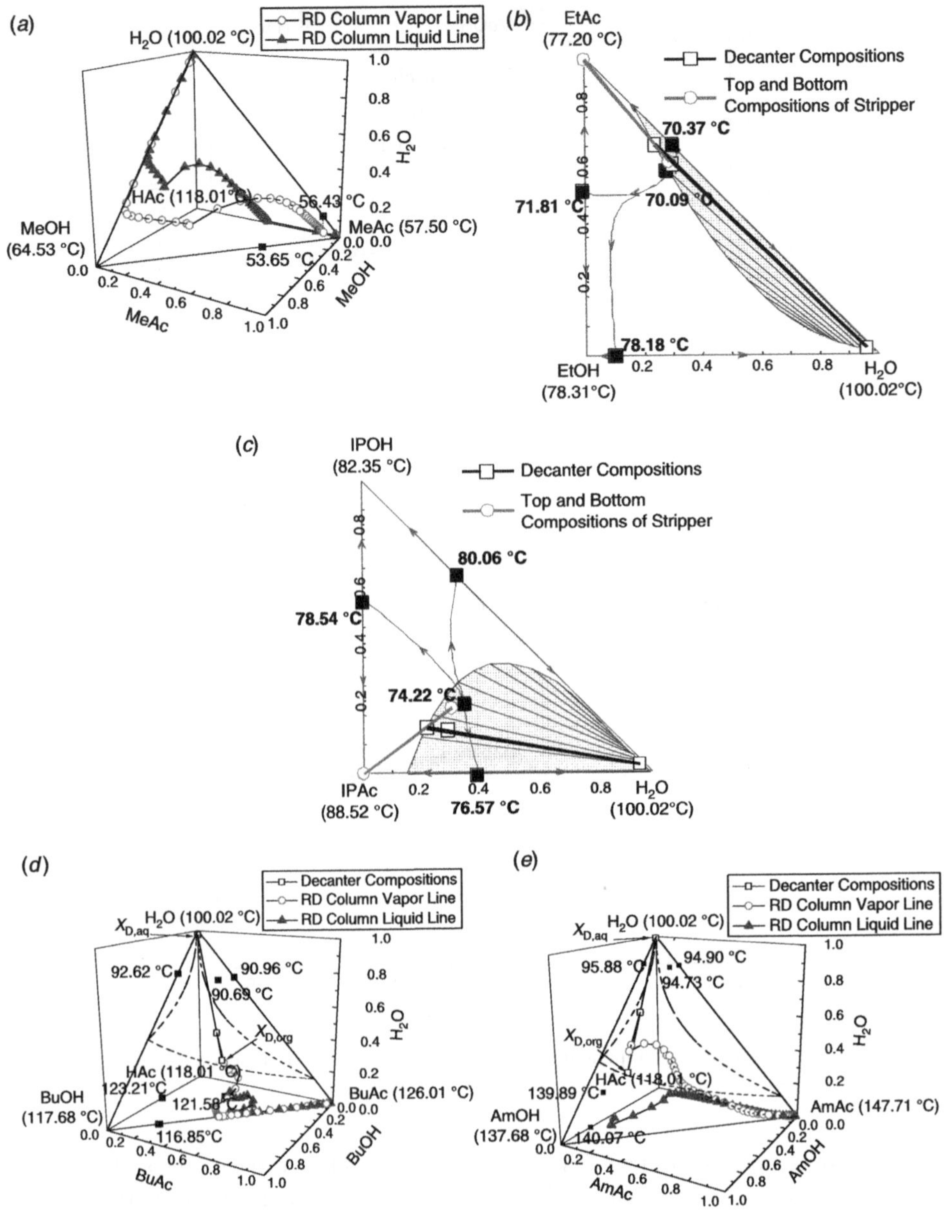

Figure 7.10 Composition profiles and material balances lines for the five esterification systems.

composition, and a favorable reaction temperature. The temperature profile in the reactive distillation column shows monotonic behavior that is dominated by the acid profile, and a similar temperature is observed in the stripper.

Very similar composition profiles are observed for the IPAc system and, again, most of the conversion takes place in the column base of the reactive distillation column (Fig. 7.13). The separation in the stripper is quite straightforward, and high purity IPAc can be achieved

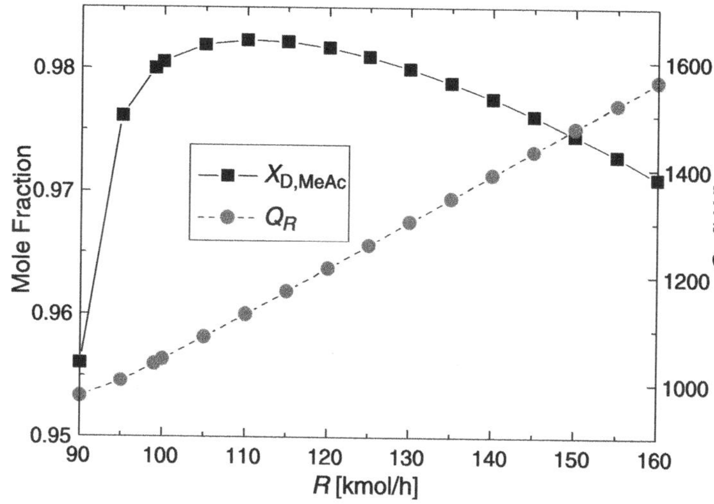

Figure 7.11 Nonmonotonic behavior of MeAc reactive distillation column.

with little difficulty. Monotonically decreasing temperature profiles can also be seen for the IPAc system. The material balance lines in Figure 7.10 show how separation is achieved around the decanter.

7.4.3 Type III: BuAc and AmAc

For the BuAc system, acetic acid has the second highest normal boiling temperature and therefore the acid is reacting away in the lower part of the reactive zone, as shown in the upper graph in Figure 7.14. The composition profile in the stripping section clearly shows that the stripper is performing the BuOH/BuAc separation. Because of a large LL envelope, only four rectifying trays are needed to drive the condensate of the top vapor into the LL zone. We also observed that significant conversion occurs in the upper section of the reactive zone where two feeds are located nearby. The concentration effects of these two reactants enhance the reaction in that region. The lower graph in Figure 7.14 shows an almost monotonic temperature because the two reactants are close boiling components with a 1 °C temperature difference in the normal boiling points. The vapor and liquid composition profile in a three-dimensional plot (Fig. 7.10) clearly shows the separation characteristic in the stripping section (along the BuOH/BuAc edge) as well as in the LL envelope.

For the AmAc system, AmOH has the second highest normal boiling point temperature (which is different from the ranking for the BuAc system). In theory, we should consume most of the alcohol in the reactive zone. Because we do not have stringent product specifications on the alcohol, a small amount of AmOH is allowed to leave the reactive zone. Still, the stripping section is performing the separation between HAc/AmAc, as demonstrated in the upper graph in Figure 7.15. Because the AmAc system has a larger LL envelope compared to the BuAc system, only two rectifying trays are required to drive the overhead condensate into the LL zone. A major portion of the reaction takes place in the upper section of the reactive zone where the two feeds are located. However, fewer reactive trays are required to react away the heavy reactant, compared to the BuAc system. A

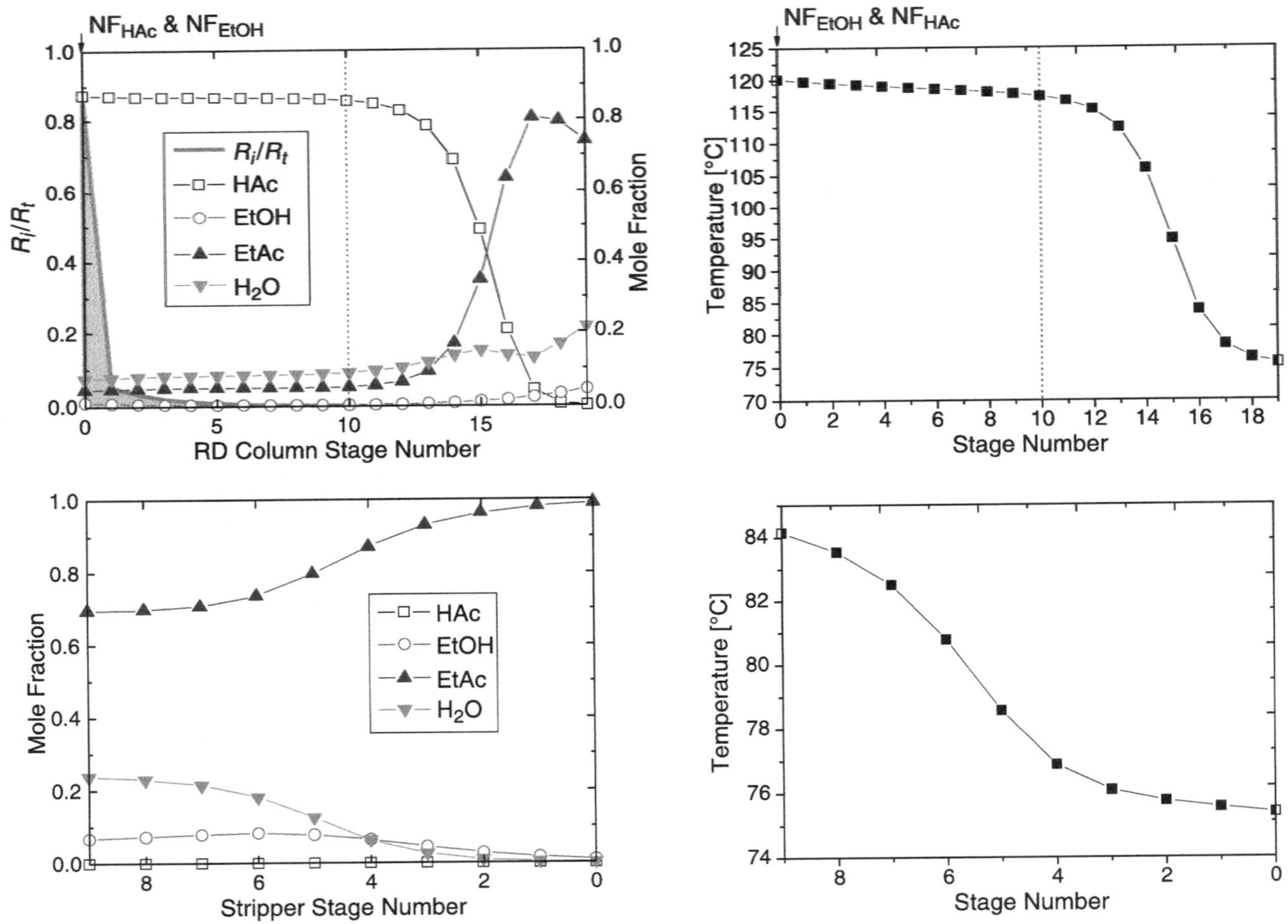

Figure 7.12 Composition and temperature profiles of the EtAc system and fraction of total conversion.

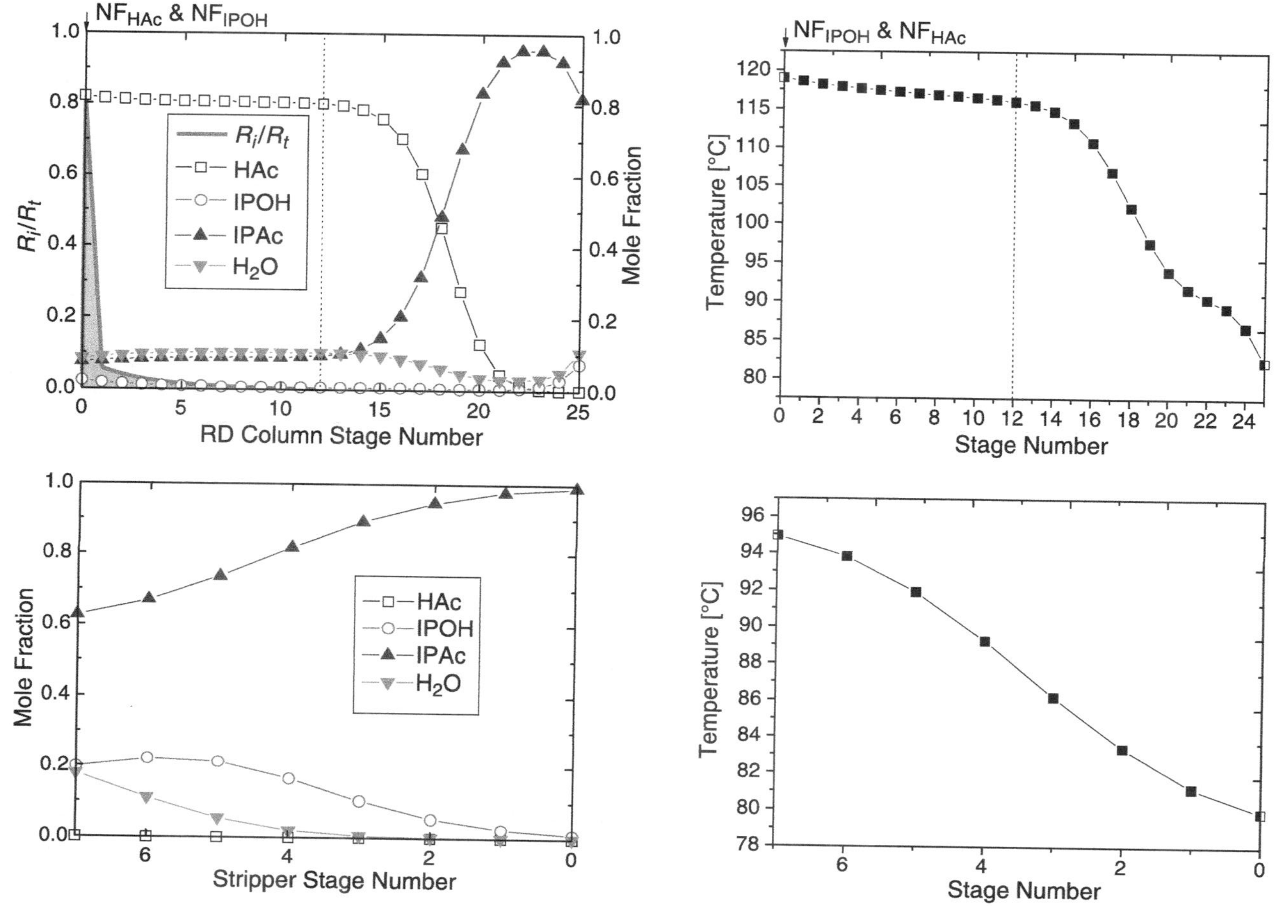

Figure 7.13 Composition and temperature profiles of the IPAc system and fraction of total conversion.

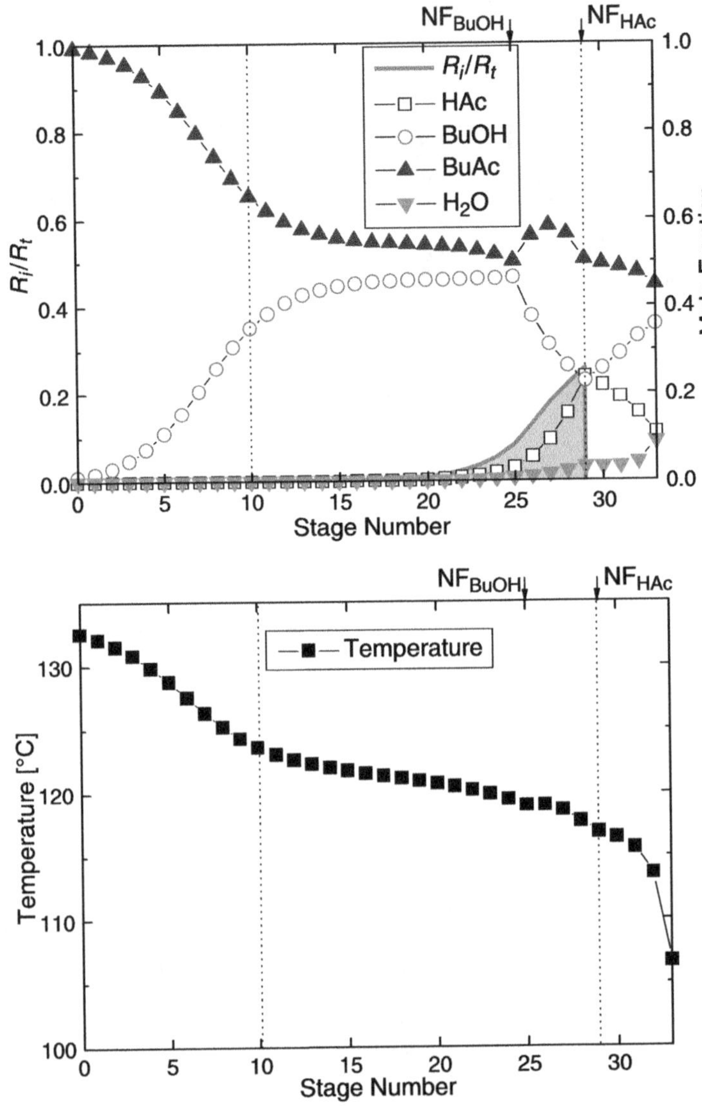

Figure 7.14 Composition and temperature profiles for the BuAc system.

nonmonotonic temperature profile is observed as a result of a large boiling point temperature difference (~20 °C) between these two reactants. Figure 7.10 shows a different characteristic in the stripping section where the compositions are distributed along the HAc/AmAc edge.

Before leaving this section, it should be pointed out that the type III flowsheet is probably the most general configuration for acetic acid esterification. In addition to BuAc and AmAc systems, the production of *n*-propyl acetate, isobutyl acetate, isoamyl acetate, and *n*-hexyl acetate also fall into this category when economic factors are taken into consideration.

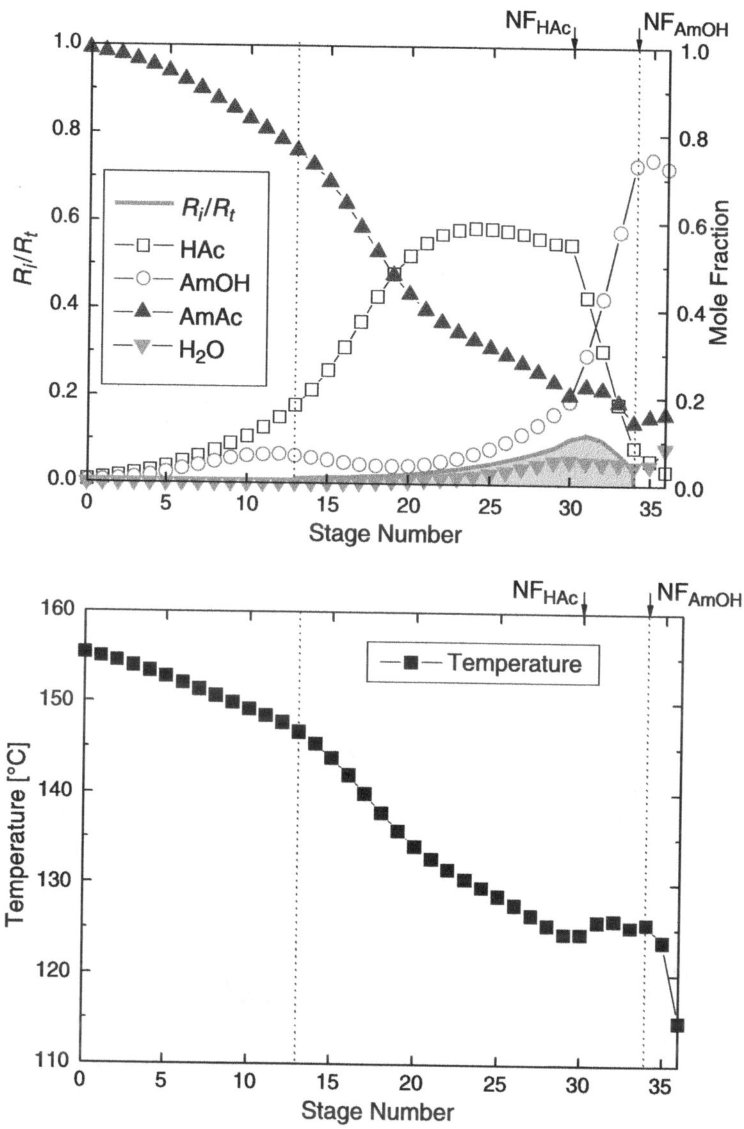

Figure 7.15 Composition and temperature profiles for the AmAc system.

7.5 DISCUSSION

The analyses of the esterification of acetic acid with five alcohols ranging from MeOH to AmOH (C_1–C_5) are intended to gain insight into the design of reactive distillation by varying the chemical species discretely. As the carbon number in the alcohol increases, the flowsheet changes from type I to type II and then to type III (Fig. 7.2). The determinant factors in the flowsheet selection are the ranking of the pure component/azeotrope temperatures (Table 7.4) and the size of the LL envelope (Fig. 7.1). This implies that the structure of the flowsheet can be determined once the VLLE data become available. The flowsheets may look different in the arrangement, but they all include the following units: a stripping section, a reactive zone, and a rectifying section. Once the flowsheet structure is determined, the design can be carried out in a sequential manner by minimizing the TAC. It is interesting that most of the dominant design variables are associated with the feed: the feed tray locations (for type I and III) and the feed ratio (for type II). This can be understood because the reactants' composition distribution is important for kinetically controlled reactive distillation. We also observed that the function of the reactive zone goes beyond providing necessary conversion. The reactive section also facilitates separation by reacting away the heavy reactant toward the lower part of the reactive zone and by consuming most of the light reactant toward the upper part.

The TAC comparison in Figure 7.16 shows that the type II (EtAc and IPAc) flowsheet is the most expensive process, followed by the type III flowsheet (BuAc and AmAc). The type I flowsheet is the most economical process. This is because the type II flowsheet boils up both products to the top of the reactive distillation column (Fig. 7.2) and recycles an almost ternary azeotropic composition back to the decanter from the stripper. Thus, this flowsheet is energy intensive and requires significant capital investment in the heat transfer area. The difference in the TACs between EtAc and IPAc comes from the VLLE advantage of the IPAc system, which has a larger LL envelope (Fig. 7.1). The type III flowsheet, despite not boiling up both products to the top, needs to boil up some acetate/alcohol along with H_2O to get into the LL envelope and then recycles the organic phase *totally* back into the column. We would expect larger energy consumption for this type of system compared to systems without a decanter. Again, the difference between BuAc and AmAc is that the AmAc system has a larger LL envelope (Fig. 7.1). Finally, the MeAc system (type I) only boils up components that are necessary to achieve desired product purity and therefore much smaller energy consumption is expected. Another reason for the MeAc to have the lowest TAC is that the purity level of the acetate typically seen in the literature is 0.98 (mol fraction), compared to 0.99 in the other four cases. The reason for the difference is that higher purity levels at both ends (e.g., $x_{D,MeAc} = x_{B,H_2O} = 0.99$) are not feasible for

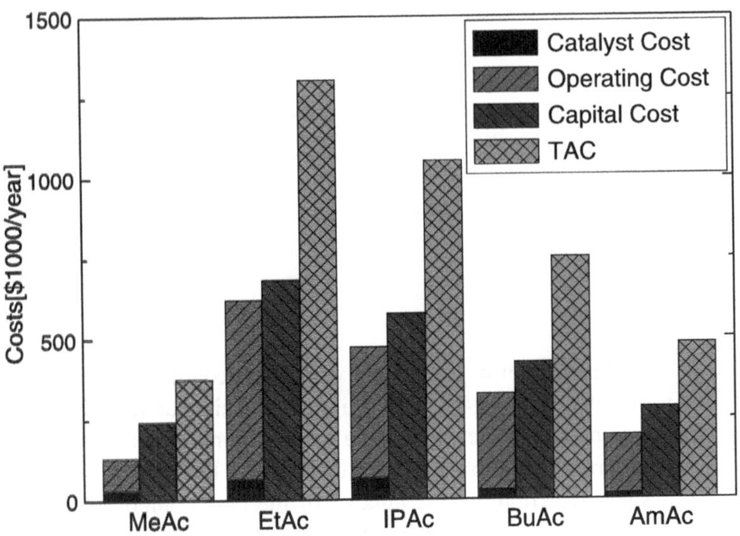

Figure 7.16 TAC (based on 50 kmol/h production rate) for five esterification systems.

the MeAc case. However, the purity level difference will not lead to such a significant reduction in the TAC.

7.6 CONCLUSION

Acetic acid esterifications using reactive distillation with five different alcohols, ranking from methanol to amyl alcohol (C_1–C_5), were explored in this chapter. First, the *qualitative* relationship between phase equilibria and possible process flowsheets was established. Three process flowsheets (types I–III) were identified, which led to relatively pure products. Two important factors in the flowsheet determination were the ranking of the pure component/azeotrope boiling point temperature and the size and location of the LL phase zone. Second, a systematic design procedure was devised to optimize the *quantitative* design based on the TAC. This sequential design procedure overcame the fragility of the simulation algorithm even with the state-of-the-art process simulator for reactive distillation. The dominant design variables were also identified: feed locations for the type I flowsheet, the feed ratio and number of trays in the rectifying section for the type II flowsheet, and feed locations for the type III flowsheet. Third, the characteristics of the optimally designed reactive distillation for these three types of flowsheets were investigated and physical explanations were given. Fourth and finally, the TACs of different flowsheets were compared and the economic potential was ranked and explained.

CHAPTER 8

DESIGN OF TAME REACTIVE DISTILLATION SYSTEMS

In this chapter we take a look at an important example of a reactive distillation column operating in a plantwide environment. The reactive column is part of a multiunit process that includes other columns for recovery of one of the reactants. The process may give the impression that the reactive column is not operating in neat mode because of the need for reactant recovery. We will show that this is really not the case. The recovery of reactant is made necessary by the presence of azeotropes that unavoidably remove one of the reactants from the reactive column.

The example is the production of TAME, which is used as a high-octane gasoline blending component. With the phase-out of lead and the reduction of aromatics in gasoline, the production of various oxygen-containing compounds such as alcohols and esters has grown in importance. The largest volume component in the past was MTBE, but it is being phased out because of groundwater contamination problems. Therefore, TAME is becoming more important. It is produced by the reaction of methanol with unsaturated C5 isoamylenes (2-methyl-1-butene and 2-methyl-2-butene). The liquid-phase reversible reactions considered are

$$\begin{aligned} &\text{2M1B} + \text{MeOH} \Leftrightarrow \text{TAME} \\ &\text{2M2B} + \text{MeOH} \Leftrightarrow \text{TAME} \\ &\text{2M1B} \Leftrightarrow \text{2M2B} \end{aligned} \tag{8.1}$$

This system is fundamentally a ternary system with inerts. The heavy component is TAME, which leaves the reactive distillation column in the bottoms.

The process flowsheet has a prereactor, a reactive distillation column, and a methanol recovery section. A methanol recovery section is required because the inert C5 components coming in with the reactive isoamylenes in the C5 fresh feed form azeotropes with methanol. The result is that a significant amount of methanol is present in the distillate from the

---

Reactive Distillation Design and Control. By William L. Luyben and Cheng-Ching Yu

reactive column. This requires that more methanol must be fed into the reactive column than just that amount required by the stoichiometry of the reactions.

The designs of two different types of methanol recovery sections are compared. In the first, pressure-swing azeotropic distillation is used. In the second, extractive distillation is used.

The material in this chapter is based on articles by Subawalla and Fair,[1] AlArfaj and Luyben,[2] and Luyben.[3] Rigorous steady-state simulation of the process is performed using Aspen Plus. All columns use rigorous RadFrac models. Details of how to use Aspen Plus for simulating conventional as well as reactive distillation columns are provided in work by Luyben.[4]

8.1 CHEMICAL KINETICS AND PHASE EQUILIBRIUM

8.1.1 Chemical Kinetics

The three reactions are first order in their respective reactants and products. Concentration units are mole fractions.

R1: $$\Re_1 = M_{cat}(k_{F1}x_{2M1B}x_{MeOH} - k_{B1}x_{TAME}) \quad (8.2)$$

R2: $$\Re_2 = M_{cat}(k_{F2}x_{2M2B}x_{MeOH} - k_{B2}x_{TAME}) \quad (8.3)$$

R3: $$\Re_3 = M_{cat}(k_{F3}x_{2M1B} - k_{B3}x_{2M2B}) \quad (8.4)$$

Kinetic parameters for the three liquid-phase reversible reactions are given in Table 8.1. These reaction rates are given in units of kilomoles per second per kilogram of catalyst and are converted to the Aspen required units of kilomoles per second per cubic meter by using a catalyst bulk density of 900 kg/m^3. The reactive stages in the column each contain 1100 kg of catalyst. This corresponds to 1.22 m^3 on each tray, which gives a weir height of 0.055 m for a reactive column with a diameter of 5.5 m.

TABLE 8.1 Reaction Kinetics for TAME Reactions

Reaction	A_F (kmol s^{-1} kg^{-1})	E_F (kJ/mol)	A_B (kmol s^{-1} kg^{-1})	E_B (kJ/mol)	ΔH_{RX} (kJ/mol)
R1	1.3263×10^8	76.103737	2.3535×10^{11}	110.540899	−34.44
R2	1.3718×10^{11}	98.2302176	1.5414×10^{14}	124.993965	−26.76
R3	2.7187×10^{10}	96.5226384	4.2933×10^{10}	104.196053	−7.67

[1]H. Subawalla and J. R. Fair, Design guidelines for solid-catalyzed reactive distillation systems, *Ind. Eng. Chem. Res.* **38**, 3693 (1999).
[2]M. A. Al-Arfaj and W. L. Luyben, Plantwide control for TAME production using reactive distillation, *AIChE J.* **50**, 1462 (2004).
[3]W. L. Luyben, Comparison of pressure-swing and extractive-distillation methods for methanol recovery systems in the TAME reactive-distillation process, *Ind. Eng. Chem. Res.* **44**, 5715–5725 (2005).
[4]W. L. Luyben, *Distillation Design and Control Using Aspen Simulation*, Wiley, Hoboken, NJ, 2006.

8.1.2 Phase Equilibrium Using Aspen Plus

The phase equilibrium of this system is complex because of the existence of azeotropes. The inert components in the C5 feedstream include isopentane, *n*-pentane, 1-pentene, and 2-pentene. Essentially all of these inerts go out the top of the reactive distillation column. To illustrate the vapor–liquid equilibrium issues involved in the separation, we consider the ternary system iC_5, methanol, and TAME.

Pure Component and Binary Information. The normal boiling points of these three components are 301, 338, and 359 K, respectively. The reactive column operates at 4 bar, at which the boiling points are 348, 377, and 413 K, respectively. This pure component information can be easily obtained in Aspen Plus by going to the top tool bar and clicking *Tools*, *Conceptual Design*, and *TernaryMap Analysis*. The window shown in Figure 8.1 opens in which components, pressures, and a physical property model (UNIFAC) are specified. Then click *Pure Components* in the *Output* list, as shown in Figure 8.2. To find the azeotropes, click *Azeotropes* on the list (Fig. 8.3). Compositions and temperatures of the azeotropic mixtures are listed for the two binary azeotropes at 4 bar.

iC_5/methanol: 339 K, 25.01 mol% methanol
methanol/TAME: 376 K, 84.43 mol% methanol

Note that the UNIFAC physical property package predicts that the iC_5/methanol azeotrope is heterogeneous (two liquid phases).

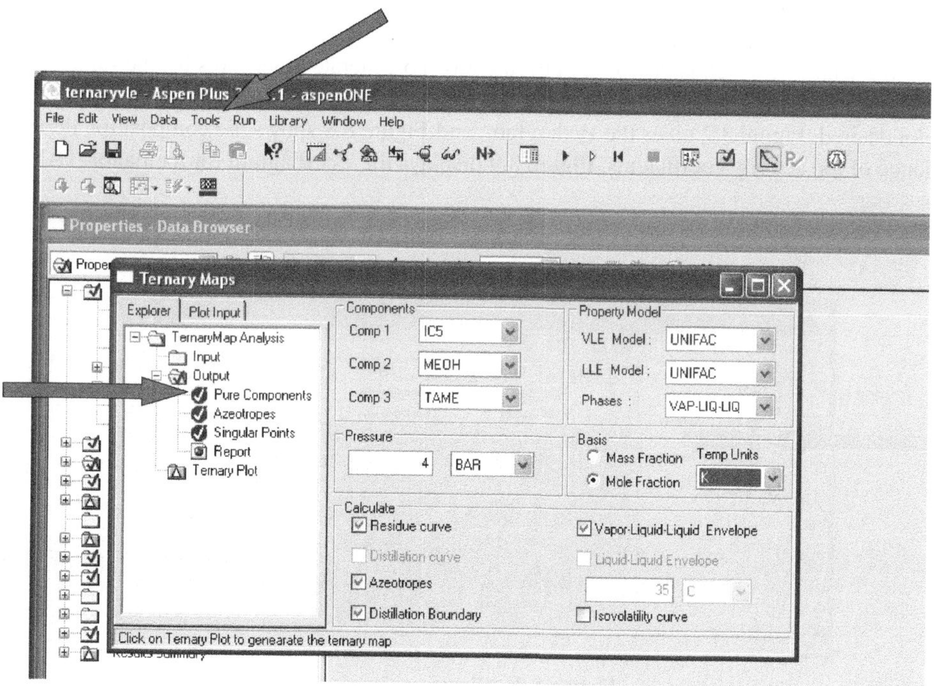

Figure 8.1 Select components and pressure.

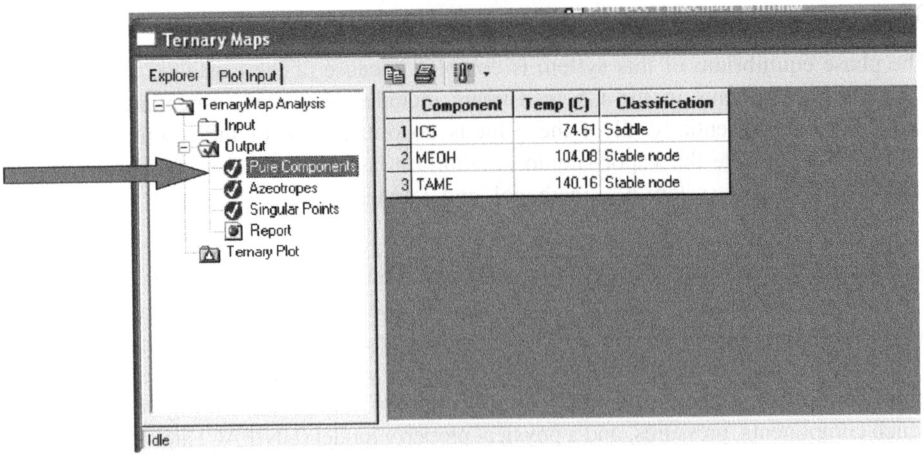

Figure 8.2 Pure component vapor pressures.

The boiling point of the first azeotrope (339 K) is lower than the boiling point of the lightest component iC_5 (348 K). This means that the overhead from the column will have a composition close to the azeotropic composition.

Various types of phase diagrams can be generated in Aspen Plus. Binary *Txy* diagrams are produced by going to the top tool bar and clicking *Tools*, *Analysis*, *Property*, and *Binary*. The window shown in Figure 8.4 opens in which components and pressures are specified. Clicking the *Go* button at the bottom produces the *Txy* diagram shown in Figure 8.5 for iC_5/methanol. A detailed table of results is given in the *Binary Analysis Results* window, which is provided in Figure 8.6. To generate other plots for the same components, click the *Plot Wizard* button at the bottom of this window and select the type of plot desired. Figure 8.7 gives the *xy* diagram, and Figure 8.8 shows the composition dependence of the activity coefficients. Figure 8.9 displays the *Txy* diagram for methanol/TAME at 4 bar.

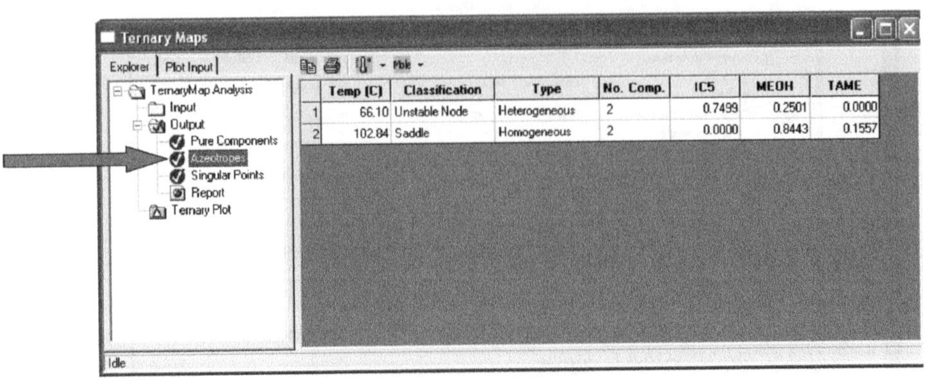

Figure 8.3 Binary azeotropes.

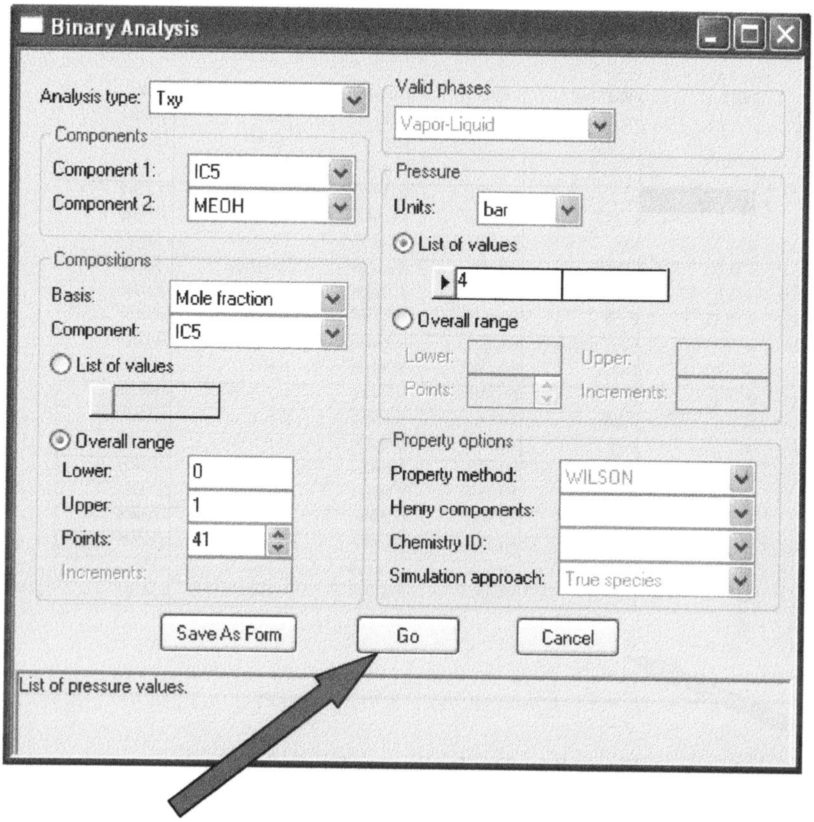

Figure 8.4 Binary *Txy* diagrams.

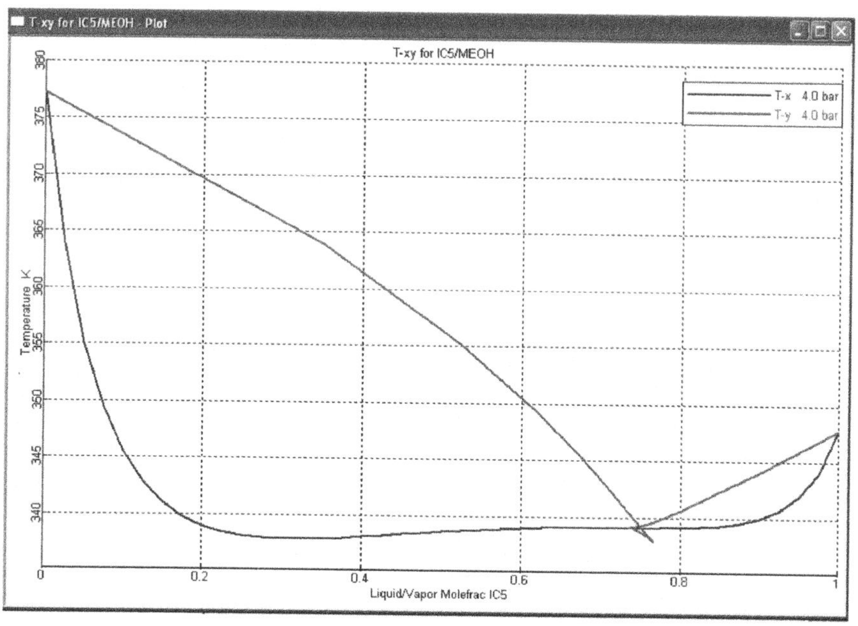

Figure 8.5 The *Txy* diagram for iC_5/methanol.

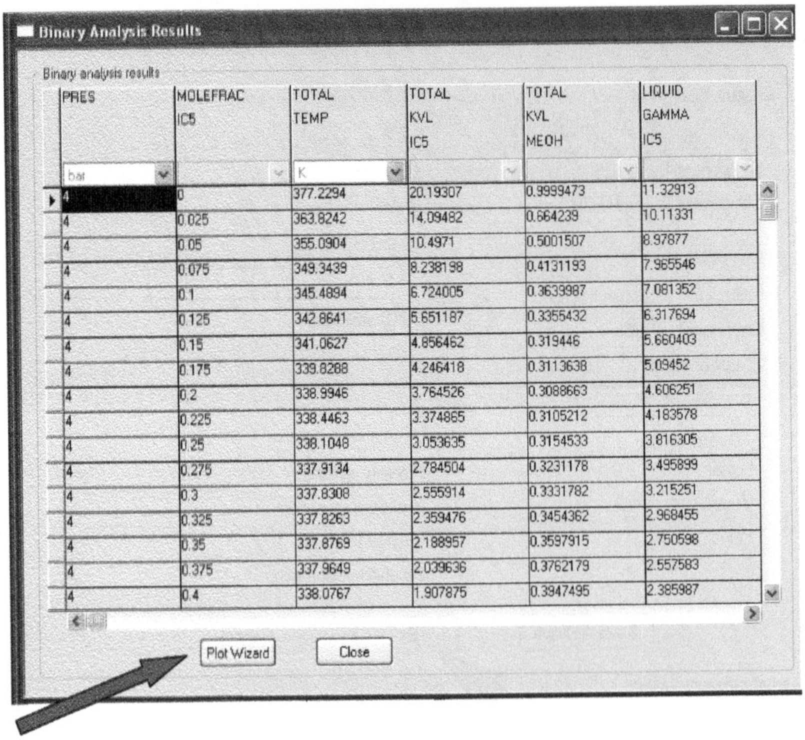
Binary Analysis Results

Binary analysis results

PRES	MOLEFRAC IC5	TOTAL TEMP	TOTAL KVL IC5	TOTAL KVL MEOH	LIQUID GAMMA IC5
bar		K			
4	0	377.2294	20.19307	0.9999473	11.32913
4	0.025	363.8242	14.09482	0.664239	10.11331
4	0.05	355.0904	10.4971	0.5001507	8.97877
4	0.075	349.3439	8.238198	0.4131193	7.965546
4	0.1	345.4894	6.724005	0.3639987	7.081352
4	0.125	342.8641	5.651187	0.3355432	6.317694
4	0.15	341.0627	4.856462	0.319446	5.660403
4	0.175	339.8288	4.246418	0.3113638	5.09452
4	0.2	338.9946	3.764526	0.3088663	4.606251
4	0.225	338.4463	3.374865	0.3105212	4.183578
4	0.25	338.1048	3.053635	0.3154533	3.816305
4	0.275	337.9134	2.784504	0.3231178	3.495899
4	0.3	337.8308	2.555914	0.3331782	3.215251
4	0.325	337.8263	2.359476	0.3454362	2.968455
4	0.35	337.8769	2.188957	0.3597915	2.750598
4	0.375	337.9649	2.039636	0.3762179	2.557583
4	0.4	338.0767	1.907875	0.3947495	2.385987

Figure 8.6 The *Txy* diagram for iC_5/methanol.

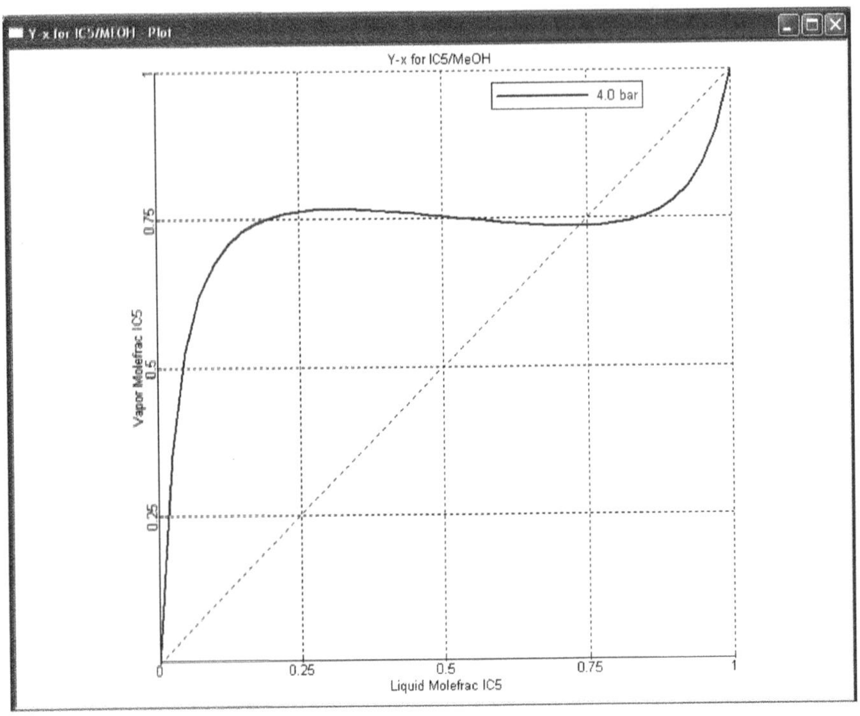

Figure 8.7 The *xy* diagram for iC_5/methanol.

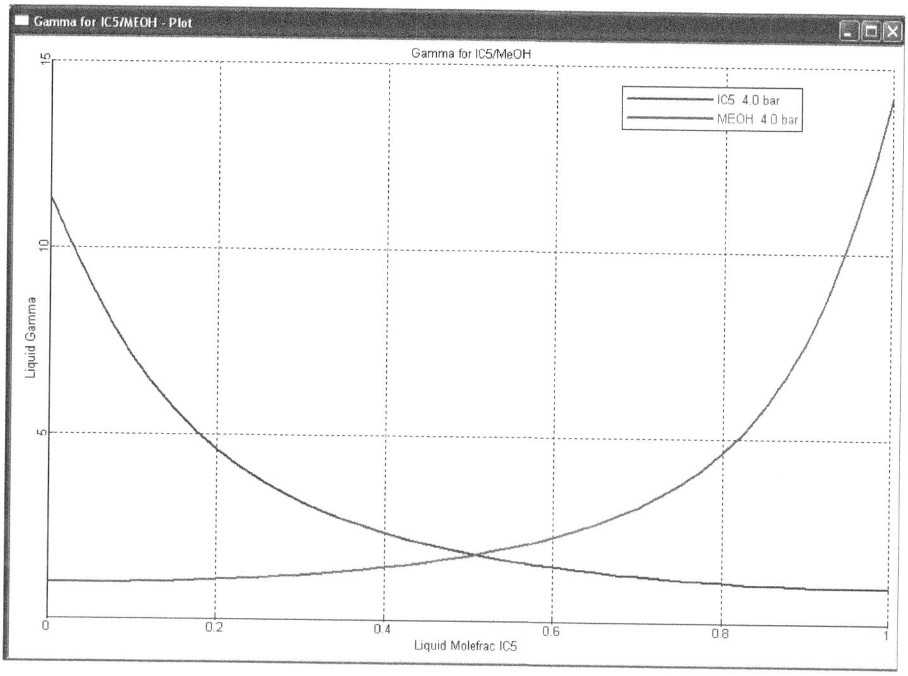

Figure 8.8 Activity coefficients for iC_5/methanol.

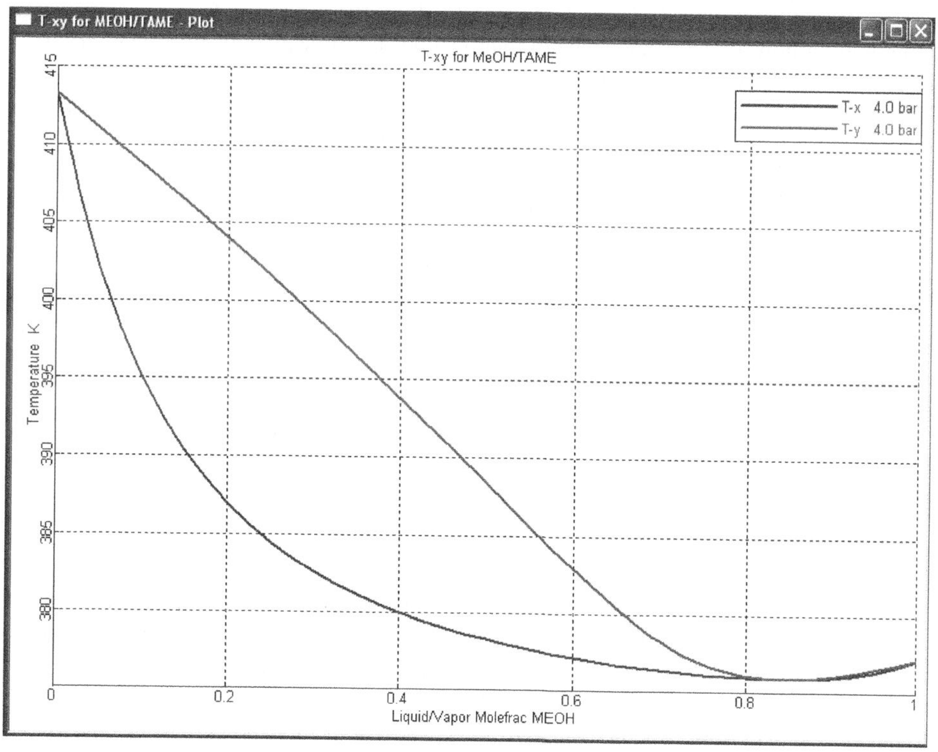

Figure 8.9 The *Txy* diagram for methanol/TAME.

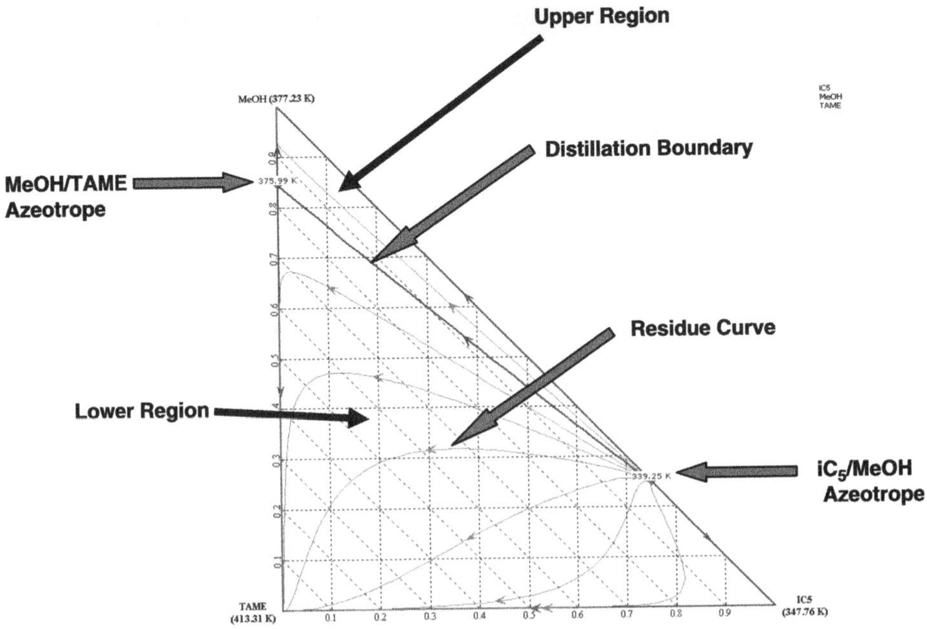

Figure 8.10 Ternary diagram at 4 bar.

Ternary Diagrams. Ternary diagrams are very useful in analyzing three-component systems and in helping to design columns, particularly for systems with complex vapor--liquid equilibrium. We illustrate some of their features in this section using the ternary iC_5/methanol/TAME system at 4 bar.

Figure 8.10 illustrates the ternary diagram for this system. The two binary azeotropes produce a distillation boundary. Residue curves are shown, which indicate that a column operating in the lower region will have a bottoms product that is mostly TAME and a distillate product that is somewhere near the iC_5/methanol azeotrope. A column operating in the upper region will have a bottoms product that is mostly methanol and a distillate product that is somewhere near the iC_5/methanol azeotrope.

We can locate various streams on this ternary diagram. For example, plotting a point with coordinates $z_{MeOH} = 0.3$ and $z_{iC5} = 0.4$ will indicate a feedstream to the column with these compositions. In the same way, the distillate and bottoms points can be located on the diagram. Because of the "ternary mixing rule," the feed coordinates must lie on a straight line connecting the distillate and bottoms coordinate points.

8.1.3 Conceptual Design

To generate ternary plots and to use them for design, Aspen Split is used. This software is imbedded in Aspen Plus and can be accessed by going to the toolbar and clicking *Library* and *References*. The window shown in Figure 8.11 opens in which the *Aspen Split* box should be checked. A new page tab will appear at the bottom of the process flow diagram next to those of the standard unit operation models, which is shown in

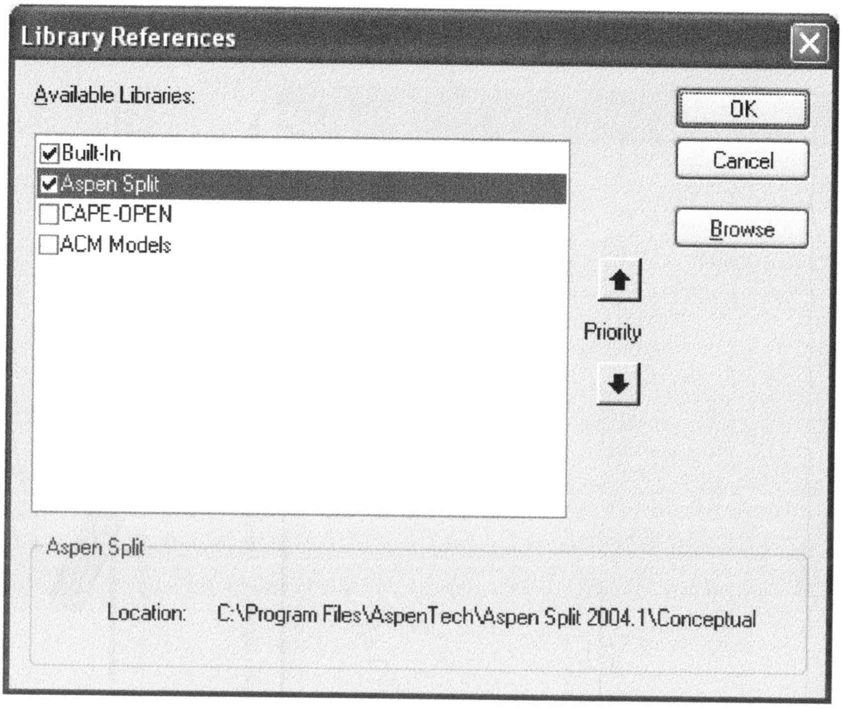

Figure 8.11 Accessing the Aspen Split library.

Figure 8.12. Selecting this page tab reveals the icon *ConSep*. Dragging and dropping this icon onto the process flowsheet window places a block there, which we labeled "Ternary." Feed and product streams are inserted in the normal way using *Material STREAMS*. The conditions of the feed are specified, as shown in Figure 8.13. The feed is 40 mol% iC_5, 30 mol% methanol, and 30 mol% TAME in this illustrative example.

Next we open the Ternary block and select *Inputs* (Fig. 8.14). The components, pressure, and physical property package are specified. The *Design* mode is selected on the right side of the window. Then, the blue "*N*" (next) button is clicked to run the program. The window shown in Figure 8.15 opens up showing the *Setup* view. Selecting *Analysis* opens the window in Figure 8.16 in which the three boxes are checked to generate residue curves and distillation boundaries and to show the locations of azeotropes.

Selecting *Specifications* opens the window in Figure 8.17. There are 4 degrees of freedom. One of these is the reflux ratio, which is specified to be 5. That leaves three compositions that can be specified. Let us assume that we are trying to produce fairly pure TAME out the bottom of the column, so we specify that the bottoms composition is 0.1 mol% iC_5 and 1 mol% methanol. Likewise, we want to lose very little TAME in the distillate, so we specify that the distillate is 0.1 mol% TAME. These are preliminary guesses that may or may not be attainable.

Finally, the *Calculate* button at the top of the window is clicked. If the overall component balances are not feasible with the compositions selected and with the given feed, an error message is displayed. If the balances are feasible, the window shown in Figure 8.18 opens. The program performs tray to tray calculations, starting from each

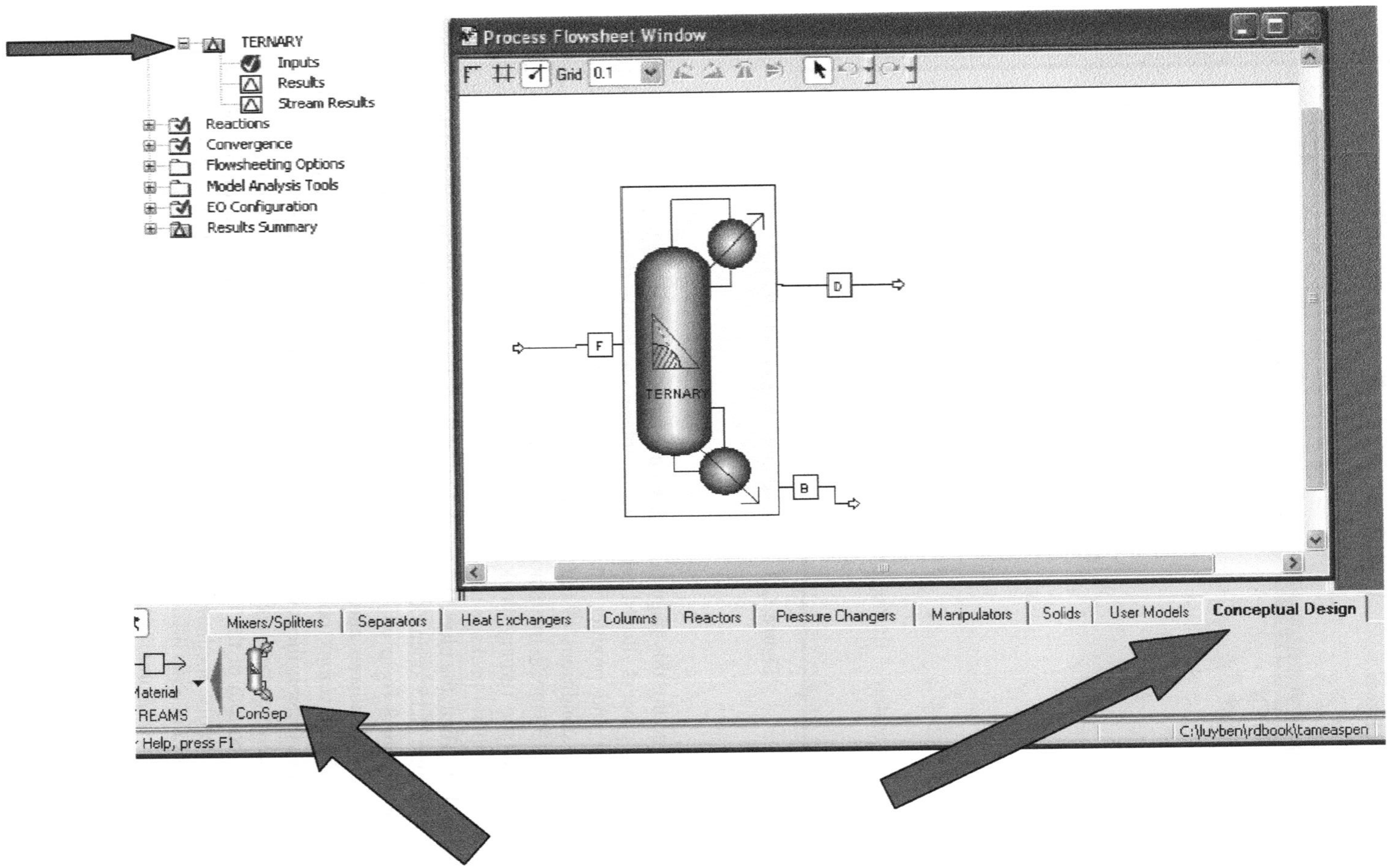

Figure 8.12 Conceptual design page tab.

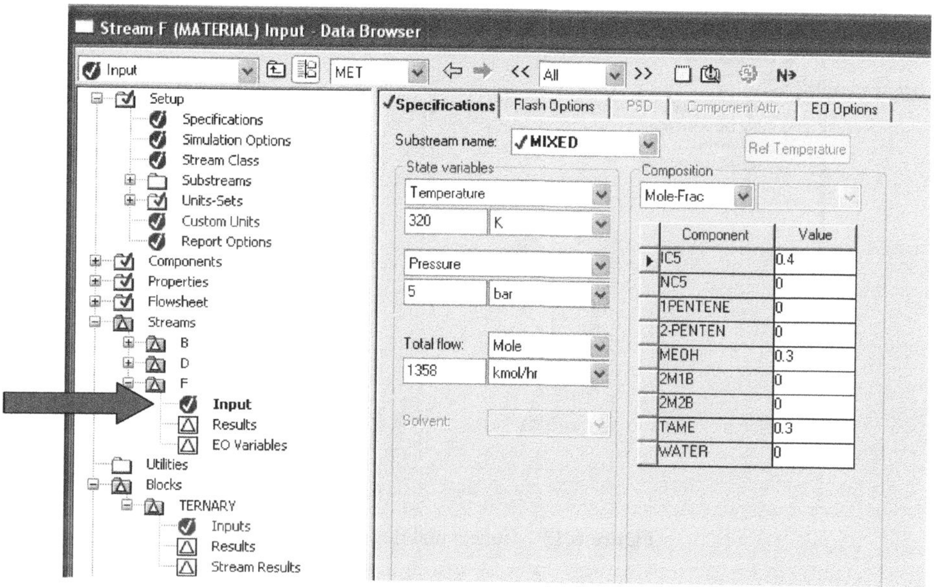

Figure 8.13 Specifying feedstream.

end of the column, using the reflux ratio specified. One trajectory of liquid compositions from tray to tray moves up the stripping section, and the other moves down the rectifying section. If the two trajectories intersect, the design is feasible.

As Figure 8.18 demonstrates, the compositions that were selected cannot be attained. The problem is clearly shown in the ternary diagram. The bottoms and feed points are in the lower region, but the distillate is in the top region. This separation is infeasible.

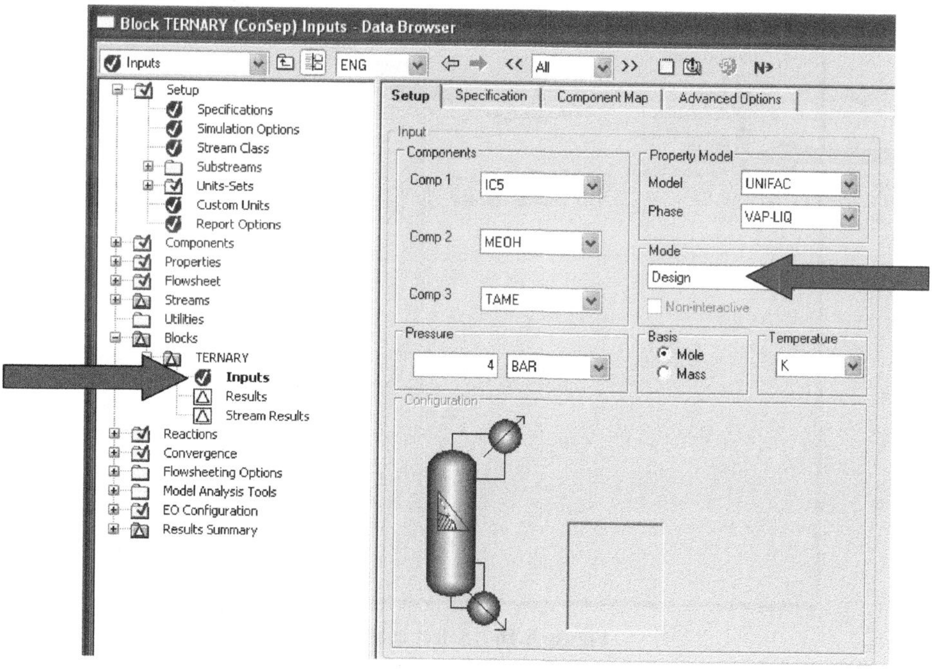

Figure 8.14 Conceptual design.

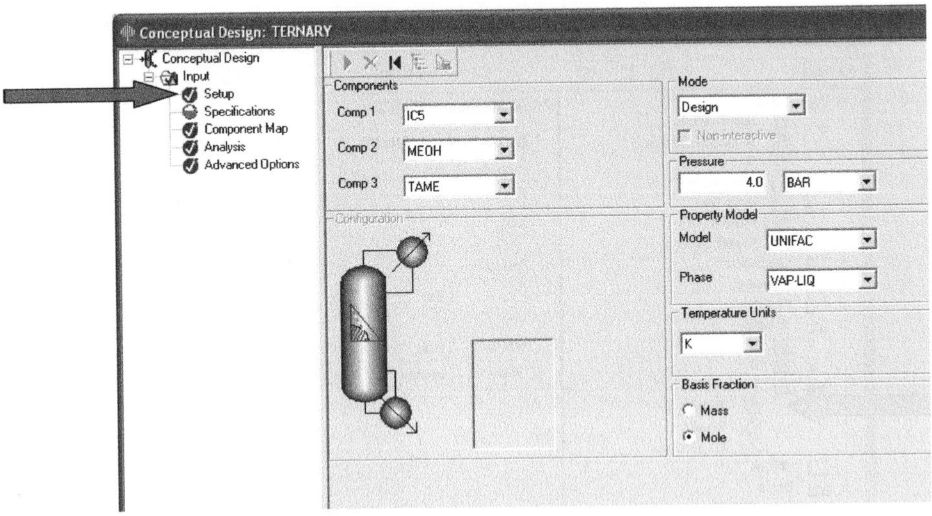

Figure 8.15 Conceptual design setup.

The distillate composition must be specified to be in the lower region where the feed and bottoms are located. Figure 8.19 reveals that the distillate point is still in the upper region when the specified distillate composition is increased to 1 mol% TAME. Figure 8.20 gives the results when the specified distillate composition is increased to 7 mol% TAME and the

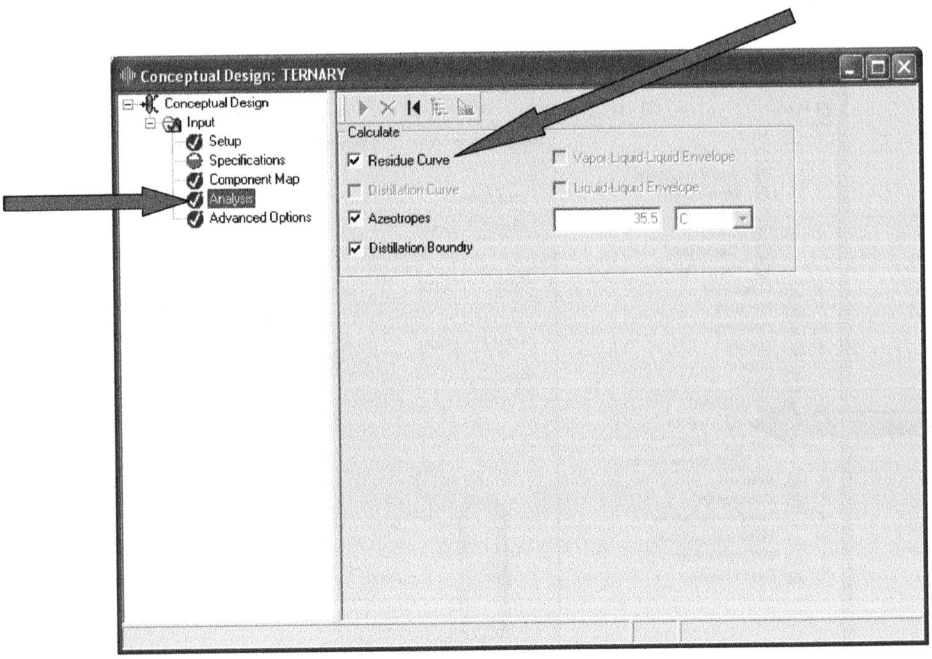

Figure 8.16 Select analysis.

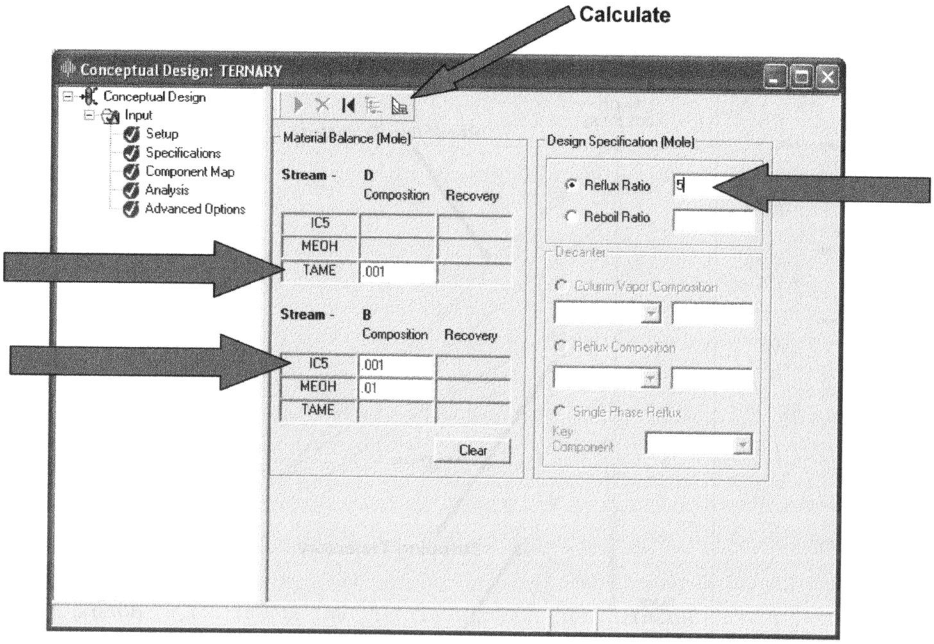

Figure 8.17 Specifications.

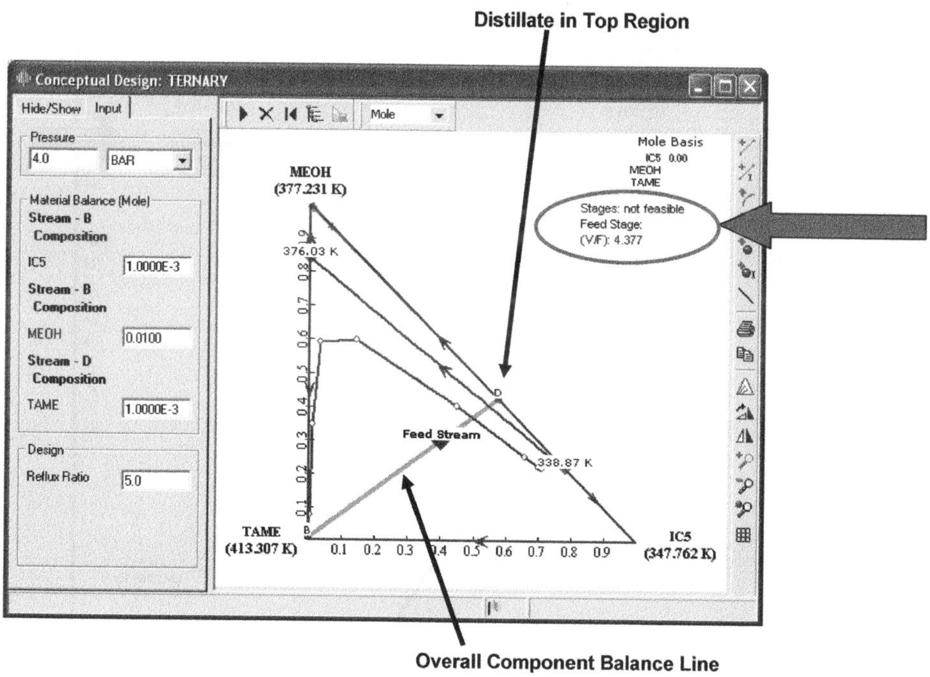

Figure 8.18 Ternary diagram.

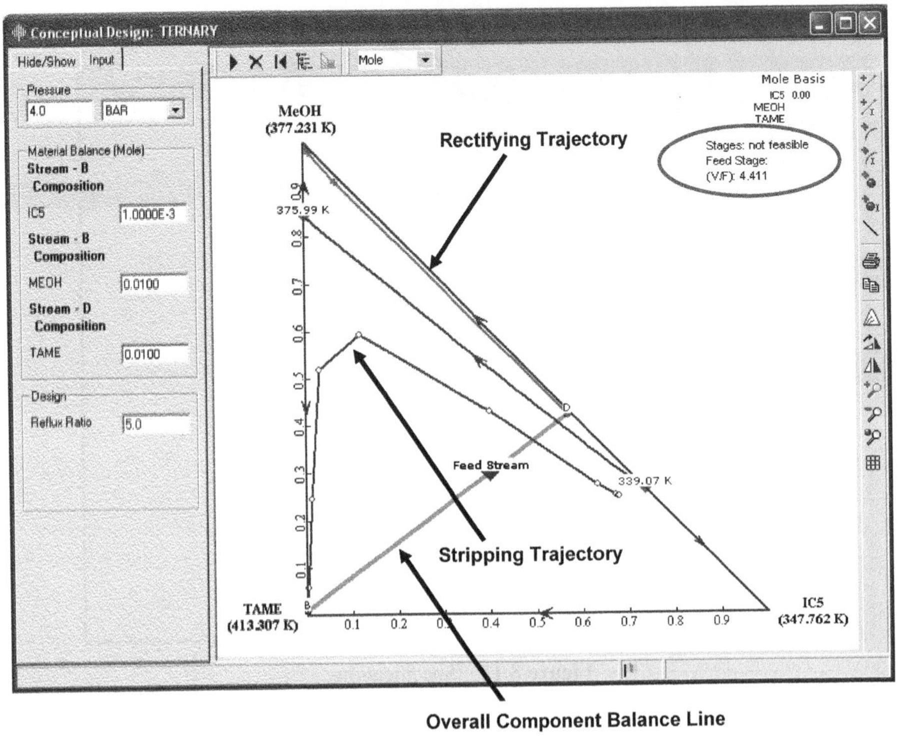

Figure 8.19 Distillate composition changed.

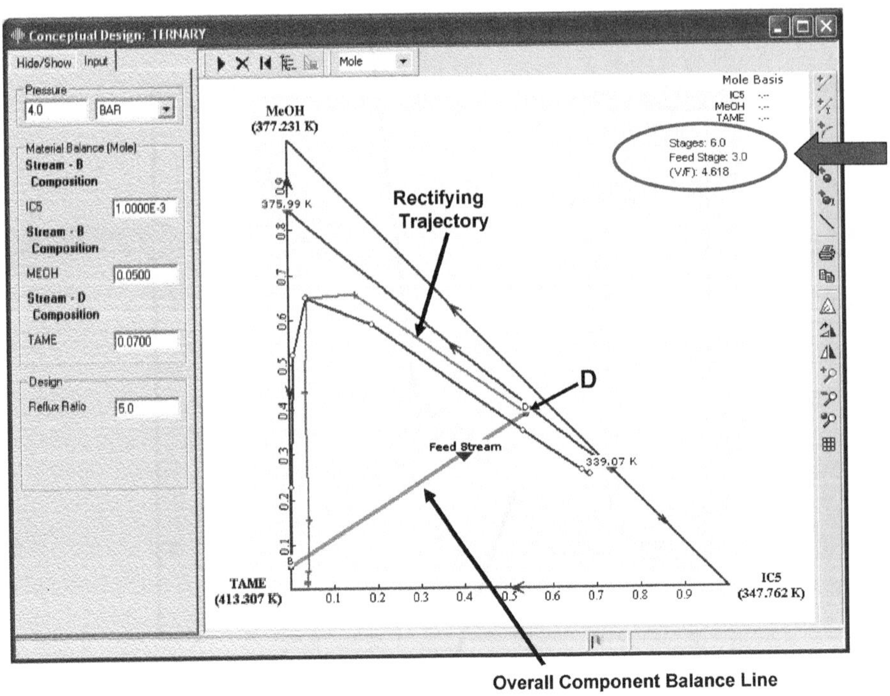

Figure 8.20 Compositions changed.

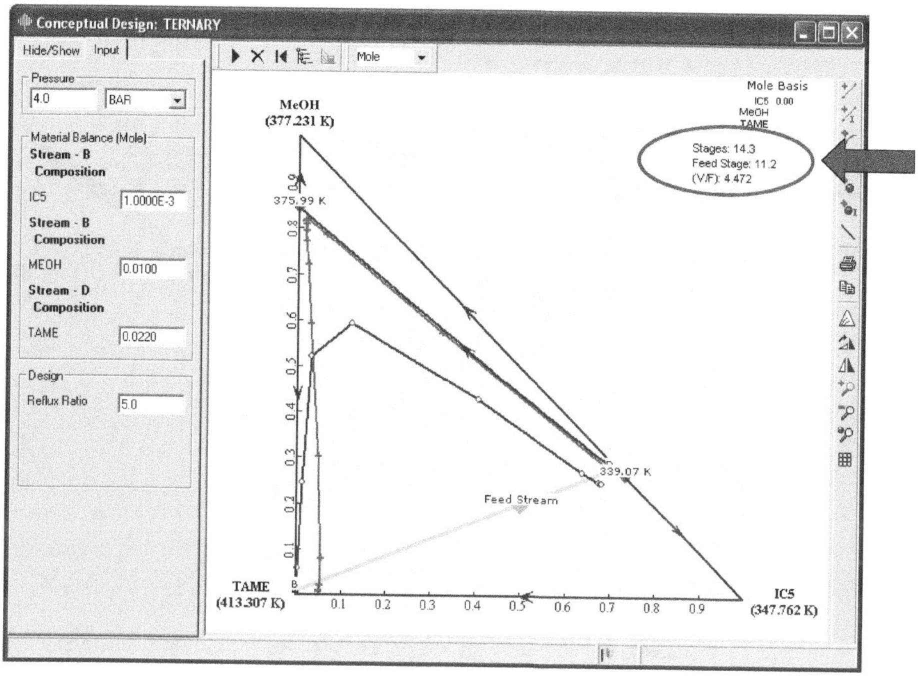

Figure 8.21 Feed composition changed.

bottoms composition is changed to 5 mol% methanol. Now the distillate point is just inside the lower region and very near the distillation boundary. With a reflux ratio of 5, the number of stages required is six.

The picture obtained from the ternary diagram clearly shows that a distillate that contains very little TAME cannot be attained with the specified feed composition. Suppose the feed contains less methanol. The feed point moves down in the diagram, which also moves the distillate point down. The distillation boundary is closer to the zero TAME hypotenuse of the triangle. Thus, a distillate with a lower composition of TAME can be attained. In Figure 8.21 the feed composition is changed from 30 to 20 mol% methanol, with an appropriate change in the composition of iC_5. The distillate composition is specified to be 2.2 mol% TAME and the bottoms 0.1 mol% iC_5 and 1 mol% methanol. A feasible design is obtained, which requires 14.3 stages. Figure 8.22 shows the same diagram with the residue curve added. These are obtained by clicking the third button on the right-hand side of the window, locating the cursor somewhere on the diagram and clicking. A residue curve is drawn through this point.

Keep in mind that the analysis applies to a nonreactive column that is simply separating a ternary mixture. However, the example indicates that adding an excess of methanol in the reactive column may result in the loss of TAME in the overhead. The TAME reactive distillation column is therefore operated in a "pseudoneat" mode. Enough methanol must be added for the reaction plus that required for the C5/methanol azeotropes. The numerical example given in the next section illustrates this point.

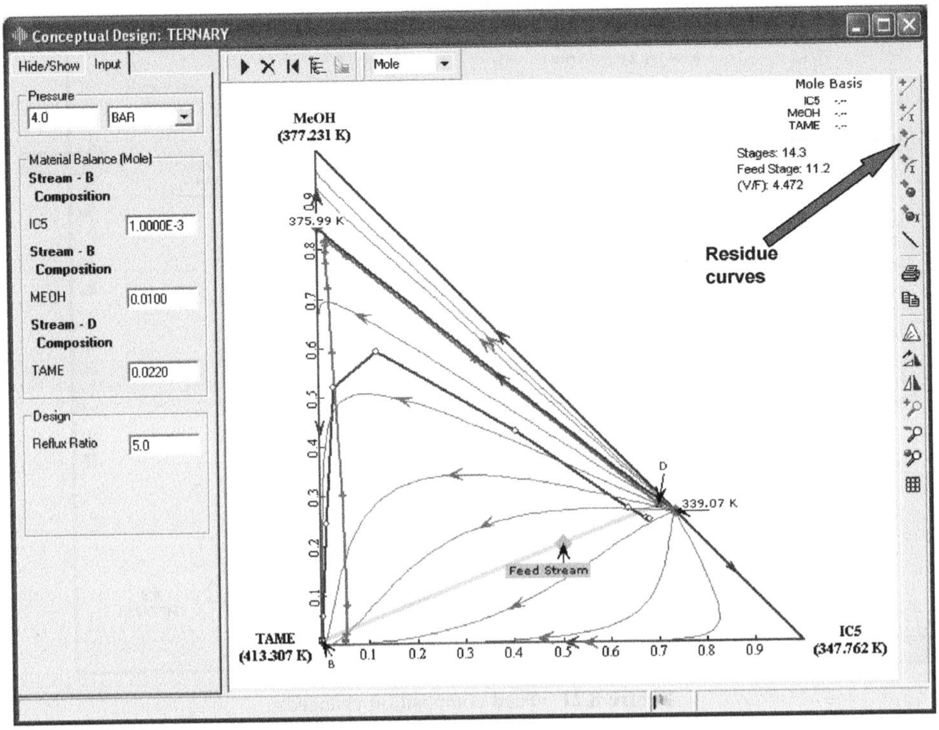

Figure 8.22 Feed composition changed.

8.2 COMPONENT BALANCES

Using the flowsheet numbers given in Table 8.2, there are 809 kmol/h of all of the C5 components in the distillate product (790 kmol/h inerts, 19 kmol/h of unreacted isoamylenes) that must go out the top of the column. The iC_5/methanol azeotrope is

TABLE 8.2 Stream Information for Prereactor and Column C1

	Fresh Methanol (kmol/h)	Feed (kmol/h)	MeOH Reactor (kmol/h)	Reactor Effluent (kmol/h)	MeOH C1 (kmol/h)	B1 (kmol/h)	D1 (kmol/h)
MeOH	232	—	315	188	235	—	316
2M1B	—	85.6	—	13.7	—	0.01	3.31
2M2B	—	165	—	112	—	0.60	14.1
TAME	—	—	—	125	—	232	—
nC_5	—	88.4	—	88.4	—	0.22	88.2
iC_5	—	501	—	501	—	0.17	501
1-Pentene	—	38.1	—	38.1	—	0.18	37.9
2-Pentene	—	162	—	162	—	0.76	161
Total	232	1040	315	1228	235	233	1122
T (K)	325	343	351	355	351	415	341
P (atm)	17	10	7	6	4	4.29	3.95

0.25 mol fraction methanol. The actual overhead composition is $316/1122 = 0.2817$ mol fraction methanol, so the overhead is close to the azeotropes of the C5 components with methanol.

The amount of TAME produced is 232 kmol/h, which consumes 232 kmol/h of methanol. Note that this is equal to the fresh feed of methanol into the plant. However, the total amount of methanol fed to the column/prereactor must be

$$232 + 316 = 548 \text{ kmol/h methanol}$$

The large amount of methanol in the overhead must be recovered and recycled. In a later section of this chapter two types of methanol recovery flowsheets are developed and their TACs are compared.

8.3 PREREACTOR AND REACTIVE COLUMN

There are two fresh feedstreams. One is methanol. The other is a mixture of C5 components that contains reactive isoamylenes plus other C5 paraffins, naphthenes, and olefins. This C5 stream typically comes from a petroleum refinery light-ends unit that separates light hydrocarbon components generated in catalytic cracking into various streams. Because the boiling points of all of the C5 components are quite similar, it is uneconomical to separate out the isoamylene reactants from the other C5s. The C5 fresh feedstream contains 24 mol% reactive isoamylenes. The remaining components are pentanes and pentene (largely isopentane), which are inert in the TAME reaction.

8.3.1 Base Case Design of Reactive Column

Subawalla and Fair[1] presented a design of a prereactor and reactive distillation column. Figure 8.23 gives the flowsheet with a pressure-swing methanol recovery section. Figure 8.24 displays the flowsheet with an extractive distillation methanol recovery section. Note that the prereactor and reactive column are identical in both flowsheets.

The prereactor is a cooled tubular reactor containing 9544 kg of catalyst. All of the C5 fresh feed is fed to the prereactor along with a portion of the total methanol. The C5 fresh feed flowrate is 1040 kmol/h. The fresh feed of methanol coming into the process is 232 kmol/h, but there is a large methanol recycle stream from the recovery section. The flowrate of the recycle methanol is 318 kmol/h. The total methanol (550 kmol/h) is split between the prereactor (315 kmol/h) and the reactive distillation column (235 kmol/h).

The reactor effluent is fed into a 36-stage reactive distillation column (C1) on stage 28. Aspen notation of numbering stages from the top is used, with stage 1 being the reflux drum. The Subawalla and Fair design[1] has catalyst present on stages 7–23. The reactor effluent is fed five trays below the reactive zone. A methanol stream is fed at the bottom of the reactive zone (stage 23). The flowrate of the methanol feed to the reactive column is 235 kmol/h.

Figure 8.25 gives composition profiles in reactive column C1. Note the low concentration of methanol in the lower part of the column and the fairly high concentration of one of the reactants (2M2B) in the same zone. This is below the reactive trays, which are stages 7–23. Figure 8.26 shows the temperature profile. The reflux ratio is 4, which gives a bottoms purity of 99.2 mol% TAME and a distillate impurity of 0.1 ppm TAME.

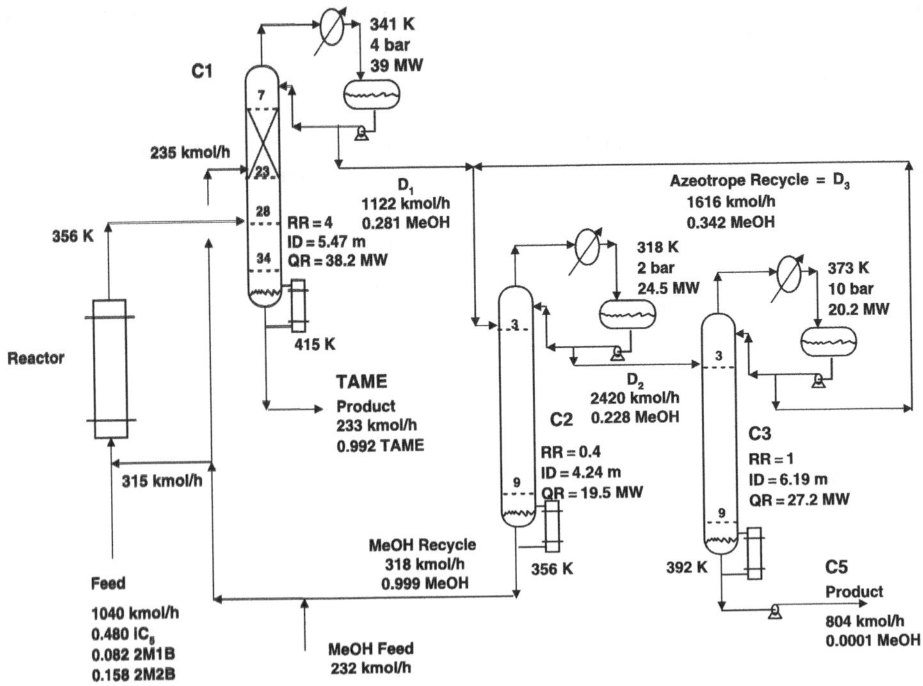

Figure 8.23 The TAME process with pressure-swing methanol recovery.

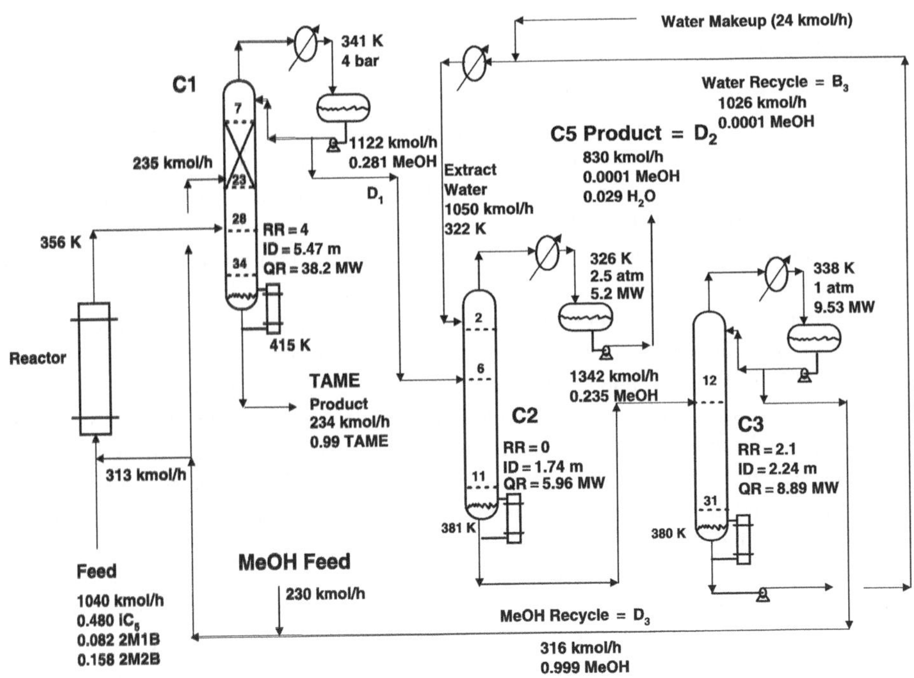

Figure 8.24 The TAME process with extractive distillation methanol recovery.

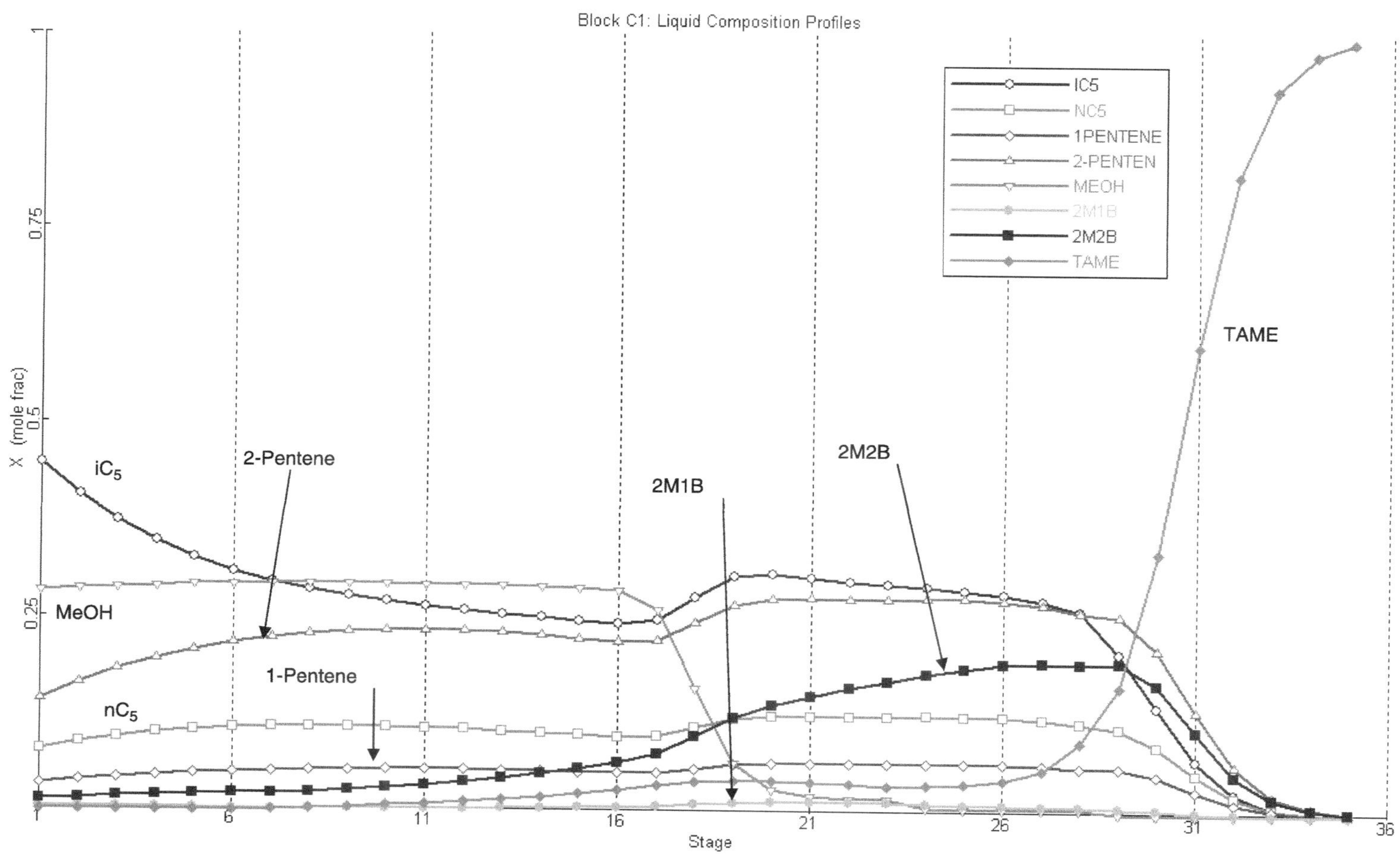

Figure 8.25 Composition profiles in reactive column.

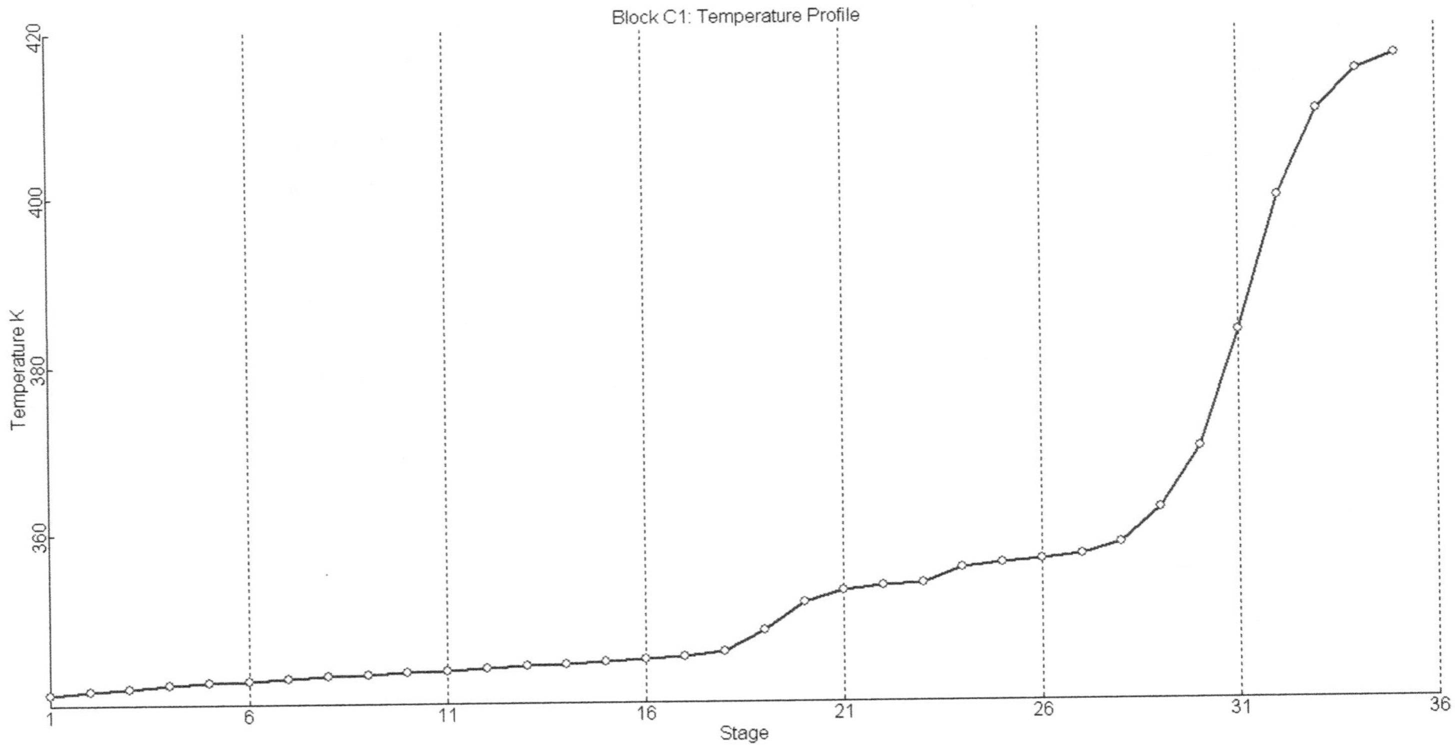

Figure 8.26 Temperature profile in reactive column.

Reboiler heat input and condenser heat removal are 38.2 and 39 MW, respectively. The operating pressure is 4 bar, and the column diameter is 5.5 m. The overall conversion of 2M1B and 2M2B in the C5 fresh feed is 92.4%.

Distillate D_1 has a composition of methanol (28 mol% methanol) that is near the azeotrope at 4 bar. It is fed at a rate of 1122 kmol/h to the methanol recovery columns section. Table 8.2 gives information about important streams associated with the prereactor and reactive distillation column.

8.3.2 Effect of Design Parameters on Reactive Column

It is interesting to explore reactive columns that are different from the one recommended by Subawalla and Fair.[1] In this section several design parameters are changed, and their impacts on the steady-state performance of the reactive column are determined.

The reactive column is simulated in Aspen Plus using RadFrac with reactive trays. The convergence of this very nonlinear, highly nonideal column can be quite difficult. Using the *Estimates* feature in the column block helps the convergence problem (see Figs. 8.27 and 8.28). These estimates are copied from temperatures and compositions found in the *Profile* list under the C1 column block once a converged case is attained. The *Azeotropic* convergence method is selected on the *Setup* page tab of the column block for reactive column C1, as shown in Figure 8.29.

In order to obtain the same basic operation while design parameters are changed, two "design spec/vary" functions are set up in column block C1.

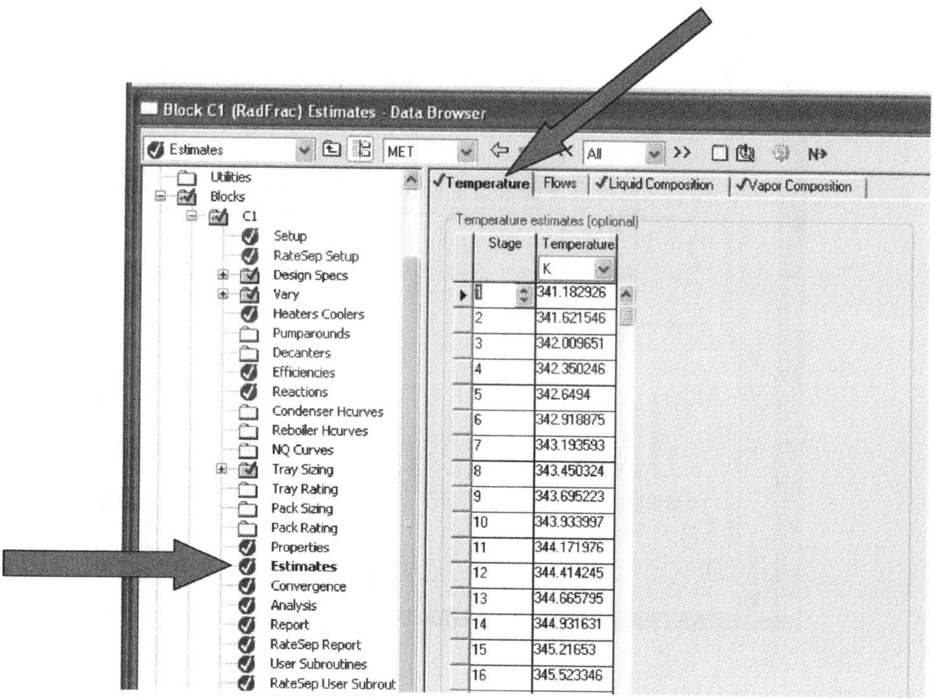

Figure 8.27 Temperature estimates.

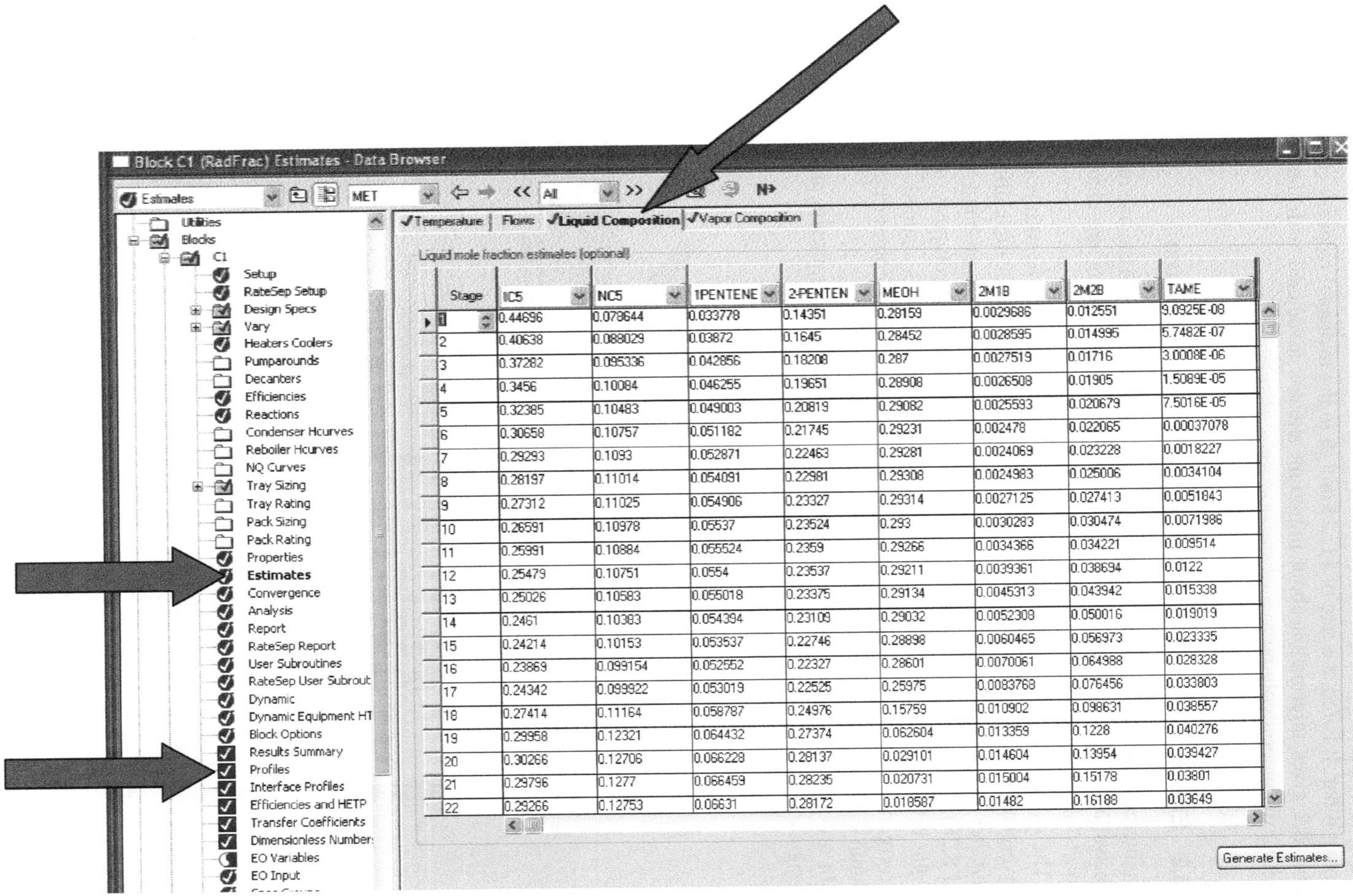

Figure 8.28 Composition estimates.

1. The first maintains the composition of TAME in the bottoms at 99.14 mol% by varying the bottoms flowrate.
2. The second maintains the composition of TAME in the distillate at 1×10^{-7} mole fraction by varying the reflux ratio.

Figure 8.29 shows that the two *Operating Specifications* are bottoms rate and reflux ratio. Figure 8.30 illustrates setting up one of the *Design Spec* functions using the three page tabs (*Specifications*, *Components*, and *Feed/Product Streams*). Figure 8.31 shows the selection of the reflux ratio as the *Vary*.

Effect of Pressure. The base case pressure is 4 bar. Table 8.3 gives results for TAME production, reactant losses, and energy consumption over a very narrow range of pressures. These results clearly show the extremely high sensitivity to pressure. A small reduction in pressure from 4 to 3.8 bar produces a very significant reduction (27%) in energy consumption (36.82 to 26.94 MW). The amount of TAME produced increases by 3% from (234.2)(0.9914) = 232.2 kmol/h in the base case to (242.2)(0.9914) = 240.1 kmol/h when the column pressure is 3.85 bar. The increase in TAME comes from the decrease in the losses of reactants in the distillate. The total of the two reactants losses drops from 17.53 to 9.71 kmol/h.

Further decreases in pressure below 3.8 bar begin to increase energy consumption and reactant losses in the distillate. Increasing pressure from the base case produces very rapid increases in energy consumption and reactant losses, and TAME production drops sharply. Figure 8.32 provides the temperature profiles at two different pressures, and Figure 8.33 shows composition profiles. Note the large changes in the position of the methanol and reactant profiles over the small range of pressures.

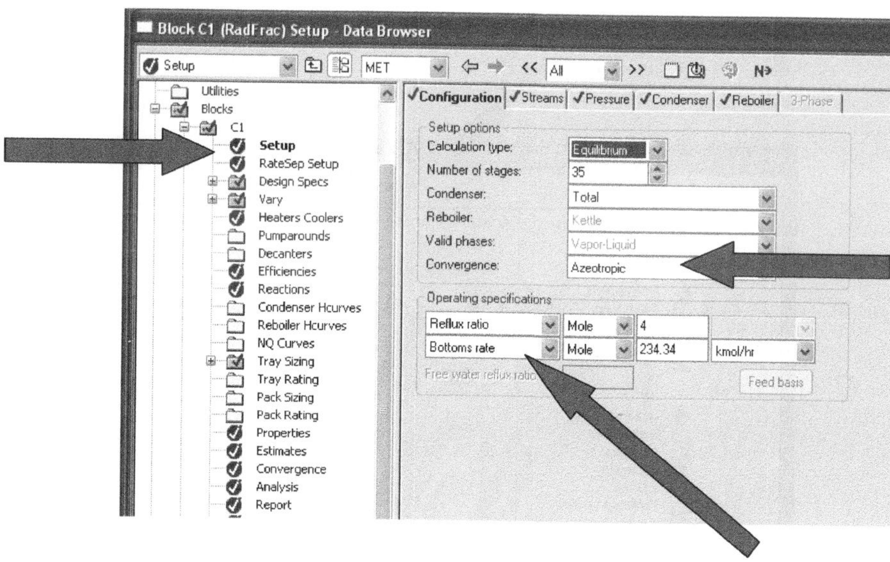

Figure 8.29 Manipulated variables *B* and RR.

The TAME reactive column is very sensitive to pressure. This example illustrates that reactive distillation columns are much more affected by pressure than are conventional columns.

Effect of Number of Reactive Trays. In the base case, stages 7–23 contain catalyst, so the reactive zone has 17 trays. Reactor effluent is fed on stage 28 and methanol on stage 23 (the lowest reactive stage). There are 6 rectifying trays and 11 stripping trays.

Table 8.4 gives results over a range of values of the number of reactive trays. The numbers of stripping and rectifying trays are held constant, as are the pressure and the design specifications on the concentration of TAME in the bottoms and distillate.

Adding more reactive trays produces a small decrease in energy consumption, and reducing the number from the base case of 17 results in increases in energy consumption. If only 10 reactive stages are used, energy consumption increases quite significantly.

Note that the effects of reactive stages on the losses of the two reactants are different. The 1M2B reactant losses decrease as fewer reactive trays are used until N_{RX} drops to 10.

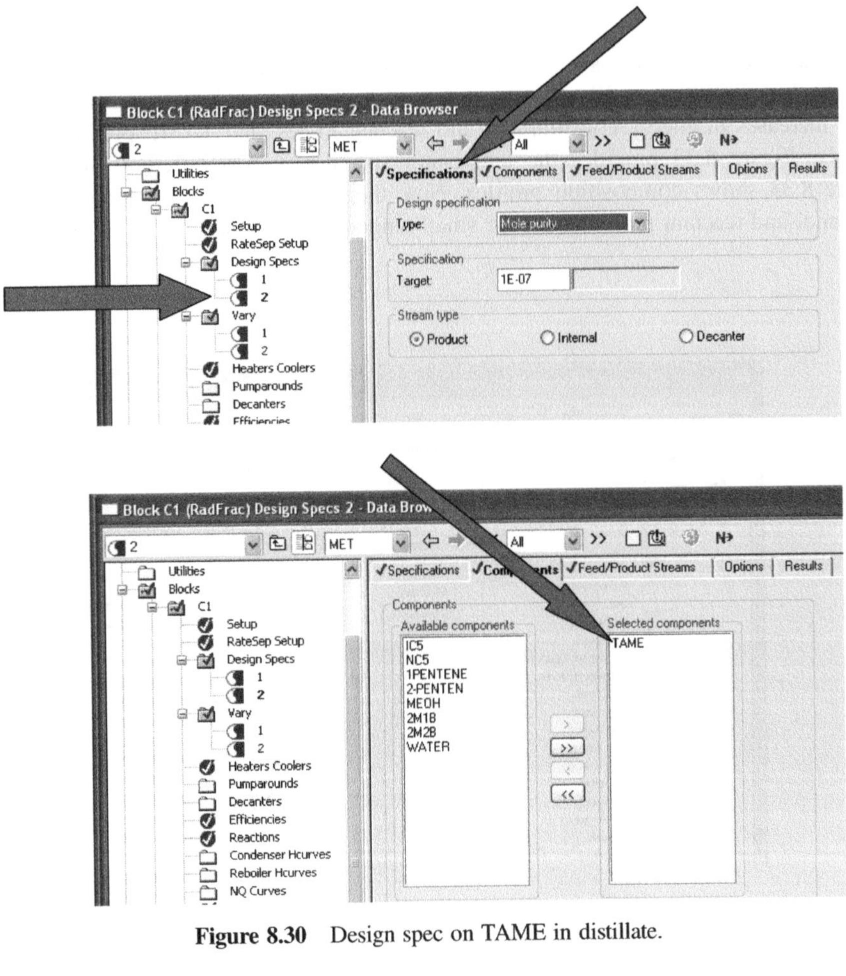

Figure 8.30 Design spec on TAME in distillate.

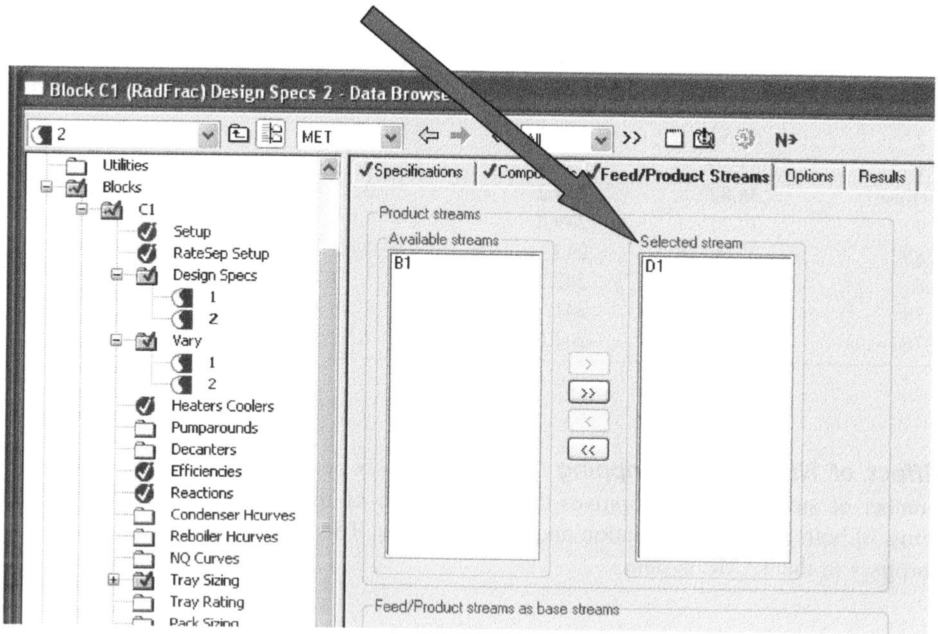

Figure 8.30 (*Continued*)

However, the 1M1B reactant losses increase as N_{RX} is reduced over the whole range of tray numbers.

These results show that the TAME reactive distillation column is fairly insensitive to the number of reactive stages.

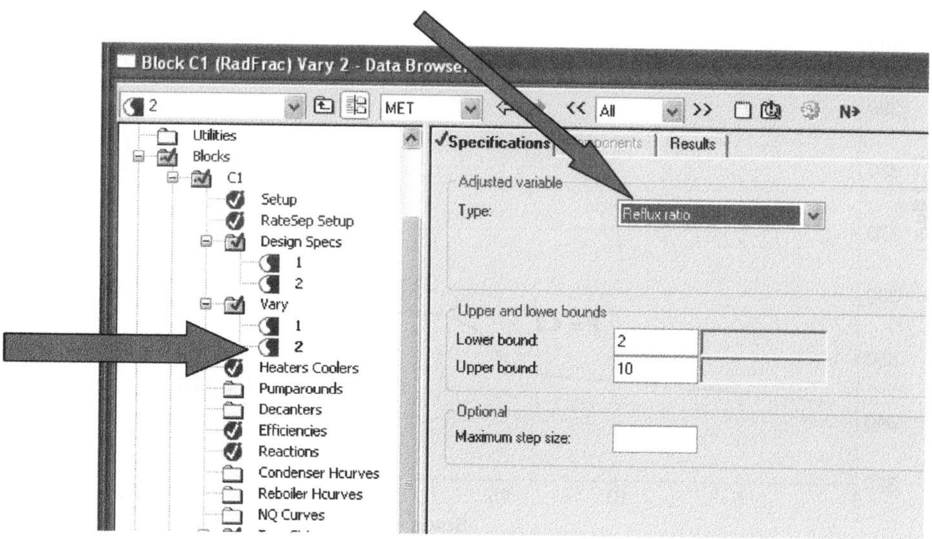

Figure 8.31 Vary reflux ratio.

TABLE 8.3 Effect of Pressure

Pressure (bar)	Reboiler Heat Input (MW)	Bottoms (kmol/h)	1M2B Losses in Distillate (kmol/h)	1M1B Losses in Distillate (kmol/h)
4.2	51.53	224.1	6.68	20.90
4.1	43.86	229.1	4.85	17.74
4 (Base)	**36.82**	**234.2**	**3.27**	**14.26**
3.9	39.27	239.4	1.96	**10.43**
3.85	27.09	242.1	1.42	8.38
3.8	26.94	242.2	1.40	8.31
3.78	26.97	241.76	1.51	9.04
3.76	27.00	241.61	1.55	9.30

Effect of Number of Stripping Trays. Table 8.5 demonstrates that increasing the number of stripping trays improves the performance of the TAME reactive column in terms of both energy consumption and reactant losses. This design parameter presents no surprises in the TAME system.

Effect of Holdup on Reactive Trays. In all of the ideal cases considered in previous chapters, increasing the holdup on reactive trays improves performance. This corresponds to our intuition. However, in the TAME reactive column the effect of M_{RX} is unexpectedly different as Table 8.6 shows. *Increasing* the reactive tray holdup *increases* the energy

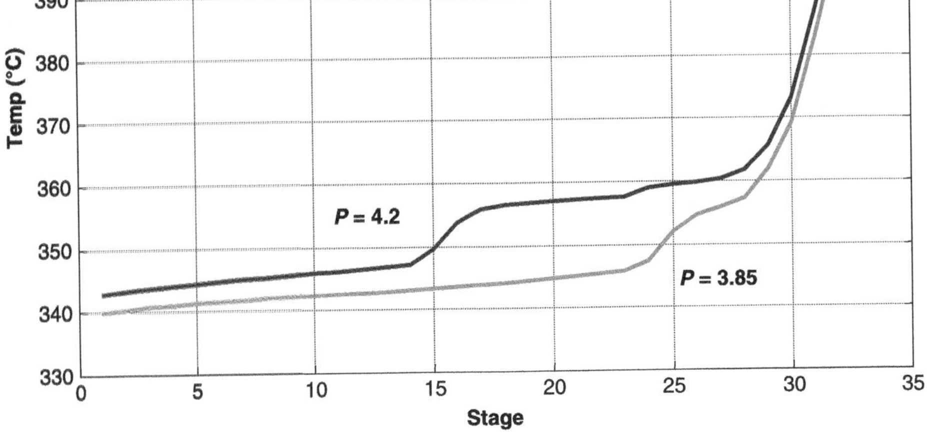

Figure 8.32 Temperature profiles.

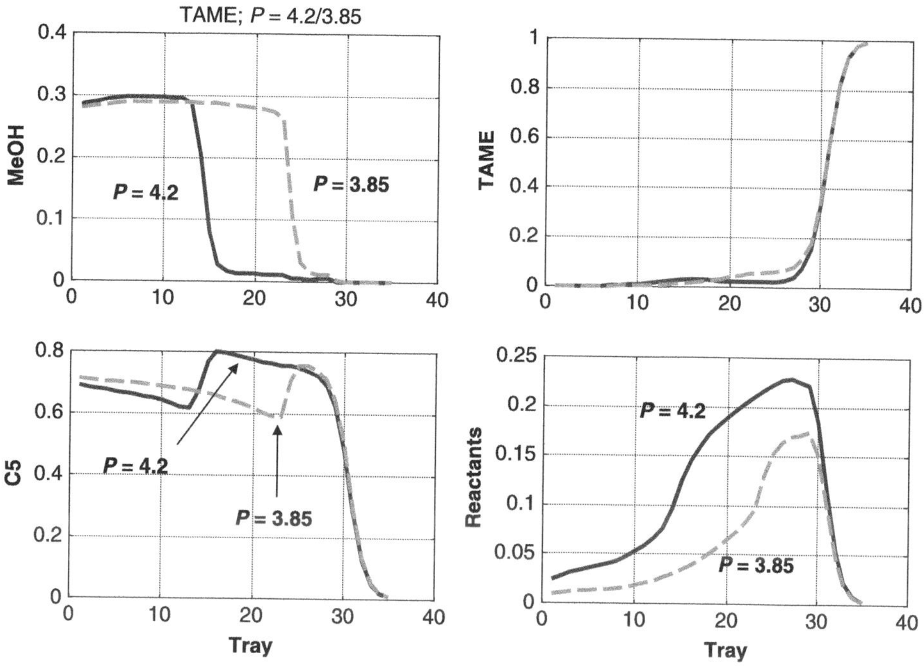

Figure 8.33 Composition profiles.

TABLE 8.4 Effect of Number of Reactive Trays N_{RX}

N_{RX}	N_T	Reboiler Heat Input (MW)	Bottoms (kmol/h)	1M2B Losses in Distillate (kmol/h)	1M1B Losses in Distillate (kmol/h)
32	50	36.79	234.25	3.45	14.20
27	45	36.78	234.25	3.25	14.21
22	40	36.80	234.25	3.25	14.22
17 (Base)	**35**	**36.82**	**234.2**	**3.27**	**14.26**
14	32	36.90	234.21	3.13	14.47
12	30	36.97	234.17	2.97	14.71
11	29	37.13	233.98	2.92	14.94
10	28	39.78	231.43	3.42	17.25

TABLE 8.5 Effect of Stripping Trays N_S

N_S	N_T	Reboiler Heat Input (MW)	Bottoms (kmol/h)	1M2B Losses in Distillate (kmol/h)	1M1B Losses in Distillate (kmol/h)
6 (Base)	**35**	**36.82**	**234.25**	**3.27**	**14.26**
8	37	36.77	234.26	3.25	14.22
11	40	36.71	234.26	3.23	14.16
13	42	36.67	234.27	3.22	14.13
16	45	36.62	234.27	3.21	14.08

TABLE 8.6 Effect of Reactive Tray Holdup M_{RX}

M_{RX}	Reboiler Heat Input (MW)	Bottoms (kmol/h)	1M2B Losses in Distillate (kmol/h)	1M1B Losses in Distillate (kmol/h)
2.44	51.27	234.90	3.81	13.55
2	46.27	234.35	3.67	13.73
1.22 (Base)	**36.82**	**234.25**	**3.27**	**14.26**
1	33.93	234.20	3.08	14.51
0.8	31.14	234.14	2.84	14.83
0.75	30.42	234.12	2.77	14.92
0.73	30.12	234.11	2.74	14.96

consumption of the TAME reactive column. This counterintuitive effect illustrates the complexity of the reactive distillation process.

As holdup is decreased, reactant 1M2B losses decrease slightly, and reactant 1M1B losses increase somewhat more rapidly. The net effect is a small decrease in TAME production (the bottoms flowrate decreases).

Figure 8.34 provides the temperature profiles for three different reactive tray holdups. Temperatures increase in the reactive zone (stages 7–23) as holdup increases. One would expect that the concentration of the 2M2B and 2M1B reactants would be lower in the reactive zone as the holdup increases. The lower right graph in Figure 8.35 shows that exactly the opposite occurs. The concentration of the C5 reactants increases. The upper left graph in Figure 8.35 shows that there is a corresponding decrease in the concentration of methanol.

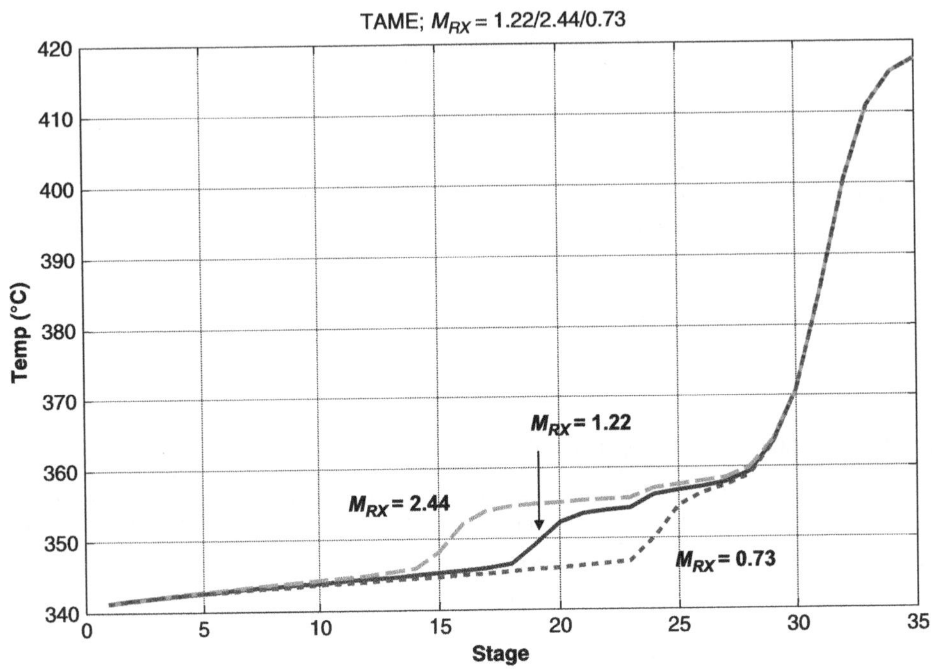

Figure 8.34 Temperature profiles.

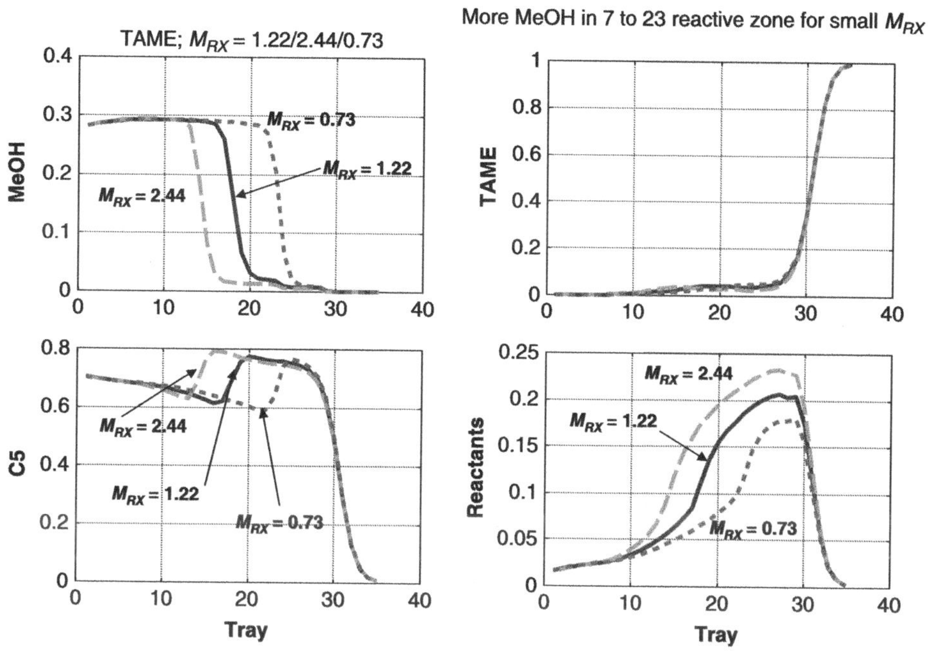

Figure 8.35 Composition profiles.

Effect of Methanol Feed Tray Location. In the Subawalla and Fair design, the methanol is introduced on the bottom tray in the reactive zone (stage 23). Table 8.7 gives results when the methanol is fed higher in the column. There is a gradual decrease in energy consumption as the methanol feed tray is moved up the column. The losses of 1M1B increase slightly, as does the bottoms flowrate.

However, if the methanol is fed on the top tray (stage 2), there are dramatic and unexpected drops in both energy and reactant losses.

The complexity of this reactive distillation system is clearly demonstrated by these counterintuitive results.

Effect of Prereactor Effluent Feed Tray Location. In the Subawalla and Fair design, the effluent from the prereactor is introduced on stage 28, which is five stages below the reactive zone. Table 8.8 shows the effect of raising this feed tray higher in the

TABLE 8.7 Effect of Methanol Feed Tray Location M_{MeOH}

M_{MeOH}	Reboiler Heat Input (MW)	Bottoms (kmol/h)	1M2B Losses in Distillate (kmol/h)	1M1B Losses in Distillate (kmol/h)
23 (Base)	**36.82**	**234.25**	**3.27**	**14.26**
18	36.81	234.25	3.26	14.26
10	36.79	234.24	3.24	14.27
4	34.24	234.20	3.07	14.47
3	33.44	234.14	3.03	14.57
2	28.09	239.00	1.92	10.87

TABLE 8.8 Effect of Prereactor Effluent Tray Location N_{Eff}

N_{Eff}	Reboiler Heat Input (MW)	Bottoms (kmol/h)	1M2B Losses in Distillate (kmol/h)	1M1B Losses in Distillate (kmol/h)
28 (Base)	**36.82**	**234.25**	**3.27**	**14.26**
25	36.78	234.26	3.25	14.23
20	36.75	234.26	3.23	14.22
17	36.88	234.26	3.26	14.32
15	37.08	234.26	3.26	14.53
14	37.26	234.26	3.27	14.72

column. The optimum location, in terms of both energy consumption and reactant losses, is stage 20. This is eight trays higher in the column than the base case design.

The results presented in this section indicate that the Subawalla and Fair design[1] can be significantly improved. Little can be done with pressure because of the extreme sensitivity. However, adjusting tray holdups and feed tray locations could yield columns that are more energy efficient and produce higher yields of TAME.

8.4 PRESSURE-SWING METHANOL SEPARATION SECTION

In this section the design of the pressure-swing methanol recovery system is developed. The flowsheet is shown in Figure 8.23. Information about important streams is given in Table 8.9. The methanol recovery column C2 operates at a pressure of 2 bar so that the reflux-drum temperature (318 K) is high enough for the use of cooling water in the condenser. The separation is a fairly easy one, so using only 10 stages and a reflux ratio of 0.4 yield a bottoms purity of 99.9 mol% methanol. The bottoms (318 kmol/h) is mixed with a fresh feed of pure methanol (230 kmol/h), and the total is split between the methanol fed to the prereactor and the methanol fed to the reactive column.

TABLE 8.9 Stream Information for Columns C2 and C3 in Pressure-Swing Process

	B2 (kmol/h)	D2 (kmol/h)	B3 (kmol/h)	D3 (kmol/h)
MeOH	230	—	313	188
2M1B	—	85.6	—	13.7
2M2B	—	165	—	112
TAME	—	—	—	125
nC_5	—	88.4	—	88.4
iC_5	—	501	—	501
1-Pentene	—	38.1	—	38.1
2-Pentene	—	162	—	162
Total	230	1040	313	1228
T (K)	325	343	351	355
P (atm)	17	10	7	6

The distillate D_2 composition (22.8 mol% methanol) is near the azeotropic composition at the 2 bar pressure. Reboiler heat input and condenser heat removal are 19.5 and 24.5 MW, respectively. The column diameter is 4.2 m.

Distillate D_2 is fed to the C5 recovery column C3. This column operates at a pressure of 10 bar, which shifts the azeotropic composition so that the distillate stream from column D_3 has a composition of 34.2 mol% methanol. Higher and lower pressures were explored to see their effect on the economics. The 10 bar pressure seems to be about the optimum because going above this pressure does not shift the azeotrope significantly and raises the base temperature, which would require higher temperature energy input.

The separation is a fairly easy one, so using only 10 stages and a reflux ratio of 1 yield a bottoms impurity of 0.01 mol% methanol. This bottoms stream B_3 is the C5 product stream containing the inert isopentane. The reflux drum temperature is 373 K at this high pressure, which means that some heat integration between C2 and C3 may be economical (the base of C2 is at 356 K and the reflux drum of C3 is at 373 K). This possibility is not considered.

The distillate is recycled back to C2 at a flowrate of 1616 kmol/h. Reboiler heat input and condenser heat removal are 27.2 and 20.2 MW, respectively. The column diameter is 6.2 m.

As any user of flowsheet simulators knows, the convergence of steady-state simulators when recycle streams are present can be very difficult. Such is the case with both of these processes because they both involve two recycle streams. An alternative approach is to use a dynamic simulation to converge the process flowsheet to a steady state. The process conditions shown in Figures 8.23 and 8.24 are obtained in this way. This procedure is discussed in work by Luyben.[5]

8.5 EXTRACTIVE DISTILLATION METHANOL SEPARATION SECTION

An alternative flowsheet is shown in Figure 8.24 in which water is used as an extractive agent in an extractive distillation column to remove the methanol from the distillate stream coming from the reactive distillation column. A second column separates the methanol/water mixture coming from the base of the extraction column and recycles both methanol and water back to upstream units in the process.

The reactor and column C1 are identical to those used in the pressure-swing process. The methanol-containing distillate D_1 from the top of the reactive column is fed to stage 6 of a 12-stage extraction column. Water is fed on the top tray at a rate of 1050 kmol/h and a temperature of 322 K, which is achieved by using a cooler (heat removal = 1.24 MW). The column is a simple stripper with no reflux. The column operates at 2.5 atm so that cooling water can be used in the condenser (reflux-drum temperature = 326 K). Reboiler heat input is 5.96 MW. The overhead vapor is condensed and it is the C5 product stream.

This column is designed by specifying a very small loss of methanol in the overhead vapor (0.01% of the methanol fed to the column) and finding the minimum flowrate of extraction water that achieves this specification. Adding more than 10 trays or using reflux did not affect the recovery of methanol.

[5]W. L. Luyben, Use of dynamic simulation to converge complex process flowsheets, *Chem. Eng. Ed.* **38**, 2 (2004).

TABLE 8.10 Stream Information for Columns C2 and C3

	B2 (kmol/h)	D2 (kmol/h)	B3 (kmol/h)	D3 (kmol/h)	Water Makeup (kmol/h)	Extract Water to C2 (kmol/h)
MeOH	316	0.03	1.03	315	—	—
2M1B	—	3.31	—	—	—	—
2M2B	—	14.1	—	—	—	—
TAME	—	—	—	—	—	—
nC_5	—	88.4	—	—	—	—
iC_5	—	501	—	—	—	—
1-Pentene	—	38.0	—	—	—	—
2-Pentene	—	161	—	—	—	—
Water	1025	—	1026	0.32	24	1050
Total	230	830	1027	315	24	1050
T (K)	325	326	379	338	325	322
P (atm)	17	2.5	1.2	1.0	7	2.5

The bottoms is essentially a binary methanol/water (23.5 mol% methanol) mix, which is fed on stage 12 of a 32-stage column operating at atmospheric pressure. The number of trays in the second column was optimized by determining the TAC of columns over a range of tray numbers. Reboiler heat input and condenser heat removal are 8.89 and 9.53 MW, respectively. The column diameter is 2.24 m.

A reflux ratio of 2.1 produces 316 kmol/h of high purity methanol in the distillate (99.9 mol%) and 1026 kmol/h of high purity water in the bottoms (99.9 mol%). The methanol is combined with 230 kmol/h of fresh methanol feed, and the total is split between the methanol feedstreams to the reactor and to the reactive column. The water is combined with a small water makeup stream, cooled, and recycled back to extractive column C2.

Some makeup water is needed because a small amount of water goes overhead in the vapor from C2 (2.9 mol% water). The solubility of water in pentanes is quite small, so the reflux drum of column C2 would form two liquid phases (not shown in Fig. 8.2). The aqueous phase would be 19.9 kmol/h and 99.9 mol% water. The organic phase would be 809 kmol/h and 0.5 mol% water.

Table 8.10 provides molar flowrates and conditions for the important streams in the separation section of this flowsheet.

8.6 ECONOMIC COMPARISON

Table 8.11 gives the basis for equipment sizing and the costs used for equipment capital investment and energy. Design parameters, capital costs, and energy costs for the separation sections of both processes are listed in Table 8.12. The reboiler heat inputs in the two columns of the extractive distillation process are about 30% of those in the pressure-swing process. This reduces column diameters and heat exchanger areas, so the capital cost is also much smaller (about 40% lower).

These economics clearly indicate a large advantage for the extractive distillation process over the pressure-swing process from the perspective of steady-state economics.

TABLE 8.11 Basis of Economics

Condensers
Heat transfer coefficient = 0.852 kW/K m^2
Differential temperature = 13.9 K
Capital cost = 7296 (area)$^{0.65}$
Area (m^2)
Reboilers
Heat transfer coefficient = 0.568 kW/K-m^2
Differential temperature = 34.8 K
Capital cost = 7296 (area)$^{0.65}$
Area (m^2)
Column vessel capital cost = 17,640 (D)$^{1.066}$ (L)$^{0.802}$
Diameter and length (m)
Energy cost = \$4.7 per 10^6 kJ
$\text{TAC} = \dfrac{\text{capital cost}}{\text{payback period}} + \text{energy cost}$
Payback period = 3 years

TABLE 8.12 Design Parameters and Economic Comparison

		Pressure-Swing Process	Extractive Distillation Process
C2	Diameter (m)	4.2	1.74
	Stages	10	12
	QC (MW)	24.5	5.20
	AC (m^2)	2065	438
	QR (MW)	19.5	5.96
	AR (m^2)	985	301
	QHX (MW)	—	1.24
	AHX (m^2)	—	104
	Shell cost (10^6 \$)	0.339	0.158
	HX cost (10^6 \$)	1.69	0.828
	Energy cost (10^6 \$/year)	2.89	0.985
	Capital (10^6 \$)	2.02	1.21
C3	Diameter (m)	6.2	2.2
	Stages	10	32
	QC (MW)	20.2	9.53
	AC (m^2)	1703	803
	QR (MW)	27.2	9.05
	AR (m^2)	1374	457
	Shell cost (10^6 \$)	0.508	0.496
	HX cost (10^6 \$)	0.304	0.954
	Energy cost (10^6 \$/year)	4.03	1.34
	Capital (10^6 \$)	2.23	1.45
Total capital (10^6 \$)		4.25	2.44
Total energy (10^6 \$/year)		6.92	2.22
TAC (10^6 \$/year)		**8.34**	**3.04**

8.7 CONCLUSION

In this chapter we presented an important example of a reactive distillation column operating in a plantwide context. The reactive column was coupled with a two-column reactant recovery section.

The effects of some design parameters in this real chemical system were different than those found in the ideal systems. The complexity and counterintuitive performance of this reactive distillation system illustrated the challenges associated with the design and control of these very challenging and interesting systems.

CHAPTER 9

DESIGN OF MTBE AND ETBE REACTIVE DISTILLATION COLUMNS

In this chapter we examine two reactive distillation column systems that are used for the production of real chemical components. The two systems are quite similar and are basically ternary systems with inerts that have characteristics similar to those discussed in Chapter 5 for ideal components.

The first system is the production of MTBE from the reaction of methanol with isobutene. The second is the production of ETBE from the reaction of ethanol with isobutene.

The simulation tool used to explore these real systems is Aspen Plus, which we would expect to be convenient because it has built-in physical property packages and models of reactive distillation column. However, as we will discuss in more detail in later sections of this chapter, the current version of Aspen Plus has some important limitations that make using it less user friendly than we would like.

9.1 MTBE PROCESS

The MTBE reactive distillation process was patented several decades ago, and the process was widely used in the petroleum industry. Many reactive columns were installed around the world to produce MTBE, which was blended into gasoline. This process was probably the largest application of reactive distillation in terms of the number of columns and total production capacity. Because MTBE presents groundwater contamination problems, it is gradually being phased out of use in gasoline.

The reactive distillation column is essentially a ternary system with inerts. The liquid-phase reversible reaction is

$$\text{isobutene} + \text{MeOH} \Leftrightarrow \text{MTBE} \tag{9.1}$$

Reactive Distillation Design and Control. By William L. Luyben and Cheng-Ching Yu

The heavy component is MTBE, which leaves the reactive distillation column in the bottoms. The isobutene feed is contained in a mixed C4 stream from an upstream refinery unit. This stream contains a number of other C4 hydrocarbons because of the difficulty of separating the various components with very similar relative volatilities. In the numerical example, we assume that *n*-butene is the chemically inert component. Most of this inert component leaves in the distillate stream.

9.1.1 Phase Equilibrium

After the isobutene is consumed by the reaction, the resulting mixture of *n*-butene, methanol, and MTBE has vapor–liquid equilibrium properties that are nonideal with two binary azeotropes. At a pressure of 11 bar, which is the operating pressure of the column, the pure component boiling points are 76.1, 140.6, and 152.8 °C, respectively.

The two binary azeotropes at 11 bar are homogeneous and minimum boiling. The first is 89.3 mol% *n*-butene and 10.7 mol% methanol at 73.2 °C. The second is 42.3 mol% MTBE and 57.7 mol% methanol at 132.4 °C. The UNIQUAC physical properties model is used in the numerical example presented in this chapter.

9.1.2 Reaction Kinetics

The reaction is reported to be equilibrium limited, not kinetically limited, and most authors assume chemical equilibrium on each reactive tray. For example, Huang and Wang[1] use the following equation for the temperature dependence of the chemical equilibrium constant:

$$\ln K_{EQ} = 16.33 - \frac{6820}{T} \tag{9.2}$$

However, a kinetic model supplied by Aspen Technology is used in this chapter. It uses a kinetic expression with concentrations in terms of activities and calculates a chemical equilibrium constant using a more complex function of temperature.

9.1.3 Aspen Plus Simulation Issues

Simulation of reactive distillation using the standard models in Aspen Plus has some significant problems. The reactions can be specified to be either kinetic or equilibrium. In the former case, the choices of concentration units are limited to mole fraction, molarity, or partial pressure. Unfortunately, *activity* is not on this list. This is a problem because activities are frequently used to describe kinetic data.

In addition, the reaction rate expression is limited to a power law in reactive distillation. Reaction rate expressions such as Langmuir–Hinshelwood–Hougen–Watson (LHHW) cannot be used. These restrictions make the use of Aspen Technology simulations tools for reactive distillation somewhat inconvenient.

In the MTBE case in which equilibrium can be assumed, this problem would seem to be of no consequence. In the steady-state design using Aspen Plus, the chemical equilibrium model can be used. However, a serious limitation arises when one attempts to export the file

[1]K. Huang and S. J. Wang, Design and control of a methyl tertiary butyl ether (MTBE) decomposition reactive distillation column, *IEC Res.* **46**, 2508–2519 (2007).

to Aspen Dynamic to conduct control studies. The reactions *must be* kinetic in Aspen Dynamics.

One solution to this problem is to develop a special user-generated subroutine for the reaction. Aspen Technology has such a subroutine in its library for MTBE (Program Files\AspenTech\AMSystem 2004.1\Help\ADExamples.pdf). It is written in Fortran and is called RAMTBE.f. We will use this subroutine in both the steady-state design in this chapter and in the dynamic control discussed in Chapter 15.

In the ETBE case considered later in this chapter, a user-supplied kinetic subroutine had to be developed. The efforts of Bobby Hung of the National Taiwan University are gratefully acknowledged in its development.

9.1.4 Setting up the Aspen Plus Simulation

The first step is to set up the reaction. Figure 9.1 shows how the stoichiometry of the reaction is specified by opening the *Specifications* page tab under the *Reactions* item in the *Data Browser* window. Opening the *Kinetics* page tab gives the view shown in Figure 9.2 in which the user-supplied subroutine is selected and the reaction is specified to be in the liquid phase. Then the name of the subroutine is entered by opening the *Subroutine* page tab, which is displayed in Figure 9.3.

The reactive distillation block is called T1, and selecting the *Reactions* item opens the window illustrated in Figure 9.4. The location of the reactive stages is specified on the *Specifications* page tab (stages 4–10). The holdup on the reactive stage is supposedly specified as shown in Figure 9.5. As we will discuss later, this does not seem to work with the user subroutine.

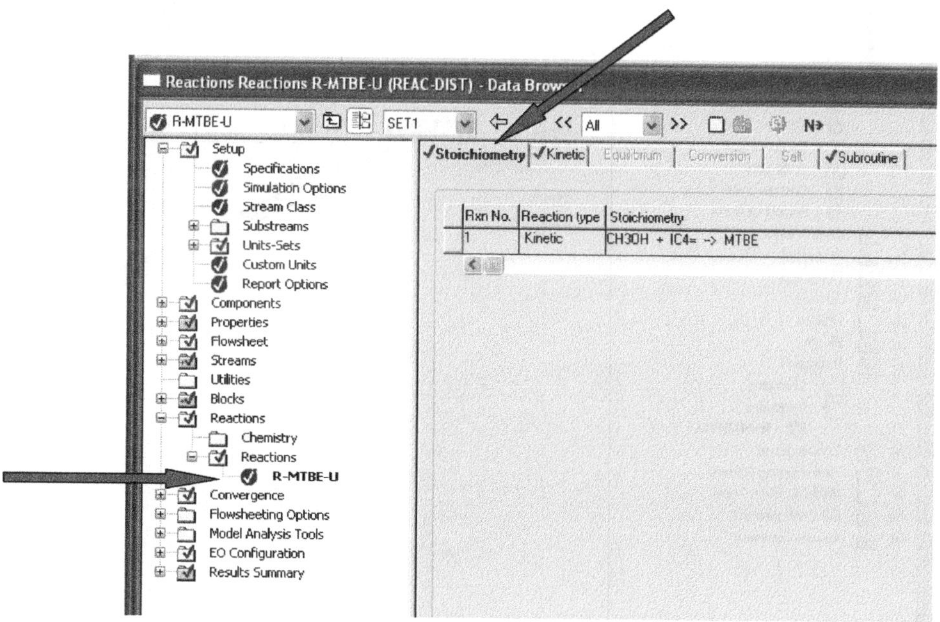

Figure 9.1 MTBE reaction stoichiometry.

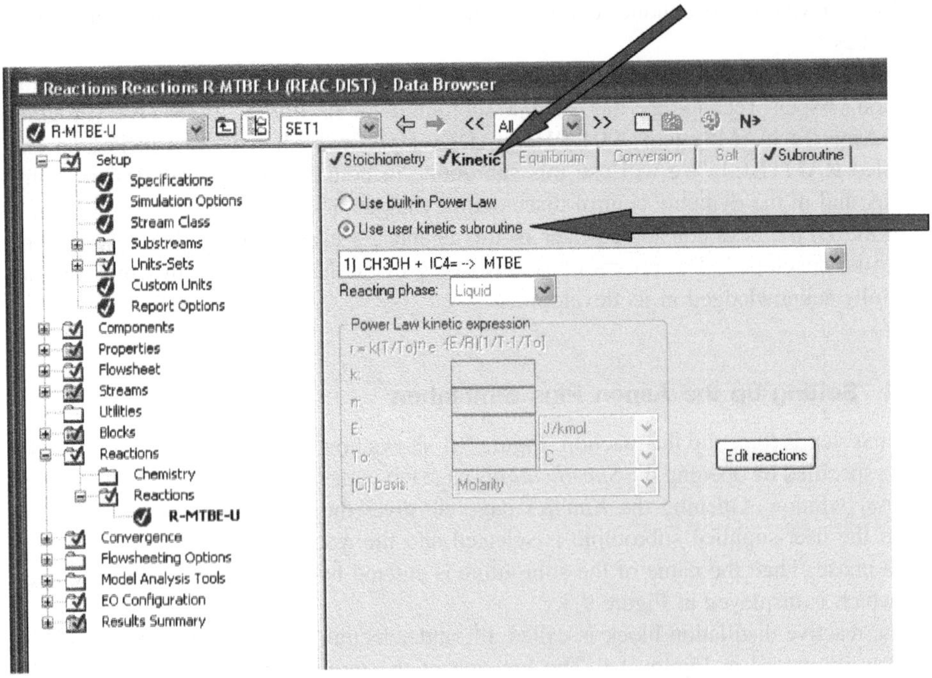

Figure 9.2 Specifying user kinetic subroutine.

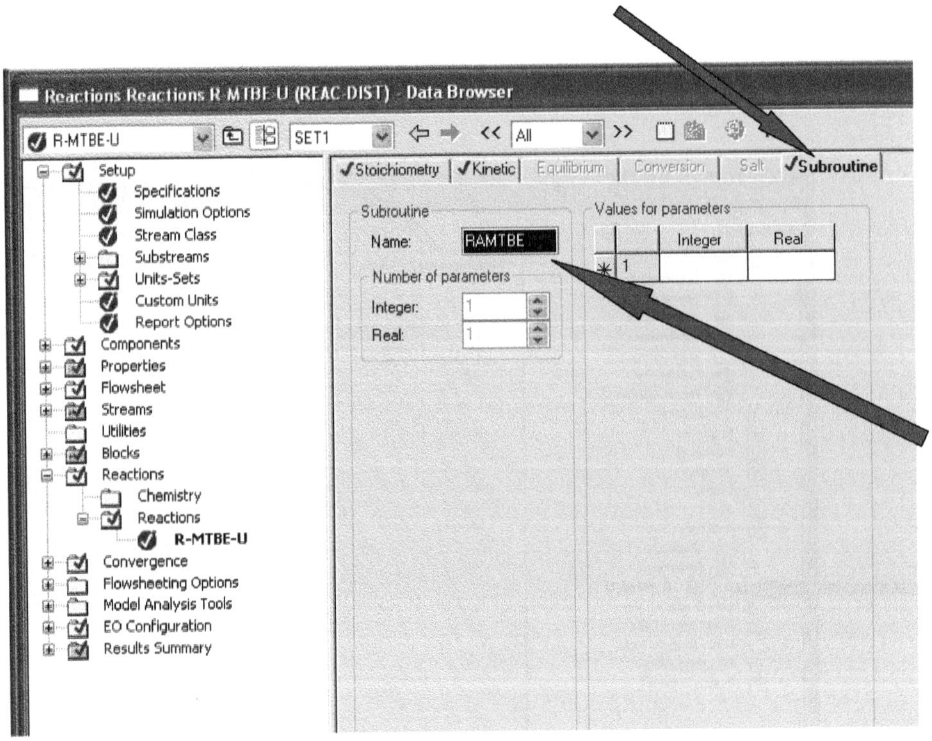

Figure 9.3 Specifying RAMTBE subroutine.

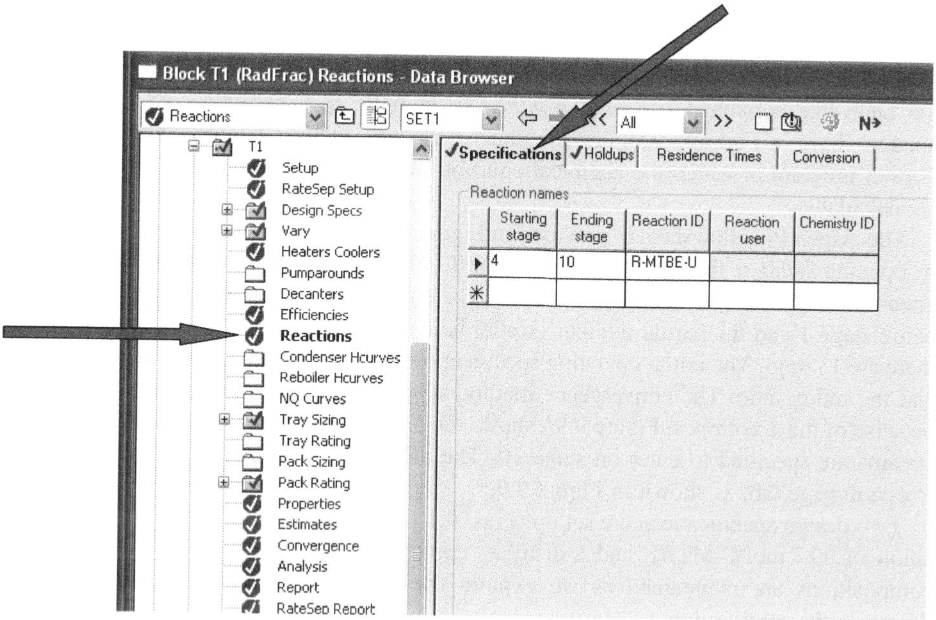

Figure 9.4 Specifying reactive stages.

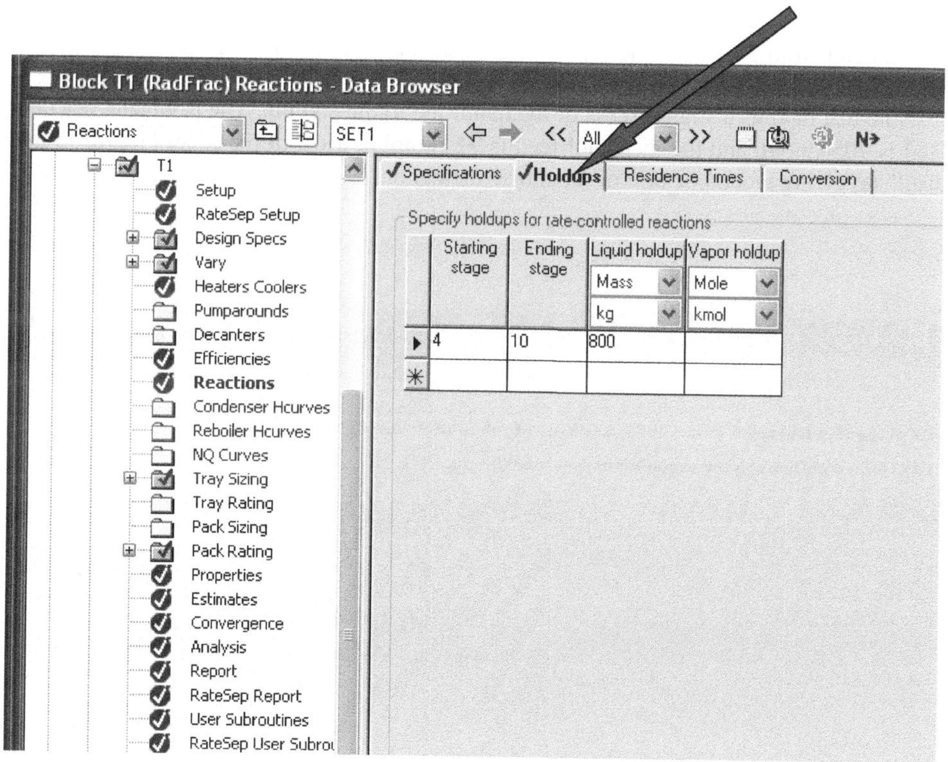

Figure 9.5 Specifying reactive stage catalyst holdup.

The difference between using a user-supplied subroutine and one of the built-in functions is the need to compile the subroutine. Figure 9.6 demonstrates how this is done. Instead of going to *Aspen User Interface*, you go to *Aspen Simulation Engine* and use the DOS-like command *aspcomp ramtbe.f* to compile the subroutine. Of course, you need to have a Fortran compiler on your computer. Figure 9.7 lists the RAMTBE.f Fortran program in which the chemical equilibrium constant, activities, and reaction rates are calculated.

The Aspen Plus flowsheet is provided in Figure 9.8. The column configuration is set up by opening *Setup* on the list of items under the T1 block. Figure 9.9*a* shows the window that opens with the *Configuration* page tab open. The column has 17 stages with the reflux drum being stage 1 and the partial reboiler (kettle) being stage 17, using Aspen notation. Thus, there are 15 trays. The initial operating specifications are selected to be the bottoms flowrate and the reflux ratio. The convergence method is selected to be *Strongly non-ideal liquid* because of the azeotropes. Figure 9.9*b* shows the *Streams* page tab on which the two feedstreams are specified to enter on stage 10. The pressure of the column is entered on the *Pressure* page tab, as shown in Figure 9.9*c*.

Two design specifications are set up (Figs. 9.10 and 9.11) to achieve a bottoms composition of 99.7 mol% MTBE and a distillate composition of 0.0421 mol% MTBE. These compositions are maintained as we explore the effects of various parameters on the design in the next section.

Figure 9.12 gives the process flowsheet with many important variables, Figure 9.13 gives detailed stream information, and Figure 9.14 gives temperature and composition profiles. Notice that the isobutene composition is low throughout the column. There is a peak methanol concentration of about 25 mol% on stage 10. Temperatures in the column run from 347 K at the top to 425 K at the bottom with an operating pressure of 11 bar.

Note that there is a slight excess of methanol fed to the column. The isobutene reactant feed is 705 kmol/h and the methanol feed is 775 kmol/h. Most of the excess leaves in the distillate, giving a composition of 6.6 mol% methanol.

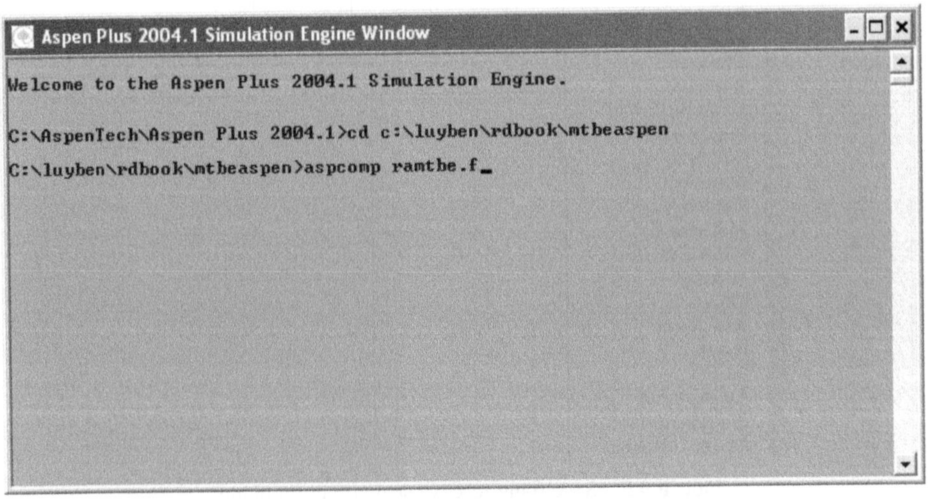

Figure 9.6 Compiling Fortran subroutine.

9.1.5 Effect of Design Parameters

The Aspen Plus file is used to explore the effects of several parameters. Convergence issues prevented a detailed study over a wide range of parameter values. This is one of the problems with commercial software. If the program does not converge, you are given little guidance as to what to do.

Pressure. The pressure used in the Aspen example is 11 bar, and this pressure was also used by Huang and Wang.[1] Table 9.1 gives results for different pressures with distillate and

```
      SUBROUTINE RAMTBE (NSTAGE, NCOMP,   NR,      NRL,      NRV,
     2                   T,      TLIQ,    TVAP,    P,        VF,
     3                   F,      X,       Y,       IDX,      NBOPST,
     4                   KDIAG,  STOIC,   IHLBAS,  HLDLIQ,   TIMLIQ,
     5                   IHVBAS, HLDVAP,  TIMVAP,  NINT,     INT,
     6                   NREAL,  REAL,    RATES,   RATEL,    RATEV,
     7                   NINTB,  INTB,    NREALB,  REALB,    NIWORK,
     8                   IWORK,  NWORK,   WORK)
      IMPLICIT NONE
C
C     DECLARE VARIABLES USED IN DIMENSIONING
C
      INTEGER NCOMP, NR,    NRL,    NRV,    NINT,
     +        NINTB, NREALB,NIWORK,NWORK, N_COMP
C
C     DECLARE PARAMETERS & VARIABLES USED IN PARAMETERS
C
      INTEGER K_MEOH,K_IC4, K_NC4, K_MTBE
      PARAMETER(K_MEOH=1)
      PARAMETER(K_IC4=2)
      PARAMETER(K_NC4=3)
      PARAMETER(K_MTBE=4)
      PARAMETER(N_COMP=4)
C
C     DECLARE ARGUMENTS
C
      INTEGER IDX(NCOMP),    NBOPST(6),     INT(NINT),
     +        INTB(NINTB),   IWORK(NIWORK),NSTAGE,
     +        KDIAG, IHLBAS,IHVBAS,NREAL, KPHI,
     +        KER,    L_GAMMA,      J
      REAL*8 X(NCOMP,3),    Y(NCOMP),
     +       STOIC(NCOMP,NR),       RATES(NCOMP),
     +       RATEL(NRL),    RATEV(NRV),
     +       REALB(NREALB),WORK(NWORK),  B(1),   T,
     +       TLIQ,   TVAP,   P,      VF,      F
      REAL*8 HLDLIQ,TIMLIQ,HLDVAP,TIMVAP,TZERO,
     +       FT
C     DECLARE SYSTEM FUNCTIONS
      REAL*8 DLOG
C     DECLARE LOCAL VARIABLES
      INTEGER IMISS, IDBG
      REAL*8 REAL(NREAL),   RMISS, C1,    C2,    C3,
     +       C4,     C5,    C6,    DKA,   DKR,
     +       Q,      RATE,  RATNET
      REAL*8 PHI(N_COMP)
      REAL*8 DPHI(N_COMP)
      REAL*8 ACTIV(N_COMP)
```

Figure 9.7 RAMTBE Fortran subroutine.

```
C
#include "ppexec_user.cmn"
      EQUIVALENCE (RMISS, USER_RUMISS)
      EQUIVALENCE (IMISS, USER_IUMISS)
#include "dms_maxwrt.cmn"
#include "dms_ipoff3.cmn"
#include "dms_plex.cmn"
      EQUIVALENCE(B(1),IB(1))
C      DATA STATEMENTS
      DATA IDBG/0/
C     thermodynamic rate constant DKA
C     ===============================
 9010 FORMAT(1X,3(G13.6,1X))
 9000 FORMAT(' fugly failed at T=',G12.5,' P=',G12.5,' ker=',I4)
 9020 FORMAT(' compo ',I3,' mole-frac=',G12.5,' activity=',G12.5)
 9030 FORMAT(' stage=',I4,' spec-rate=',G12.5,' net-rate=',G12.5)
C
C     BEGIN EXECUTABLE CODE
C
      TZERO=298.15D+00
      C1=-1493.D+00
      C2=-77.4D+00
      C3=0.508D+00
      C4=-0.913D-03
      C5=1.11D-06
      C6=-0.628D-09
      FT=C1*(1.D+00/T-1.D+00/TZERO)
     *  +C2*DLOG(T/TZERO)
     *  +C3*(T-TZERO)
     *  +C4*(T*T-TZERO*TZERO)
     *  +C5*(T**3-TZERO**3)
     *  +C6*(T**4-TZERO**4)

      DKA=284D+00*DEXP(FT)  <----- Chemical equilibrium constant
C     reaction rate constant
C     ======================
      DKR=3.67D+12*DEXP(-11110.D+00/T)

      IF(IDBG.GE.1)THEN
        WRITE(MAXWRT_MAXBUF(1),9010) FT,DKA,DKR
        CALL DMS_WRTTRM(1)
      ENDIF
C     calculation of components activities
C     ====================================
C     calculate only fugacity coefficient
      KPHI=1
C     fugacity coefficient of components in the mixture
      CALL PPMON_FUGLY(T,P,X(1,1)
     +      ,Y,NCOMP,IDX,NBOPST,KDIAG,KPHI,PHI,DPHI,KER)
      IF(KER.NE.0)THEN
        WRITE(MAXWRT_MAXBUF(1),9000) T,P,KER
        CALL DMS_WRTTRM(1)
      ENDIF
C     set offset to get activity coefficients
```

Figure 9.7 (*Continued*).

```
C       (see vol5, p 11-11 and asp$sor search for 'GAMMAL')
        L_GAMMA=IPOFF3_IPOFF3(24)
C       calculate activities for plex data
        DO J=1,NCOMP
          ACTIV(J)=DEXP(B(L_GAMMA+J))*X(J,1)          Activities
        END DO
        IF(IDBG.GE.1)THEN
          DO J=1,NCOMP
            WRITE(MAXWRT_MAXBUF(1),9020) J,X(J,1),ACTIV(J)
            CALL DMS_WRTTRM(1)
          END DO
        ENDIF
C       reaction rate
C       =============
C       RATE : mol/s/kgcata
C       Q    : equiv/kgcata
C       DKR  : mol/s/equiv
        Q=4.9D+00
        RATE=Q*DKR*
     *        (ACTIV(K_IC4)/ACTIV(K_MEOH)
     *       -ACTIV(K_MTBE)/DKA/ACTIV(K_MEOH)**2)          Reaction rate
c
c change catalyst
c
c        realb(1)=600.
        RATNET=RATE*REALB(1)*1.D-03
        RATES(K_IC4)=-RATNET
        RATES(K_MEOH)=-RATNET
        RATES(K_MTBE)=RATNET
        RATES(K_NC4)=0.D+00
        IF(IDBG.GE.1)THEN
          WRITE(MAXWRT_MAXBUF(1),9030) NSTAGE,RATE,RATNET
          CALL DMS_WRTTRM(1)
        ENDIF
        RETURN
#undef P_MAX3
        END
```

Figure 9.7 (*Continued*).

bottoms compositions held at their specified values. The minimum energy consumption occurs at 11.2 bar. The production of MTBE also increases slightly as pressure is increased, so operation at a pressure slightly higher than 11.2 bar may be more economical despite the slightly higher energy cost. The increase in MTBE production occurs because less of the isobutene reactant is lost in the distillate. The losses in the bottoms are small, but they actually increase with increasing pressure.

Reactive Tray Holdup. Changing this specification in the reactor block had no effect. Even a zero holdup gave the same solution. We tried changing the holdup in the Fortran subroutine and recompiling, but the column would not converge.

Therefore, this parameter could not be explored because of the limitations and bugs in Aspen Plus.

Feed Stage Location. In the base case, both C4 and methanol are fed onto stage 10, which is at the bottom of the reactive stages (stages 4–10). The effect of changing the

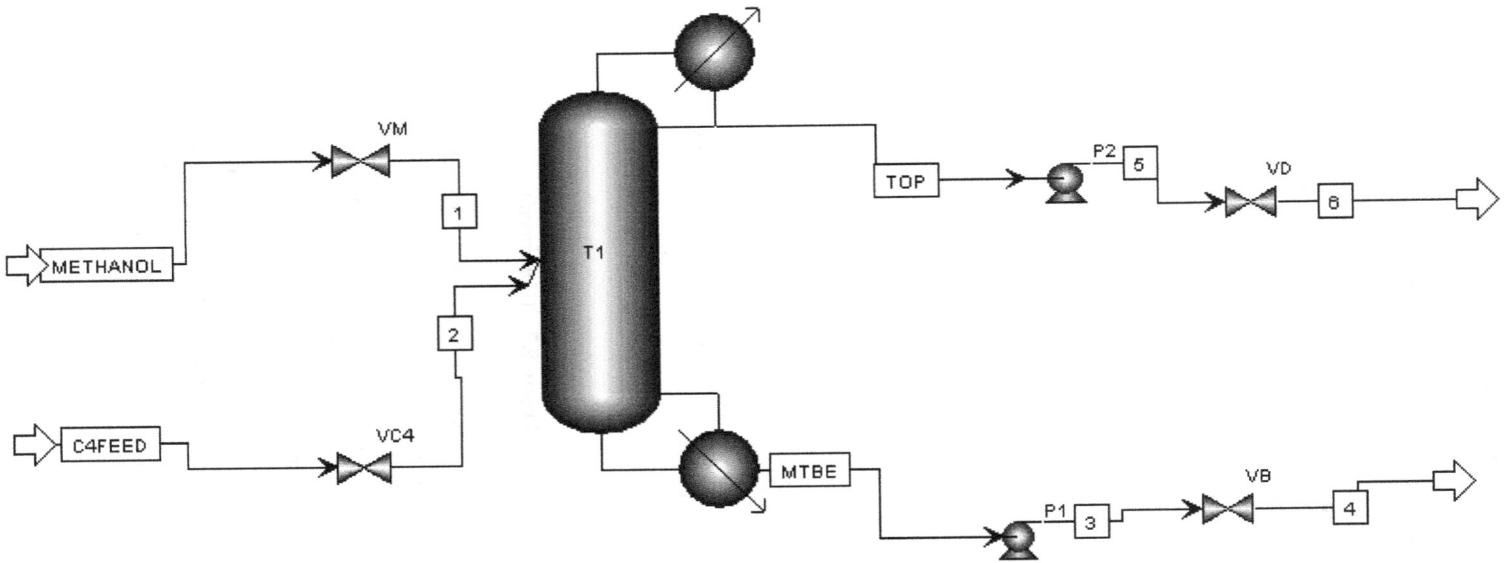

Figure 9.8 Aspen Plus flowsheet.

location of each of the two feedstreams was explored. The simulation would not converge when the C4 feed was fed to any stage other than stage 10. The effect of moving the methanol feed point is shown in Table 9.2. Moving up one stage has two advantages: energy consumption decreases and MTBE production increases. Moving still higher in the column gives poorer results.

Stripping Trays. In the base case there are six stripping trays (stages 11–16). Table 9.3 give the results for increasing this number. The program would not converge if fewer than six stripping trays were used.

(*a*)

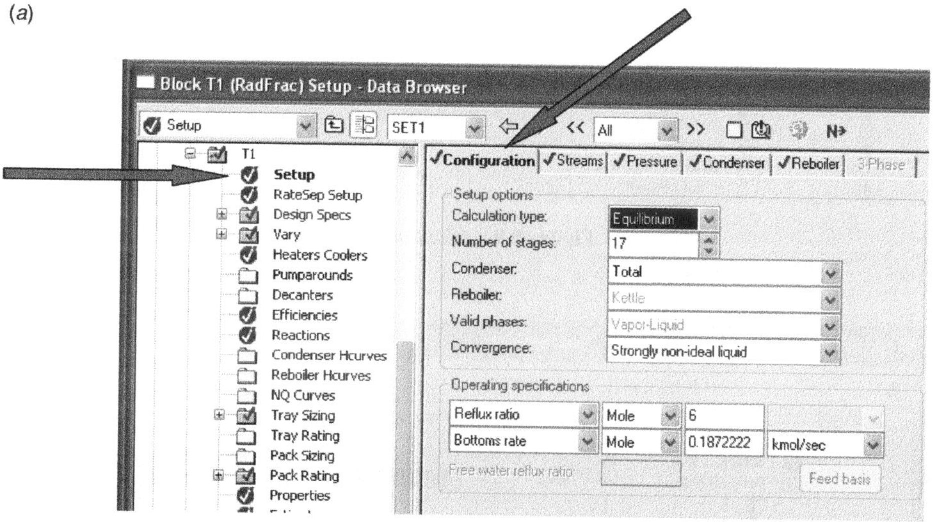

(*b*)

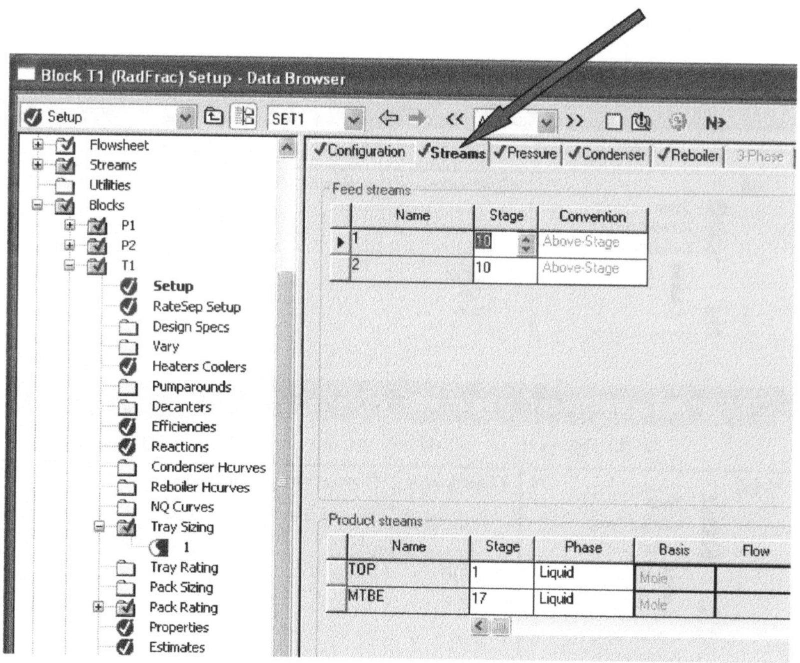

Figure 9.9 (*a*) Configuration, (*b*) streams, and (*c*) pressure.

(*c*)

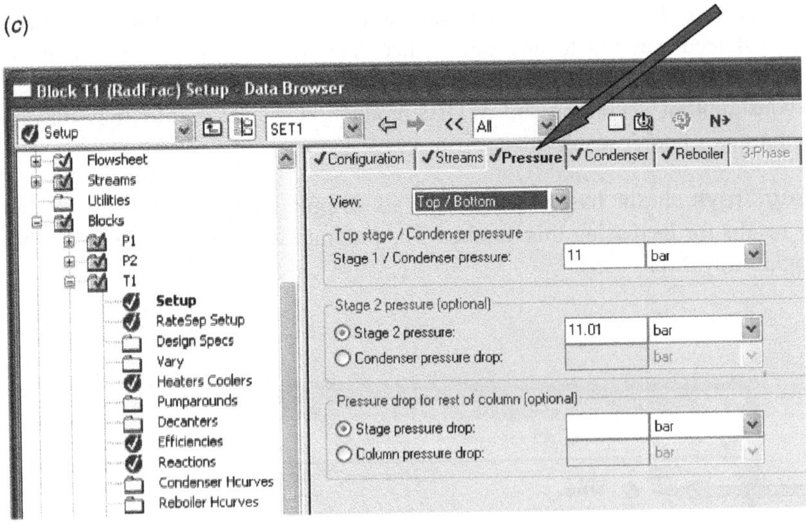

Figure 9.9 (*Continued*).

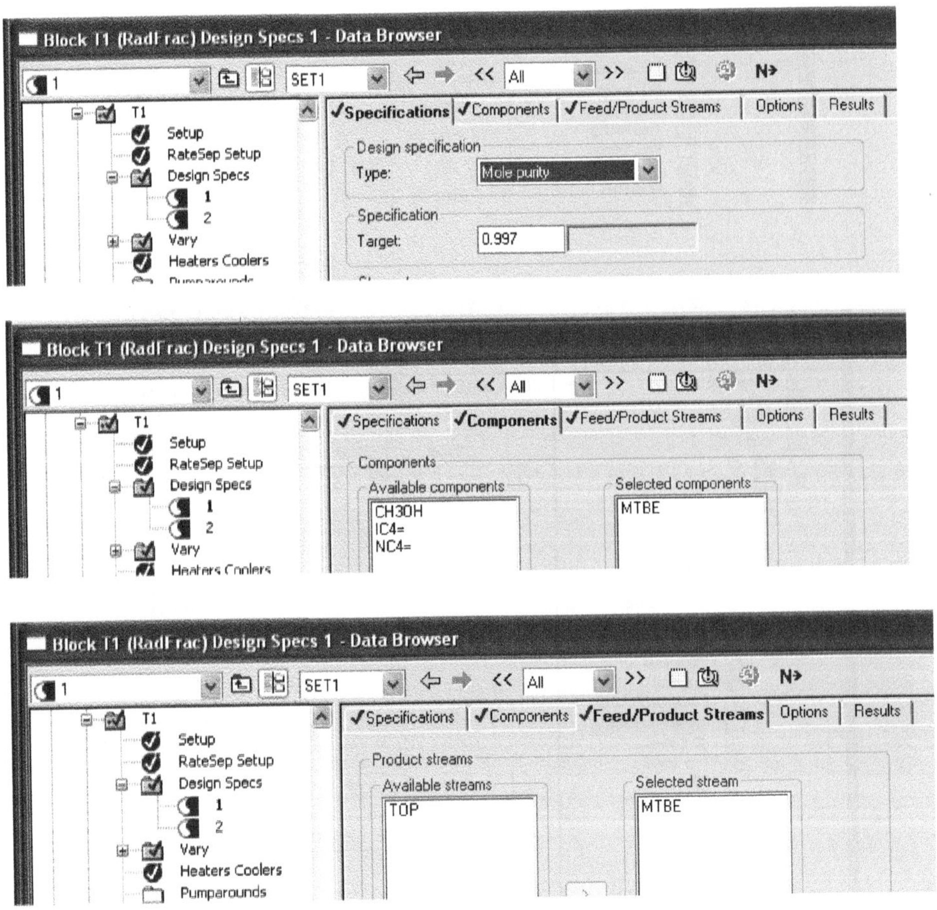

Figure 9.10 Design spec for bottoms.

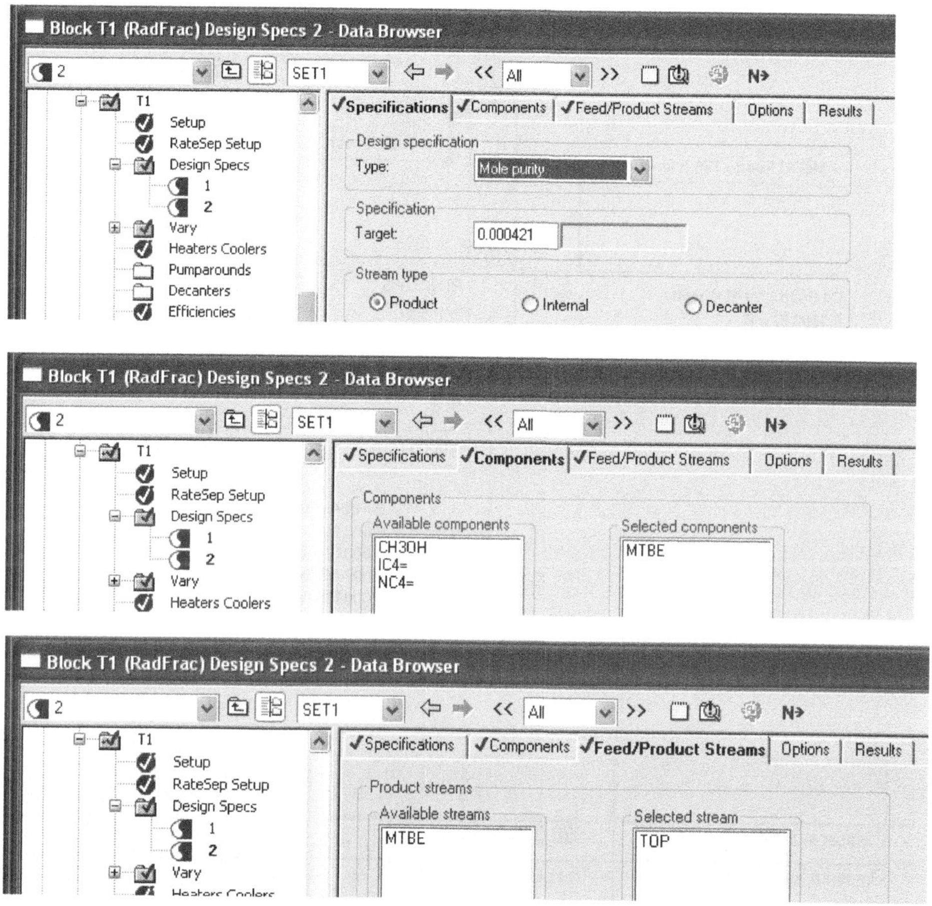

Figure 9.11 Design spec for distillate.

Adding more stripping trays reduces energy consumption and increases MTBE production. However, the improvements become increasingly smaller above about nine trays.

Other Parameters. Attempts to explore the number of reactive stages were unsuccessful because the program would not converge. The same occurred when the number of rectifying stages was changed.

Aspen Plus sometimes has trouble converging standard conventional distillation columns, particularly with design specifications. Convergence problems with reactive distillation columns are encountered even more frequently.

9.1.6 Chemical Equilibrium Model

In the previous sections we utilized the kinetic subroutine supplied by Aspen Technology, which uses a kinetic expression for the reaction rate. It is interesting to compare this column design with the results when chemical equilibrium is assumed on the reactive trays. This is done in Aspen Plus by simply specifying that the reaction is *Equilibrium* on the *Edit Reactions* window, as shown in the upper picture of Figure 9.15. The lower picture shows that the chemical equilibrium constant is calculated from the Gibbs free energies.

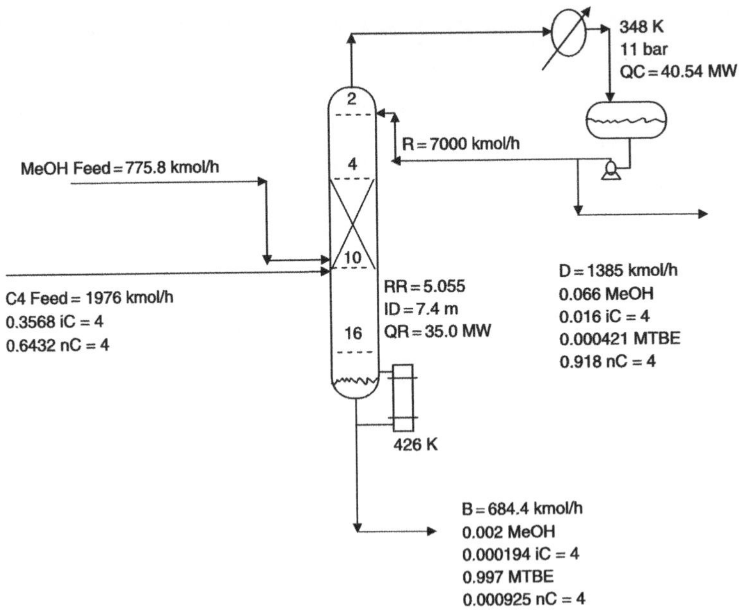

Figure 9.12 Steady-state conditions with kinetic model.

	METHANOl	C4FEED	TOP	MTBE
Temperature K	320.0	350.0	347.4	425.7
Pressure atm	15.000	15.000	10.856	10.859
Vapor Frac	0.000	0.000	0.000	0.000
Mole Flow kmol/h	775.800	1976.400	1384.774	684.448
Mass Flow kg/h	24858.308	110890.903	75512.081	60237.130
Volume Flow L/min	536.350	3586.137	2372.866	1762.028
Enthalpy MMkcal/h	-43.827	-9.262	-9.485	-48.303
Mole Flow kmol/h				
CH3OH	775.800		91.535	1.287
IC4=		705.180	22.069	0.133
NC4=		1271.220	1270.587	0.633
MTBE			0.583	682.395
Mole Frac				
CH3OH	1.000		0.066	0.002
IC4=		0.357	0.016	194 PPM
NC4=		0.643	0.918	925 PPM
MTBE			421 PPM	0.997

Figure 9.13 Steady-state stream information.

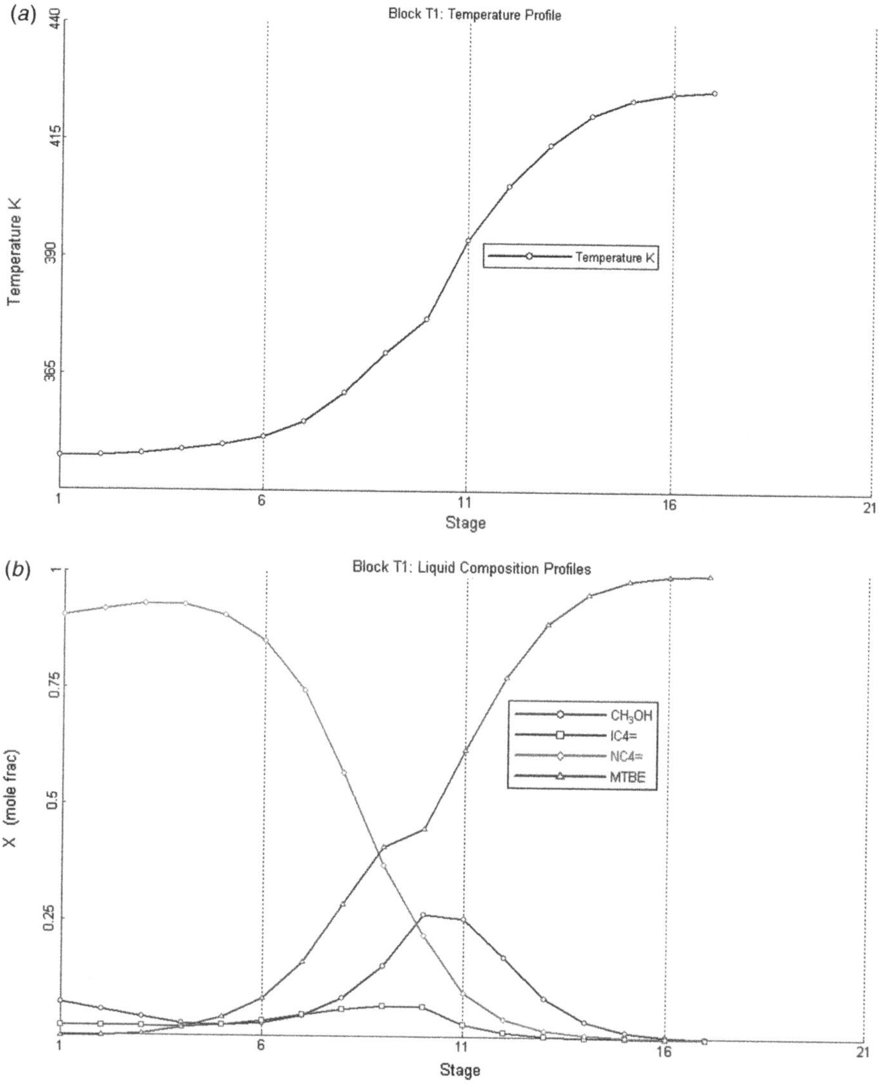

Figure 9.14 (*a*) Temperature profile in MTBE reactive column and (*b*) composition profiles with kinetic model.

TABLE 9.1 Pressure Effect Results for MTBE

Pressure (bar)	Reboiler Heat Input (MW)	Reflux Ratio	MTBE Recovered in Bottoms (kmol/h)	Isobutene Lost in Distillate and Bottoms (kmol/h)
10.85	37.97	5.392	675.46	29.03 + 0.10
10.9	36.49	5.222	678.84	25.64 + 0.11
11	**35.04**	**5.056**	**682.39**	**22.07 + 0.13**
11.2	34.16	4.975	685.62	17.84 + 0.17
11.4	34.46	5.049	687.4	17.19 + 0.20
11.7	36.36	5.369	688.26	16.11 + 0.23
12	40.05	5.954	688.29	16.06 + 0.25

TABLE 9.2 Methanol Feed Stage Results for MTBE

Methanol Feed Stage	Reboiler Heat Input (MW)	Reflux Ratio	MTBE Recovered in Bottoms (kmol/h)	Isobutene Lost in Distillate and Bottoms (kmol/h)
10	**35.04**	**5.056**	**682.39**	**22.07 + 0.13**
9	31.14	4.501	685.08	19.19 + 0.34
8	37.71	5.346	674.58	29.58 + 0.42
7	34.38	4.569	650.32	53.77 + 0.48

TABLE 9.3 Stripping Tray Results for MTBE

No. of Stripping Stages	Reboiler Heat Input (MW)	Reflux Ratio	MTBE Recovered in Bottoms (kmol/h)	Isobutene Lost in Distillate and Bottoms (kmol/h)
6	**35.04**	**5.056**	**682.39**	**22.07 + 0.13**
7	30.01	4.360	687.23	17.31 + 0.06
8	29.15	4.239	688.01	16.56 + 0.02
9	28.93	4.209	688.14	16.46 + 0.007
11	28.78	4.188	688.26	16.34 + 0.001
13	28.76	4.185	688.29	16.32 + 0

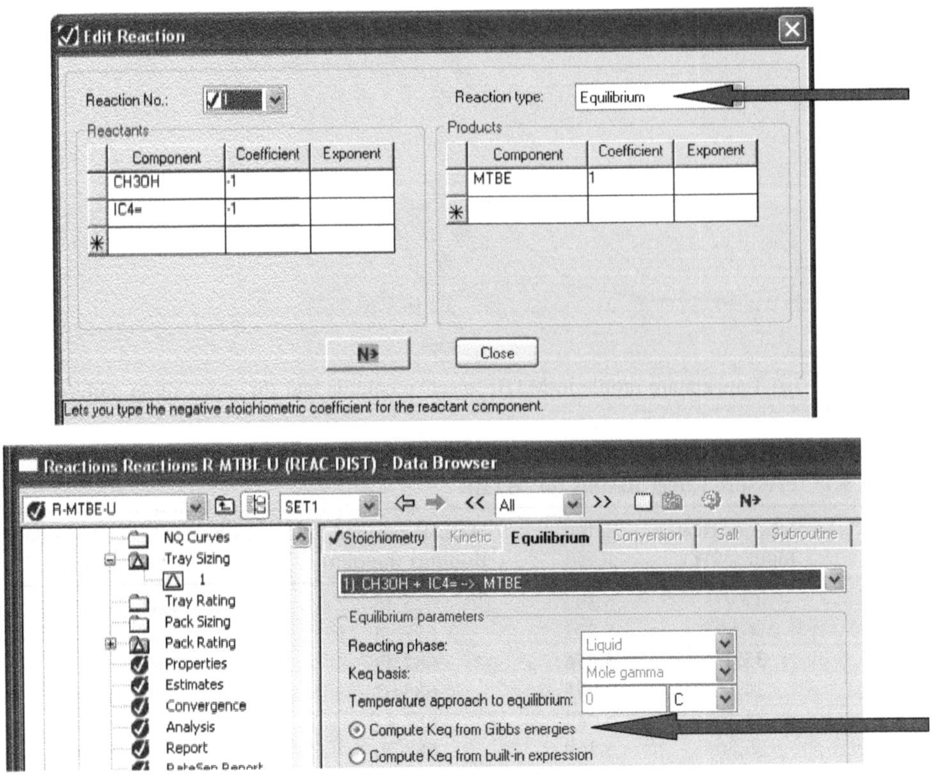

Figure 9.15 Setting up chemical equilibrium model.

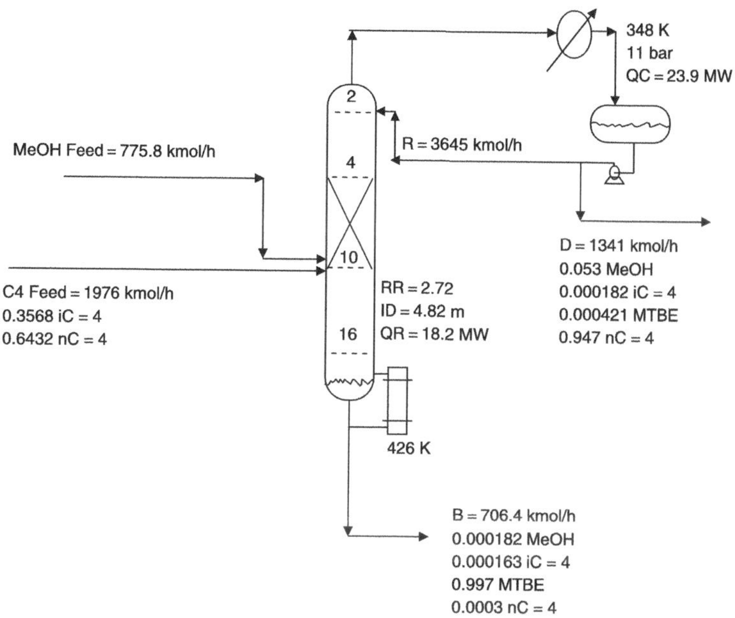

Figure 9.16 Steady-state conditions with chemical equilibrium.

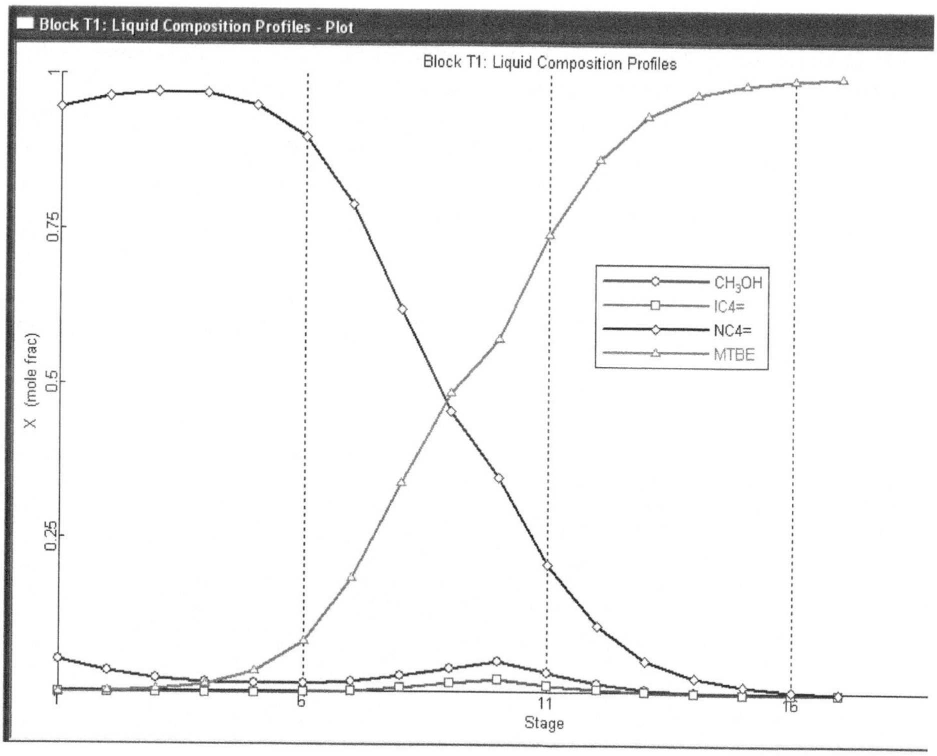

Figure 9.17 Composition profiles with chemical equilibrium.

The resulting design flowsheet is given in Figure 9.16 and should be compared with the one for the kinetic model given in Figure 9.12. The differences are striking. The reflux ratio required to achieve the same product purities drops from 5.05 to 2.74, which corresponds to a very large reduction in energy consumption (from 35.0 to 18.2 MW). The diameter of the column is reduced from 7.4 to 4.8 m.

The composition profiles are quite different between the two cases. Figure 9.17 presents the results for the equilibrium design. A comparison with Figure 9.14*b* shows that the reactant compositions are much smaller for the equilibrium model (peak methanol concentration = ~5 mol%) compared with those for the kinetic model (peak methanol concentration = ~25 mol%).

9.2 ETBE PROCESS

The reactive distillation column is essentially a ternary system with inerts. The liquid-phase reversible reaction is

$$\text{isobutene} + \text{EtOH} \Leftrightarrow \text{ETBE} \tag{9.3}$$

The heavy component is ETBE, which leaves the reactive distillation column in the bottoms. Most of the isobutene is consumed in the reaction, while the chemically inert *n*-butene in the C4 feedstream goes overhead in the distillate.

9.2.1 Kinetic Model

The kinetic equations that we used are from an article by Sneesby et al.[2] These were used by Al-Arfaj and Luyben to develop a Fortran simulation of the system.[3]

The kinetic model contains concentration terms given as activities. The overall reaction rate is not a simple power law but has absorption terms in LHHW form. This type of chemical kinetic cannot be used in the standard Aspen Plus reactive distillation model. The various formulas follow for the overall reaction rate (mol/s):

$$\mathcal{R}_{\text{ETBE}} = \frac{m_{\text{cat}} k_{\text{rate}} \left(a_{\text{iC4}} (a_{\text{EtOH}})^2 - \dfrac{a_{\text{ETBE}}}{K_{\text{EQ}} a_{\text{EtOH}}} \right)}{(1 + K_{\text{A}} a_{\text{EtOH}})^3} \tag{9.4}$$

specific reaction rate (mol s^{-1} kg^{-1}):

$$k_{\text{rate}} = 2.0606 \times 10^{12} \exp(-60{,}400/\text{RT}) \tag{9.5}$$

reaction equilibrium constant:

$$\ln K_{\text{EQ}} = 10.387 + \frac{4060.59}{T} - 2.89055 \ln T - 0.0191544T + 5.28586 \times 10^{-5} T^2 - 5.32977 \times 10^{-8} T^3 \tag{9.6}$$

[2]M. G. Sneesby, M. D. Tade, R. Datta, and T. N. Smith, ETBE synthesis via reactive distillation. 1. Steady-state simulation and design aspects, *IER Res.* **36**, 1855–1869 (1997).

[3]M. A. Al-Arfaj and W. L. Luyben, Control study of ETBE reactive distillation, *IEC Res.* **41**, 3784–3796 (2002).

absorption constant:

$$\ln K_A = -1.0707 + \frac{1323.1}{T} \tag{9.7}$$

and activity:

$$a_j = x_j \gamma_j \tag{9.8}$$

where x_j is the mole fraction of component j, γ_j is the liquid-phase activity coefficient of component j, a_j is the activity of component j, T is the temperature (K), E is the activation energy (kJ/kmol), and m_{cat} is the catalyst weight (g).

9.2.2 Process Studied

The ETBE reactive distillation column studied by Al-Arfaj and Luyben[3] is a double-feed system with pure ethanol fed to stage 6 at the top of the reactive zone and a mixed C4 stream fed to stage 20 at the bottom of the reactive zone. The C4 feed is 40 mol% isobutene and 60 mol% n-butene. The column has 4 stripping trays, 15 reactive trays, and 4 rectifying trays, which are shown in Figure 9.18.

The column operates at 7.5 atm pressure and a reflux ratio of 2.94. The catalyst holdup on the reactive trays is 1000 kg. The bottoms stream contains ETBE with a purity of

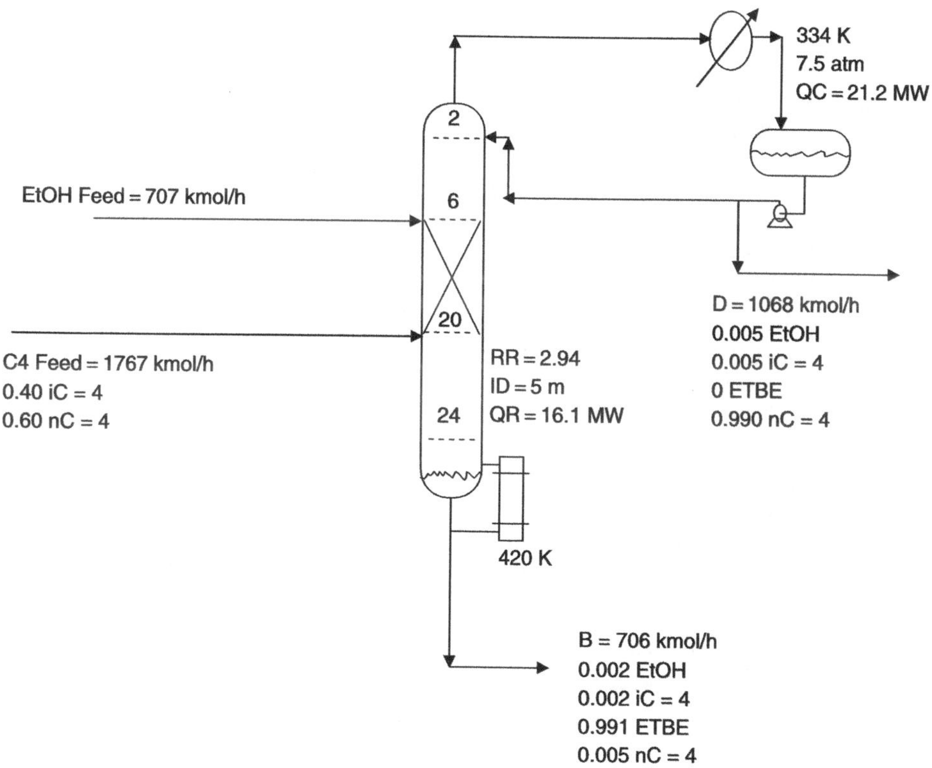

Figure 9.18 Al-Arfaj and Luyben ETBE flowsheet.[3]

```
      SUBROUTINE RAETBELB (NSTAGE, NCOMP,   NR,      NRL,      NRV,
     2                      T,        TLIQ,     TVAP,    P,        VF,
     3                      F,        X,        Y,       IDX,      NBOPST,
     4                      KDIAG,    STOIC,    IHLBAS,  HLDLIQ,   TIMLIQ,
     5                      IHVBAS,   HLDVAP,   TIMVAP,  NINT,     INT,
     6                      NREAL,    REAL,     RATES,   RATEL,    RATEV,
     7                      NINTB,    INTB,     NREALB,  REALB,    NIWORK,
     8                      IWORK,    NWORK,    WORK)
      IMPLICIT NONE
      INTEGER NCOMP, NR,    NRL,   NRV,   NINT,
     +        NINTB, NREALB,NIWORK,NWORK, N_COMP
      INTEGER K_ETOH,K_IC4, K_NC4, K_ETBE
      PARAMETER(K_ETOH=1)
      PARAMETER(K_IC4=2)
      PARAMETER(K_NC4=3)
      PARAMETER(K_ETBE=4)
      PARAMETER(N_COMP=4)
      INTEGER IDX(NCOMP),    NBOPST(6),     INT(NINT),
     +        INTB(NINTB),   IWORK(NIWORK),NSTAGE,
     +        KDIAG, IHLBAS,IHVBAS,NREAL, KPHI,
     +        KER,   L_GAMMA,      J
      REAL*8 X(NCOMP,3),    Y(NCOMP),
     +       STOIC(NCOMP,NR),      RATES(NCOMP),
     +       RATEL(NRL),    RATEV(NRV),
     +       REALB(NREALB),WORK(NWORK),  B(1),   T,
     +       TLIQ,  TVAP,  P,      VF,     F
      REAL*8 HLDLIQ,TIMLIQ,HLDVAP,TIMVAP,TZERO,
     +       FT
      REAL*8 DLOG
      INTEGER IMISS, IDBG
      REAL*8 REAL(NREAL),  RMISS, C1,     C2,     C3,
     +       C4,    C5,    C6,    DKA,    DKR,
     +       Q,     RATE,  RATNET, KETBE,  KA, KRATE
      REAL*8 PHI(N_COMP)
      REAL*8 DPHI(N_COMP)
      REAL*8 ACTIV(N_COMP)
#include "ppexec_user.cmn"
      EQUIVALENCE (RMISS, USER_RUMISS)
      EQUIVALENCE (IMISS, USER_IUMISS)
#include "dms_maxwrt.cmn"
#include "dms_ipoff3.cmn"
#include "dms_plex.cmn"
      EQUIVALENCE(B(1),IB(1))
      DATA IDBG/0/
 9010 FORMAT(1X,3(G13.6,1X))
 9000 FORMAT(' fugly failed at T=',G12.5,' P=',G12.5,' ker=',I4)
 9020 FORMAT(' compo ',I3,' mole-frac=',G12.5,' activity=',G12.5)
 9030 FORMAT(' stage=',I4,' spec-rate=',G12.5,' net-rate=',G12.5)
C
C     BEGIN EXECUTABLE CODE
```

Figure 9.19 User subroutine for ETBE.

```
C
      KETBE=DEXP(10.387D0+4060.59D0/T-2.89055D0*DLOG(T)-0.0191544D0*T+
     &       5.28586D-5*T**2-5.32977D-8*T**3)
      KA=DEXP(-1.0707D0+1323.1D0/T)
      KRATE=(2.0606D12*DEXP(-60.4D3/8.314D0/T))
      IF(IDBG.GE.1)THEN
        WRITE(MAXWRT_MAXBUF(1),9010) FT,DKA,DKR
        CALL DMS_WRTTRM(1)
      ENDIF
      KPHI=1
C     fugacity coefficient of components in the mixture
      CALL PPMON_FUGLY(T,P,X(1,1)
     +      ,Y,NCOMP,IDX,NBOPST,KDIAG,KPHI,PHI,DPHI,KER)
      IF(KER.NE.0)THEN
        WRITE(MAXWRT_MAXBUF(1),9000) T,P,KER
        CALL DMS_WRTTRM(1)
      ENDIF
      L_GAMMA=IPOFF3_IPOFF3(24)
      DO J=1,NCOMP
        ACTIV(J)=DEXP(B(L_GAMMA+J))*X(J,1)
      END DO
      IF(IDBG.GE.1)THEN
        DO J=1,NCOMP
          WRITE(MAXWRT_MAXBUF(1),9020) J,X(J,1),ACTIV(J)
          CALL DMS_WRTTRM(1)
        END DO
      ENDIF
      RATE=REALB(1)*KRATE*(ACTIV(K_ETOH))**2.d0*
     &      (ACTIV(K_IC4)-ACTIV(K_ETBE)/KETBE/ACTIV(K_ETOH))
      RATE=(RATE/(1.D0+KA*ACTIV(K_ETOH))**3.d0)/1.d3
      RATES(K_IC4)=-RATE
      RATES(K_ETOH)=-RATE
      RATES(K_ETBE)=RATE
      RATES(K_NC4)=0.D+00
      IF(IDBG.GE.1)THEN
        WRITE(MAXWRT_MAXBUF(1),9030) NSTAGE,RATE,RATNET
        CALL DMS_WRTTRM(1)
      ENDIF
      RETURN
#undef P_MAX3
      END
```

Figure 9.19 *(Continued)*.

99.1 mol%. The conversion of the isobutene fed is

$$\text{conversion} = \frac{\text{ETBE in bottoms}}{\text{isobutene in C4 feed}} = \frac{(706\ \text{kmol/h})(0.991)}{(1767\ \text{kmol/h})(0.40)} = 0.99 \qquad (9.9)$$

9.2.3 User Subroutine for ETBE

The kinetics given above were used by Bobby Hung to develop a kinetic subroutine RAETBE.f, and this was run in Aspen Plus. Figure 9.19 lists the Fortran subroutine.

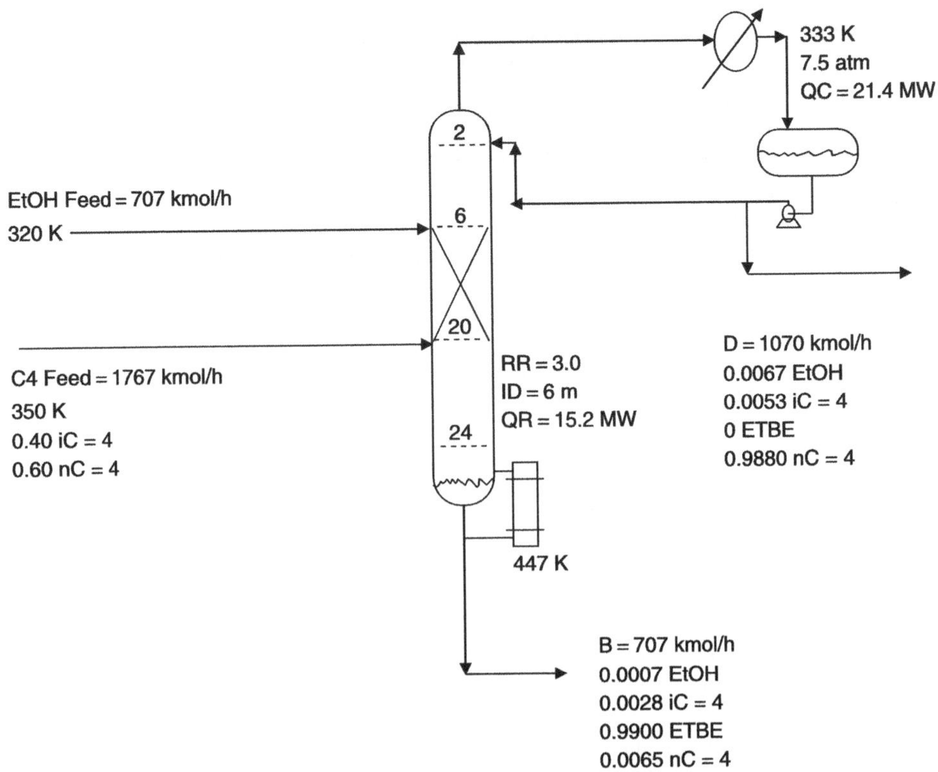

Figure 9.20 Aspen Plus steady-state conditions for ETBE.

Figure 9.20 provides the flowsheet results from the Aspen Plus simulation, which are in good agreement with the results of Al-Arfaj and Luyben.[3] Figures 9.21 and 9.22 give composition and temperature profiles.

There is a peak isobutene concentration of 22.5 mol% at stage 20. This is the location where the C4 fresh feedstream is introduced. The reaction zone (stages 6–20) has a fairly high concentration of ethanol in the region between stage 16 and stage 6.

9.2.4 Chemical Equilibrium Model

Because it is reported that the ETBE reaction is equilibrium limited, not kinetically limited, we might expect that the chemical equilibrium option available in Aspen Plus could be used to look at the steady state. It cannot be used in Aspen Dynamics, but at least some useful steady-state design information might be obtained. In Section 9.19.1 a chemical equilibrium model was explored for the MTBE system and found to give a design requiring less energy than that needed when using the kinetic model. These results are what we would expect.

The equilibrium model is easy to generate, as discussed in Section 9.1.6. The reaction is changed from *kinetic* to *equilibrium* and the equilibrium constant is calculated from Gibbs free energies.

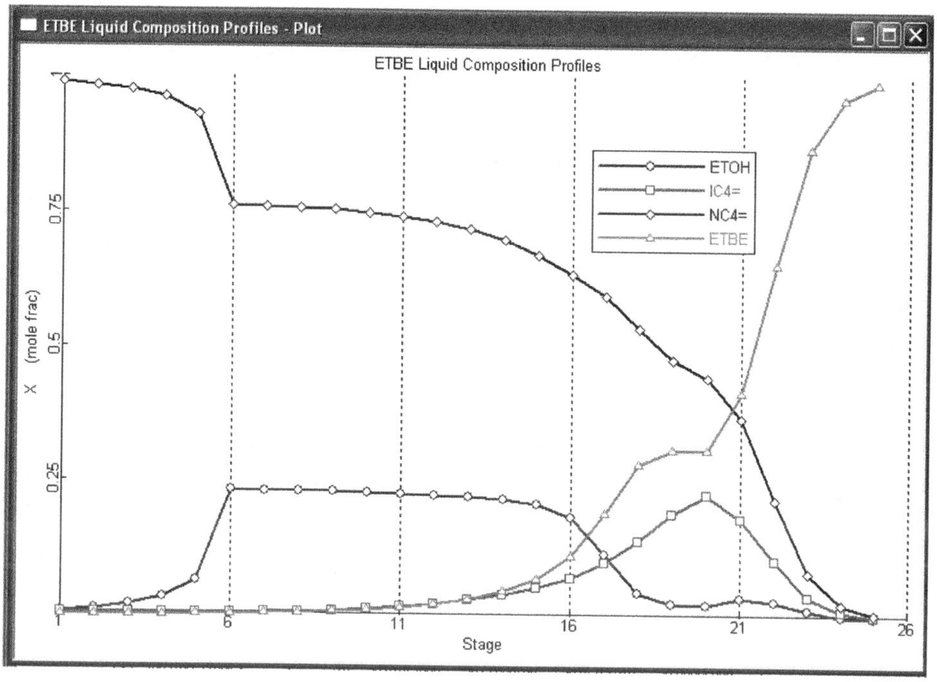

Figure 9.21 Aspen Plus composition profiles for ETBE.

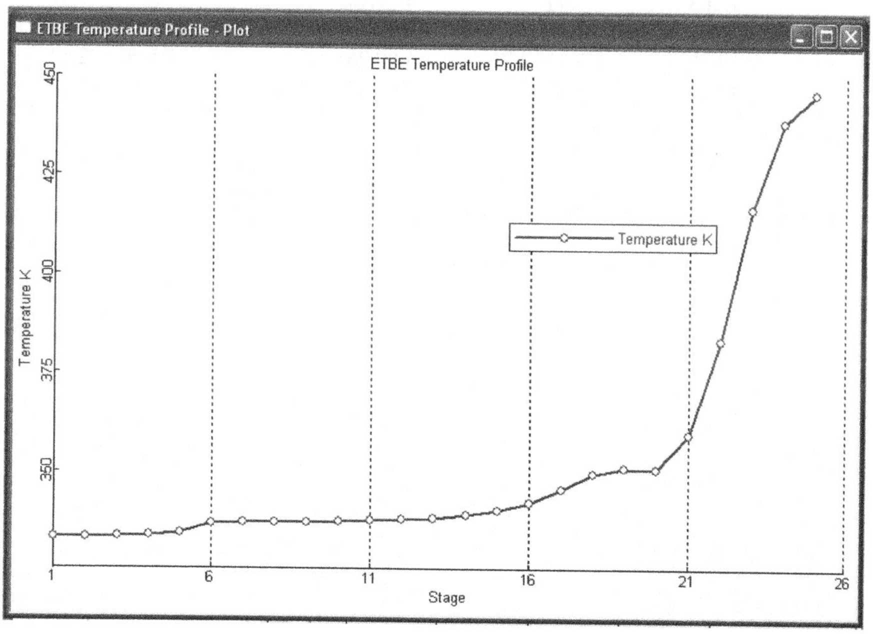

Figure 9.22 Aspen Plus temperature profile for ETBE.

Thus, the same idea was applied to the ETBE simulation. However, the results were quite unexpected. The conversion dropped to less than 50%, and the concentrations of both reactants in the entire reaction zone were quite high. We are at a loss to explain these results.

9.2.5 Effects of Design Parameters

Attempts were made to explore the impact of several parameters on the design of the ETBE system. Convergence issues and frequent Fortran system errors severely limited this investigation.

Two design specifications were set up to hold the bottoms purity at 99 mol% ETBE and the distillate ethanol impurity at 0.7 mol% by manipulating the bottoms flowrate and reflux ratio.

Pressure. The base case design pressure is 7.5 atm. Table 9.4 gives results over a small range of pressures around this value. Decreasing pressure reduces energy consumption, but the production of ETBE decreases somewhat because of more losses of reactant in the distillate. The program would not converge for pressures lower than 7.2 atm.

TABLE 9.4 Pressure Results for ETBE

Pressure (atm)	Reboiler Heat Input (MW)	Reflux Ratio	ETBE Recovered in Bottoms (kmol/h)	Isobutene Lost in Distillate and Bottoms (kmol/h)
7.2	12.46	2.474	698.09	7.00 + 1.91
7.3	13.26	2.622	698.69	6.25 + 2.06
7.4	14.12	2.770	698.95	5.94 + 2.11
7.5	**15.03**	**2.947**	**699.10**	**5.77 + 2.13**
7.6	16.01	3.129	699.19	5.67 + 2.14
7.7	17.07	3.324	699.26	5.60 + 2.14
7.8	18.20	3.536	699.30	5.57 + 2.13

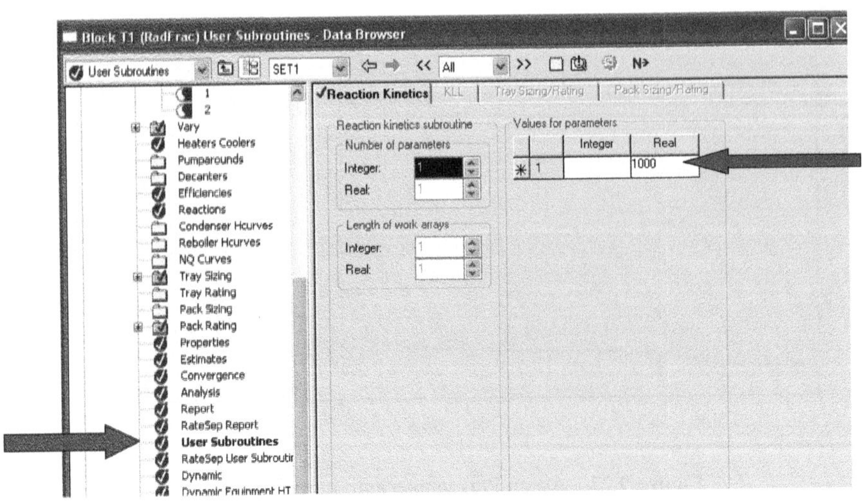

Figure 9.23 Changing holdup on reactive trays.

Number of Stripping Trays. The base case has 4 stripping trays and requires a reflux ratio of 2.947. The program would not converge when fewer stripping trays were used. Increasing the number over the range from 4 to 12 resulted in a decrease in the reflux ratio to 2.885. Thus, the number of stripping trays has little effect as long as there is more than the minimum.

Fresh Feed Location for C4. In the base case the C4 feed is introduced at the bottom of the reactive zone on stage 20. Feeding this stream higher in the column reduced the reflux ratio only slightly. The reflux ratio is 2.947 at stage 20 and 2.900 at stage 15. The program would not converge for higher locations.

Other Parameters. We were unsuccessful in trying to change the reactive tray holdup (Fig. 9.23), the number of reactive trays, the number of stripping trays, or the location of the ethanol fresh feed. The program either would not converge or produced Fortran errors and shut down.

9.3 CONCLUSION

The design of two real reactive distillation systems was explored. Both the MTBE and ETBE systems are basically ternary systems with inerts. The control of these two systems will be studied in Chapter 15.

PART IV

CONTROL OF IDEAL SYSTEMS

In Parts I and II we explored the steady-state designs of several ideal hypothetical systems. The following three chapters examine the control of these systems. Chapter 10 considers the four-component quaternary system with the reaction A + B $\Leftrightarrow$ C + D under conditions of neat operation. Chapter 11 looks at control of two-column flowsheets when an excess of one of the reactants is used. Chapter 12 studies the ternary system A + B $\Leftrightarrow$ C, with and without inerts, and the ternary system A $\Leftrightarrow$ B + C. We will illustrate that the chemistry and resulting process structure have important effects on the control structure needed for effective control of reactive distillation columns.

CHAPTER 10

CONTROL OF QUATERNARY REACTIVE DISTILLATION COLUMNS

Several different control structures have been proposed for reactive distillation columns. The appropriate control structure depends on the flowsheet and on the type of reactions occurring in the column. If two reactants are involved and if it is desirable to operate the process without any excess of reactant, it is necessary to manage the fresh feedstreams so that the stoichiometry is exactly balanced. A composition analyzer that measures an internal composition in the column is sometimes required. However, if two products are produced, it may be possible to avoid the use of an analyzer by using two temperatures in the column to adjust the two feedstreams. This type of structure was proposed by Roat et al.[1] for the ideal reaction $A + B \Leftrightarrow C + D$ in one of the earliest articles dealing with reactive distillation control. We call this control structure the "Eastman structure."

In this chapter we explore the effectiveness of this two-temperature control structure for various column designs (number of reactive stages) to quantify the impact of design on controllability. We also discuss the issues of the selection of the trays whose temperatures are to be controlled and the tuning of the two interacting temperature controllers. Disturbances in production rate and fresh feed compositions are made to examine the rangeability of this control structure. One of the main conclusions is that the locations of the temperature control trays should be made such that the two temperature controllers both have direct action (an increase in temperature increases feed), which requires negative open-loop process gains for both loops.

[1]S. Roat, J. Downs, E. Vogel, and J. Doss, Integration of rigorous dynamic modeling and control system synthesis for distillation columns, In J. D. Birdwell and J. R. Cockette, Editors, *Chemical Process Control—CPC III*, Elsevier, New York, 1986.

The material in this chapter is based on an article by Kaymak and Luyben.[2]

10.1 INTRODUCTION

Several types of controllers have been studied in the reactive distillation literature, ranging from simple proportional-integral (PI) controllers to complex nonlinear model predictive controllers. Even if we limit ourselves to PI controllers, a variety of alternative control structures has been studied.

In addition, a variety of chemical systems and a number of types of flowsheets have been investigated. For example, some of the common types of reactions are the following:

1. Reactions with two reactants and two products
2. Reactions with two reactants and one product
3. Reactions with one reactant and two products

The flowsheet for the third type of chemistry is usually a single reactive column with reactant fed in and the two products removed in the distillate and bottoms streams. The control of this type of system is usually quite simple because the feed is a single reactant. Unlike the two-reactant chemistry, there is no need to precisely adjust the two feedstreams to balance the stoichiometry. This system is considered in Chapter 12.

However, the flowsheets for the first and second types of chemistry come in two flavors. The flowsheet can consist of either a single reactive column or a two-column system with a reactive distillation column followed by a recovery column and recycle of excess reactant back to the reactive column. The type of flowsheet depends on whether we want to operate the reactive distillation column neat (i.e., no excess of either reactant). The "excess reactant" flowsheet has higher capital and operating costs, but its control is easier. The control of this system is considered in Chapter 11.

The "neat" flowsheet is more difficult to control because the two reactants must be fed in exactly the correct amounts to satisfy the stoichiometry *down to the last molecule*. This places a heavy demand on the control structure and requires some type of feedback of information from within the process to indicate the accumulation or depletion of at least one of the reactants. In some chemical systems the only way this can be achieved is by the use of a composition analyzer.

Al-Arfaj[3] studied a number of types of reactions and a number of control structures. An empirical finding from the many case studies is that a composition analyzer may not be required in the two-reactant case if there are *two* products produced.

If only *one* product is produced, the temperature profile does not display enough information to enable the use of inferential control for component inventory control purposes. There is only one product leaving the column, and information can be extracted from a limited region where the concentration of this product becomes large. Industrial examples of this type of system are MTBE, ETBE, and TAME.

[2] D. B. Kaymak and W. L. Luyben, Evaluation of a two-temperature control structure for a two-reactant/two-product type of reactive distillation column, *Chem. Eng. Sci.* **61**, 4432–4450 (2006).

[3] M. A. Al-Arfaj, Control of reactive distillation systems, PhD Thesis, Lehigh University, 2002.

However, if *two* products are produced, the temperature information may be rich enough to infer compositions with sufficient accuracy so that direct composition measurement is not absolutely necessary for neat operation. The ideal case considered by Roat et al.[1] (A + B ⇔ C + D) and the production of methyl acetate (methanol + acetic acid ⇔ methyl acetate + water) are examples of this type of chemistry. In this chapter, we consider the control of the ideal quaternary system. In Chapter 13 the control of the methyl acetate system and other similar real systems are explored.

To summarize, the process under study is a reactive distillation column, which is operated without an excess of one of the reactants, in which there are two reactant feedstreams and two products are produced.

10.2 STEADY-STATE DESIGN

The numerical case that was studied assumes that the relative volatilities are constant at 2, the product purities are 95 mol%, and the $(K_{EQ})_{366}$ is 2. Figure 10.1 gives the steady-state flowsheet considered for these dynamic control studies.

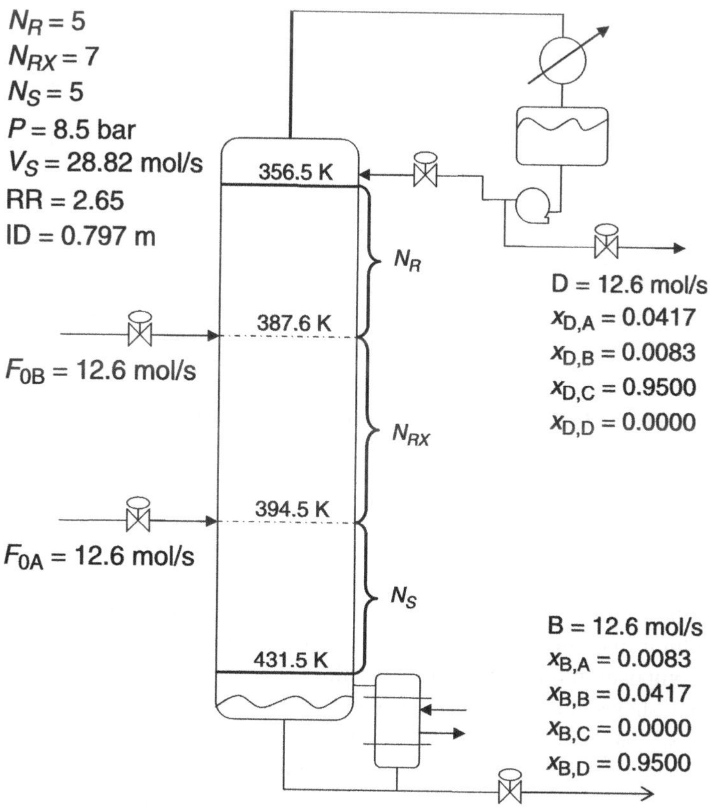

Figure 10.1 Steady-state flowsheet.

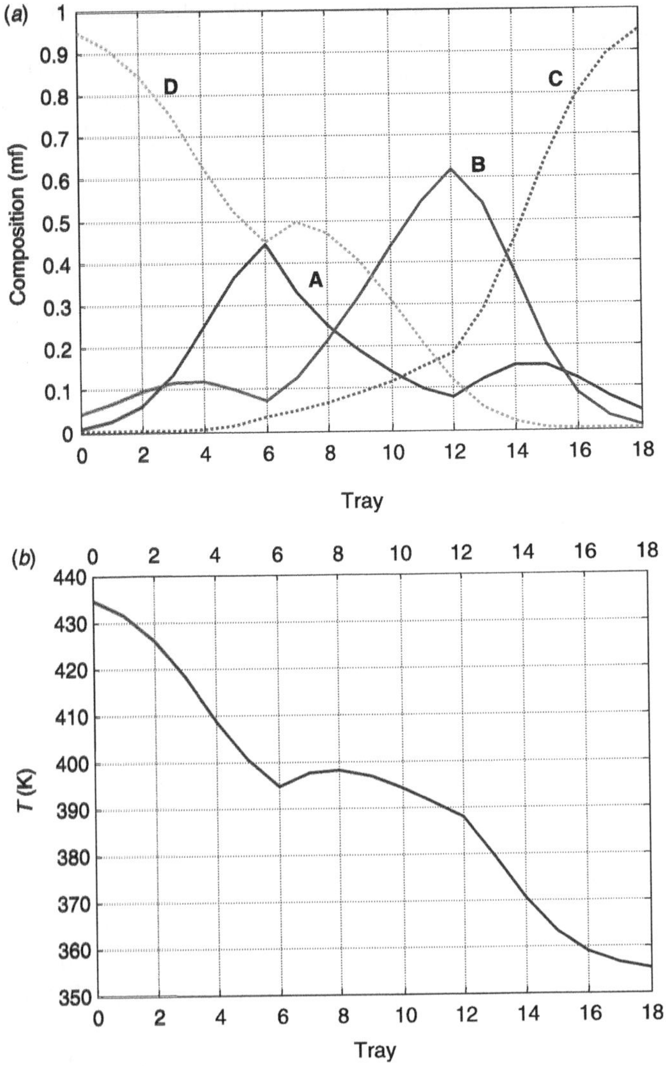

Figure 10.2 Steady-state composition and temperature profiles.

The column has a total of 17 trays: 5 stripping trays, 7 reactive trays, and 5 rectifying trays. The column pressure is 8.5 bar, and the reflux ratio is 2.65. Figure 10.2 gives the steady-state composition and temperature profiles.

Figure 10.3 shows the effects of the number of reactive trays on the TAC for the pressure $P = 8.5$ bar with the number of separation stages $N_R = N_S = 5$. The optimum number of reactive trays balances capital and energy costs. Designs that incorporate more reactive trays suffer an economic penalty. However, as demonstrated later in this chapter, these designs may result in greatly improved dynamic controllability.

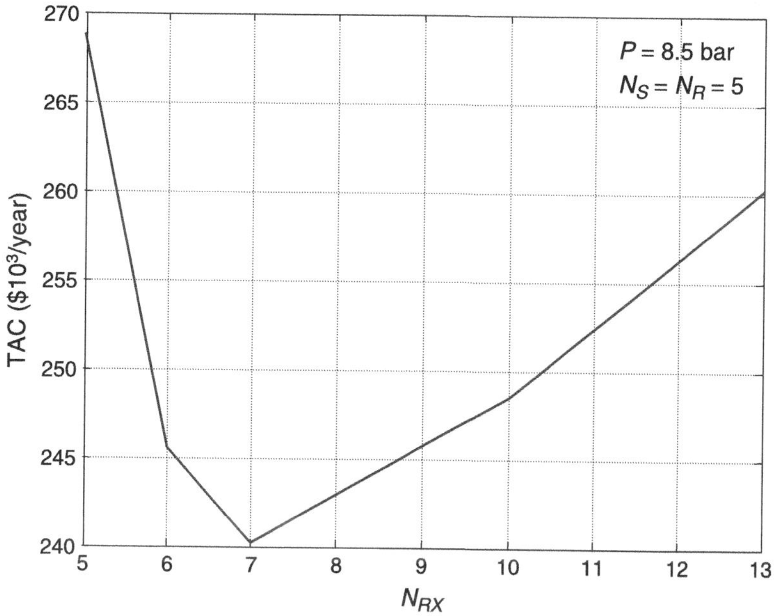

Figure 10.3 Effect of number of reactive trays on TAC.

10.3 CONTROL STRUCTURES

Because composition analyzers are expensive, require high maintenance, and introduce deadtime into the control loop, it is desirable to use inferential temperature measurements instead of direct composition measurements whenever possible.

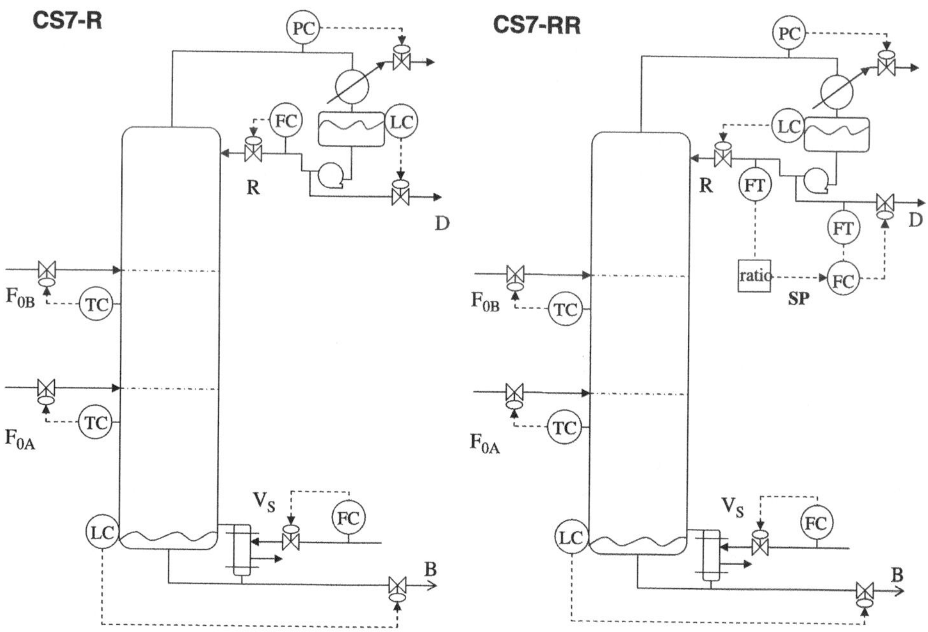

Figure 10.4 Two alternatives for Eastman control structure.

In the Eastman control structure, two PI temperature controllers are used to maintain two tray temperatures in the column by manipulating the two fresh feedstreams. Production rate changes are achieved by changing the vapor boilup. The base level is controlled by manipulating the bottoms flowrate.

There are two alternatives for this control structure: a constant reflux flowrate and a constant reflux ratio. The left system in Figure 10.4 shows the control structure (CS7-R) where the reflux flowrate is fixed and the reflux drum level is controlled by manipulating the distillate flowrate. The right system in Figure 10.4 shows the alternative version (CS7-RR) where the reflux drum level is controlled by the reflux flowrate and the distillate flowrate is adjusted to give a constant RR.

Both control structures are single-input, single-output structures with PI controllers for temperature control and P controllers for level control. The relay-feedback method is used to obtain the ultimate gains and ultimate periods. The valves are designed to be half open at steady state. Two 60-s first-order measurement lags are used in both temperature loops. The temperature controllers are tuned using the Tyreus–Luyben tuning method. The column pressure is assumed to be constant, which is achieved by manipulating the condenser heat duty.

10.4 SELECTION OF CONTROL TRAY LOCATION

The selection of the trays on which to control temperature is the main issue in this structure. From the rigorous steady-state simulations, the gain matrix between the inputs (two fresh feed flowrates) and the outputs (the temperatures on all trays) is calculated numerically. The procedure will be discussed in detail. Using the steady-state gains, the singular value decomposition (SVD) method is used to indicate the most sensitive trays to be controlled. Some engineering judgment must be exercised in this selection. The steady-state gains between the tray temperature and the two feeds can be positive in some sections of the column and negative in others. Thus, the action of the two temperature controllers (direct or reverse acting) can be different, depending on what trays are selected.

One of the principles we develop in this chapter is that the trays should be selected so that both controllers have direct action (an increase in temperature produces an increase in feed flowrate). This means that the controller gains must be negative, which requires that the open-loop process steady-state gains between tray temperatures and feed flowrates should be negative.

Figure 10.5 gives values for steady-state gains (K_{F0A} and K_{F0B}) between tray temperatures and feed flowrates for both the CS7-R and CS7-RR control structures. Note that these gains are not normalized and are in units of Kelvin per (mole per second).

The steady-state gains should correspond to a linearized version of the process. These are the gains that should be used for controller design because the closed-loop system will be held near this steady-state condition by the controllers. Determining the gains numerically requires that very small changes should be made in the inputs. The procedure is to calculate the steady-state gains for decreasing magnitudes of input changes. When the gains stop changing as the increment is decreased, we can assume that the true linear gains have been found. The increment size had to be reduced to $\sim$0.005%.

Figure 10.5 provides the steady-state gains and their related SVD results for both the CS7-R structure (dashed lines) and the CS7-RR structure (solid lines). Note that the steady-state gains between tray temperatures and the F_{0A} feed are always negative (an increase in feed flowrate decreases tray temperature) for both structures and for all trays. The steady-state gains between tray temperatures and the F_{0B} feed are positive for all trays with the CS7-R structure. However, there is a small region with the CS7-RR structure where the gains have small negative values.

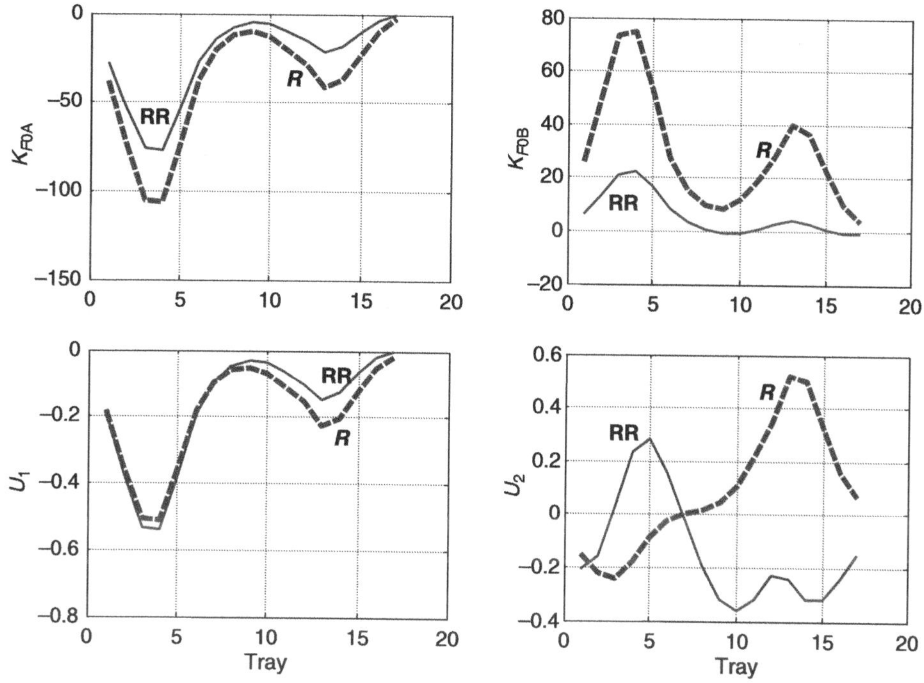

Figure 10.5 Gains and SVD analysis.

These occur between trays 8 and 10. Remember that these results are for the 5/7/5 design ($N_S = 5$, $N_{RX} = 7$, and $N_R = 5$). The effect of changing the number of reactive trays on these steady-state gains is discussed in a later section.

Looking at the U_1 parameter in Figure 10.5, the SVD analysis suggests that tray 4 is the most sensitive temperature measurement for both control structures. The temperature of this tray could be controlled by manipulating the fresh feedstream F_{0A}, which is fed at the bottom of reactive zone. Looking at the U_2 parameter in Figure 10.5, different trays should be selected for the two control structures. For the CS7-R structure, tray 13 is the most sensitive and it is tray 10 for the CS7-RR structure.

There is also a difference in the actions of the control loops of these control structures. The two temperature control loops of the CS7-R structure have opposite actions. The tray 4 temperature controller will be direct acting because the T_4/F_{0A} steady-state gain is negative, but the tray 13 temperature controller will be reverse acting because the T_{13}/F_{0B} steady-state gain is positive. However, the T_{10}/F_{0B} temperature controller in the CS7-RR structure has a negative steady-state gain on tray 10, so both temperature controllers in this structure will have direct action.

10.5 CLOSED-LOOP PERFORMANCE

10.5.1 CS7-R Structure

Figure 10.6 shows the response of the CS7-R structure for $\pm 1\%$ changes in the vapor boilup. The temperatures go back to their setpoints for both positive and negative disturbances. However, even these small disturbances result in large changes in the purities of the product streams $x_{D,C}$ and $x_{B,D}$.

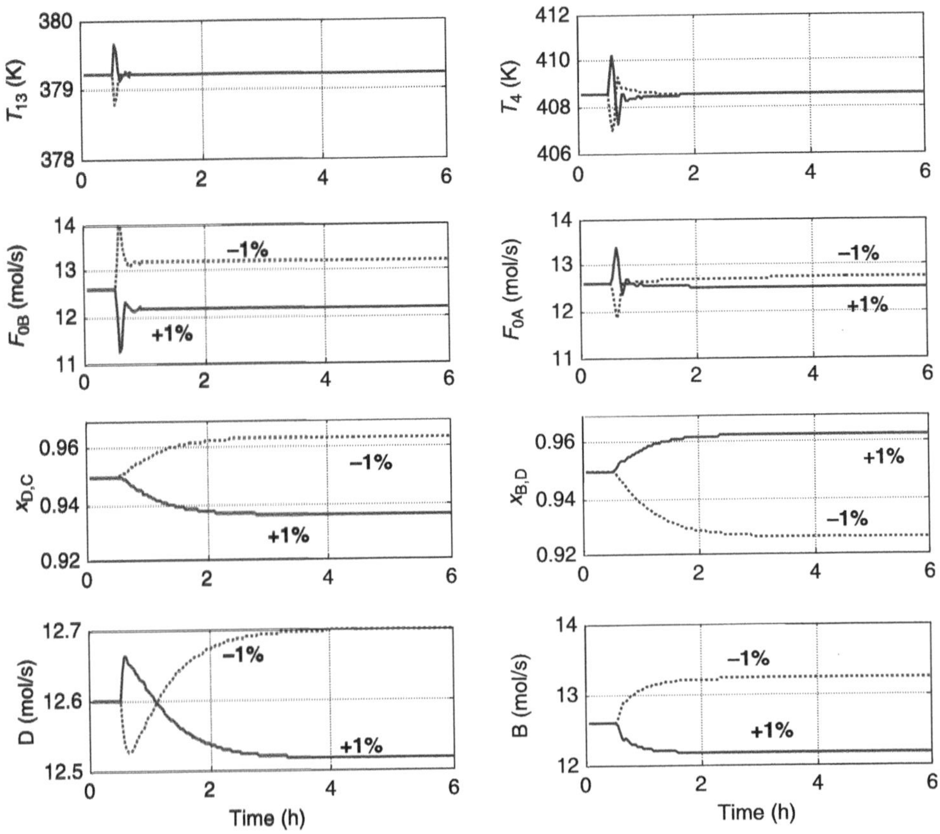

Figure 10.6 Control performance for CS7-R.

The final effects of increasing vapor boilup are decreases in both manipulated variables (feed flowrates) and product streams. Because the two temperature controllers in the CS7-R structure have opposite actions, the initial responses of the two controllers are different. An increase in vapor boilup produces an increase in the temperature on both control trays (trays 4 and 13). The tray 4 temperature controller has direct action, so it increases the F_{0A} feed flowrate. However, the tray 13 temperature controller has reverse action, so it decreases the F_{0B} feed flowrate. Not enough reactant B is added in a timely enough manner to prevent light reactant A from going to the upper section of the column. Once this breakthrough occurs, the column moves to a different operating condition in which the impurity of A in the distillate stream increases and product purities are lower.

We can conclude that the CS7-R control structure provides good dynamically stable control, but it does not maintain conversion and product purities.

10.5.2 CS7-RR Structure

Figure 10.7 shows the responses of the CS7-RR structure for +1% and +5% changes in vapor boilup. The responses are very oscillatory because of the very small steady-state gain of tray 10, which is the tray recommended by SVD calculations.

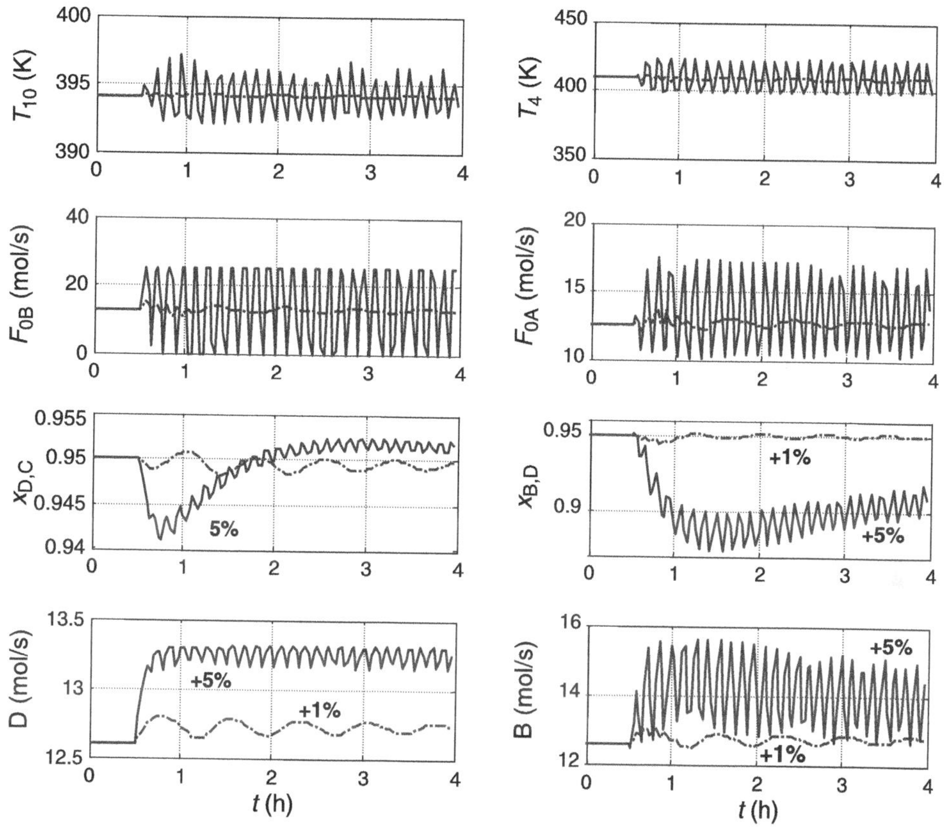

Figure 10.7 Control performance for CS7-RR.

However, note that the product purities remain fairly close to the desired 95% level, even for the +5% change in vapor boilup. This suggests that the CS7-RR structure should perform better than the CS7-R structure if the dynamics can be improved. The next section demonstrates that this can be achieved by using a suboptimal steady-state design that has more reactive trays.

10.6 USING MORE REACTIVE TRAYS

The economical optimum number of reactive trays is seven in this system. However, the results demonstrate that this column has poor dynamics performance. Adding more reactive trays increases capital cost, but can it improve dynamic controllability? We now demonstrate that it can indeed make control better. This is a good example of the ever present tradeoff between design and control that occurs in chemical processes.

10.6.1 Steady-State Design

Table 10.1 provides three designs using 7, 10, and 13 reactive trays. The pressure and the number of stripping and rectifying trays are held constant. Note that the 5/7/5 design has the lowest vapor boilup. This is why it is the optimum economic design.

TABLE 10.1 Steady-State Results of Different Design Cases (Ideal Case)

	7	N_{RX} 10	13
Design parameters			
N_S	5	5	5
N_{RX}	7	10	13
N_R	5	5	5
P (bar)	8.50	8.50	8.50
V_S (mol/s)	28.82	29.34	30.48
R (mol/s)	33.45	33.97	35.12
ID (m)	0.797	0.802	0.812
Design temperatures (K)			
Base	434.7	433.4	432.3
Reactive zone bottom	434.5	392.2	391.5
Reactive zone top	387.6	391.8	392.9
Reflux drum	355.2	355.3	355.4

Because the dynamics of the 5/10/5 and 5/13/5 designs are similar, only the results of the case with 10 reactive trays will be discussed. We found that adding more reactive trays did not solve the problem of a drop in product purities for the CS7-R structure. Thus, we consider only the CS7-RR structure.

10.6.2 SVD Analysis

Figure 10.8 shows how the steady-state gains and the SVD results change as more reactive trays are added. The curves for the different cases look similar superficially, but closer inspection reveals important differences. Although there is a small shift in the location of the most sensitive tray for fresh feedstream F_{0A} with an increase in the number of reactive trays, it stays in the stripping section for all three design cases. The magnitude of steady-state gains for input F_{0A} (K_{F0A}) decreases, but this can be compensated for by larger controller gains.

However, the differences between optimum (5/7/5) and suboptimum (5/10/5 and 5/13/5) designs regarding K_{F0B} and U_2 are more important. The trays in the rectifying section and upper part of the reactive zone have negative steady-state gains when more trays are added to the reactive zone. Their magnitudes are bigger compared to the very small negative gain of the optimum design ($N_{RX} = 7$) at tray 10. Another difference is that the most sensitive trays from the SVD analysis shift to the rectifying section with an increase in the number of reactive trays. Note that there are also other sensitive regions (second negative peaks that are smaller) in reactive zones for the 5/10/5 and 5/13/5 designs.

The steady-state gains and the SVD analysis of the 5/10/5 case suggest using tray 18 in the rectifying section by manipulating fresh feed flowrate F_{0B} and the temperature of tray 2 in the stripping section is controlled by manipulating fresh feed flowrate F_{0A}. Figure 10.9 gives relay-feedback test results for both loops using these trays. As given in Table 10.2, the controller gain and reset times calculated from the relay-feedback test data have reasonable values for tray 2. However, the controller for tray 18 has a reset time of 347 min. This occurs because the T_{18}/F_{0B} open-loop transfer function has an inverse response. The shape of the tray 18 temperature response shown in Figure 10.9 (shark's tooth appearance) is characteristic of a process with an inverse response. This

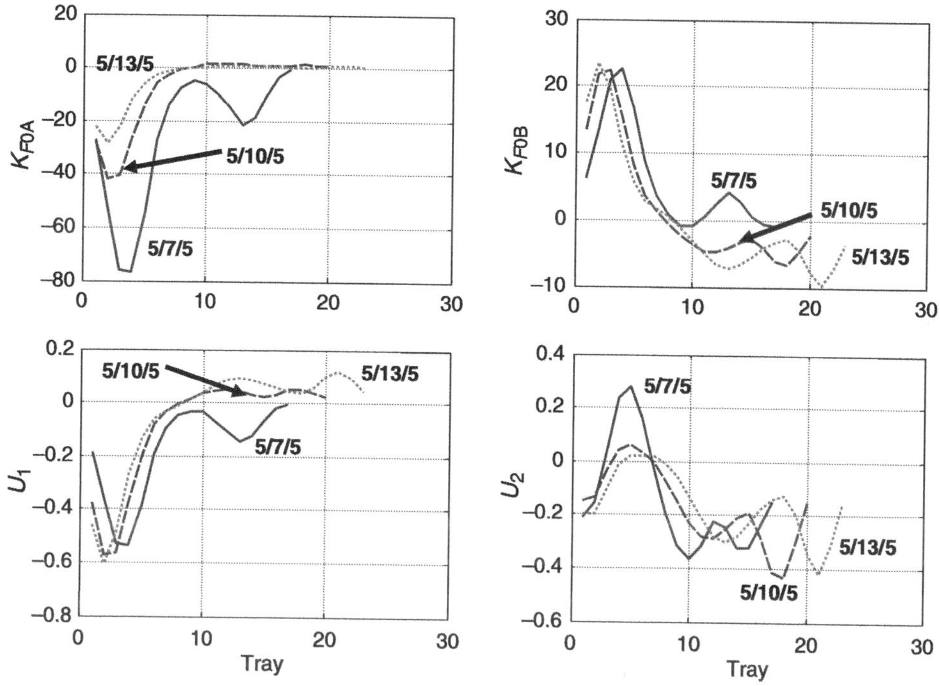

Figure 10.8 Gains and SVD analysis for other values of N_{RX}.

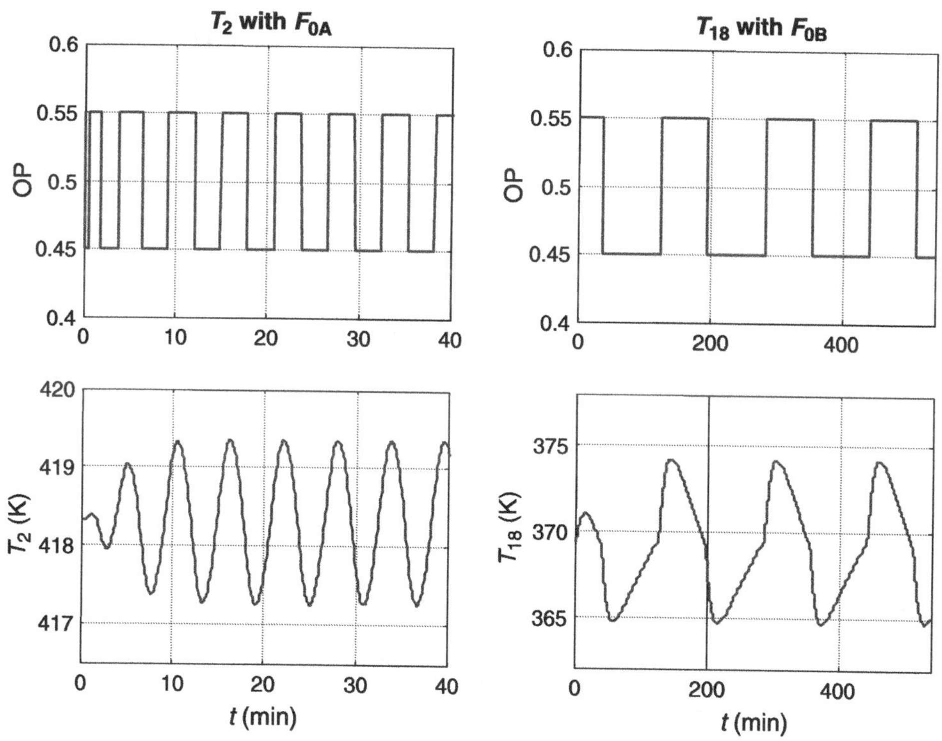

Figure 10.9 Relay-feedback tests for T_2 and T_{18}.

TABLE 10.2 Tuning Parameters for CS7-RR

Design	Pairing	K_C	τ_I (min)
5/7/5	$F_{OA}-T_4$	0.86	12.83
	$F_{OB}-T_{10}$	40.23	8.43
5/10/5	$F_{OA}-T_2$	0.95	12.83
	$F_{OB}-T_{12}$	8.78	16.87
	$F_{OB}-T_{18}$	0.20	347.60
5/13/5	$F_{OA}-T_2$	1.00	12.10
	$F_{OB}-T_{13}$	7.46	16.87
	$F_{OB}-T_{21}$	0.21	371.07

Controller gains are dimensionless using temperature transmitter spans of 50 K and valve sizes twice the steady-state flowrates.

was confirmed by a step test. Therefore, the T_{18}/F_{OB} pairing is not a good choice for the controllability of the system.

As mentioned earlier, there is another region in the 5/10/5 column that is sensitive to the changes in fresh feed flowrate F_{OB}. This is located at tray 12 in the reactive zone where the steady-state gain (Fig. 10.8) is fairly large. Figure 10.10 shows the relay-feedback test and step test results for tray 12. The ultimate gain and ultimate period from the relay-feedback test look reasonable (Table 10.2), and the step test shows no inverse response behavior for

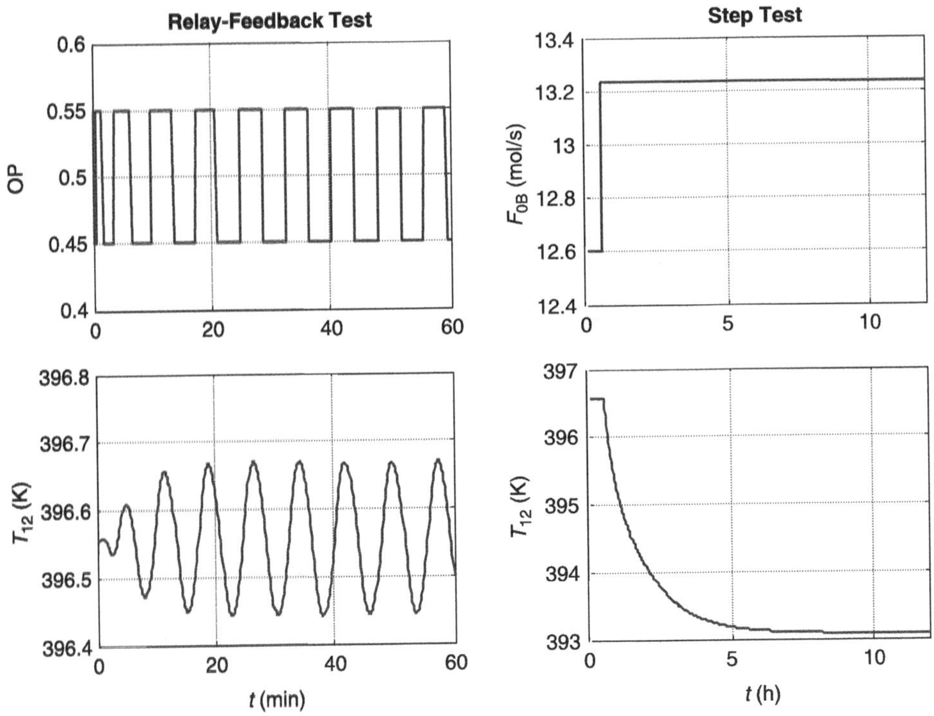

Figure 10.10 Relay-feedback test and step test for T_{12}.

T_{12}/F_{0B} pairing. Thus, the closed-loop dynamics of the system should be improved by using tray 12 instead of tray 18 in the 5/10/5 design.

10.6.3 Dynamic Performance of CS7-RR

Figure 10.11 shows the effectiveness of this modified CS7-RR structure. The disturbances are 10% increases and decreases in vapor boilup. The system is quite stable and achieves a new steady state in about 2 h. Product purities remain close to 95%.

There are two reasons for the improvement in control. The first has already been discussed: the elimination of inverse response. The second is equally important: the two temperature controllers in the CS7-RR structure have the same action (direct). This means that when both control loops see a positive vapor boilup disturbance, which increases both tray temperatures, the two controllers will increase both fresh feeds. This helps to maintain the delicate stoichiometric balance between the reactants that is essential for neat operation of a reactive distillation column. Because a reactive distillation column acts like a pure integrator with respect to the reactants, this similar initial response is very important for CS7-RR, where feedstreams are used as manipulated variables.

Conversely, a negative change in V_S causes a decrease in both temperatures, which results in an initial decrease in both fresh feed flowrates F_{0A} and F_{0B}. This keeps the stoichiometric ratio close to the original value and prevents a breakthrough of reactants. Thus,

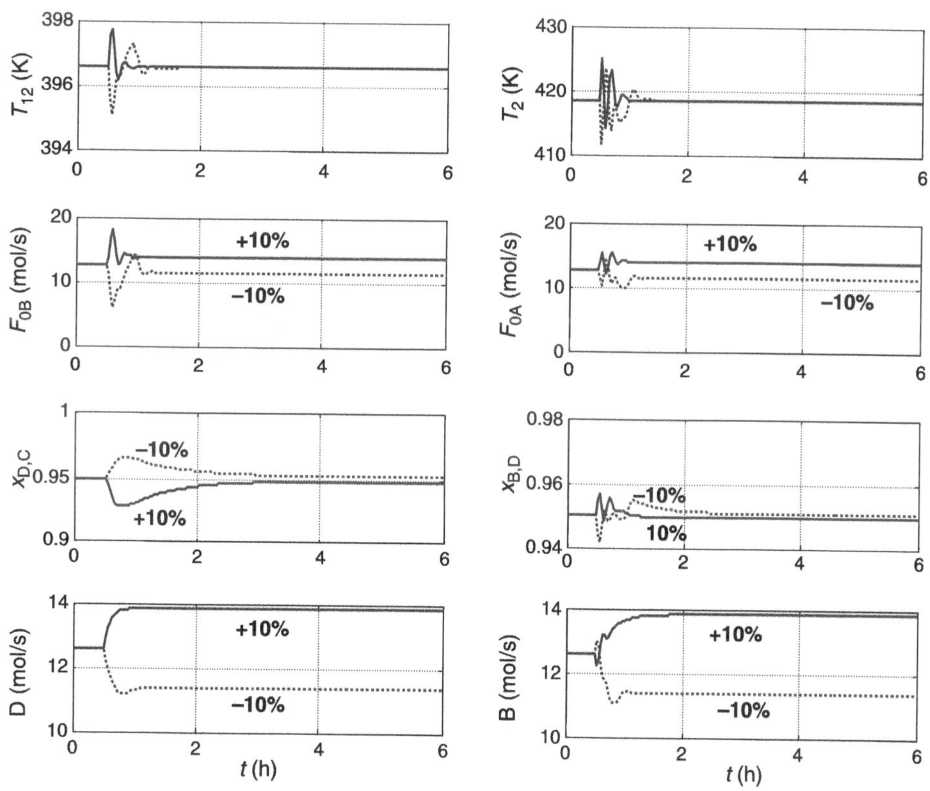

Figure 10.11 Performance using T_2 and T_{12}.

the system ends up at lower feed and product stream flowrates but with desired product purities.

10.7 INCREASING HOLDUP ON REACTIVE TRAYS

Thus far in this chapter, the holdup per tray is kept constant at 1000 mol when the number of reactive trays is increased in an attempt to improve the dynamics of the CS7-RR structure. This means that the suboptimal design (5/10/5) has a total of 10,000 mol of reactive holdup and the optimal design (5/7/5) has a total of 7000 mol. The question that arises is what if the total holdup of the optimal design is kept the same and just distributed over a larger number of trays?

To see what effect this has, the molar holdup of the 5/10/5 case is decreased from 1000 to 700 mol/tray. Figure 10.12 shows the steady-state gains (K_{F0A} and K_{F0B}) between tray temperatures and feed flowrates for these two cases. The upper plot shows the case where the suboptimal design has 1000 mol/tray. The lower plot shows the case where the total holdup is kept the same as the optimum case, so each tray has 700 mol. The shapes of the steady-state gain curves of these two cases are similar in terms of signs and magnitudes

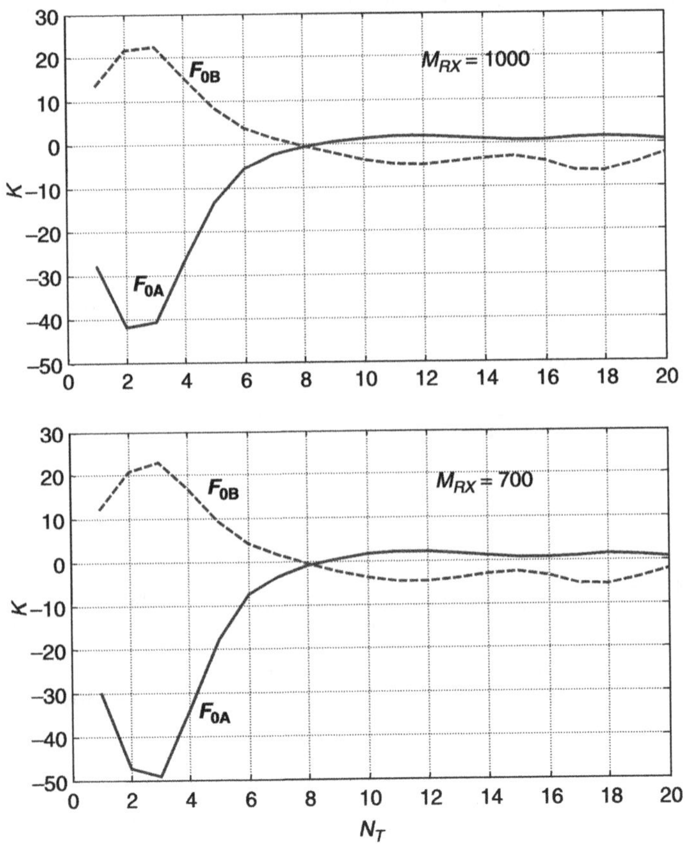

Figure 10.12 Gains for $N_{RX} = 10$.

for any region of the column. These cases also have similar SVD results and controller parameter (K_C and τ_I) values. Therefore, these two cases show similar control behavior (robustness and rangeability) with just slight differences, even though they have different amounts of reactive holdup. Thus, these results indicate that the controllability problems are truly caused by the number of reactive trays instead of the total amount of holdup.

To support this conclusion, the optimum design case (5/7/5) with larger holdup per tray was also studied. Figure 10.13 shows the steady-state gains (K_{F0A} and K_{F0B}) for two cases. The upper plot shows the original optimal design where each tray has 1000 mol of holdup. The lower plot shows the case where the holdup per tray is increased to 1500 mol. The shapes of the steady-state gain curves for these two cases are similar in terms of signs and magnitudes for any region of the column. The most sensitive tray available to pair with F_{0B} is tray 10 with small negative steady-state gains for both cases. Thus, they give similar results that display poor dynamic controllability. For example, both cases have the same type of oscillatory responses for +5% changes in production rates. Both processes also shut down with a −3% change in production rate. Thus, these results confirm that the number of reactive trays affects the controllability more than modest amounts of additional holdup. Very large increases in holdup may improve controllability but may not be physically possible because of hydraulic constraints.

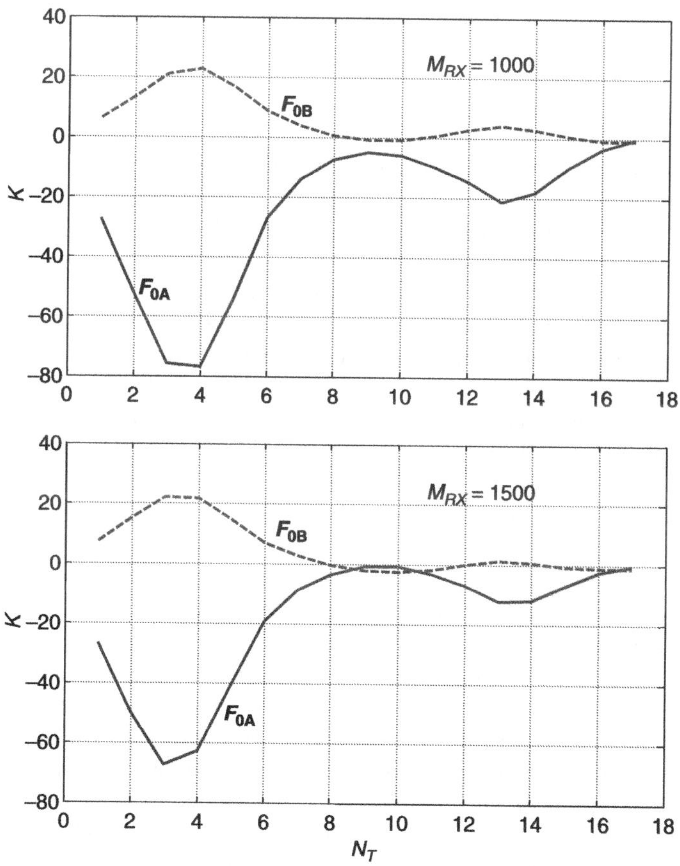

Figure 10.13 Gains for $N_{RX} = 7$.

10.8 RANGEABILITY

It is important to evaluate the region of disturbances for which the control system is able to provide stable effective control and maintains reasonable conversions and product purities. We call this property the rangeability of the control structure.

The first type of disturbance studied is a range of plus or minus step changes in vapor boilup (the production rate handle). Results show that very large positive changes of up to +50% can be handled by the CS7-RR control structure. The purities of both products are maintained within 1% of the desired 95% specification.

However, the robustness of the system for decreases in vapor boilup is more limited. When vapor boilup decreases, tray temperatures throughout the column decrease. Both temperature controllers decrease the fresh feeds. A negative 10% change can be handled, but a 12% decrease shuts off the F_{0B} stream completely and the system shuts down after a short time.

The second type of disturbance is a step change in the composition of the fresh feeds. For a change in composition of fresh feed F_{0A} from pure component A to a mixture of components A and B, the system is dynamically stable, but disturbances larger than 4% of B in this feedstream result in a decrease in bottoms purity below 94%. The distillate purity is above specification.

Because component B is heavier than component A, when putting some B in the F_{0A} feedstream, the temperatures throughout the column increase initially. The temperature controllers increase both fresh feeds because they both have direct action. This introduces even more reactant B into the system. The final result is a higher concentration of reactant B in the bottoms (for the same two tray temperatures and the same vapor boilup).

For a change in the composition of fresh feed F_{0B} from pure component B to a mixture of components A and B, the system is dynamically stable for changes up to 8% A in this feedstream, but the system shuts down for 10% composition changes. In the stable region both the distillate and bottoms purities are above specification.

The details for the 5/10/5 case were presented, and the results are the same qualitatively for the 5/13/5 case. Similar procedures are applied to design the control structure of the 5/13/5 case. The SVD analysis suggests the use of tray 21 paired with fresh feed flowrate F_{0B} and tray 2 paired with fresh feed flowrate F_{0A}. Controller gains and reset times calculated from the relay-feedback test data give reasonable results for tray 2. However, as a step test demonstrated, tray 21 has an inverse response. Therefore, this T_{21}/F_{0B} pairing is not a good choice for the controllability of the system. Similar to the 5/10/5 case, there is another region in the column that is sensitive to changes in fresh feed flowrate F_{0B}: tray 13 in the reactive zone. The ultimate gain and ultimate period from the relay-feedback test look reasonable for this tray, and a step test shows that there is no inverse response behavior for the T_{13}/F_{0B} pairing. Table 10.2 gives the controller parameters of control structure CS7-RR for all design cases.

Figure 10.14 shows the operability regions for designs using control structure CS7-RR with 10 and 13 reactive trays. The borders shown in the figure are set by either system shutdowns or a minimum purity of either product that drops below 94%. The upper boundaries and the left-hand boundaries shown in Figure 10.14 are all set by shutdowns, and the lower boundaries are set by low product purities.

Both designs can easily handle large positive changes in vapor boilup. The total operability region increases as the number of reactive trays is increased. A column with more reactive trays can handle larger negative changes in vapor boilup (left-hand boundary). It also handles bigger changes in the feed composition of F_{0B} (top boundary).

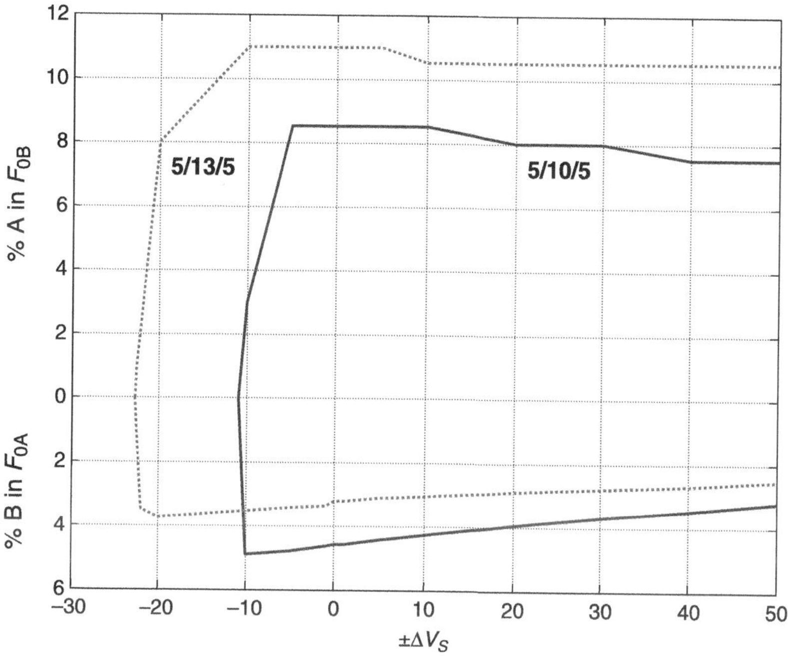

Figure 10.14 Rangeability with A and B impurities in feeds.

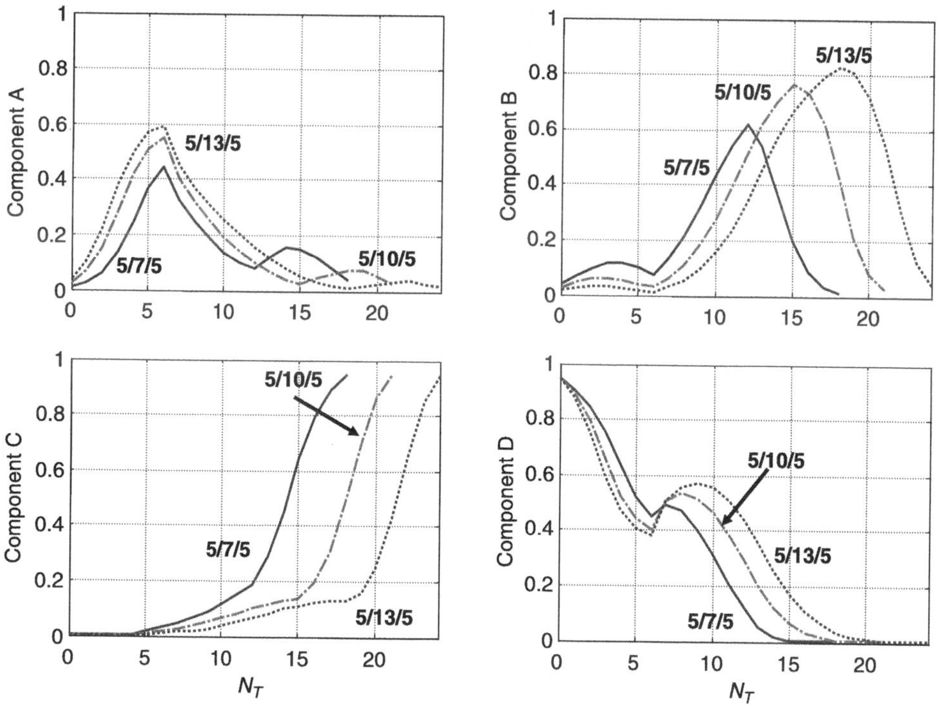

Figure 10.15 Composition profiles.

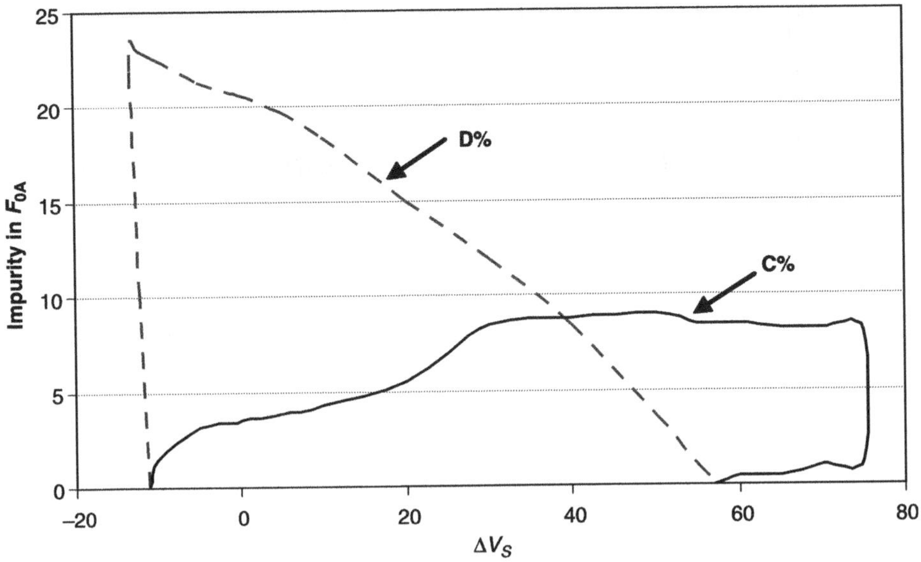

Figure 10.16 Rangeability with C and D in F_{0A}.

Both designs can handle big changes in the feed composition of F_{0A} without shutting down. However, the product purities drop faster for the disturbance in the feed composition of F_{0A} for the 5/13/5 design. This makes the acceptable change in F_{0A} composition smaller for the 5/13/5 design than for the 5/10/5 design.

We have no explanation for this counterintuitive result. It may be caused by differences in the shapes of the composition profiles among the various numbers of reactive trays. Figure 10.15 gives composition profiles for the three cases.

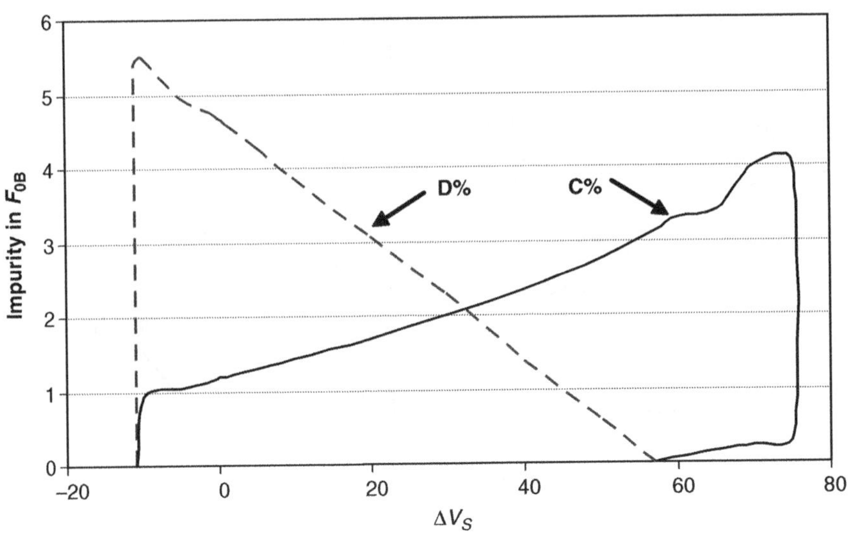

Figure 10.17 Rangeability with C and D in F_{0B}.

Two changes occur as more reactive trays are added:

1. The peaks in reactant A and B concentrations increase as the number of reactive trays is increased. This is counterintuitive because we would expect the presence of more catalyst to promote more reaction and lower the reactant concentrations. However, this can be explained by observing that the concentration of one reactant increases while the concentration of the other reactant decreases.
2. The impurities in the two products change from being essentially the adjacent volatility component (mostly B impurity in the bottoms with product D and mostly A impurity in the distillate with product C) to being a mixture of both reactants at both ends.

We cannot explain these unexpected results, but they do help to explain why the optimum number of reactive trays is 7. Recall that vapor boilup increases as we increase N_{RX} from 7 to 10 to 13 (see Table 10.1). This occurs because the larger reactant concentrations at the ends of the reactive zone require more energy to keep these impurities out of the product streams. We have no physical explanation of why these changes make the column more sensitive to disturbances in the composition of F_{0A}.

Two other types of impurities in the fresh feeds are also possible. There may be some of the product components (C or D) in the fresh feedstreams. The rangeability of the system for the 5/10/5 case is shown in Figures 10.16 and 10.17. In Figure 10.16 either C or D are impurities in fresh feed F_{0A}. The system can handle higher concentrations of D because the F_{0A} stream is fed near the bottom of the column from which D is removed. The impurity levels can be fairly large.

However, Figure 10.17 shows that the impurities levels are much lower for F_{0B}, particularly for D impurities because this heavy product has to work itself all the way down through the column from the high feed location of F_{0B}.

10.9 CONCLUSION

The dynamic controllability of the ideal quaternary two-reactant, two-product system was explored. Adding more reactive trays improved dynamic controllability. The two-temperature control structure provided fairly effective control of the quaternary system.

Two changes occur as more reactive trays are added:

CHAPTER 11

CONTROL OF EXCESS REACTANT SYSTEMS

In Chapter 4 we compared the economics of two alternative reactive distillation processes for the quaternary system with the reaction A + B ⇔ C + D. The first was a single reactive column operating in neat mode. The second process was a two-column flowsheet in which the reactive column was fed with an excess of one of the reactants and the excess was recovered in a second distillation column. The one-column system was shown to be significantly less expensive.

However, we mentioned that the one-column process may be more difficult to control. There are several reasons for this concern. The one-column system requires a control system that can achieve the necessary precise balancing of the fresh feedstreams so that there is no gradual buildup of one of the reactants, which could cause a loss of conversion and product purities. Some method for measuring or inferring the amount of at least one of the reactants inside the process must be available. A direct composition measurement may be necessary, which has major disadvantages compared to the use of simple and reliable temperature measurements. The one-column system also has the inherent disadvantage of having fewer manipulated variables to achieve control and attenuate disturbances.

11.1 CONTROL DEGREES OF FREEDOM

The number of manipulated variables is called the control degrees of freedom, which is equal to the number of control valves. In a conventional distillation column system there are six control valves: feed, condenser cooling water, reboiler steam, reflux, distillate, and bottoms. One control degree of freedom must be used to control throughput. This is usually the feed, but it can be a product stream in an "on-demand" control structure. One control degree of freedom must be used to control pressure (typically condenser

Reactive Distillation Design and Control. By William L. Luyben and Cheng-Ching Yu

duty but sometimes a vapor distillate or reboiler heat input is used). Two control degrees of freedom must be used to control the two liquid levels (typically distillate and bottoms flowrates). That leaves two (*and only two*) optional manipulated variables that can be used to achieve two other control objectives: reflux and steam. Ideally, these control objectives are the impurity levels in the distillate and bottoms products: the concentration of the heavy key component in the distillate and the concentration of the light key component in the bottoms. In practice, the two optional control objectives are very often the temperature on some tray in the column and the reflux/feed ratio (or the reflux ratio).

In contrast, a quaternary reactive distillation column has seven control valves, not six. The additional manipulated variable is the second feedstream. The other six are the same as in a conventional distillation column. Thus, a one-column reactive distillation column operating in neat mode has 7 control degrees of freedom. If we subtract the degrees of freedom required for setting throughput, maintaining two levels and holding pressure, the reactive distillation column has 3 optional control degrees of freedom left.

In the two-temperature control structure studied in Chapter 10, the 7 control degrees of freedom are allocated as follows:

1. Throughput set by vapor boilup
2. Pressure controlled by condenser duty
3. Base level controlled by bottoms flowrate
4. Reflux-drum level controlled by distillate flowrate
5. Reflux ratio maintained by measuring distillate flowrate and adjusting reflux flowrate
6. Temperature on some tray in the column controlled by one of the fresh feeds
7. Another temperature in the column controlled by the other fresh feed

In this structure, the 3 optional degrees of freedom are consumed by the reflux ratio and two temperatures.

A two-column process with a reactive column and a recovery column has seven control valves in the first column and five in the second column, giving a total of 12 control degrees of freedom. There are only 5 degrees of freedom in the second column because its feed comes from the first column. For example, if there is an excess of reactant B fed to the first reactive column, the bottoms from the first column is the feed to the second recovery column.

The two-column system has two pressures and four levels to control. This leaves $12 - 6 = 6$ control degrees of freedom remaining. Throughput is set by 1 control degree of freedom somewhere in the system. This leaves 5 optional degrees of freedom in the two-column system. In the control structure discussed in this chapter for the two-column system, these 5 optional degrees of freedom are consumed by one total feed, two reflux ratios, and two temperatures (one in each column).

Thus, the two-column system has more optional manipulated variables (five) than the single reactive distillation column (three). These variables are available to achieve control objectives. This inherent advantage should theoretically result in improved control because there are more handles to adjust.

We present quantitative comparisons of the dynamic controllability of these two flowsheets in this chapter. Two different control structures are considered for the one-column neat system and one for the two-column excess reactant system.

11.2 SINGLE REACTIVE COLUMN CONTROL STRUCTURES

Two different control structures are explored for the single reactive column process. The first is the two-temperature structure considered in Chapter 10. This system uses only temperatures and does not require a direct composition measurement. The second control structure uses a composition analyzer to measure a composition of one of the reactants on a selected tray in the column to adjust one of the fresh feeds.

The system is the same base case considered in Chapter 4 with five stripping, nine reactive, and five rectifying trays, operating at 8 bar with 1000 mol on reactive trays and 356 mol on separation trays. The column diameter is 0.805 m, and the reflux ratio is 2.66. Figure 11.1 presents the flowsheet with stream information and equipment sizes. The production rates are 12.6 mol/s of both products with purities of 95 mol%. The corresponding conversion is 95%. Figure 11.2 and Figure 11.3 give the temperature and composition profiles, respectively. The peaks in the concentration profiles of reactants A and B occur at their respective feed trays: tray 6 for A and tray 14 for B. The temperature profile is flat in the reactive section but changes fairly rapidly from tray to tray in both the stripping and rectifying sections.

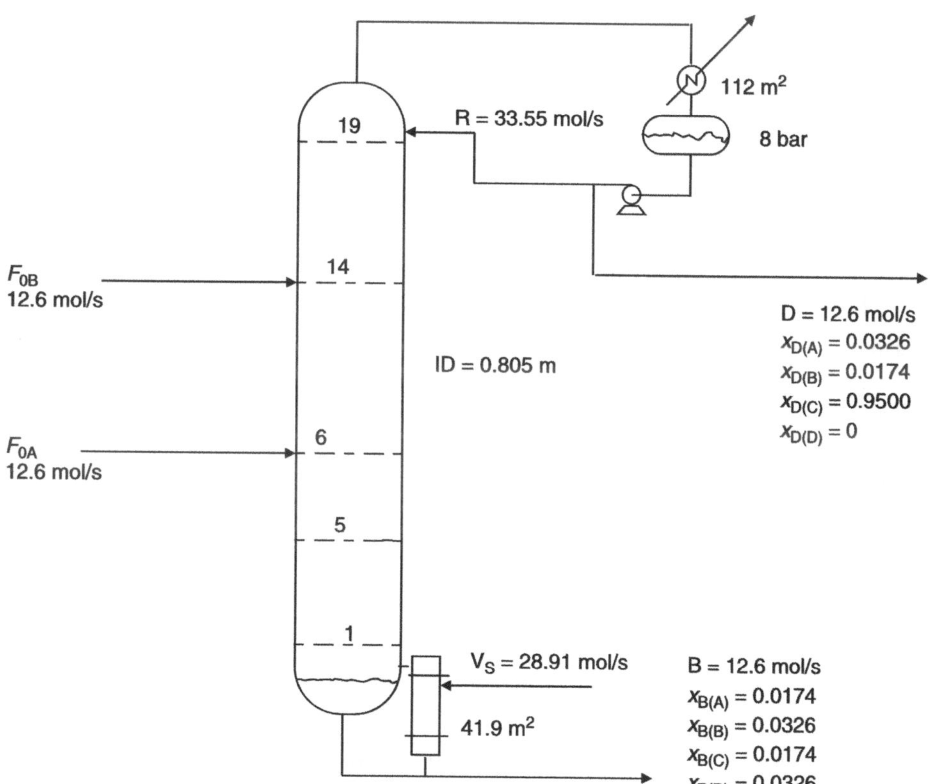

Figure 11.1 Reactive column with neat operation.

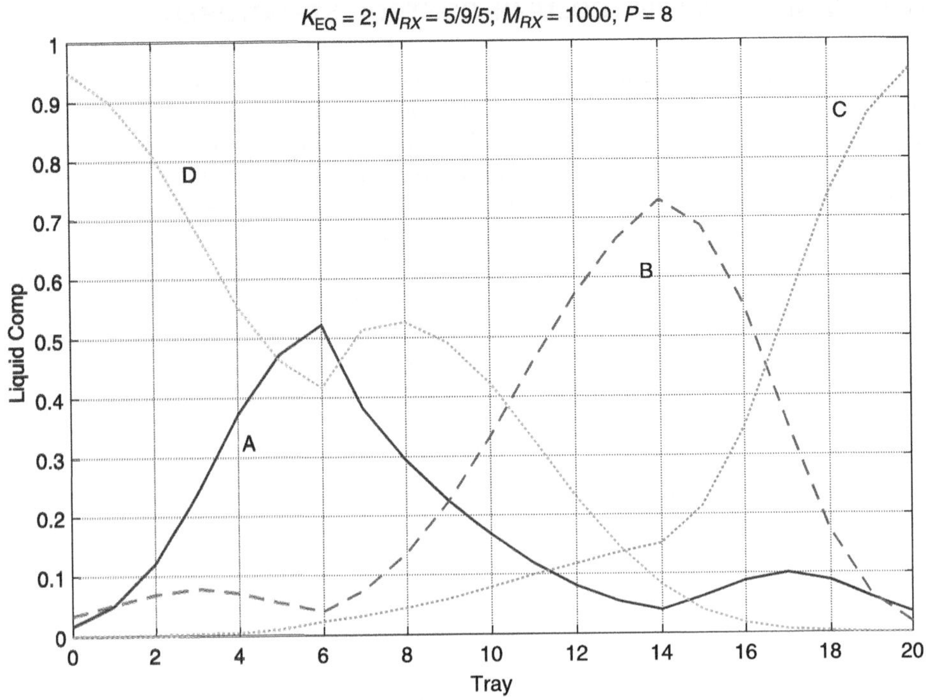

Figure 11.2 Composition profiles for neat operation.

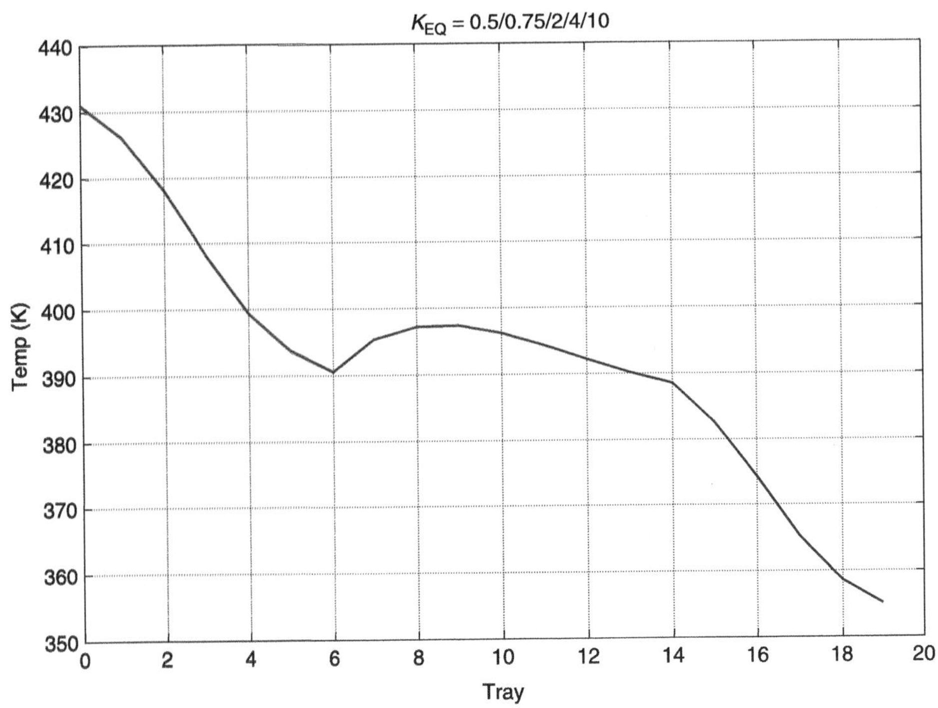

Figure 11.3 Temperature profile for neat operation.

11.2.1 Two-Temperature Control Structure

Figure 11.4 shows the single reactive column operating in neat mode with two temperatures controlled by manipulating the two fresh feeds. The reflux ratio is controlled. Vapor boilup is the production rate handle. Liquid levels are controlled by product removal rates. Column pressure is controlled condenser cooling water in all of the control structures and is not shown in the figures.

As discussed in Chapter 10, the key issue in this control structure is the selection of the control trays. Steady-state gains between tray temperatures and the two fresh feed inputs F_{0A} and F_{0B} are calculated numerically by making very small changes (0.5%) in one of these input variables and converging to a new steady state. The procedure is repeated for the second input variable. Figure 11.5 give these gains for all trays.

The gains are large and *negative* for changes in F_{0A} in the lower part of the column: an increase in the lighter reactant lowers temperatures in the stripping section and in the lower part of the reactive zone. However, they are positive and very small in the upper part of the column. These results suggest that the tray 2 temperature could be controlled by manipulating fresh feed F_{0A} with a direct acting controller (an increase in temperature increases feed).

The gains are large and *positive* for changes in F_{0B} in the lower part of the column: an increase in the heavier reactant raises temperatures in the stripping section and in the lower part of the reactive zone. However, they are negative and small in the upper part of the column.

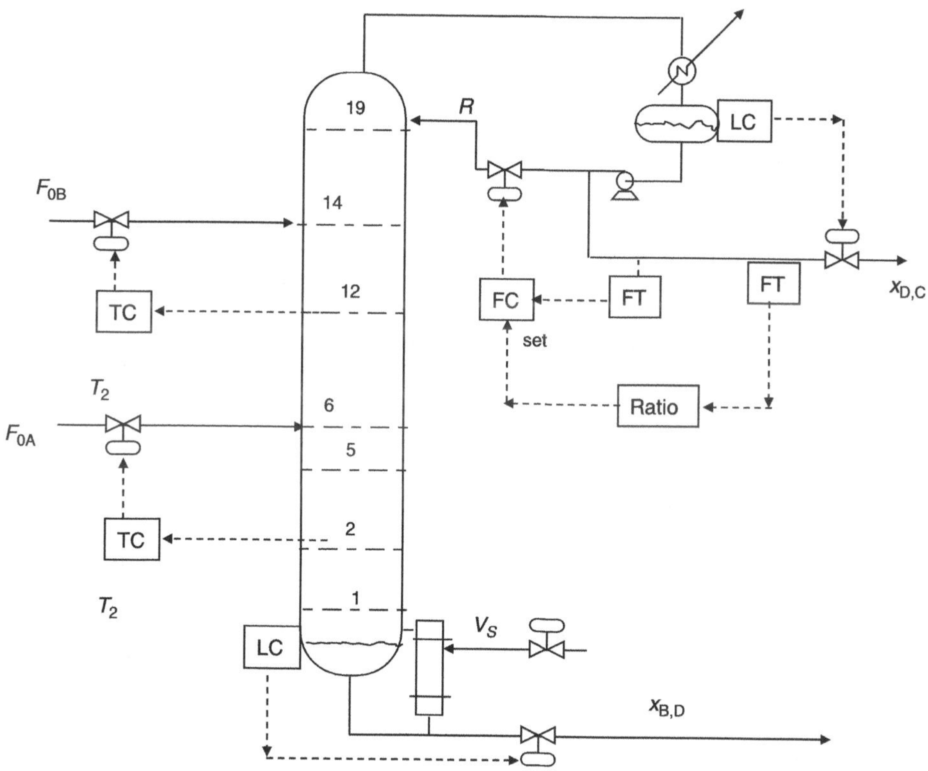

Figure 11.4 Control of T_2/T_{12} for neat operation.

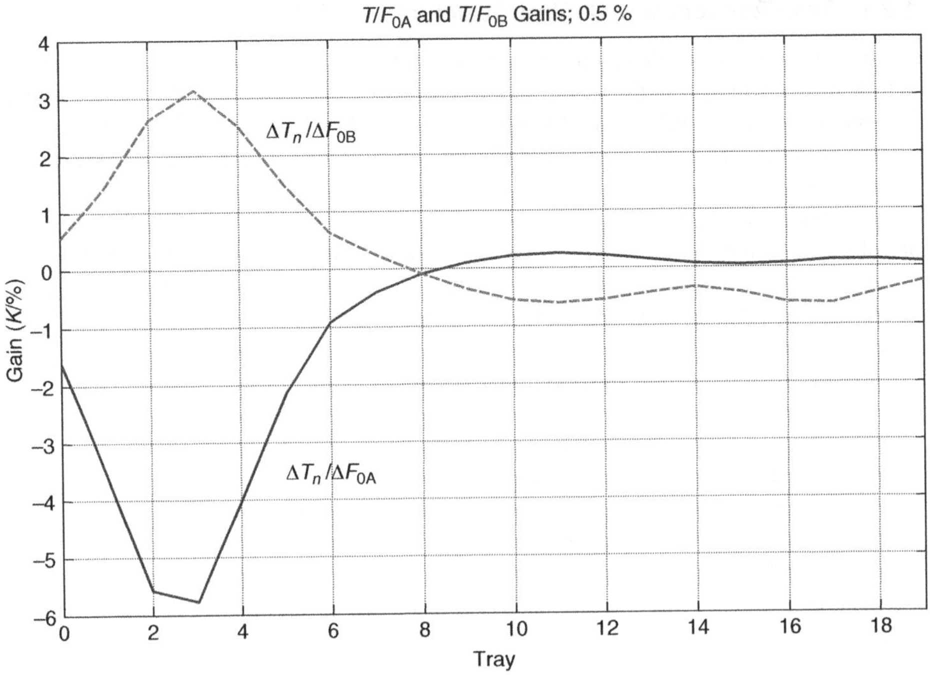

Figure 11.5 Gains for neat operation.

The results from Chapter 10 suggest that the two temperature controllers should have the same action (both direct acting). Therefore, a tray somewhere in the region where the gain between temperature and F_{0B} is negative should be selected. Tray 12 exhibits a negative gain, but it is small.

The SVD analysis provided in Figure 11.6 suggests that there are sensitive regions in the lower part of the column around trays 2 and 3 and in the upper part of the column around trays 11–16. We select trays 2 and 12 to control with the two fresh feeds. Both controllers are direct acting controllers (an increase in temperature increases feed).

Relay-feedback tests are run individually on both temperature loops with the other loop on manual. Three 1-min temperature measurement lags are included. Temperature transmitter spans are 50 K. Valves are half open at steady state. Controller tuning constants are given in Table 11.1. Tyreus–Luyben tuning constants are used with some reductions in controller gains to give less oscillatory behavior.

All of the systems and control structures are subjected to a number of disturbances in throughput and feed compositions. Step disturbance are made at 10 min.

Figures 11.7 and 11.8 gives responses to positive and negative 20% changes in vapor boilup, the throughput handle in this control structure. These disturbances are handled well by the two-temperature control structure. Stable base-level regulatory control is achieved. The increase in V_S results in increases in both fresh feeds, and the distillate and bottoms streams increase. Product purities $x_{D(C)}$ and $x_{B(D)}$ are maintained fairly close to their desired values. Product purities drop slightly below their specifications for the increase in V_S but rise above specifications for the decrease in throughput. Reflux increases because of the reflux ratio control structure.

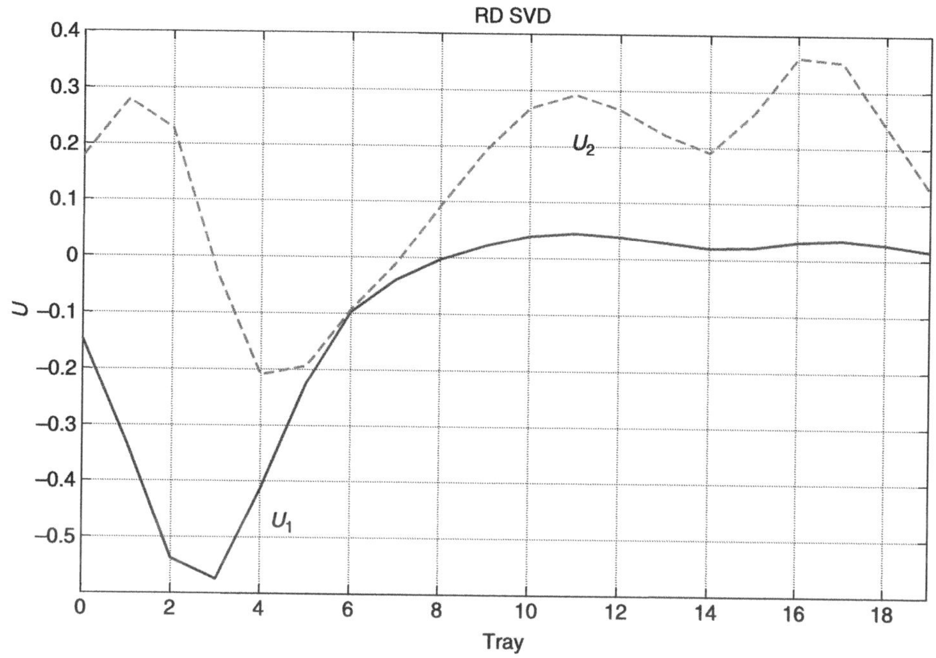

Figure 11.6 The SVD analysis.

Figures 11.9–11.14 demonstrate the responses of the system with the two-temperature control structure for a variety of feed composition changes. Instead of feeding pure reactant A in the F_{0A} fresh feedstream and feeding pure reactant B in the F_{0B} fresh feedstream, the composition of these streams $z_{0A(j)}$ and $z_{0B(j)}$ are changed in various ways. In Figure 11.9

TABLE 11.1 Controller Tuning Parameters

	Reactive Two-Temp. Control Structure	Column Internal Composition Control Structure	Two-Column Reactive Column	Process Recovery Column
Controlled variable	T_2	T_3	T_{17}	T_{18}
Manipulated variable	F_{0A}	V_S	V_{S1}	V_{S2}
Setpoint (K)	407.4	408.0	372.3	338.3
K_U	4.2	1.6	1.8	12.7
P_U (min)	3.0	2.3	2.5	5.0
K_C	0.5	0.25	0.57	4.0
τ_I (min)	6.6	5	5.5	11
Controlled variable	T_{12}	$x_{(6,A)}$		
Manipulated variable	F_{0B}	F_{0A}/F_{0B}		
Setpoint (K)	392.2	0.5224		
K_U	120	2.9		
P_U (min)	2.2	2.1		
K_C	10	0.45		
τ_I (min)	50	5		

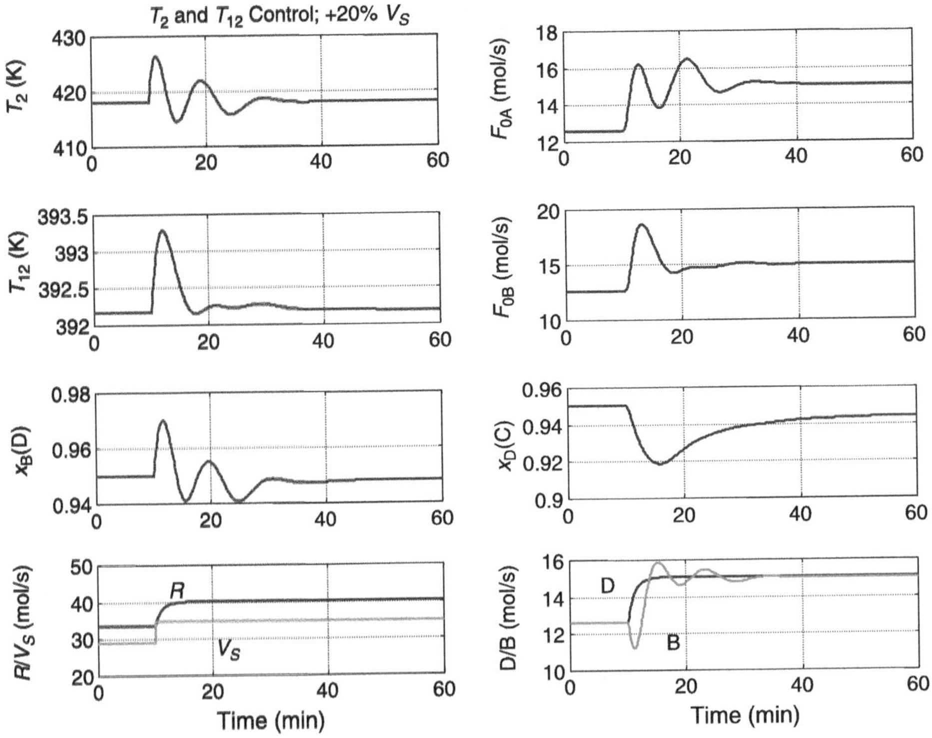

Figure 11.7 Control of T_2/T_{12} at $+20\%$ V_S.

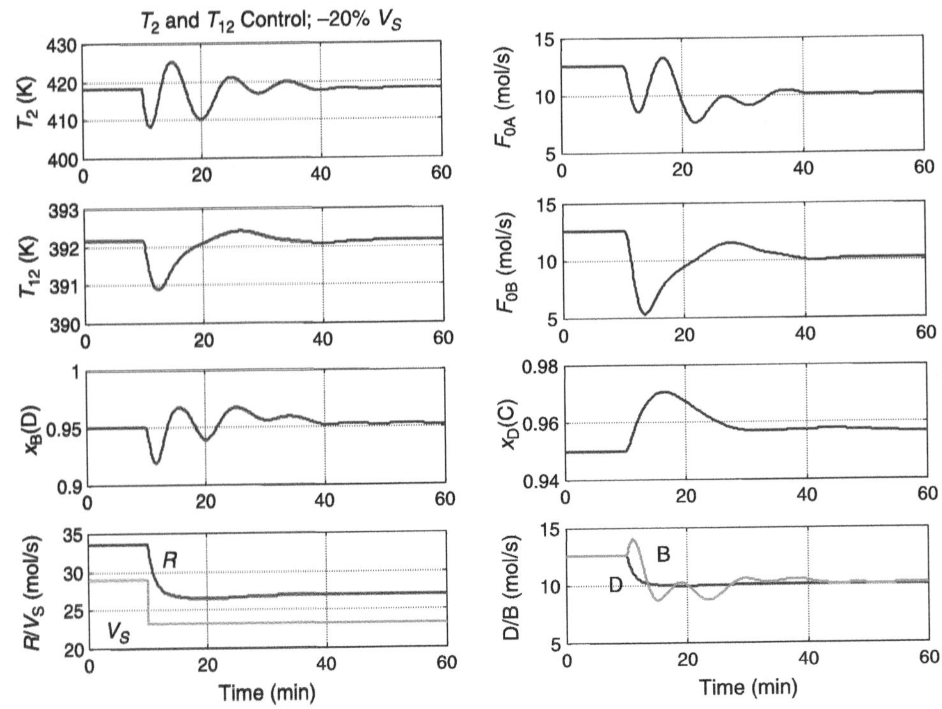

Figure 11.8 Control of T_2/T_{12} at -20% V_S.

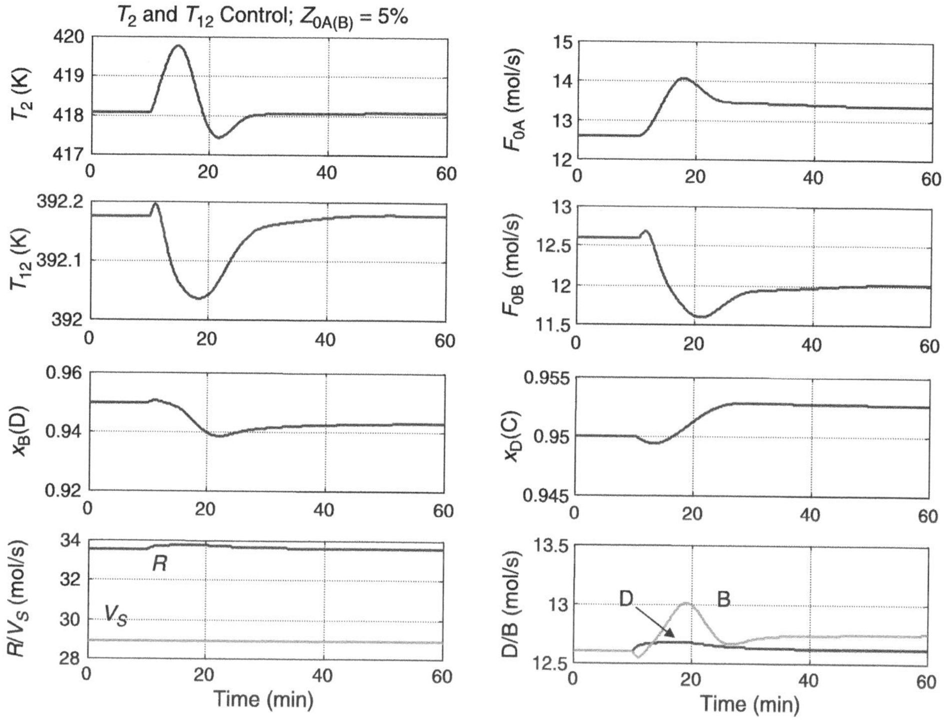

Figure 11.9 Control of T_2/T_{12} at $z_{0A(B)} = 0.05$.

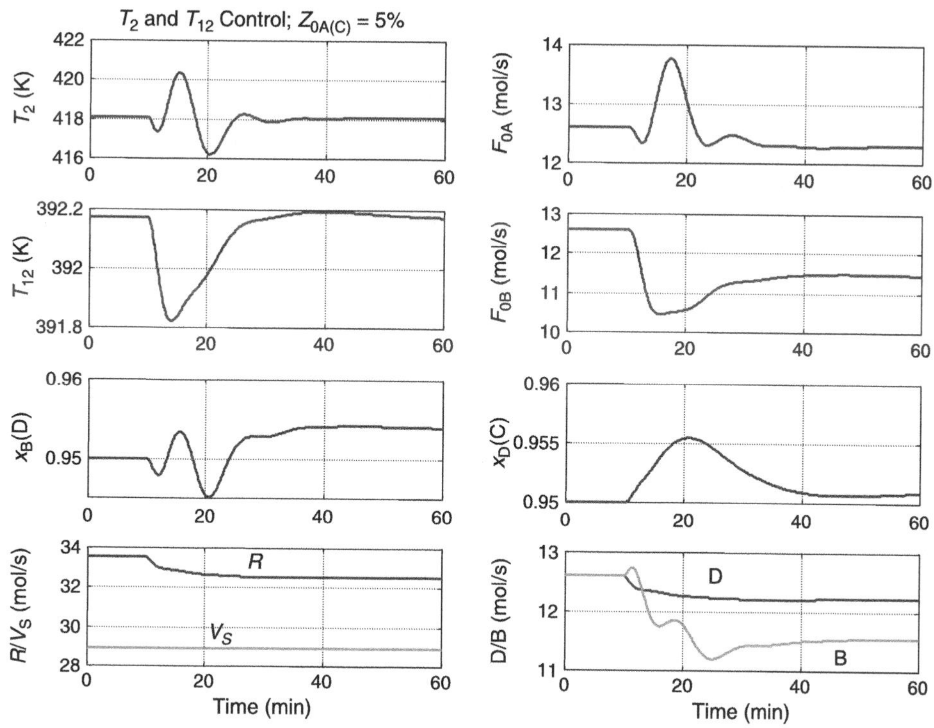

Figure 11.10 Control of T_2/T_{12} at $z_{0A(C)} = 0.05$.

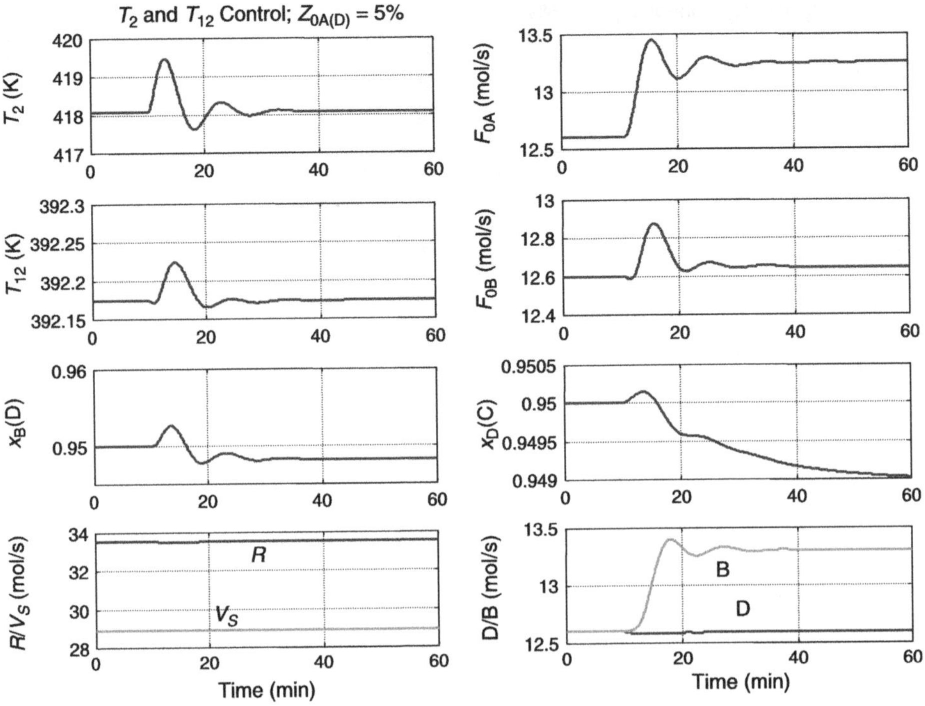

Figure 11.11 Control of T_2/T_{12} at $z_{0A(D)} = 0.05$.

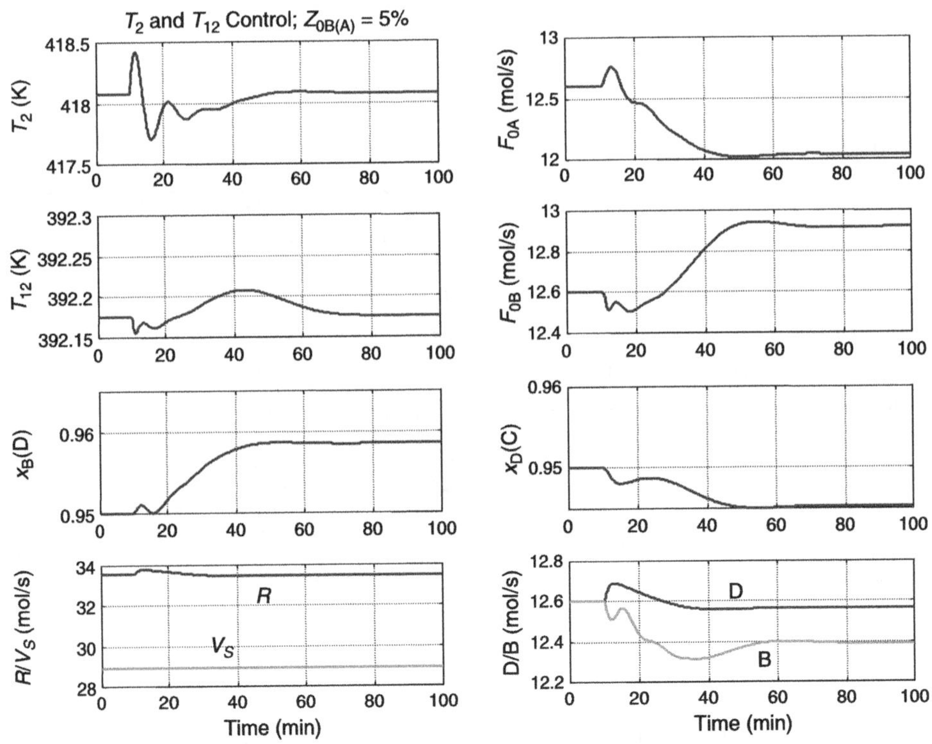

Figure 11.12 Control of T_2/T_{12} at $z_{0B(A)} = 0.05$.

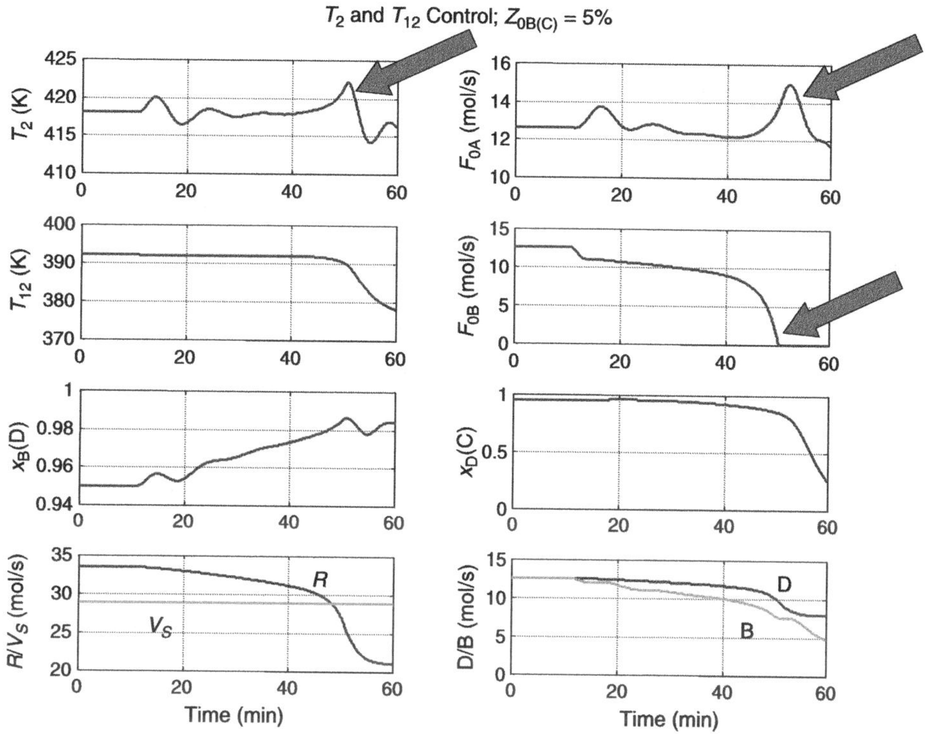

Figure 11.13 Control of T_2/T_{12} at $z_{0B(C)} = 0.05$.

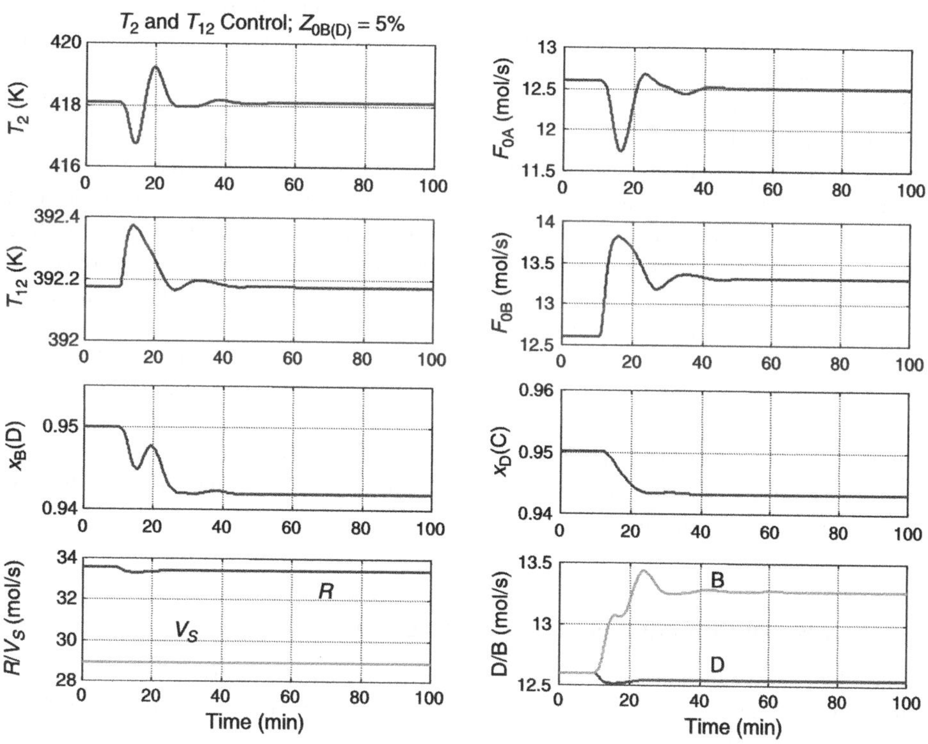

Figure 11.14 Control of T_2/T_{12} at $z_{0B(D)} = 0.05$.

the composition of F_{0A} is changed from pure A to contain 5 mol% B ($z_{0A(A)} = 0.95$, $z_{0A(B)} = 0.05$). The control scheme handles this disturbance well, but the purity of the bottoms decreases somewhat. The fresh feed of B is decreased and the fresh feed of A is increased. The net result is a small increase in the bottoms product flowrate because of its slightly lower purity.

In Figure 11.10 the composition of F_{0A} is changed from pure A to contain 5 mol% C ($z_{0A(A)} = 0.95$, $z_{0A(C)} = 0.05$). The control scheme handles this disturbance well, with the purities of both products improving. The fresh feed of B is decreased, and the fresh feed of A is decreased slightly. The flowrates of both products decrease with the larger reduction in the bottoms flowrate because the C in the feed leaves in the distillate.

The final z_{0A} disturbance adds some D to fresh feed F_{0A}. Figure 11.11 shows that this disturbance is handled well. As expected, the production of D increases (bottoms flowrate rises). Product purities drop slightly.

Similar feed composition disturbances are made in fresh feed F_{0B}. In Figure 11.12 the composition of F_{0B} is changed from pure B to 5 mol% A ($z_{0B(B)} = 0.95$, $z_{0B(A)} = 0.05$). The control scheme handles this disturbance well. The flowrate of F_{0B} increases, and the flowrate of F_{0A} decreases.

However, the two-temperature control structure fails when the feed composition disturbance is the introduction of some C in the F_{0B} fresh feed ($z_{0B(B)} = 0.95$, $z_{0B(C)} = 0.05$). Figure 11.13 shows that a shutdown occurs in about 1 h. The tray 12 temperature controller tries to hold the temperature as it begins to fall at about 45 min by cutting back on F_{0B}. However, the tray 2 temperature has increased, so the controller increases F_{0A}. Adding more light reactant tends to drop the temperature on tray 12 even more. The system crashes when the F_{0B} feed is cut off completely. The distillate purity drops drastically because there is no B being fed to produce it.

In Figure 11.14 the composition of F_{0B} is changed from pure B to 5 mol% D ($z_{0B(B)} = 0.95$, $z_{0B(D)} = 0.05$). The control scheme handles this disturbance well with a small decrease in product purities.

In summary, the two-temperature control structure, which does not use any composition measurement, handles all of these disturbance well except when some of the lightest component C is present in the fresh feed of B. Remember the F_{0B} feed is trying to control the tray 12 temperature, and it is a direct acting controller. Thus, when the light C component enters near the top of the column in the F_{0B} stream, it drops the temperature and the temperature controller decreases F_{0B}. However, the decrease in F_{0B} produces an increase in the temperatures in the lower part of the column (remember the T/F_{0B} gains are positive in the lower part of the column). As a result, the tray 2 temperature increases and the second temperature controller then increases F_{0A} (see upper left graph, Fig. 11.13). This puts more light material in the column and drives the tray 12 temperature even lower. The result is a system shutdown.

In the next section we study an alternative control structure that uses an internal composition measurement. Will this alternative control scheme be able to handle these disturbances without shutting down?

11.2.2 Internal Composition Control Structure

Figure 11.15 shows an alternative control structure for the single reactive column operating in neat mode. Instead of controlling two temperatures by manipulating the two fresh feedstreams and setting the production rate by vapor boilup (Fig. 11.4), this alternative scheme

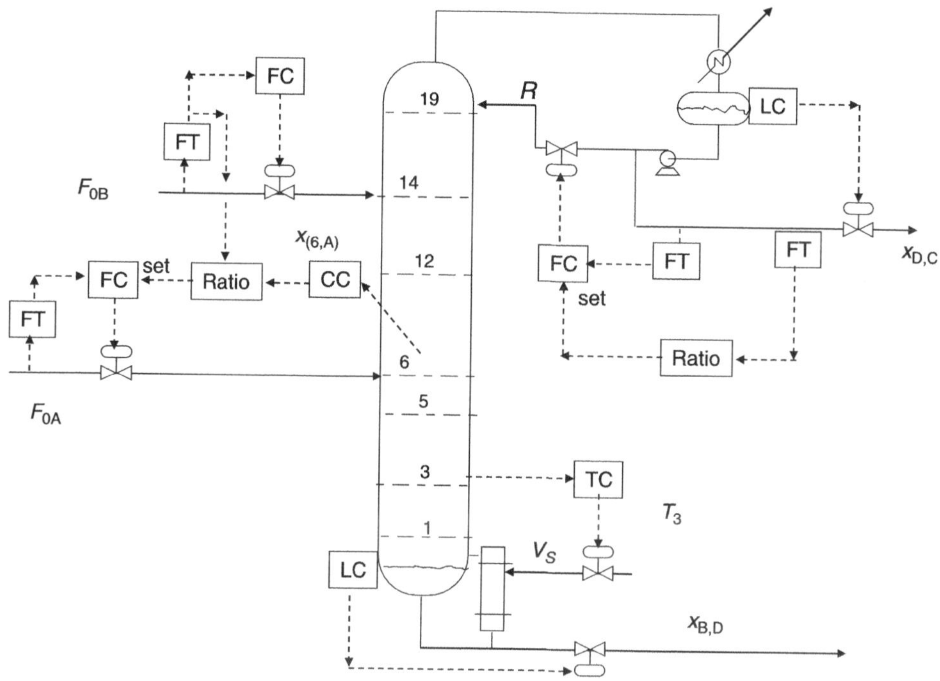

Figure 11.15 Control of $T_3/x_{(6,A)}$ for neat operation.

sets the production rate by controlling the flow of the fresh feed of B (F_{0B}), controlling the temperature on tray 3 by manipulating vapor boilup, and controlling the composition of A on tray 6 by manipulating the ratio of F_{0A} to F_{0B}. This control structure directly detects if A is building up or being depleted in the column and makes appropriate adjustment of the fresh feed of A. It has the advantage of providing a direct handle on throughput, but it has the disadvantage of requiring composition measurement, which can be more expensive and less reliable than temperature measurement.

The temperature and composition loops are tuned in the same way as previously discussed. Three 1-min lags are used in both loops. The composition transmitter span is 20 mol%. Tray 6 is selected because this is where the peak in the composition curve of A occurs (see Fig. 11.2). Tuning parameters are given in Table 11.1.

The same disturbances used in the previous control structure are imposed on the system, except the throughput change is now a step in F_{0B}. Figure 11.16 gives the response for a 20% increase in F_{0B} when the tray 6 composition controller sets the flowrate of F_{0A} directly without the ratio in service. This response should be compared with that given in Figure 11.17 with the ratio in service. There are smaller deviations in $x_{(6,A)}$ and the bottoms product returns more quickly to near its desired value. Figure 11.18 shows the response for a 20% decrease in throughput. These throughput changes are handled well by this control structure.

Figures 11.19–11.24 demonstrate the responses to all of the feed composition disturbances. Adding some impurity of B, C, or D in fresh feed F_{0A} does not result in an unstable system (Figs 11.19–11.21). The same result is seen for adding impurities of A, C, or D in fresh feed F_{0B} (Figs 11.22–11.24).

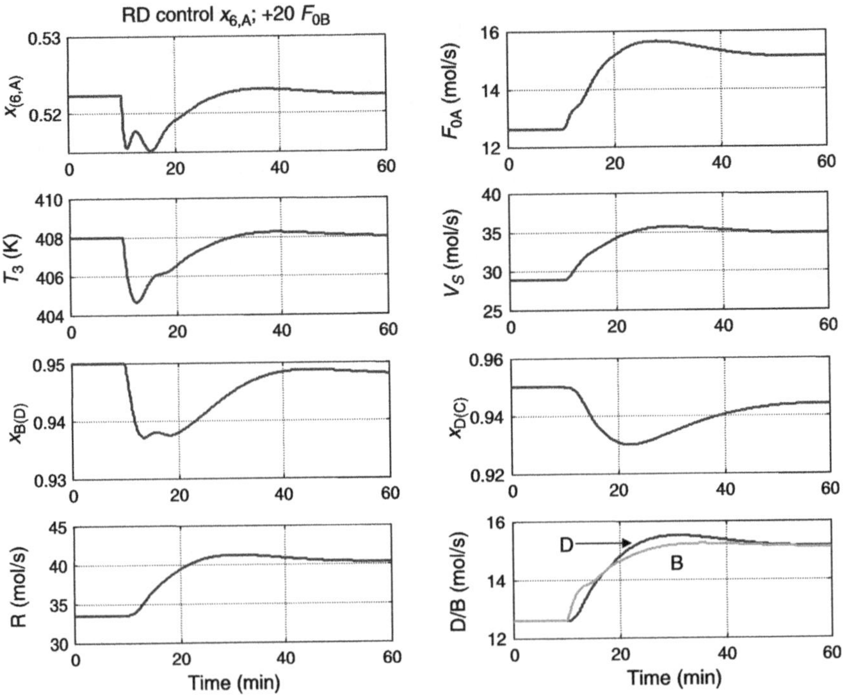

Figure 11.16 Control of $T_3/x_{(6,A)}$ without ratio for $+20\%$ F_{0B}.

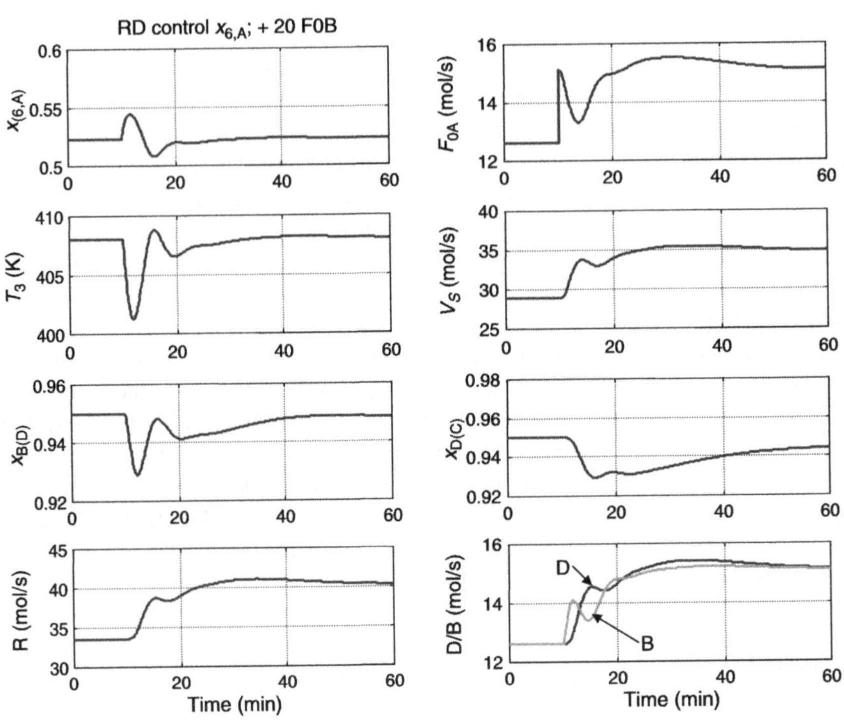

Figure 11.17 Control of $T_3/x_{(6,A)}$ with ratio for $+20\%$ F_{0B}.

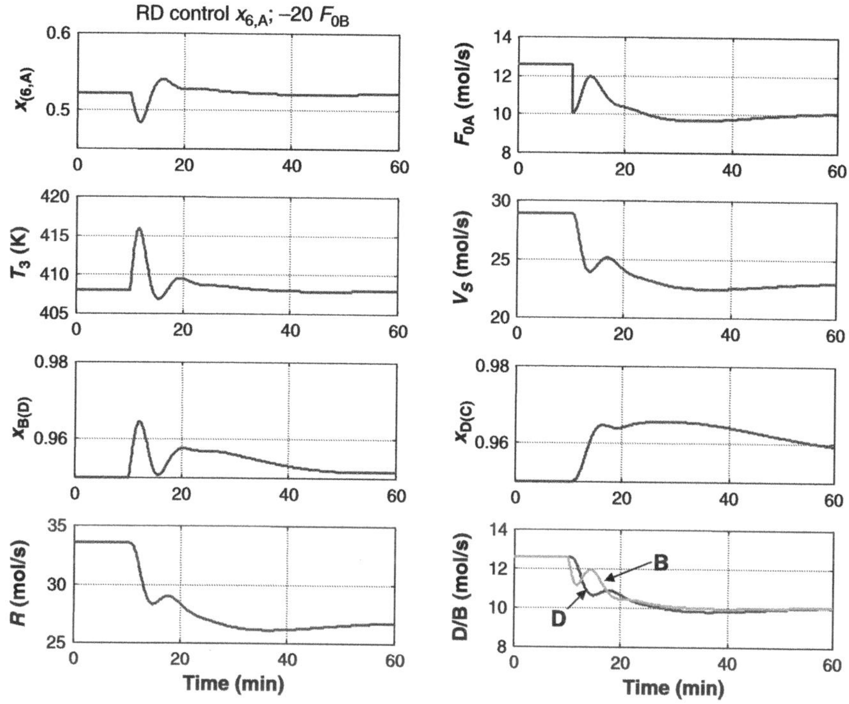

Figure 11.18 Control of $T_3/x_{(6,A)}$ with ratio for -20% F_{0B}.

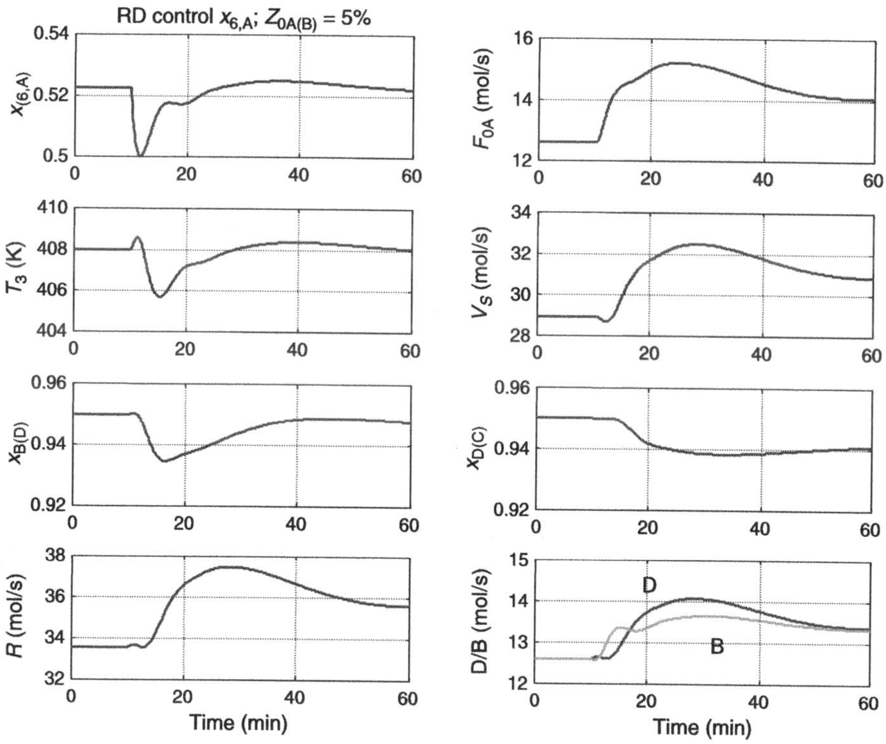

Figure 11.19 Control of $T_3/x_{(6,A)}$ for $z_{0A(B)} = 0.05$.

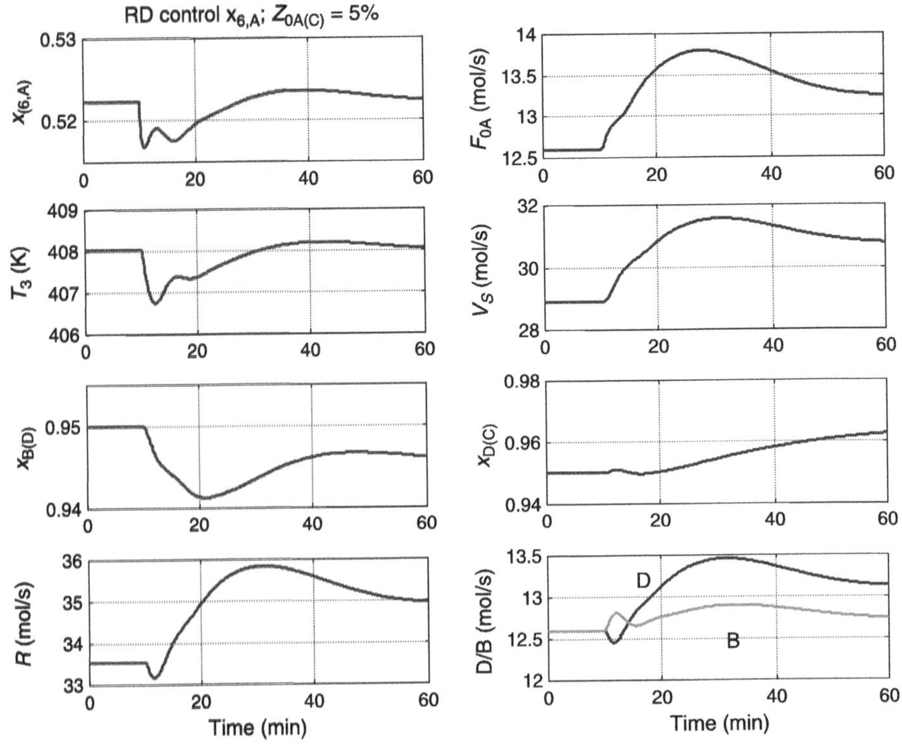

Figure 11.20 Control of $T_3/x_{(6,A)}$ for $z_{0A(C)} = 0.05$.

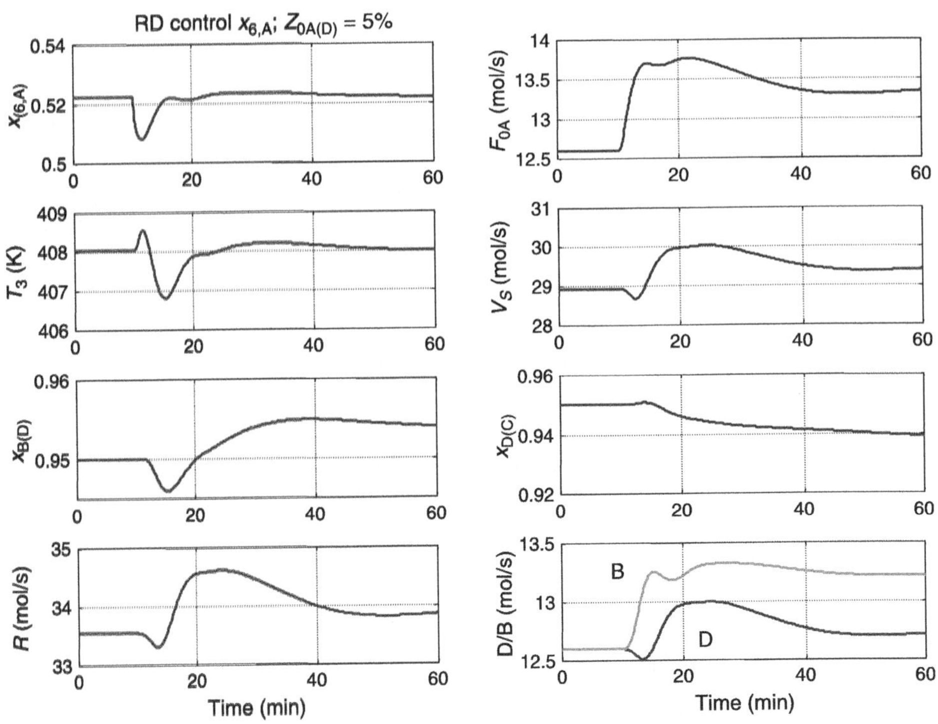

Figure 11.21 Control of $T_3/x_{(6,A)}$ for $z_{0A(D)} = 0.05$.

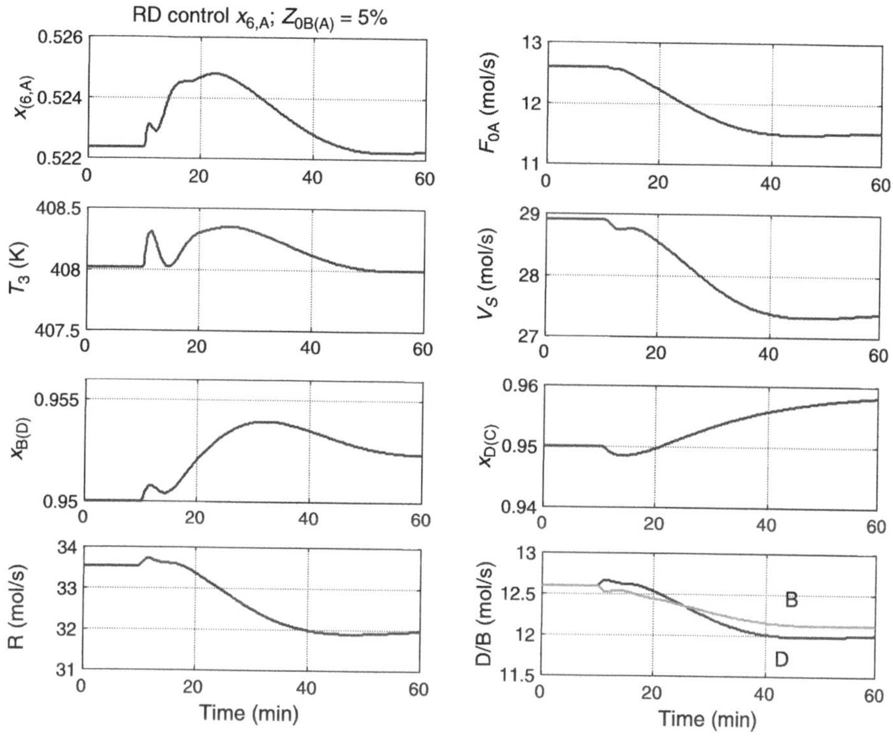

Figure 11.22 Control of $T_3/x_{(6,A)}$ for $z_{0B(A)} = 0.05$.

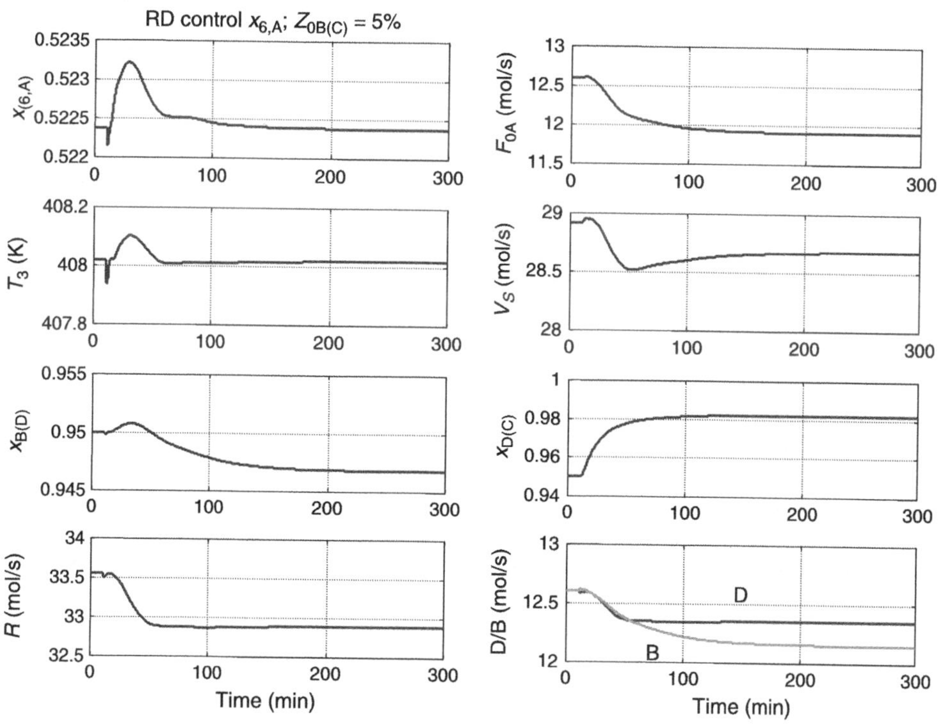

Figure 11.23 Control of $T_3/x_{(6,A)}$ for $z_{0B(C)} = 0.05$.

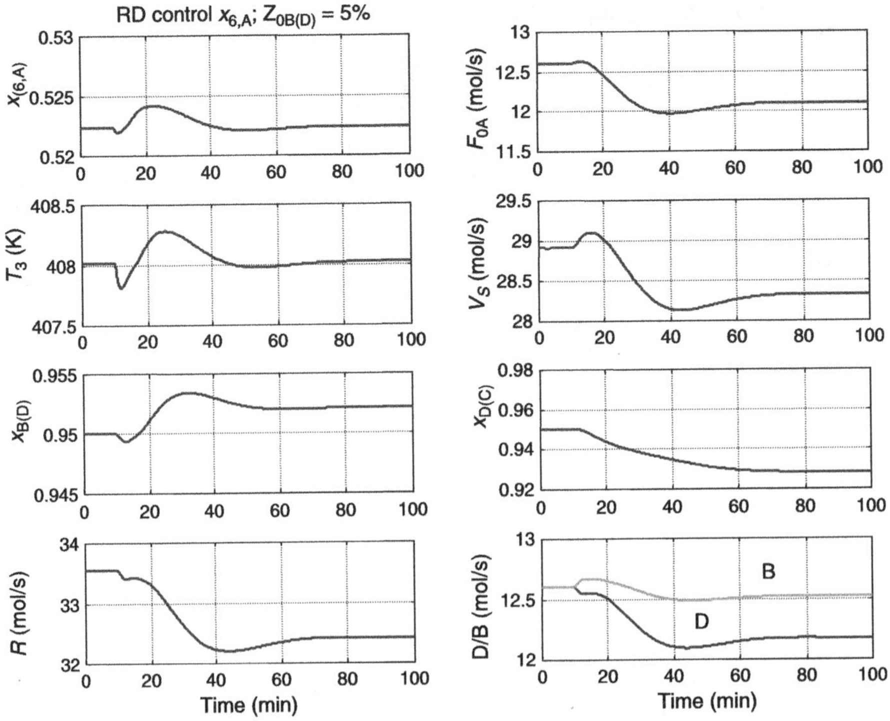

Figure 11.24 Control of $T_3/x_{(6,A)}$ for $z_{0B(D)} = 0.05$.

Note that the $z_{0B(C)} = 0.05$ disturbance (Fig. 11.23), which crashed the two-temperature control scheme, causes no problem in the internal composition control structure and that the time scale has been increased to 300 min to make sure that there is long-term instability. The internal composition controller detects the increase in reactant A composition on tray 6 because there is less B coming into the column in the F_{0A} feed. The composition controller reduces the flowrate of F_{0A}. The temperature on tray 3 increases as less D is produced because less B is entering the column. The temperature controller cuts back on the vapor boilup, and the reflux ratio cuts back on the reflux. The final steady state produces somewhat lower purity bottoms and less of it. Distillate purity is higher with only a slightly lower flowrate because of the C coming in with the F_{0B} feed.

11.3 CONTROL OF TWO-COLUMN SYSTEM

Now that we have demonstrated that the one-column neat reactive distillation system can be controlled but may require a composition measurement to handle all types of disturbances, we want to see how the two-column system with an excess of reactant handles disturbances. The case considered is the 20% excess of B fed to the reactive column. Figure 11.25 contains the flowsheet of the reactive column/recovery column system. The two product streams are distillate D_1 from the reactive column containing product C and bottoms B_2 from the recovery column containing product D. The distillate D_3 from recovery column recycles the excess B back to the reactive column. Figure 11.26 gives composition profiles in the reactive column, and Figure 11.27 gives composition profiles in the recovery column.

Temperature profiles for both columns are shown in Figure 11.28. Note that the reactive column has a fairly sharp temperature break in the rectifying section where the concentration of the lightest component, product C, is building up. A tray temperature in this region should give a fairly accurate estimation of the purity of the distillate. In the recovery column, there is a moderate change in temperatures from tray to tray in the upper part of the column where the concentration of the heavier reactant component B is building up. Because the bottoms stream is the product, we would prefer to use a temperature near the base. However, the temperature profile in the lower part of the column is very flat, so using a tray temperature in this section may not give effective control. If precise control of the bottoms purity is required, it might be necessary to use a composition controller in a composition/temperature cascade structure to change the setpoint of the tray temperature controller.

11.3.1 Two-Temperature Control

Two temperatures are controlled in the structure used for this two-column system. There is no direct measurement of composition. If the recovery column does its job and prevents B from escaping out the bottom of the column, the flowrate of distillate recycle stream D_2 is an indication of the amount of excess B that is leaving the reactive column and entering the separation column.

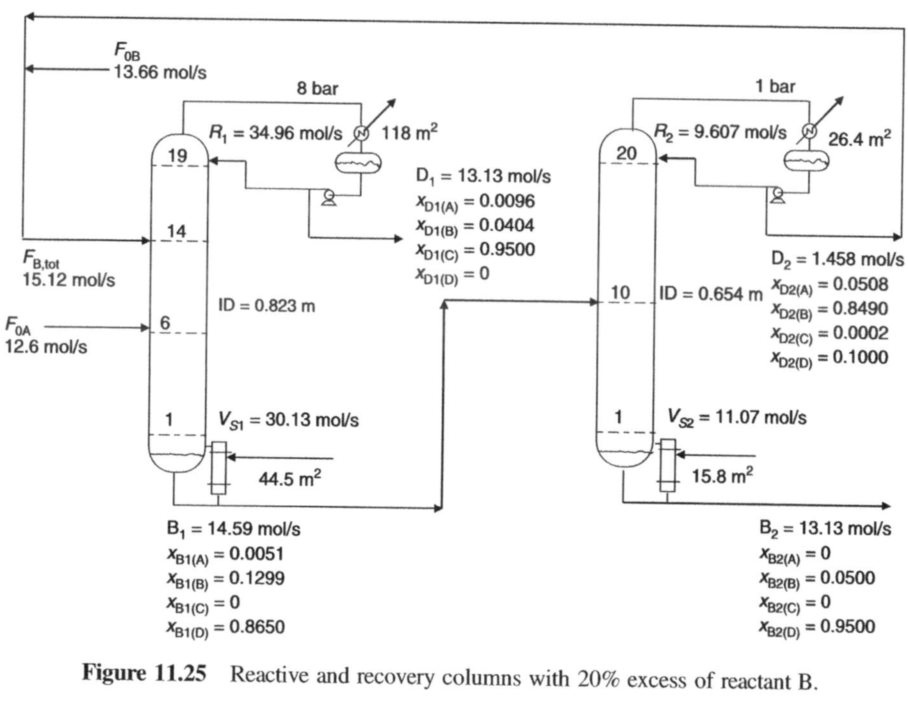

Figure 11.25 Reactive and recovery columns with 20% excess of reactant B.

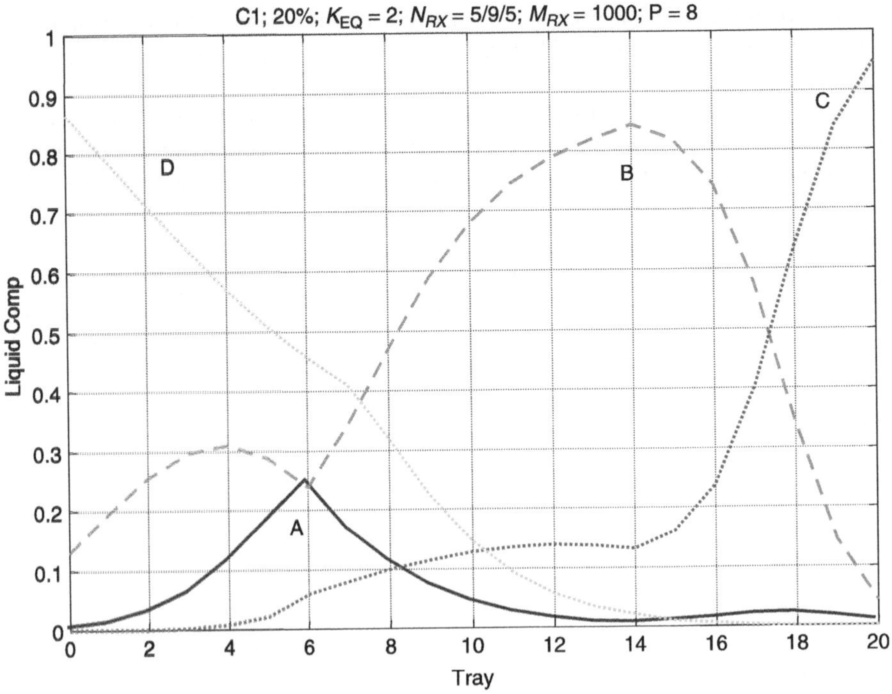

Figure 11.26 Reactive column with 20% excess of reactant B.

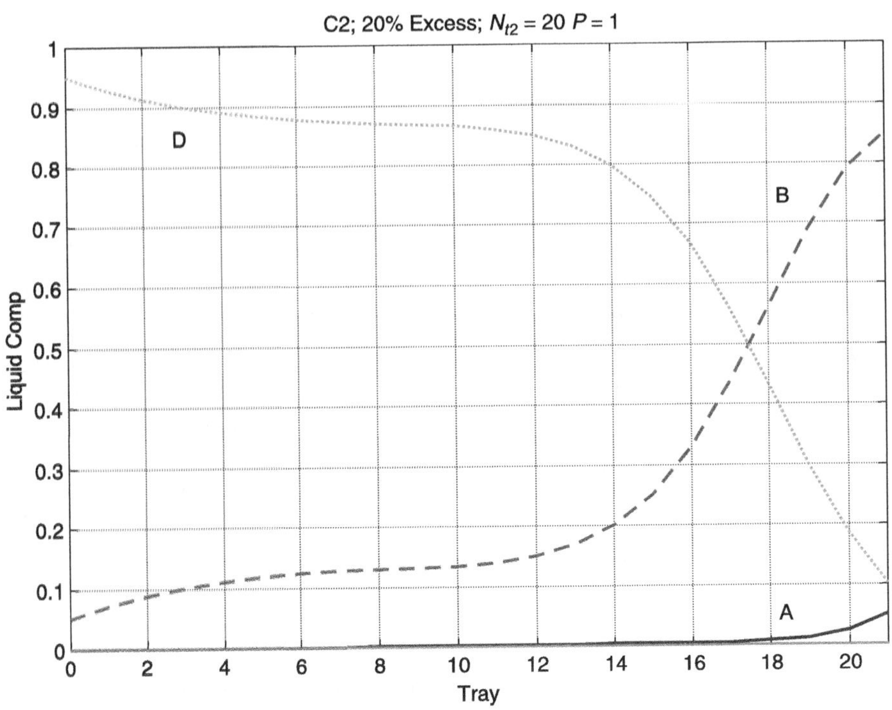

Figure 11.27 Recovery column with 20% excess of reactant B.

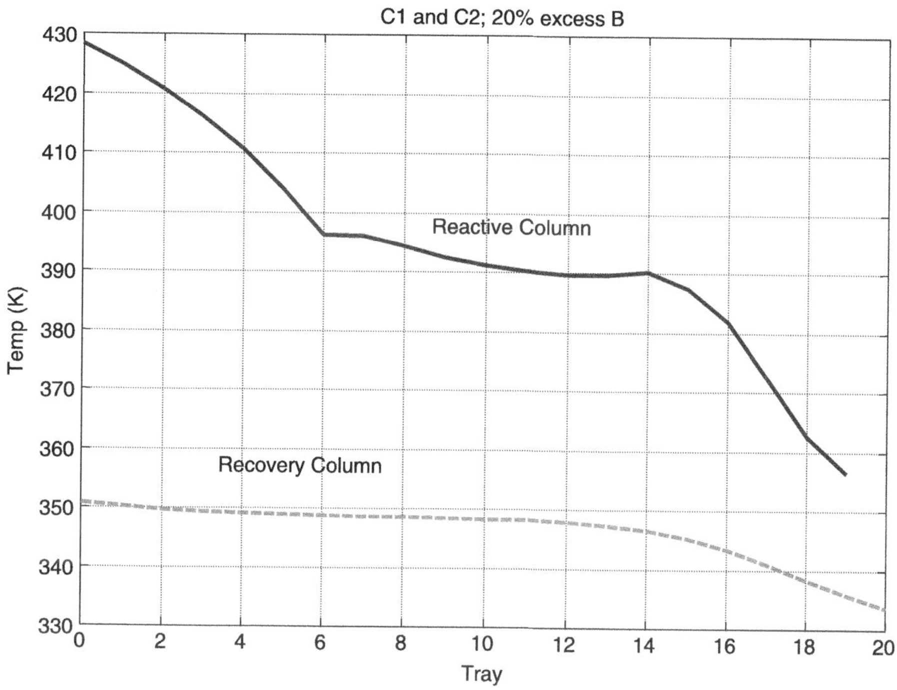

Figure 11.28 Reactive and recovery columns with 20% excess of reactant B.

Figure 11.29 shows the control structure used in this two-column system. The various loops are detailed in the following:

1. Throughput is set by the flow controller on the fresh feed of A (F_{0A}).
2. The total flow to the top of reactive zone $F_{B,tot}$ is the sum of the fresh feed of B (F_{0B}) and distillate D_2. This total stream is flow controlled by manipulating the

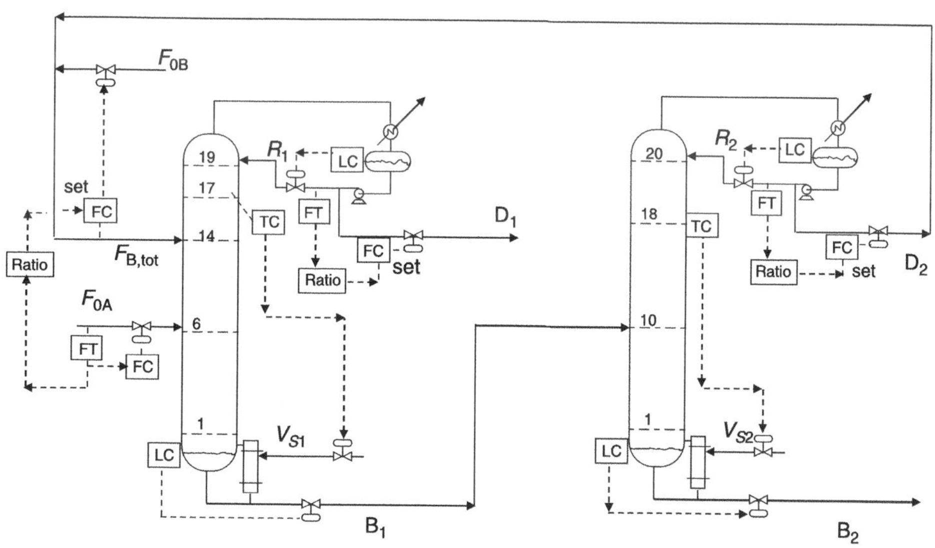

Figure 11.29 Reactive and recovery columns with excess of reactant B.

flowrate of fresh feed F_{0B}. The setpoint of this total flowrate controller is ratioed to flowrate F_{0A}.

3. Pressures in both columns are controlled by condenser duties (not shown in Fig. 11.29).
4. Base levels are controlled by bottoms flowrates.
5. Reflux-drum levels are controlled by reflux flowrates.
6. Reflux ratios are maintained by measuring distillate flowrates and adjusting reflux flowrates.
7. The temperature on tray 17 in the reactive column is controlled by manipulating vapor boilup V_{S1} in the reactive column.
8. The temperature on tray 18 in the recovery column is controlled by manipulating vapor boilup V_{S2} in the recovery column.

The ratio is set to give a 20% excess of B. For the numerical case considered in this example, the ratio is 1.2, that is, $F_{B,tot} = 1.2\ F_{0A}$.

The same disturbances used in the previous studies of the single reactive column are imposed on the system. The throughput change is now a step in F_{0A}. Figure 11.30 gives the response for a 20% increase in F_{0A}. The first thing to note is that the time scale has been increased from that used for the one-column case. The system with recycle has significantly larger time constants than the single column. It takes 2–3 h for the transients to die

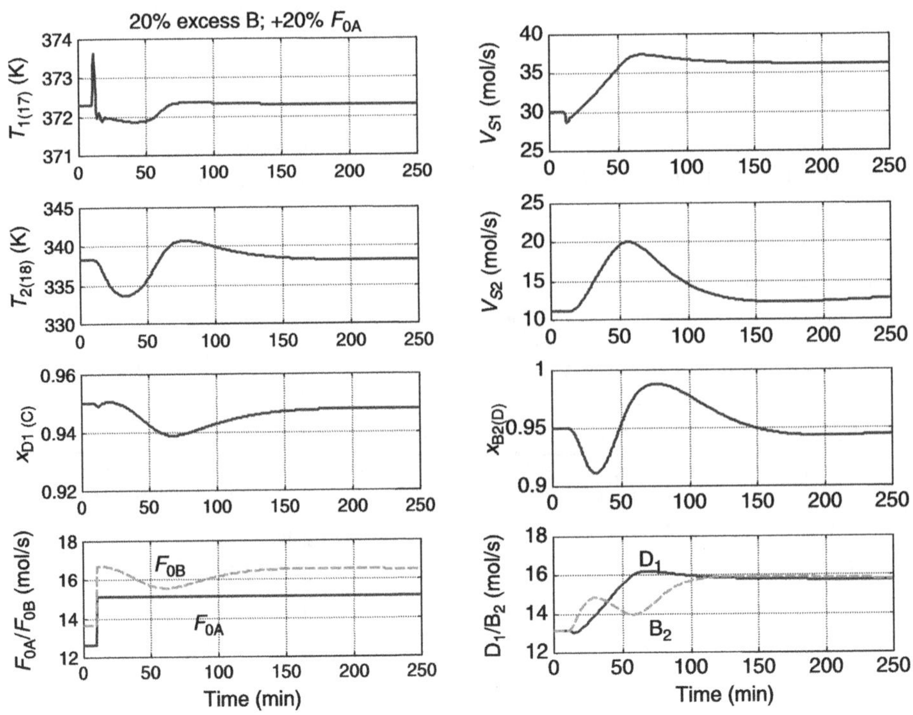

Figure 11.30 Reactive and recovery columns for +20% F_{0A}.

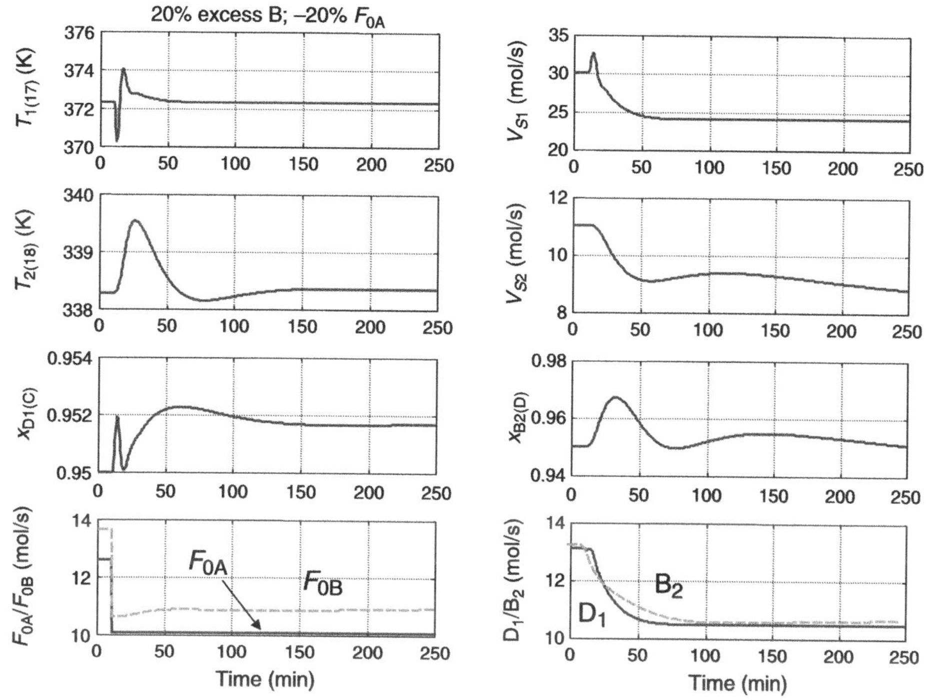

Figure 11.31 Reactive and recovery columns for –20% F_{0A}.

out in the two-column system with recycle compared to about 1 h for the one-column system.

Figures 11.30 and 11.31 show that positive and negative 20% step changes in throughput (F_{0A}) are handled well. Product purities drop very slightly for an increase and rise slightly for a decrease. The largest deviation in product purity is a drop to about 92 mol% D in the bottoms of the recovery column during the transient. This deviation could be reduced by using a vapor boilup to feed ratio on this column to anticipate that more flow to the column requires more vapor boilup.

Figures 11.32–11.38 give the response of the two-column system to the same sequence of feed composition disturbances applied in the one-column cases. In Figure 11.32 some B impurity is added to fresh feed F_{0A} ($z_{0A(A)} = 0.95$, $z_{0A(B)} = 0.05$). The larger amount of B coming into the reactive column produces an increase in the bottoms B_1 fed to the recovery column. The vapor boilup in recovery column V_{S2} increases to twice its normal steady-state value, at which point the control valve is wide open. Control of tray 18 in the recovery column is lost. Despite hitting this constraint, product purities are not adversely affected. In Figure 11.33 the maximum value of V_{S2} is increased to 4 times the steady-state value, which removes the constraint. Now both temperatures are controlled, and the B_2 product stream becomes very pure in D.

In Figure 11.34 some C impurity is added to the fresh feed F_{0A} ($z_{0A(A)} = 0.95$, $z_{0A(C)} = 0.05$). The disturbance is handled well, but a large increase in V_{S2} is required and the B_2 product stream becomes very pure in D. The responses to these two disturbances illustrated that controlling a temperature in the upper part of the recovery column does not hold the bottoms composition very close to its specification. A bottoms composition

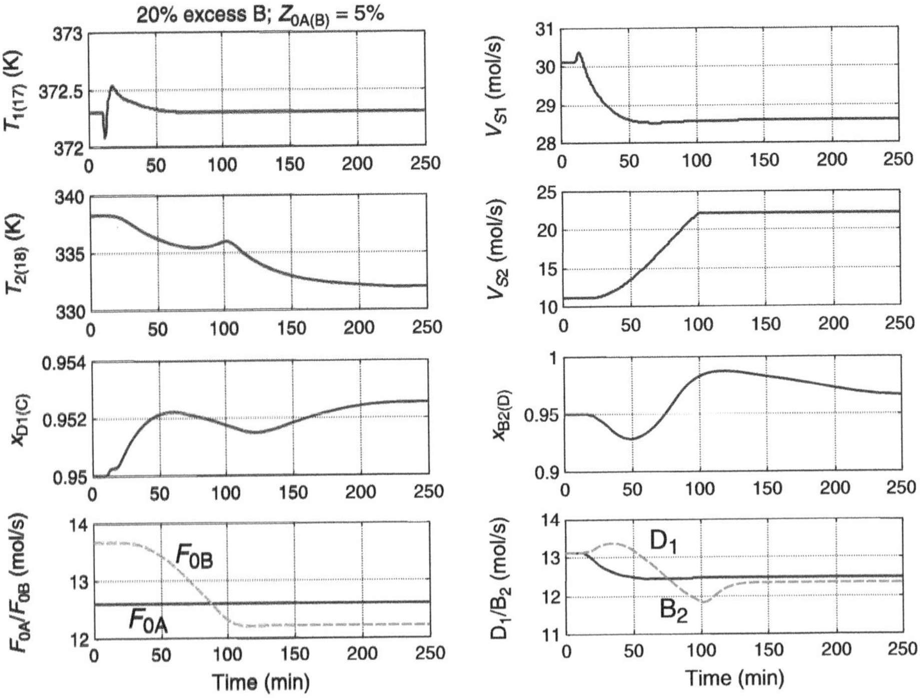

Figure 11.32 Reactive and recovery columns for $z_{0A(B)} = 0.05$.

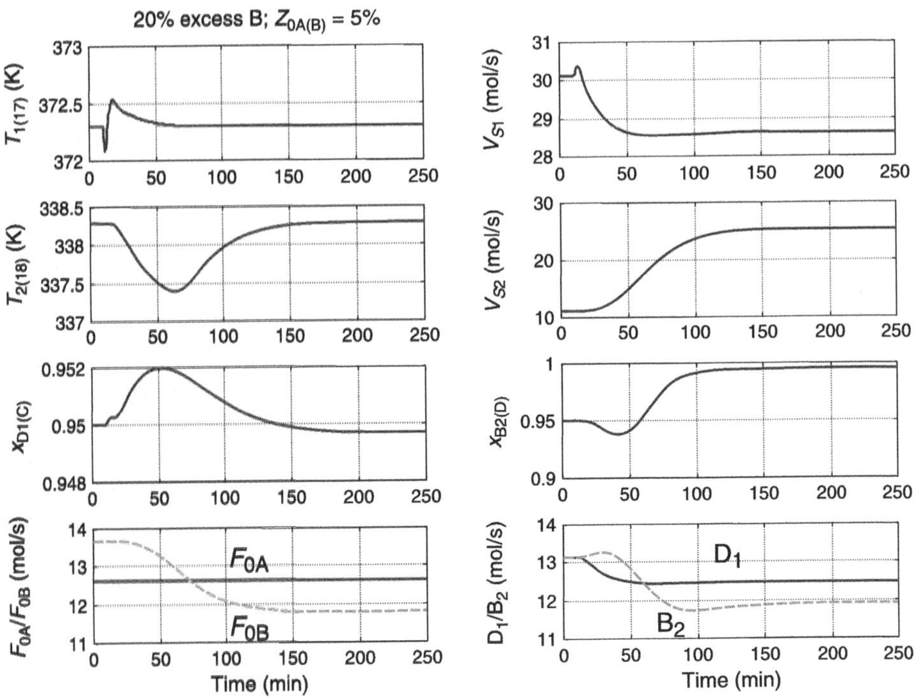

Figure 11.33 Range of V_{S2} doubled for $z_{0A(B)} = 0.05$.

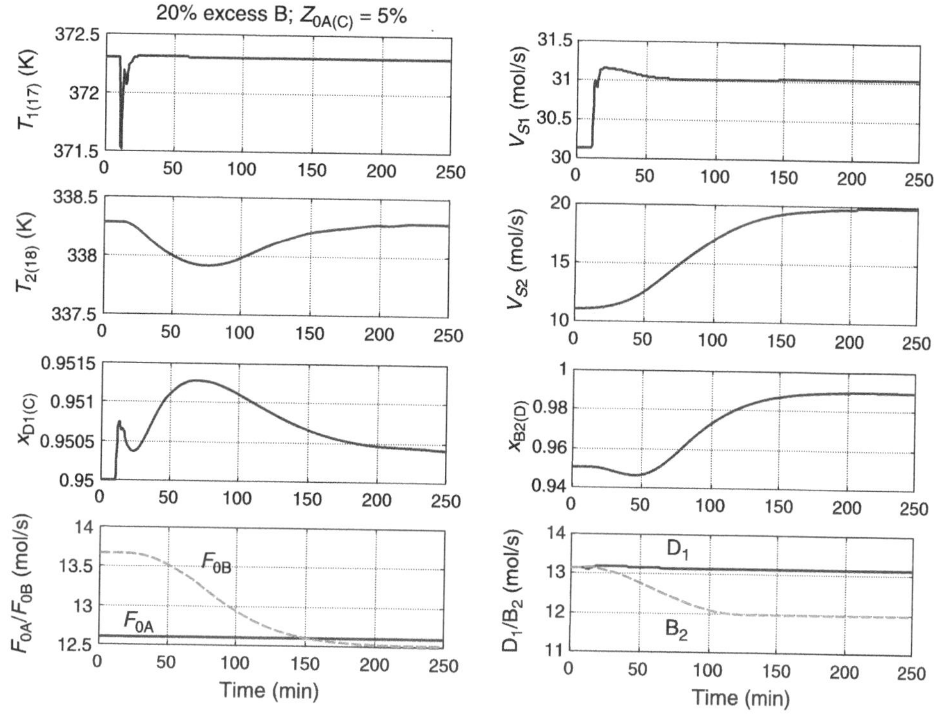

Figure 11.34 Reactive and recovery columns for $z_{0A(C)} = 0.05$.

controller would be required to achieve this objective, and this is discussed in the next section.

Figures 11.35–11.38 show similar results for the other feed composition disturbances. Stable base-level regulatory control is achieved, but the purity of the B_2 product stream varies quite significantly.

11.3.2 Temperature/Composition Cascade Control

The results presented in the previous section show that using a temperature near the top of the recovery column does not maintain the composition of the bottoms product stream for some of the feed composition disturbances. In this section a composition controller is added that changes the setpoint of the tray 18 temperature controller. Figure 11.39 shows the revised control scheme. The composition controller setpoint is 95 mol% D.

In the cascade structure, the secondary loop does not need integral action, so the tray 18 temperature controller is made proportional only with a gain of 4. Three 3-min lags are inserted in the composition loop. The temperature transmitter span is 50 K, so the composition controller output span is 50 K. The composition transmitter span is 20 mol% D. With the tray 18 temperature controller on automatic, a relay-feedback test gives an ultimate gain of 0.16 and an ultimate period of 4.5 min. The tuning parameters of the composition controller are $K_C = 0.05$ and $\tau_I = 10$ min.

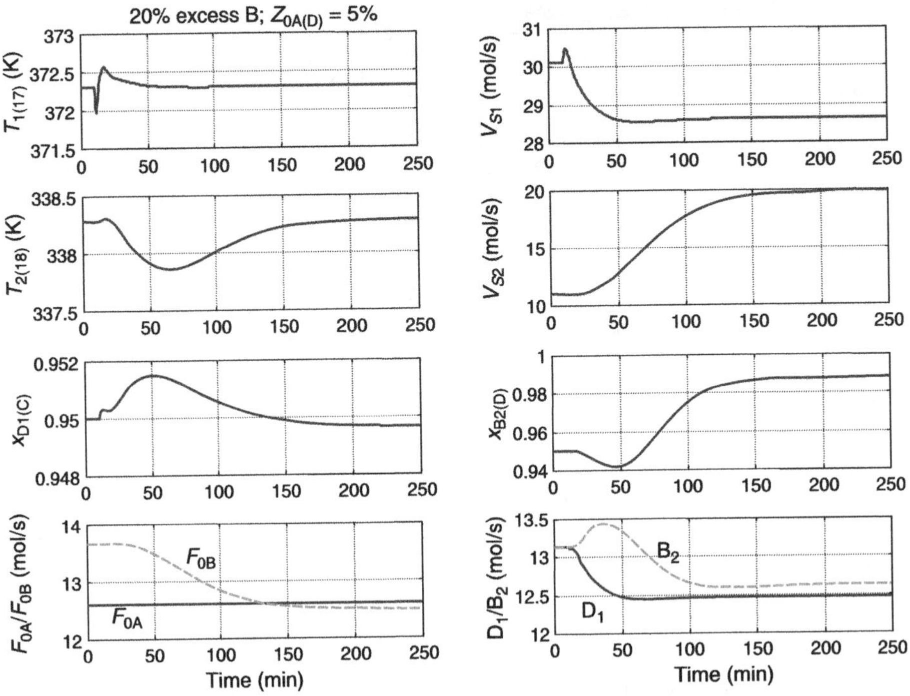

Figure 11.35 Reactive and recovery columns for $z_{0A(D)} = 0.05$.

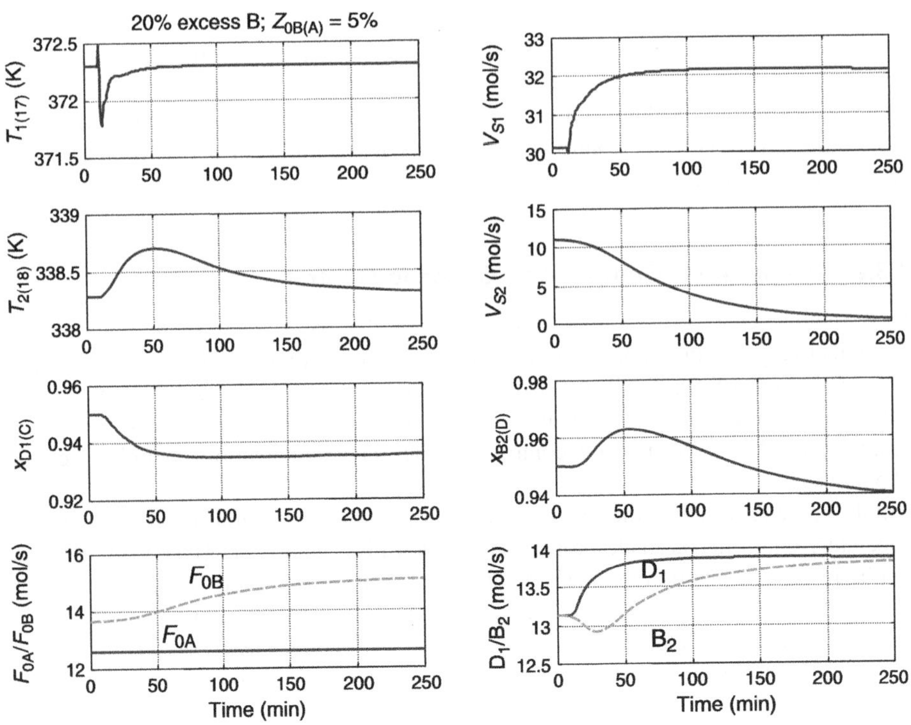

Figure 11.36 Reactive and recovery columns for $z_{0B(A)} = 0.05$.

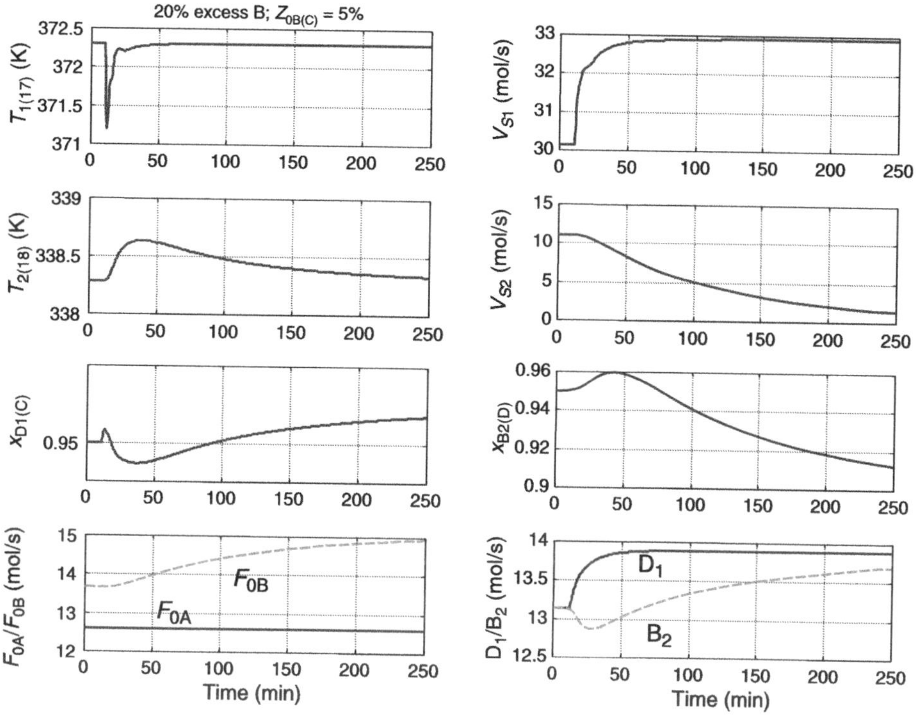

Figure 11.37 Reactive and recovery columns for $z_{0B(C)} = 0.05$.

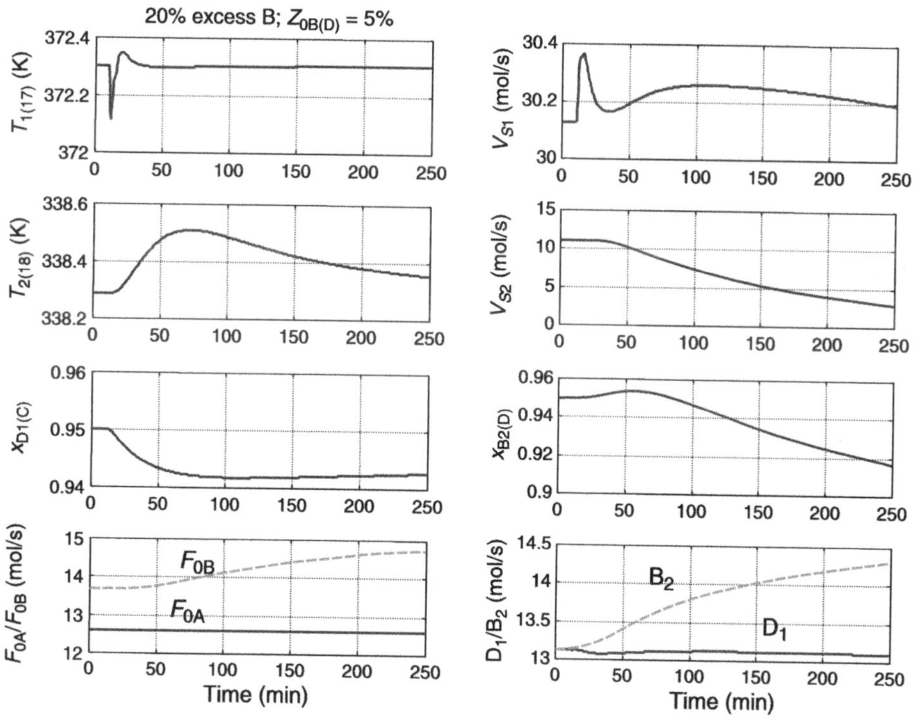

Figure 11.38 Reactive and recovery columns for $z_{0B(D)} = 0.05$.

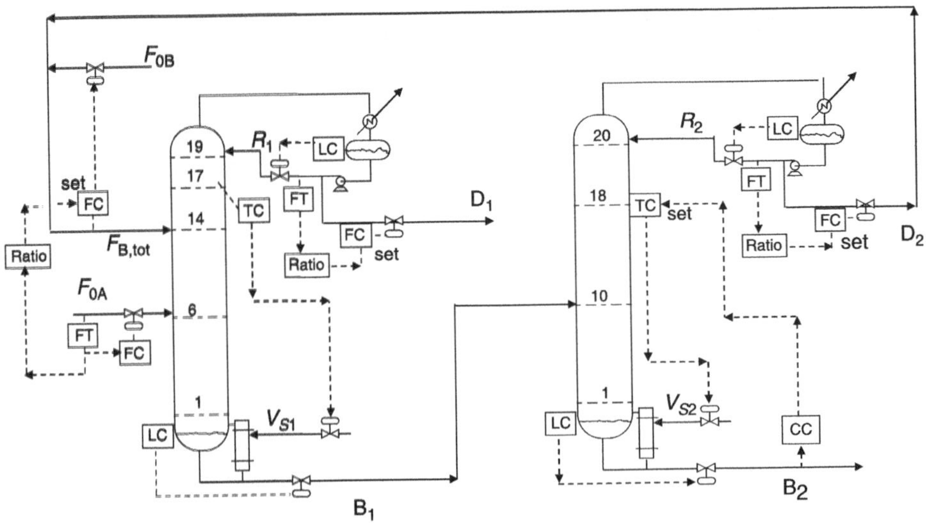

Figure 11.39 Recovery column with composition/temperature cascade.

Figures 11.40–11.47 provide the response of the system for all of the disturbances imposed on the previous systems. Stable base-level regulatory control is achieved in all cases, with the composition of the bottoms from the recovery column driven back to its specification. Figures 11.40 and 11.41 show the responses to step changes in the throughput

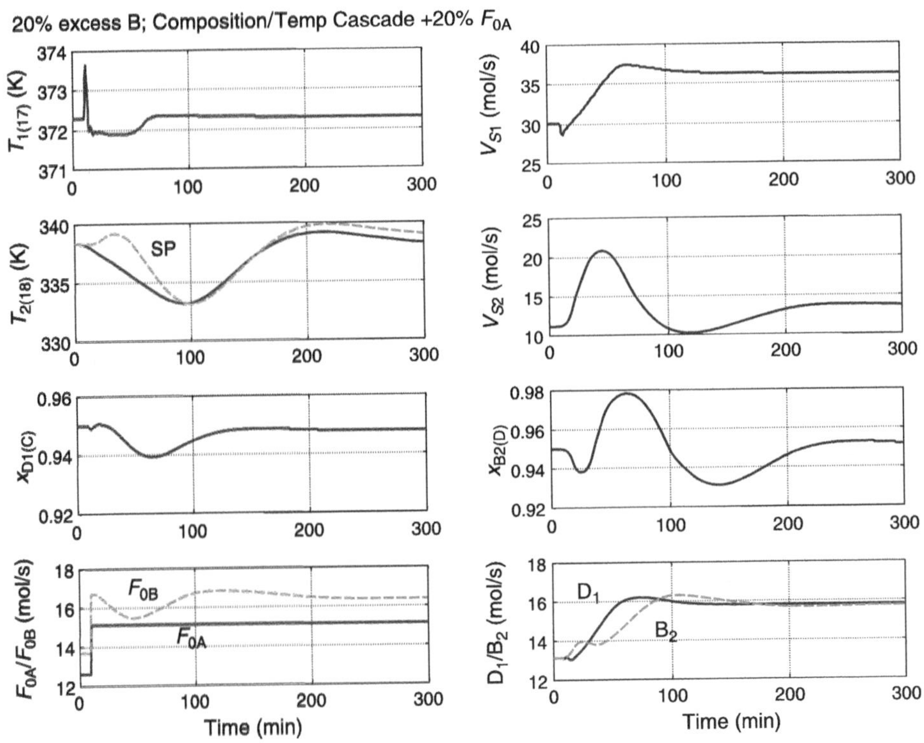

Figure 11.40 Composition/temperature cascade for $+20\%$ F_{0A}.

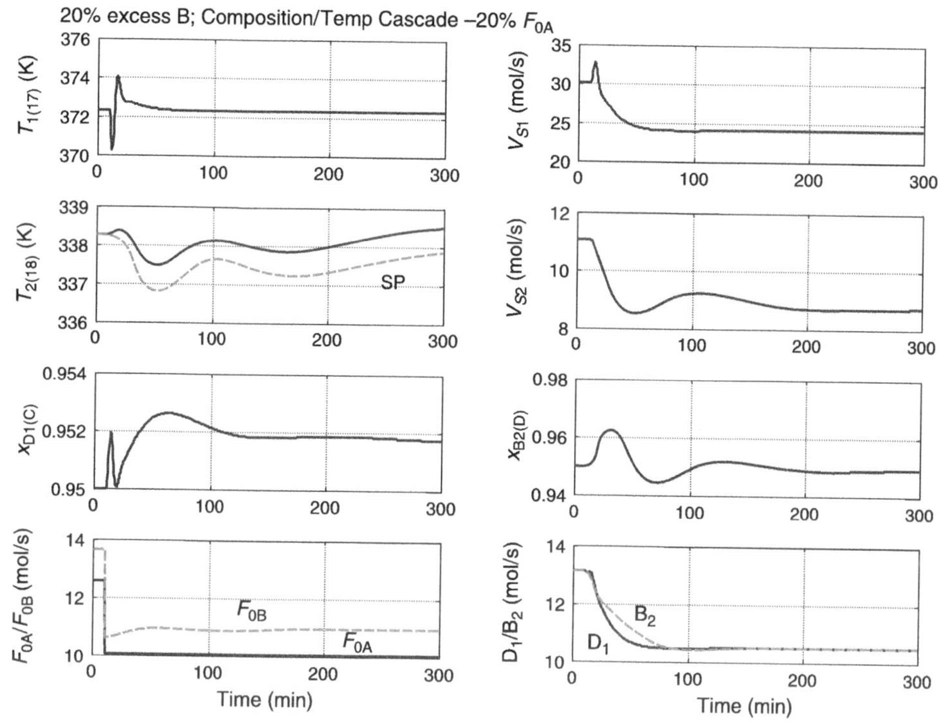

Figure 11.41 Composition/temperature cascade for -20% F_{0A}.

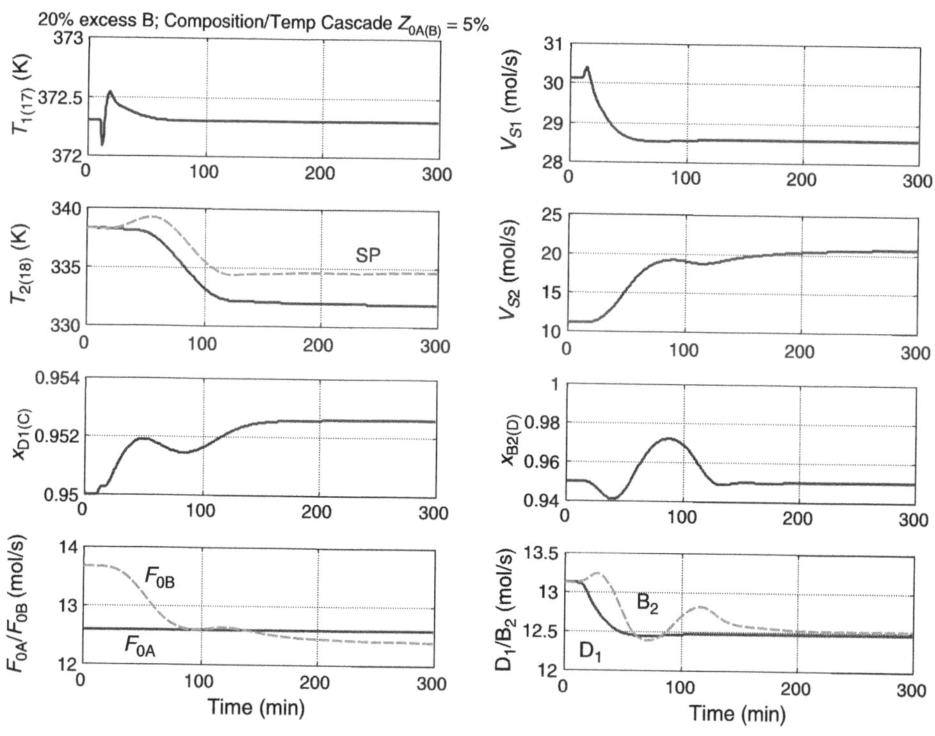

Figure 11.42 Composition/temperature cascade for $z_{0A(B)} = 0.05$.

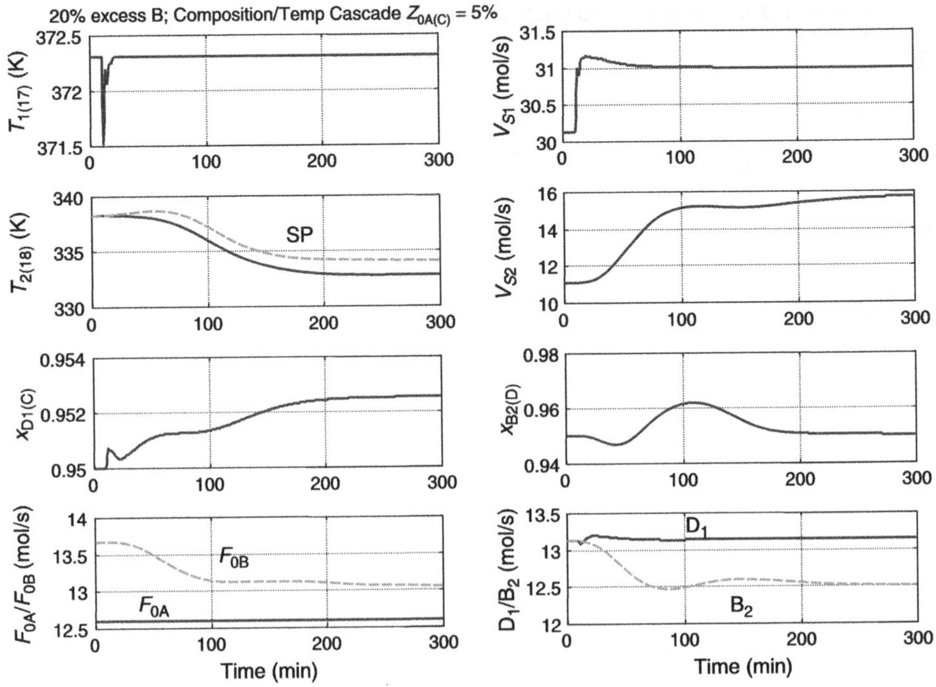

Figure 11.43 Composition/temperature cascade for $z_{0A(C)} = 0.05$.

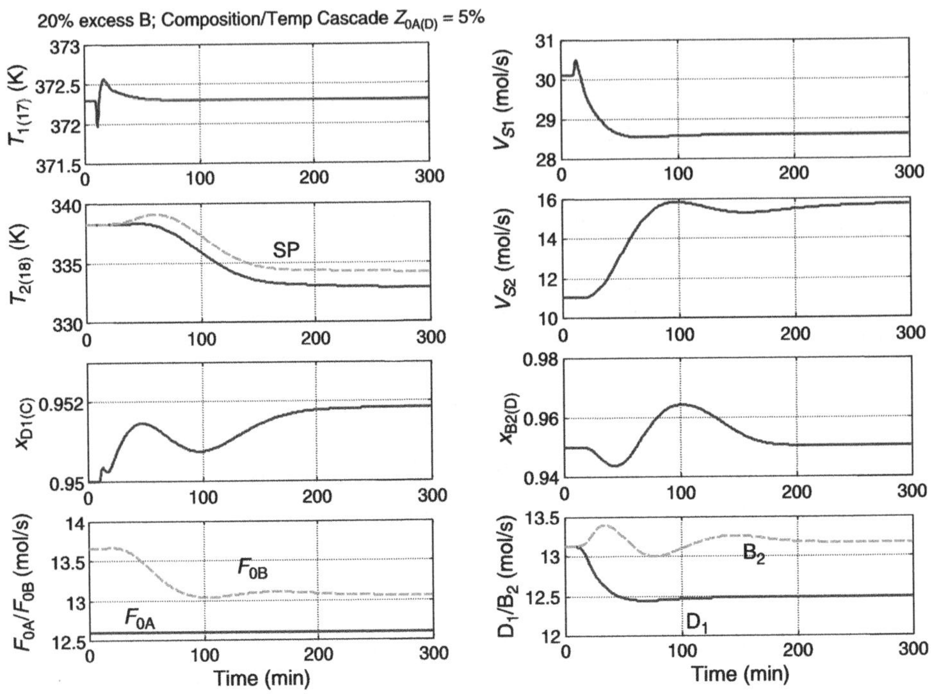

Figure 11.44 Composition/temperature cascade for $z_{0A(D)} = 0.05$.

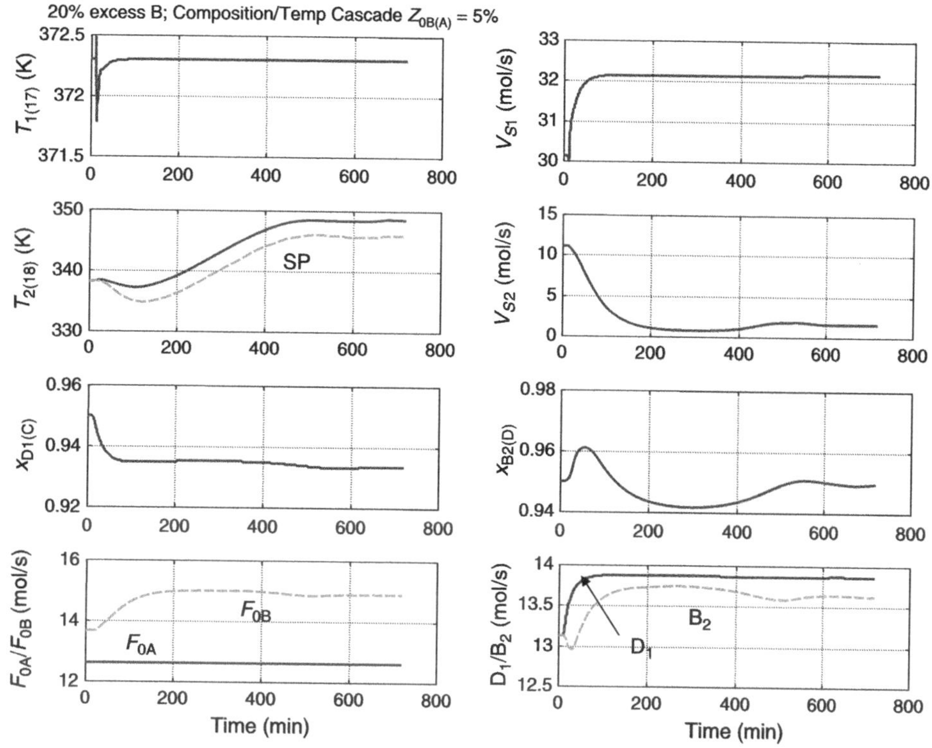

Figure 11.45 Composition/temperature cascade for $z_{0B(A)} = 0.05$.

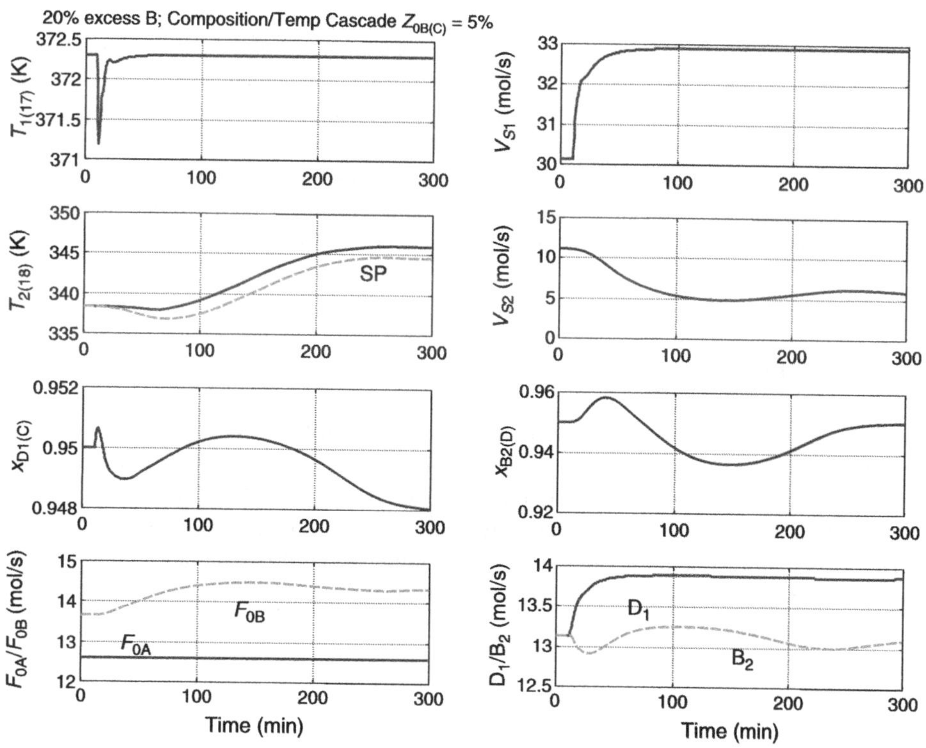

Figure 11.46 Composition/temperature cascade for $z_{0B(C)} = 0.05$.

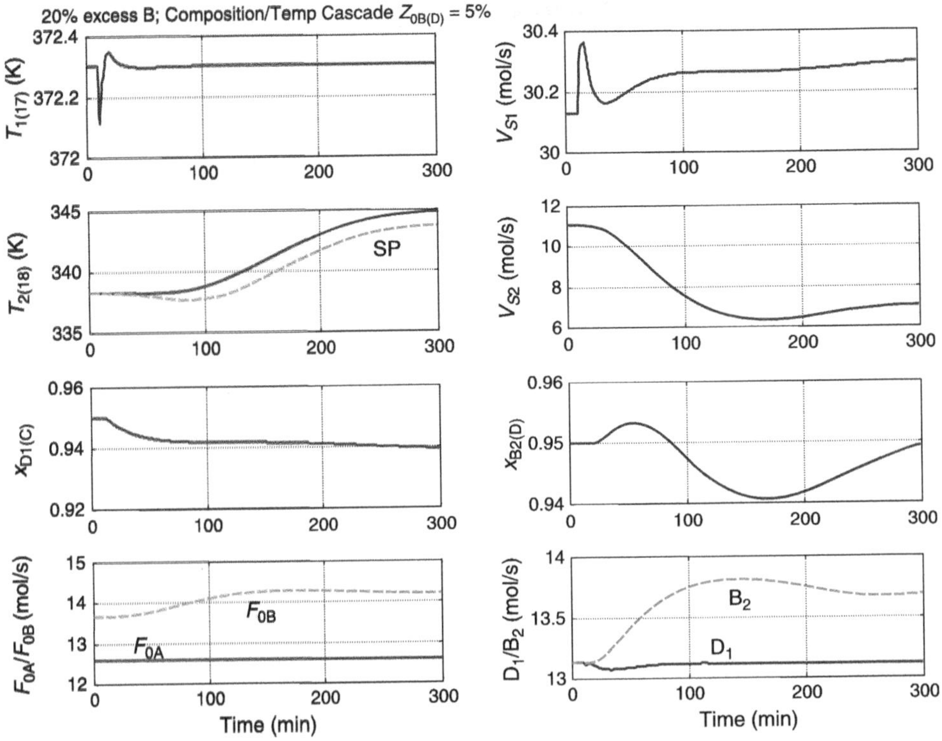

Figure 11.47 Composition/temperature cascade for $z_{0B(D)} = 0.05$.

handle F_{0A}. Figures 11.42–11.45 give the responses to changes in the composition $z_{0A(j)}$. Note that the setpoint of the tray 18 temperature controller is changed to different values in order to drive the bottoms composition to the desired 95 mol% D. Note also that there is an offset between the setpoint and the temperature because of the proportional-only control.

Figures 11.45–11.47 provide the responses to changes in the composition $z_{0B(j)}$. The setpoint of the tray 18 temperature controller is changed to different values in order to drive the bottoms composition to the desired 95 mol% D. Note the time scale in Figure 11.45. It takes about 10 h to come to a new steady state.

11.4 CONCLUSION

The one-column and two-column systems are both controllable using several types of control structures. The two-temperature control scheme for the neat reactive column handles most disturbances, but it cannot handle one type of feed composition disturbance. The use of an internal composition measurement provides more robust control.

The two-column system provides stable operation but does not maintain tight control of product composition for some feed composition disturbances. A cascade composition/temperature structure may be required.

CHAPTER 12

CONTROL OF TERNARY REACTIVE DISTILLATION COLUMNS

Control systems were explored in Chapters 10 and 11 for quaternary reactive distillation systems with the classical two-reactant, two-product, reversible A + B ⇔ C + D reaction. In this chapter we study the control of several types of ternary systems. Their steady-state designs were discussed in Chapters 5 and 6.

Three types of chemical systems are considered. In the first, the reaction is A + B ⇔ C, which is reversible, liquid phase, and exothermic. There are two reactants but only one product. There are no inerts in the feed. In the second system, inerts are added to the feed. In the third system, the reaction is A ⇔ B + C, which has a single feed but two products.

12.1 TERNARY SYSTEM WITHOUT INERTS

12.1.1 Column Configuration

With a one-product reaction system without inerts, the column has only a bottoms product because product component C is heavier than reactant components A and B. Figure 12.1 shows the flowsheet. Steady-state conditions and design parameters are given in Table 12.1. The steady state is the base case considered in Chapter 5. Figures 12.2 and 12.3 give composition and temperature profiles, respectively.

Column diameter is calculated from the sizing relationships given in Chapter 3. A liquid height of 0.05 m is assumed on the stripping trays, giving a tray holdup of 650 mol on these trays. The holdup on the reactive trays is 1000 mol. The holdups in the column base and reflux drum are sized to give a 5-min residence time when 50% full, based on the total liquid entering. In the reflux drum this is the reflux flowrate, which is equal to the overhead

Reactive Distillation Design and Control. By William L. Luyben and Cheng-Ching Yu

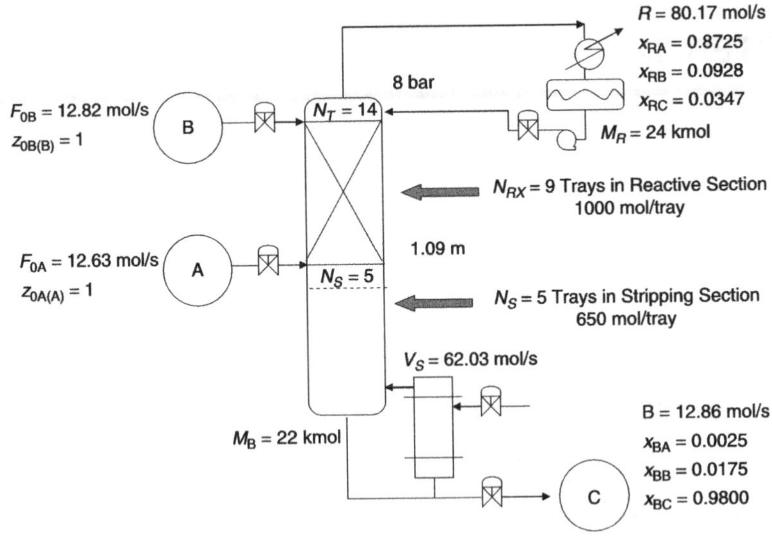

Figure 12.1 Ternary reactive distillation column without inerts.

TABLE 12.1 Steady-State Conditions and Design Parameters for Ternary System Without Inerts

Fresh feed flowrate		
A (mol/s)	12.63	
B (mol/s)	12.82	
Distillate flowrate (mol/s)	0	
Bottoms flowrate (mol/s)	12.857	
Vapor boilup (mol/s)	62.03	
Reflux flowrate (mol/s)	80.17	
Overhead vapor flowrate (mol/s)	80.17	
Stripping trays	5	
Reactive trays	9	
Rectifying trays	0	
Liquid holdup		
On reactive trays (mol)	1000	
On separation trays (mol)	650	
In base (kmol)	22	
In reflux drum (kmol)	24	
Column ID (m)	1.09	
Pressure (bar)	8	
Vapor Pressure Constants	A_j	B_j
A	12.34	3862
B	11.65	3862
C	10.96	3862
Product Composition (Mole Fraction)	Reflux	Bottoms
A	0.8725	0.0025
B	0.0928	0.0175
C	0.0347	0.9800

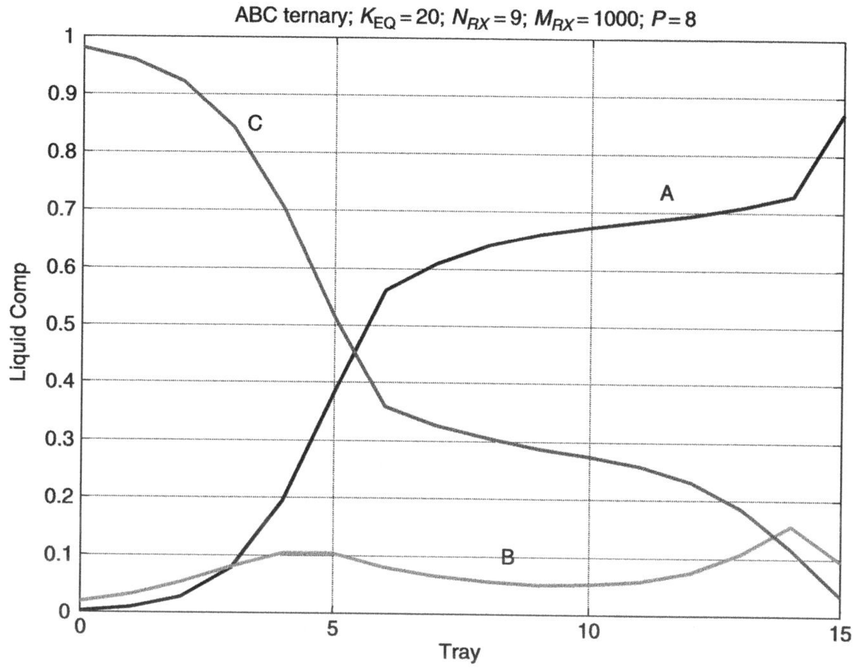

Figure 12.2 Composition profiles.

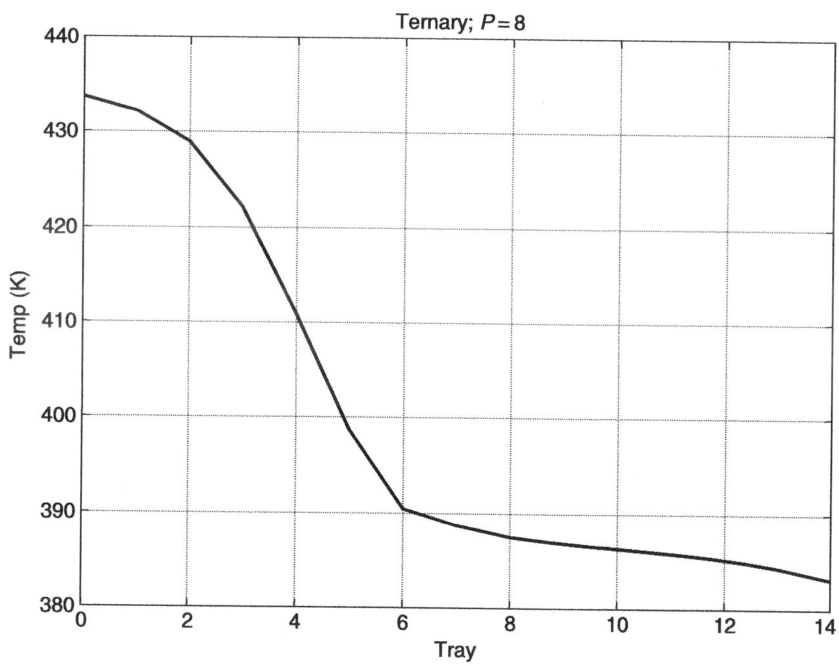

Figure 12.3 Temperature profile.

vapor flowrate because there is no distillate. In the column base, the total liquid entering is the liquid coming from tray 1, which is the bottoms plus the vapor boilup.

The following sections explore three alternative control structures. The primary objective is to maintain the specified purity of the bottoms product at 98 mol% C in the face of disturbances in production rate and feed compositions.

12.1.2 Control Structure CS1

Figure 12.4 shows a control structure that uses two direct composition measurements. Bottoms product purity (mol% C) is measured and controlled by vapor boilup. The composition of A on tray 5 is measured and controlled by the flowrate of fresh feed F_{0A}. Throughput is set by flow controlling fresh feed F_{0B}. Reflux-drum level is controlled by manipulating the reflux flowrate, and base level is controlled by manipulating the bottoms flowrate.

Tray 5 is selected because this is the location where the composition of A is changing rapidly from tray to tray. The steady-state composition is 38 mol% A. Two 0.5-min first-order composition measurement lags are inserted in both composition control loops. Composition transmitter spans are 20 mol%. All valves are 50% open at steady state.

The controllers are tuned by running relay-feedback tests. Values of ultimate gains and periods are provided in Table 12.2. Tyreus–Luyben tuning is used in most cases, but some loops are detuned to give larger closed-loop damping coefficients.

Results for the CS1 control structure are given in Figures 12.5–12.8. In Figure 12.5 the disturbance is a positive 20% step change in the throughput handle, V_S, at 10 min. The

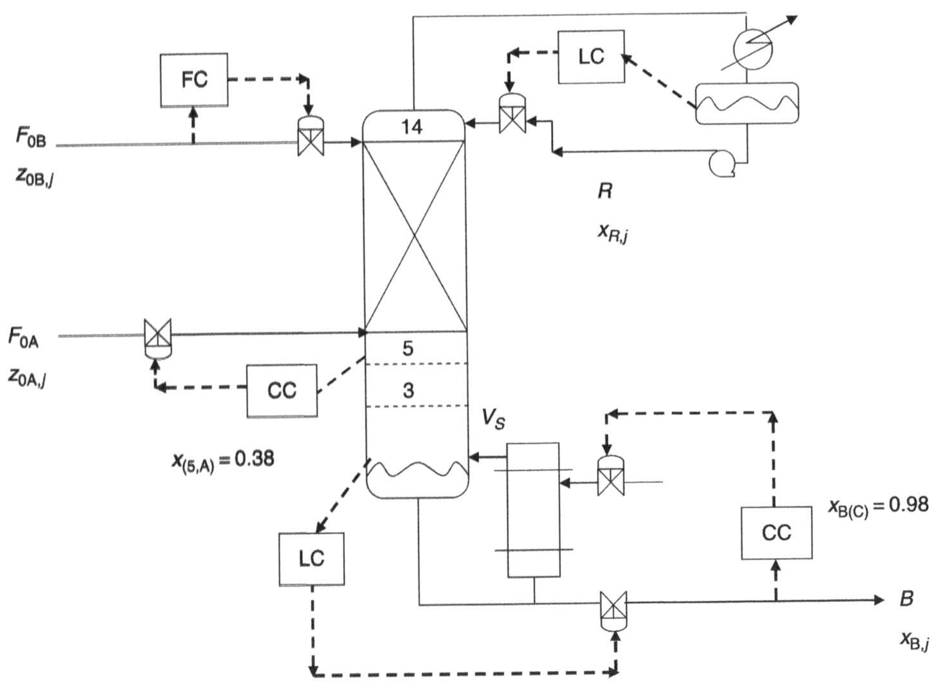

Figure 12.4 Control structure CS1.

TABLE 12.2 Controller Tuning Parameters for Ternary Without Inerts

CS1		CS2		CS3	
Controlled Variable $x_{B(C)}$	Manipulated Variable V_S	Controlled Variable Temp. Tray 3 (K)	Manipulated Variable V_S	Controlled Variable Temp. Tray 3 (K)	Manipulated Variable F_{0A}
K_U	9.1		2.6		12
P_U (min)	5.0		2.0		2.7
K_C	1.0		0.8		3.6
τ_I (min)	11		4.4		6.0
Setpoint	0.98		422 (K)		422.1 (K)
$x_{(5,A)}$	F_{0A}	$x_{(5,A)}$	F_{0A}	Temp. Tray 5 (K)	F_{0B}
K_U	1.8		1.8		2.6
P_U (min)	3.0		3.0		17
K_C	0.5		0.5		0.83
τ_I (min)	6.6		6.6		37
Setpoint	0.38		0.38		398.8 (K)

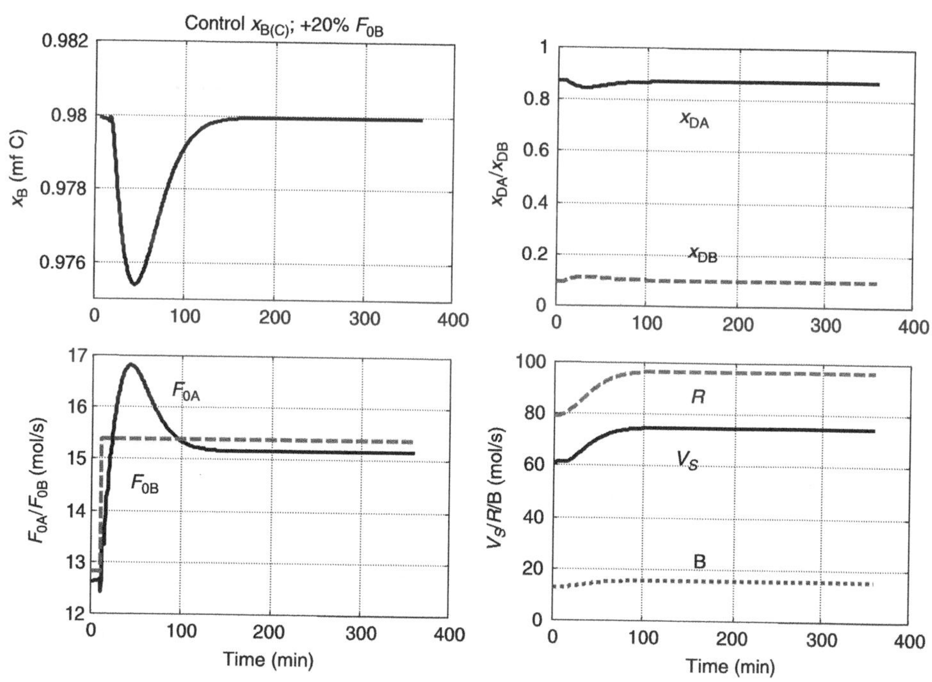

Figure 12.5 Control structure CS1 with +20% ΔF_{0B}.

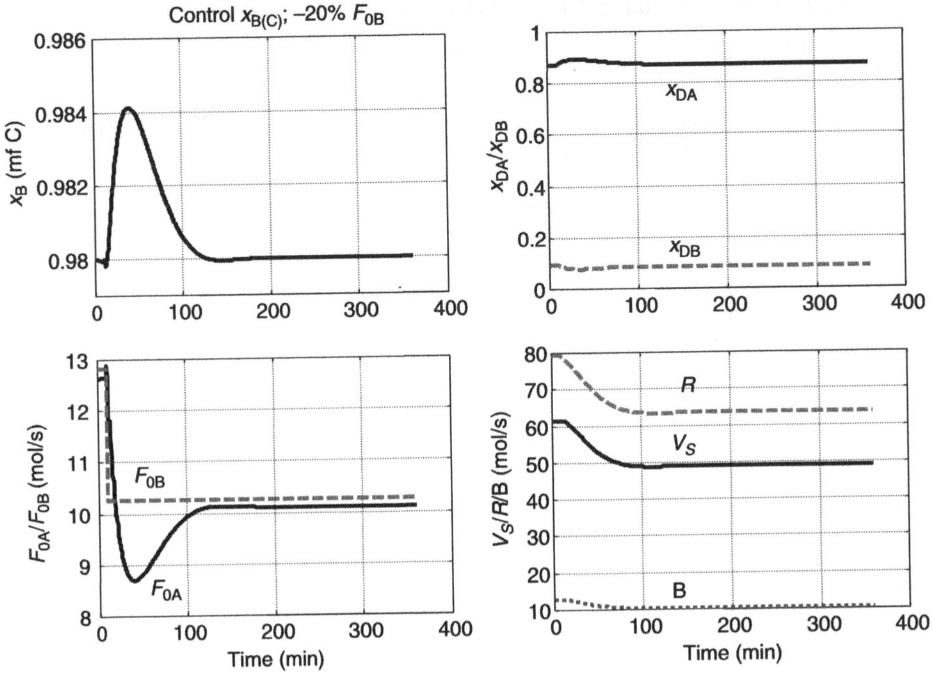

Figure 12.6 Control structure CS1 with −20% ΔF_{0B}.

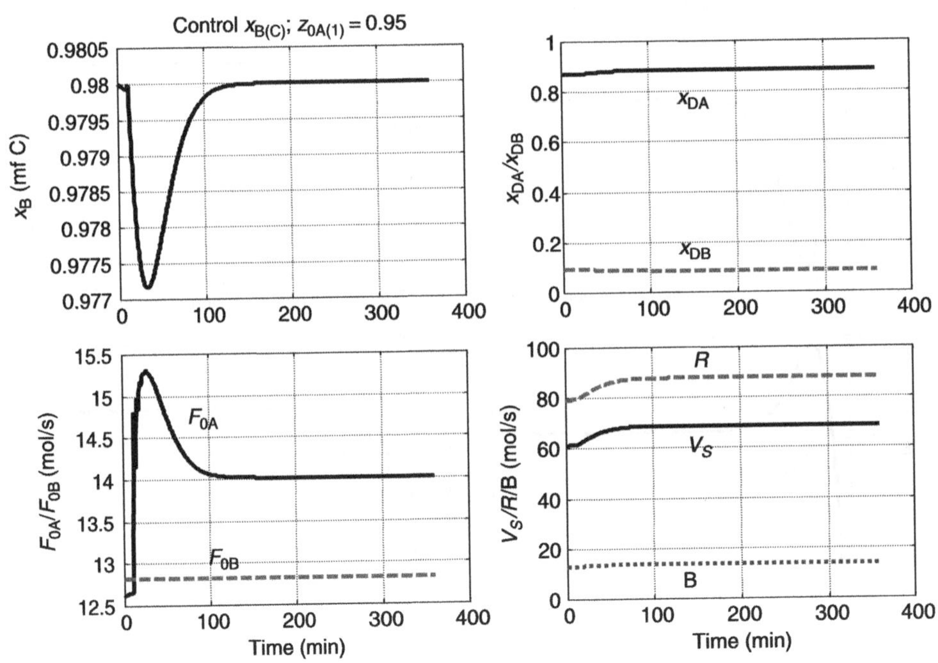

Figure 12.7 Control structure CS1 with $z_{0A(A)} = 0.95$ and $z_{0A(B)} = 0.05$.

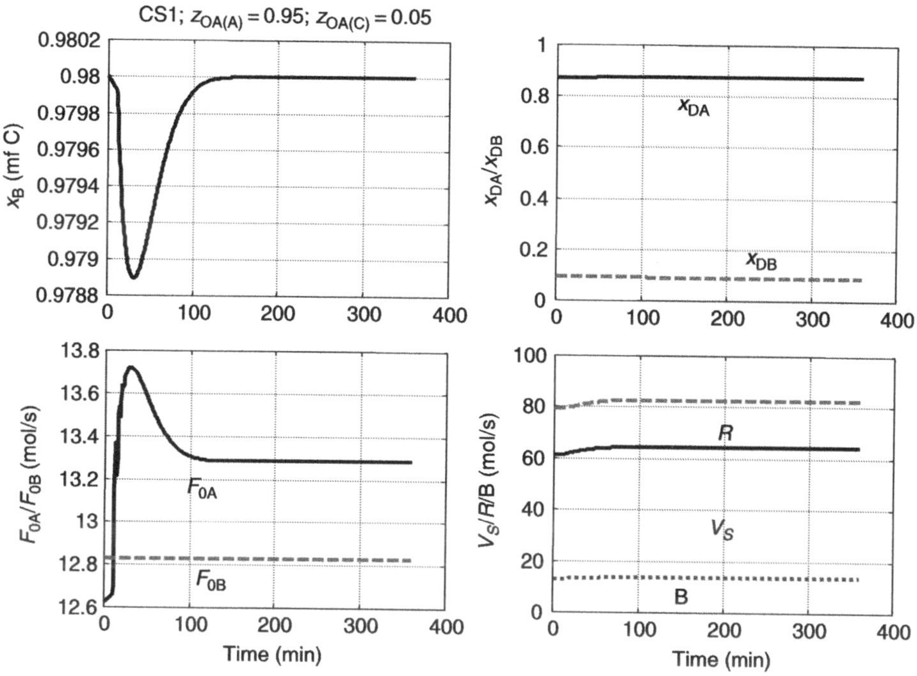

Figure 12.8 Control structure CS1 with $z_{0A(A)} = 0.95$ and $z_{0A(C)} = 0.05$.

bottoms product purity drops to about 97.5 mol% C during the transient but recovers back to its specified value of 98 mol% C in about 2 h. The fresh feed flowrate F_{0A} of A is increased by the $x_{(5,A)}$ controller to bring in the required amount of A to react with the increased input of B. More vapor boilup V_S and reflux R are required. Reflux composition x_{Dj} changes very little.

Figure 12.6 gives results for a negative 20% step change in V_S. Control performance is essentially the same as for the positive disturbance. Figure 12.7 shows what happens when the composition z_{0A} of fresh feed F_{0A} is changed from pure A ($z_{0A(A)} = 1$) to a mixture of A and B ($z_{0A(A)} = 0.95$, $z_{0A(B)} = 0.05$). The $x_{(5,A)}$ controller increases the F_{0A} flowrate to compensate for the lower amount of A in that stream. Because more B is coming into the system, more product C is produced and the bottoms flowrate increases. More vapor boilup and reflux are required.

Figure 12.8 shows what happens when composition z_{0A} of fresh feed F_{0A} is changed from pure A ($z_{0A(A)} = 1$) to a mixture of A and C ($z_{0A(A)} = 0.95$, $z_{0A(C)} = 0.05$). The $x_{(5,A)}$ controller increases the F_{0A} flowrate to compensate for the lower amount of A in that stream. Because the same amount of B is coming into the system, the same amount of product C is produced in the reaction, but the small additional amount of C in the feed produces a small increase in the bottoms flowrate.

The CS1 control structure successfully handles these fairly large disturbances. However, it has the disadvantage of requiring two on-line composition measurements. In the following sections we explore the use of temperature measurements instead of composition measurements. Compositions are much more difficult and expensive to measure than temperatures, so it is highly desirable to find control schemes that provide effective control using only temperatures when this is possible, which is not always the case.

12.1.3 Control Structure CS2

Figure 12.9 shows a control structure in which, instead of measuring the bottoms composition, the temperature on tray 3 is controlled by manipulating vapor boilup. The temperature profile shown in Figure 12.3 is changing rapidly from tray to tray at this location, so it should provide a good indication of the composition of heavy component C at a tray near the location from which the product is removed. The setpoint of this $T_{(3)}$ temperature controller is 422 K. Relay-feedback testing yields the tuning parameters given in Table 12.2. The span of the temperature transmitter is 50 K.

Figures 12.10–12.13 show the performance of this CS2 control structure for the same four disturbances considered in the previous section. Because the bottoms composition is not measured directly and we are only holding a temperature constant, there is no guarantee that the composition of the bottoms will be held exactly at its specification. However, for the throughput disturbances (changes in F_{0B}), the bottoms purity is maintained quite close to its specification. There is more of an offset in the bottoms composition for the two feed composition disturbances, but this small drop in product purity may be quite acceptable. Thus, the use of a temperature in the stripping section may permit elimination of the need for a direct composition measurement of the bottoms.

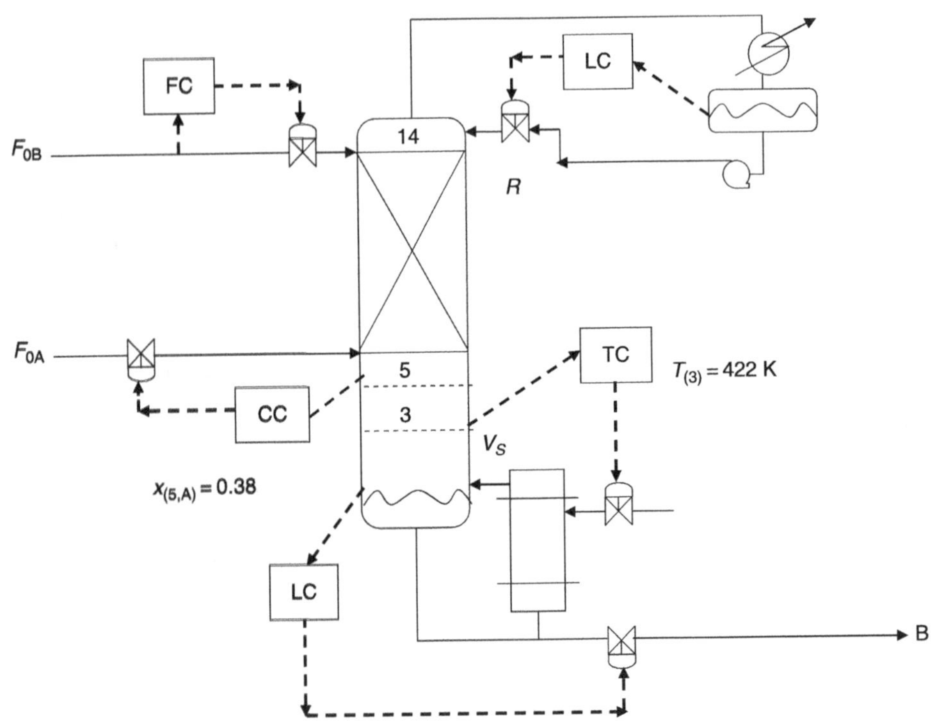

Figure 12.9 Control structure CS2.

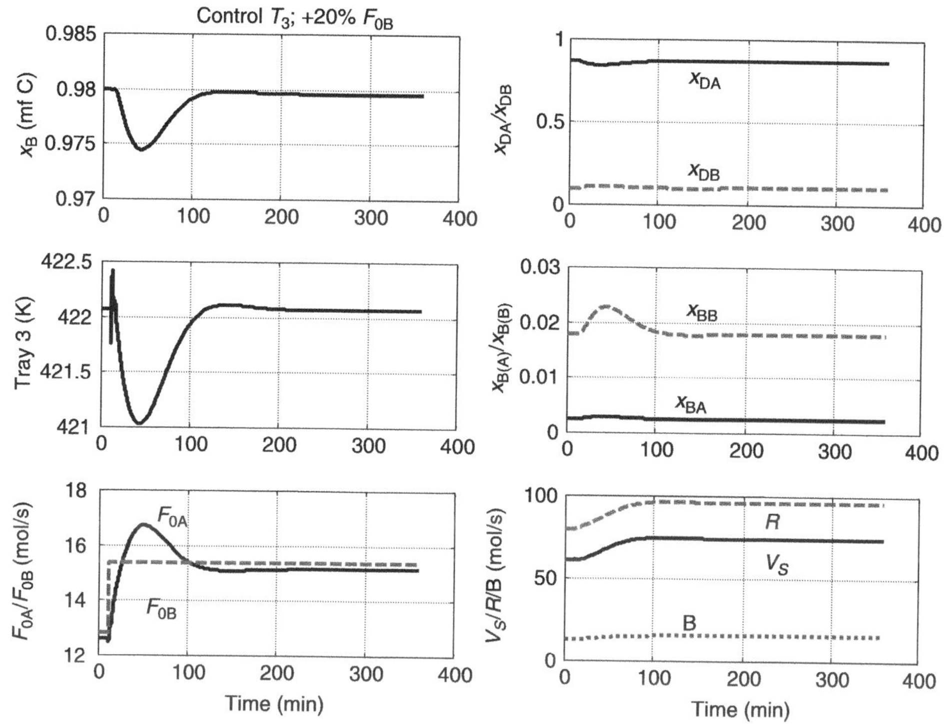

Figure 12.10 Control structure CS2 with +20% ΔF_{0B}.

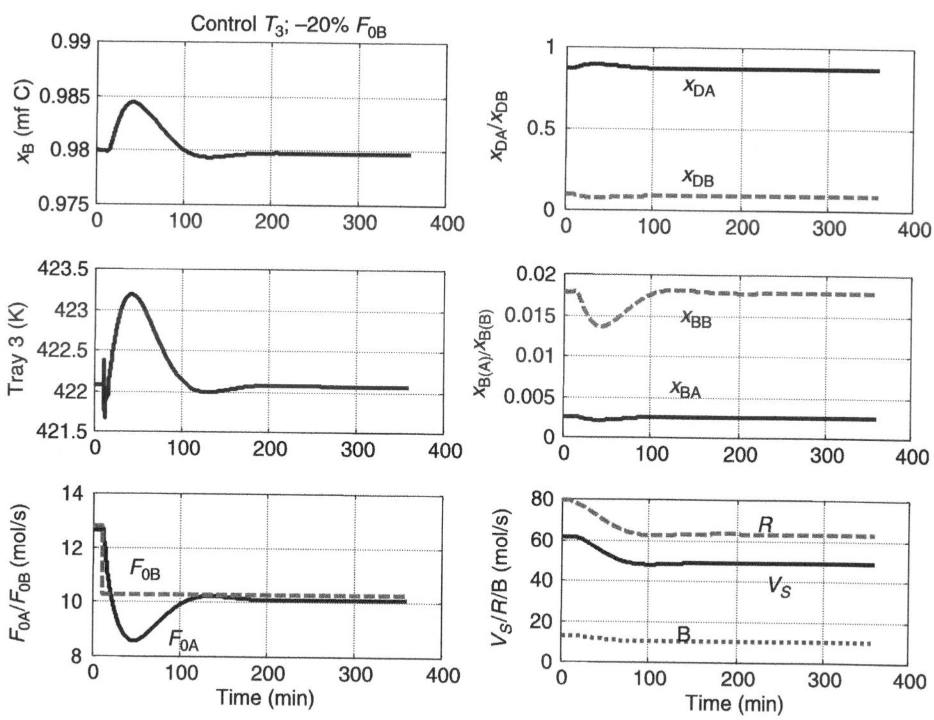

Figure 12.11 Control structure CS2 with –20% ΔF_{0B}.

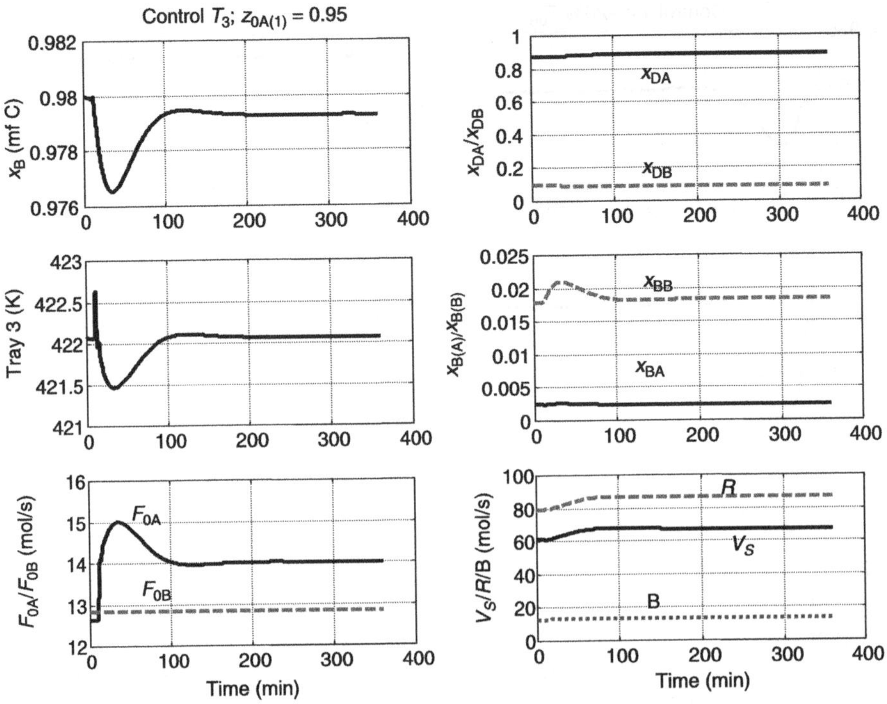

Figure 12.12 Control structure CS2 with $z_{0A(A)} = 0.95$ and $z_{0A(B)} = 0.05$.

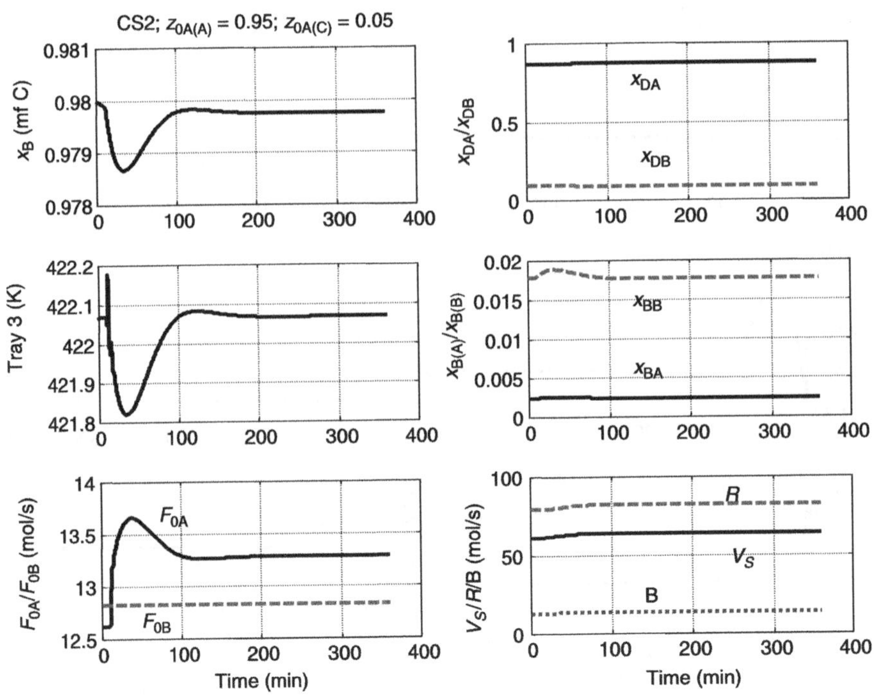

Figure 12.13 Control structure CS2 with $z_{0A(A)} = 0.95$ and $z_{0A(C)} = 0.05$.

12.1.4 Control Structure CS3

In Chapter 10 we investigated a two-temperature control structure for the quaternary, two-product system. We demonstrated that an internal composition measurement is not required in that system to provide the extremely precise balancing of the stoichiometry of the reaction, that is, feeding exactly the right amount of reactants so that no excess builds up in the column. Will a similar two-temperature control structure be effective in the ternary system without inerts? This structure is shown in Figure 12.14. The two fresh feeds are manipulated to control the temperatures on two trays.

The first issue is to find what trays to control. In Chapter 10 we found the steady-state gains and used SVD to help in the selection of control trays. The same approach is applied in the ternary system.

Using Steady-State Gains. Figure 12.15 shows the steady-state changes in tray temperatures throughout the column for several small changes in vapor boilup. These are obtained by making small changes in vapor boilup from its steady-state value and converging to a new steady state. The level controllers are on automatic during this procedure. The ΔV_S must be quite small to obtain the true linear gains (gain values do not change as the size of the ΔV_S is decreased). These results show that tray 4 is the most sensitive for vapor boilup changes.

The same procedure is used for the other two inputs into the system: fresh feed flowrates F_{0A} and F_{0B}. Results are provided in Figure 12.16. Tray 3 is the most sensitive to changes in

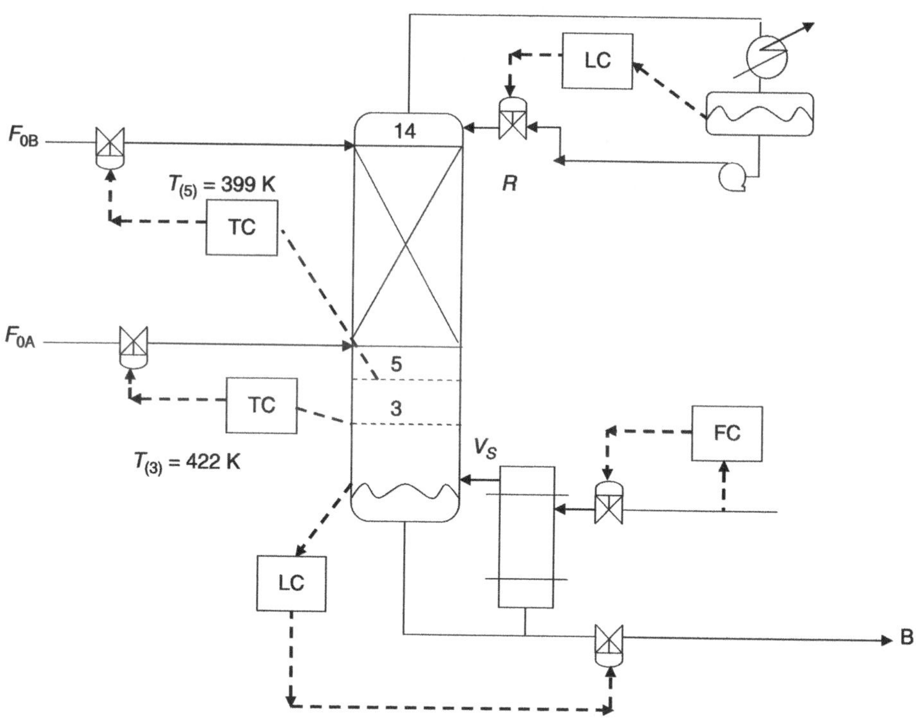

Figure 12.14 Control structure CS3.

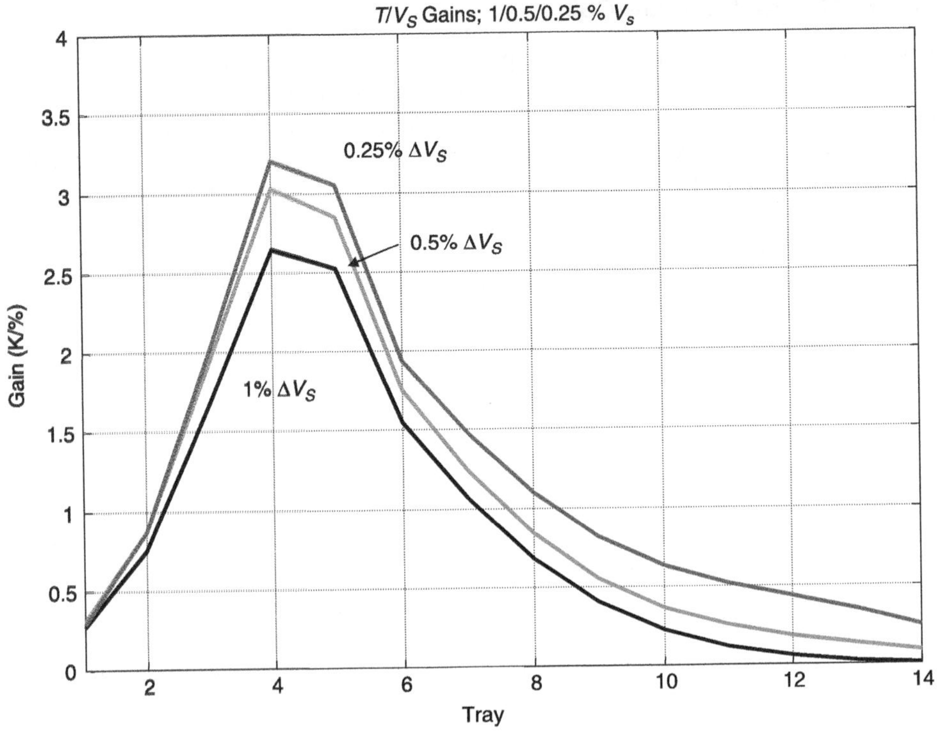

Figure 12.15 Steady-state gains for different ΔV_S values.

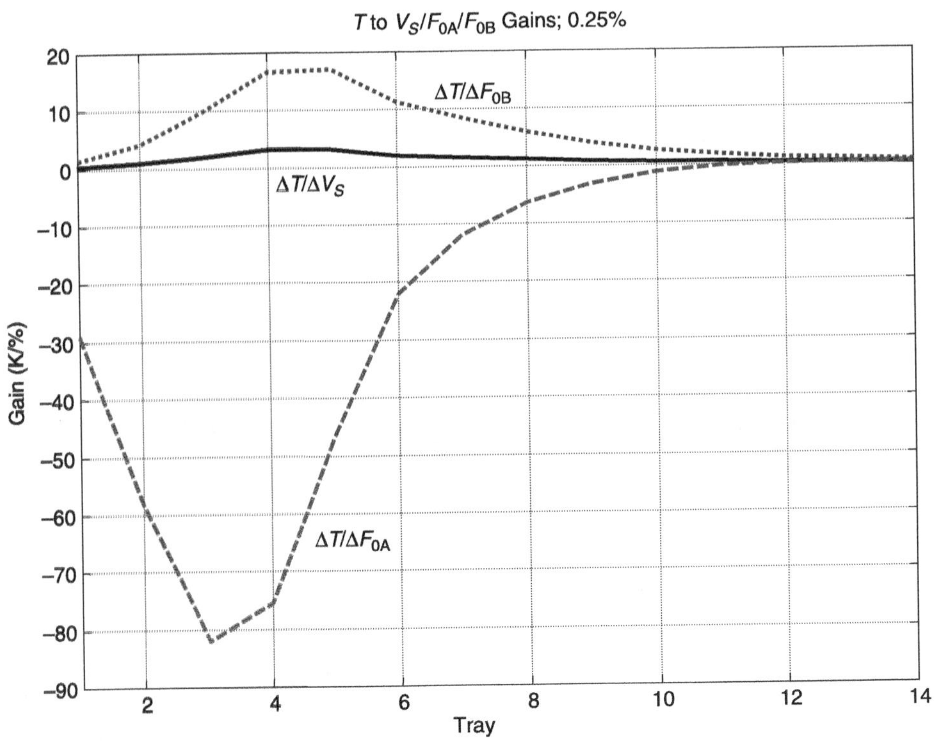

Figure 12.16 Steady-state gains.

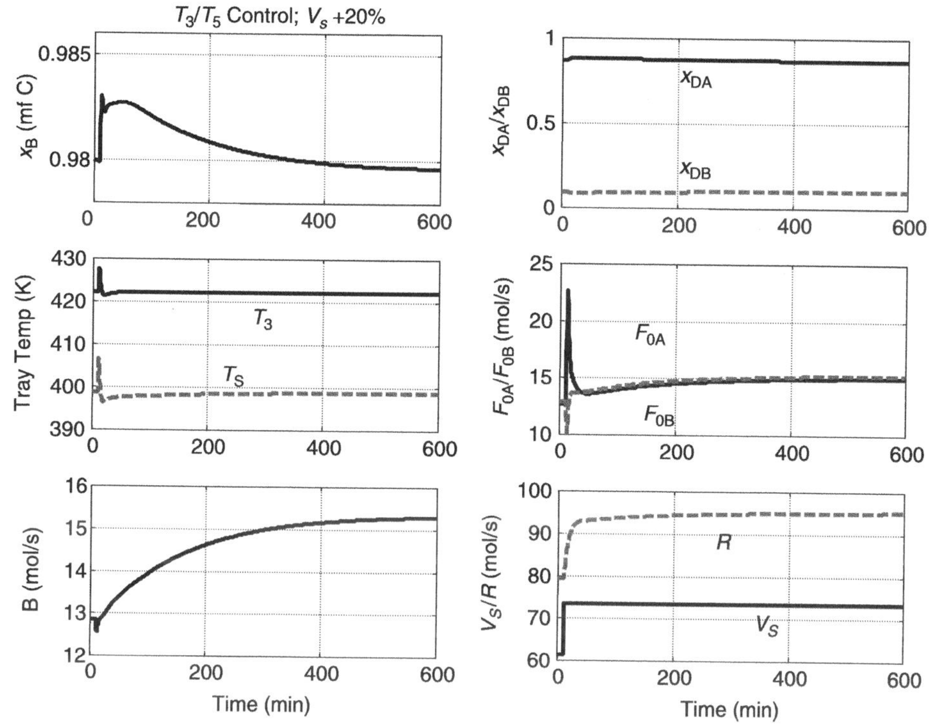

Figure 12.17 Control structure CS3 with +20% V_S.

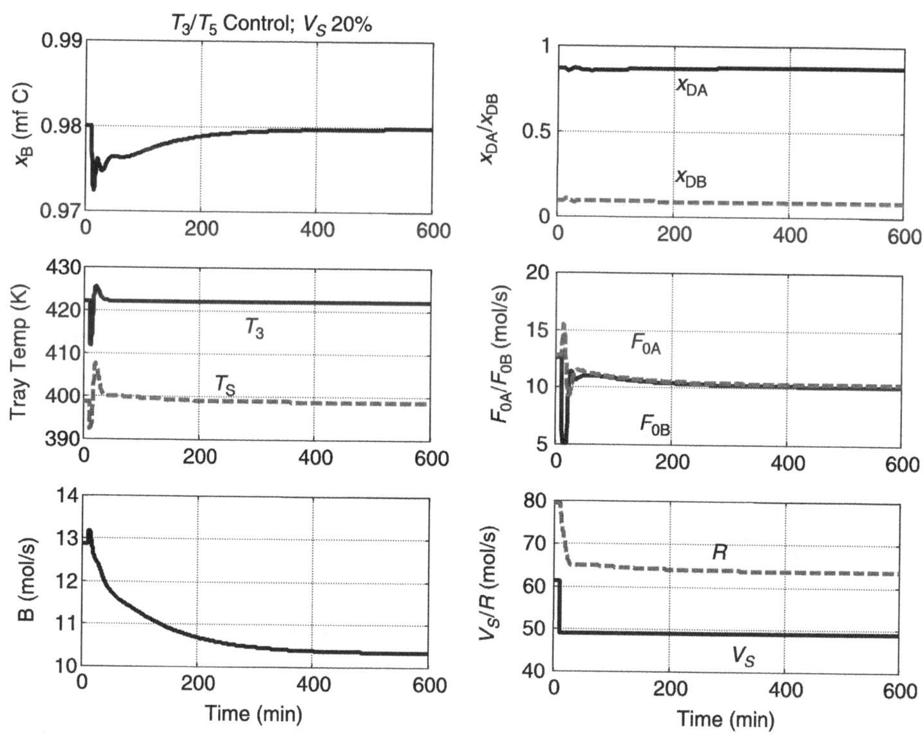

Figure 12.18 Control structure CS3 with −20% V_S.

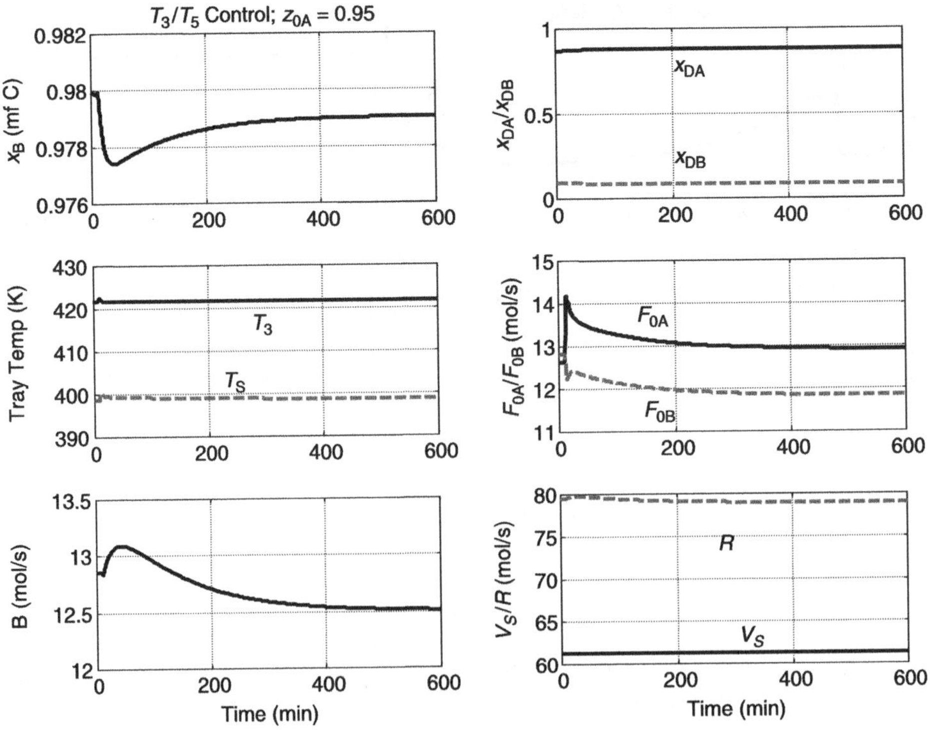

Figure 12.19 Control structure CS3 with $z_{0A(A)} = 0.95$ and $z_{0A(B)} = 0.05$.

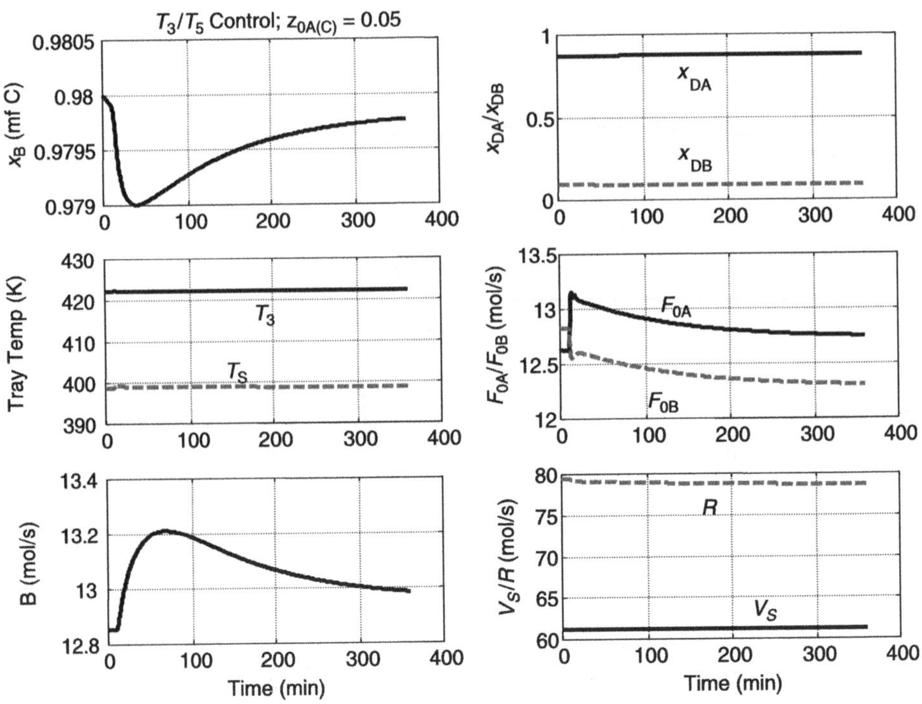

Figure 12.20 Control structure CS3 with $z_{0A(A)} = 0.95$ and $z_{0A(C)} = 0.05$.

F_{0A}, and tray 5 is the most sensitive to changes in F_{0B}. Therefore, these locations are selected, as shown in Figure 12.14. The two singular values for this control structure are $\sigma_1 = 142$ and $\sigma_2 = 17$, giving a condition number of 8.4, which indicates that this control structure should perform adequately.

Figures 12.17–12.20 demonstrate that this control structure provides quite acceptable control in the face of all four of the disturbances. Stable base-level regulatory control is attained. The bottoms purity is maintained close to the desired value for all of the disturbances.

Using SVD Analysis. Figure 12.21 shows the U vectors from the SVD analysis. The most sensitive locations are trays 3 and 5, which are those suggested by the steady-state gains. However, there is a third location at tray 12 that is indicated as sensitive. This is unexpected because the gains in this region, as shown in Figure 12.6, are all quite small. However, the $\Delta T/\Delta F_{0B}$ gain at tray 12 is the largest of the three. Thus, a control structure is evaluated in which the tray 5 temperature is controlled by manipulating F_{0A} and tray 12 temperature is controlled by manipulating F_{0B}.

The results given in Figures 12.22–12.25 demonstrate that this selection of control trays does provide stable base-level regulatory control. However, the purity of the bottoms product is not held as close to its specification as with the original choice of trays 3 and 5. This is particularly true for the changes in feed composition.

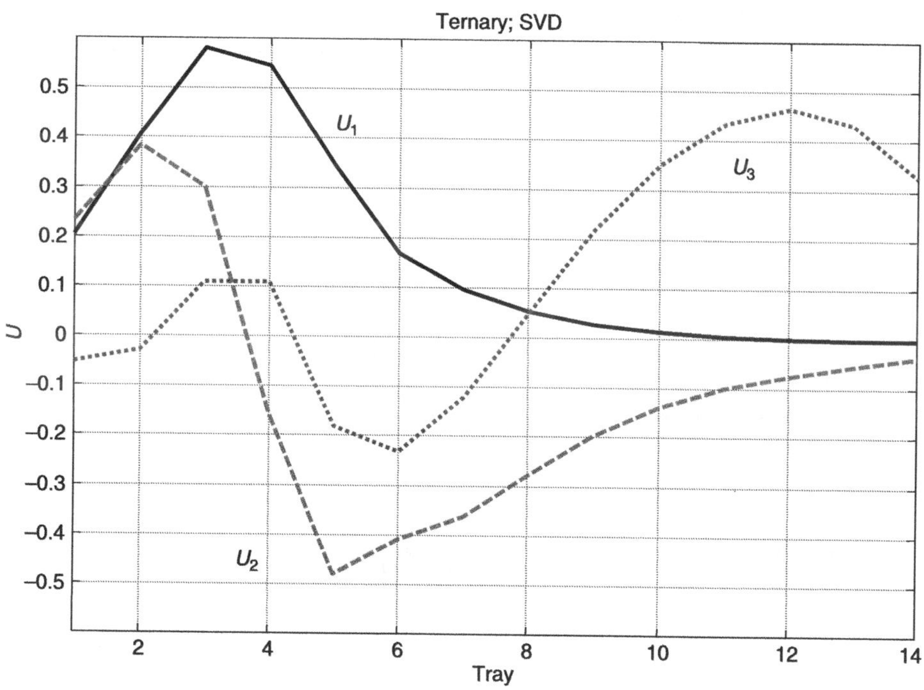

Figure 12.21 The SVD analysis.

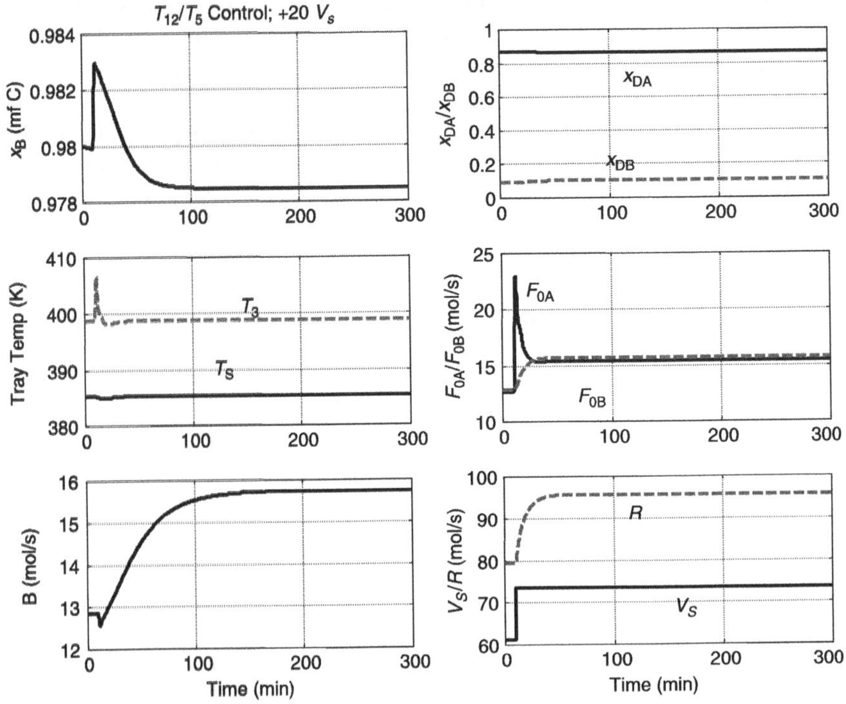

Figure 12.22 Control structure CS3 with +20% V_S for T_5 and T_{12}.

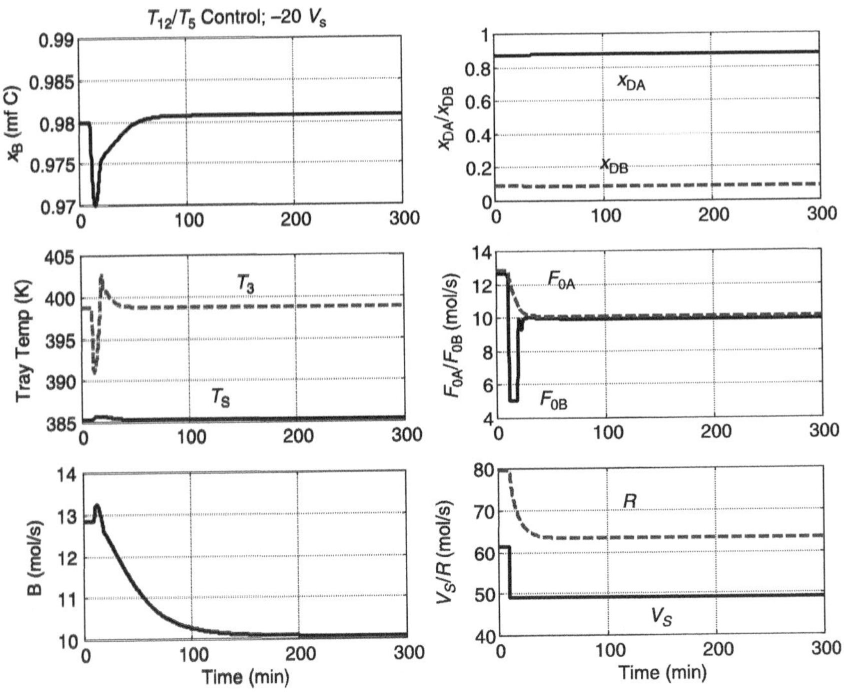

Figure 12.23 Control structure CS3 with −20% V_S for T_5 and T_{12}.

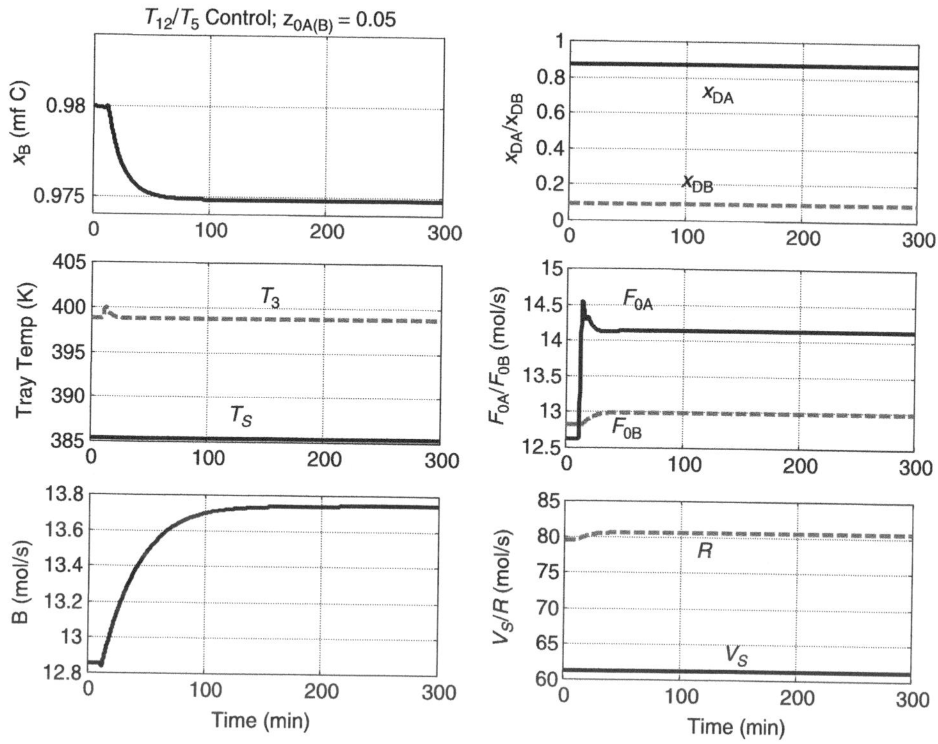

Figure 12.24 Control structure CS3 with $z_{0A(B)} = 0.05$ for T_5 and T_{12}.

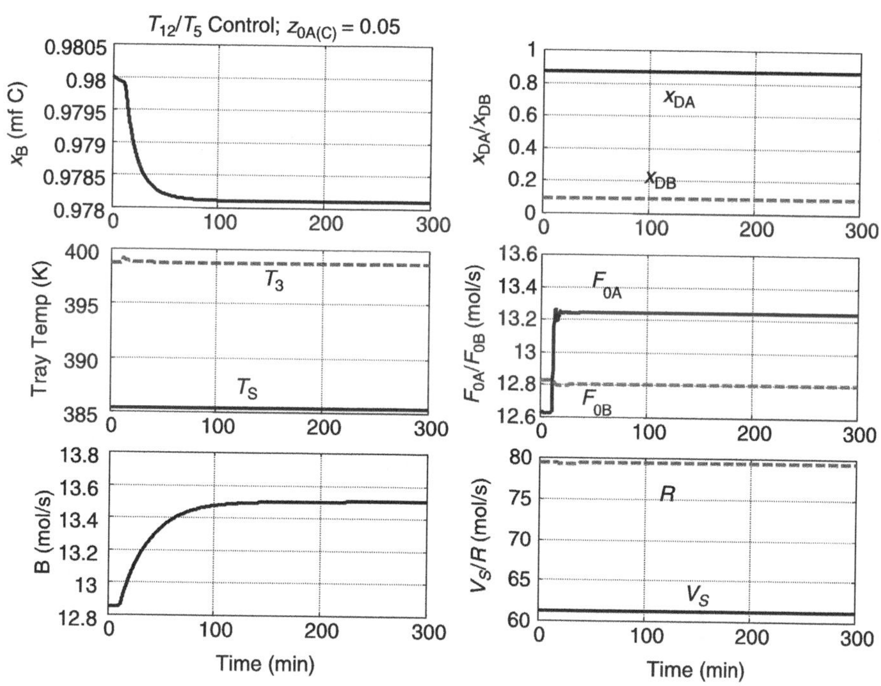

Figure 12.25 Control structure CS3 with $z_{0A(C)} = 0.05$ for T_5 and T_{12}.

12.2 TERNARY SYSTEM WITH INERTS

12.2.1 Column Configuration

With a one-product reaction system without inerts, the column has only reactive and stripping trays and there is no distillate product. The addition of light inerts in the feed requires a column with all three zones (rectifying, reactive and stripping), and there are both distillate and bottoms product streams. Reactant C leaves in the bottoms, and the distillate should be mostly the light inert component I. Remember that the volatility of I is the same as light reactant A in the example designed in Chapter 5. Refer to Table 12.3 for variable values at steady state and design parameters. The flowsheet is provided in Figure 12.26.

Figure 12.27 gives composition profiles, and Figure 12.28 gives the temperature profile. There are 5 stripping trays, 15 reactive trays, and 5 rectifying trays. The column diameter (1.1 m) is calculated from the sizing relationships given in Chapter 3. A liquid height of 0.05 m is assumed on the stripping and rectifying trays, giving a tray holdup of 670 mol on these trays. The holdup on the reactive trays is 2000 mol, which corresponds to a liquid height of 0.15 m. The holdups in the column base (23.2 kmol) and reflux drum (24.8 kmol) are sized to give 5 min of residence time when 50% full, based on the total liquid entering. In the reflux drum, this is the sum of the reflux and distillate flowrates.

Three alternative control structures are explored in the following sections. The primary objectives in the face of disturbances in production rate and feed compositions are to maintain the specified purity of the bottoms product at 98 mol% C and to avoid losses of reactants in the distillate.

12.2.2 Control Structure CS1

Figure 12.29 shows a control structure that uses two direct composition measurements. The bottoms product purity (mol% C) is measured and controlled by vapor boilup.

TABLE 12.3 Controller Tuning Parameters for Ternary with Inerts

CS1		CS2		CS3	
Controlled Variable $x_{B(C)}$	Manipulated Variable V_S	Controlled Variable Temp. Tray 3 (K)	Manipulated Variable V_S	Controlled Variable Temp. Tray 5 (K)	Manipulated Variable F_{0A}
K_U	4.0		1.6		5.3
P_U (min)	6.0		2.5		3.0
K_C	0.5		0.125		1.6
τ_I (min)	13		5.5		6.6
Setpoint	0.98		418.5 (K)		339.4 (K)
$x_{(6,A)}$	F_{0A}	$x_{(5,A)}$	F_{0A}	Temp. Tray 13 (K)	F_{0B}
K_U	11		11		9.1
P_U (min)	3.0		3.0		55
K_C	2.0		2.0		2.8
τ_I (min)	6.6		6.6		100
Setpoint	0.30		0.30		387.5 (K)

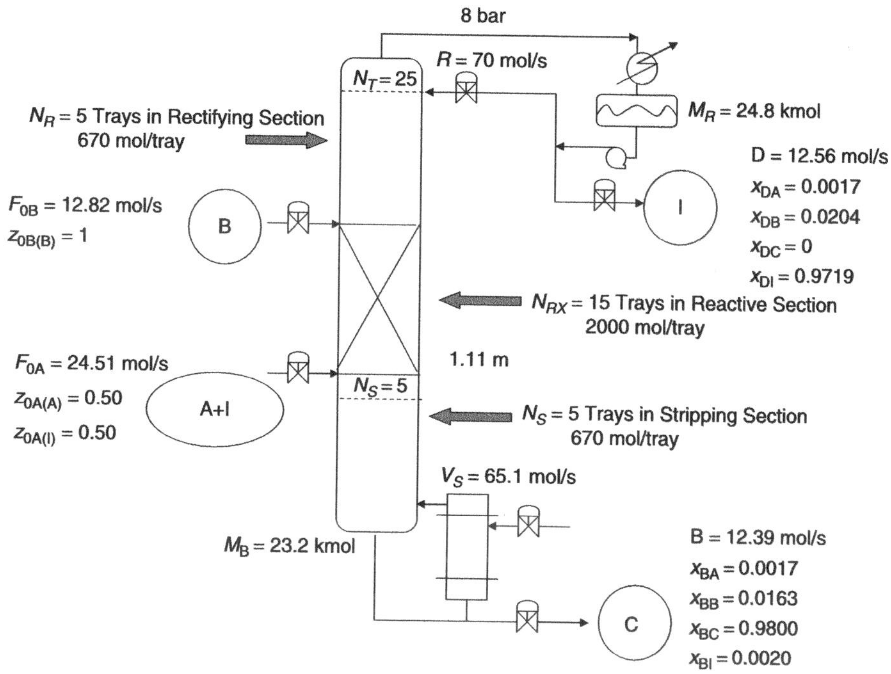

Figure 12.26 Ternary reactive distillation column with inerts.

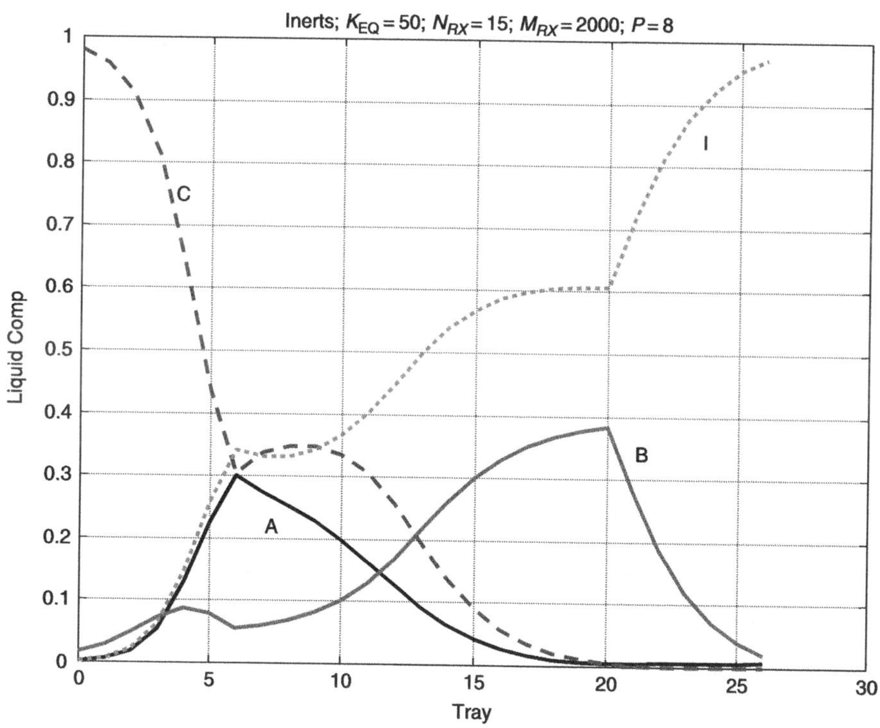

Figure 12.27 Ternary with inerts composition profiles.

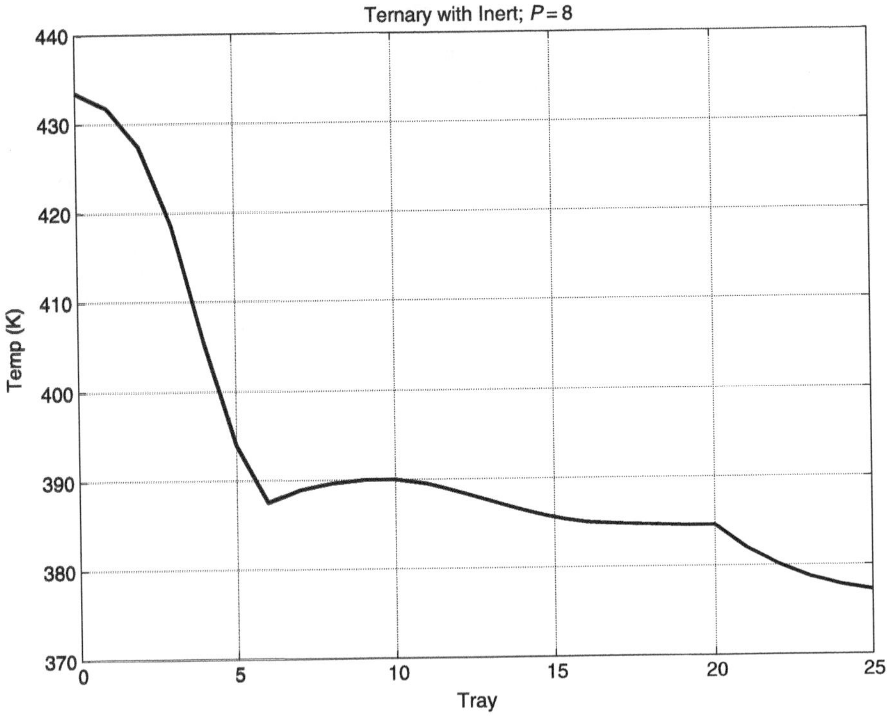

Figure 12.28 Ternary with inerts temperature profile.

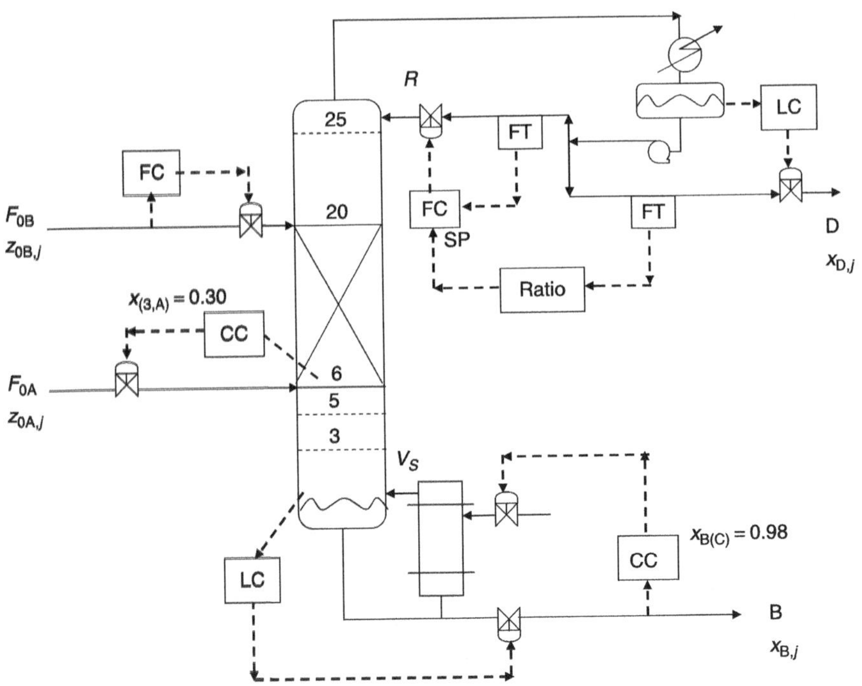

Figure 12.29 Ternary with inerts for control structure CS1-RR.

The composition of A on tray 6 (30 mol% A) is measured and controlled by the flowrate of fresh feed F_{0A}. Throughput is set by flow controlling fresh feed F_{0B}. The reflux-drum level is controlled by manipulating the distillate flowrate, and the base level is controlled by manipulating the bottoms flowrate.

Two alternative structures are evaluated with this control scheme. In the first, the reflux flowrate is simply fixed. In the second (Fig. 12.29), the reflux ratio is controlled. This is achieved by measuring the distillate flowrate, multiplying this signal in the "ratio" element by the desired reflux ratio. The output signal from the ratio changes the setpoint of the reflux flow controller. The steady-state reflux ratio is $70/12.58 = 5.58$. The bottoms composition is 98 mol% C. The distillate composition is 97.19 mol% I, which means that there are only small losses of reactants A and B in the distillate.

Two 0.5-min first-order composition measurement lags are inserted in both composition control loops. Composition transmitter spans are 20 mol%. All valves are 50% open at steady state. The controllers are tuned by running relay-feedback tests. The values of ultimate gains and periods are given in Table 12.3. Tyreus–Luyben tuning is used in most cases, but some loops are detuned to give larger closed-loop damping coefficients.

Figure 12.30 gives the response of the system with the fixed reflux flowrate to a +20% change in the production rate handle F_{0B}. The bottoms composition is maintained at its specified value, but the bottom left graph shows that the inert composition of the distillate drops drastically to about 87 mol% I. This means large amount of the reactants are being lost out the top of the column. The distillate flowrate increases, as does fresh feed F_{0A} and vapor boilup. The production of product C only increases by 8% (bottoms flowrate

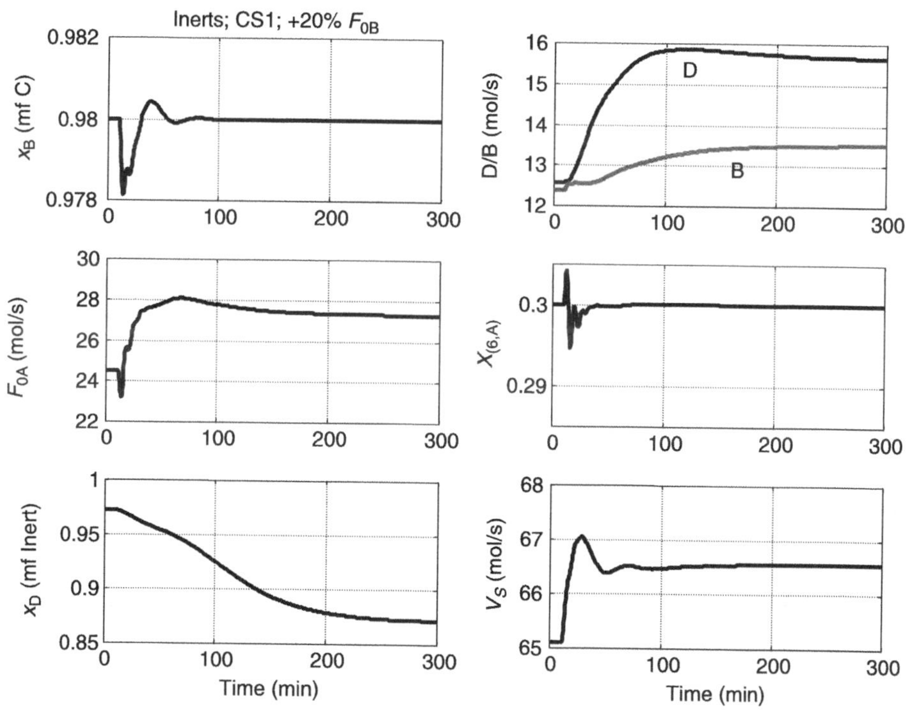

Figure 12.30 Inerts CS1 with fixed reflux for +20% F_{0B}.

increases from 12.39 to 13.5 mol/s) despite the 20% increase in the amount of B being fed. The losses of both reactants are very large.

Implementing the reflux ratio strategy improves the situation, as shown in Figure 12.31. The distillate inert composition only drops to about 96.8 mol%. The bottoms flowrate increases to 14.9 mol/s, which should be compared with the 13.5 mol/s seen in the fixed reflux control structure. Figures 12.32–12.35 give the response to other disturbances in production rate and feed compositions.

In Figure 12.33 fresh feed F_{0A} contains less inert, so the distillate rate decreases and less F_{0A} is fed for a fixed value of F_{0B}. In Figure 12.34 more inert is included in fresh feed F_{0A}, so the distillate rate and F_{0A} both increase. Production rate, as reflected in the bottoms flowrate, stays constant at the new steady state. The feed composition disturbance in Figure 12.35 is changing F_{0B} from pure B to 5 mol% A. The bottoms purity is not affected, but distillate drops about 1 mol% I. Because less B is fed, less F_{0A} is fed and both bottoms and distillate flowrates decrease.

12.2.3 Control Structure CS2

Figure 12.36 shows a control structure in which the temperature on tray 3 is controlled instead of direct composition control of the bottoms composition. Figures 12.37–12.41 show that this "inferential" control scheme does a fairly good job of maintaining product purity and keeps reactant losses low for throughput changes. However, for feed composition disturbances, both the distillate and bottoms compositions move away fairly significantly

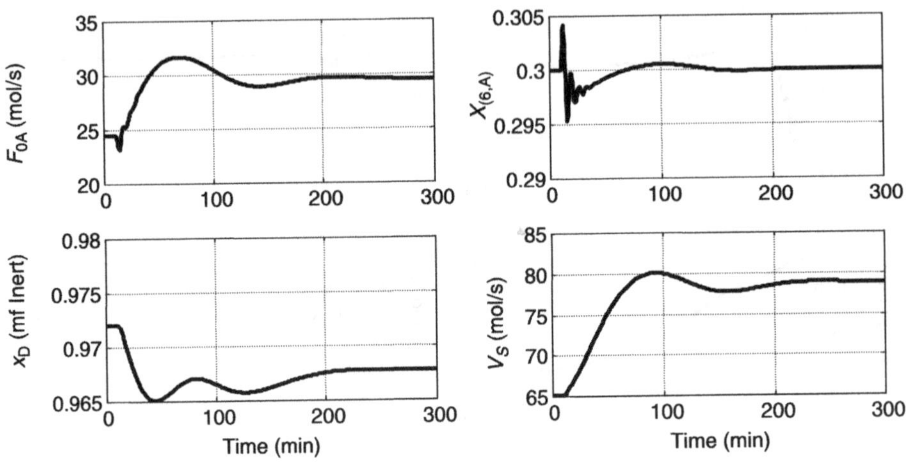

Figure 12.31 Inerts CS1-RR for +20% F_{0B}.

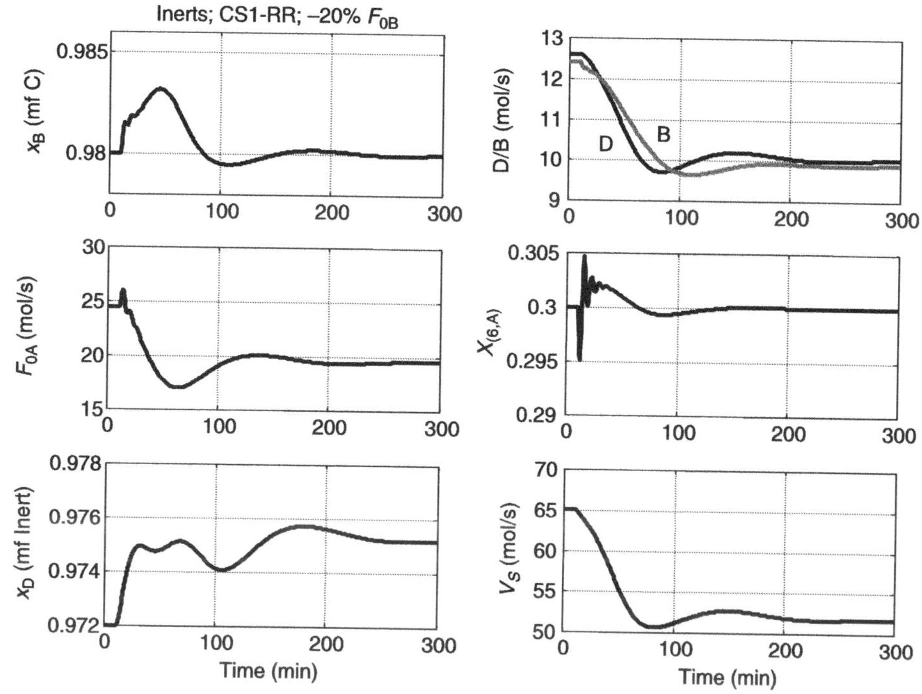

Figure 12.32 Inerts CS1-RR for –20% F_{0B}.

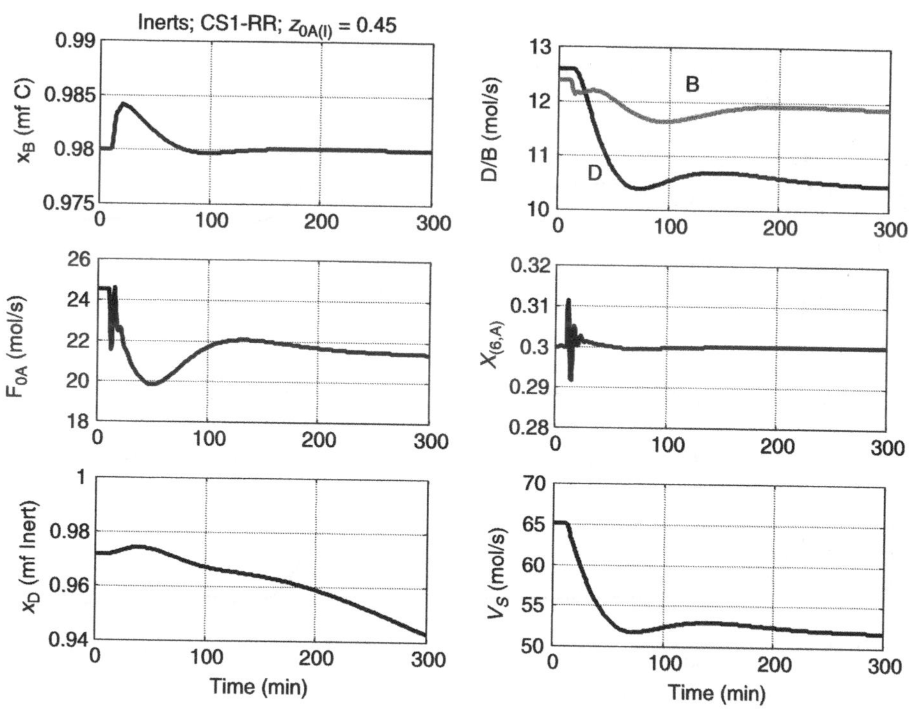

Figure 12.33 Inerts CS1-RR for $z_{0A(A)} = 0.55$ and $z_{0A(I)} = 0.45$.

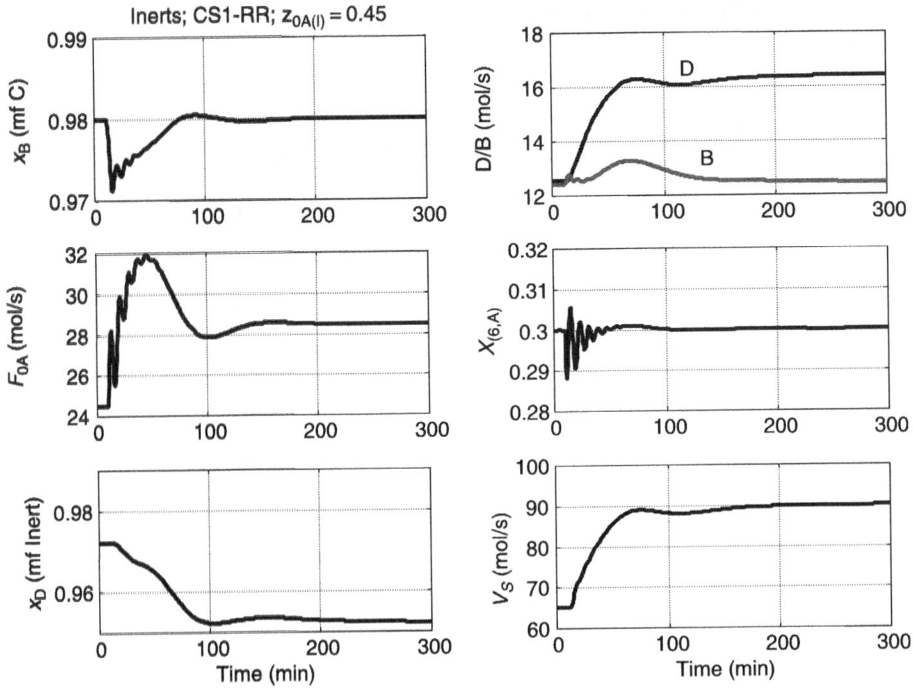

Figure 12.34 Inerts CS1-RR for $z_{0A(A)} = 0.45$ and $z_{0A(I)} = 0.55$.

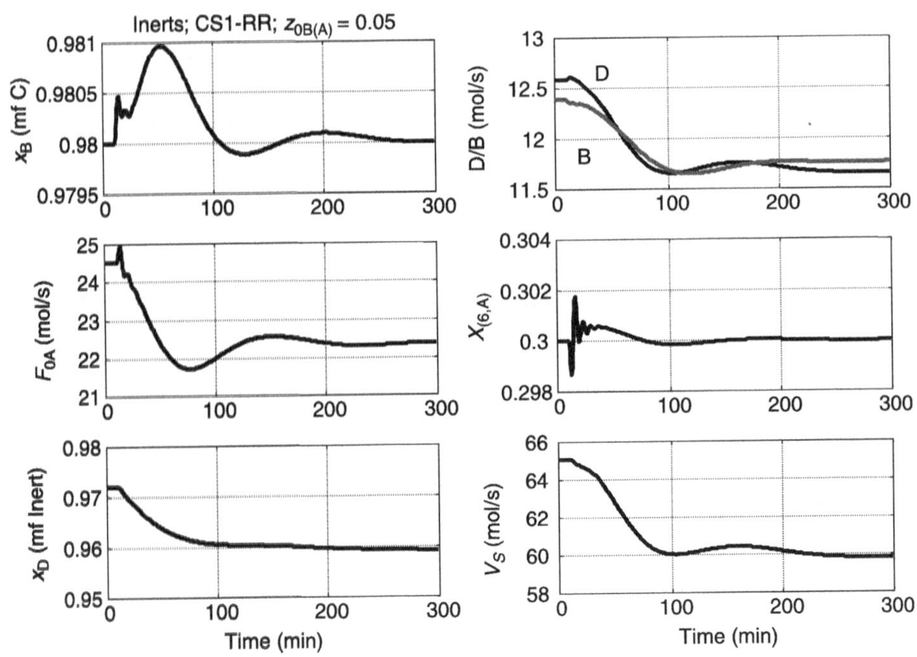

Figure 12.35 Inerts CS1-RR for $z_{0B(A)} = 0.05$.

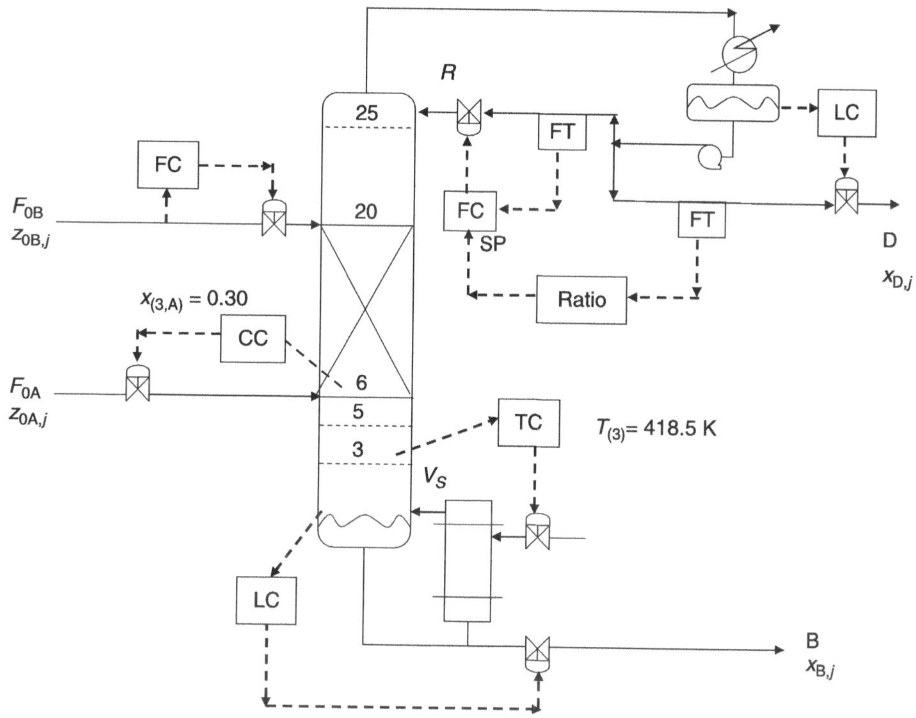

Figure 12.36 Ternary with inerts for control structure CS2-RR.

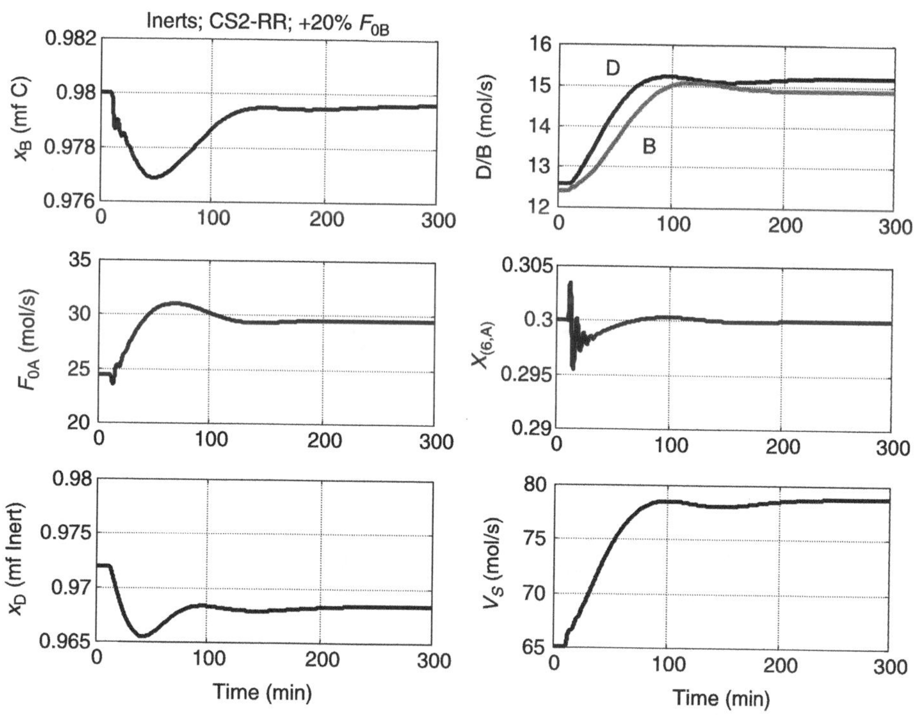

Figure 12.37 Inerts CS2-RR for +20% F_{0B}.

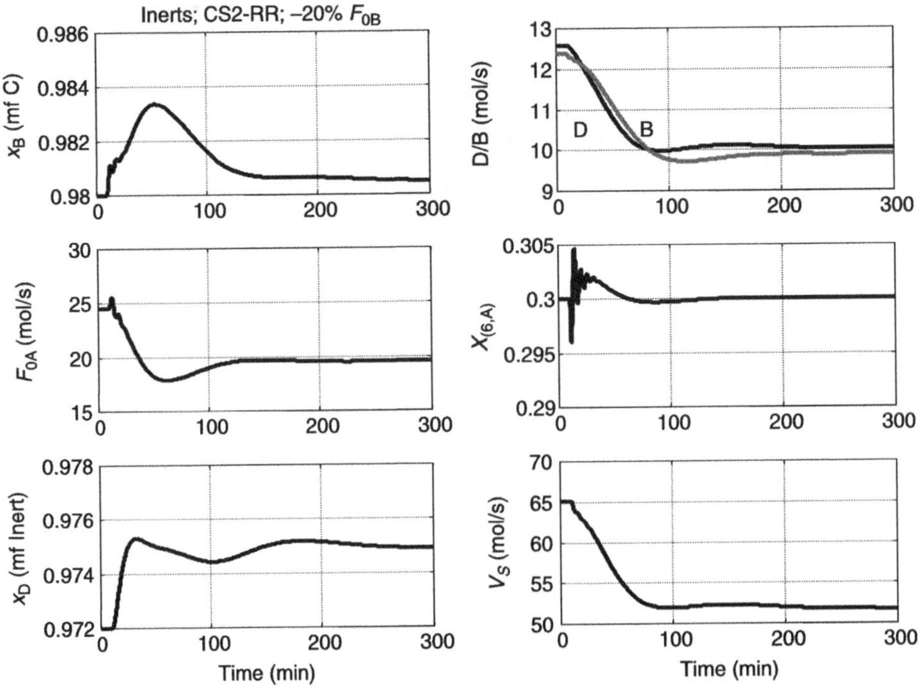

Figure 12.38 Inerts CS2-RR for -20% F_{0B}.

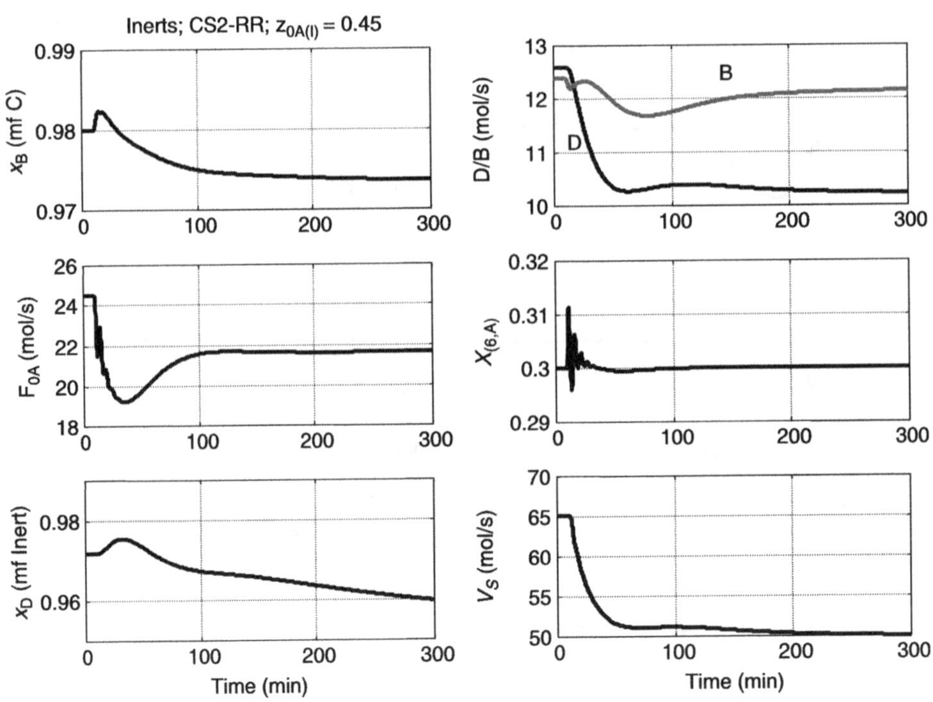

Figure 12.39 Inerts CS2-RR for $z_{0A(I)} = 0.45$ and $z_{0A(A)} = 0.55$.

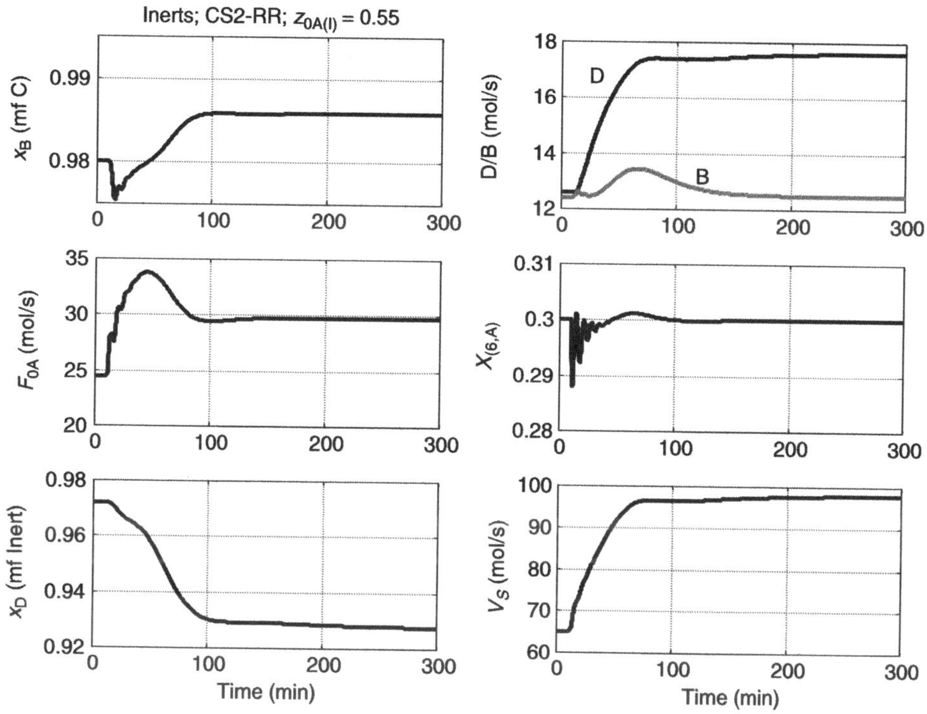

Figure 12.40 Inerts CS2-RR for $z_{0A(I)} = 0.55$ and $z_{0A(A)} = 0.45$.

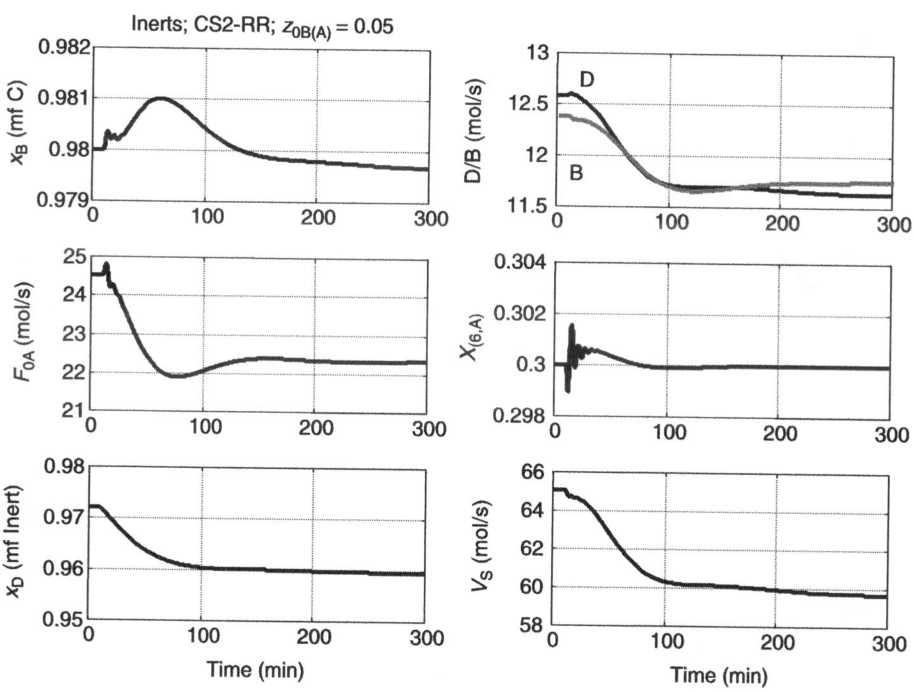

Figure 12.41 Inerts CS2-RR for $z_{0B(A)} = 0.05$.

from the desired values. This is particularly true when the amount of inert in the feed increases. Distillate composition drops below 93 mol% I (Fig. 12.40).

12.2.4 Control Structure CS3

The final control structure studied is the two-temperature scheme that eliminates the need for composition measurements. This control structure works quite well in both the quaternary system and in the ternary system without inerts. Will it work as well in the ternary system with inerts?

To determine what control trays to select, steady-state gains are calculated and SVD analysis is performed. Figure 12.42 gives the steady-state gains for all trays in the column. These results indicate that the effects of both F_{0A} and F_{0B} are largest around trays 3 and 4. These two locations are too close together to provide independent temperature measurements. Calculating the two U vectors in the SVD analysis for the two inputs F_{0A} and F_{0B} gives the results provided in Figure 12.43. The two sensitive locations are indicated as tray 3 and tray 5. These are also very close together, but the singular values of the gain matrix are 219 and 76, which gives a reasonable condition number.

This control structure was tested without success. The temperature on tray 3 was controlled by manipulating F_{0A}. The process gain is negative, so the controller gain should

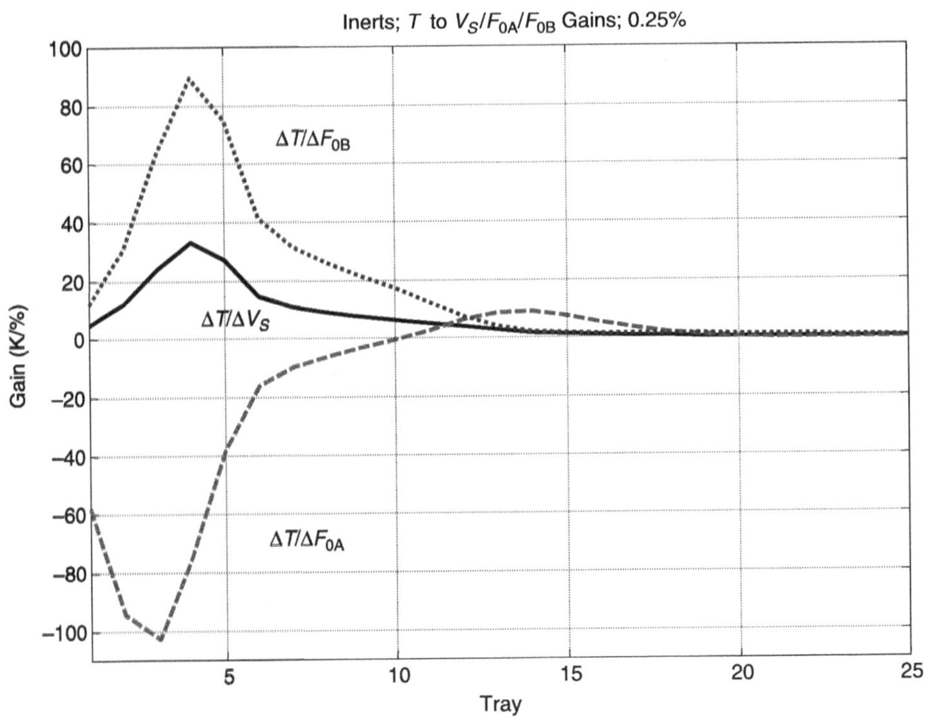

Figure 12.42 Inerts steady-state gains.

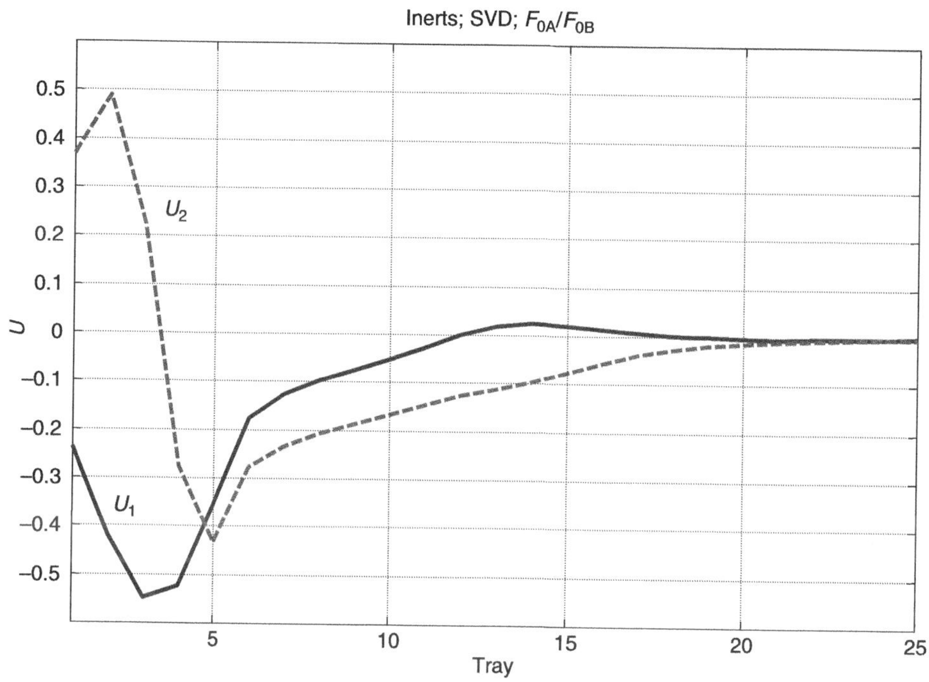

Figure 12.43 Inerts SVD analysis.

be negative (direct acting: an increase in temperature increases F_{0A}). This loop was tuned and worked well when the other temperature loop was on manual.

The second temperature on tray 5 was controlled by manipulating F_{0B}. The process gain is positive, so the controller gain should be positive (reverse acting: an increase in temperature decreases F_{0A}). This loop did not work, as illustrated in Figure 12.44. The tray 3 temperature controller is on manual, that is, F_{0A} is fixed. No disturbance is made. After about 10 min, the temperature on tray 5 begins to drop. The tray 5 temperature controller increases F_{0B} up to its limit (twice the steady-state value), but the temperature does not recover. We were unable to make this control structure work using trays 3 and 5.

There is a slight hump in the U_1 curve at tray 13 shown in Figure 12.43. We tested a control structure that used F_{0A} to control the tray 5 temperature and F_{0B} to control the tray 13 temperature. These two loops were successfully tuned, but the tray 13 controller had a very large ultimate period (55 min). Table 12.3 gives the controller tuning parameters.

The performance of this control structure is given in Figure 12.45 for a +20% change in V_S. The system is stable, but the bottoms purity and distillate purity are not well controlled. Figure 12.46 shows that the response to a 20% decrease in V_S is even worse in terms of product purities. The bottoms composition drops to 76 mol% C and the distillate composition drops slightly.

These results demonstrate that a two-temperature control scheme does not provide effective control of the ternary system with inerts. The two-temperature control structure works for the quaternary system and for the ternary without inerts, but not for the ternary with

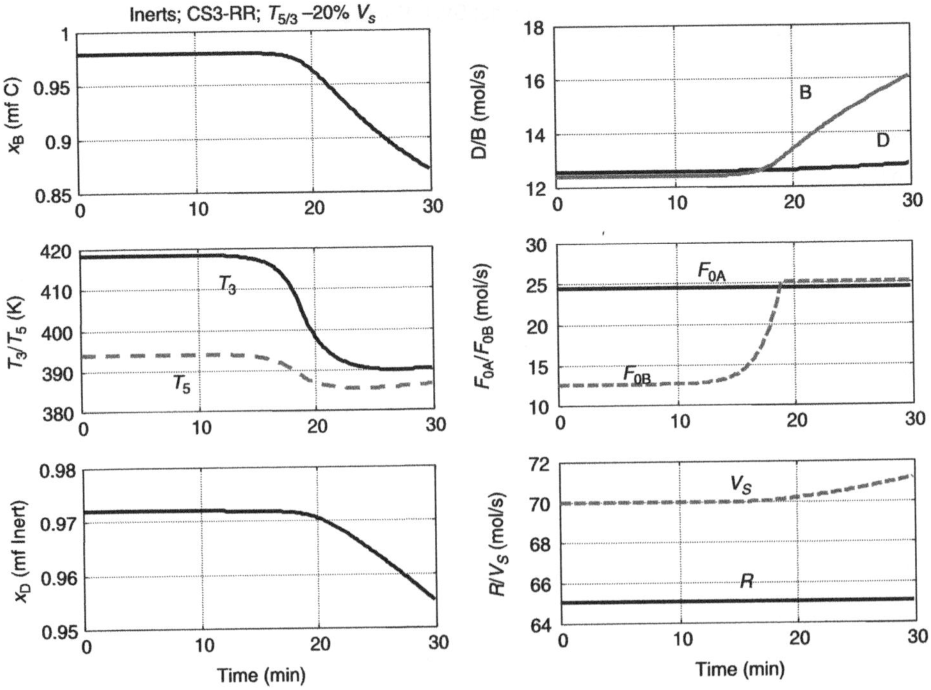

Figure 12.44 Inerts CS3-RR for T_5 to F_{0B} loop.

inerts. The same conclusion was found by Al-Arfaj and Luyben for the ETBE system[1] and by Wang et al. for the MTBE system.[2]

12.2.5 Conclusion for Ternary A + B ⇔ C System

Two different ternary systems were explored. Both have two reactants and one product. However, one has an additional component coming in with one of the feeds that is not involved in the reaction. It is inert as far as the chemistry is concerned. However, it dilutes the concentrations of the reactants and the products, so it does affect the overall reaction rates.

Of more importance, the presence of the inert component has a major impact on both the structure of the column (two outlet streams are required instead of one) and the vapor–liquid phase equilibrium. As this chapter illustrates, it also has a profound effect on the control structure required for adequate control.

[1]M. A. Al-Arfaj and W. L. Luyben, Control study of ETBE reactive distillation, *Ind. Eng. Chem. Res.* **41**, 3784–3796 (2002).

[2]S. J. Wang, D. S. H. Wong, and E. K. Lee, Effect of interaction multiplicity on control system design for a MTBE reactive distillation column, *J. Process Control* **13**, 503–515 (2003).

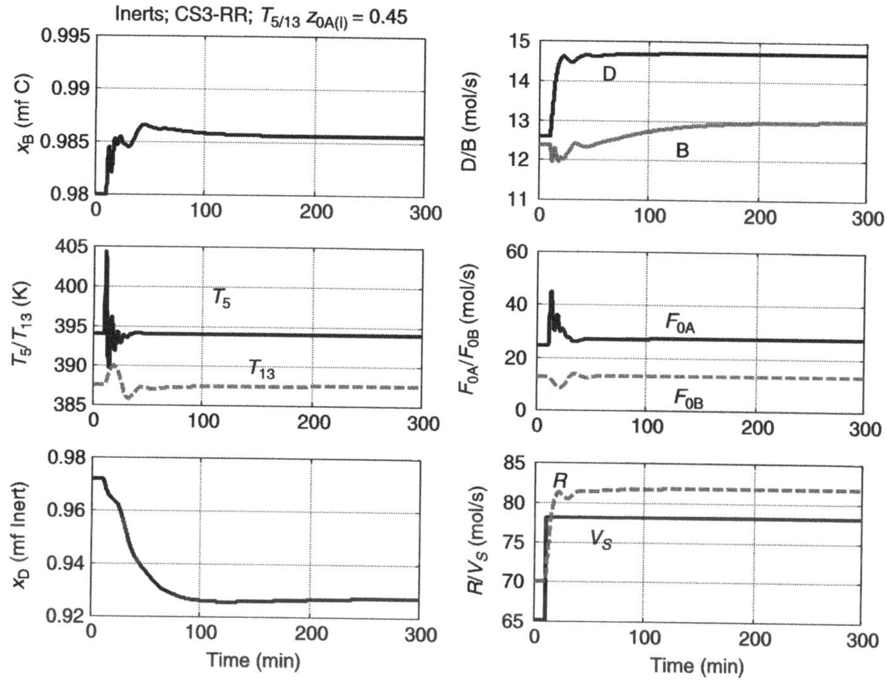

Figure 12.45 Inerts CS3-RR for T_5/T_{13} for $+20\%$ V_S.

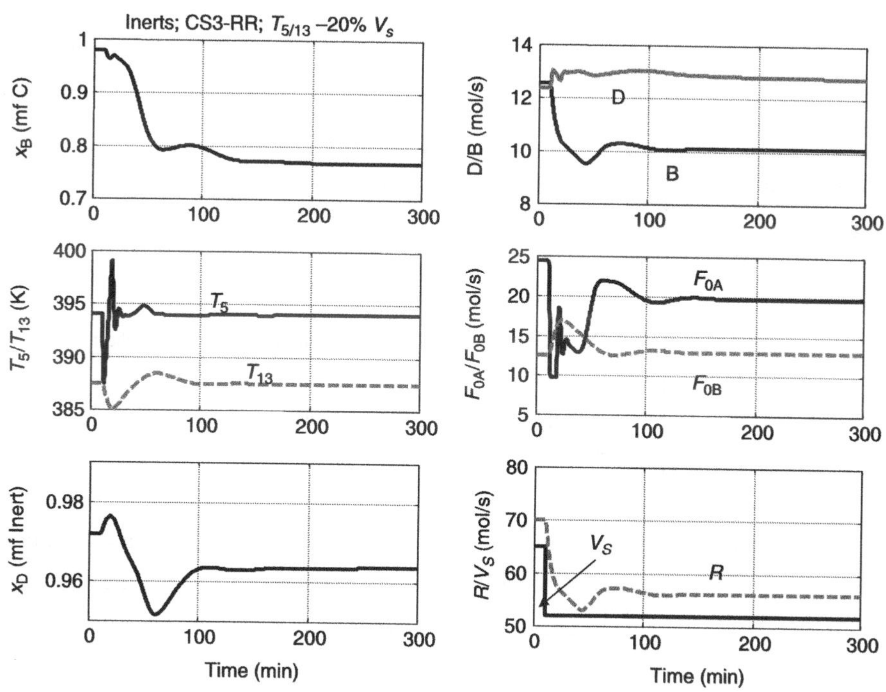

Figure 12.46 Inerts CS3-RR for T_5/T_{13} for -20% V_S.

12.3 TERNARY A ⇔ B + C SYSTEM: INTERMEDIATE-BOILING REACTANT

12.3.1 Column Configuration

For the decomposition reaction A ⇔ B + C, we consider the case when the reactant is the intermediate boiler, that is, IK ⇔ LK + HK. With a one-reactant, two-product reaction system, the column has a bottoms product and a top product because the two products B and C are the lightest and heaviest boilers, respectively. Figure 12.47 provides the flowsheet. Steady-state conditions and design parameters are given in Table 12.4. This is the optimal design in terms of TAC as discussed in Chapter 6. Figure 12.48 contains composition profiles and the fraction of total conversion on tray i (R_i/R_t) in the reactive zone. Figure 12.49 gives the temperature profile.

The column diameter is calculated from the sizing relationships given in Chapter 6, which gives a diameter of 0.92 m. A liquid height of 0.05 m is assumed for each tray. Tray holdup on the reactive trays is 909 mol. The holdups in the column base and reflux drum are sized to give a 5-min residence time when the liquid level is held at 50%.

Three alternative control structures are explored in the following sections. The objective is to maintain the specified purities of the top product at 98 mol% B and the bottoms product at 98 mol% C. The load disturbances that are considered are production rate changes and feed composition variations.

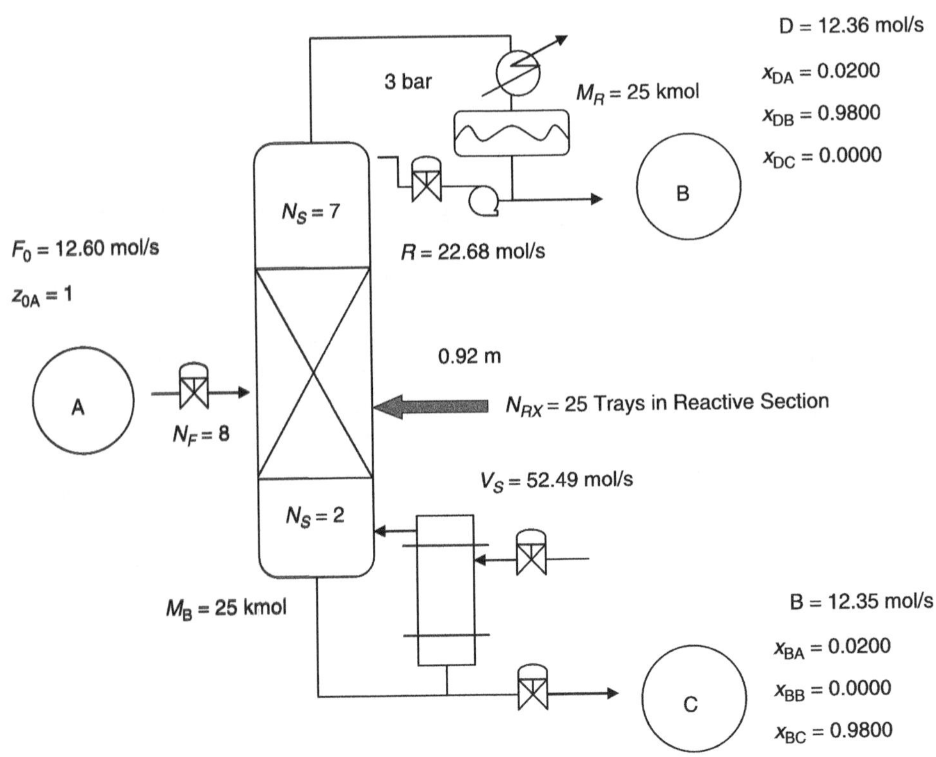

Figure 12.47 Ternary reactive distillation column (IK ⇔ LK + HK).

TABLE 12.4 Steady-State Conditions and Design Parameters for TAC Optimum Case

Fresh feed flowrate of *A* (mol/s)	12.60	
Distillate flowrate (mol/s)	12.36	
Bottoms flowrate (mol/s)	12.35	
Vapor boilup (mol/s)	52.49	
Reflux flowrate (mol/s)	22.68	
Stripping trays	2	
Reactive trays	25	
Rectifying trays	7	
Liquid holdup on reactive trays (mol)	909.3	
Pressure (bar)	3	
Product Composition (Mole Fraction)	Distillate	Bottoms
A	0.0200	0.0200
B	0.9800	0.0000
C	0.0000	0.9800

There is only one reactant in the ternary decomposition reaction. Unlike the quaternary two-feed, two-product systems and ternary synthesis reactions with two feeds and one product, the stoichiometric balance of the reactants is no longer a problem. The control strategy will focus on holding product purities at setpoints. Thus, the control problem is similar to that of conventional distillation column.

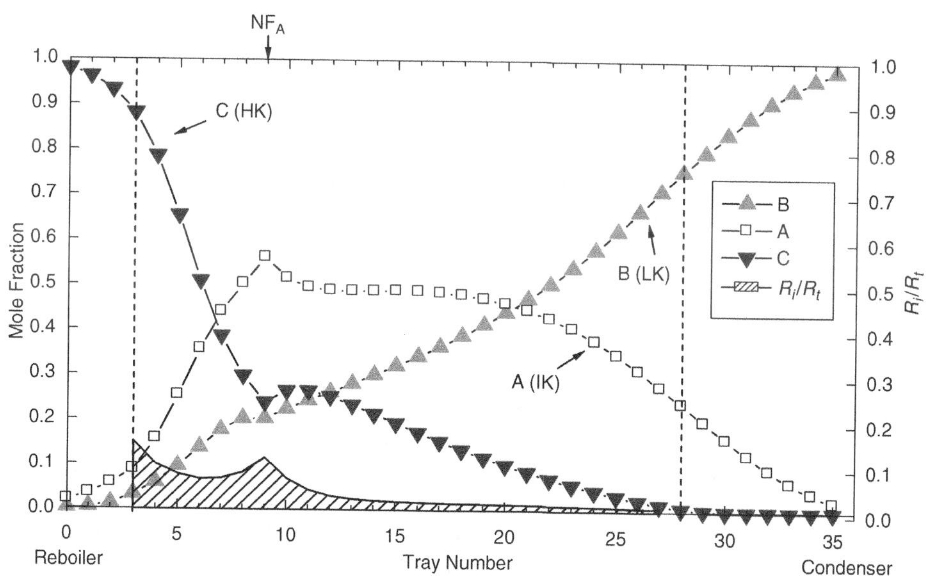

Figure 12.48 Composition profile for A ⇔ B + C with reactive zone marked with NF_A and the fraction of total conversion (R_i/R_t) shown in shaded area.

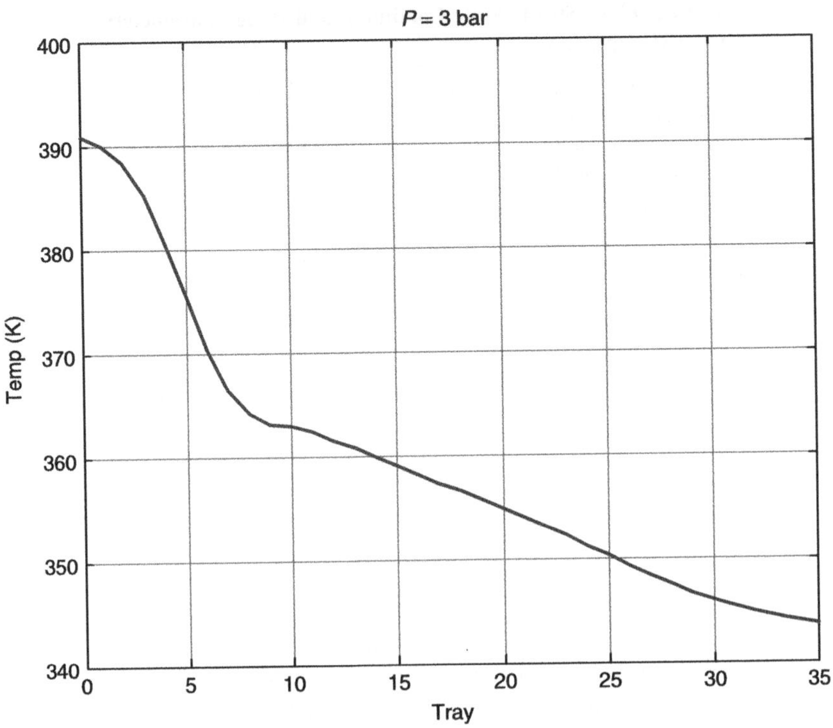

Figure 12.49 Temperature profile.

12.3.2 Control Structure CS1

A two-temperature control structure is considered first. Selecting the vapor boilup and reflux ratio as manipulated variables, the control problem then is to find the best locations for the two temperature control trays. Figure 12.50 shows the steady-state changes in tray temperatures throughout the column for several small changes in vapor boilup and reflux ratio.

We used SVD to select the control trays. Figure 12.51 shows the U vectors from the SVD analysis. The most sensitive locations are trays 26 and 5. Figure 12.52 shows the control structure. The reflux-drum level is controlled by manipulating the distillate flowrate, and the base level is controlled by manipulating the bottoms flowrate. A third-order measurement lag with a time constant of 10 s is assumed in each temperature controller. The controllers are tuned by running relay-feedback tests. Ultimate gains and periods values are given in Table 12.5. Tyreus–Luyben tuning is used in most cases, but some loops are detuned to give larger closed-loop damping coefficients.

Results for the CS1 control structure are presented in Figures 12.53–12.55. In Figure 12.53 the disturbances are positive and negative 20% step changes in the feed flowrate at a time of 100 min. The temperature dynamics are very fast and steady state can be reached in a few minutes. The composition dynamics are a little slower and settle in 50 min. However, 0.1 mol% steady-state offsets in product purities are observed. Figure 12.54 shows what happens when composition z_0 of fresh feed F_0 is changed from pure reactant A ($z_{0A} = 1$) to a mixture of A and B ($z_{0A} = 0.95$, $z_{0B} = 0.05$). When light

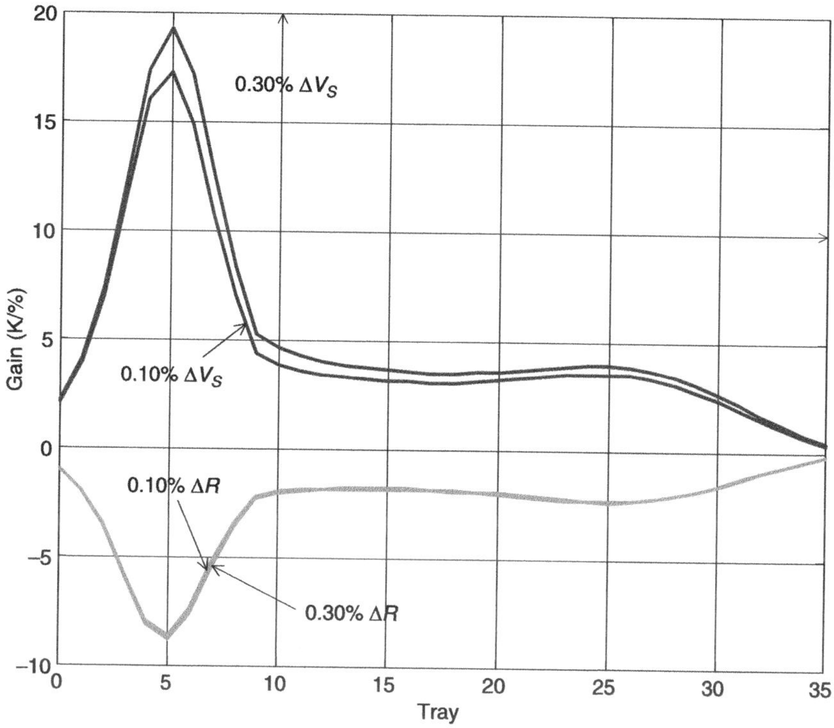

Figure 12.50 Steady-state gains for different ΔV_S and ΔR values.

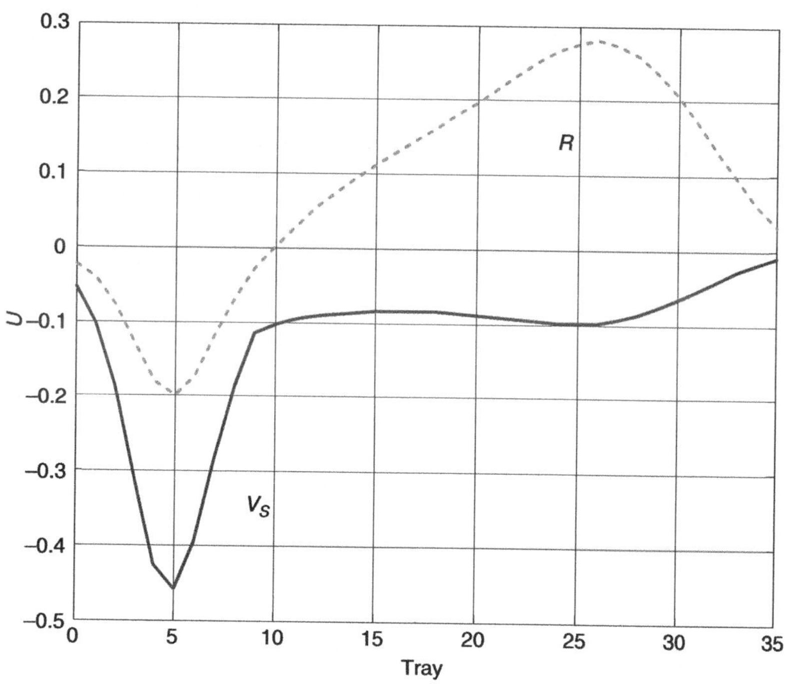

Figure 12.51 The U matrix from SVD analysis.

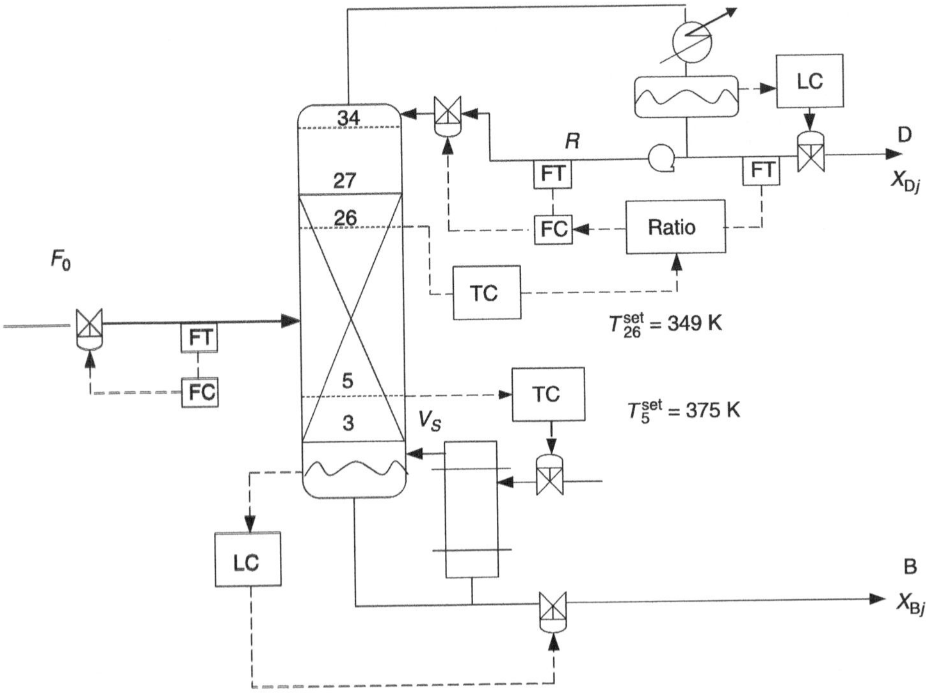

Figure 12.52 Control structure CS1.

product B is fed into the column, the reflux ratio decreases and the boilup rate decreases to maintain the column temperatures. This will lead to lower top product purity and purer heavy product in the bottoms. Figure 12.55 demonstrates what happens when composition z_0 of the fresh feed is changed from pure A ($z_{0A} = 1$) to a mixture of A and C ($z_{0A} = 0.95$,

TABLE 12.5 Controller Tuning Parameters for Ternary Decomposition Reaction: IK ⇔ LK + HK

CS1		CS2		CS3	
Controlled Variable	Manipulated Variable	Controlled Variable	Manipulated Variable	Controlled Variable	Manipulated Variable
T_{26}	RR	$x_{D,B}$	RR	$x_{D,B}$	T_{26}^{set}
K_U	104.4		9.8		0.2
P_U (min)	1.7		30.0		14.7
K_C	27.8		1.6		0.01
τ_I (min)	4.3		120.0		29.3
Setpoint	349 (K)		0.98		0.98
T_5	V_S	$x_{B,C}$	V_S	$x_{B,C}$	T_5^{set}
K_U	8.0		1.4		0.5
P_U (min)	1.5		25.4		15.5
K_C	2.6		0.24		0.04
τ_I (min)	2.9		102.0		31.0
Setpoint	375 (K)		0.98		0.98

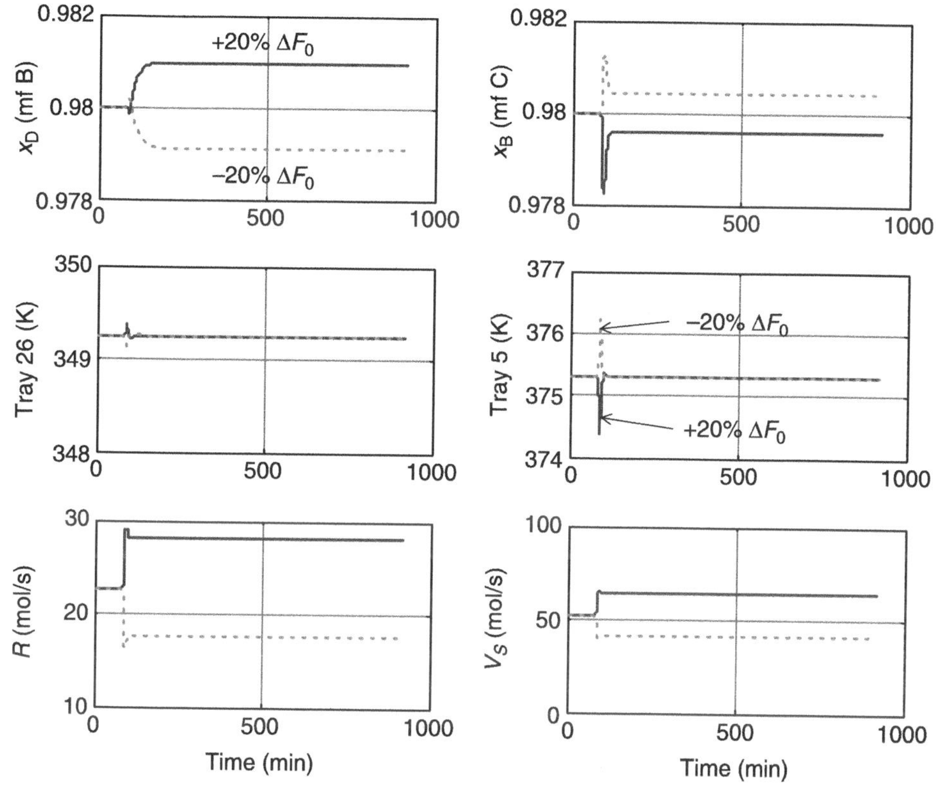

Figure 12.53 Control structure CS1 with $\pm 20\%$ ΔF_0.

$z_{0C} = 0.05$). Similarly, the reflux flowrate and vapor boilup rate are both decreased. However, there are opposite effects on product composition. Nonetheless, fast dynamics responses are observed.

12.3.3 Control Structure CS2

To eliminate steady-state offsets in product compositions, dual composition control is implemented. Figure 12.56 shows that the distillate composition is controlled by changing the reflux ratio, and the bottoms composition is maintained by adjusting vapor boilup. Because the composition analyzer has a slower response, a measurement dead time of 4 min is assumed in the simulation. Relay-feedback tests are performed to find settings for the PI controllers. Table 12.5 shows that because of the dead time in the composition measurement, the reset times for the top and bottoms loops now become 120 and 102 min, compared to 4.3 and 2.9 min in the case of temperature control.

Results for the CS2 control structure are given in Figure 12.57. For feedflow disturbances, it takes longer than 10 h to reach steady state while keeping both product purities on spec. For feed composition disturbances, the dynamics are similar to that of feedflow changes. In summary, CS2 has the advantage of eliminating steady-state offsets, but the system needs more time to reach steady state and produces larger product purity variations in the transient state.

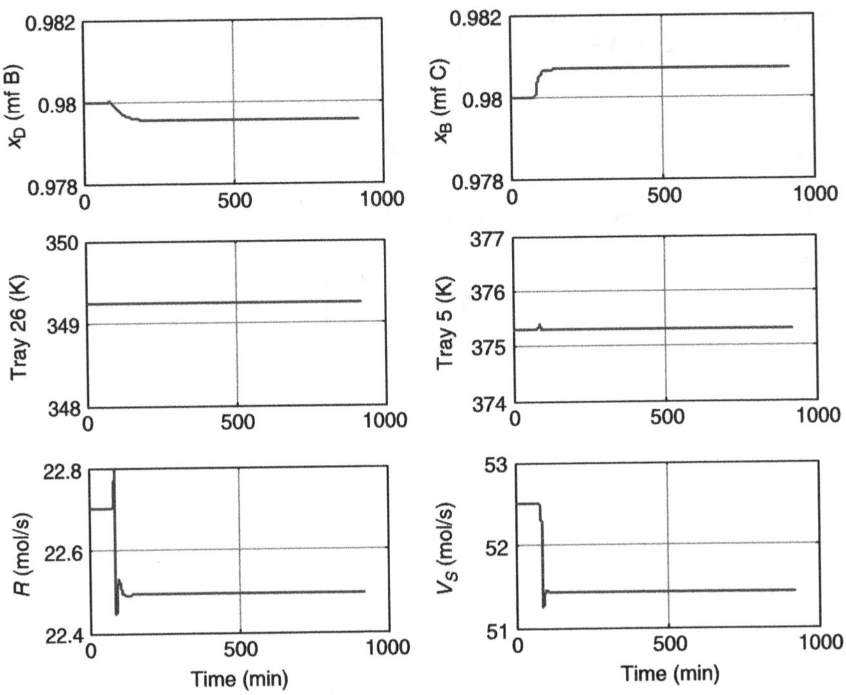

Figure 12.54 Control structure CS1 with $z_{0A} = 0.95$ and $z_{0B} = 0.05$.

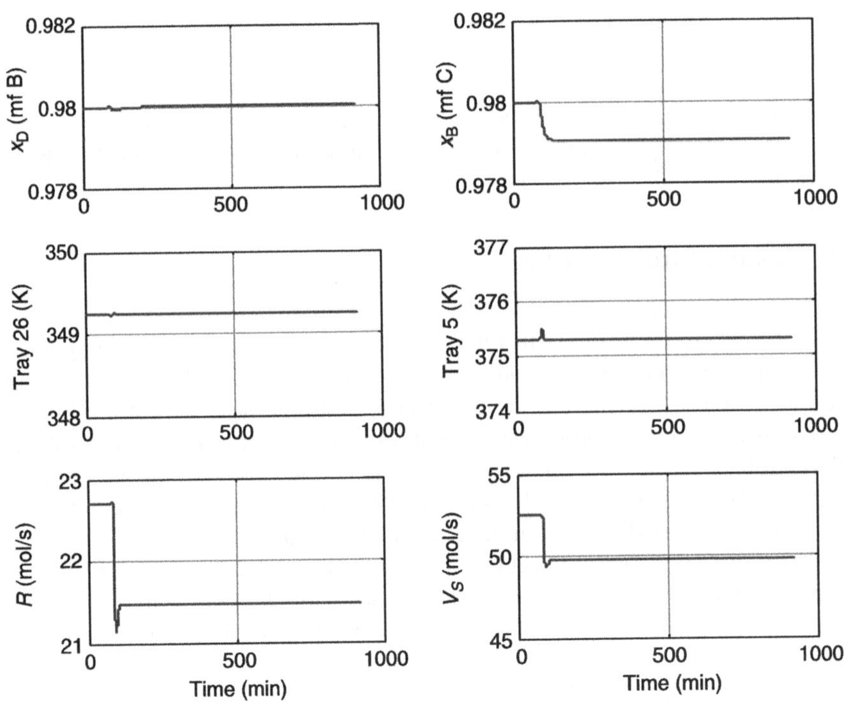

Figure 12.55 Control structure CS1 with $z_{0A} = 0.95$ and $z_{0C} = 0.05$.

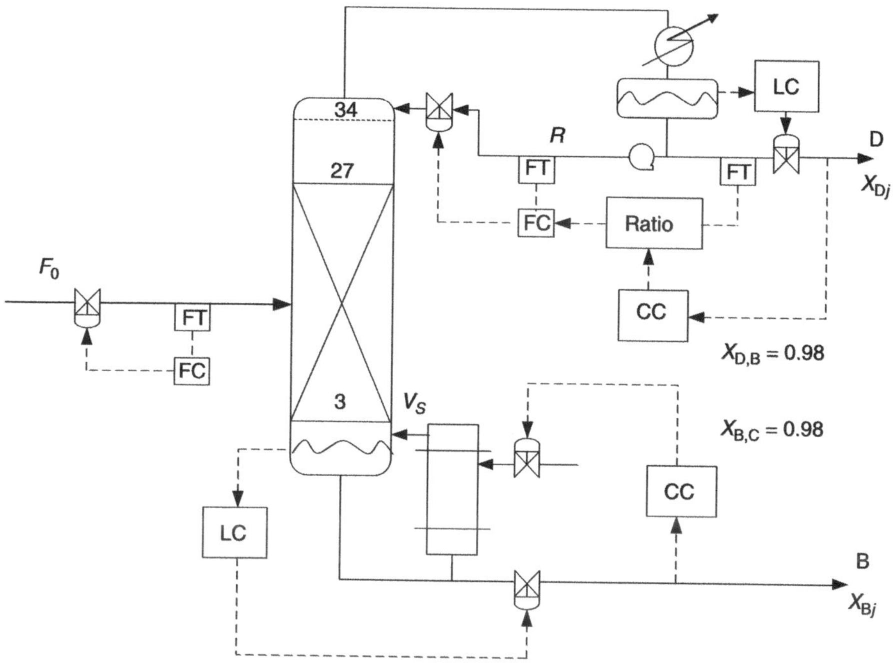

Figure 12.56 Control structure CS2.

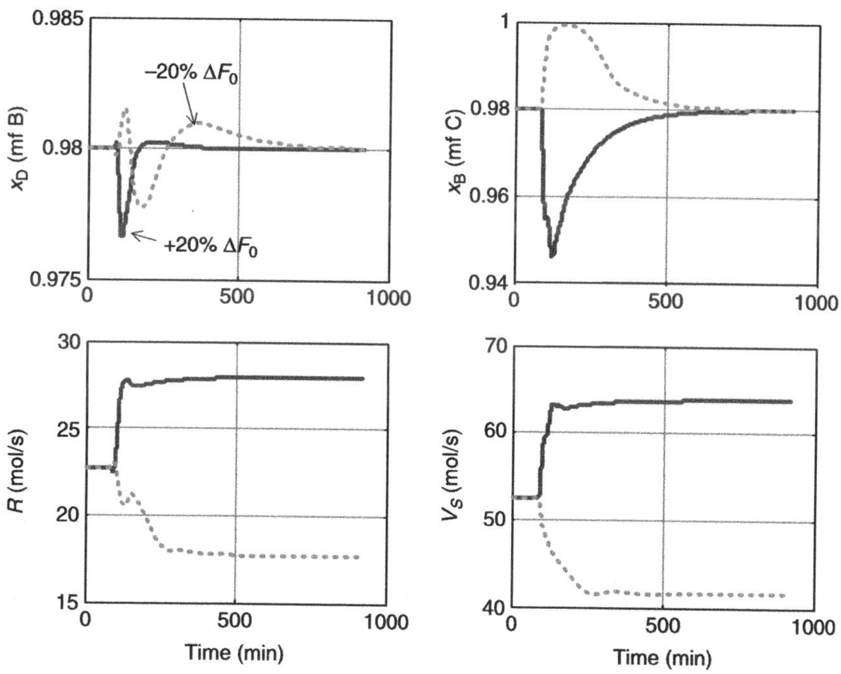

Figure 12.57 Control structure CS2 with $\pm 20\%\ \Delta F_0$.

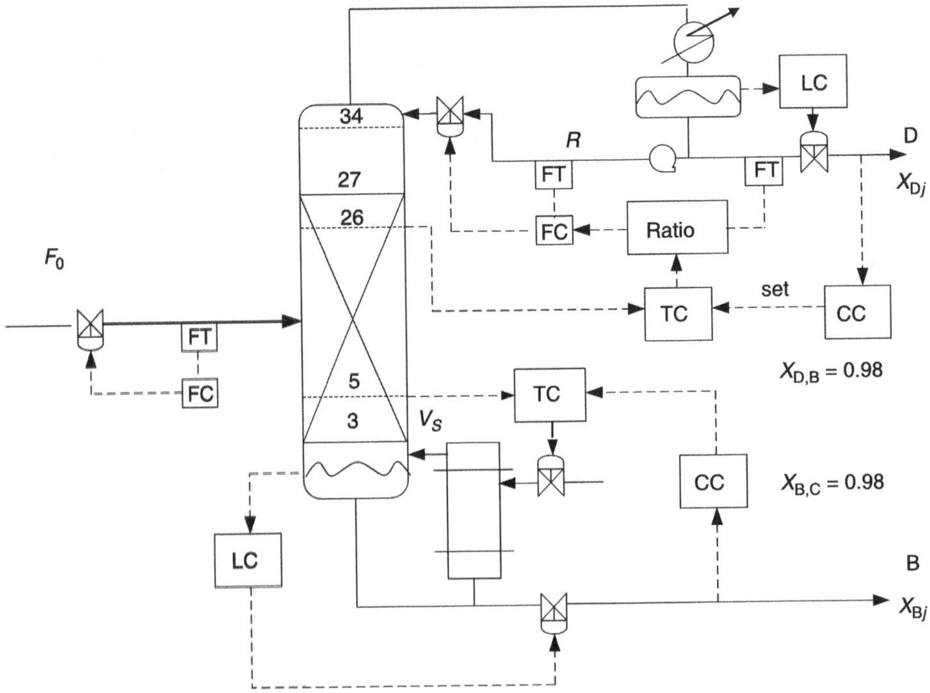

Figure 12.58 Control structure CS3.

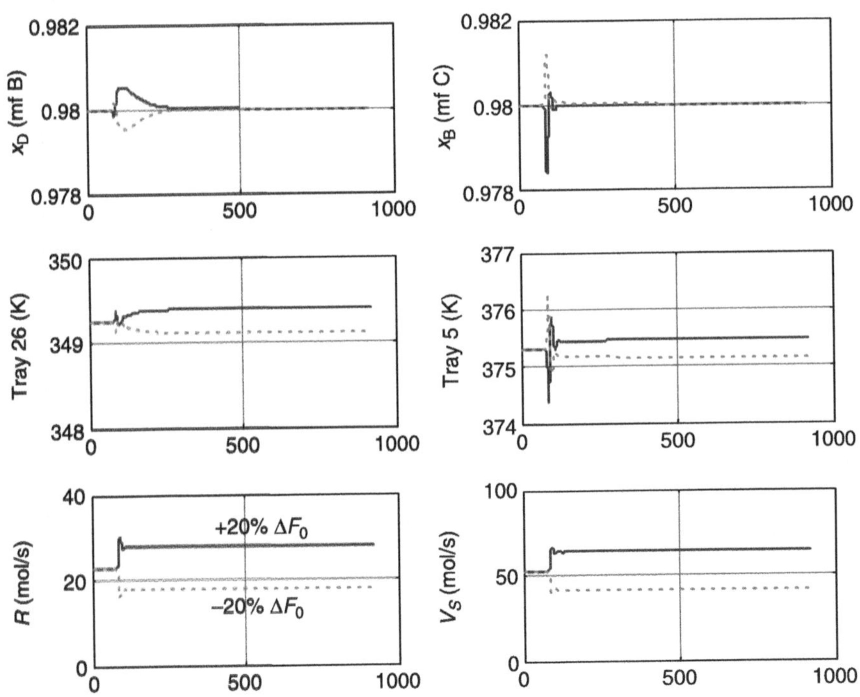

Figure 12.59 Control structure CS3 with $\pm 20\%$ ΔF_0.

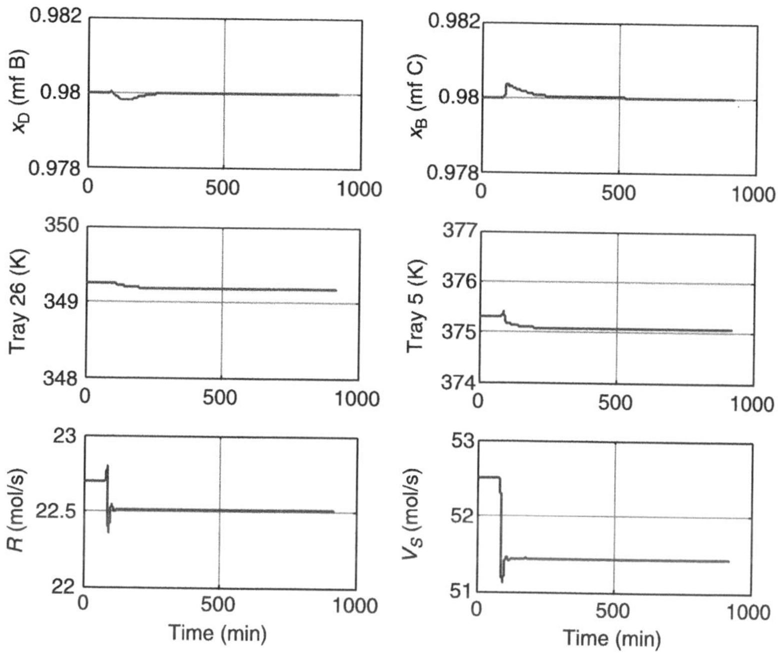

Figure 12.60 Control structure CS3 with $z_{0A} = 0.95$ and $z_{0B} = 0.05$.

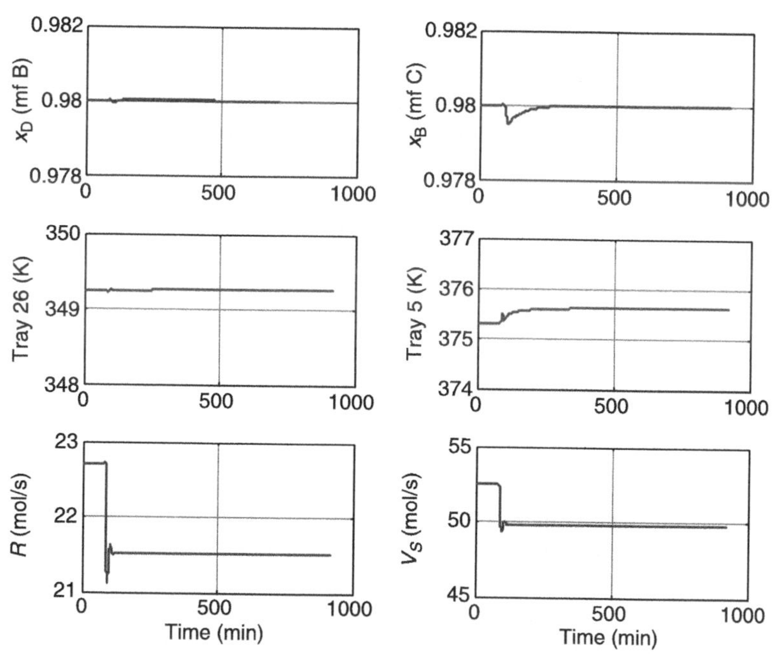

Figure 12.61 Control structure CS3 with $z_{0A} = 0.95$ and $z_{0C} = 0.05$.

12.3.4 Control Structure CS3

To maintain product purities while having tight control performance, a temperature/composition cascade control structure is explored. Figure 12.58 shows that the setpoint of the tray 26 temperature is adjusted by the output of a composition controller to keep x_D at its setpoint. Similarly, the setpoint of tray 5 is changed by the output of the bottoms composition controller. Using the same controller settings from CS1 in the temperature loops, relay-feedback tests are used to obtain settings for the composition controllers. The ultimate gains and periods values for both temperature and composition loops are given in Table 12.5.

Results for the CS3 control structure are given in Figures 12.59–12.61. The dynamics are almost as fast as that of the temperature control, CS1, while eliminating steady-state errors. The results clearly indicate the advantage of temperature/composition cascade control for reactive distillation systems when offset-free composition control is required.

12.4 TERNARY A $\Leftrightarrow$ B + C SYSTEM: HEAVY REACTANT WITH TWO-COLUMN CONFIGURATION

12.4.1 Column Configuration

The reaction HK $\Leftrightarrow$ LK + IK is a more general case in ternary decomposition reactions. Steady-state design was explored in Chapter 6. There are two design options: a two-column configuration and a one-column reactive distillation configuration. We have shown that, in terms of steady-state economics, these two configurations have their own applicable range (Fig. 6.23). The two-column configuration is favored when the chemical equilibrium constant is small and the one-column reactive distillation configuration is preferred when the chemical equilibrium constant is large. Now the question is, do these two configurations give comparable dynamic behavior?

We first explore the two-column configuration as shown in Figure 12.62. In the first distillation column, the heavy reactant (A) is fed to the column base of the reactive distillation column while the light and intermediate-boiling products (B and C) are withdrawn from the top. The binary mixture is then sent into a distillation column for further separation. Figure 12.62 gives the process flowsheet of the optimum design in terms of TAC. Steady-state conditions and design parameters are given in Table 12.6. Figures 12.63 and 12.64 provide the composition and temperature profiles of the reactive distillation column. Figures 12.65 and 12.66 show the composition and temperature profiles of the second distillation column.

12.4.2 Control Structure CS1

As discussed in the previous section, temperature control gives much faster responses compared to composition control. In this two-column configuration, temperature control is used in the reactive distillation column followed by temperature or composition control for the product column (second column). Because we are dealing with binary separation in the second column, two-temperature control is considered first. Figure 12.67 gives the temperature control schemes. For the reactive distillation column, the most sensitive tray, tray 9 ($T_{1,9}$), is chosen for temperature control. The temperature is maintained at 385 K by

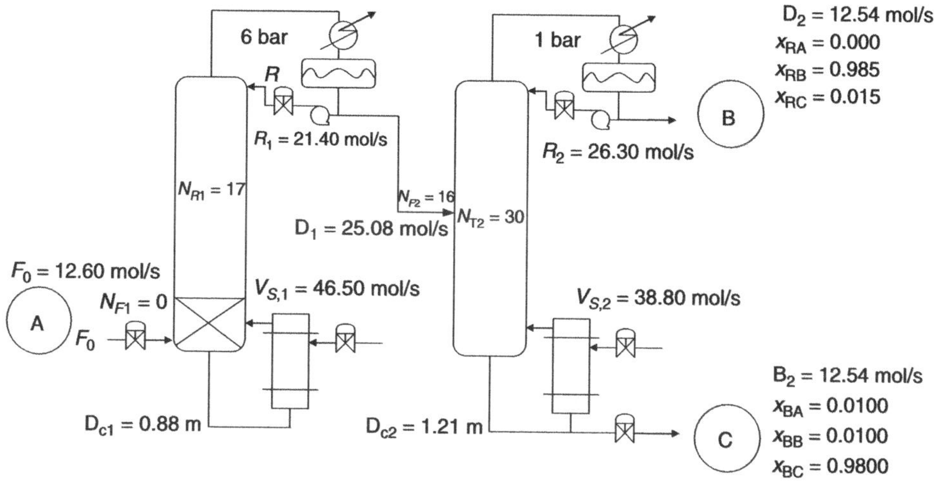

Figure 12.62 Two-column configuration for decomposition reaction (HK ⇔ IK + LK).

manipulating the reflux ratio. For the distillation column, trays 23 and 8 ($T_{2,23}$ and $T_{2,8}$) are chosen as control trays by using SVD analysis. Relay-feedback tests are performed to find controller settings for these three temperature control loops. Table 12.7 gives controlled and manipulated variables and controller parameters.

The performance of this control structure is given in Figure 12.68 for $\pm 20\%$ changes in the feed flowrate. Relative fast transients can be achieved, and little deviation in the top and bottoms compositions of column 2 are observed. Specifically, the deviations in product compositions are all less than 0.01%. Because we have two columns in series, the product compositions settle in ~7 h and the temperatures return to setpoints in ~3 h. Figure 12.69 shows what happens when composition z_0 of fresh feed F_0 is changed from pure A ($z_{0A} = 1$) to a mixture of A and B ($z_{0A} = 0.95$, $z_{0B} = 0.05$). Figure 12.70 shows what happens when composition z_0 of fresh feed F_0 is changed from pure A ($z_{0A} = 1$) to a mixture of A and C ($z_{0A} = 0.95$, $z_{0C} = 0.05$). As shown in Figures 12.69 and 12.70, the process is well controlled using temperature control. For feed composition disturbances, the steady-state offset in product composition is kept below 0.21%, while having fast temperature responses.

12.4.3 Control Structure CS2

We can eliminate composition offsets altogether by implementing two-composition control in the second distillation column. The control scheme is shown in Figure 12.71. The temperature control of the reactive distillation column remains unchanged. However, two-temperature control of column 2 is replaced by two-composition control. Recall that a 4-min analyzer dead time is assumed for composition measurements. Top product purity is maintained at 98.5% B by manipulating the reflux ratio, and the bottoms product composition is controlled at 98% C by adjusting the vapor boilup. Relay-feedback tests are carried out to find ultimate gains and ultimate periods, followed by Tyreus–Luyben tuning with possible detuning. The tuning parameters are given in Table 12.7.

Figure 12.72 shows that much slower responses result for composition control. Product composition responses also show large deviations during transients as well as asymmetric

TABLE 12.6 Steady-State Conditions and Design Parameters for Two-Column Configuration

Column 1 (Reactive Distillation Column)		
Fresh feed flowrate of A F_{0A} (mol/s)	12.60	
Distillate flowrate D_1 (mol/s)	25.08	
Vapor boilup $V_{S,1}$ (mol/s)	46.50	
Reflux flowrate R_1 (mol/s)	21.40	
Reactive trays $N_{RX,1}$	0	
Rectifying trays $N_{R,1}$	17	
$(K_{EQ})_{366}$	0.2866	
Liquid holdup on reactive trays M_{RX} (mol)	891.0	
Pressure P_1 (bar)	6	
Product Composition (Mole Fraction)	Distillate x_{Dj}	
A	0.0050	
B	0.4975	
C	0.4975	
Column 2 (Distillation Column)		
Distillate flowrate D_2 (mol/s)	12.54	
Bottoms flowrate B_2 (mol/s)	12.54	
Vapor boilup $V_{S,2}$ (mol/s)	38.80	
Reflux flowrate R_2 (mol/s)	26.30	
Stripping trays $N_{S,2}$	15	
Rectifying trays $N_{R,2}$	15	
Liquid holdup M_2 (mol)	1653.0	
Pressure P_2 (bar)	1	
Product Composition (Mole fraction)	Distillate x_{Dj}	Bottoms x_{Bj}
A	0.0000	0.0100
B	0.9850	0.0100
C	0.0150	0.9800

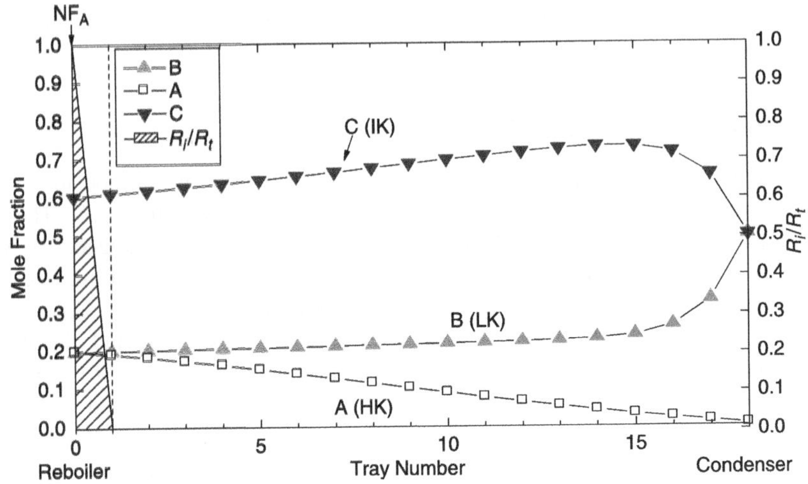

Figure 12.63 Composition profile of reactive distillation column with reactive zone placed in column base, feedtray NF$_A$ and fraction of total conversion also given.

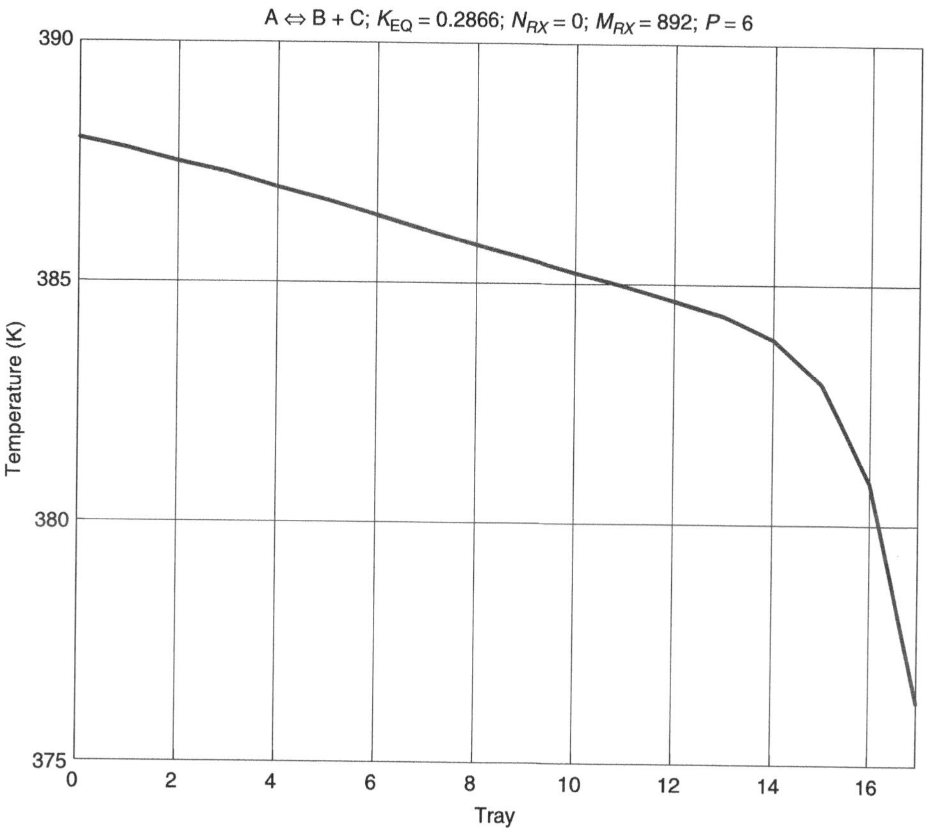

Figure 12.64 Temperature profile of reactive distillation column.

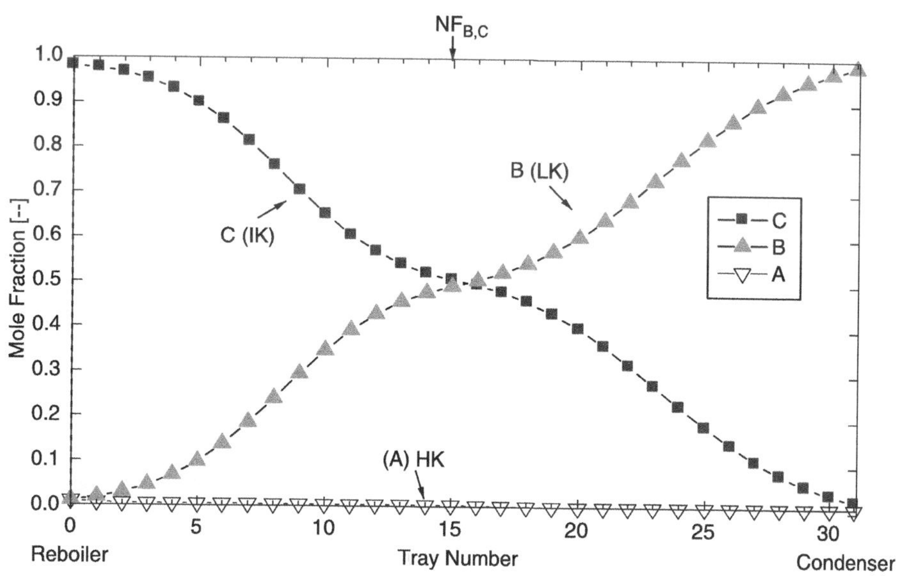

Figure 12.65 Composition profile of distillation column.

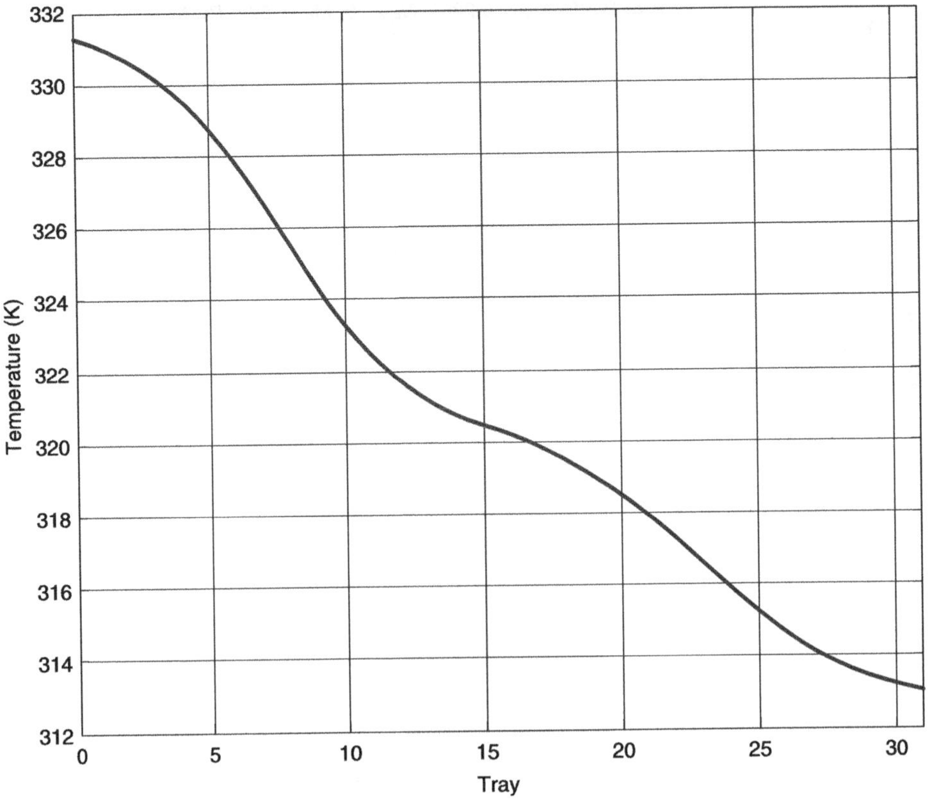

Figure 12.66 Temperature profile of distillation column.

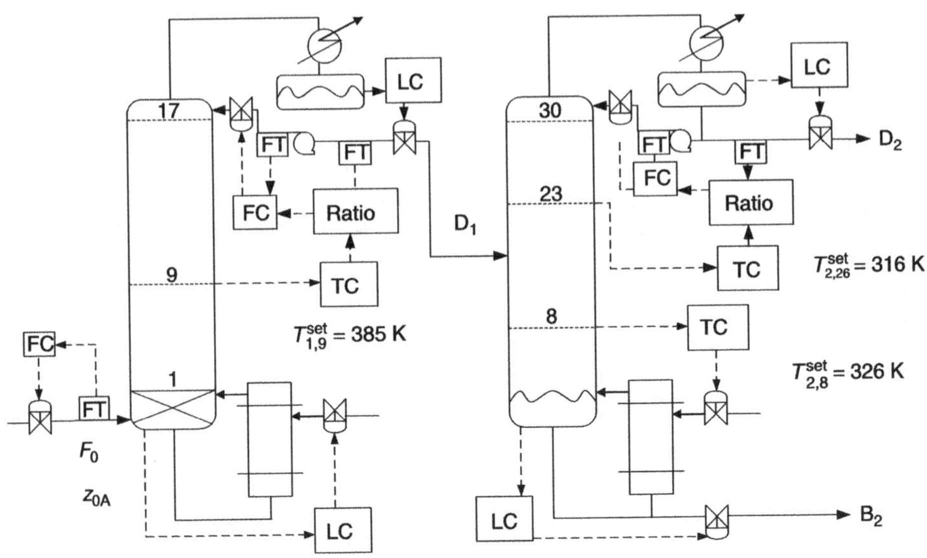

Figure 12.67 Control structure CS1.

TABLE 12.7 Controller Tuning Parameters for HK ⇔ IK + LK with Two-Column Configuration

CS1		CS2	
Column 1 (Reactive Distillation Column)			
Controlled Variable	Manipulated Variable	Controlled Variable	Manipulated Variable
$T_{1,9}$	RR_1	$T_{1,9}$	RR_1
K_U	36.4	K_U	36.4
P_U (min)	1.7	P_U (min)	1.7
K_C	12.1	K_C	12.1
τ_I (min)	3.5	τ_I (min)	3.5
Setpoint	385 (K)	Setpoint	385 (K)
Column 2 (Distillation Column)			
$T_{2,23}$	RR_2	$x_{D2,B}$	RR_2
K_U	220.3	K_U	12.7
P_U (min)	1.7	P_U (min)	44.9
K_C	47.0	K_C	4.2
τ_I (min)	5.4	τ_I (min)	89.8
Setpoint	316 (K)	Setpoint	0.985
$T_{2,8}$	$V_{s,2}$	$x_{B2,C}$	$V_{s,2}$
K_U	51	K_U	0.6
P_U (min)	2.2	P_U (min)	35.8
K_C	17	K_C	0.2
τ_I (min)	4.3	τ_I (min)	71.6
Setpoint	326 (K)	Setpoint	0.98

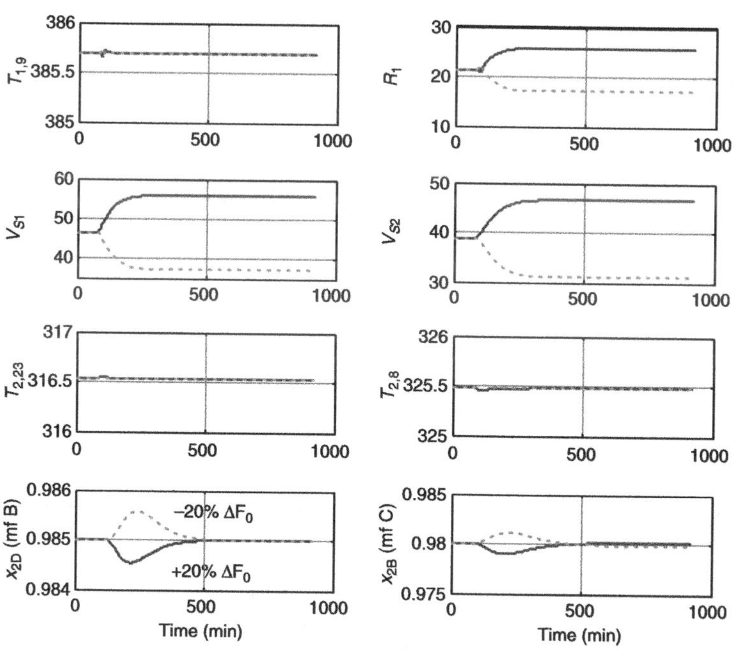

Figure 12.68 Control structure CS1 with $\pm 20\%\ \Delta F_0$.

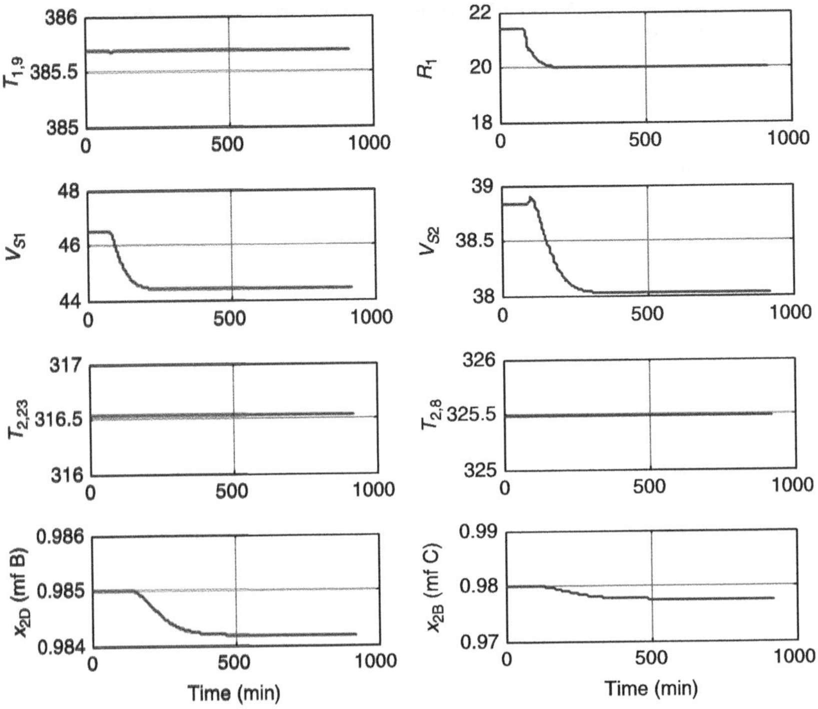

Figure 12.69 Control structure CS1 with $z_{0A} = 0.95$ and $z_{0B} = 0.05$.

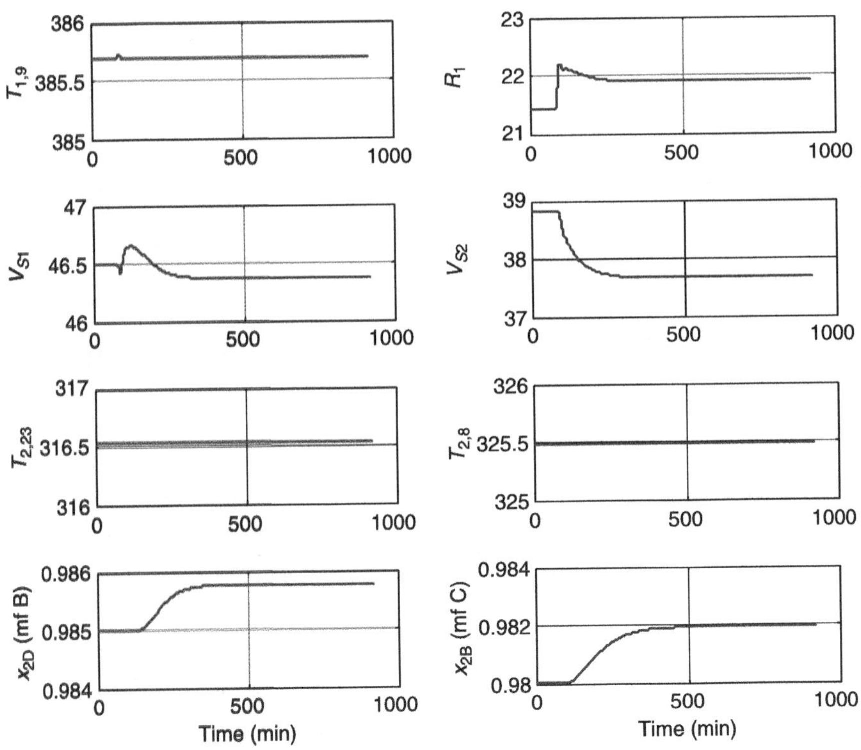

Figure 12.70 Control structure CS1 with $z_{0A} = 0.95$ and $z_{0C} = 0.05$.

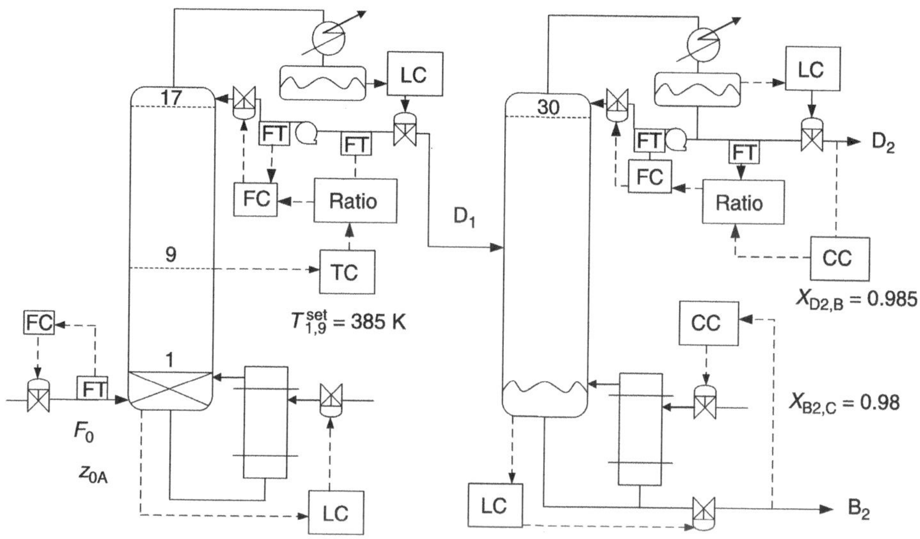

Figure 12.71 Control structure CS2.

behavior for positive and negative feedflow changes. Similar to feedflow disturbances, sluggish responses are also observed for feed composition disturbances. For this two-column configuration, temperature control gives acceptable performance compared to that of composition control (in the product column).

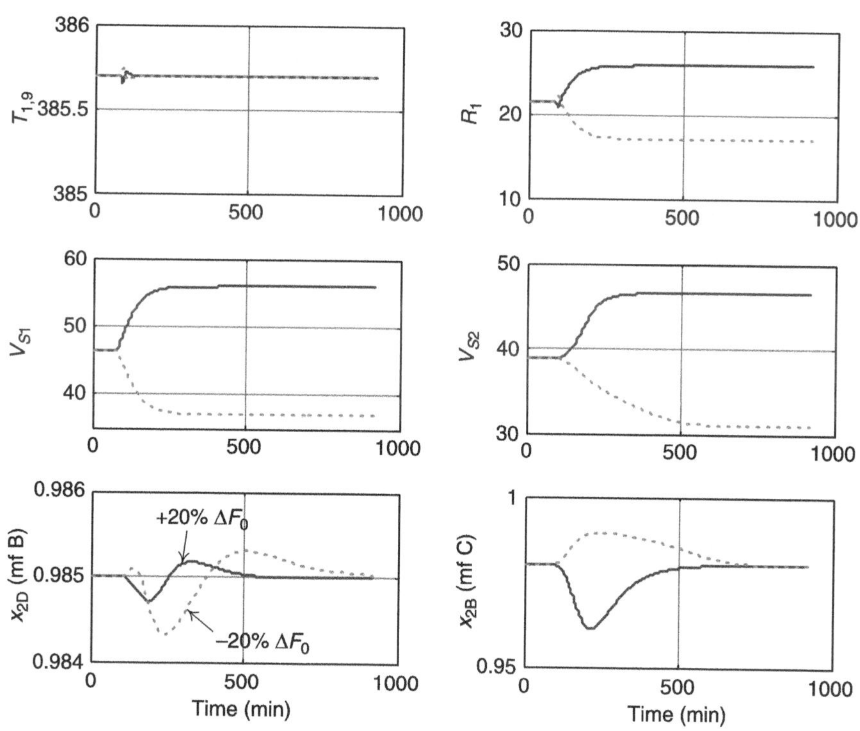

Figure 12.72 Control structure CS2 with $\pm 20\%$ ΔF_0.

12.5 TERNARY A $\Leftrightarrow$ B + C SYSTEM: HEAVY REACTANT WITH ONE-COLUMN CONFIGURATION

12.5.1 Column Configuration

Now we explore the control structure and dynamic performance of the one-column configuration for the same reaction, HK $\Leftrightarrow$ LK + IK. The design for this configuration is based on overcoming the unfavorable boiling point ranking reaching the reactive azeotrope at the bottom of the column. If we consume most of the heavy reactant by the time it reaches the bottoms of the reactive distillation column, the intermediate-boiling product can be withdrawn from the column base at the desired purity. This can only be achieved for systems with a large chemical equilibrium constant as discussed in Chapter 6. In previous sections, the two-column configuration was shown to be controllable with simple temperature control. Will the single column configuration be dynamically controllable, especially when reaching the reactive azeotrope at one end of the reactive distillation column?

Table 12.8 gives the steady-state conditions and design parameters. For keeping the composition near chemical equilibrium, the entire lower section of the column contains reactive trays and the column base has 10 times the amount of catalyst used on each reactive tray. Figure 12.73 shows the flowsheet. Compared to the two-column configuration, this is a very tall column with 90 trays, versus 17 trays in the reactive distillation column. Figure 12.74 gives composition profiles, and Figure 12.75 gives the temperature profile.

Three control structures are considered here: one-temperature control, two-temperature control, and temperature/composition cascade control. The control objective is to keep the product purity at 98 mol% at both ends.

12.5.2 Control Structure CS1

Because we have a reactive azeotrope toward the bottoms of the column, it serves as a composition control mechanism in the lower part of the column. Figure 12.76 shows a control

TABLE 12.8 Steady-State Conditions and Design Parameters for One-Column Configuration

Fresh feed flowrate of A F_0 (mol/s)	12.60	
Distillate flowrate D (mol/s)	12.60	
Bottom flowrate B (mol/s)	12.45	
Vapor boilup V_S (mol/s)	73.90	
Reflux flowrate R (mol/s)	44.10	
Reactive trays N_{RX}	79	
Rectifying trays N_R	11	
Feed tray location N_F	79	
$(K_{EQ})_{366}$	0.2866	
Liquid holdup on reactive tray M_{RX} (mol)	1067.6	
Pressure P (bar)	6	
Product Composition (Mole Fraction)	Distillate x_{Dj}	Bottoms x_{Bj}
A	0.0000	0.0080
B	0.9800	0.0120
C	0.0200	0.9800

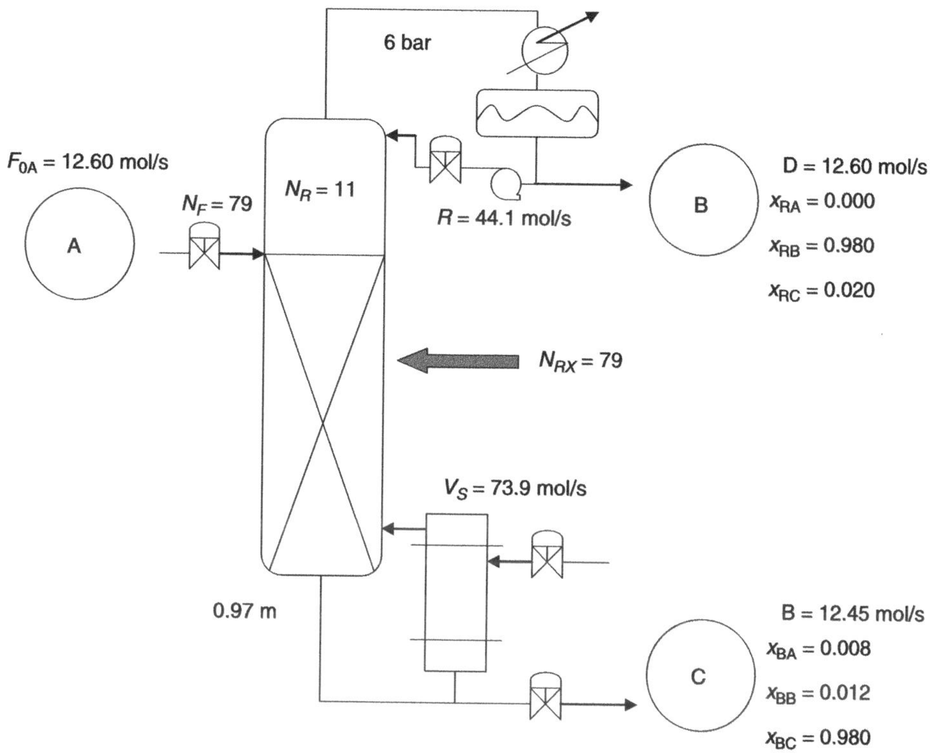

Figure 12.73 One-column configuration for decomposition reaction (HK ⇔ LK + IK).

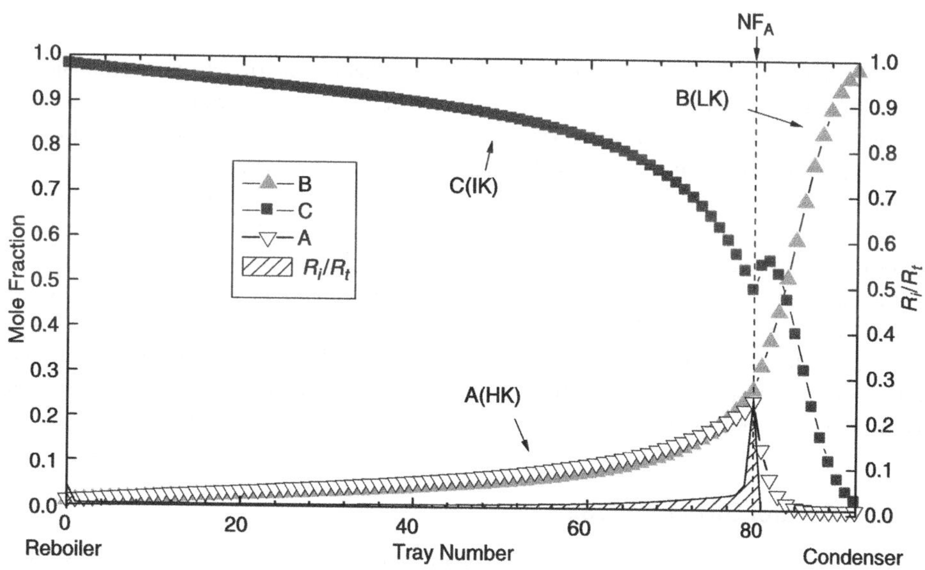

Figure 12.74 Composition profile for HK ⇔ LK + IK with reactive zone marked with NF$_A$ and the fraction of total conversion (R_i/R_t) shown in shaded area.

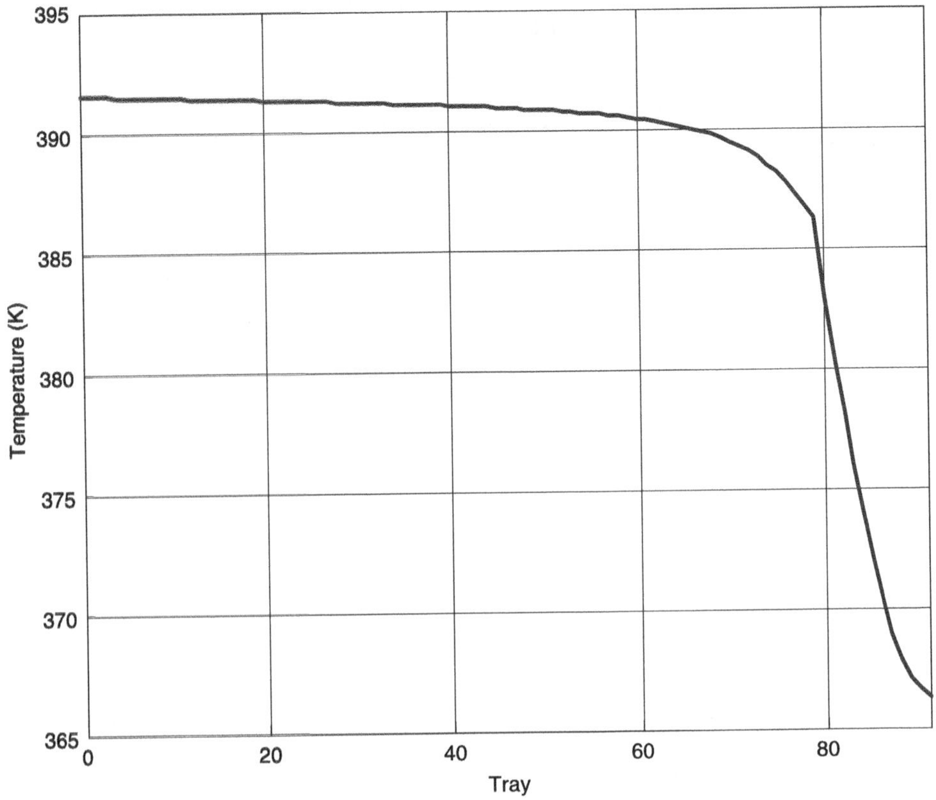

Figure 12.75 Temperature profile.

structure where the tray 85 temperature is controlled using the reflux ratio and the vapor boilup is ratioed to the feedflow. Figure 12.77 shows that almost offset free (composition) control can be achieved for feedflow disturbances and fast responses can also be observed. Figure 12.78 demonstrates what happens when the feed composition is changed from pure A ($z_{0A(A)} = 1$) to a mixture of A and B ($z_{0A(A)} = 0.95$, $z_{0A(B)} = 0.05$). Because less A is coming into the system, more B (LK) and C (IK) are going toward the top, with the vapor boilup kept constant. Increased IK C leads to a slight increase in the temperatures in the top of the column, and the temperature controller increases the reflux ratio to maintain the tray 85 temperature. Figure 12.79 shows similar behavior when the feed composition is changed from pure A ($z_{0A(A)} = 1$) to a mixture of A and C ($z_{0A(A)} = 0.95$, $z_{0A(C)} = 0.05$). Because product C is the intermediate boiler, which could appear both in the top and bottom of the column, the reflux ratio will increase to separate products C and B. For all disturbances, the steady-state offsets in product compositions are less than 0.3%.

12.5.3 Control Structure CS2

Using the reflux ratio and vapor boilup as the manipulated variables, the two-temperature control structure is shown in Figure 12.80. Two temperature control trays, tray 85 and tray 5, are selected based on SVD analysis. Using relay-feedback tests, values of ultimate

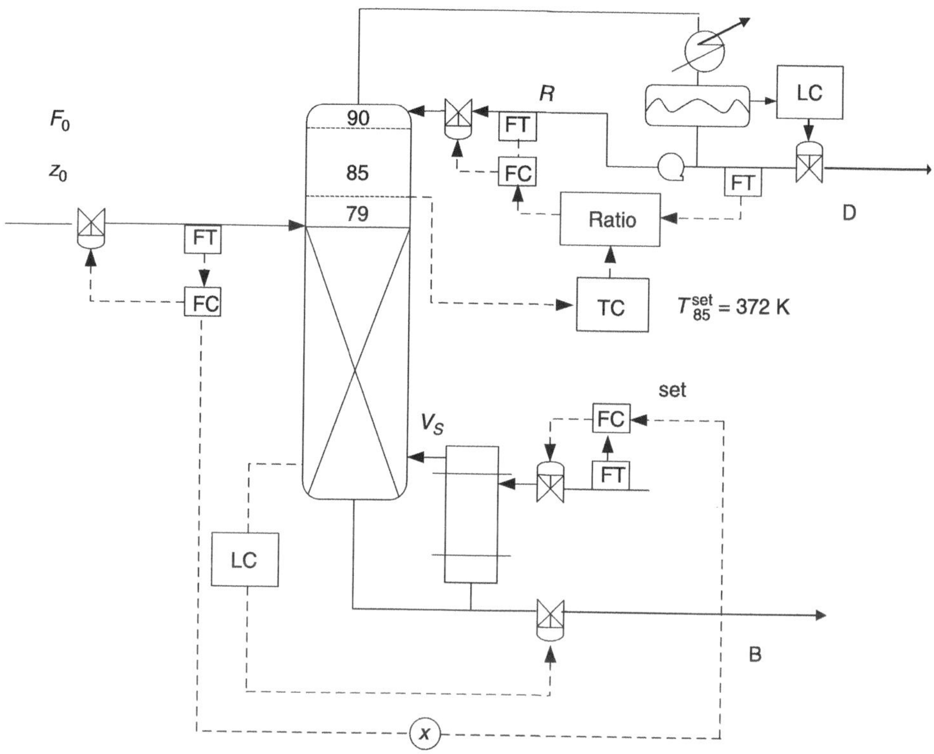

Figure 12.76 Control structure CS1.

gains and periods are obtained and provided in Table 12.9. Tyreus–Luyben tuning is used, but some loops are detuned for less oscillatory responses.

Figures 12.81 illustrates the performance of this CS2 control structure under the feedflow disturbances. Tight control can be achieved with a settling time of ~2 h. Figure 12.82 shows what happens when the feed composition is changed from pure A ($z_{0A(A)} = 1$) to a mixture of A and B ($z_{0A(A)} = 0.95$, $z_{0A(B)} = 0.05$). Because less A is coming into the system, less vapor boilup and reflux are required. Figure 12.83 shows what happens when the feed composition is changed from pure A ($z_{0A(A)} = 1$) to a mixture of A and C ($z_{0A(A)} = 0.95$, $z_{0A(C)} = 0.05$). Because product C is the intermediate-boiling key, which could appear both in the top and bottom of the column, the reflux ratio will increase to separate product C and product B. For all disturbances, the steady-state error in product composition is less than 0.2%, which is mostly in the top product composition. Compared to CS1, the energy consumption of two-temperature control is less, especially when product B is coming into the system from the feed (Figs. 12.78 and 12.82).

12.5.4 Control Structure CS3

The composition offsets can be eliminated using a temperature/composition cascade as suggested in the case of IK ⇔ LK + HK. Will the parallel cascade control also be effective for this single column configuration with heavy reactant, HK ⇔ IK + LK?

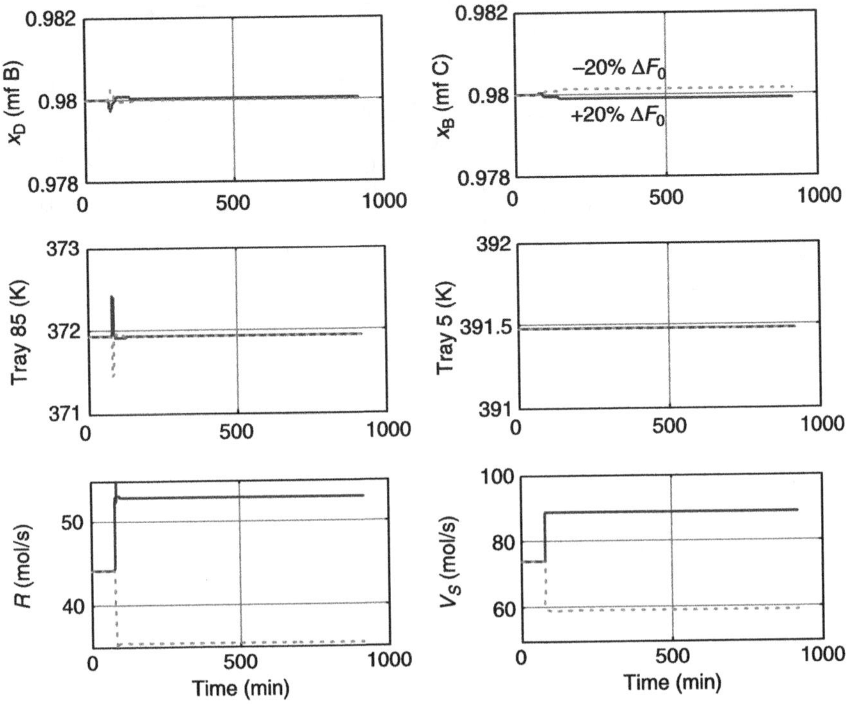

Figure 12.77 Control structure CS1 with $\pm 20\%\ \Delta F_0$.

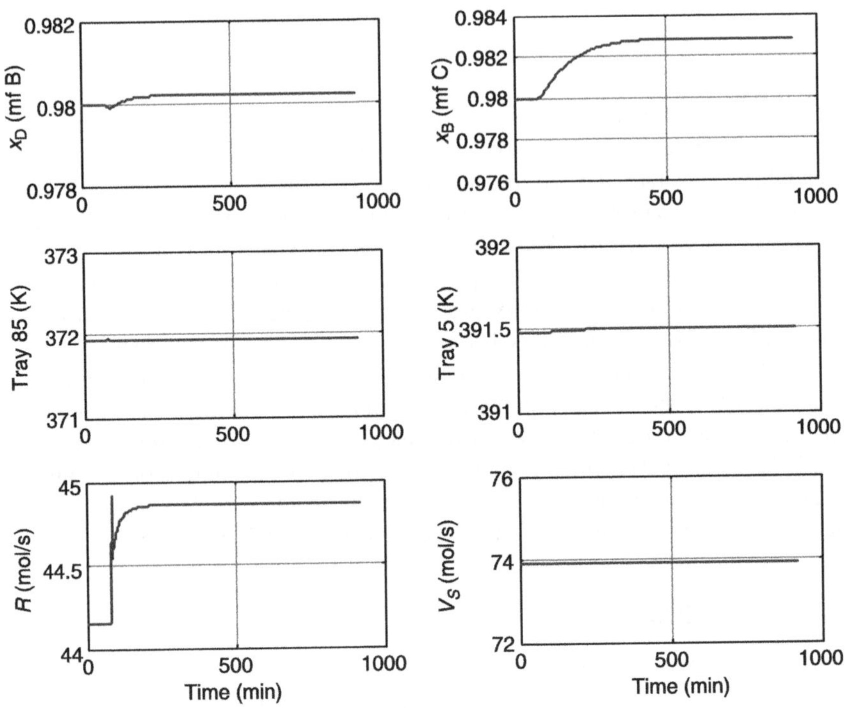

Figure 12.78 Control structure CS1 with $z_{0A} = 0.95$ and $z_{0B} = 0.05$.

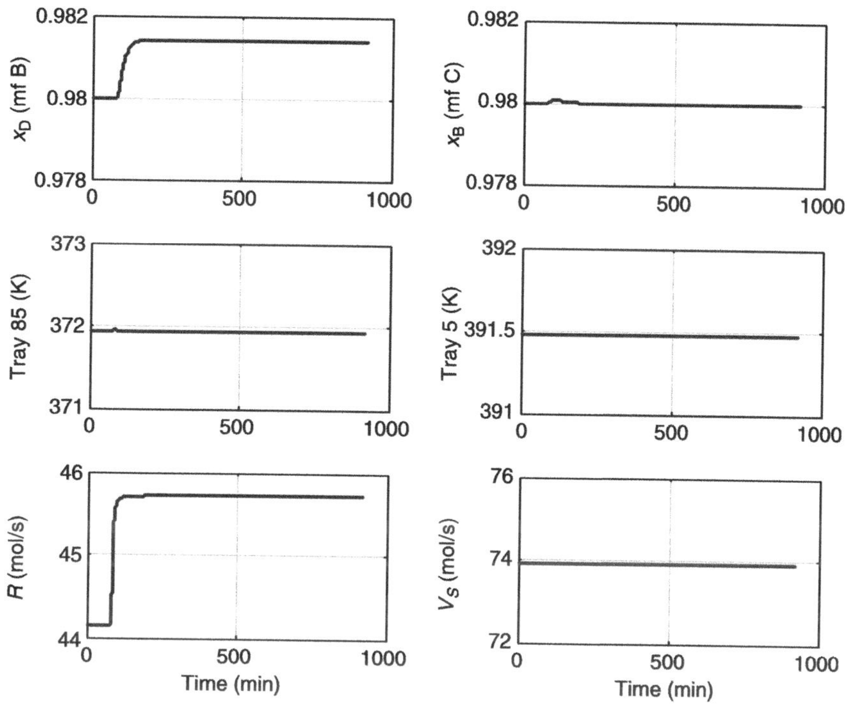

Figure 12.79 Control structure CS1 with $z_{0A} = 0.95$ and $z_{0C} = 0.05$.

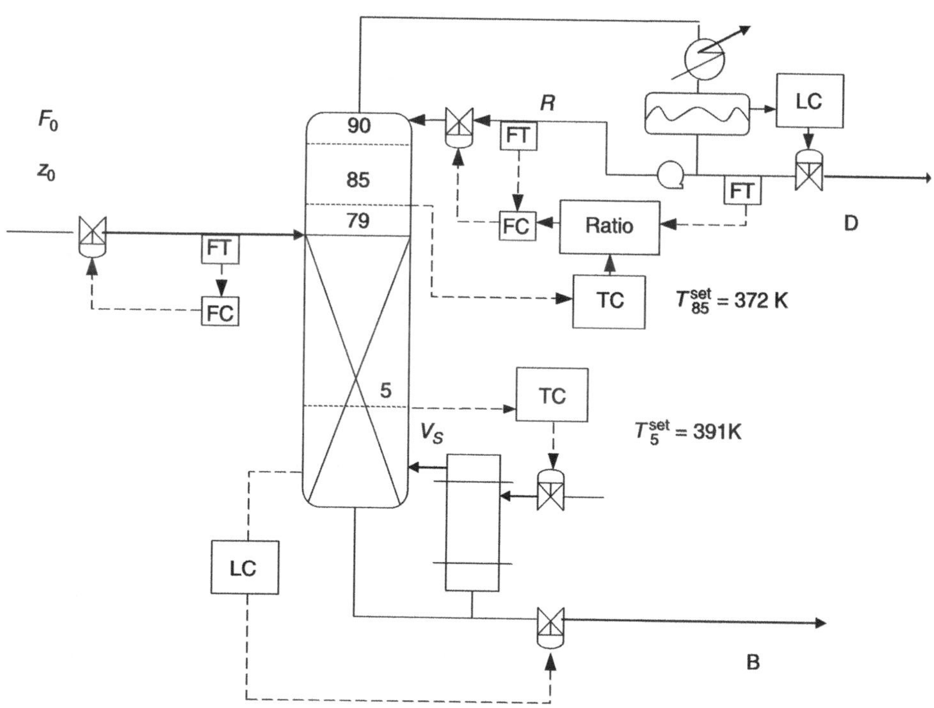

Figure 12.80 Control structure CS2.

TABLE 12.9 Controller Tuning Parameters for Ternary Decomposition Reaction: HK ⇔ LK + IK

CS1		CS2		CS3	
Controlled Variable T_{85}	Manipulated Variable RR	Controlled Variable T_{85}	Manipulated Variable RR	Controlled Variable x_D	Manipulated Variable T_{85}^{set}
K_U	37.5		37.5		0.3
P_U (min)	1.58		1.58		14.7
K_C	12.5		12.5		0.01
τ_I (min)	3.1		3.1		29.5
Setpoint	372 (K)	Setpoint	372 (K)		0.98
		T_5	V_S	x_B	T_5^{set}
		K_U	4681		0.05
		P_U (min)	1.2		22.2
		K_C	1560		0.002
		τ_I (min)	2.4		44.3
		Setpoint	391 (K)		0.98

With the same temperature controllers (CS2) implemented, composition measurements are made. The setpoint of the tray 85 temperature is adjusted to maintain top product purity and the setpoint of the tray 5 temperature is changed to keep the bottoms product composition as shown in Figure 12.84. The ultimate gains and ultimate periods values are given in

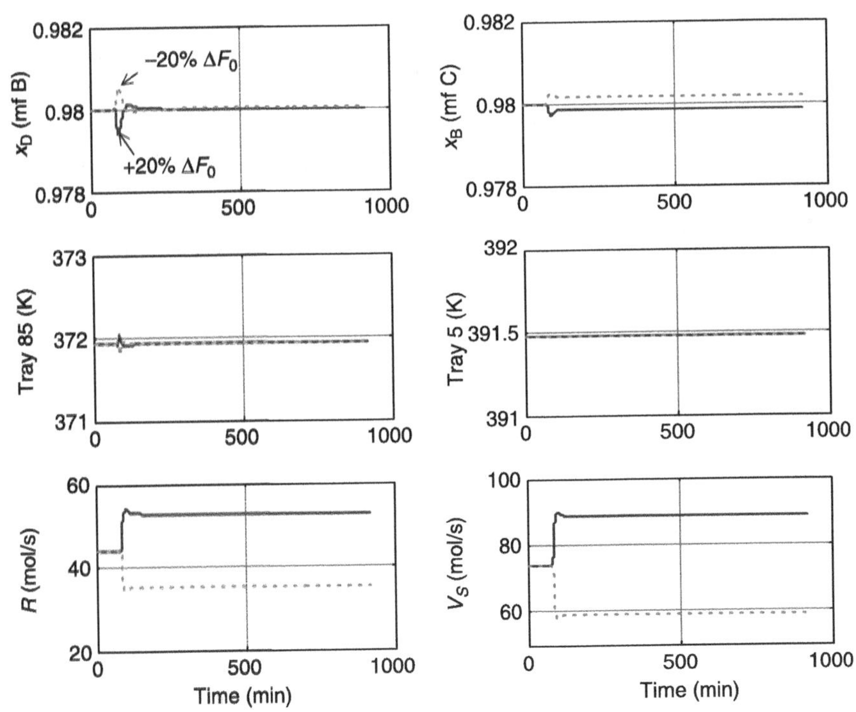

Figure 12.81 Control structure CS2 with $\pm 20\%$ ΔF_0.

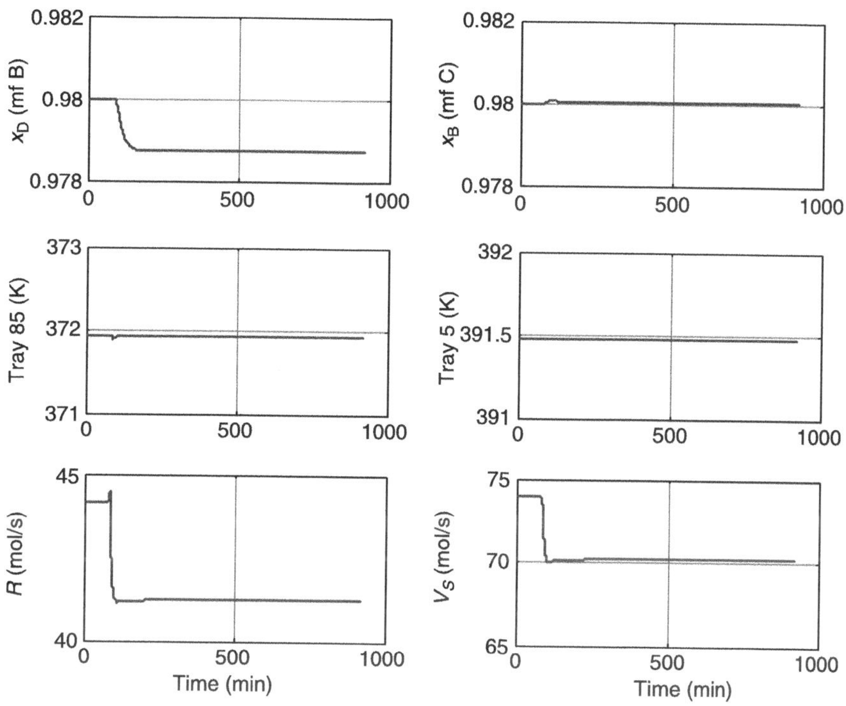

Figure 12.82 Control structure CS2 with $z_{0A} = 0.95$ and $z_{0B} = 0.05$.

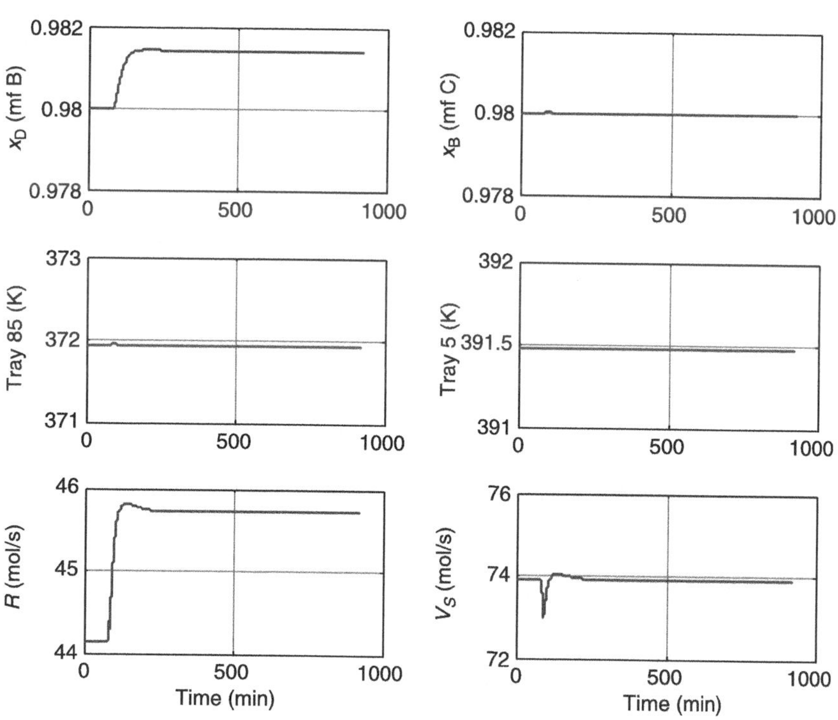

Figure 12.83 Control structure CS2 with $z_{0A} = 0.95$ and $z_{0C} = 0.05$.

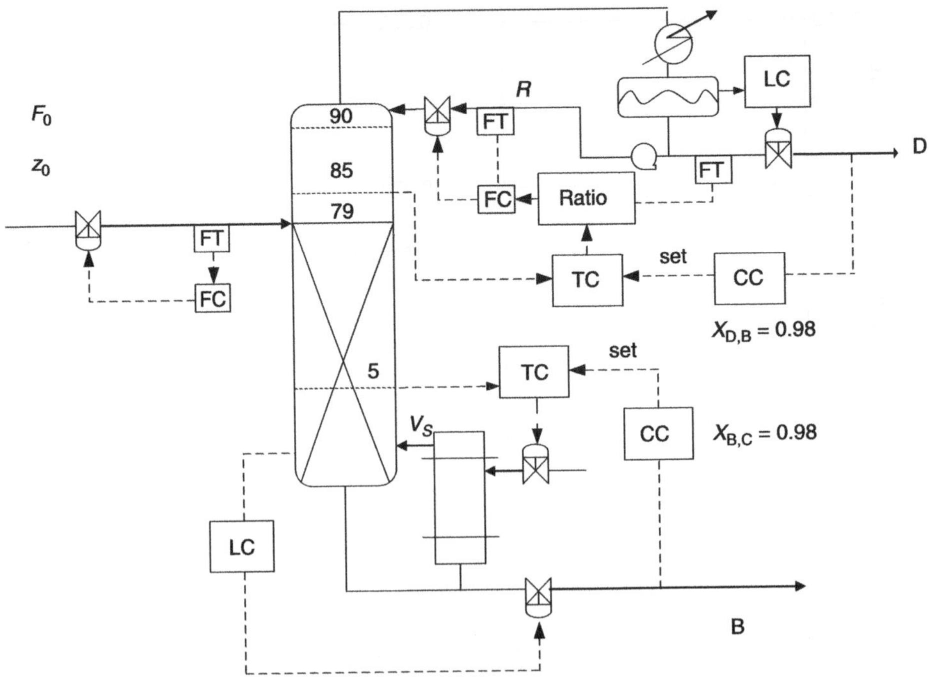

Figure 12.84 Control structure CS3.

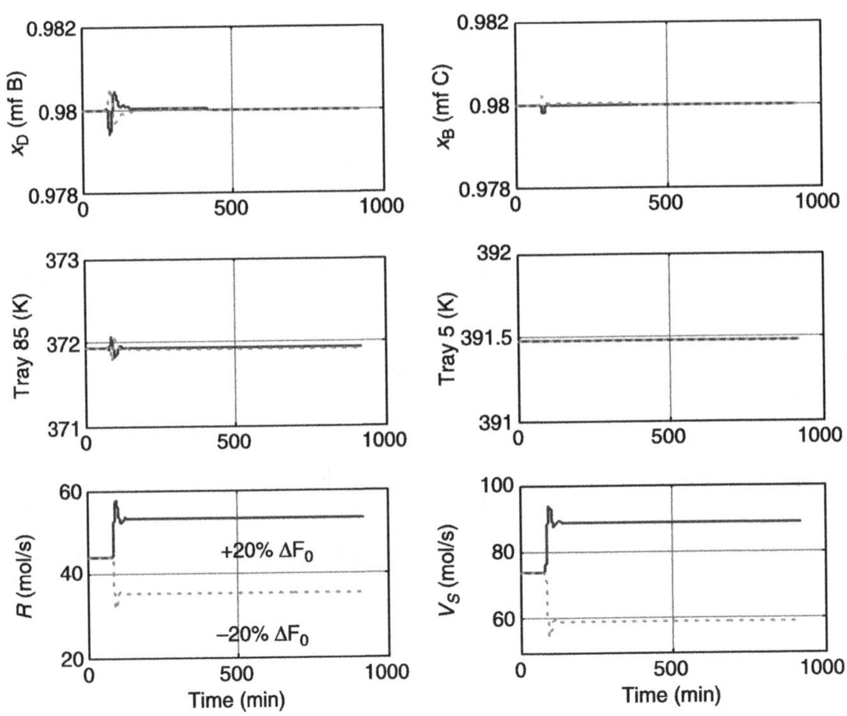

Figure 12.85 Control structure CS3 with $\pm 20\%$ ΔF_0.

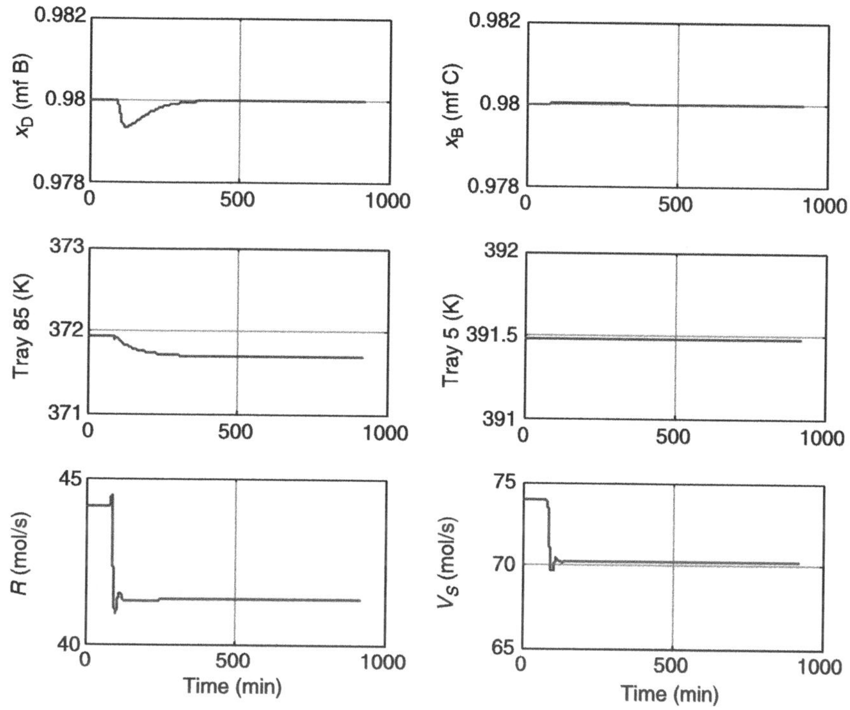

Figure 12.86 Control structure CS3 with $z_{0A} = 0.95$ and $z_{0B} = 0.05$.

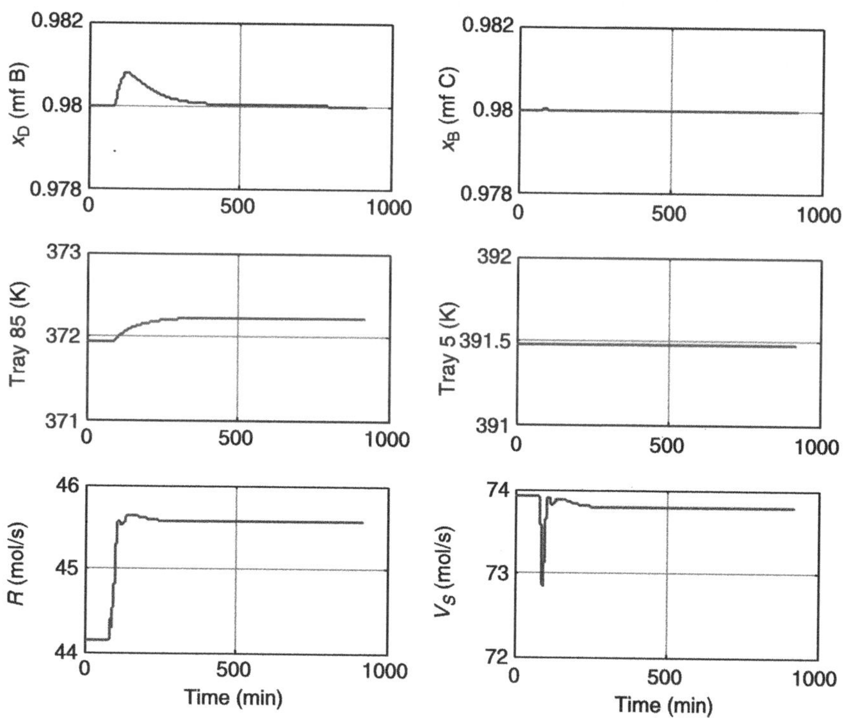

Figure 12.87 Control structure CS3 with $z_{0A} = 0.95$ and $z_{0C} = 0.05$.

Table 12.9. The results show that the ultimate periods of the composition controllers are 1 order of magnitude larger than that of the temperature controllers.

For feedflow disturbances, Figure 12.85 reveals that tight composition control can be achieved in an offset-free manner. Because of the almost offset-free temperature control, little adjustment in the temperature setpoints is made. For feed composition disturbances (Figs. 12.86 and 12.87), the speed of the responses is similar to that of temperature control (Figs. 12.83 and 12.84), except that both product compositions return to their setpoints.

12.5.5 Conclusion for Ternary A $\Leftrightarrow$ B + C System

For ternary decomposition reactions, control of reactive distillation systems was explored for two different boiling point rankings: IK $\Leftrightarrow$ LK + HK and HK $\Leftrightarrow$ LK + IK. For nonlinear reactive distillation systems, fast measurement (temperature measurement) is essential for tight control and control performance deteriorates if significant measurement delay is present. Thus, parallel cascade control offers an attractive alternative for the control of reactive distillation systems. Compared to the A + B $\Leftrightarrow$ C + D and A + B $\Leftrightarrow$ C systems, the decomposition reaction system is relatively easy to control because stoichiometric balance is no longer an issue. Following systematic control structure design, good control performance can be achieved. For IK $\Leftrightarrow$ LK + HK, the most intuitive control structure could handle this process. Both temperature and temperature/composition cascade controls work well. For HK $\Leftrightarrow$ LK + IK, we have two process alternatives: a two-column configuration and a one-column reactive distillation configuration. The results show that both process configurations give reasonable dynamic performance as long as prompt process measurements are available. Thus, parallel cascade control is a candidate control structure. Moreover, single-end control will work well for the one-column configuration because we have a reactive azeotrope toward the column base.

PART V

CONTROL OF REAL SYSTEMS

CHAPTER 13

CONTROL OF REACTIVE DISTILLATIONS FOR ACETIC ACID ESTERIFICATION

Steady-state design of the esterifications of acetic acid with five different alcohols (C_1–C_5) was explored in Chapter 7. This chapter explores the control of these five reactive distillation systems using three different flowsheets. The degree of process nonlinearity is computed quantitatively based on the fraction of "sign reversal" for all tray temperatures or based on Allgower's nonlinearity measure.[1,2] These measures provide useful information about potential problems in closed-loop control. Next, a systematic design procedure is proposed to devise control structures for all three types of flowsheets for these five esterification systems. The simulation results reveal that reasonable control can be achieved for all five systems with different degrees of asymmetry in closed-loop responses as predicted by the nonlinearity measures. Two-temperature control and one-temperature-one-composition control are studied. Simulation results clearly show that simple decentralized control provides a workable solution for highly nonlinear reactive distillation columns with various flowsheet configurations.

13.1 PROCESS CHARACTERISTICS

13.1.1 Process Studies

In Chapter 7 we showed that the esterification of acetic acid with different types of alcohols (C_1–C_5) can be classified into three types of flowsheets (types I–III), which were shown in

[1]P. H. Menold, F. Allgower, and P. K. Pearson, Nonlinear structure identification of chemical processes, *Comput. Chem. Eng*. **21**, S137–S142 (1997).
[2]T. Schweickhardt and F. Allgower, Quantitative nonlinearity assessment: An introduction to nonlinearity measure, In P. Seferlis and M. C. Georgiadis, Editors, *Integration of Process Design and Control*, Elsevier, Amsterdam, 2004.

Figure 7.2. The gradual change in the process configuration is the direct consequences of two factors: increased immiscibility (Fig. 7.3) and the shift of the boiling point ranking of the two products (water and acetate). Table 7.5 gives the optimized design for the production of methyl acetate (MeAc, type I), ethyl acetate (EtAc, type II), isopropyl acetate (IPAc, type II), butyl acetate (BuAc, type III), and amyl acetate (AmAc, type III) systems.

Steady-state analysis indicates that the type I and type III systems are more economical than the type II system. This chapter explores the dynamic controllability of these three flowsheets. Of more importance, we want to devise a systematic approach to the control of these three types of reactive distillations. All of the results are obtained from steady-state and dynamic simulations using Aspen Plus and Aspen Dynamics.

13.1.2 Quantitative Analysis

Manipulated Inputs. Before delving into a detailed quantitative analysis, we need to identify the manipulated variables for these three different types of processes. As pointed out by Luyben,[3] it is important to maintain stoichiometric balance for neat reactive distillation. Al-Arfaj and Luyben[4] chose to use one of the feedrates. In this chapter the feed ratio is used as one manipulated variable. In addition to holding the stoichiometric balance, we need to control two product compositions using two manipulated variables. However, for reactions such as $A + B \Leftrightarrow C + D$, if the conversion is properly maintained and the product flowrates are equally distributed, one-end composition control will do a fairly good job.

Following Al-Arfaj and Luyben,[4] for the type I flowsheet of MeAc production we chose to control the bottoms composition using vapor boilup while fixing the reflux ratio.

For the type II flowsheet, the product composition of water from the first column (reactive distillation column) is determined by the LL equilibrium, so no composition control is necessary. However, the reflux ratio of the reactive distillation column is fixed. The acetate product is withdrawn from the bottoms of the stripper and the composition is controlled by manipulating the vapor boilup as shown in Figure 13.1.

Similar to the type II flowsheet, a decanter is used for the type III flowsheet to separate the water from the overhead condensate and therefore composition control is not necessary. The organic phase is totally refluxed back to the column. However, the bottoms acetate composition is controlled by changing the reboiler duty. Note that all of the control structures mentioned in this section use temperature control. In summary, the manipulated variables are

Type I: feed ratio and reboiler duty (fixed reflux ratio)

Type II: feed ratio and reboiler duty in the stripper (fixed organic reflux ratio in reactive distillation column)

Type III: feed ratio and reboiler duty (organic phase totally refluxed)

Nonlinearity and Output Multiplicity. Once the manipulated variables are determined, we can evaluate the process nonlinearity for these three different types of flowsheets. The tray temperatures are treated as the state variables. The manipulated variables are the heat input Q_R and feed ratio FR. The upper and lower bounds of the steady-state

[3]W. L. Luyben, Economic and dynamic impact of the use of excess reactant in reactive distillation systems, *Ind. Eng. Chem. Res.* **39**, 2935–2946 (2000).

[4]M. A. Al-Arfaj and W. L. Luyben, Comparative control study of ideal and methyl acetate reactive distillation, *Chem. Eng. Sci.* **57**, 5039–5050 (2002).

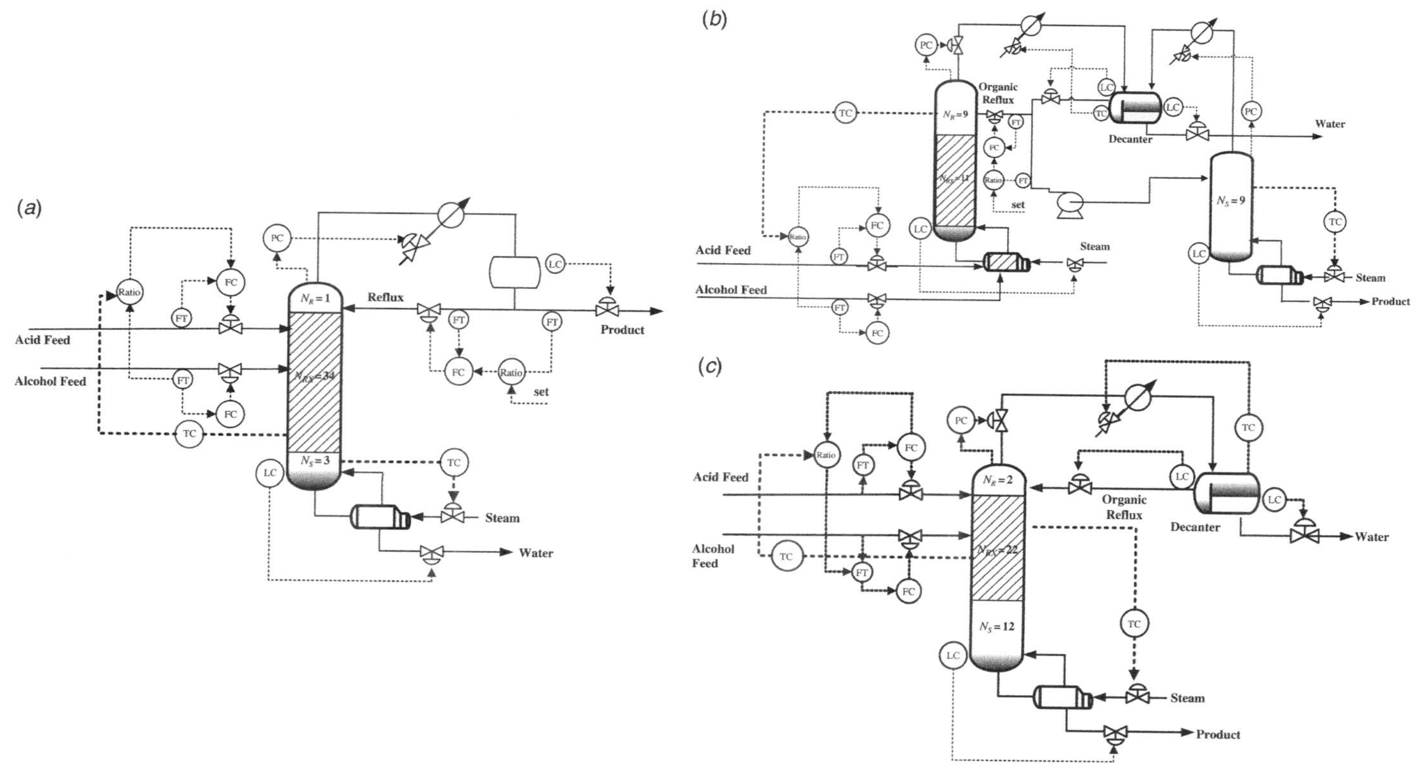

Figure 13.1 Process flowsheets and temperature control configurations for (*a*) type I, (*b*) type II, and (*c*) type III systems.

gains between the tray temperatures and the manipulated variables (Q_R and FR) are obtained for a range of input variations. We made -5% to $+5\%$ changes in the heat input and -1% to $+1\%$ changes in the feed ratio. Note that, for a truly linear system, the upper and lower bounds should coincide with each other. Figure 13.2 clearly shows that the reactive distillation columns exhibit strong nonlinearity for all five systems but with different degrees of severity. Moreover, the sign reversal is also observed for all five systems for both Q_R or FR changes. The sign reversal indicates that the steady-state gain of a specific tray temperature changes sign as the magnitude of the same manipulated variables varies.

The results presented here are rather unconventional because chemical processes are known to be quite nonlinear but not to this degree in such a consistent manner. Two measures are used to differentiate the degree of nonlinearity for these three types of reactive distillation systems. One obvious choice is the fraction of trays that exhibit sign reversal.

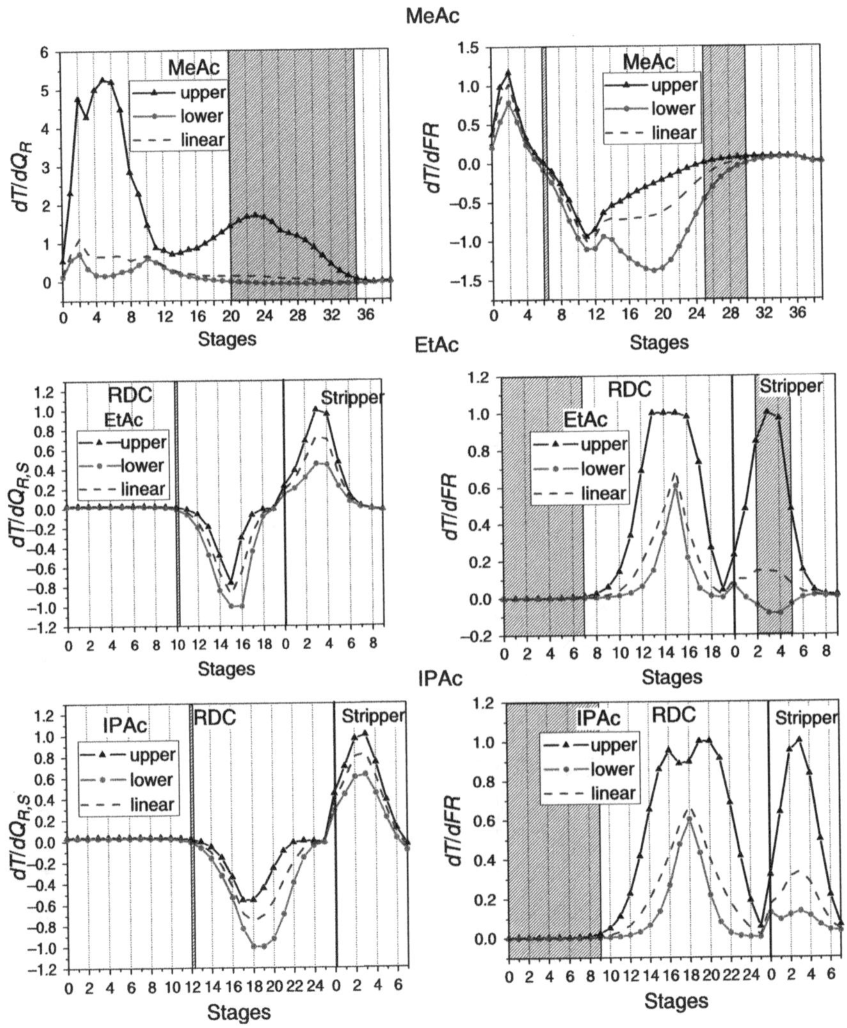

Figure 13.2 Upper and lower bounds of steady-state gains of all tray temperatures for $\pm 5\%$ reboiler duty and $\pm 1\%$ feed ratio changes. Sign reversal indicated as shaded areas for all five systems.

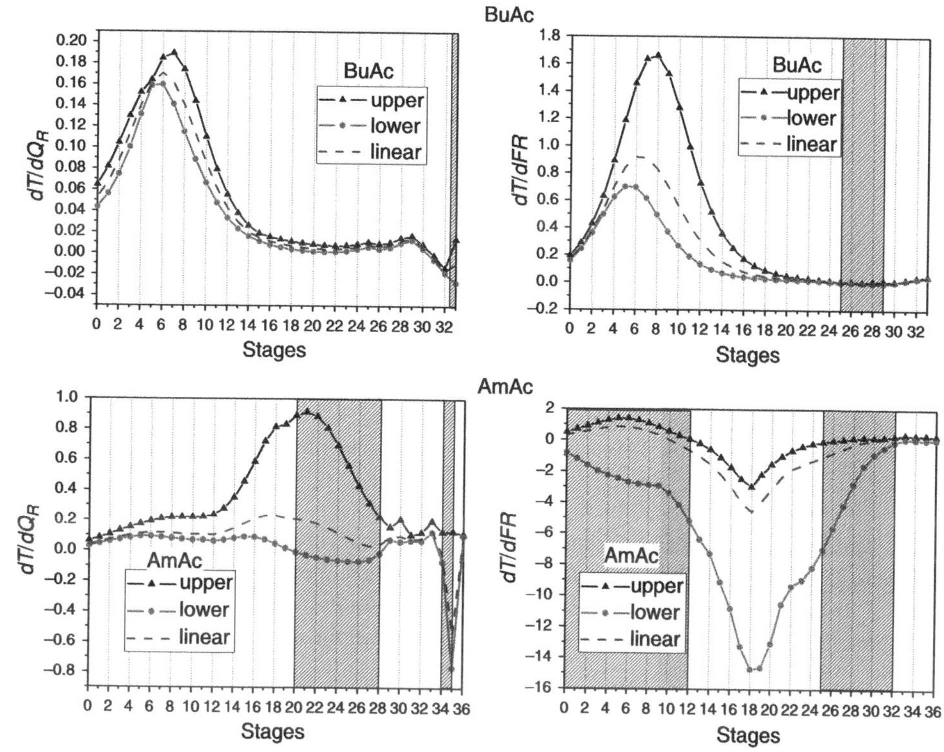

Figure 13.2 *(Continued).*

In this aspect, the AmAc system (Fig. 13.2) has more than half of the trays showing sign reversal, followed by the MeAc system (type I) in which almost half of the tray temperatures exhibit sign reversal. The type II system (EtAc and IPAc) shows that almost one-third of the tray temperatures exhibit sign reversal. The BuAc system is the system with the least sign changes in the tray temperatures. Table 13.1 summarizes the fraction of sign changes for all five systems.

The second nonlinearity indicator is ϕ^N that was first proposed by Allgower[1] for general dynamic systems and further studied by Hernjak and Doyle[5] for systems under feedback control. Schweickhardt and Allgower[2] give an updated summary on the nonlinearity measure. In this chapter we only consider the steady-state aspect (viewing it as the nonlinearity measure for a static function). Schweickhardt and Allgower's approach uses the 2-norm to compute ϕ^N, and each manipulated variable is considered separately. The measure is defined as

$$\phi^N = \inf_{G \in \mathrm{G}} \sup_{u \in \mathrm{U}} \frac{\|G(\boldsymbol{u}) - N(\boldsymbol{u})\|_2}{\|N(\boldsymbol{u})\|_2} \tag{13.1}$$

where G is a linear operator, N is a nonlinear static function, U is the input set, and G is the set of all linear operators. Physically, this can be viewed as the relative deviation of a

[5]N. Hernjak and F. J. Doyle III, Correlation of process nonlinearity with closed-loop disturbance rejection, *Ind. Eng. Chem. Res.* **42**, 4611–4619 (2003).

TABLE 13.1 Fractions of Sign Reversal and Nonlinearity Measures for All Five Esterification Systems

Flowsheet Type	System	Fraction of Sign Reversal			Nonlinearity Measure[a]			Overall Assessment[d]
		Q_R	FR	Overall[b]	Q_R	FR	Overall[c]	
I	MeAc	0.40	0.18	0.43	0.84	0.44	0.67	H
II	EtAc	0.03	0.40	0.43	0.34	0.70	0.55	M
	IPAc	0.03	0.29	0.32	0.32	0.60	0.48	M
III	BuAc	0.03	0.15	0.18	0.16	0.46	0.34	L
	AmAc	0.30	0.57	0.76	0.78	0.79	0.79	H

[a]Schweickhardt and Allgower.[2]
[b]Delete overlapping (from each input) trays.
[c]Taken as the 2-norm of two inputs divided by 2.
[d]High if the average value is >0.5, medium if the average value is between 0.3 and 0.5, and low if the average value is <0.3.

steady-state gain of a linear transfer function to the nonlinear function in a normalized sense. The measure ϕ^N ranges from 0 to 1 with $\phi^N = 0$ indicating a linear system and ϕ^N increasing toward 1 as the nonlinearity becomes more severe. For static functions with the upper and lower bounds available, the solution to the optimization problem is simply

$$\frac{\|\bar{G} - G_+\|_2}{\|G_+\|_2} = \frac{\|\bar{G} - G_-\|_2}{\|G_-\|_2} \tag{13.2}$$

where G_+ is the upper bound of $N(\boldsymbol{u})/\boldsymbol{u}$ and G_- is the lower bound of $N(\boldsymbol{u})/\boldsymbol{u}$. By solving Eq. (13.2), the linear approximation $\bar{G}$ can be obtained as shown in Figure 13.2. Consequently, the nonlinear measure can be computed as

$$\phi^N = \frac{\|\bar{G} - G_+\|_2}{\|G_+\|_2} = \frac{\|\bar{G} - G_-\|_2}{\|G_-\|_2} \tag{13.3}$$

The vector G corresponds to tray temperatures throughout the column. Because we treat two manipulated inputs separately, two ϕ^N values are available for a given system. Table 13.1 gives the nonlinearity measures for all five systems with two different inputs. Qualitatively, the results of ϕ^N are consistent with the fractions of sign reversal shown in Table 13.1. The 2-norm is employed for overall nonlinearity assessment based on ϕ^N, and the results are consistent with the previous sign reversal analysis. The ranking of the processes from linear to nonlinear becomes the following: BuAc (type III) → IPAc (type II) → EtAc (type II) → MeAc (type I) → AmAc (type III).

In addition to the tray temperatures, we are also interested in the behavior of product composition for a range of input changes. As pointed out by Doherty and Malone,[6] Al-Arfaj and Luyben,[4] and Lin et al.,[7] the MeAc system exhibits "input multiplicity" where more than one set of inputs give the same outputs. Figure 13.3 clearly shows that both the heat input and the feed ratio result in input multiplicity for the compositions of

[6]M. F. Doherty and M. F. Malone, Reactive distillation, *Ind. Eng. Chem. Res.* **39**, 3953 (2000).
[7]Y. D. Lin, H. P. Huang, and C. C. Yu, Relay feedback tests for highly nonlinear processes: Reactive distillation, *Ind. Eng. Chem. Res.* **45**, 4081–4092 (2006).

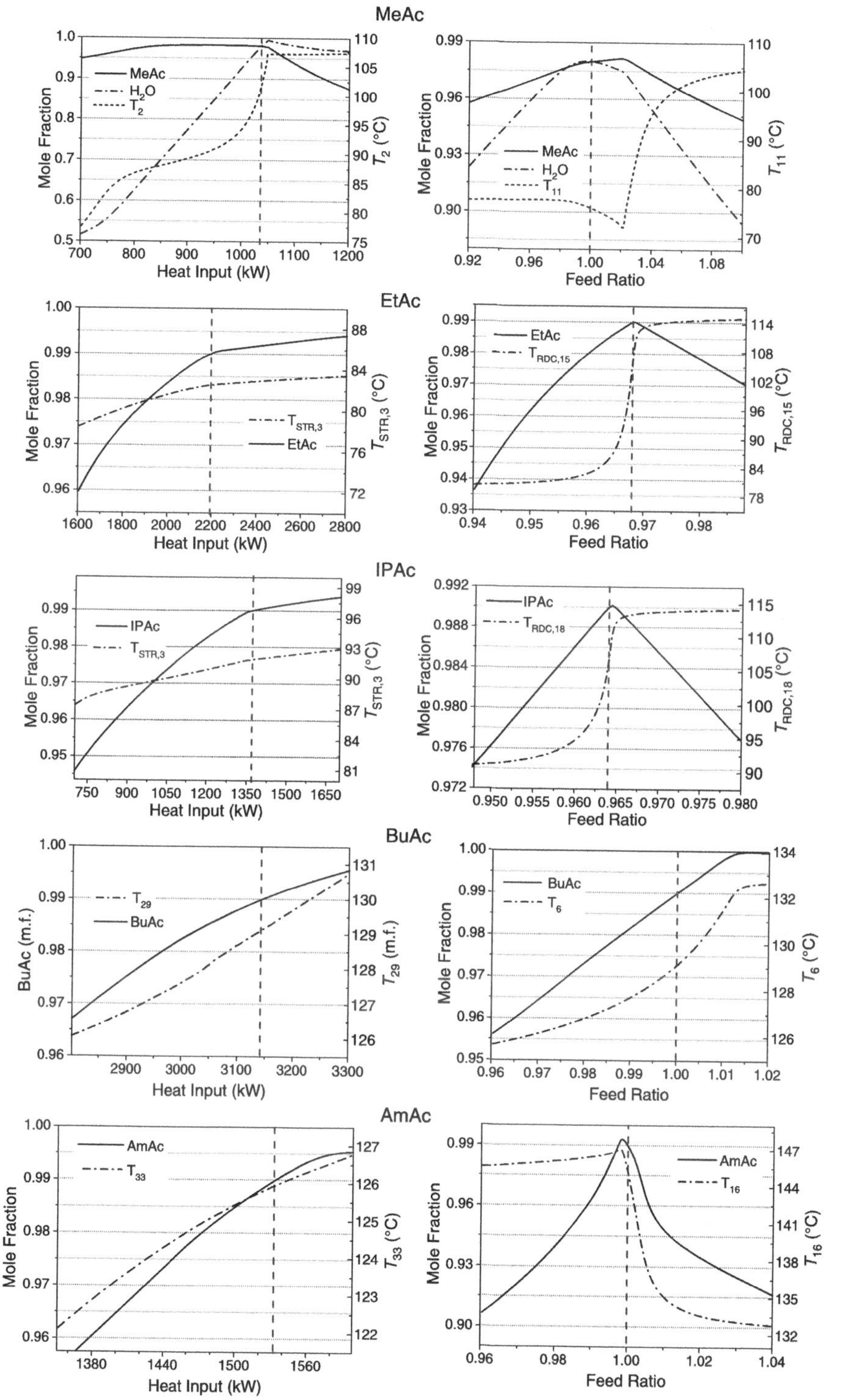

Figure 13.3 Trends of product compositions and temperature responses for a range of changes in the manipulated variables (heat input and feed ratio). (- - -) Nominal design indicated.

the two products, MeAc and H_2O. The vertical dashed line in Figure 13.3 indicates the nominal steady state. From a separation perspective, an increase in the heat input will enrich the bottoms product, but Figure 13.3 indicates that the H_2O composition goes through an increase followed by a decrease. Similarly, the top product composition, MeAc, also shows nonmonotonic behavior as the heat input changes. These unexpected results are also observed for the feed ratio change (Fig. 13.3). If a tray temperature T_{11} is used, instead of composition, the multiplicity problem can be alleviated slightly but cannot be eliminated completely.

For the type II flowsheet, the input multiplicity only occurs between the feed ratio and the acetate composition, which is illustrated in Figure 13.3. Because the feeds are introduced into the reactive distillation column and the product EtAc or IPAc is withdrawn from the stripper, this multiplicity will not affect process operation. Moreover, it is possible to find a tray temperature such that the input multiplicity can be eliminated (e.g., $T_{RDC,15}$ for EtAc and $T_{RDC,18}$ for IPAc, where RDC stands for the reactive distillation column), as shown in Figure 13.3.

The two type III systems show quite different process characteristics. Similar to the previous nonlinearity analysis, the BuAc system does not reveal any multiplicity; in contrast, the AmAc system shows input multiplicity for the feed ratio variation. If a temperature T_{16} is used instead, the input multiplicity cannot be eliminated completely.

Multiplicity analysis indicates that composition control of the MeAc system can be difficult, and this is also true for the AmAc system in which the input multiplicity cannot be completely eliminated. As for the type II flowsheets (EtAc and IPAc), the input multiplicity can be overcome by using the temperature control as shown in Figure 13.3. Similar to the previous analysis, the BuAc system does not show any potential problem in control.

The analysis presented here clearly indicates that the reactive distillation systems, regardless of the types of flowsheet, exhibit severe nonlinearities that include significant sign reversal, extremely large values of Allgower's nonlinearity measure (ϕ^N), and input multiplicity. Under these circumstances, control structure design becomes important.

As shown in Figure 13.3 for the type II flowsheet, the input multiplicity can be avoided by controlling a tray temperature instead of product composition control. However, the problem cannot be avoided for the type I flowsheet for MeAc production. The problem then becomes, what kind of controllers can provide adequate control for such a nonlinear process?

13.2 CONTROL STRUCTURE DESIGN

In this section a systematic approach is proposed to design the control structures for these three types of reactive distillation flowsheets. Because all five reactive distillation systems (Table 7.5) have almost equal molar feedflows (neat flowsheet), the stoichiometric balance has to be maintained.[3] Here we adjust the feed ratio to prevent accumulation of unreacted reactants attributable to stoichiometric imbalance. The next issue is, how many product compositions or inferred product purities should be controlled? For the esterification reactions with $A + B \Leftrightarrow C + D$ with a neat flowsheet, controlling one-end product purity implied a similar purity level on the other end, provided the product flowrates are equal. Thus, a single-end composition (or temperature control) is preferred. This leads to 2×2 multivariable control, as opposed to a 3×3 multiple-input-multiple-output system. The next problem is consideration of robustness. Because of input multiplicities and potential sign reversals, the uncertainty associated with a linear model can be significant. Instead of inverting all four transfer function models (for a 2×2 system) for controller design, a decentralized control is used in which only information on diagonal elements is employed for initial design. In other words, the uncertainty associated with the process is so large that we want to minimize the exercise of model inversion. Thus, decentralized control is preferred. In summary, the following principles are recommended:

1. Maintain the stoichiometric balance using the feed ratio.
2. Control only one-end composition (or temperature).
3. Use decentralized control to maintain robust stability.

This leads to the following design procedure for temperature control of reactive distillation systems:

1. With the feed ratio as one manipulated variable, select a second manipulated variable. Typically, the other manipulated input is the heat input or the reflux ratio.
2. Use the nonsquare relative gain (NRG)[8] to select temperature control trays. The larger row sums of the NRG indicate a potential temperature control tray. Note that the temperatures with sign reversal (Fig. 13.3) cannot be used as controlled variables.
3. Use the relative gain array (RGA)[9] for variable pairing once the inputs and outputs are determined.
4. Perform a sequential relay-feedback test to find the ultimate gain (K_u) and ultimate period (P_u).
5. Use Tyreus–Luyben tuning to set the tuning constants for the PI controllers. A simple version is $K_c = K_u/3$ and $\tau_I = 2P_u$.

In step 2, the row sum of the NRG is used instead of the more familiar SVD. The reason for this is that the row sum of the ith row is equal to the square of the 2-norm of the ith row of the U matrix from the SVD (i.e., $G = U\Sigma V^T$ and ith row sum $= \|e_i U\|_2^2$, where e_i is the ith unit vector). Thus, the two approaches are the same.

13.2.1 Selection of Temperature Control Trays

Sensitivity analyses were performed on these five esterification reactive distillation systems. In order to find the steady-state gains of the tray temperature in the linear region, extremely small step changes ($\pm 0.1\%$) are made in the manipulated variables.

For the MeAc system (Fig. 13.1*a*), a step increase in the heat input (Q_R) leads to a temperature increase in the lower section of the column and a relative small temperature decrease toward the top of the column as shown in Figure 13.4. This is uncharacteristic because, for conventional distillation, we generally observe temperature increases throughout the column. For the change in the second manipulated variable (ratio of acid feed flowrates to alcohol feedrate, $F_{acid}/F_{alcohol}$), an increase in the flowrate of the

[8] J. W. Chang and C. C. Yu, The relative gain for non-square multivariable systems, *Chem. Eng. Sci.* **45**, 1309–1323 (1990).

[9] E. H. Bristol, On a new measure of interaction for multivariable process control. *IEEE Trans. Automat. Control* **AC11**, 133–134 (1966).

heavy reactant (F_{acid}) results in a temperature rise in the lower section of the column, followed by a decrease toward the middle section of the column (Fig. 13.4). The reason for the temperature increase in the lower section is the excess of the heavy reactant (HAc). Because of the lowered conversion, less water (the second highest boiler) is formed, which subsequently leads to a lower tray temperature in the midsection of the column.

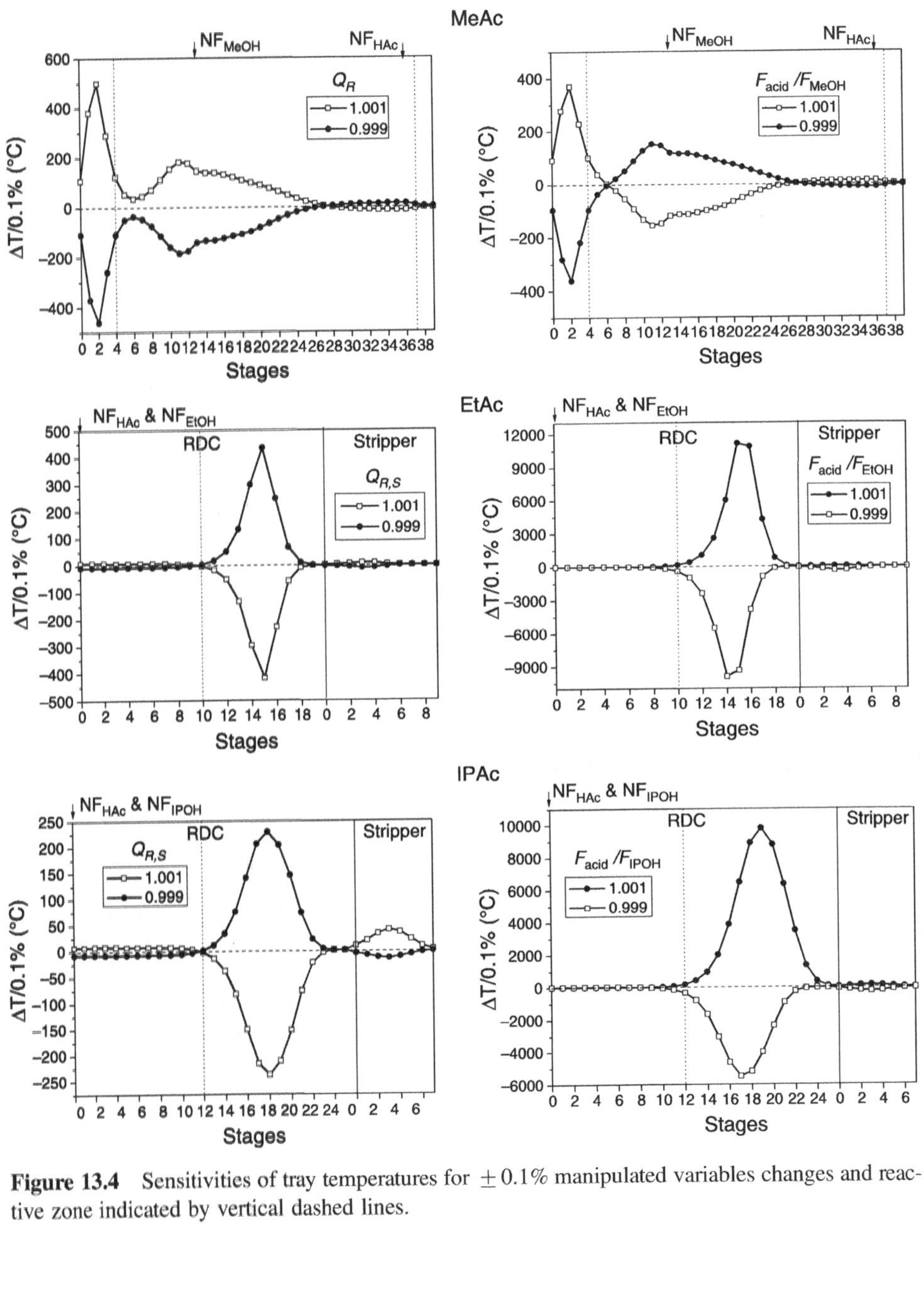

Figure 13.4 Sensitivities of tray temperatures for $\pm 0.1\%$ manipulated variables changes and reactive zone indicated by vertical dashed lines.

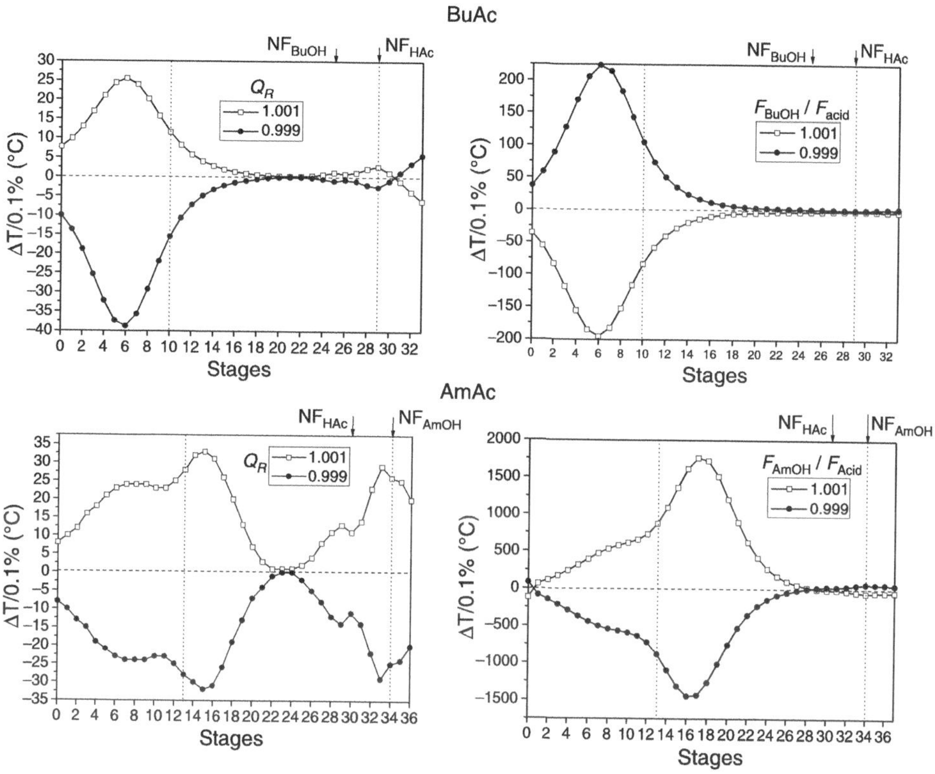

Figure 13.4 (*Continued*).

The type II flowsheet shows a different configuration where the base level of the reactive distillation column is controlled by the heat input (a dependent variable) and the heat input of the stripper ($Q_{R,S}$) is the manipulated variable (Fig. 13.1*b*). The second manipulated variable is the feed ratio, and the reflux ratio of the reactive distillation column is fixed. Note that in Figure 13.4 the temperatures in the reactive distillation column and the stripper are combined and labeled on the x axis. For the EtAc system, an *increase* in $Q_{R,S}$ leads to a *decrease* in the tray temperatures of the reactive distillation column (Fig. 13.4). This is a rather unusual phenomenon that is a direct consequence of the process configuration. An increase in $Q_{R,S}$ results in a larger vapor rate from the stripper back to the decanter, and this implies a larger recycle flowrate for the reactive distillation column. As for the acid feed ($F_{acid}/F_{alcohol}$) changes, an increase in the heavy reactant results in a reactive distillation column tray temperature rise while the stripper temperatures show little variation. The arguments also apply to the IPAc system as shown in Figure 13.4.

The type III flowsheet is quite similar to the type I flowsheet topographically except that a decanter is used to replace the reflux drum and the organic phase is under total reflux (Fig. 13.1). For the BuAc system, an increase in the heat input leads to temperature rises throughout the column except for the upper section of the column. This behavior is quite similar to that of the MeAc system. In the column design, the HAc concentration is kept low toward the lower section of the column (Chap. 7) to meet the product specification (acid in a parts per million level). Thus, an increase in the acid feedrate increases the formation of BuAc in the lower section and results in a higher tray temperature as illustrated in Figure 13.4. Because the two reactants have similar boiling points and because the feed arrangement results in the acid being consumed early in the reactive zone, the trend in the temperature decrease is not as significant as in the MeAc system. Similar behavior can also be seen for the AmAc system except that the heavy reactant is the alcohol (AmOH) instead of the acid (HAc), as seen in Figure 13.4.

Chang and Yu's NRG[8] was used to find the temperature control trays. The NRG (Λ^N) is defined as

$$\Lambda^N = K_p \otimes (K_p^+)^T \tag{13.3}$$

where K_p is the steady-state gain matrix, $\otimes$ denotes the element-by-element multiplication, superscript $+$ is the pseudoinverse, and superscript T is the transpose. The largest row sum of the NRG is selected as the temperature control trays. Figure 13.5 shows the row sums for all five systems. The controlled variables are

1. T_2 and T_{11} for MeAc
2. $T_{\mathrm{RDC},15}$ and $T_{\mathrm{STR},3}$ for EtAc
3. $T_{\mathrm{RDC},18}$ and $T_{\mathrm{STR},3}$ for IPAc
4. T_6 and T_{29} for BuAc
5. T_{16} and T_{33} for AmAc

where STR is the stripper. Because NRG is a linear analysis, the nonlinearity measure for each individual tray (ϕ) is also computed (Fig. 13.5) to evaluate potential conflict between linear and nonlinearity analyses. Note that the ϕ shown here is obtained by taking the 2-norm of individual nonlinearity measure from each input and then dividing by 2. The nonlinear results are generally in good agreement with that of the linear one, except for the BuAc case where the trays close to the column top are not used to avoid poor mixing. It is also important to ensure that the selected temperature control trays do not exhibit sign changes when manipulated inputs are varied (Fig. 13.2). The shaded areas in Figure 13.2 indicate that for these five systems the temperature control trays do not exhibit sign reversal in the ranges ($\pm 5\%$ for heat input, $\pm 1\%$ for feed ratio) of manipulated variable variations. Note that if the NRG selected temperature falls within the sign reversal area, an alternative temperature should be sought.

13.2.2 Control Structure and Controller Design

Because we chose to use the decentralized control for highly nonlinear reactive distillation systems, the next step is to find the variable pairing for the controlled and manipulated variables. Table 13.2 (fourth column) gives the steady-state gain matrices from the linear analysis. The RGA[9] is used for variable pairings. The selected pairings (fifth column, Table 13.2) give diagonal RGAs ranging from 0.53 to 1.04.

For the MeAc system, the heat input is used to control T_2 and T_{11} is maintained by changing the feed ratio (i.e., T_2–Q_R and T_{11}–FR), which looks reasonable from a dynamic perspective (note that we are counting the tray numbers from the bottoms up).

For the type II flowsheets (EtAc and IPAc), a stripper temperature is maintained using the heat input to the stripper and a temperature in the reactive distillation column is controlled using the ratio of fresh feeds into the reactive distillation column (i.e., $T_{\mathrm{STR},3}$–$Q_{R,S}$ and $T_{\mathrm{RDC},15}$–FR for EtAc, $T_{\mathrm{STR},3}$–$Q_{R,S}$ and $T_{\mathrm{RDC},18}$–FR for IPAc).

The control structures for the type III flowsheet may look odd from a conventional distillation control perspective, but they have been frequently seen for reactive distillation. Here, the heat input is paired with an upper section tray temperature and a lower section tray temperature is controlled by the feed ratio (e.g., T_{29}–Q_R and T_6–FR for BuAc, T_{33}–Q_R and T_{16}–FR for AmAc). The relay-feedback test is used to find the ultimate gain K_u and the ultimate period P_u followed by the Tyreus–Luyben PI tuning rule. The

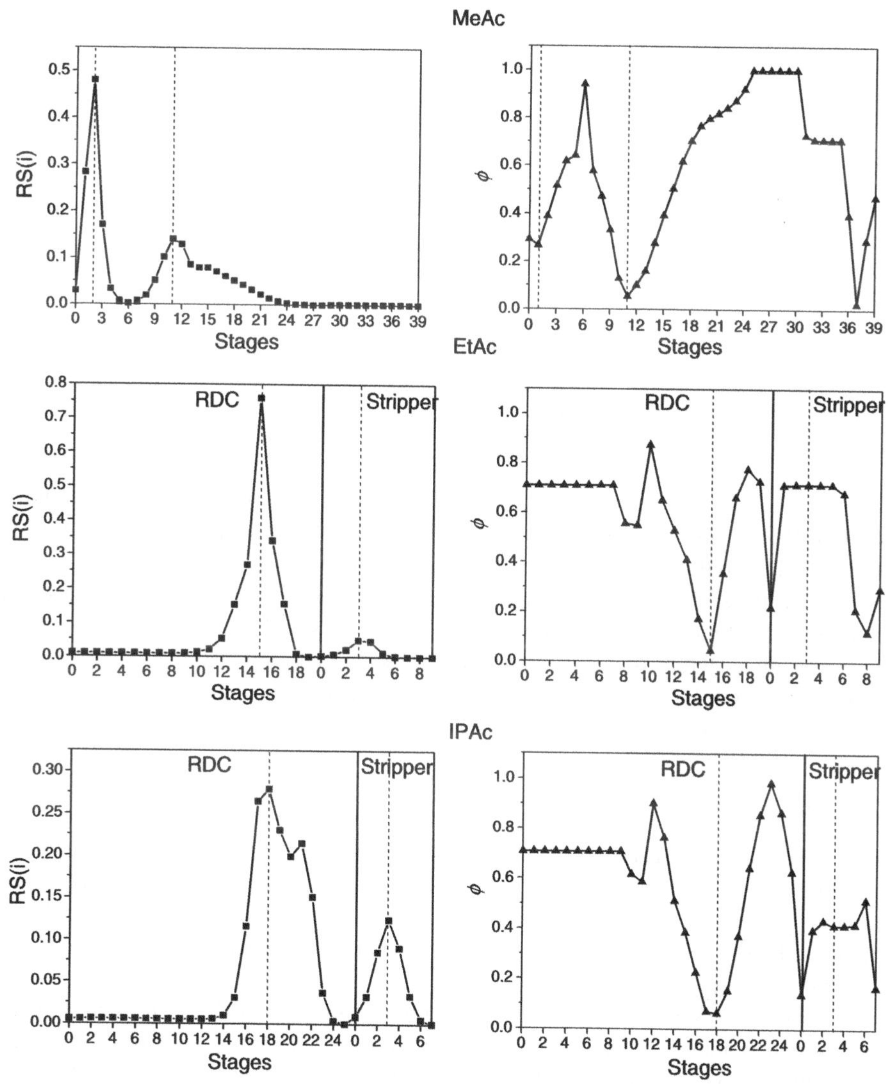

Figure 13.5 Row sums and nonlinearity measures (ϕ) of individual tray temperatures for all five systems.

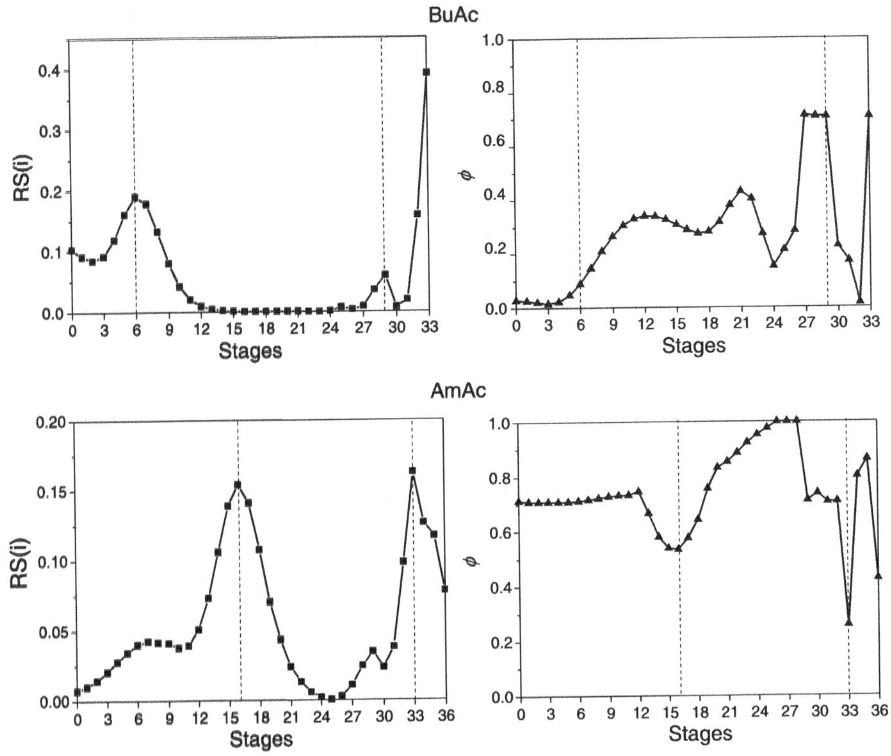

Figure 13.5 (*Continued*).

identification-tuning step is carried out sequentially to find the controller settings for the PI controllers. We find the sequential relay-feedback autotuning procedure is very effective for these five highly nonlinear processes. Table 13.2 summarizes the settings for all five reactive systems with two PI loops. Note that the large reset time is associated with the feed ratio loop and the reset time for the heat input loop is relatively small. This implies that we have two dynamically different speeds of response. The heat input loop takes care of the disturbance initially followed by a gradual effort to maintain the stoichiometric balance (to prevent long-term accumulation).

13.2.3 Performance

Feedflow and feed ratio disturbances are used to evaluate the dynamic performance of the temperature control structures for these five esterification systems. Recall that these reactive distillation systems are highly nonlinear (Fig. 13.2) with significant sign reversal and input multiplicity. Figure 13.6 shows that the simple PI temperature control actually works quite well for all five systems.

For the MeAc system, two temperature control trays (T_2 and T_{11}) show asymmetric responses and somewhat oscillatory behavior is observed for T_{11} (Fig. 13.6*a*). The product composition, $X_{D,\text{acetate}}$ in particular, does not settle down 15 h after the 20% feed flowrate change is introduced. Asymmetrical responses are observed for most of the process variables and steady-state offsets (~0.002 mf; Fig. 13.6*a*). For the type II

TABLE 13.2 Controlled Variables, Manipulated Variables, Process Gain Matrices, RGA, and Tuning Parameters for Five Esterification Systems Under Temperature Control

	Controlled Variables	Manipulated Variables	Steady-State Gain[a,b]	RGA	Tuning Parameter
MeAc	T_2 T_{11}	F_{acid}/F_{MeOH} Q_R	$\begin{bmatrix} T_{11} \\ T_2 \end{bmatrix} = \begin{bmatrix} 1.839 & -1.522 \\ 4.802 & 3.659 \end{bmatrix} \begin{bmatrix} Q_R \\ F_{acid}/F_{MeOH} \end{bmatrix}$	$\begin{matrix} & Q_R & F_{acid}/F_{MeOH} & \\ \Lambda = & 0.479 & 0.520 & T_{11} \\ & 0.520 & 0.479 & T_2 \end{matrix}$	$Q_R - T_2$: $K_c = 0.464$, $\tau_I = 0.333$ (h) $F_{acid}/F_{MeOH} - T_{11}$: $K_c = 1.237$, $\tau_I = 1.98$ (h)
EtAc	$T_{STR,3}$ $T_{RDC,15}$	F_{acid}/F_{EtOH} $Q_{R,S}$	$\begin{bmatrix} T_{STR,3} \\ T_{RDC,15} \end{bmatrix} = \begin{bmatrix} 0.087 & 1.967 \\ -4.26 & 102.57 \end{bmatrix} \begin{bmatrix} Q_{R,S} \\ F_{acid}/F_{EtOH} \end{bmatrix}$	$\begin{matrix} & Q_{R,S} & F_{acid}/F_{EtOH} & \\ \Lambda = & 0.517 & 0.482 & T_{STR,3} \\ & 0.482 & 0.517 & T_{RDC,15} \end{matrix}$	$Q_{R,S} - T_{STR,3}$: $K_c = 22.65$, $\tau_I = 0.018$ (h) $F_{acid}/F_{EtOH} - T_{RDC,15}$: $K_c = 0.755$, $\tau_I = 2.069$ (h)
IPAc	$T_{STR,3}$ $T_{RDC,18}$	F_{acid}/F_{IPOH} $Q_{R,S}$	$\begin{bmatrix} T_{STR,3} \\ T_{RDC,18} \end{bmatrix} = \begin{bmatrix} 0.227 & 1.574 \\ -2.336 & 70.439 \end{bmatrix} \begin{bmatrix} Q_{R,S} \\ F_{acid}/F_{IPOH} \end{bmatrix}$	$\begin{matrix} & Q_{R,S} & F_{acid}/F_{IPOH} & \\ \Lambda = & 0.812 & 0.187 & T_{STR,3} \\ & 0.187 & 0.812 & T_{RDC,18} \end{matrix}$	$Q_{R,S} - T_{STR,3}$: $K_c = 25.5$, $\tau_I = 0.08$ (h) $F_{acid}/F_{IPOH} - T_{RDC,18}$: $K_c = 3.92$, $\tau_I = 2.66$ (h)
BuAc	T_{29} T_6	F_{BuOH}/F_{acid} Q_R	$\begin{bmatrix} T_{29} \\ T_6 \end{bmatrix} = \begin{bmatrix} 0.026 & -0.007 \\ 0.299 & -2.127 \end{bmatrix} \begin{bmatrix} Q_R \\ F_{BuOH}/F_{acid} \end{bmatrix}$	$\begin{matrix} & Q_R & F_{BuOH}/F_{acid} & \\ \Lambda = & 1.041 & -0.041 & T_{29} \\ & -0.041 & 1.041 & T_6 \end{matrix}$	$Q_R - T_{29}$: $K_c = 38.86$, $\tau_I = 0.06$ (h) $F_{BuOH}/F_{acid} - T_6$: $K_c = 5.16$, $\tau_I = 0.9$ (h)
AmAc	T_{33} T_{16}	F_{AmOH}/F_{acid} Q_R	$\begin{bmatrix} T_{33} \\ T_{16} \end{bmatrix} = \begin{bmatrix} 0.29 & -0.059 \\ 0.31 & 15.995 \end{bmatrix} \begin{bmatrix} Q_R \\ F_{AmOH}/F_{acid} \end{bmatrix}$	$\begin{matrix} & Q_R & F_{AmOH}/F_{acid} & \\ \Lambda = & 0.961 & 0.038 & T_{33} \\ & 0.038 & 0.961 & T_{16} \end{matrix}$	$Q_R - T_{33}$: $K_c = 29.72$, $\tau_I = 0.08$ (h) $F_{AmOH}/F_{acid} - T_{16}$: $K_c = 9.1$, $\tau_I = 1.2$ (h)

[a]Transmitter span: twice the steady-state value of the temperature (°C).
[b]Valve gains: twice the steady-state value for Q_R ($Q_{R,S}$) and F.

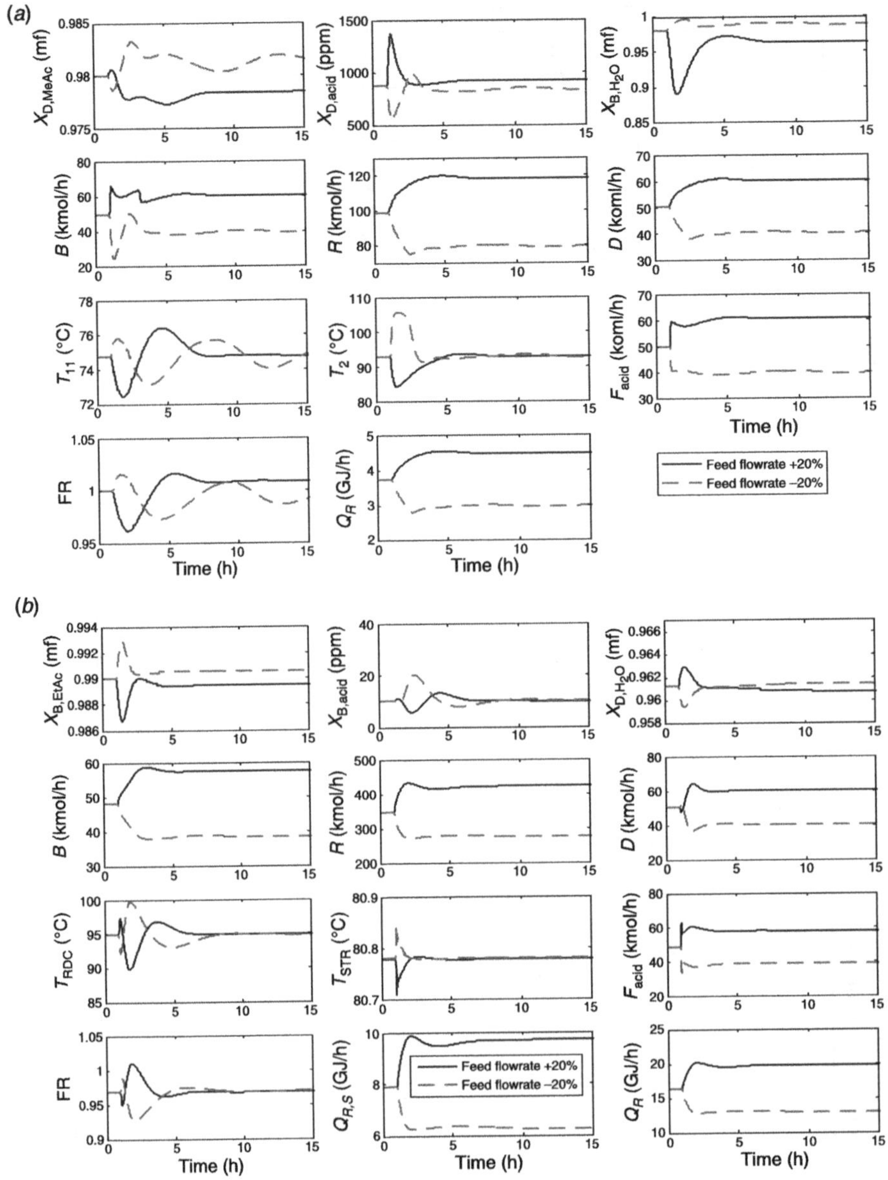

Figure 13.6 Temperature control responses for $\pm 20\%$ production rate changes for (*a*) MeAc, (*b*) EtAc, (*c*) IPAc, (*d*), BuAc, and (*e*) AmAc systems.

flowsheets, faster responses and much more symmetrical responses can be obtained for the controlled temperatures as well as major product compositions as shown in Figures 13.6*b* and 13.6*c*. The product composition settles in <10 h and much smaller offsets in the acetate composition can be achieved (~0.001 mf for EtAc, nil for IPAc). For the type III flowsheets, symmetrical responses in the temperature control trays can be seen in Figures 13.6*d* and 13.6*e*. However, two different composition dynamics are observed, which can

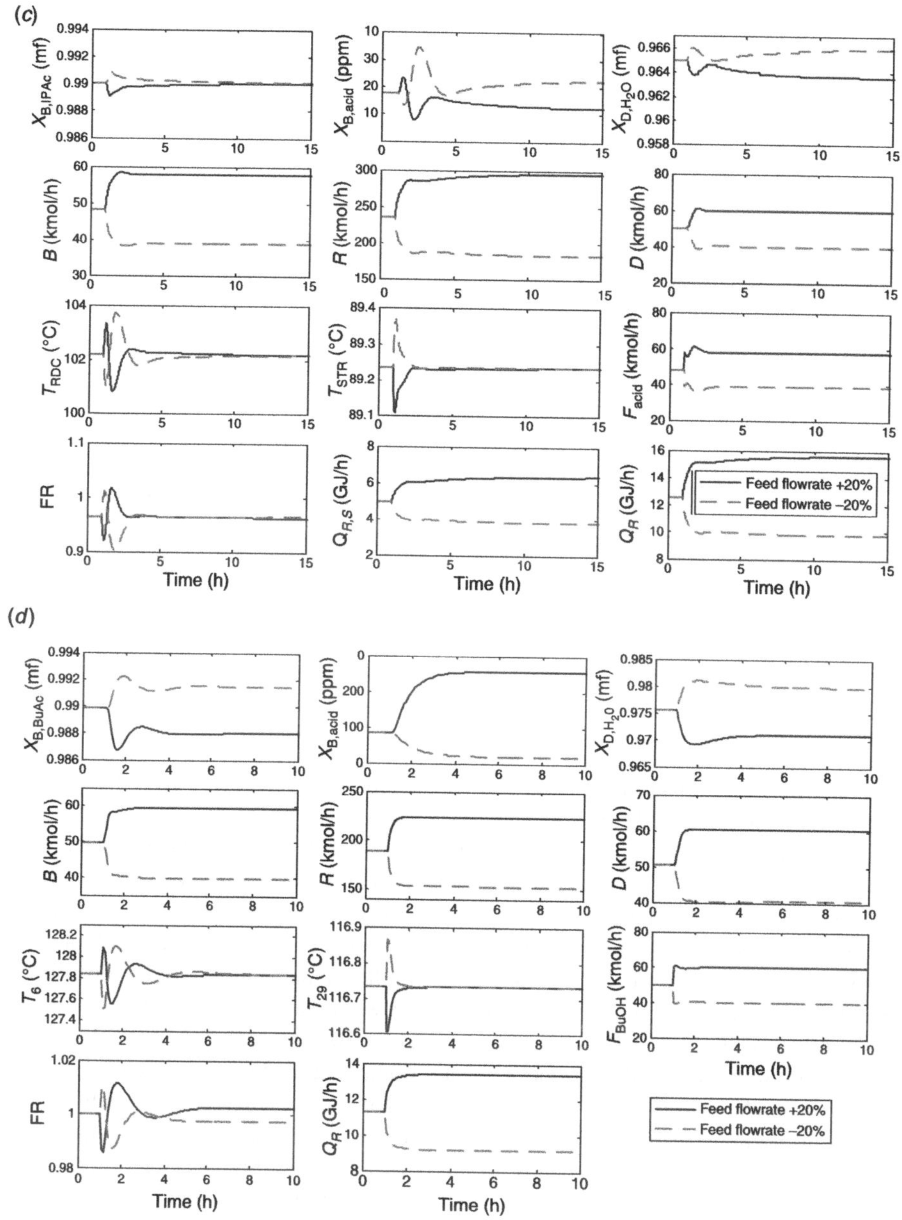

Figure 13.6 (*Continued*).

be foreseen from nonlinear analysis (Table 13.1). For the BuAc system, the product composition dynamics settles in <5 h (the fastest response) and symmetrical responses can also been seen for most of the process variables except the trace acid concentration (Fig. 13.6*d*). Steady-state offsets are also observed, but the error is around 0.002 mf for ±20% feedflow changes. For the AmAc system the dynamics are fast, but the responses are asymmetrical

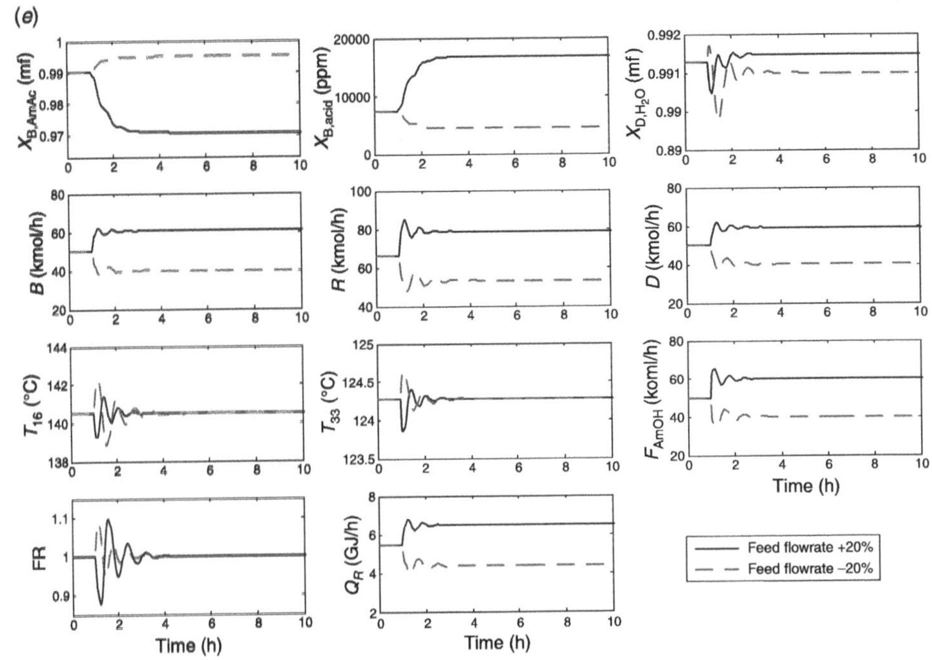

Figure 13.6 (*Continued*).

especially for the product compositions ($X_{B,acetate}$ and X_{D,H_2O}). A large steady-state offset (~0.02 mf) is also found for the acetate composition.

The second disturbance that we explored was ±5% feed ratio changes. This can be viewed as steplike flow measurement errors (e.g., bias). Figure 13.7 shows the control performance for ±5% feed ratio disturbances. Generally, offset-free composition control can be achieved by holding two temperatures at their setpoints. As demonstrated in Figure 13.7, the speed of responses for each system is quite similar to that of the feedflow disturbances. The MeAc system takes more than 15 h to settle while showing significant nonlinearity (Fig. 13.7*a*). The EtAc and IPAc systems take ~10 h to return to steady state with almost negligible error in the product (acetate) composition (Figs. 13.7*b* and 13.7*c*). The BuAc and AmAc systems give the fastest dynamics and the product compositions return to setpoint in approximately 5 h (Figs. 13.7*d* and 13.7*e*). For the AmAc system, the feed ratio disturbance is much better handled compared to the feedflow changes (Fig. 13.6*e*).

Despite strong nonlinearity (Figs. 13.3 and 13.4), workable temperature control of these reactive distillation systems can be obtained using a systematic design procedure with rather simple control structures. It should be emphasized that the selections of controlled and manipulated variables (control structure design) play a crucial role for these highly nonlinear processes and that decentralized control provides a better structure to cope with steady-state gain variations than model-inversion-based control structures.

Table 13.3 provides the control performance qualitatively for these five systems. As expected, the MeAc system (type I flowsheet) exhibits relatively poor control performance as can be seen early from qualitative argument or quantitative nonlinearity measures. The closed-loop behaviors of the EtAc and IPAc (type II flowsheet) systems are not quite as nonlinear as the steady-state measures predict. Relatively fast and symmetrical dynamics can be obtained, which is shown in Figures 13.6*b*, 13.6*c*, 13.7*b*, and 13.7*c*. One reason for this is that the type II flowsheet has a two-column configuration with the columns separated by a decanter with typically a 20-min holdup. In other words, a large surge tank is placed between these two units, which serves to dynamically decouple them. As predicted, the

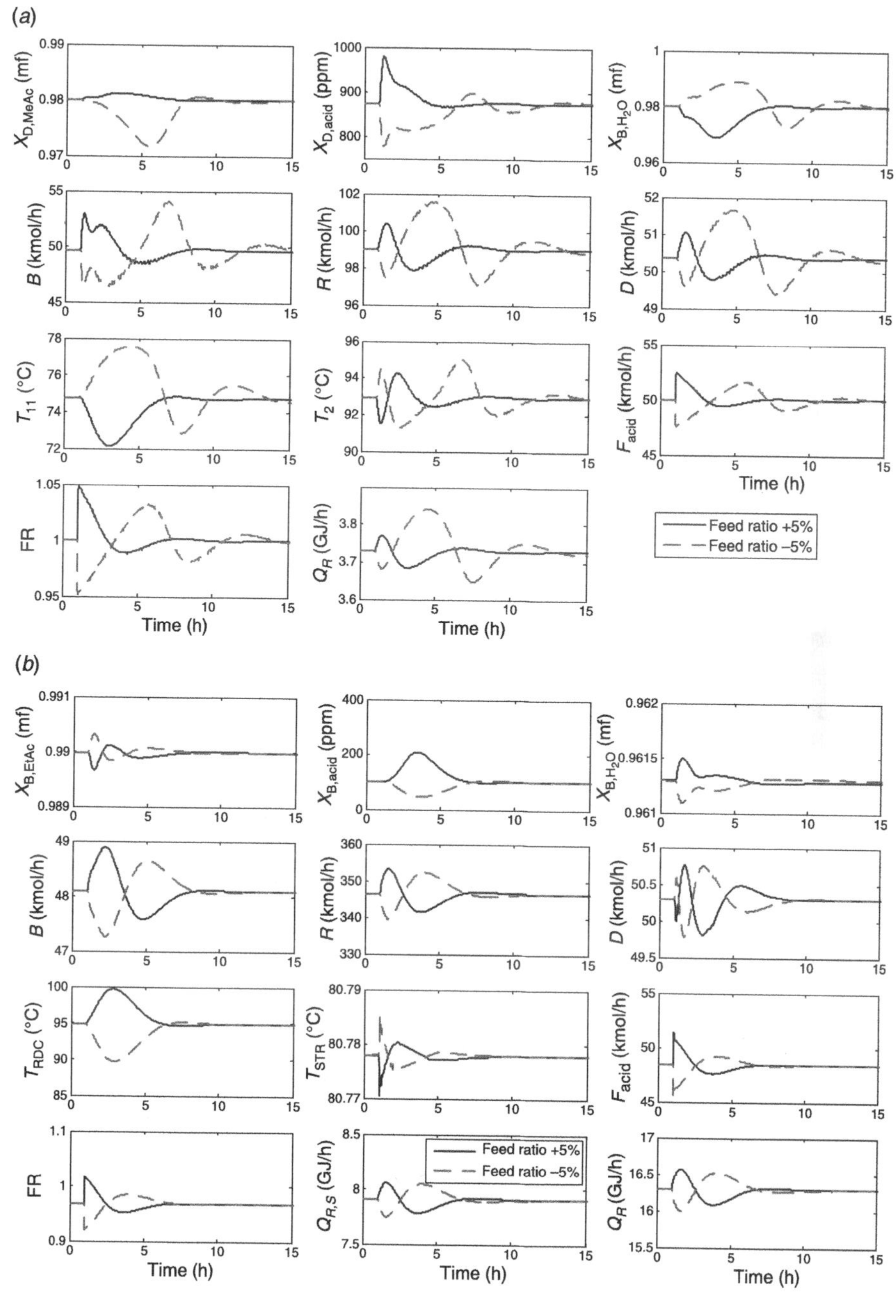

Figure 13.7 Temperature control responses for ±5% feed ratio changes for (*a*) MeAc, (*b*) EtAc, (*c*) IPAc, (*d*) BuAc, and (*e*) AmAc systems.

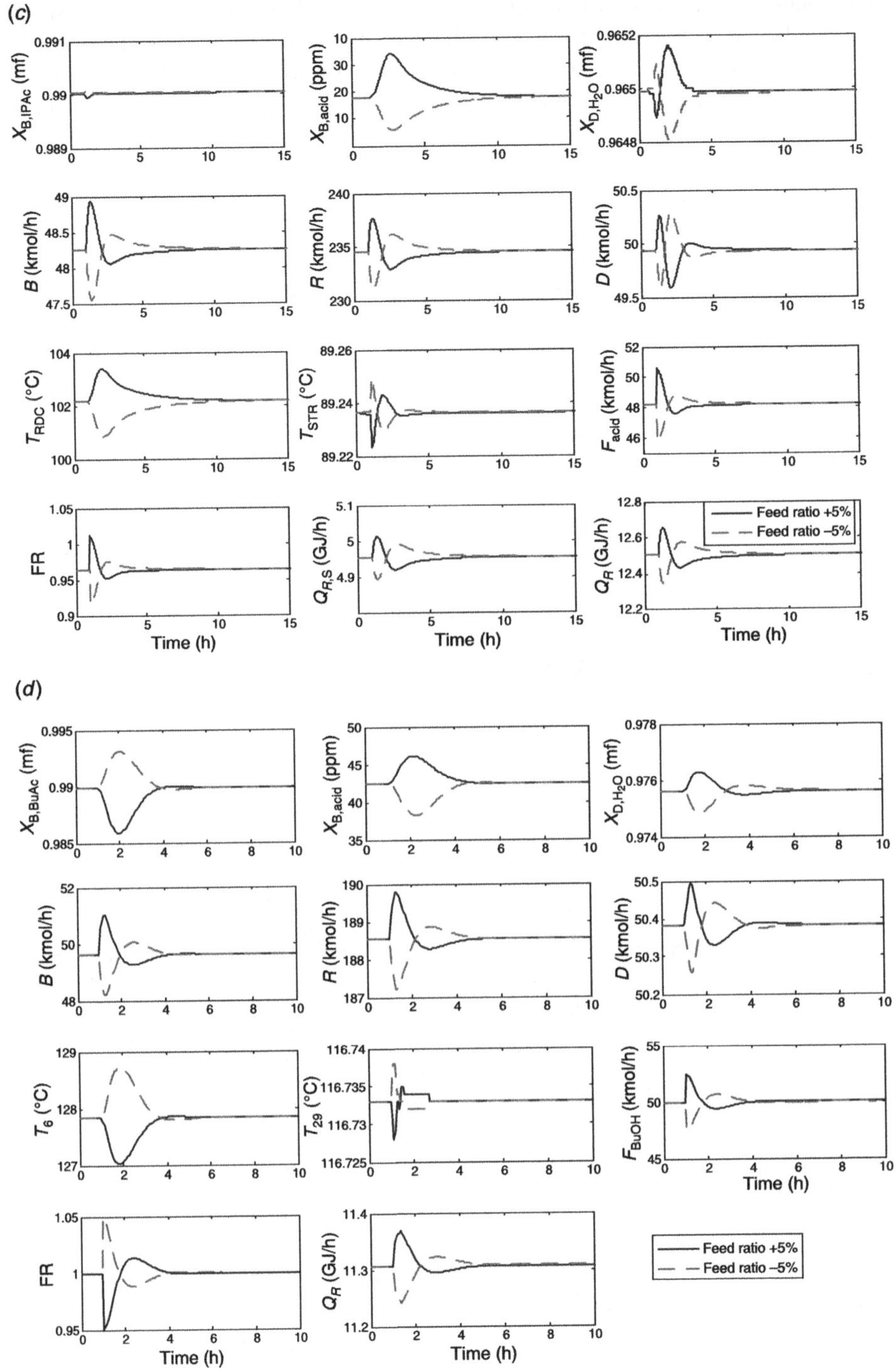

Figure 13.7 (*Continued*).

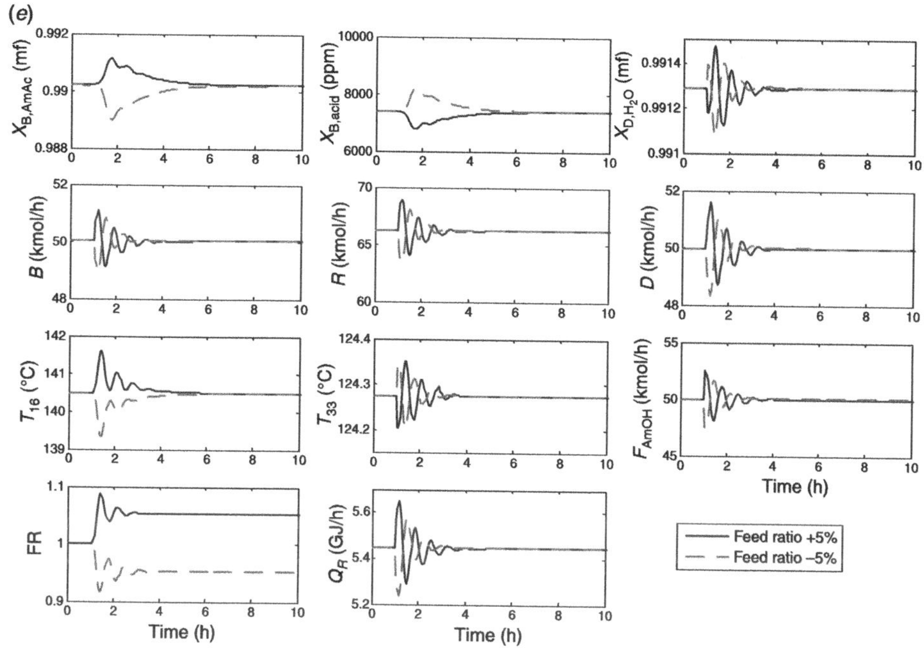

Figure 13.7 (*Continued*).

BuAc system should be an easy one to control, which the closed-loop responses confirmed (Figs. 13.6*d* and 13.7*d*).

In contrast, the other type III flowsheet, the AmAc system, gives significantly different closed-loop performance. Less symmetrical responses are observed as predicted by the nonlinearity measures, and a large composition offset results for feed flowrate changes (Fig. 13.6*e*). However, similar to the BuAc system, the type III flowsheet generally gives fast closed-loop dynamics. Table 13.3 summarizes the performance, settling time, and steady-state errors in product composition for these five temperature controlled reactive distillation systems. The results are for ±20% feedflow changes. Note that the control performance could be improved by incorporating a feedforward control for the measured

TABLE 13.3 Dynamic Performance for Both Control Structures Under ±20% Feed Flow Changes

Flowsheet Type		Two-Temp. Control		One-Temp., One-Comp. Control	
Dynamic Perform. for Product	System	Settling Time (h)	Offset (mf)	Settling Time (h)	Peak Errors (mf)
I	MeAc	~15	~0.002	~10	~0.05
II	EtAc	~10	~0.001	~5	~0.005
	IPAc	~10	~0.001	~5	~0.005
III	BuAc	~5	~0.002	<5	~0.002
	AmAc	~5	~0.02	~10	~0.02

disturbance. For example, we could ratio the feed flowrate to the reboiler duty for better control performance for the MeAc system.

Finally, it is interesting to note that the two systems (MeAc and AmAc) with lower TAC give relatively poor closed-loop responses as evidenced by either slow dynamics or large steady-state offsets. The TACs for these five designs are given in Table 7.5.

13.2.4 Alternative Temperature Control Structures

For the two-temperature control structures studied thus far, one of the fresh feeds is fixed, which provides a direct handle on the production rate. The ratio to the second fresh feed (feed ratio FR) is adjusted to control a tray temperature and subsequently the stoichiometric balance. A second tray temperature is controlled by manipulating the reboiler heat duty. This is called the Q-FR control structure (Fig. 13.8*a*). Roat et al.[10] propose a different two-temperature control structure for the control of the MeAc system. Two tray temperatures in the column are maintained using two fresh feedstreams. Production rate changes are achieved by adjusting the vapor boilup (Fig. 13.8*c*). We call this an F-F control structure, which has been studied by Kaymak and Luyben.[11,12] A third control structure is similar to the F-F control structure but, instead of manipulating both feedstreams, one fresh feedstream is used to maintain a tray temperature and the ratio to the second fresh feed is adjusted to control a second tray temperature, which we call the F-FR control structure shown in Figure 13.8b. The purpose here is to compare the effectiveness of these three alternative control structures. A comparison will be made for dynamic responses for production rate changes for the MeAc, BuAc, and AmAc systems.

Following the same procedure used for temperature control tray selection, two temperature control trays are selected for MeAc, BuAc, and AmAc systems, which are provided in Table 13.4. The variable pairing for the decentralized control is determined from RGAs. Finally, Tyreus–Luyben tuning is used to initialize PI controller settings followed by possible detuning to achieve similar damping for these alternative control structures. Table 13.4 summarizes the control structures and PI controller parameters for these three systems with different two-temperature control structures. Similar to the Q-FR control structure case (Table 13.2), we find a much larger reset time (τ_I) associated with the FR loop for the F-FR control structure compared to the reset times in the F-F control structure.

For the MeAc system, the F-FR and F-F control structures give slightly better performance for a 20% production rate increase compared to that of the Q-FR control structure, as shown in Figure 13.9. The temperatures of the Q-FR control structure take ~6 h to settle while the F-FR and F-F structures need only ~3 h. In terms of peak composition error, the F-FR shows the least errors. However, the steady-state offsets are the same for all three control structures because the temperature control trays are the same for all three cases.

[10] S. Roat, J. Downs, E. Vogel, and J. Doss, Integration of rigorous dynamic modeling and control system synthesis for distillation columns, In J. D. Birdwelland J. R. Cockette, Editors, *Chemical Process Control—CPC III*, Elsevier, New York, 1986, pp. 99–138.

[11] D. Kaymak and W. L. Luyben, Comparison of two types of two-temperature control structures for reactive distillation columns, *Ind. Eng. Chem. Res.* **44**, 4625–4640 (2005).

[12] D. Kaymak and W. L. Luyben, Evaluation of a two-temperature control structure for a two-reactant/two-product type of reactive distillation column, *Ind. Eng. Chem. Res.* **61**, 4432–4450 (2006).

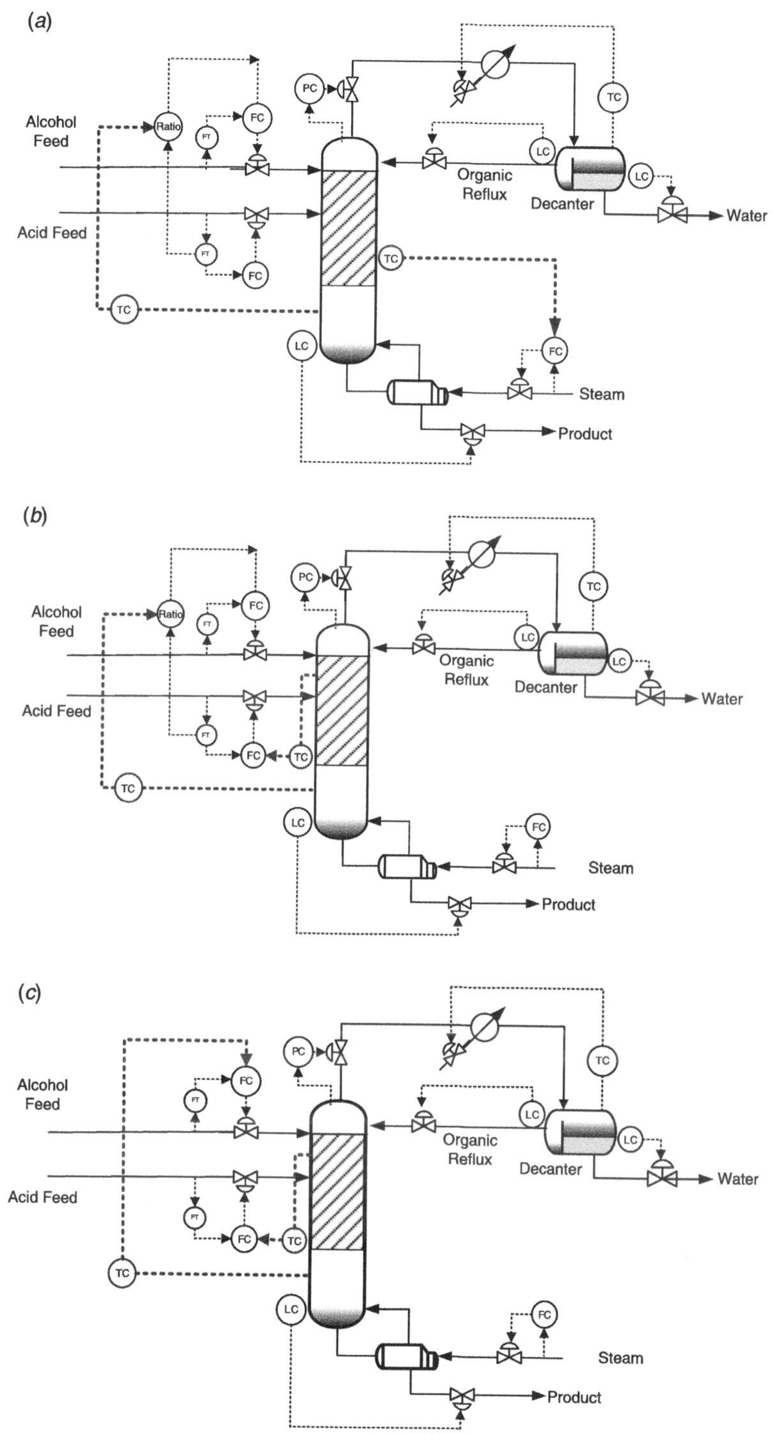

Figure 13.8 Alternative temperature control structures using manipulated variables: (*a*) Q-FR, (*b*) F-FR, and (*c*) F-F.

TABLE 13.4 Controlled Variables, Manipulated Variables, Process Gain Matrices, RGA, and Tuning Parameters for Three Esterification Systems with Two Different Two-Temperature Control Structures

	Controlled Variables	Manipulated Variables	Steady-State Gain	RGA	Tuning Parameter
MeAc	T_2 T_{11}	F_{acid}/F_{MeOH} F_{MeOH}	$\begin{bmatrix} T_{11} \\ T_2 \end{bmatrix} = \begin{bmatrix} -1.769 & -2.217 \\ -4.883 & 1.852 \end{bmatrix} \begin{bmatrix} F_{MeOH} \\ F_{acid}/F_{MeOH} \end{bmatrix}$	$\begin{matrix} & F_{MeOH} & F_{acid}/F_{MeOH} & \\ \Lambda = & \begin{bmatrix} 0.233 \\ 0.767 \end{bmatrix} & \begin{matrix} 0.767 \\ 0.233 \end{matrix} & \begin{matrix} T_{11} \\ T_2 \end{matrix} \end{matrix}$	$F_{MeOH}-T_2$: $K_c = 1.368$, $\tau_I = 0.836$ (h) $F_{acid}/F_{MeOH}-T_{11}$: $K_c = 6.33$, $\tau_I = 1.1$ (h)
MeAc	T_2 T_{11}	F_{acid} F_{MeOH}	$\begin{bmatrix} T_{11} \\ T_2 \end{bmatrix} = \begin{bmatrix} 0.450 & -2.217 \\ -6.748 & 1.852 \end{bmatrix} \begin{bmatrix} F_{MeOH} \\ F_{acid} \end{bmatrix}$	$\begin{matrix} & F_{MeOH} & F_{acid} & \\ \Lambda = & -0.06 & 1.06 & T_{11} \\ & 1.06 & -0.06 & T_2 \end{matrix}$	$F_{MeOH}-T_2$: $K_c = 6.28$, $\tau_I = 0.42$ (h) $F_{acid}-T_{11}$: $K_c = 2.65$, $\tau_I = 0.89$ (h)
BuAc	T_{28} T_6	F_{BuOH}/F_{acid} F_{acid}	$\begin{bmatrix} T_{29} \\ T_6 \end{bmatrix} = \begin{bmatrix} -0.039 & -0.008 \\ -0.545 & -2.283 \end{bmatrix} \begin{bmatrix} F_{acid} \\ F_{BuOH}/F_{acid} \end{bmatrix}$	$\begin{matrix} & F_{acid} & F_{BuOH}/F_{acid} & \\ \Lambda = & 1.052 & -0.052 & T_{29} \\ & -0.052 & 1.052 & T_6 \end{matrix}$	$F_{acid}-T_{29}$: $K_c = 17.75$, $\tau_I = 0.6$ (h) $F_{BuOH}/F_{acid}-T_6$: $K_c = 22.26$, $\tau_I = 3.3$ (h)
BuAc	T_{28} T_6	F_{BuOH} F_{acid}	$\begin{bmatrix} T_{29} \\ T_6 \end{bmatrix} = \begin{bmatrix} -0.013 & -0.008 \\ 1.978 & -2.283 \end{bmatrix} \begin{bmatrix} F_{acid} \\ F_{BuOH} \end{bmatrix}$	$\begin{matrix} & F_{acid} & F_{BuOH} & \\ \Lambda = & 0.657 & 0.342 & T_{29} \\ & 0.342 & 0.657 & T_6 \end{matrix}$	$F_{acid}-T_{29}$: $K_c = 33.75$, $\tau_I = 0.968$ (h) $F_{BuOH}-T_6$: $K_c = 26.25$, $\tau_I = 0.88$ (h)
AmAc	T_{33} T_{16}	F_{AmOH}/F_{acid} F_{acid}	$\begin{bmatrix} T_{33} \\ T_{16} \end{bmatrix} = \begin{bmatrix} -0.28 & -0.595 \\ -0.235 & 15.995 \end{bmatrix} \begin{bmatrix} F_{acid} \\ F_{AmOH}/F_{acid} \end{bmatrix}$	$\begin{matrix} & F_{acid} & F_{AmOH}/F_{acid} & \\ \Lambda = & 0.969 & 0.030 & T_{33} \\ & 0.030 & 0.969 & T_{16} \end{matrix}$	$F_{acid}-T_{33}$: $K_c = 7.99$, $\tau_I = 0.96$ (h) $F_{AmOH}/F_{acid}-T_{16}$: $K_c = 2.33$, $\tau_I = 3.36$ (h)
AmAc	T_{33} T_{16}	F_{AmOH} F_{acid}	$\begin{bmatrix} T_{33} \\ T_{16} \end{bmatrix} = \begin{bmatrix} 0.310 & -0.596 \\ -15.848 & 15.992 \end{bmatrix} \begin{bmatrix} F_{acid} \\ F_{AmOH} \end{bmatrix}$	$\begin{matrix} & F_{acid} & F_{AmOH} & \\ \Lambda = & -1.101 & 2.101 & T_{33} \\ & 2.101 & -1.101 & T_{16} \end{matrix}$	$F_{acid}-T_{16}$: $K_c = 2.6$, $\tau_I = 0.22$ (h) $F_{AmOH}-T_{33}$: $K_c = 28.37$, $\tau_I = 0.88$ (h)

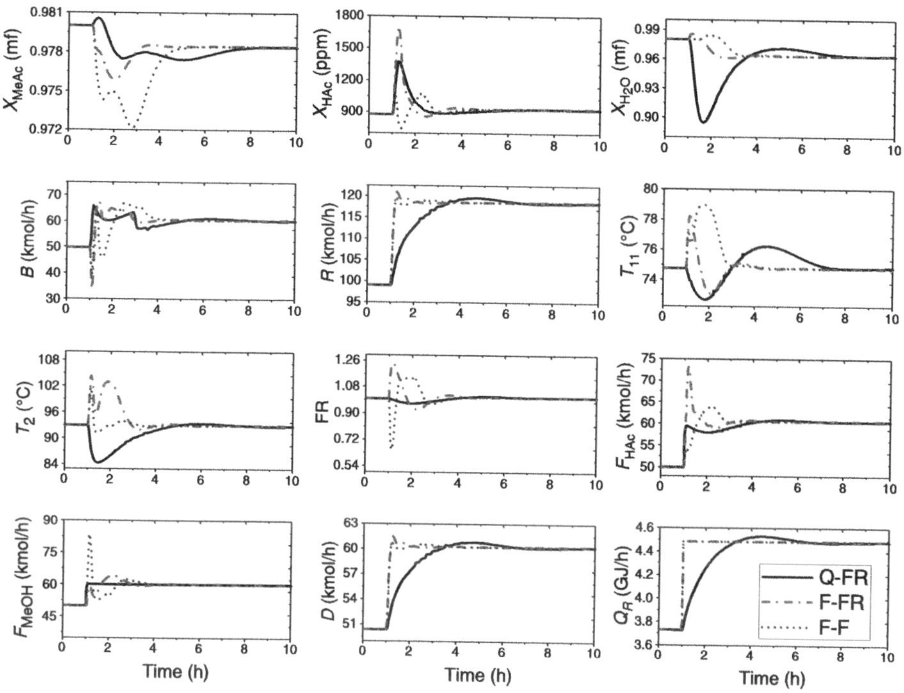

Figure 13.9 Temperature control responses for +20% production rate increase for MeAc system using three different control structures.

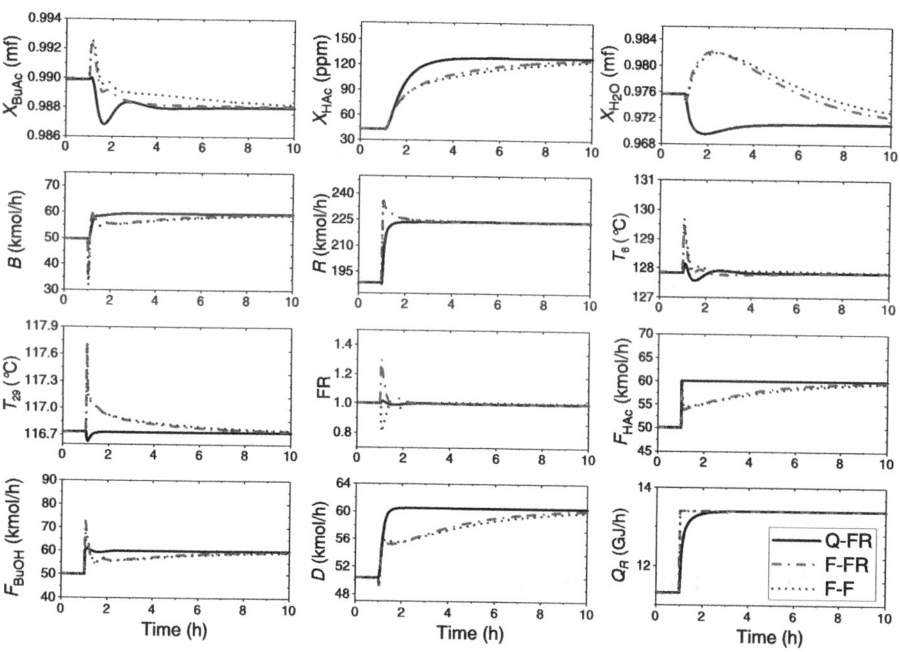

Figure 13.10 Temperature control responses for +20% production rate increase for BuAc system using three different control structures.

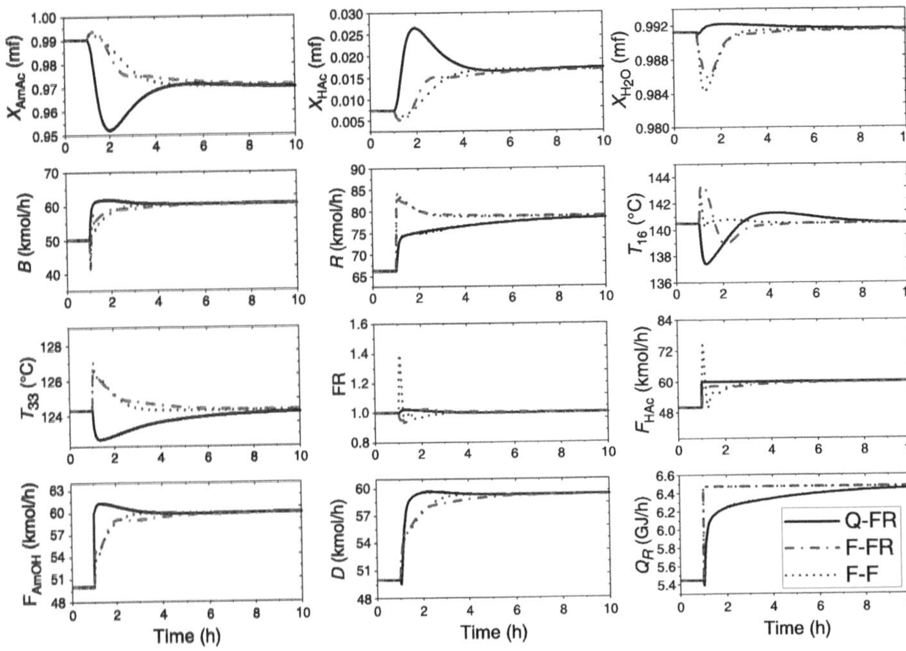

Figure 13.11 Temperature control responses for +20% production rate increase for AmAc system using three different control structures.

For the BuAc system, the Q-FR control structure outperforms the other two structures for a 20% production rate increase (Fig. 13.10). The top product composition (X_{D,H_2O}) takes more than 10 h to settle for the F-FR and F-F control structures. The reason for this is that it takes a long time for T_{28} to come back to its setpoint when two fresh feedstreams are manipulated.

For the AmAc system, the load responses of the F-FR and F-F control structures are about the same, while the Q-FR control structure exhibits a larger peak error in the acetate product composition, as shown in Figure 13.11. The peak errors of the two controlled temperatures (T_{16} and T_{33}) are about the same for all three control structures, but the settling time of the Q-FR structure is twice that of the other two structures (8 vs. 4 h, Fig. 13.11).

In summary, the performance of alternative temperature control structures depends on the system studied. Despite having the advantage of direct production rate handling, the Q-FR control structure performs better only for the BuAc system. The F-FR and F-F control structures give comparable performance, and they outperform the Q-FR control structure for the MeAc and AmAc systems.

13.3 EXTENSION TO COMPOSITION CONTROL

Because of the steady-state offsets when using temperature control, it may be necessary to use offset-free composition control in acetic acid esterification. Because of the nature of the reaction (the same amount of C and D is produced in the A + B ⇔ C + D reaction), the number of composition loops can be kept to a minimal level. We choose to control the composition of the bottoms product. This is the acetate product in all of the systems except the MeAc system. The reason for this is that the water purities in the EtAc, IPAc, BuAc, and AmAc systems are determined by tie lines from LL equilibria, a natural composition control imposed by LL equilibrium. As for the MeAc system, the acetate is a saddle as opposed to water, which is a stable node (Fig. 7.3*a*). Thus, we choose to control the water purity instead of the acetate product composition.

In the previous section, two temperature control trays are selected from the NRG analysis. Because the bottoms product composition is one of the controlled variables, the other controlled variable is the temperature further away from the acetate withdrawal ends. The fourth column of Table 13.5 gives the steady-state gain matrices of these five 2×2 systems. One temperature and one composition are controlled with two manipulated variables: heat input and feed ratio.

Next, the RGAs are computed for variable pairings. The results show that for type I (MeAc) and type II (EtAc and IPAc) flowsheets, the composition is controlled with the heat inputs and the feed ratio is used for temperature control. However, for the type III flowsheets (BuAc and AmAc), the acetate composition is maintained by changing the feed ratio and the tray temperature is controlled using the heat input (Table 13.5). The variable pairing in Table 13.5 indicates that severe nonlinearity will be encountered for the MeAc system and possibly also the AmAc systems. The multiplicity analysis in Figure 13.4 reveals that both input–output pairs (X_{D,H_2O}–Q_R and T_{11}–FR) show input multiplicity for the MeAc system. For the AmAc system, the $X_{B,AmAc}$–FR pair also exhibits input multiplicity. The accompanying difficult control problem cannot be prevented if we choose to control product composition for the MeAc system.

Once the controller structure is determined, sequential relay-feedback tests are performed to find the PI controller settings shown in Table 13.5. Similar to the settings for temperature control (Table 13.2), we have large reset times associated with the feed ratio and much smaller reset times for the heat input. Again, the control systems are designed in a systematic manner with minimal complexity in the design steps, identification, tuning, and controller types.

The performance of the one-temperature, one-composition control is evaluated next. For the composition measurement, a 4-min analyzer dead time is assumed. In general, the dynamic behaviors of these five systems with composition control are quite similar to those observed with temperature control (Fig. 13.8). For example, the MeAc and AmAc systems show significant nonlinearity for feed flowrate disturbances. In contrast, relatively symmetric responses are observed for the EtAc, IPAc, and BuAc systems. Certainly, the steady-state offset in the acetate composition can be eliminated when a composition loop is in place. For the MeAc system, the peak error in MeAc purity with composition control is 20 times greater than that with the temperature control (Figs. 13.6*a* and 13.12*a*) for a 20% feedrate decrease. Figure 13.12*a* shows that initially a small deviation in the acetate composition (dashed line) results in slow responses in the vapor boilup (should be decreasing). Thus, a low base level results in shutting down of the bottoms flow 1 h into the production rate change, which keeps the acetate not too far away from the spec. This slow response in reboiler duty eventually results in a significant deviation in product composition, as can be seen in Figure 13.12*a*.

TABLE 13.5 Controlled Variables, Manipulated Variables, Process Gain Matrices, RGA, and Tuning Parameters for Five Esterification Systems with One-Temperature, One-Composition Control

	Controlled Variables	Manipulated Variables	Steady-State Gain[a,b]	RGA	Tuning Parameter
MeAc	X_{B,H_2O} T_{11}	F_{acid}/F_{MeOH} Q_R	$\begin{bmatrix} X_{B,H_2O} \\ T_{11} \end{bmatrix} = \begin{bmatrix} 1.281 & -0.168 \\ 0.874 & -1.520 \end{bmatrix} \begin{bmatrix} Q_R \\ F_{acid}/F_{MeOH} \end{bmatrix}$	$\Lambda = \begin{matrix} Q_R & F_{acid}/F_{MeOH} & \\ 1.082 & -0.082 & X_{B,H_2O} \\ -0.082 & 1.082 & T_{11} \end{matrix}$	Q_R–X_{D,H_2O}: $K_c = 1.8$, $\tau_I = 0.316$ (h) F_{acid}/F_{MeOH}–T_{11}: $K_c = 0.67$, $\tau_I = 4.2$ (h)
EtAc	$X_{B,EtAc}$ $T_{RDC,15}$	F_{acid}/F_{EtOH} Q_R	$\begin{bmatrix} X_{B,EtAc} \\ T_{RDC,15} \end{bmatrix} = \begin{bmatrix} 0.068 & 0.572 \\ -4.26 & 102.57 \end{bmatrix} \begin{bmatrix} Q_{R,S} \\ F_{acid}/F_{EtOH} \end{bmatrix}$	$\Lambda = \begin{matrix} Q_{R,S} & F_{acid}/F_{EtOH} & \\ 0.742 & 0.258 & X_{B,EtAc} \\ 0.258 & 0.742 & T_{RDC,15} \end{matrix}$	$Q_{R,S}$–$X_{B,EtAc}$: $K_c = 14.05$, $\tau_I = 0.64$ (h) F_{acid}/F_{EtOH}–$T_{RDC,15}$: $K_c = 0.755$, $\tau_I = 2.069$ (h)
IPAc	$X_{B,IPAc}$ $T_{RDC,18}$	F_{acid}/F_{IPOH} Q_R	$\begin{bmatrix} X_{B,IPAc} \\ T_{RDC,18} \end{bmatrix} = \begin{bmatrix} 0.1514 & 0.7013 \\ -2.336 & 70.439 \end{bmatrix} \begin{bmatrix} Q_{R,S} \\ F_{acid}/F_{IPOH} \end{bmatrix}$	$\Lambda = \begin{matrix} Q_{R,S} & F_{acid}/F_{IPOH} & \\ 0.866 & 0.133 & X_{B,IPAc} \\ 0.133 & 0.866 & T_{RDC,18} \end{matrix}$	$Q_{R,S}$–$X_{B,IPAc}$: $K_c = 13.8$, $\tau_I = 0.32$ (h) F_{acid}/F_{IPOH}–$T_{RDC,18}$: $K_c = 3.92$, $\tau_I = 2.66$ (h)
BuAc	T_{29} $X_{B,BuAc}$	F_{BuOH}/F_{acid} Q_R	$\begin{bmatrix} X_{B,BuAc} \\ T_6 \end{bmatrix} = \begin{bmatrix} 2.785 & -19.043 \\ 0.026 & -0.007 \end{bmatrix} \begin{bmatrix} Q_R \\ F_{BuOH}/F_{acid} \end{bmatrix}$	$\Lambda = \begin{matrix} Q_R & F_{BuOH}/F_{acid} & \\ -0.042 & 1.042 & X_{B,BuAc} \\ 1.042 & -0.042 & T_{29} \end{matrix}$	Q_R–T_{29}: $K_c = 21.8$, $\tau_I = 0.06$ (h) F_{BuOH}/F_{acid}–$X_{B,BuAc}$: $K_c = 6.43$, $\tau_I = 5.04$ (h)
AmAc	T_{33} $X_{B,AmAc}$	F_{AmOH}/F_{acid} Q_R	$\begin{bmatrix} X_{B,AmAc} \\ T_{33} \end{bmatrix} = \begin{bmatrix} 2.337 & 55.721 \\ 0.29 & -0.595 \end{bmatrix} \begin{bmatrix} Q_R \\ F_{AmOH}/F_{acid} \end{bmatrix}$	$\Lambda = \begin{matrix} Q_R & F_{AmOH}/F_{acid} & \\ 0.0792 & 0.920 & X_{B,BuAc} \\ 0.920 & 0.079 & T_{33} \end{matrix}$	Q_R–T_{33}: $K_c = 28.9$, $\tau_I = 0.08$ (h) F_{AmOH}/F_{acid}–$X_{B,AmAc}$: $K_c = 0.92$, $\tau_I = 24.4$ (h)

[a]Transmitter span: twice the steady-state value of the temperature (°C).
[b]Valve gains: twice the steady-state value for Q_R ($Q_{R,S}$) and F.

Good disturbance rejection can be achieved for a feedrate increase. For the EtAc and IPAc systems (type II flowsheet), little difference in the speed of response, peak error, and steady-state offsets are observed (Figs. 13.6*b*, 13.6*c*, 13.12*b*, and 13.12*c*). Thus, there is little incentive to seek composition control for the type II flowsheet. For the BuAc system, good control performance is possible using composition control. The

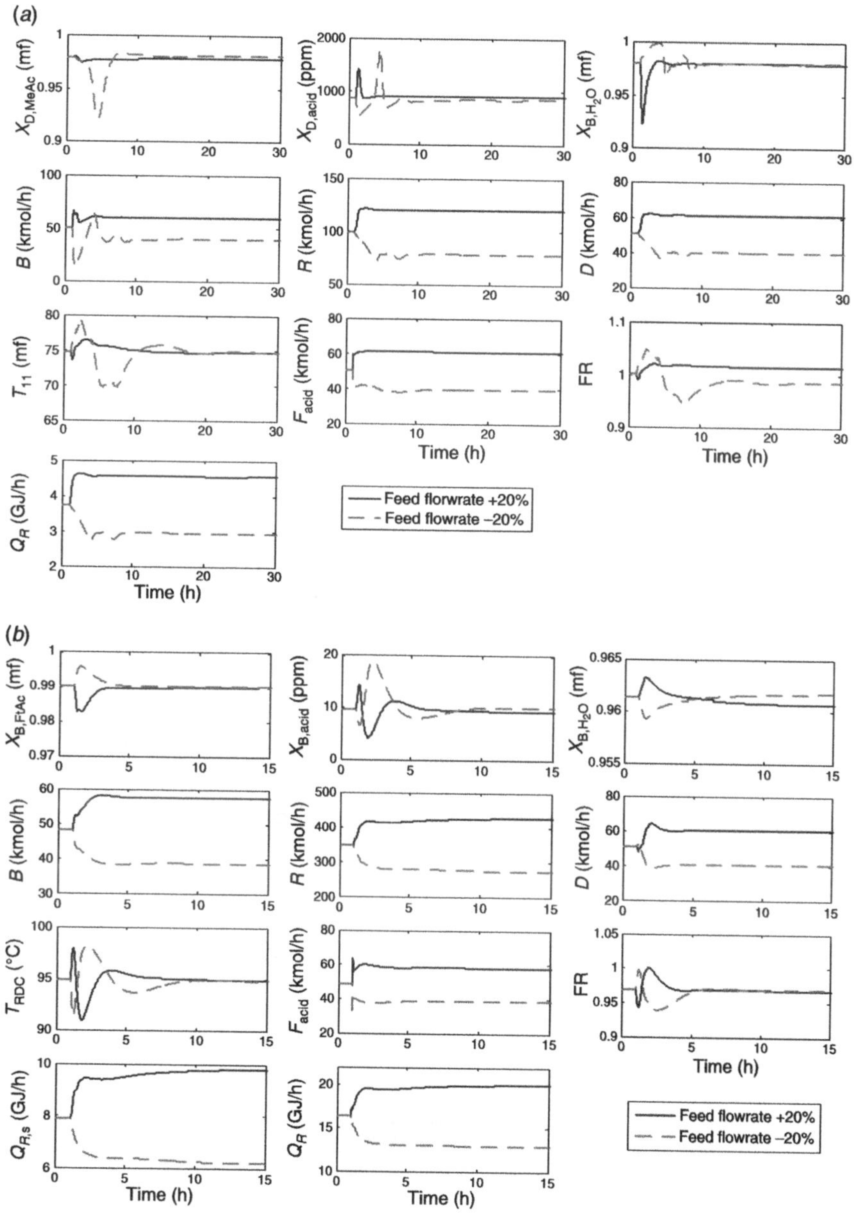

Figure 13.12 Composition control responses for $\pm 20\%$ production rate changes for (*a*) MeAc, (*b*) EtAc, (*c*) IPAc, (*d*) BuAc, and (*e*) AmAc systems

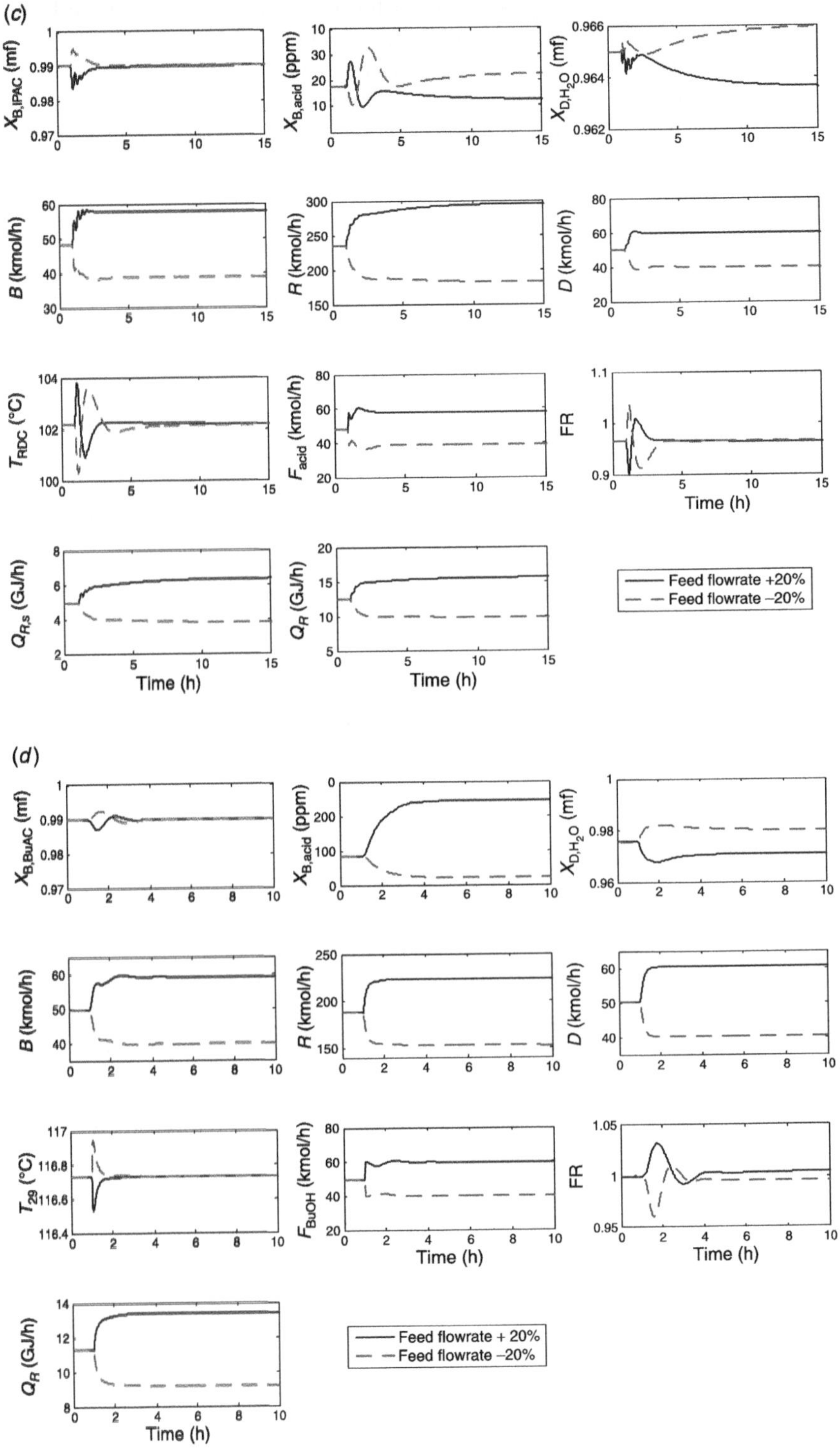

Figure 13.12 (*Continued*).

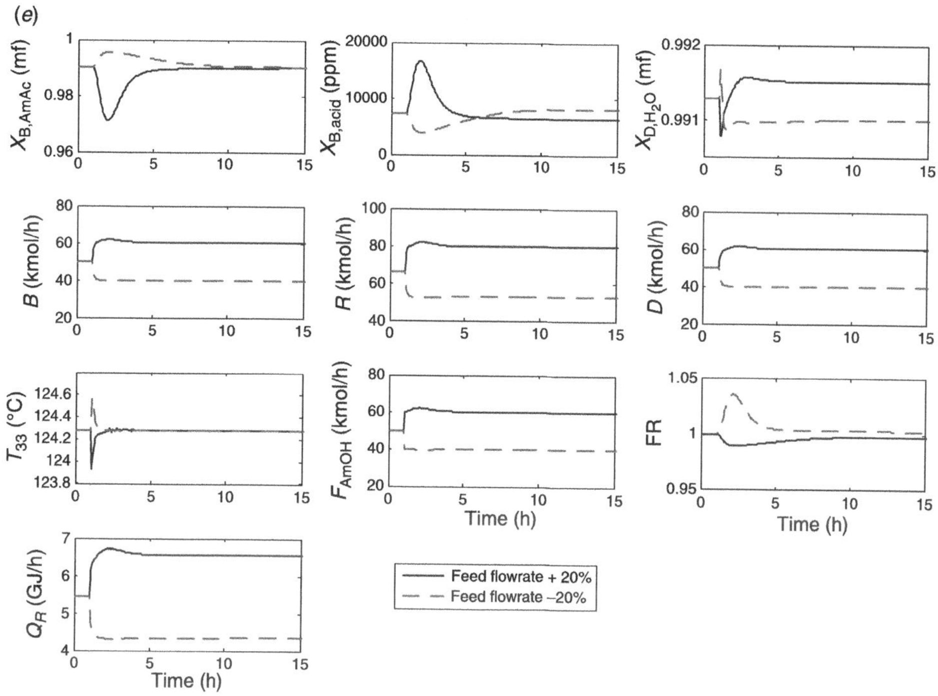

Figure 13.12 (*Continued*).

peak errors for the temperature and composition control are roughly the same, except that offset-free product composition can be obtained (Figs. 13.6*d* and 13.12*d*). Thus, composition control may be an attractive alternative for the BuAc system. For the AmAc system, the peak errors are almost the same for both temperature control and composition control structures. However, the steady-state offset can be eliminated using one composition loop. This is another likely system to consider for composition control (Figs. 13.6*e* and 13.12*e*). However, asymmetrical responses can also be seen for the AmAc system.

For feed ratio disturbances, the responses of the EtAc, IPAc and BuAc systems are symmetrical with acceptable speeds of response (Fig. 13.13). Again, oscillatory behaviors are observed for the MeAc system (Fig. 13.13*a*) and limited operability can be seen for the AmAc system (can handle only $\pm 2\%$ feed ratio changes) as shown in Figure 13.13*e*. Table 13.3 also gives the dynamic performance for one-temperature, one-composition control for all five systems.

In summary, one-temperature, one-composition control is recommended for the BuAc system, in which steady-state error can be eliminated without sacrificing closed-loop dynamics. For the AmAc system, we observed limited operability (Fig. 13.13*e*) when composition control is applied. For the MeAc system, composition control leads to a large peak error and oscillatory response. Therefore, a tradeoff has to be made between the off-spec acetate composition and poor control performance. An alternative is to use a temperature/composition cascade to solve this problem (e.g., CS3 in Figs. 12.58 and 12.84). As for the EtAc and IPAc systems, composition control provides little improvement in terms of the steady-state error.

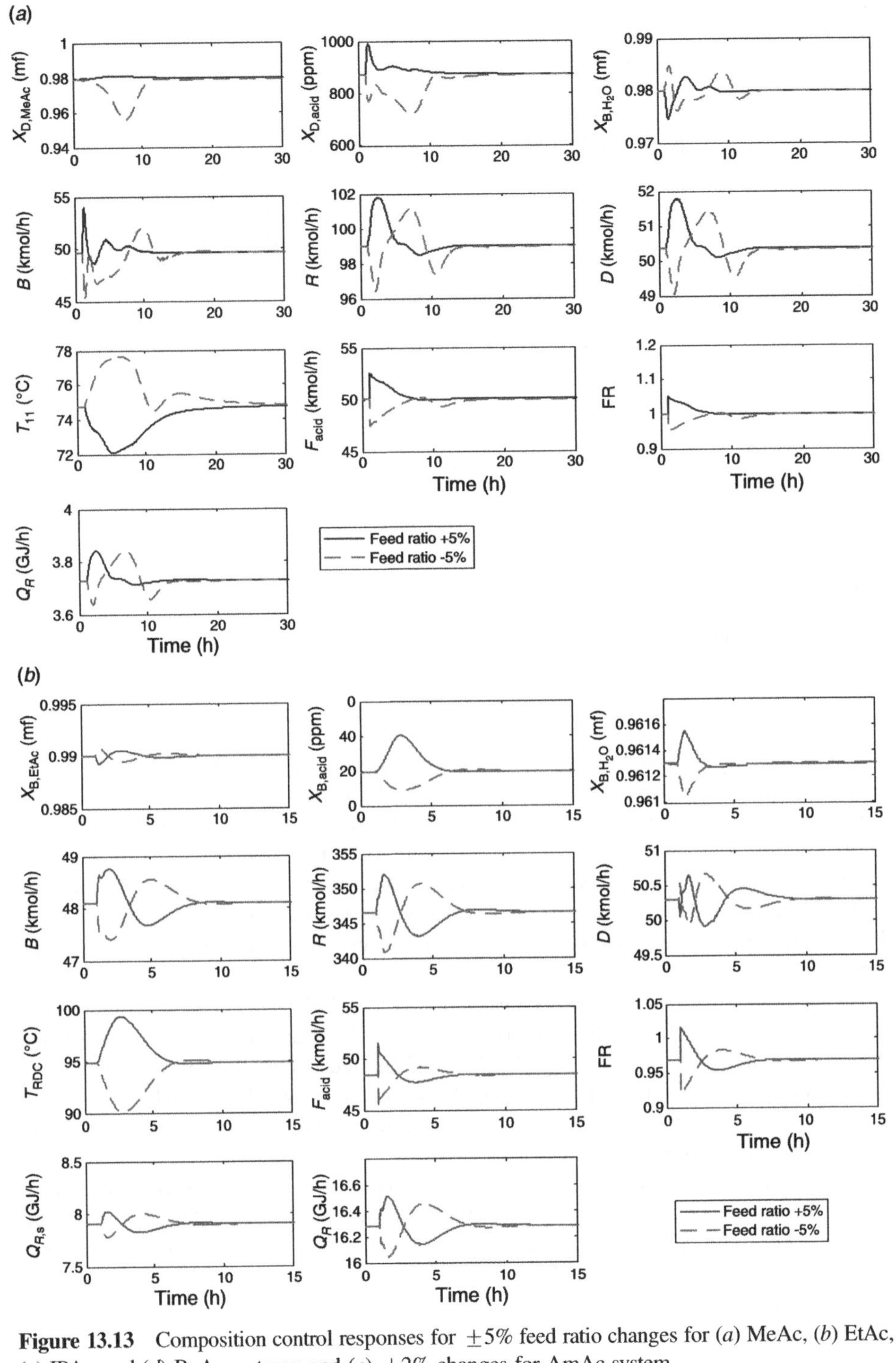

Figure 13.13 Composition control responses for ±5% feed ratio changes for (*a*) MeAc, (*b*) EtAc, (*c*) IPAc, and (*d*) BuAc systems; and (*e*) ±2% changes for AmAc system.

(c)

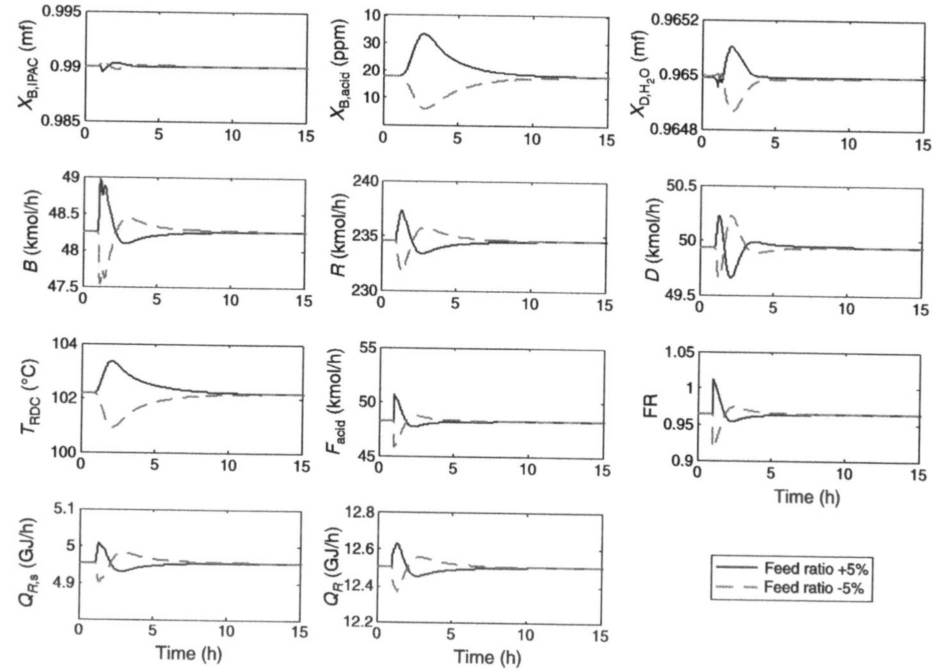

(d)

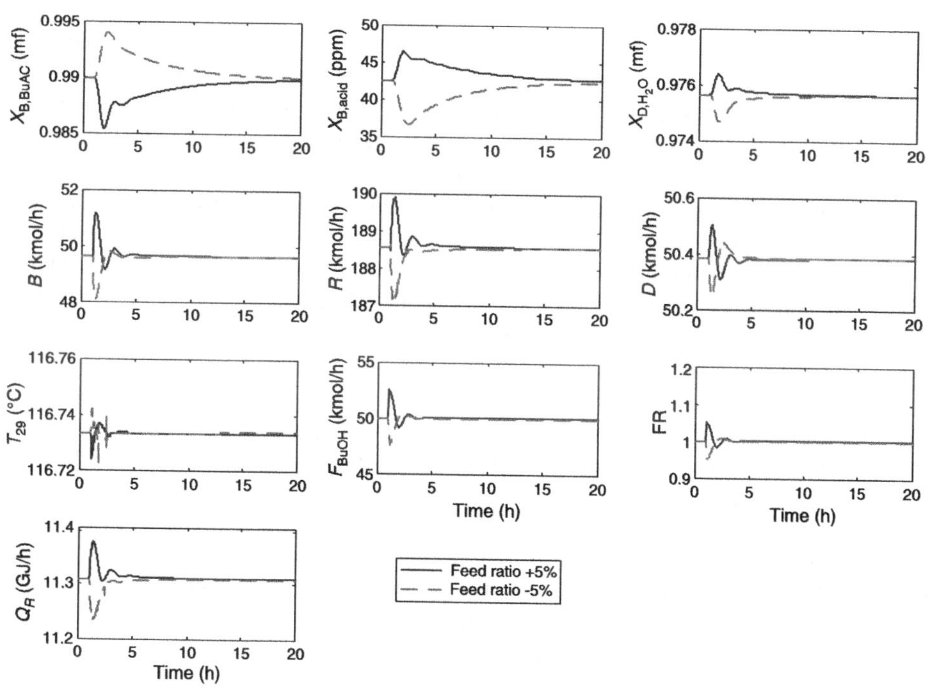

Figure 13.13 *(Continued).*

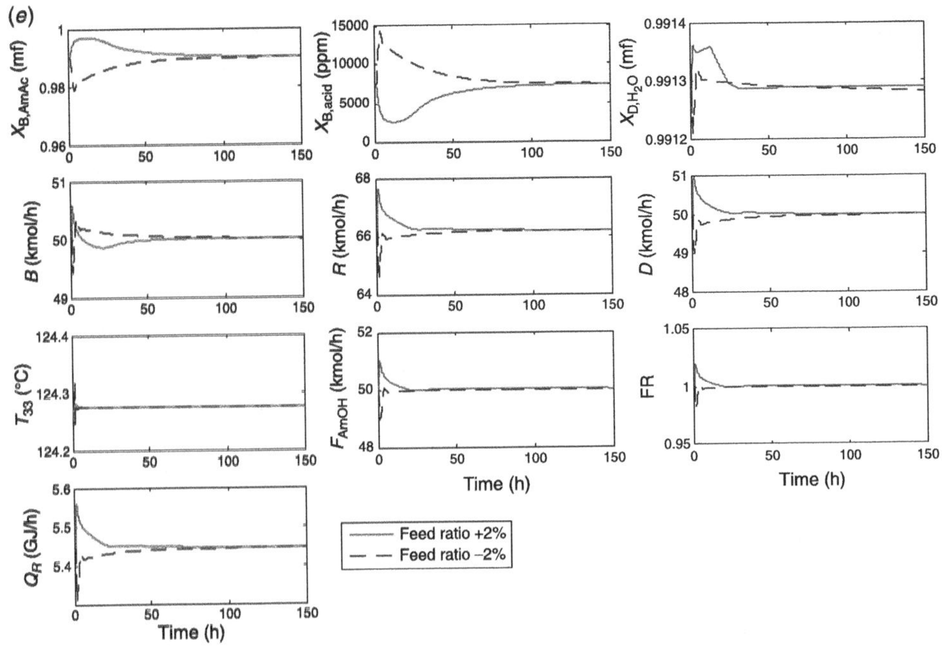

Figure 13.13 (*Continued*).

13.4 CONCLUSION

The dynamics and control for three different types of reactive distillation flowsheets (types I, II, and III) were explored in this chapter. The study covered acetic acid esterification with different alcohols ranging from C_1 (MeOH) to C_5 (AmOH). Simultaneous reaction and separation led to strong nonlinearity in all five systems. However, the degree of nonlinearity was analyzed qualitatively or computed quantitatively. The results indicated that the steady-state gains sign reversal and input multiplicity were unavoidable for all five reactive distillation systems. A systematic design procedure was proposed to devise control structures for all three types of flowsheets. The simulation results revealed that workable temperature control could be obtained for these highly nonlinear processes with simple control. Moreover, the closed-loop systems behaved as the preliminary nonlinear analyses predicted, and inherent strong nonlinearity led to asymmetrical responses, especially for the MeAc and AmAc systems.

Because of steady-state offsets in the product composition, a one-temperature, one-composition control structure was proposed to maintain on-aim product quality. This offers an alternative when the two-temperature control structure shows unacceptably large offsets in product composition.

The type II flowsheet (EtAc and IPAc) was divided into two units separated by a large decanter. This somewhat dampened disturbances and interactions between the reactive distillation column and the stripper, which subsequently led to a more controllable process. The flowsheets (BuAc and AmAc) had decanters that provided a natural one-end composition control via LL equilibrium. Interaction between top and bottoms composition control was therefore reduced.

The type I flowsheet could not escape from either the inherent nonlinearity or dynamic interactions. The nonlinearity and strong interactions led to a process that is very difficult to control.

CHAPTER 14

PLANTWIDE CONTROL OF TAME REACTIVE DISTILLATION SYSTEM

In Chapter 8 we explored the steady-state design of the TAME reactive distillation system. The reactive column is part of a multiunit process that includes other columns for recovery of the methanol reactant. The recovery is necessary because the presence of methanol/C5 azeotropes unavoidably removes methanol from the reactive column in the distillate stream. The economics of two alternative methanol recovery systems were evaluated in Chapter 8. In this chapter the dynamic control of the process is studied, and an effective plantwide control structure is developed. The process has three distillation columns: one reactive column, one extractive distillation column, and one methanol/water separation column from which methanol and water are recycled.

14.1 PROCESS STUDIED

The reaction of methanol with unsaturated C5 isoamylenes (2-methyl-1-butene and 2-methyl-2-butene) produces TAME. The liquid-phase reversible reactions are

$$\begin{aligned} &\mathrm{2M1B + MeOH \Leftarrow TAME} \\ &\mathrm{2M2B + MeOH \Leftrightarrow TAME} \\ &\mathrm{2M1B \Leftarrow 2M2B} \end{aligned} \tag{14.1}$$

This system is fundamentally a ternary system with inerts. The methanol fresh feedstream is essentially pure. The hydrocarbon feed comes from an upstream catalytic cracking unit. It is a mixed C5 stream that contains not only the two reactants (2M1B and 2M2B) but also other C5s such as isopentane, *n*-pentane, 1-pentene, and 2-pentene.

Reactive Distillation Design and Control. By William L. Luyben and Cheng-Ching Yu

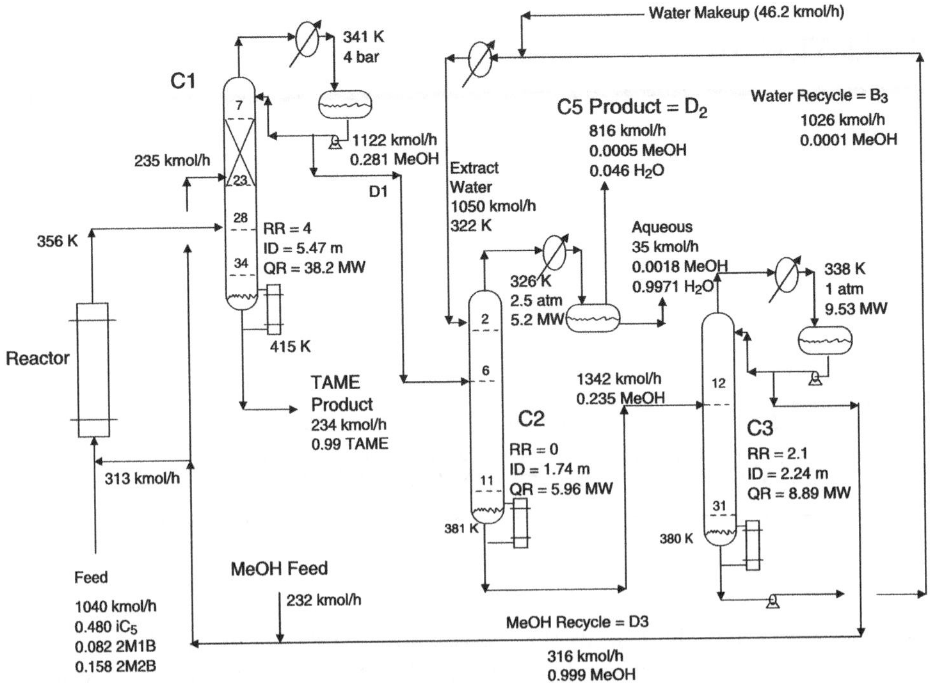

Figure 14.1 Flowsheet for TAME process with extractive distillation methanol recovery.

The heavy component is TAME, which leaves the reactive distillation column in the bottoms. The lighter C5 components leave in the distillate stream and take with them a significant amount of methanol that is tied up in the C5–methanol azeotropes. For example, the iC_5/methanol azeotrope contains 25 mol% methanol at a pressure of 4 bar with a temperature of 376 K.

The process flowsheet has a prereactor, a reactive distillation column, and a methanol recovery section. The system is simulated in Aspen Plus using UNIFAC physical properties. Figure 14.1 presents the flowsheet with steady-state design conditions and equipment sizes. This flowsheet is slightly modified from that given in Chapter 8. A decanter is used for the reflux drum in extractive distillation column C2. Water is used as the extractive agent in this column. A significant amount of water goes overhead and forms an aqueous liquid phase. The organic phase is mostly inert C5 that comes in with the fresh feedstream (816 kmol/h). The aqueous phase (35 kmol/h) is 99.71 mol% water and 0.1 mol% methanol.

14.1.1 Prereactor

The prereactor is a cooled tubular reactor containing 9544 kg of catalyst. All of the C5 fresh feed is fed to the prereactor along with a portion of the total methanol. The C5 fresh feed flowrate is 1040 kmol/h. This stream contains the reactive isoamylenes (85.6 kmol/h of 2M1B and 165 kmol/h of 2M2B). The remainder is chemically inert C5s.

The reactor is sized by assuming a catalyst bulk density of 900 kg/m^3 and reactor tubes that are 0.05 m in diameter and 5 m in length. There are 1080 tubes, giving a heat transfer area of 825 m^2. The reactor is cooled by a circulating cooling water system that is assumed to provide a constant coolant temperature throughout the shell of the reactor. An overall heat transfer coefficient of 568 W m^{-2} K^{-1} is assumed. The reactor inlet temperature is 338 K, the heat transfer rate is 0.422 MW, and the coolant temperature is 355 K. Figure 14.2 shows the temperature profile in the prereactor. The peak temperature is only 360 K.

The fresh feed of methanol coming into the process is 232 kmol/h, but there is a large methanol recycle stream from the recovery section. The flowrate of the recycle methanol is 316 kmol/h. The total methanol (548 kmol/h) is split between the prereactor (313 kmol/h) and the reactive distillation column (235 kmol/h).

14.1.2 Reactive Column C1

The reactor effluent is fed into a 36-stage reactive distillation column (C1) on stage 28. Aspen notation of numbering stages from the top is used, with stage 1 being the reflux drum. Catalyst is present on stages 7–23. The reactor effluent is fed five trays below the reactive zone. The methanol stream (235 kmol/h) is fed at the bottom of the reactive zone (stage 23).

Figure 14.3 gives composition profiles in reactive column C1. Note the low concentration of methanol in the lower part of the column and the fairly high concentration of one of the reactants (2M2B) in the same zone. This is below the reactive trays, which are stages 7–23. Figure 14.4 shows the temperature profile. The reflux ratio is 4, which gives a bottoms purity of 99.2 mol% TAME and a distillate impurity of 0.1 ppm TAME. Reboiler heat input and condenser heat removal are 38.2 and 39 MW, respectively. The operating pressure is 4 bar. The overall conversion of 2M1B and 2M2B in the C5 fresh feed is 92.4%.

Distillate D_1 has a composition of methanol (28 mol% methanol) that is near the azeotrope at 4 bar. It is fed at a rate of 1122 kmol/h to the methanol recovery columns section. Information about important streams associated with the prereactor and reactive distillation column is given in Table 8.2.

The column base and the reflux drum are sized to provide 5 min of holdup when 50% full under steady-state conditions. The column diameter is 5.5 m.

14.1.3 Extractive Column C2

Methanol-containing distillate D_1 from the top of the reactive column is fed to stage 6 of a 12-stage extraction column. Water is fed on the top tray at a rate of 1050 kmol/h and a temperature of 322 K, which is achieved by using a cooler (heat removal = 1.24 MW). The column is a simple stripper with no reflux. The column operates at 2.5 atm so that cooling water can be used in the condenser (decanter temperature = 326 K). Reboiler heat input is 5.96 MW.

The overhead vapor is condensed and fed to a decanter in which the two liquid phases are separated. The organic phase is the chemically inert C5s, and the aqueous phase is mostly water.

This column is designed by specifying a very small loss of methanol in the overhead vapor (0.01% of the methanol fed to the column) and finding the minimum flowrate of extraction water that achieves this specification. Adding more than 10 trays or using reflux did not affect the recovery of methanol. The bottoms is essentially a binary

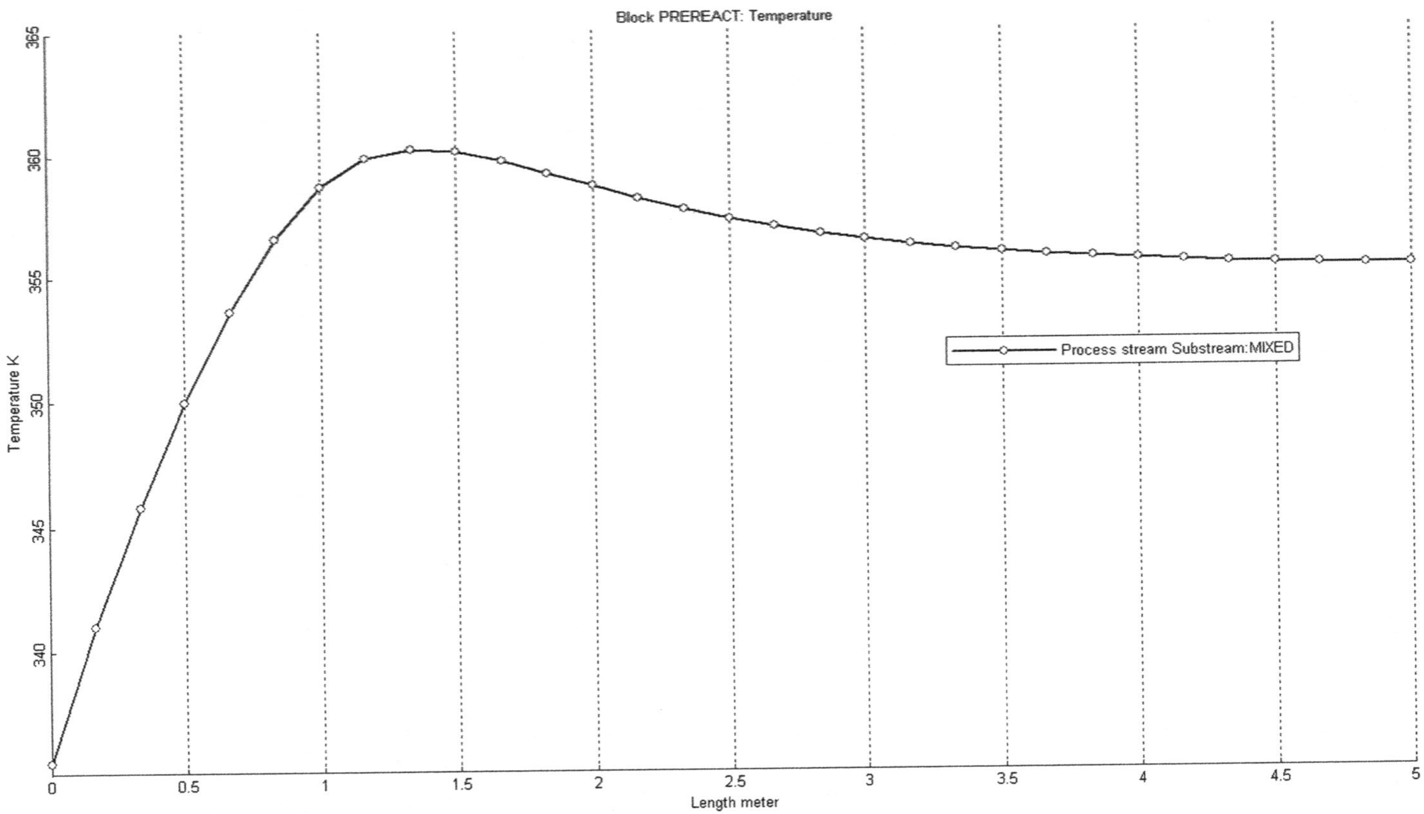

Figure 14.2 Prereactor temperature profile.

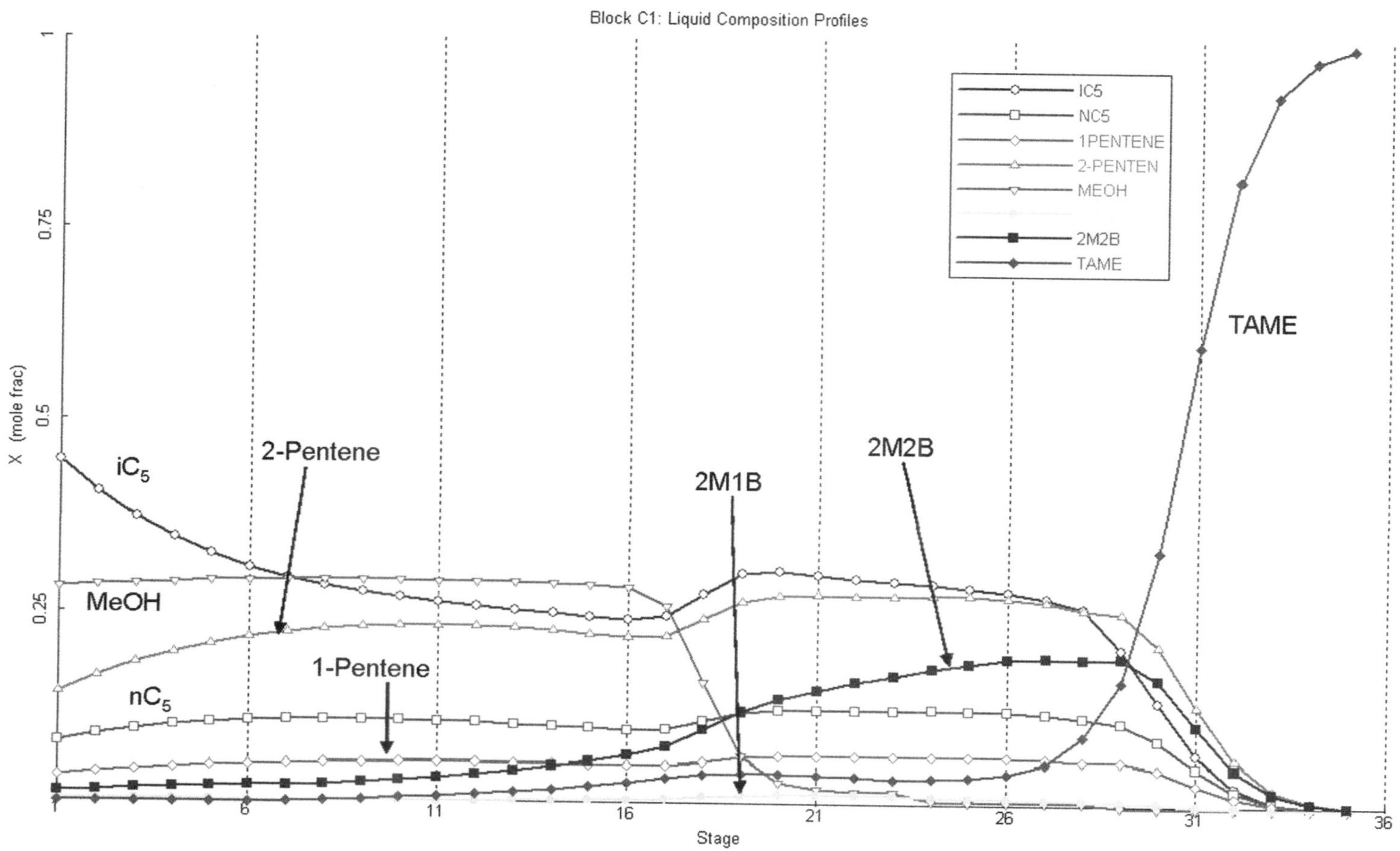

Figure 14.3 Composition profiles in reactive column.

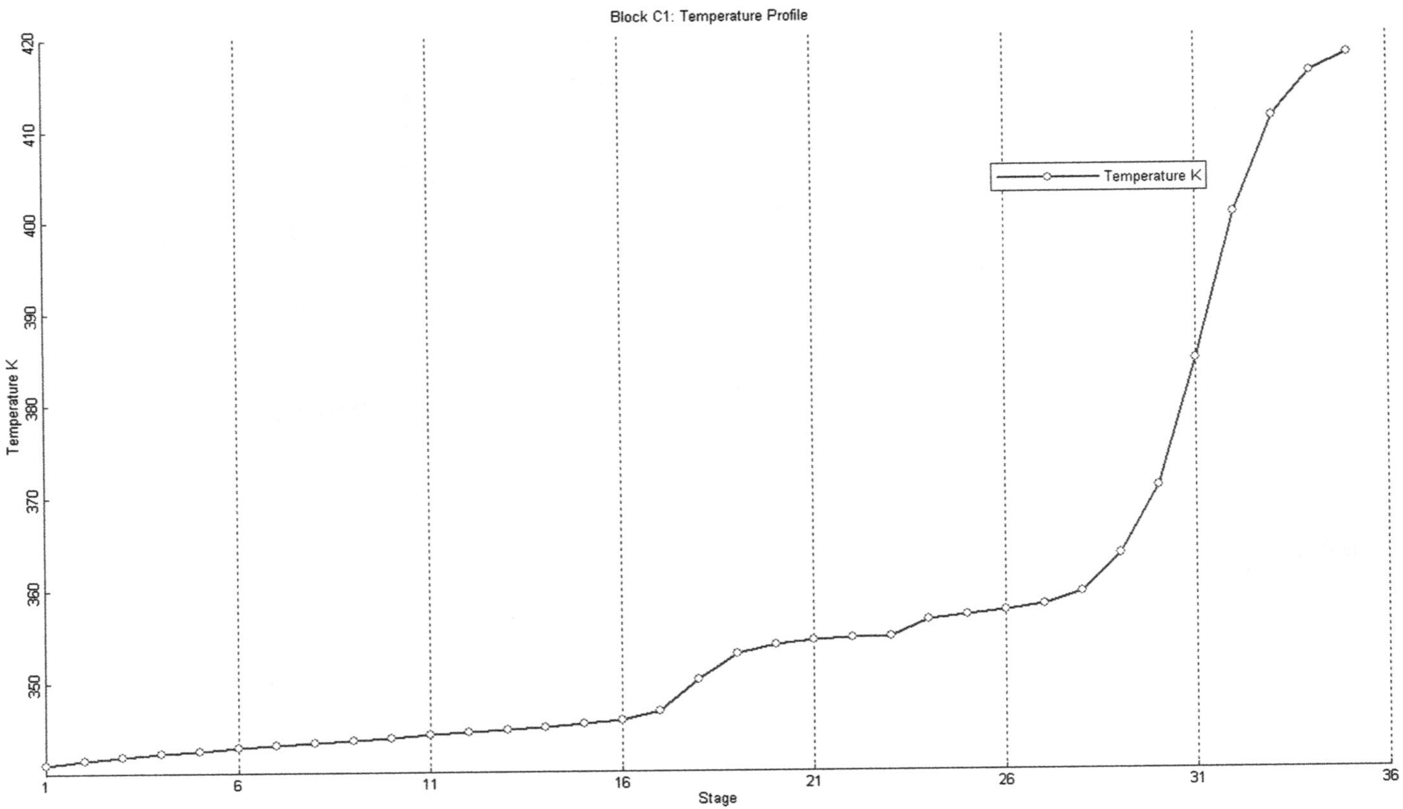

Figure 14.4 Temperature profile for C1.

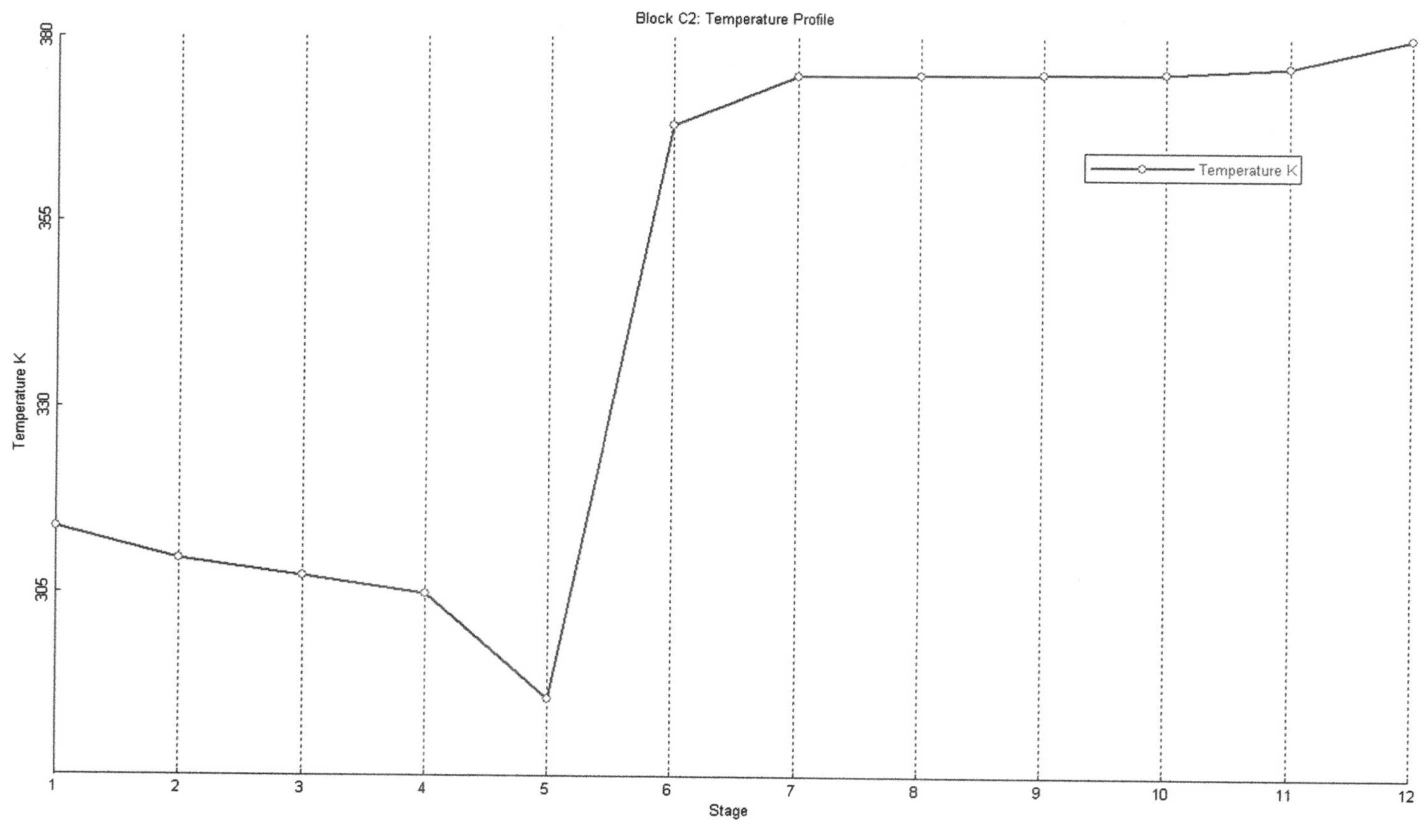

Figure 14.5 Temperature profile for C2.

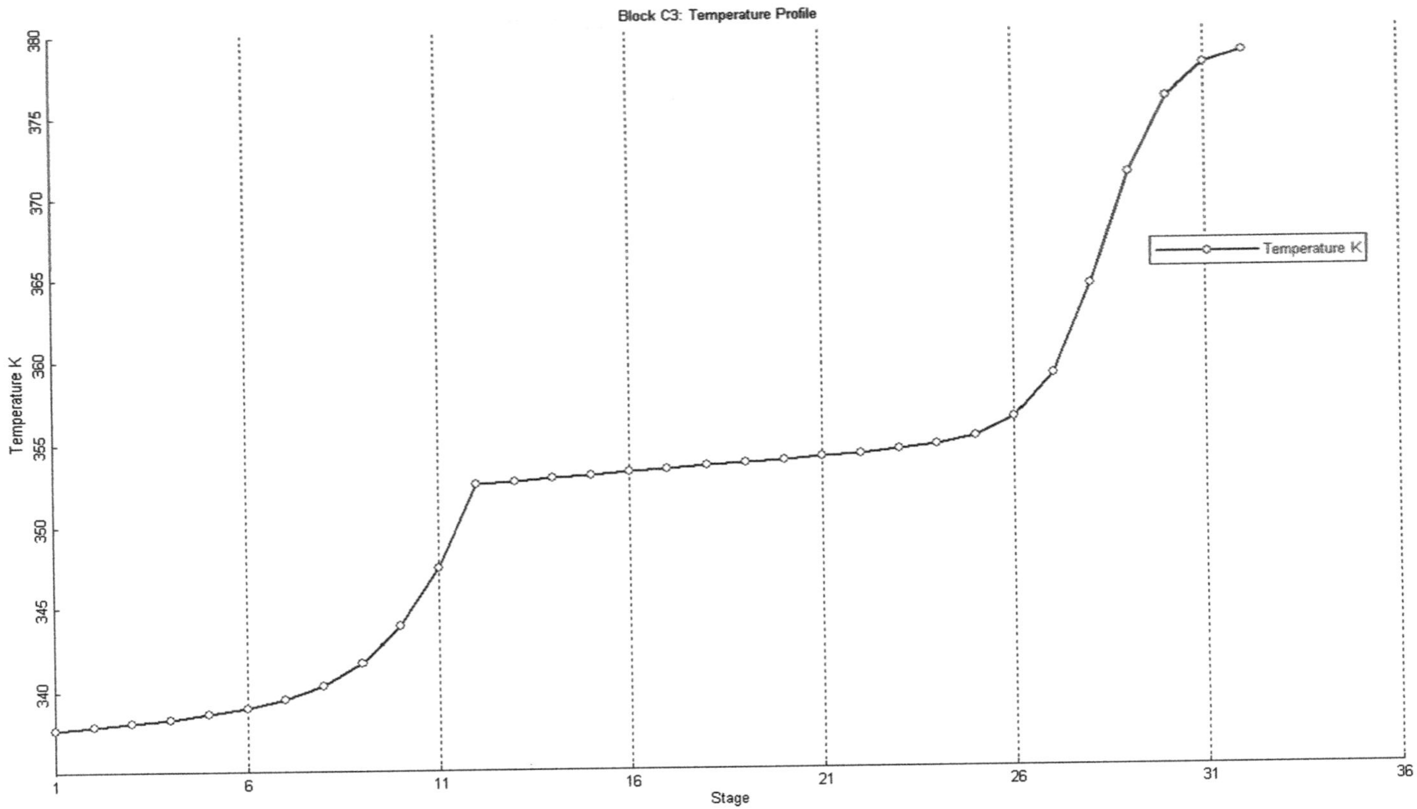

Figure 14.6 Temperature profile for C3.

methanol/water mixture (23.5 mol% methanol), which is fed to the methanol recovery column.

Figure 14.5 gives the column temperature profile, which shows a sharp temperature break over a small number of trays. There is a large temperature difference between the upper section above the feed (~300 K) and the stripping section below the feed (~370 K). This will affect the control system used in the column, as discussed in the next section.

The column base and the decanter are sized to provide 5 min of holdup when 50% full under steady-state conditions. The column diameter is 1.74 m.

14.1.4 Methanol Recovery Column C3

The feed is introduced on stage 12 of a 32-stage column operating at atmospheric pressure. The number of trays in this column was optimized by determining the TAC of columns over a range of tray numbers. Reboiler heat input and condenser heat removal are 8.89 and 9.53 MW, respectively. The reflux ratio is 2.1. High product purities are achieved in both the distillate (99.9 mol% methanol) and the bottoms (99.9 mol% water). Both of these streams are recycled.

The bottoms water stream is combined with a water makeup stream (46.2 kmol/h) and cooled before being fed as the extraction water feed to C2. The distillate methanol stream is combined with the fresh feed of methanol (232 kmol/h), and the total is split between the prereactor and the reactive distillation column. Figure 14.6 displays the temperature profile. There are two areas where the temperature changes from tray to tray are fairly large, which suggests either one-end temperature control or two-temperature control may be possible.

The column base and the reflux drum are sized to provide 5 min of holdup when 50% full under steady-state conditions. The column diameter is 2.24 m.

14.2 CONTROL STRUCTURE

In preparation for exporting the steady-state flowsheet from Aspen Plus into Aspen Dynamics, all equipment is sized. Column diameters are calculated by Aspen Tray Sizing. Pumps heads and control valve pressure drops are specified to give adequate dynamic rangeability. Typical valve pressure drops are 2 atm. The prereactor heat removal is specified to use a constant coolant temperature.

The control structure for the prereactor and reactive distillation column C1 is shown in Figure 14.7, The control structure for the extractive distillation column C2 and the methanol recovery column C3 is shown in Figure 14.8.

14.2.1 Prereactor

The fresh C5 stream containing the reactive isoamylenes and the chemically inert other C5 components is fed into the reactor on flow control. The methanol fed to the prereactor is ratioed to the fresh feed flowrate. The exit temperature of the reactor is controlled using a temperature/temperature cascade structure. The reactor effluent temperature controller changes the setpoint of the circulating cooling water temperature controller, which manipulates the cooling water makeup valve (see Fig. 14.7).

In the Aspen Dynamics simulation, a valve must be installed between the reactor and the column. In reality, no valve is necessary if the reactor pressure can be the same as the

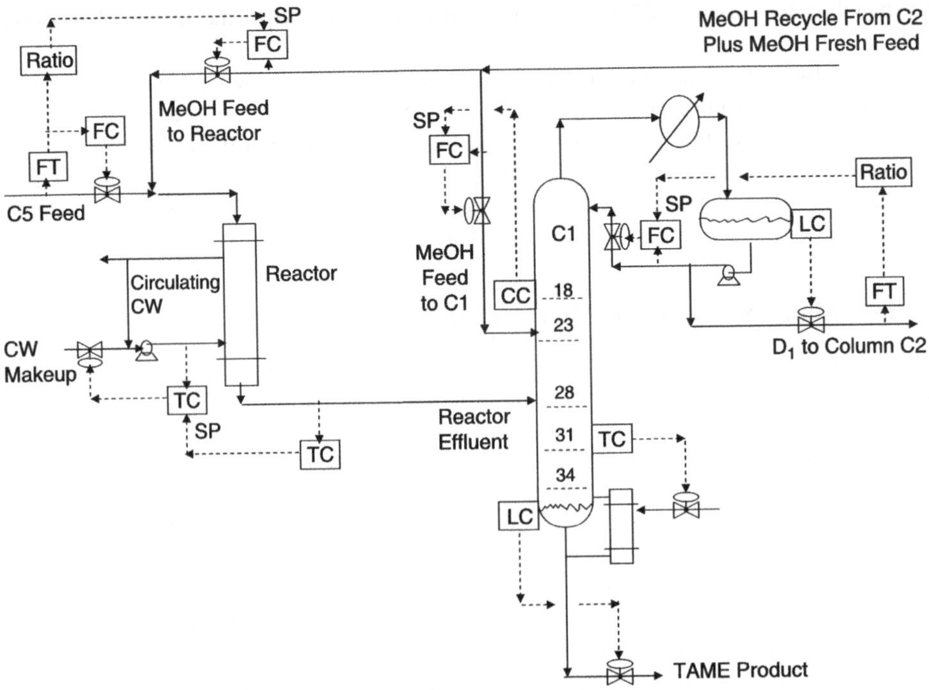

Figure 14.7 Control structure for prereactor and C1.

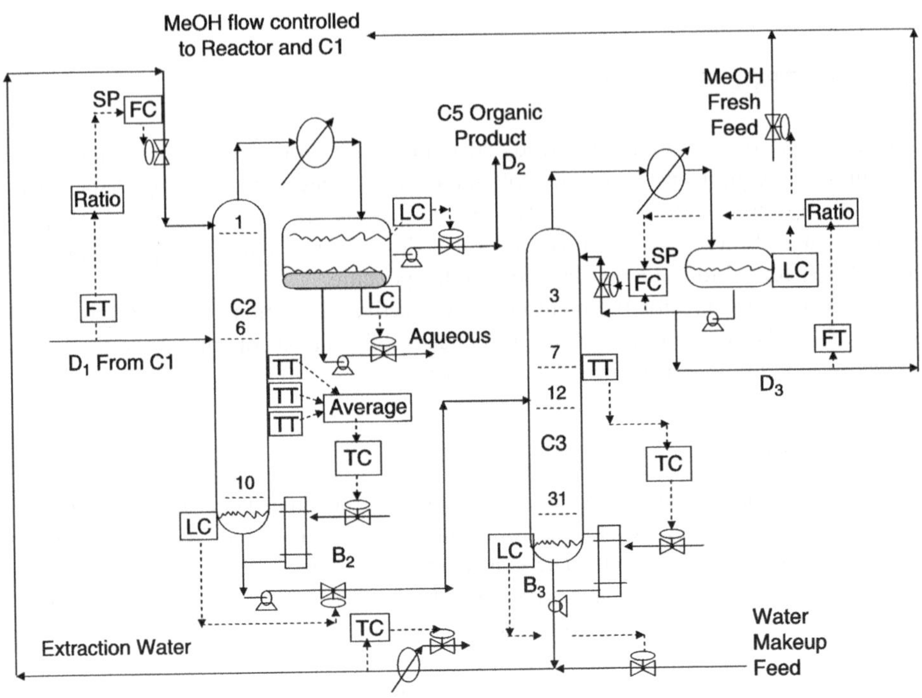

Figure 14.8 Control structure for C2 and C3.

column pressure. If it were necessary to keep the reactor pressure higher than the column pressure, a valve would be required. However, the valve in the C5 fresh feedline would have to be removed so that the first law of plumbing is not violated ("Never put two valves in series in a liquid-filled line."). The flowrate of the C5 fresh feed would have to be manipulated by using a variable speed pump instead of a valve.

14.2.2 Reactive Distillation Column C1

The temperature on stage 31 is controlled by manipulating the reboiler heat input with a 1-min dead time. Relay-feedback testing and Tyreus–Luyben settings give $K_C = 0.38$ and $\tau_I = 7.9$ min with a 100 K temperature transmitter span and a reboiler heat input range from 0 to twice the steady-state value.

The composition of methanol on stage 18 is controlled by manipulating the methanol stream to C1. A 3-min dead time is used in this composition loop. Relay-feedback testing and Tyreus–Luyben settings give $K_C = 0.73$ and $\tau_I = 37$ min with a 50 mol% composition transmitter span and a flowrate range from 0 to twice the steady-state value.

The reflux-drum level is controlled by the distillate. The reflux ratio is controlled by measuring the distillate flowrate, sending this signal to a multiplier ("Ratio") and sending the output signal to the setpoint of the reflux flow controller. The base level is controlled by manipulating the bottoms flowrate.

14.2.3 Extractive Distillation Column C2

Figure 14.8 shows the control structure for the extractive column. The flowrate of extraction water fed to the top of C2 is ratioed to the feed to this column D_1 by using a multiplier and a remotely set flow controller. Base level is controlled by manipulating the bottoms. The organic and aqueous levels in the decanter are controlled by manipulating the flowrates of the two exit streams. There is no reflux back to the column. The temperature of the extraction water is controlled by manipulating the cooling water to the cooler.

The sharp temperature profile in C2 presents control difficulties. The process gain between a single tray temperature, for example, stage 7, and reboiler heat input is very large. This means that the controller gain must be small, which results in poor load rejection. A simple solution is to measure several temperatures on trays in the column and control the average temperature.

Implementing this in Aspen Dynamics is somewhat tricky. The control devices all use metric units, so temperatures are in degrees Celsius. However, we are using temperatures in Kelvin, so we must do some conversion. The control configuration is shown in Figure 14.9. The temperatures on stages 5, 6, and 7 are connected to a *multisum* module called "sum3." These three temperatures are 304, 371, and 375 K, respectively, as shown in the top picture in Figure 14.10. You would expect the output signal to be the sum of these (1050 K). However, the output is 231 °C, which is the sum of the three inlet temperatures in degrees Celsius.

The next block is a multiplier called "average" that multiplies the input by 0.3333 to get the average temperature in degrees Celsius. The next block is another summer that adds 273 to obtain the average temperature in Kelvin. This signal is finally sent through a dead time element (1 min) and then to the average temperature controller.

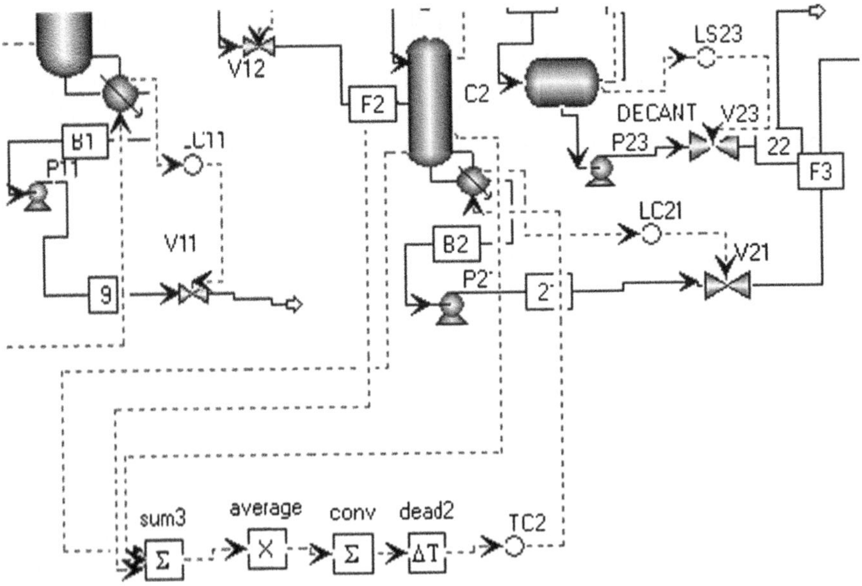

Figure 14.9 Average temperature control.

sum3.AllVariables Table

	Value	Spec	
ComponentList	Type1		
Input_(1)	304.218	Free	K
Input_(2)	371.619	Free	K
Input_(3)	374.546	Free	K
NSignal	3		
Output_	230.933	Free	

average.AllVariables Table

	Value	Spec
ComponentList	Type1	
Input1	230.933	Free
Input2	0.3333	Fixed
Output_	76.97	Free

conv.AllVariables Table

	Value	Spec
ComponentList	Type1	
Input1	76.97	Free
Input2	273.0	Fixed
Output_	349.97	Free

Figure 14.10 Parameter values in summers.

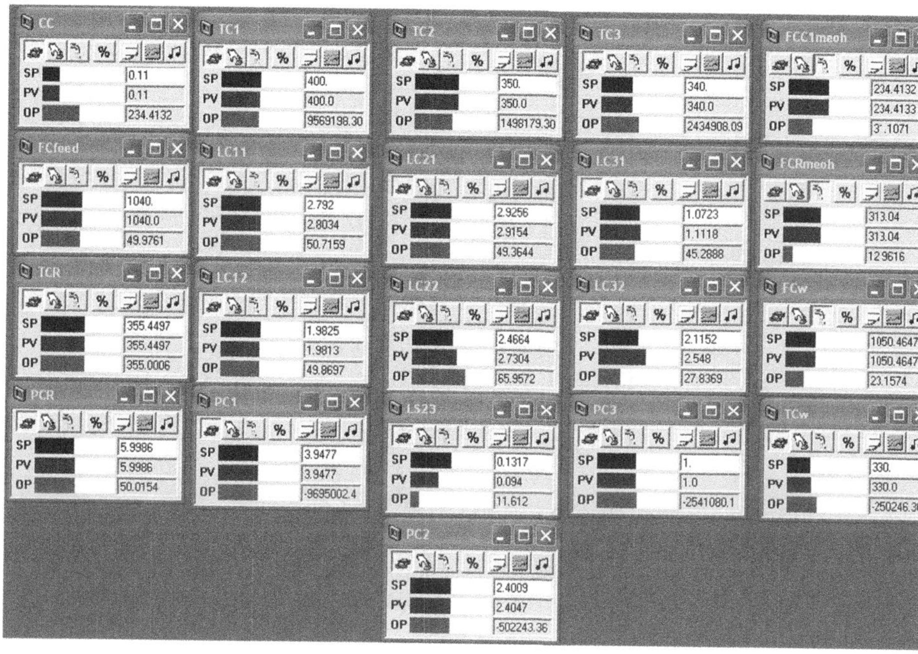

Figure 14.11 Faceplates.

Relay-feedback testing and Tyreus–Luyben settings give $K_C = 2.5$ and $\tau_I = 6.6$ min with a very large temperature transmitter span of 650 K and a reboiler heat input range from 0 to twice the steady-state value.

14.2.4 Methanol Recovery Column C3

Figure 14.8 shows the control structure used in the C3 column that is separating methanol and water. A constant reflux ratio is maintained in this column by adjusting the reflux flowrate based on the distillate flowrate. The temperature on stage 7 is controlled by the reboiler heat input.

There are two key plantwide material balance loops associated with column C3. The level in the reflux drum provides a good indication of the inventory of methanol in the system. If this level is going down, more methanol is being consumed in the reaction than is being fed into the process. Therefore, the control structure maintains the reflux-drum level in C3 by manipulating the methanol fresh feed.

Note that the flowrate of the total methanol (D_3 plus fresh methanol feed) is fixed by the two downstream flow controllers setting the flowrates to the reactor and to column C1. This means there is an immediate effect of the fresh feed flowrate on the reflux-drum level. Distillate flow D_3 changes inversely with the fresh feedflow because the downstream flowrate is fixed. Thus, the reflux-drum level changes instantaneously for changes in the flowrate of the methanol fresh feed.

At the other end of the column, the base level provides a good indication of the inventory of water in the system. Most of the extraction water just circulates between the extractive column and the recovery column. However, there is a small amount of water lost in the

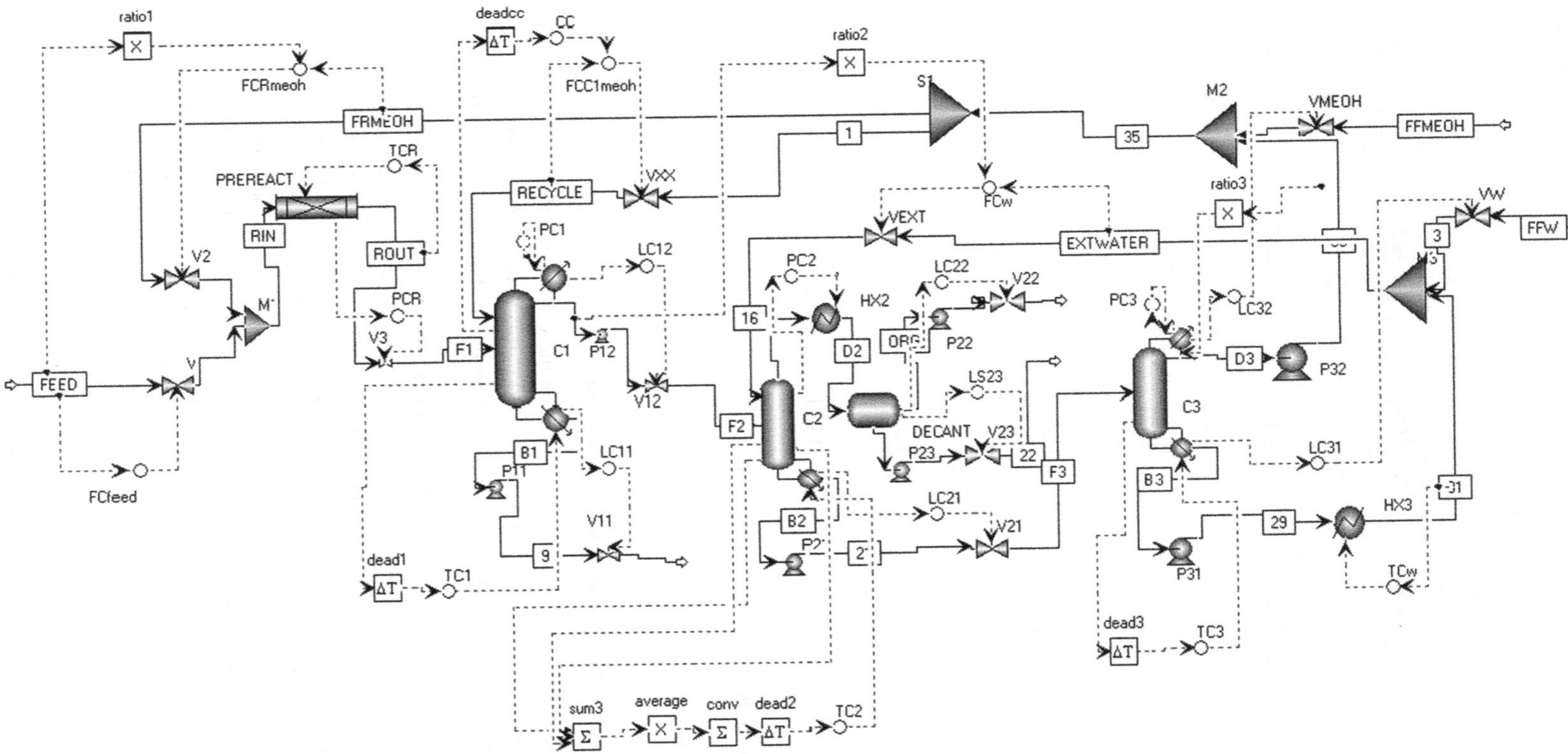

Figure 14.12 Aspen Dynamics process flow diagram.

overhead from column C2, which leaves in the aqueous stream from the decanter. A water makeup stream is used to control the liquid level in the base of C3. This makeup flow is small compared to the water circulation, so the base of column C3 must be sized to provide enough surge capacity to ride through disturbances.

Relay-feedback testing and Tyreus–Luyben settings give $K_C = 11.6$ and $\tau_I = 5.3$ min with a temperature transmitter span of 100 K and a reboiler heat input range from 0 to twice the steady-state value.

All liquid levels are controlled by proportional controllers with gains of 2. Column pressure controllers use default controller settings and manipulate condenser heat removal (not shown in Figs. 14.7 and 14.8). The faceplates for the 21 controllers in this process are shown in Figure 14.11. Note that flow controllers FCC1meoh, FCRmeoh, and FCw are on cascade because they get their setpoint signals from other controllers or multipliers. The Aspen Dynamics process flow diagram is given in Figure 14.12 with all of the control elements installed and loops closed.

14.3 RESULTS

The effectiveness of the plantwide control structure is tested by creating disturbances in the throughput (C5 feed flowrate) and feed composition. Results are shown in Figures 14.13–14.16.

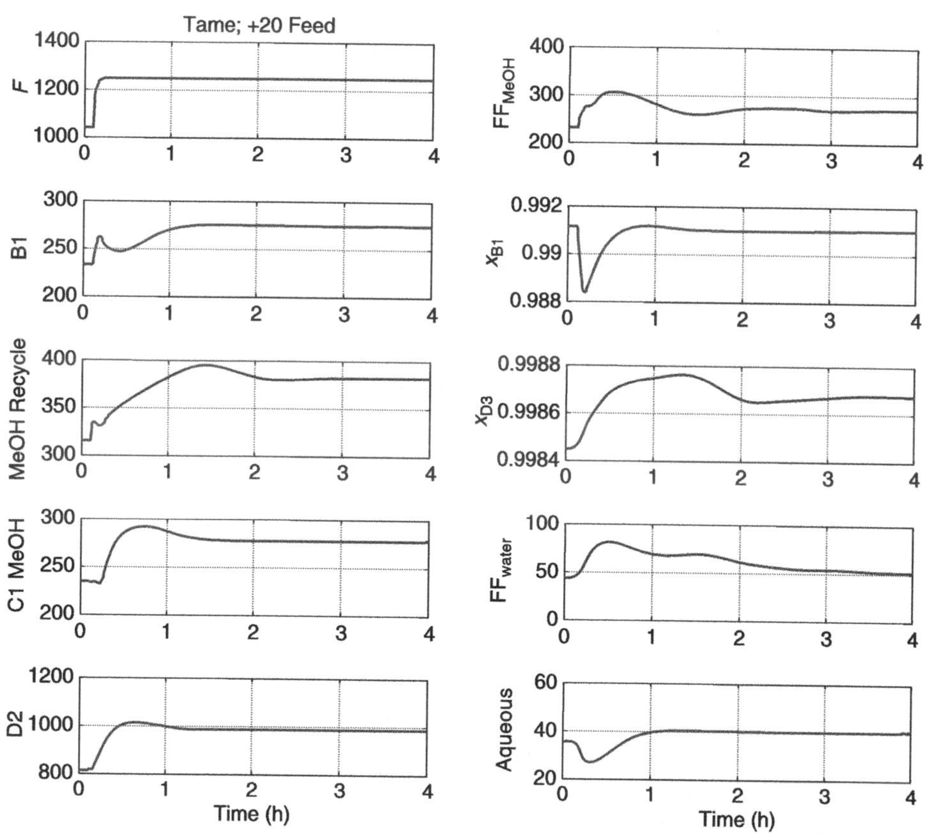

Figure 14.13 Results for 20% step increase in feed.

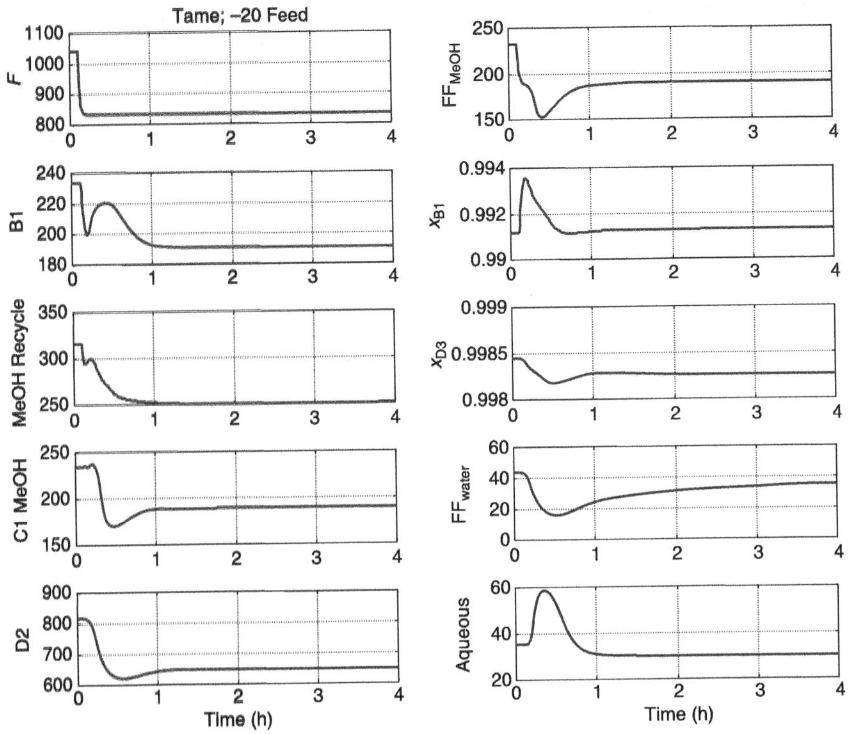

Figure 14.14 Results for 20% step decrease in feed.

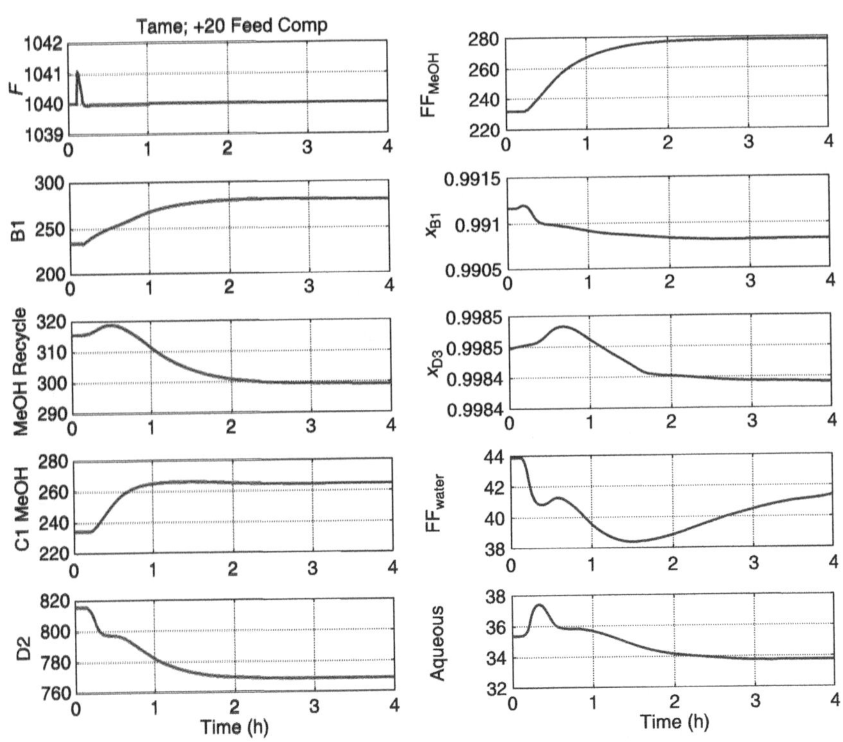

Figure 14.15 Results for 20% step increase in reactants in feed.

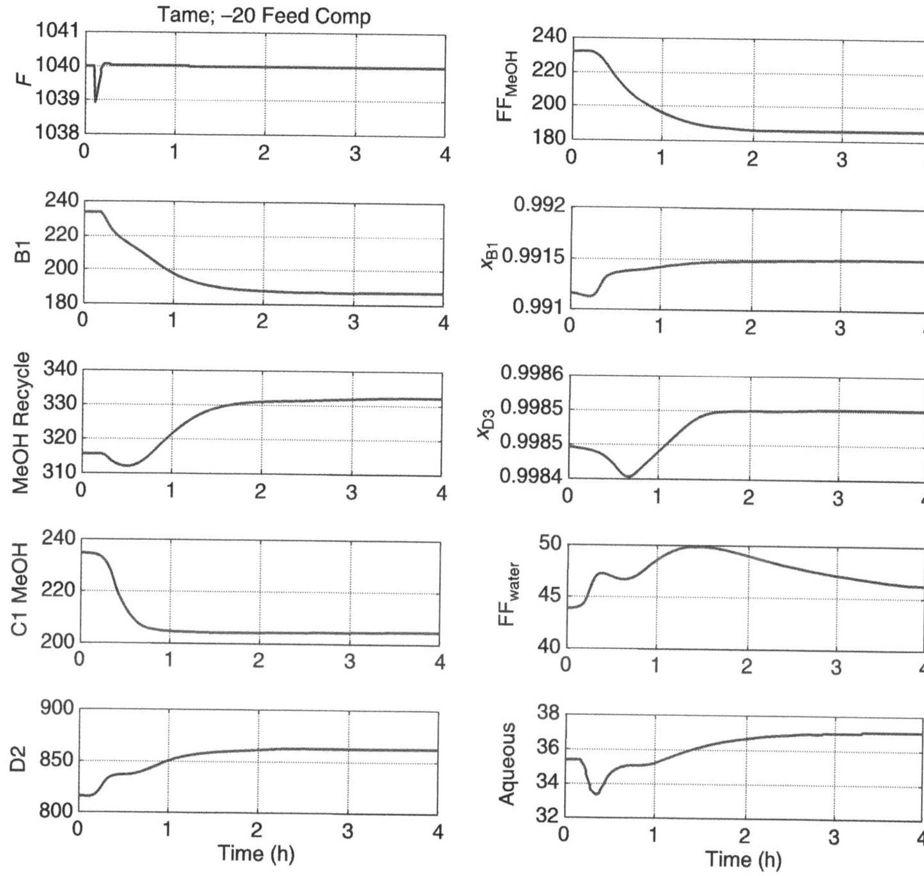

Figure 14.16 Results for 20% step decrease in reactants in feed.

A step change of +20% is made in the setpoint of the feedflow controller at 0.1 h in Figure 14.13. The production rate of TAME (stream B1) increases by 20% over a 1.5-h period. More fresh feed of methanol (FFmeoh stream) is brought into the system. The composition of the TAME product (x_{B1}) remains close to its specification. The composition controller in the reactive column increases the feed of methanol to that column (C1MeOH stream). The increased amount of inert C5s in the feed produce an increase in organic stream D2 and the aqueous stream from C2. The flowrates of the methanol recycle and the makeup water also increase. The purity of the methanol recycle x_{D3} is maintained.

In Figure 14.14 the disturbance is a 20% step decrease in feedflow. The responses are essentially the reverse of those seen for the increase. These results demonstrate that the control structure handles throughput disturbances quite well.

Feed composition disturbances are made by changing the molar flowrates of reactants 2M1B and 2M2B while decreasing the molar flowrate of the isopentane by the same amount so that the total feed flowrate remains constant. The feed composition disturbance used in Figure 14.15 is a 20% increase in the reactants (2M1B increases from 85.6 to 102.7 kmol/h and 2M2B increases from 165 to 198 kmol/h). The corresponding decrease in iC_5

is from 501 to 451 kmol/h. More TAME is produced (B1 stream), more fresh methanol is brought into the plant (FFmeoh stream), and less inert C5 has to be removed (D2 stream). The purities of the TAME and recycle methanol are maintained close to the desired values. The responses to a 20% decrease in the reactants are given in Figure 14.16. The results are essentially the inverse of those found for the increase.

14.4 CONCLUSION

The TAME reactive distillation system with a two-column methanol recovery system was successfully simulated in Aspen Dynamics. The system features two recycles (methanol and water) and three feedstreams (C5, methanol, and water). The system is essentially a ternary system with inerts, but the complex vapor–liquid equilibrium results in the formation of azeotropes that result in losses of methanol out of the top of the reactive column with the inerts. Therefore, a methanol recovery system must be included in the plant design and control.

An effective plantwide control structure was developed that handles quite large disturbances in both throughput and feed compositions. The key plantwide concepts are the detection of the inventories of methanol and water in the system and feeding in fresh feeds to maintain stable plantwide component balances.

CHAPTER 15

CONTROL OF MTBE AND ETBE REACTIVE DISTILLATION COLUMNS

In Chapter 9 we explored the steady-state designs of both the MTBE and the ETBE reactive distillation columns using Aspen Plus. In this chapter we export the files into Aspen Dynamics as pressure-driven dynamic simulations and then look at dynamics and control. The control structures evaluated on both systems are based on those developed in Chapter 12 for ternary systems with inerts.

In each of these systems, two alternative control structures are evaluated that use different production rate handles. One control scheme flow controls the C4 fresh feedstream (containing the isobutene) and brings in the alcohol fresh feedstream to control an internal alcohol composition (methanol or ethanol) on a tray in the column. The second control scheme flow controls the alcohol fresh feedstream and brings in the C4 fresh feedstream to control an internal isobutene composition on a tray in the column. The effectiveness of these alternative control structures is evaluated for disturbances in production rate and feed composition.

15.1 MTBE CONTROL

The two fresh feedstreams to the column are a pure methanol steam and a C4 stream containing the reactive isobutene and the nonreactive *n*-butene. The bottoms from the column is mostly MTBE. The distillate is mostly *n*-butene.

15.1.1 Steady State

Figure 15.1 gives the steady-state conditions and equipment parameters found in Aspen Plus. The methanol and the C4 stream are both fed on stage 10 at the bottom of the reactive

Reactive Distillation Design and Control. By William L. Luyben and Cheng-Ching Yu

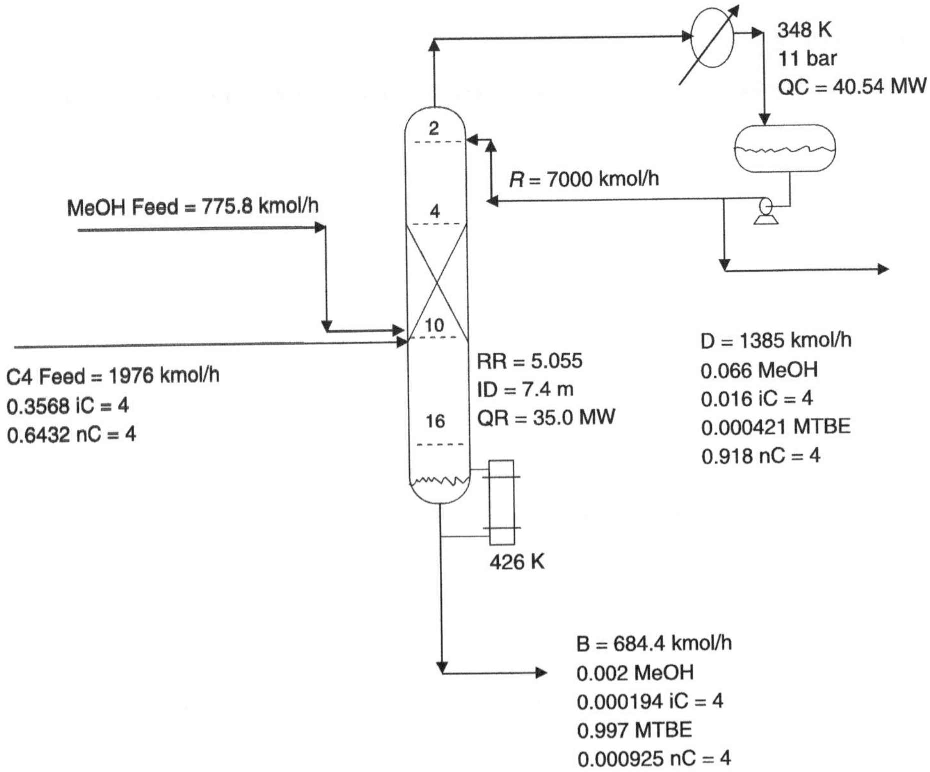

Figure 15.1 The MTBE Aspen Plus steady-state conditions.

zone (stages 4–10) of a column with 15 total trays (17 stages in Aspen notation with the reflux drum being stage 1). Figures 15.2 and 15.3 give composition and temperature profiles, respectively.

The dynamic simulation was converged to a steady-state condition after the control structure discussed in Section 15.1.2 was installed. The results (Fig. 15.4) are essentially the same as those found in Aspen Plus (Fig. 15.1). The flowrate of the C4 stream is specified to be the same (1976 kmol/h), but the methanol feed is slightly smaller (768 kmol/h in the dynamic steady state vs. 776 kmol/h in the original steady state). The bottoms MTBE flowrate and purity are the same, but the distillate flowrate is slightly smaller (1361 kmol/h in the dynamic steady state vs. 1385 kmol/h in the original steady state). The reason for this difference is the slightly higher purity of the distillate: 93.0 mol% *n*-butene in the dynamic steady state versus 91.8 mol% *n*-butene in the original steady state. The distillate contains less methanol (5.7 vs. 6.6 mol%), so less methanol needs to be fed.

15.1.2 Control Structure with C4 Feedflow Controlled

The dynamic parameters in the Aspen Plus file are inserted. The volumes of the reflux drum (144 m^3) and the column base (177 m^3) are calculated so that the holdup times are 5 min when the levels are at 50%, based on the total liquid entering. Aspen Plus *Tray Sizing* gives a diameter of 7.4 m. Aspen Plus *Packing Sizing* gives a diameter of 5.4 m, and this

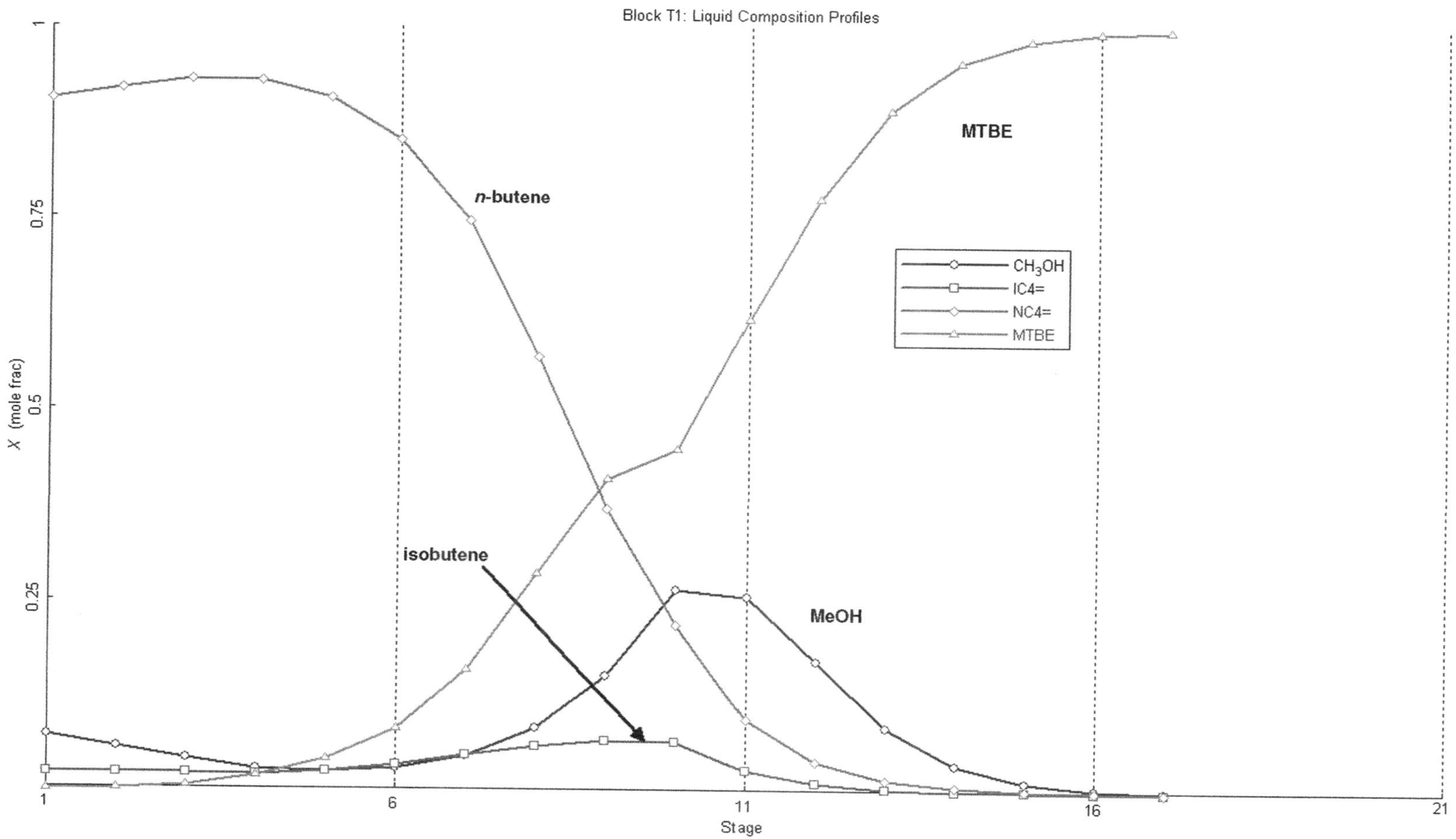

Figure 15.2 Composition profiles in MTBE reactive column.

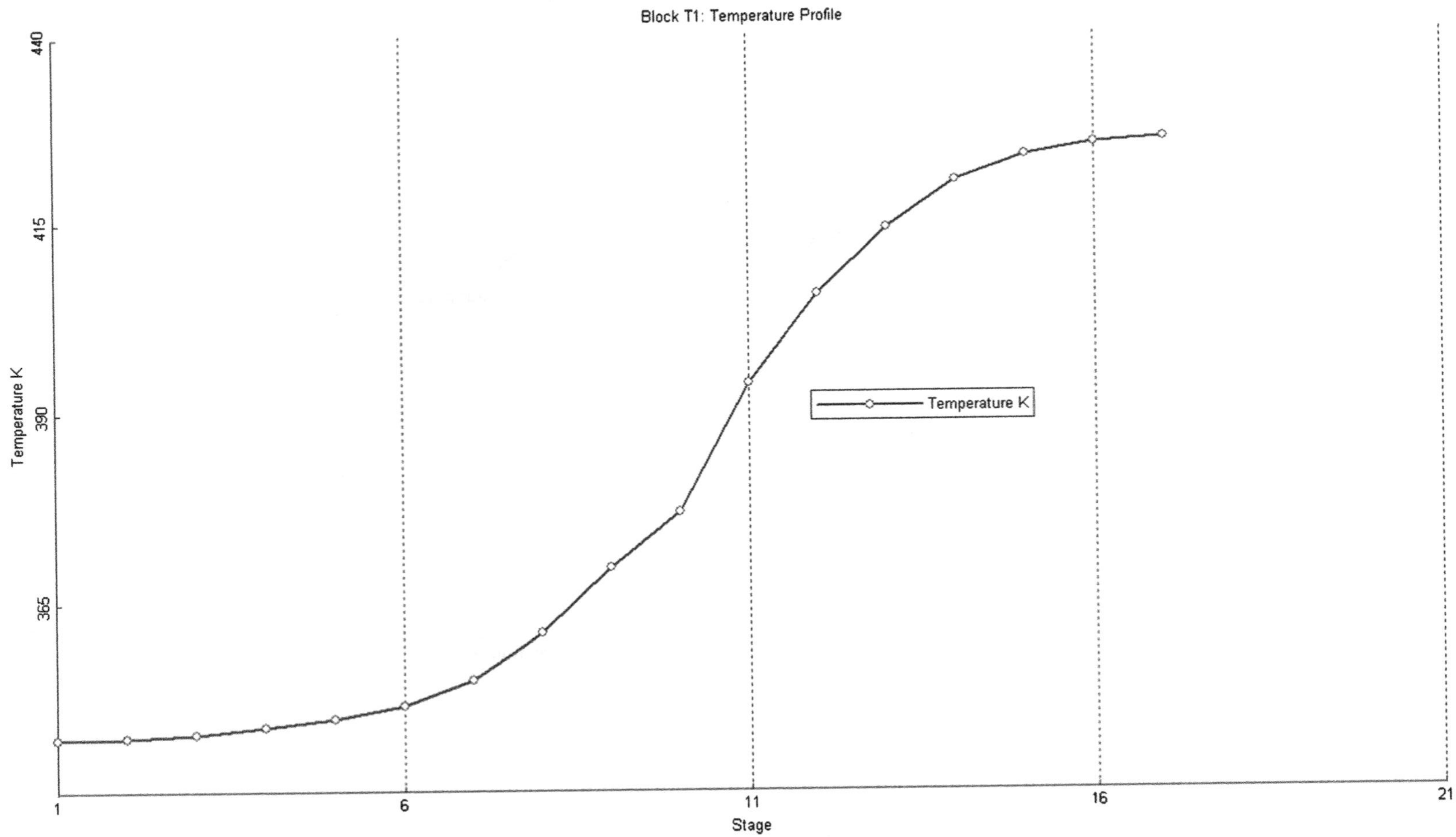

Figure 15.3 Temperature profile in MTBE reactive column.

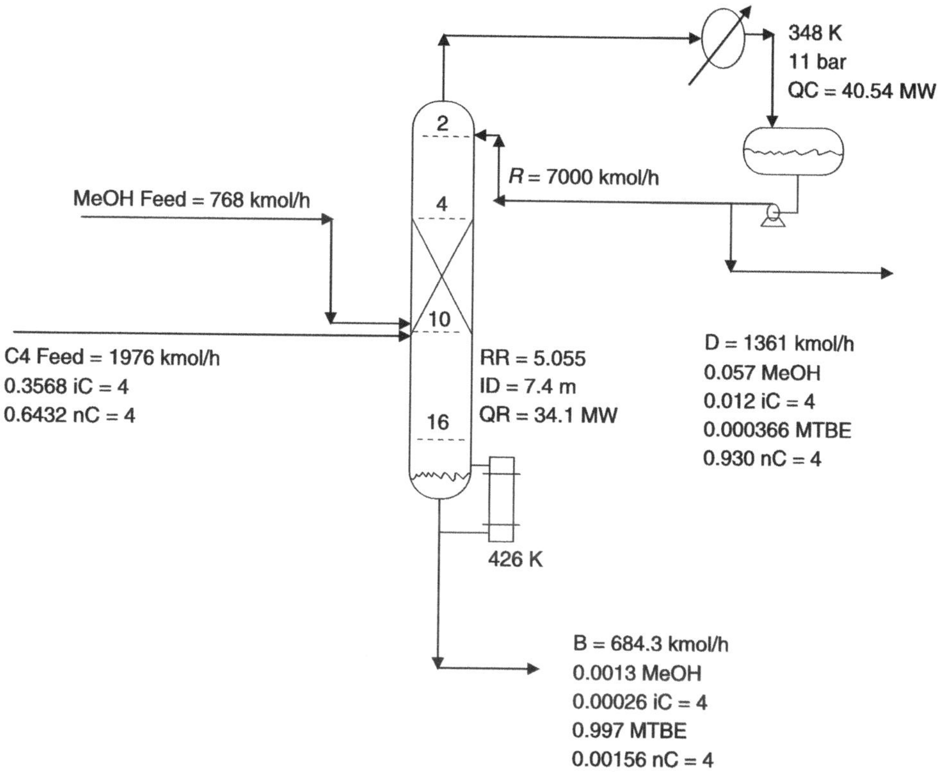

Figure 15.4 The MTBE steady state from converged dynamic simulation.

is used in the dynamic simulation. To hold 800 kg of catalyst, the liquid height is about 0.044 m on the reactive trays. Pressure drops over the four valves in the system and pump heads are set to give adequate rangeability for all of the flows (about 4 atm pressure drop at design flowrates with control valves 50% open). The file is pressure checked and exported to Aspen Dynamics.

Structure. The control structure is shown in Figure 15.5. This structure is similar to that used in the ideal ternary system with inerts, which was studied in Chapter 12. It consists of the following loops:

1. The C4 feed is flow controlled.
2. The methanol feed is ratioed to the C4 feed with the ratio set by a composition controller that controls the methanol composition on stage 10 inside the column at 18 mol% methanol. Note that the methanol flow controller is on cascade with its remote setpoint coming from the output signal of the multiplier *ratiomc4*. The two inputs to the multiplier are the flowrate of the C4 stream and the controller output signal from composition controller CC. At steady state this ratio is $768/1976 = 0.389$, so the output signal from the composition controller has this value, as shown on the faceplates given in Figure 15.6.

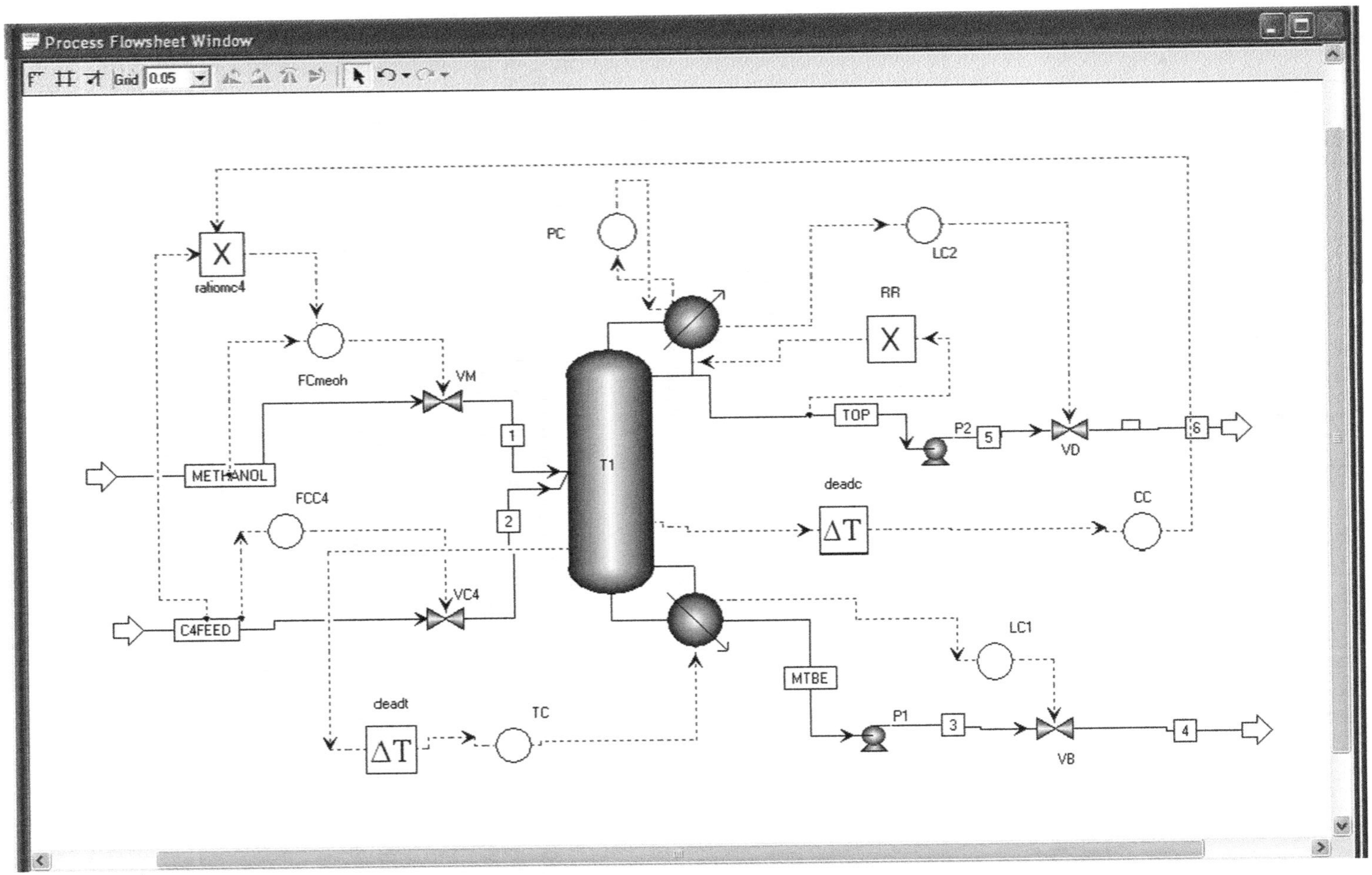

Figure 15.5 Aspen Dynamics control structure with flow control of C4.

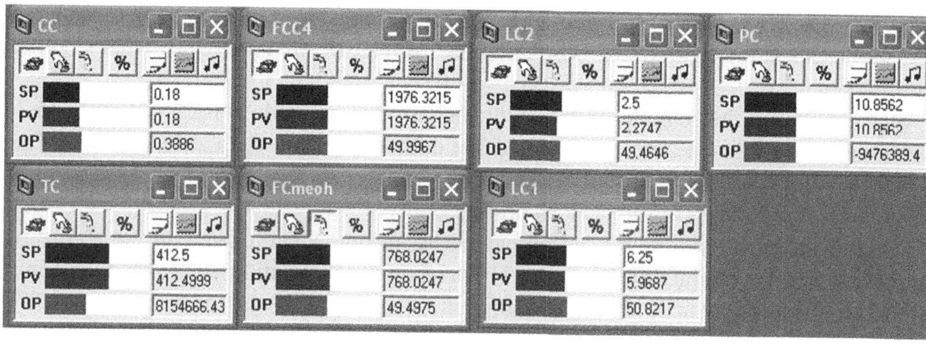

Figure 15.6 Controller faceplates MTBE with flow control of C4.

3. The temperature of stage 13 is controlled at 412.5 K by manipulating the reboiler heat input whose steady-state value is 8.15×10^6 cal/s.
4. The reflux-drum level is controlled by manipulating the distillate flow.
5. The base level is controlled by manipulating the bottoms flow.
6. The reflux ratio is controlled by measuring the mass flowrate of the distillate, sending this signal through a multiplier RR (with a second input of the desired reflux ratio of 5.055) and setting the mass flowrate of the reflux.
7. Pressure is controlled by manipulating the condenser heat removal.

Controller Tuning. The composition controller and the temperature controller are tuned by running relay-feedback tests and using Tyreus–Luyben tuning. Table 15.1 provides parameter values. The temperature controller was tuned first, with the composition controller on manual. Then the composition controller was tuned with the temperature still on automatic (sequential tuning). The level controllers are proportional only with gains of 2.

TABLE 15.1 MTBE Column Tuning Parameters

	Composition Controller with C4 Feedflow Controller	Temp. Control	Composition Controller with Methanol Feedflow Controller
Location	Stage 10	Stage 13	Stage 10
Setpoint	18 (mol% methanol)	412.5 (K)	6.98 (mol% isobutene)
Transmitter span	50 (mol%)	100 (K)	20 (mol%)
Output span	1	20×10^6 (cal/s)	5
Dead time (min)	3	2	3
Ultimate			
Gain	1.44	1.14	2.9
Period (min)	11.4	6.0	7.2
Gain	0.45	0.389	0.91
Integral time (min)	25	13.2	15.8
Output signal at steady state	0.3886	8.15×10^6 (cal/s)	3.576

Results. The effectiveness of the control system is demonstrated by subjecting the column to disturbances in throughput (the flowrate of the C4 feedstream) and in the composition of the C4 stream.

Step changes in the setpoint of the C4 flow controller are made at 0.2 h in Figure 15.7. The solid lines are for a 20% increase and the dashed lines are for a 20% decrease. The system handles these large disturbances very well. The stage 13 temperature controller holds the bottoms composition near its desired purity.

The change in C4 feedflow is accompanied by an immediate change in the methanol fresh feed because of the ratio element. The methanol composition controller then trims up the ratio to bring exactly the correct amount of methanol fresh feed to hold 18 mol% methanol on stage 10. The concentration of methanol in the distillate increases somewhat when the throughput is decreased and decreases when the throughput is increased. The concentration of the reactant isobutene in the distillate is affected in the opposite way. The system reaches a new steady state in about 2 h for the increase in throughput, but it takes about 4 h for variables to steady out for the decrease in throughput.

The disturbances in Figure 15.8 are changes in the composition of the C4 fresh feed. The solid lines represent an increase in the isobutene concentration from 35.68 to 42.82 mol%,

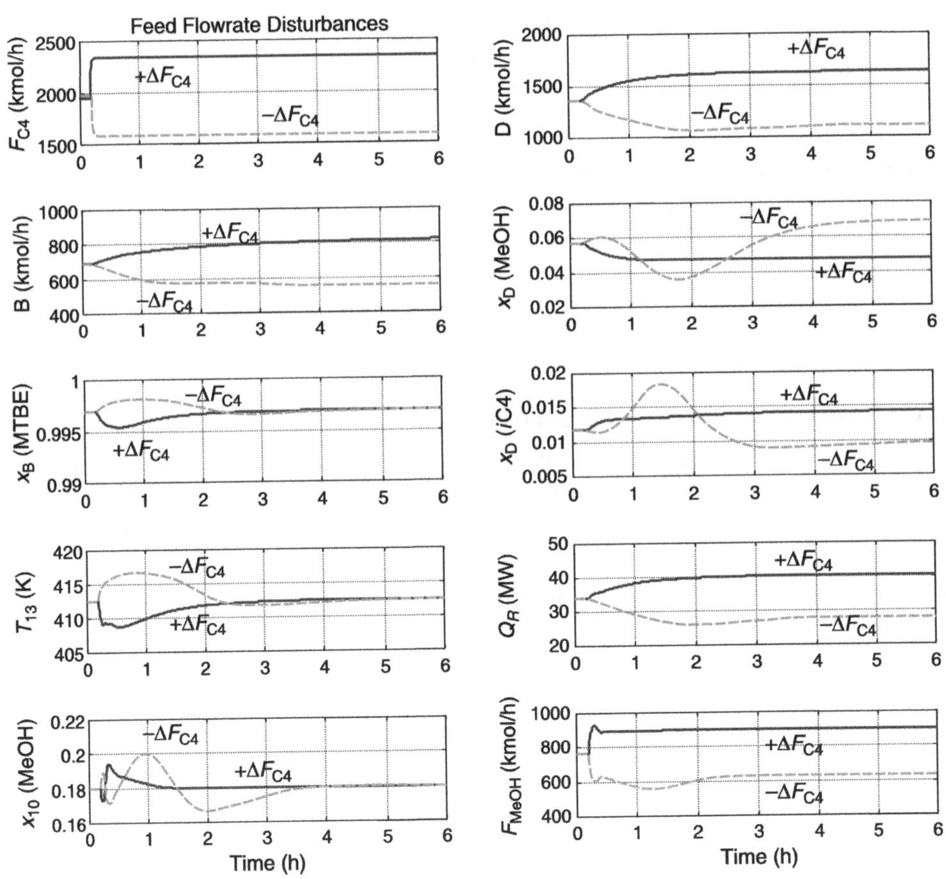

Figure 15.7 Flowrate changes for C4 with flow control of C4.

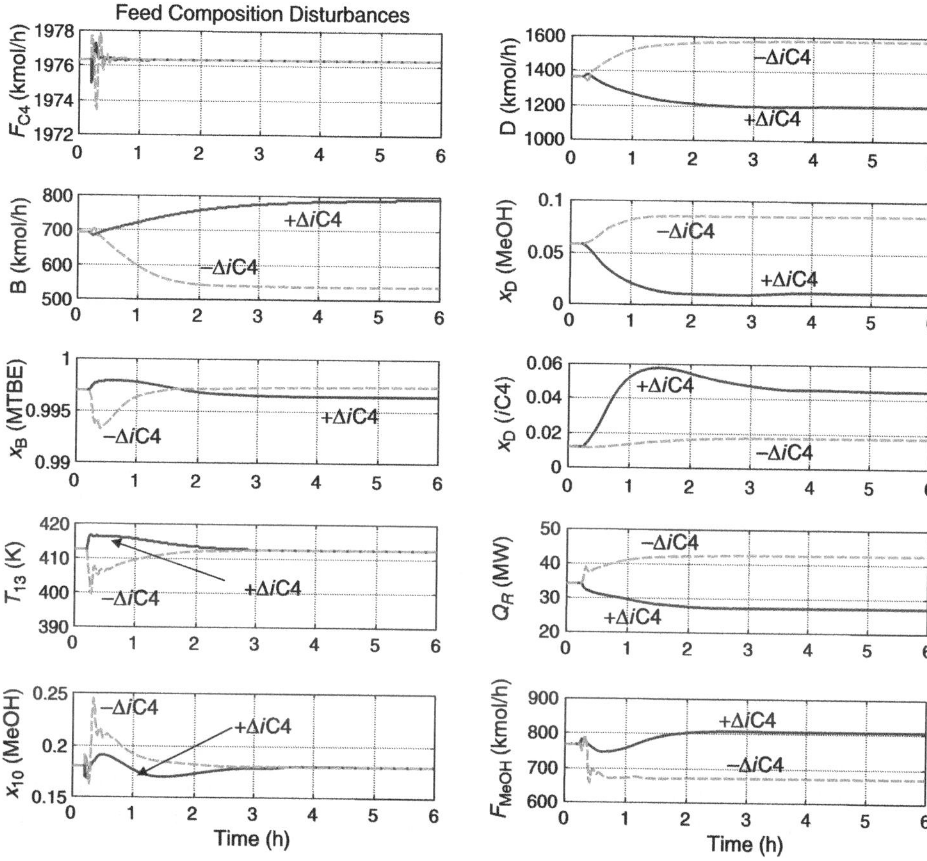

Figure 15.8 Composition changes for C4 with flow control of C4.

with a corresponding decrease in the *n*-butene concentration. The dashed lines represent a decrease in the isobutene concentration from 35.68 to 28.54 mol%, with a corresponding increase in the *n*-butene concentration. The control structure handles these large disturbances fairly well. Stable base-level regulatory control is achieved.

The MTBE purity of the bottoms stays quite close to its desired value. However, the methanol and isobutene concentrations in the distillate change very significantly. When more reactant isobutene is fed there is less inert *n*-butene, so the distillate flowrate is lower. More methanol is required in the reaction, but holding a fixed 18 mol% methanol composition on stage 10 does not increase the methanol fresh feed enough. Thus, there is less methanol in the distillate. In addition, there is more isobutene reactant lost in the distillate. Note that the bottoms product flowrate increases but not as much as it would if the isobutene conversion did not drop.

When less isobutene is fed, the effects are the reverse. The distillate flowrate increases, and the bottoms flowrate decreases. Even though less methanol is fed, the distillate methanol composition increases. There is also a slight increase in the isobutene composition of the distillate.

These results indicate that the control structure handles rate changes very well, but composition changes are less well handled in terms of maintaining conversion. The design conversion is 97.6% (705 kmol/h of isobutene fed in the C4 stream and 688 kmol/h of MTBE in the bottoms). For the increase in isobutene feed composition, the conversion decreases to 93.2% (846 kmol/h of isobutene fed in the C4 stream and 789 kmol/h of MTBE in the bottoms). For the decrease in isobutene feed composition, the conversion decreases to 95.0% (563 kmol/h of isobutene fed in the C4 stream and 535 kmol/h of MTBE in the bottoms).

15.1.3 Control Structure with Methanol Feedflow Controlled

In the alternative control structure, the throughput is set by the flowrate of the methanol fresh feed instead of the C4 fresh feed. Now the methanol feed is flow controlled, and the C4 feed is ratioed to it, as shown in Figure 15.9. The ratio is reset by a composition controller that maintains the isobutene composition at 6.98 mol% on stage 10. As shown in Figure 15.2, this is the location of the peak in the isobutene composition profile.

The rest of the control loops are identical to those used in the previous structure. Note that flow controller FCC4 is on cascade with its setpoint coming from the ratio element whose other input is the output signal of composition controller CC. Note also that this output signal is the C4-to-methanol ratio, which is 3.58 at base case conditions. Controller faceplates are given in Figure 15.10. The new tuning parameters for the composition controller are provided in Table 15.1.

Disturbances in both throughput and feed composition were used to evaluate the effectiveness of this alternative control structure. Figure 15.11 shows the responses to 20% step changes in the setpoint of the methanol feed flow controller. Solid lines are increases, and dashed lines are decreases. The increase is handled quite well. It is surprising that the decrease is not handled well. There is a significant transient drop in MTBE purity in the bottoms product (middle left graph, Fig. 15.11), which lasts for several hours. In addition and of more importance, there is a very large increase in the isobutene lost in the distillate.

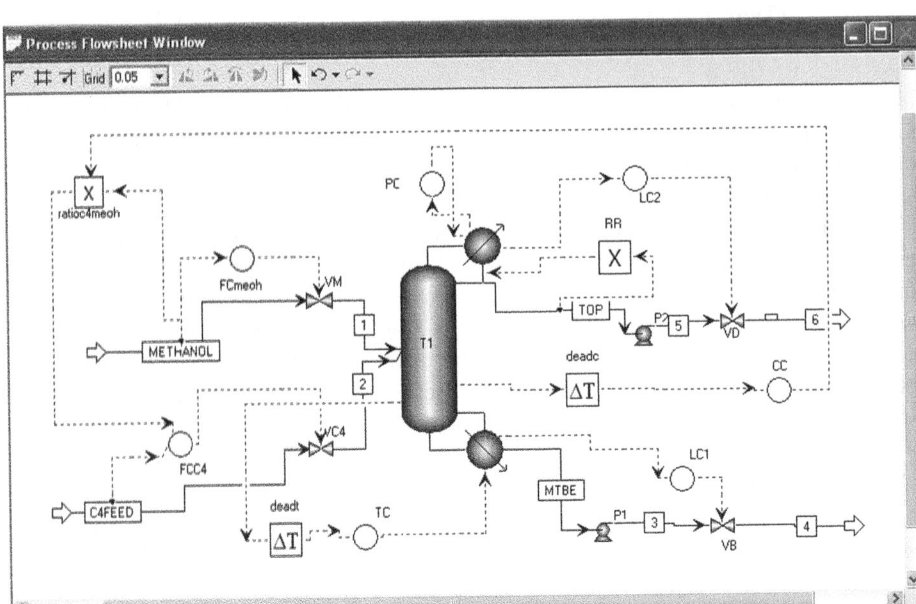

Figure 15.9 Control structure with methanol flow controlled.

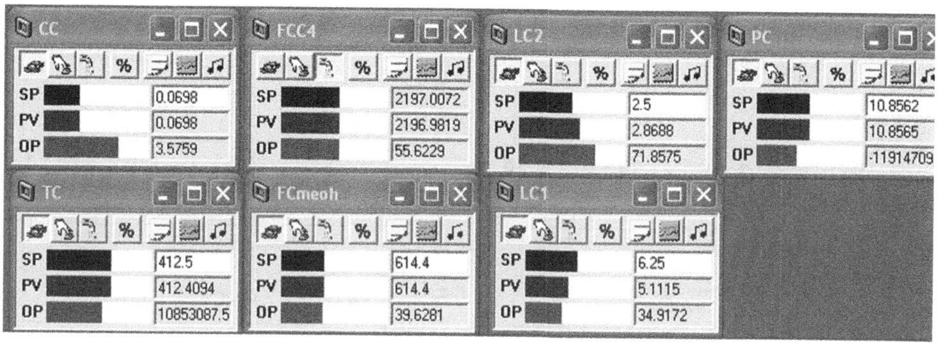

Figure 15.10 Faceplates with methanol flow controlled.

Even though the methanol feed is decreased, the C4 feed actually ends up at a new steady state that is at a higher flowrate. The distillate isobutene composition climbs to about 14 mol%.

Figure 15.12 gives the responses for changes in the composition of the C4 feedstream. An increase in isobutene concentration produces a large transient drop in MTBE purity.

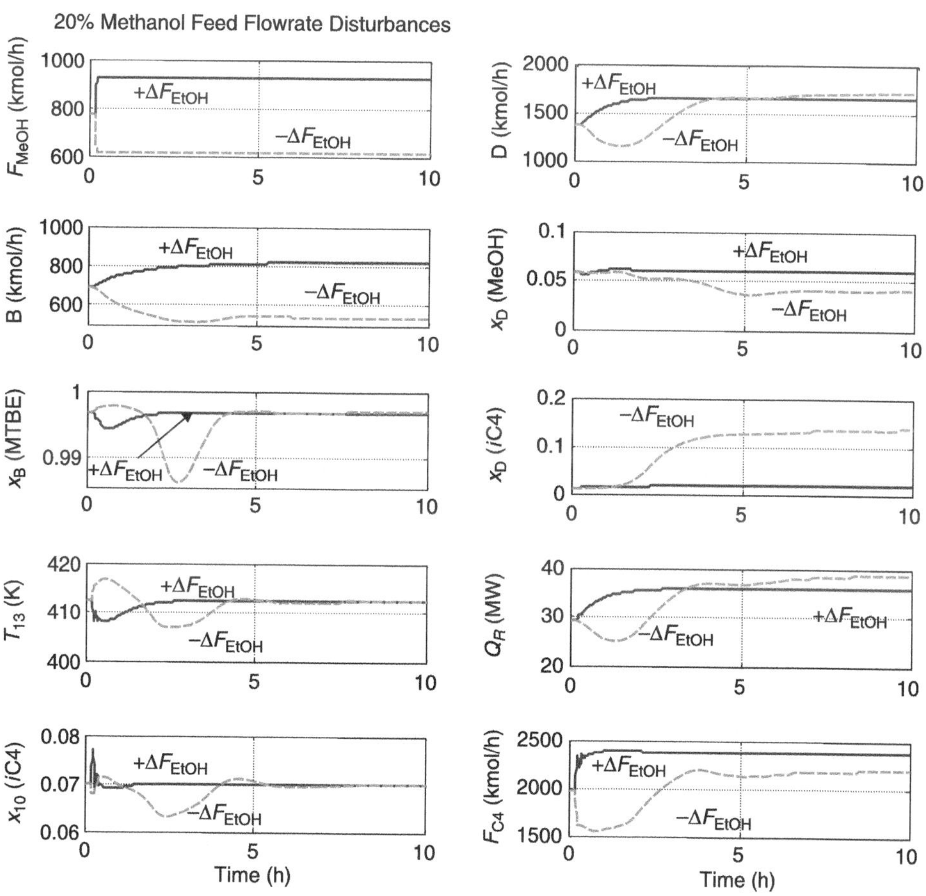

Figure 15.11 Methanol feedflow flowrate disturbances.

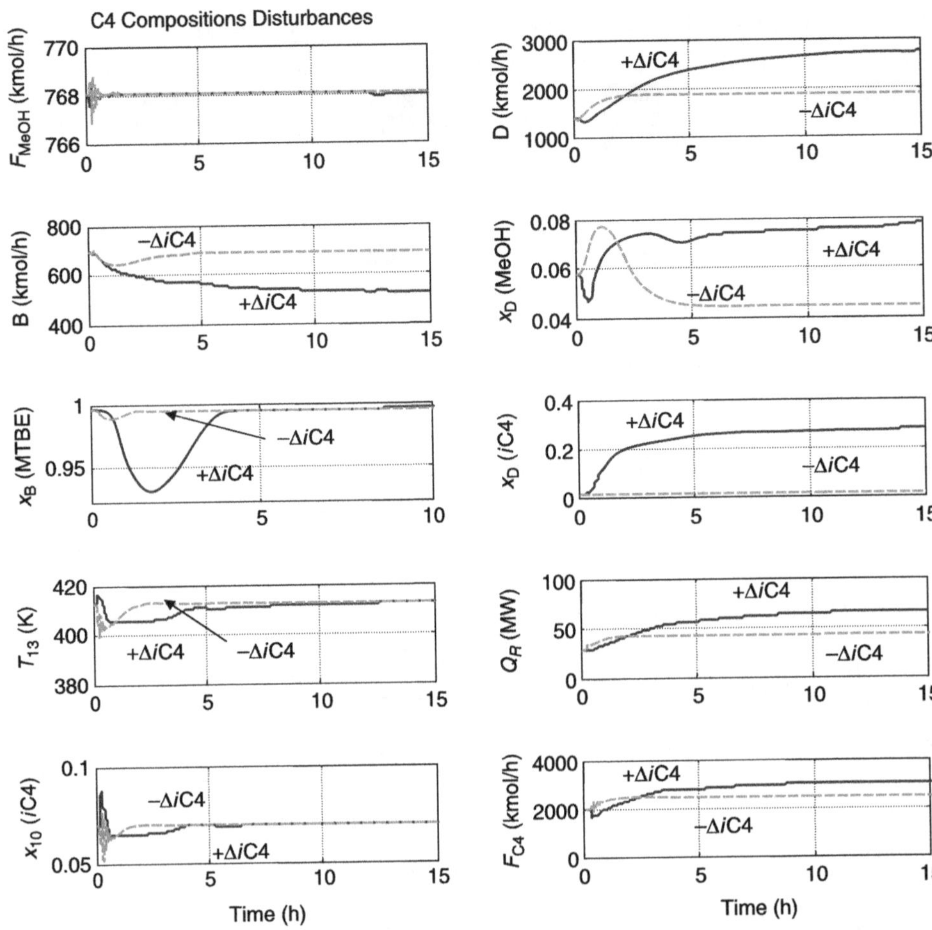

Figure 15.12 Disturbances of C4 composition with methanol feedflow controlled.

A large increase in the isobutene losses in the distillate stream also occurs. The distillate isobutene composition rises to almost 30 mol%. We would expect more MTBE to be produced and a larger bottoms flowrate when the isobutene in the feed is increased. As the second graph on the left in Figure 15.12 shows, the bottoms flowrate actually decreases. There is an increase in the distillate flowrate that is attributable to the reduction in the bottoms and to a slight increase in the flowrate of the C4 feed.

Clearly, this control structure does not provide effective control for the MTBE system. The fundamental reason for this poor performance is the very low concentration of isobutene in the column. These results indicate that it is much better to flow control the C4 feed and manipulate the methanol feed.

15.2 ETBE CONTROL

The two fresh feedstreams to the column are ethanol and a C4 stream containing reactive isobutene and nonreactive n-butene. The bottoms is mostly ETBE. The distillate is

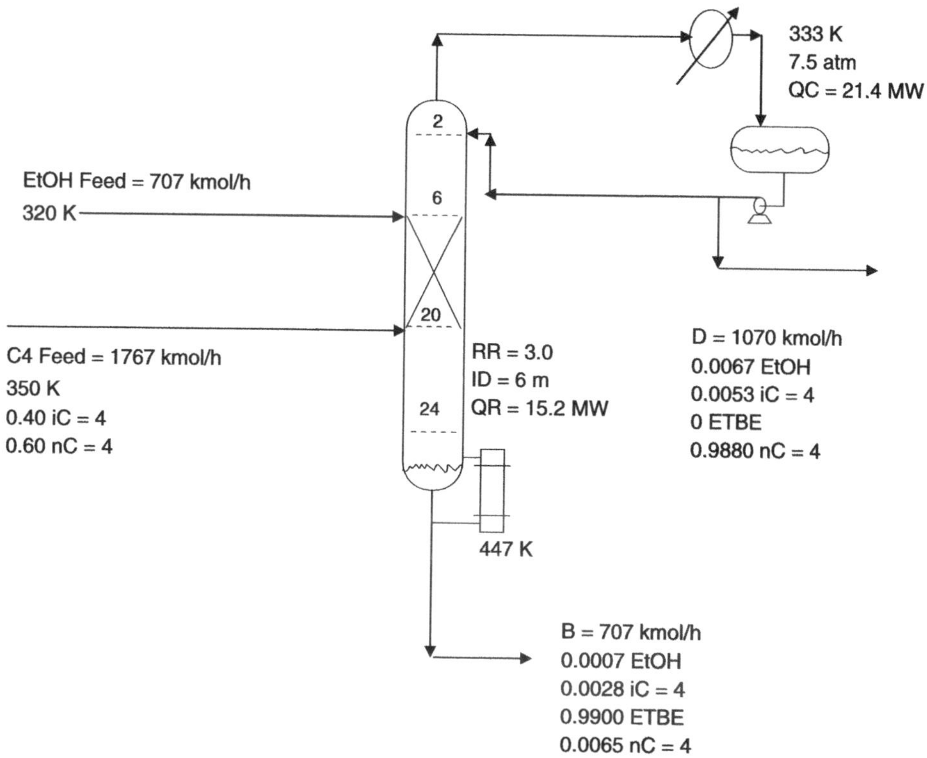

Figure 15.13 Steady-state conditions for ETBE.

mostly *n*-butene. Figure 15.13 shows the flowsheet with steady-state conditions. Figures 15.14 and 15.15 give composition and temperature profiles, respectively.

Two alternative control structures are evaluated. They differ only in which fresh feedstream is selected to set the production rate. In the first, the C4 stream is flow controlled. In the second, the ethanol feedstream is flow controlled.

15.2.1 Control Structure with Flow Control of C4 Feed

Figure 15.16 shows the control structure in which the FFC4 stream is flow controlled. Then the ethanol feed is ratioed to it, with the ratio being reset by a composition controller that holds the ethanol composition on stage 17 at 11.58 mol%. This stage is selected by looking at the composition profiles shown in Figure 15.14. The ethanol composition is changing rapidly at this location. Above stage 17 in the upper part of the reactive zone, the ethanol composition profile is flat at about 25 mol%. The isobutene compositions are low in the upper part of the reactive zone, but they begin to increase at around stage 17.

The temperature controller holds stage 22 at 383 K by manipulating the reboiler heat input. Note that the temperature profile in the ETBE system (Fig. 15.15) is much sharper than in the MTBE system (Fig. 15.3). Therefore, a larger temperature transmitter span was used.

The rest of the control structure is the same as that used for the MTBE system, as described in Section 15.1.2. The composition controller and the temperature

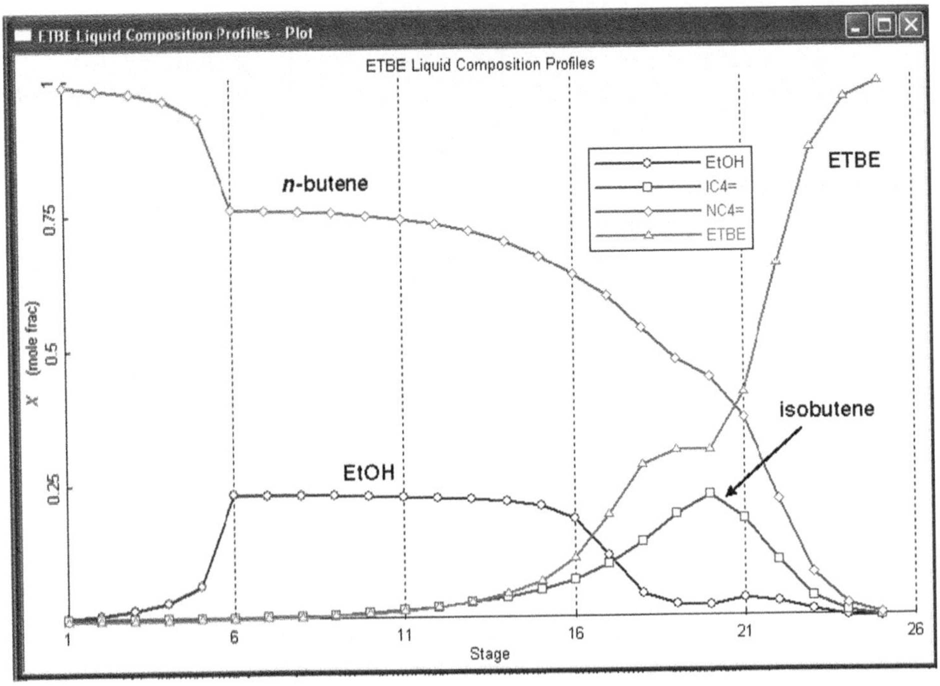

Figure 15.14 Composition profiles for ETBE.

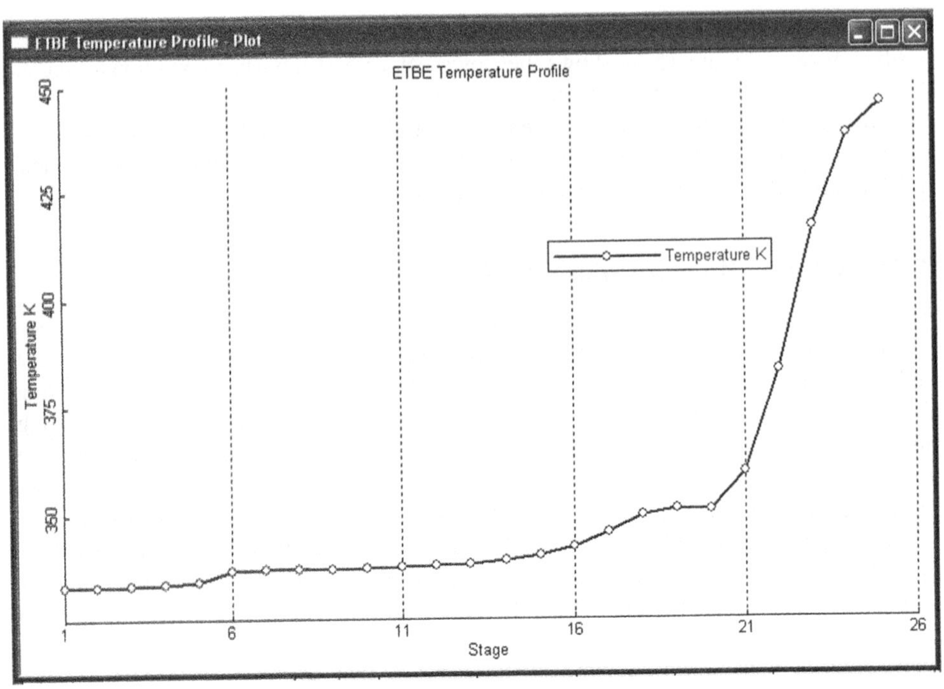

Figure 15.15 Temperature profile for ETBE.

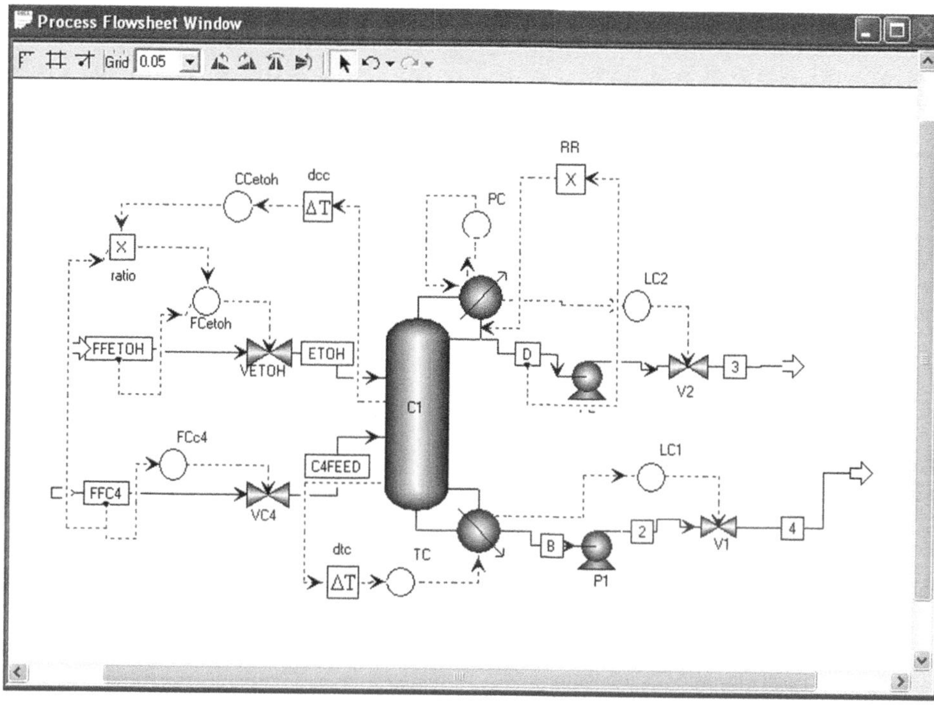

Figure 15.16 Control structure for ETBE with flow control of C4 stream.

controller are tuned by running relay-feedback tests and using Tyreus–Luyben tuning. Table 15.2 lists the parameter values. The temperature controller was tuned first, with the composition controller on manual. Then the composition controller was tuned with the temperature still on automatic (sequential tuning). The level controllers are proportional only with gains of 2. Faceplates are shown in Figure 15.17. Note that in this

TABLE 15.2 ETBE Column Tuning Parameters

	Composition Controller with Flow Control of C4 Feed	Temp. Control	Composition Controller with Flow Control of Ethanol Feed
Location	Stage 17	Stage 22	Stage 20
Setpoint	11.58 (mol% ethanol)	383.4 (K)	22.5 (mol% isobutene)
Transmitter span	25 (mol%)	200 (K)	50 (mol%)
Output span	1 (fraction)	10×10^6 (cal/s)	5
Dead time (min)	3	1	3
Ultimate			
Gain	0.27	1.56	4.26
Period (min)	16.2	4.6	11.4
Gain	0.0.84	0.49	1.33
Integral time (min)	36	10.6	25
Output signal at steady state	0.40	8.15×10^6 (cal/s)	2.5

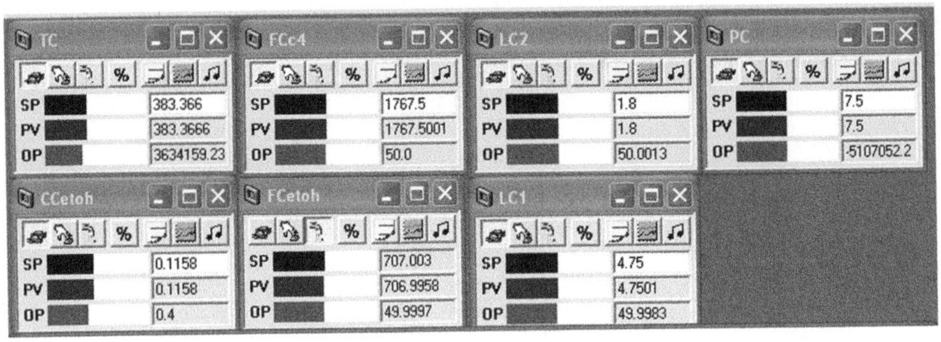

Figure 15.17 Faceplates with flow control of C4 stream.

structure the FCetoh flow controller is on cascade. Note also that the output signal of composition controller CCetoh is the ratio of the flowrates of the ethanol feed and the C4 feed ($707/1767 = 0.4$).

Results for disturbances in throughput are provided in Figure 15.18. At 0.2 h, step changes in the setpoint of the C4 feedflow controller are made. The solid lines are for

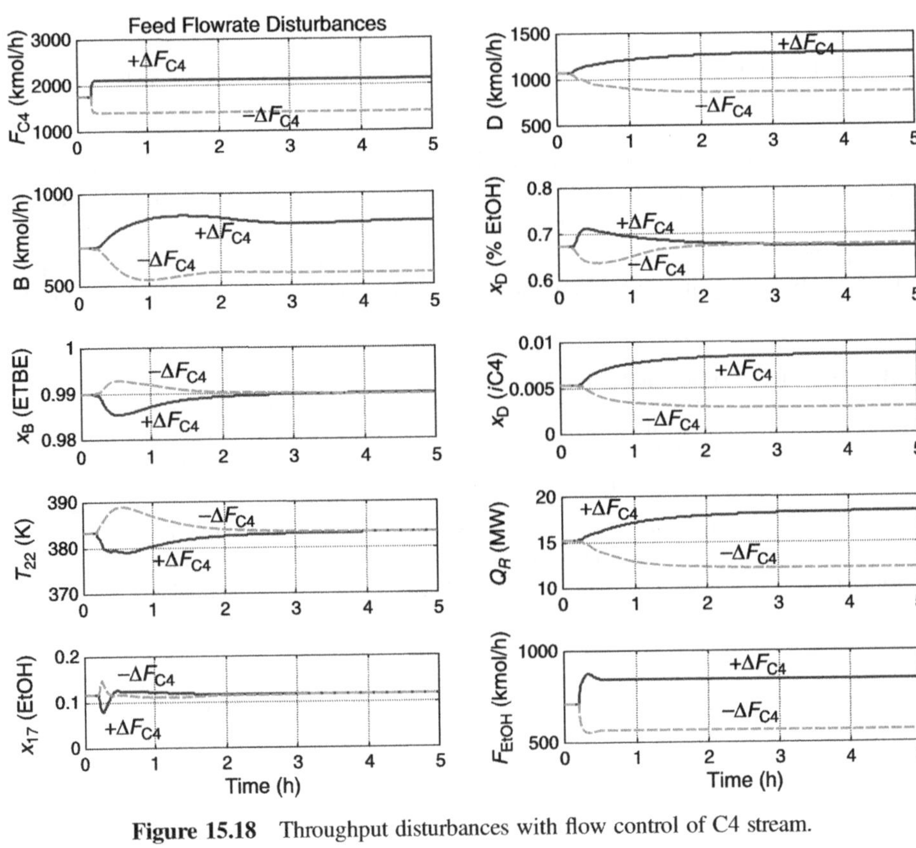

Figure 15.18 Throughput disturbances with flow control of C4 stream.

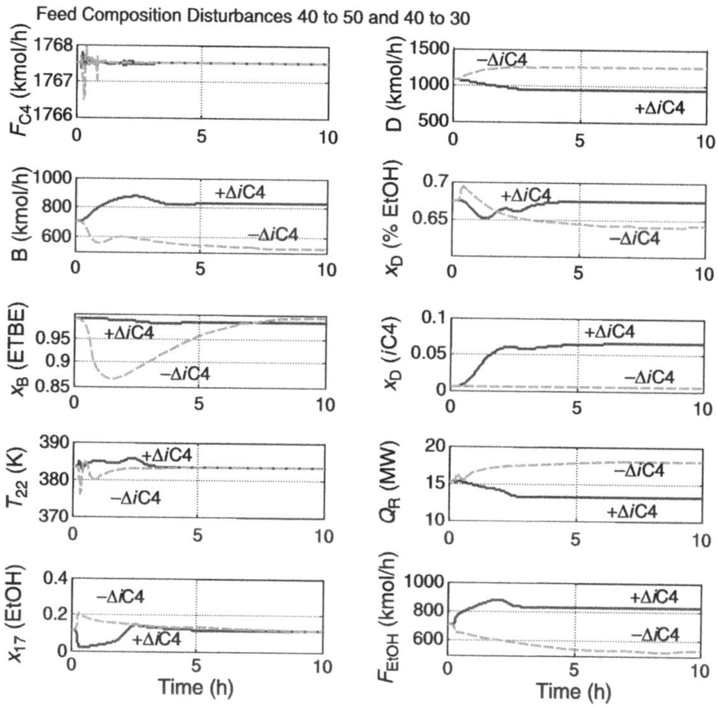

Figure 15.19 Composition disturbances with flow control of C4 stream.

+20% steps; the dashed lines are for −20% steps. The control structure handles these disturbances quite well with ETBE purity held very close to its desired value. The amount of isobutene lost in the distillate increases somewhat at higher throughputs.

In Figure 15.19 the disturbances are very large increases and decreases in the concentration of isobutene in the C4 fresh feedstream. The solid lines correspond to a change from 40 to 50 mol% isobutene, with a corresponding decrease in the *n*-butene concentration. The dashed lines correspond to a change from 40 to 30 mol% isobutene, with a corresponding increase in the *n*-butene concentration. The ETBE purity in the bottoms is held close to the desired value for the increase in isobutene in the feed, but it undergoes a large transient drop for the decrease in isobutene feed composition. More ethanol is fed and more ETBE is produced for the increase in isobutene feed composition. However, the loss of isobutene in the distillate increases, which is demonstrated in the middle right graph in Figure 15.19.

These are very large feed composition disturbances. Figure 15.20 gives results for smaller decreases. The dotted lines are results for the 10 mol% decrease, the dashed lines are results for the 7 mol% decrease, and the solid lines are results for the 5 mol% decrease. The more realistic 5 mol% disturbance is handled quite well with the bottoms ETBE purity staying close to its desired value. The decrease in isobutene and the corresponding increase in the inert *n*-butene fed into the system result in a smaller bottoms product and a larger distillate product. The composition of the distillate changes only slightly.

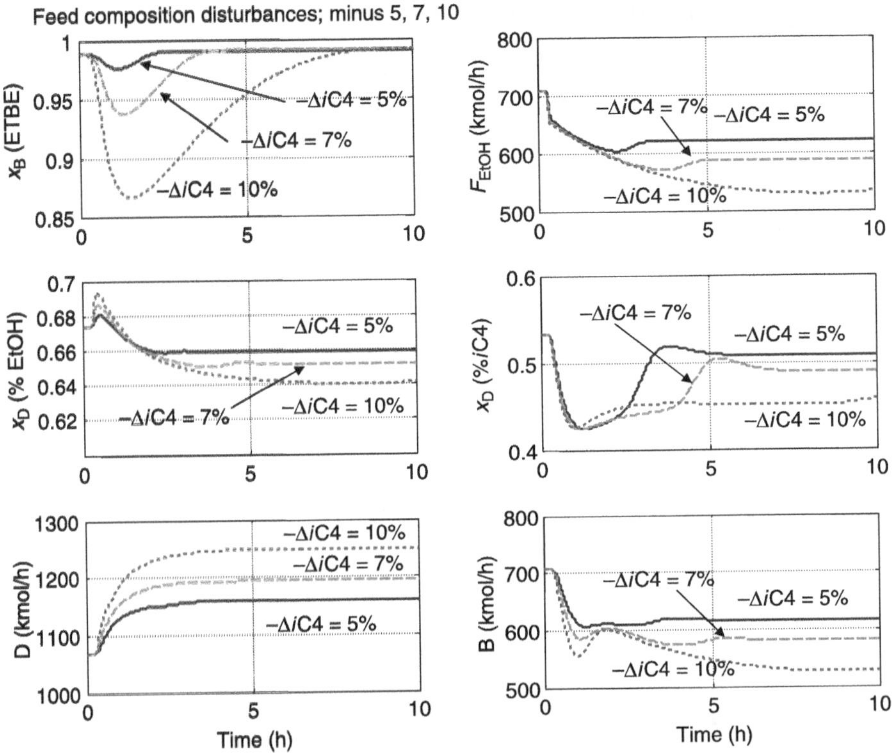

Figure 15.20 Decreases in iC_4 with flow control of C4 stream.

15.2.2 Control Structure with Flow Control of Ethanol Feed

The second control structure switches from flow controlling the C4 feedstream to flow controlling the ethanol fresh feed. Thus, the production rate handle is ethanol, not C4. In addition, the composition of isobutene is controlled on a tray inside the column instead of controlling the composition of ethanol.

Figure 15.21 shows this control structure. The FFetoh stream is flow controlled. Then the C4 feed is ratioed to it, with the ratio being reset by a composition controller that holds the isobutene composition on stage 20 at 22.5 mol%. This stage is selected by looking at the composition profiles shown in Figure 15.14. The peak in the isobutene composition occurs at stage 20.

The temperature controller is the same as in the previous control structure. The temperature on stage 22 is controlled at 383 K by manipulating the reboiler heat input. It required no retuning. However, the composition controller was retuned for the new structure. Sequential tuning was used (the temperature controller was on automatic when tuning the composition controller). The composition controller was tuned by running relay-feedback tests and using Tyreus–Luyben tuning. Table 15.2 gives the parameter values. Faceplates are shown in Figure 15.22. Note that in this structure the FCc4 flow controller is on cascade. In addition, note that the output signal of composition controller CCic4 is the ratio of the flowrates of the C4 feed and the ethanol feed ($1767/707 = 2.5$).

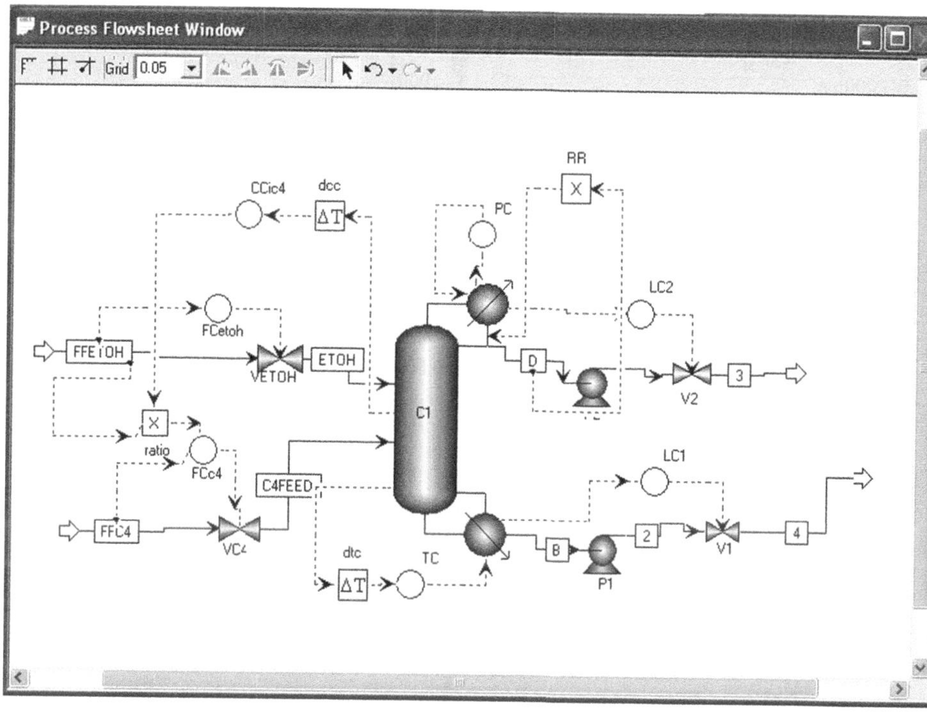

Figure 15.21 Control structure for ETBE with flow control of ethanol stream.

The results for disturbances in throughput are provided in Figure 15.23. At 0.2 h, step changes in the setpoint of the C4 feedflow controller are made. The solid lines are for +20% steps; the dashed lines are for −20% steps. The control structure handles these disturbances quite well with ETBE purity held very close to its desired value. A comparison of Figure 15.23 with Figure 15.18 shows that the two control structures handle throughput disturbances almost identically.

In Figure 15.24 the disturbances are large 10% increases and decreases in the concentration of isobutene in the C4 fresh feedstream. The solid lines correspond to a change from 40 to 50 mol% isobutene, with a corresponding decrease in the n-butene

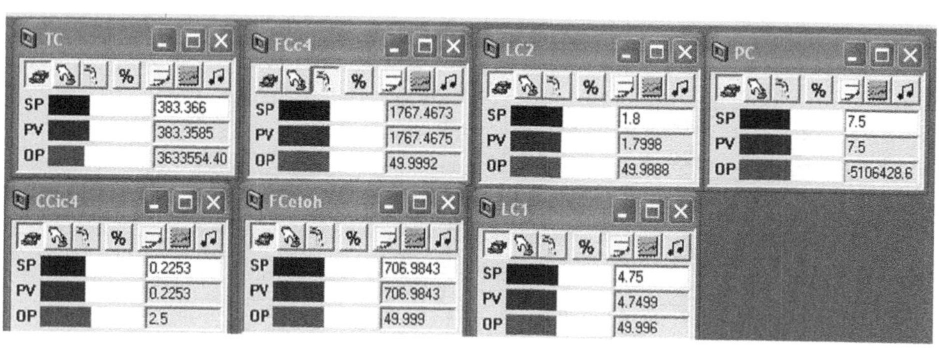

Figure 15.22 Faceplates for ETBE with flow control of ethanol stream.

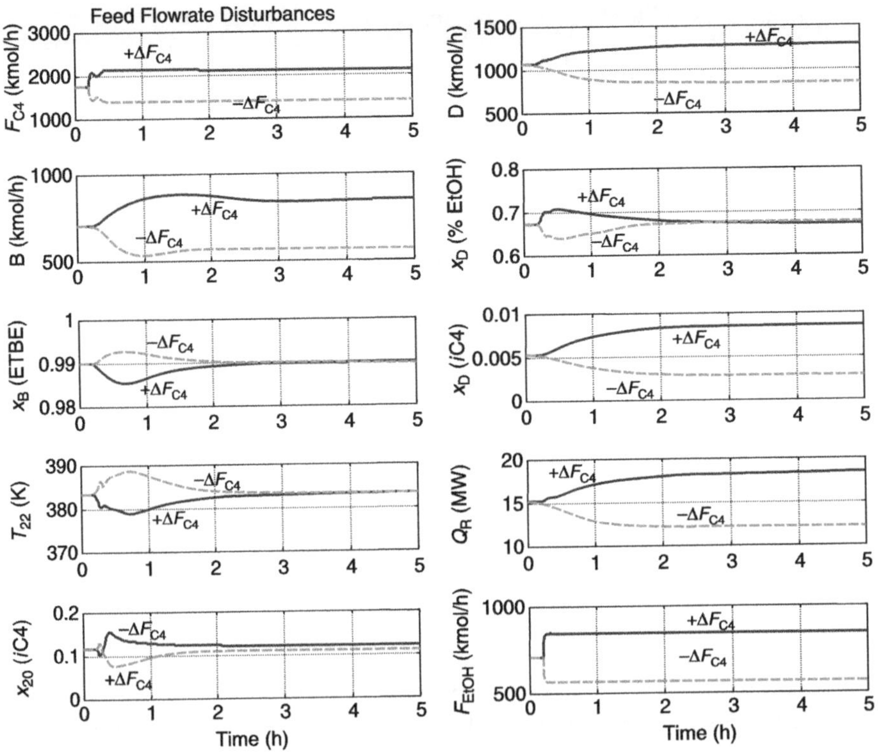

Figure 15.23 Ethanol flowrate disturbances.

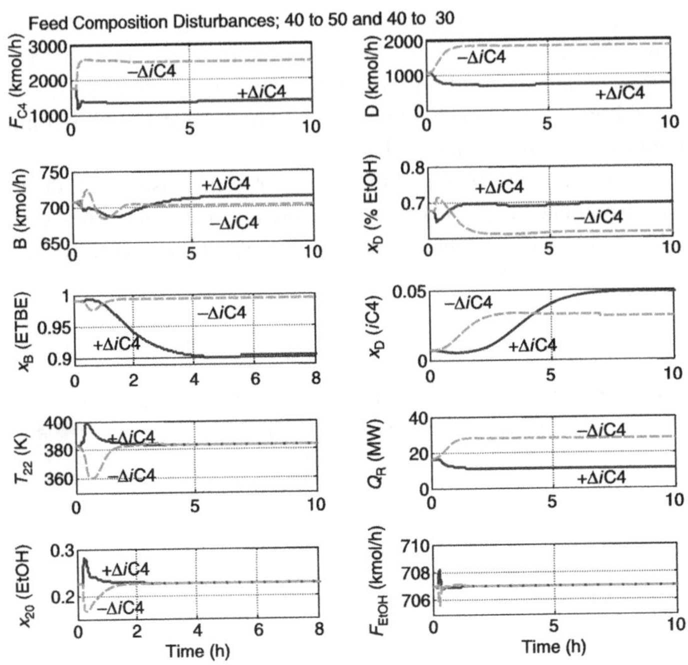

Figure 15.24 Isobutene feed composition disturbances.

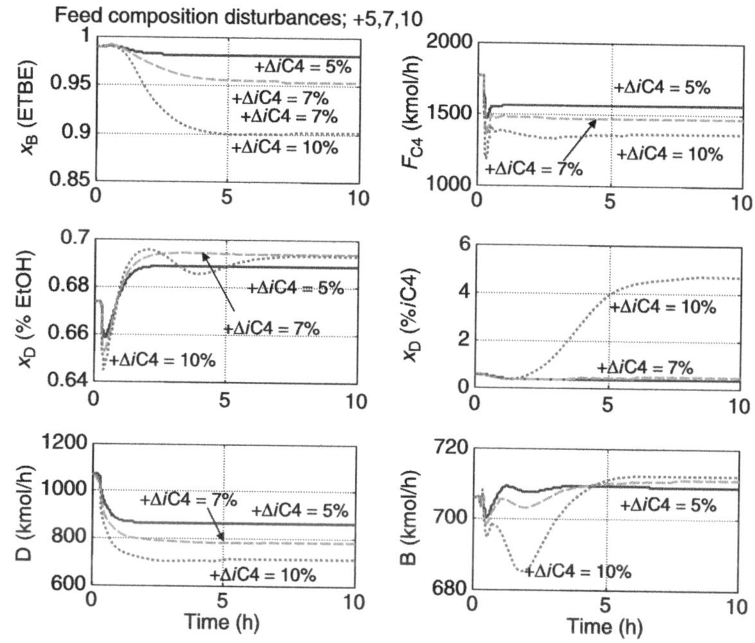

Figure 15.25 Isobutene feed composition disturbances at 5%, 7%, and 10%.

concentration. The dashed lines correspond to a change from 40 to 30 mol% isobutene, with a corresponding increase in the *n*-butene concentration. The decrease is fairly well handled, except for an increase in the loss of isobutene in the distillate. However, the increase in isobutene feed concentration results in a very large drop in ETBE purity, as shown in the middle left graph in Figure 15.24.

The responses are much improved for smaller disturbances, which is demonstrated in Figure 15.25.

PART VI

HYDRID AND NONCONVENTIONAL SYSTEMS

CHAPTER 16

DESIGN AND CONTROL OF COLUMN/ SIDE REACTOR SYSTEMS

One of the limitations for making reactive distillation economically attractive is that the temperature range suitable for reasonable chemical reaction kinetics must match the temperature range suitable for vapor–liquid equilibrium because both separation and reaction occur in a single vessel. We alluded to this limitation in several previous chapters. In particular, we examined it quantitatively in Chapter 3 where we demonstrated how the economics of reactive distillation are adversely affected when a mismatch occurs.

This temperature mismatch problem can sometimes be overcome by considering an alternative process flowsheet that features a distillation column linked with several external side reactors. The column operates at a low pressure and temperatures favorable for separation. Liquid side streams from trap-out trays at intermediate locations in the column are pumped to external reactors operating at higher pressure and temperatures that are favorable for chemical kinetics.

The results demonstrate that this configuration has better steady-state economics than conventional reactive distillation in cases where there is a temperature mismatch between reaction and separation. Effective control structures are developed and tested for this system that handle a variety of large disturbances.

Two different chemical systems are explored in this chapter. The first is the ideal two-reactant, two-product quaternary system. The second system is the real chemical system ethyl acetate.

16.1 INTRODUCTION

The applicability of reactive distillation is highly dependent on the chemical system at hand, and there are still limited applications of this process. A fundamental difference between

Reactive Distillation Design and Control. By William L. Luyben and Cheng-Ching Yu

reactive distillation and a conventional multiunit flowsheet is the selection of operating temperatures. In a conventional multiunit process, optimal temperatures for the reactor and the columns can be set independently. Conversely, this is not the case in reactive distillation because both reaction and separation occur in the same vessel operating at a single pressure. Thus, if reactive distillation is to be attractive, the temperatures that are good for reaction must match the temperatures that are good for vapor–liquid separation.

One approach to overcoming the temperature mismatch is to consider a flowsheet that features a distillation column with external side reactors. The column is operated at a pressure that is favorable while still permitting separation, which is typically as low as possible for the use of cooling water in the condenser. There is no catalyst on the trays in the column, so there is no chemical reaction occurring in the column. Several total liquid trap-out trays are located in the middle section of the column. Liquid is collected and pumped up to a sufficiently high pressure so that it remains liquid when heated to the higher temperatures that are favorable for reaction. This configuration permits the temperature for separation and the temperature for reaction to be set independently.

Figure 16.1 shows two versions of the flowsheet. The process on the left uses both heat exchangers and tubular reactors in the external loops. Preheating the feed to the reactor with the hotter reactor effluent reduces the size of the reactor. The process on the right uses only tubular reactors. The reactors must be significantly larger because of the lower inlet temperature. However, reactor vessels are relatively inexpensive compared to heat exchangers. Thus, we will limit the study to the reactor-only process.

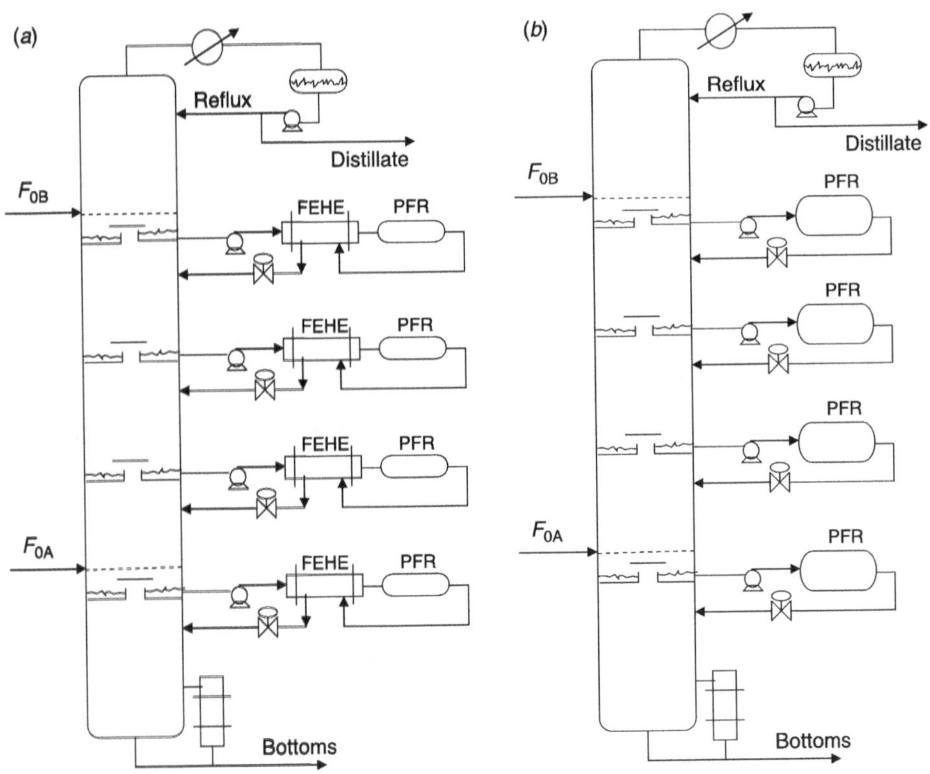

Figure 16.1 Column/side reactor process.

Both of these configurations have better economics than a reactive distillation column when there is a mismatch between favorable reaction temperatures and favorable vapor–liquid equilibrium temperatures.

16.2 DESIGN FOR QUATERNARY IDEAL SYSTEM

The process considered in this section is the ideal quaternary exothermic reversible liquid-phase reaction

$$\mathrm{A} + \mathrm{B} \Longleftrightarrow \mathrm{C} + \mathrm{D}$$

A range of temperature-dependent relative volatilities are considered, which are the same as those studied in Chapter 3. The relative volatilities between all adjacent components are assumed to be 2 at a temperature of 320 K. This temperature corresponds to a typical reflux-drum temperature when using cooling water in the condenser.

$$\alpha_{\mathrm{C}} = 8 \qquad \alpha_{\mathrm{A}} = 4 \qquad \alpha_{\mathrm{B}} = 2 \qquad \alpha_{\mathrm{D}} = 1$$

The relative volatilities are assumed to change with temperature, becoming smaller as the temperature increases. The value of the relative volatility of all adjacent components at a temperature of 390 K (α_{390}) is used as a parameter to vary the temperature dependence. The smaller the value of α_{390}, the more temperature dependent the vapor–liquid system is.

A temperature of 390 K is chosen because it gives reasonable reaction rates and chemical equilibrium constants for the numerical example under consideration. Different cases are studied for a range of α_{390} values. When $\alpha_{390} = 2$, the relative volatilities are independent of the temperature. When $\alpha_{390} = 0.95$, the adjacent component switches volatilities as the temperature approaches 390 K, so the desired separation would be infeasible.

Figure 16.2 provides two examples of different temperature-dependent vapor pressures. In the left graph the vapor pressures are constant for all temperatures. In the right graph the curves get closer as the temperature increases. The Antoine constants for the cases we considered are listed in Table 3.6.

The distillation column operates at its optimum pressure and temperature for separation (320 K in the reflux drum). The column is fed with two pure reactant fresh feedstreams: $F_{0\mathrm{A}}$ and $F_{0\mathrm{B}}$. The column has three zones. There are N_S trays and a partial reboiler in the stripping section and a rectifying section with N_R trays and a total condenser. As shown in Figure 16.1, a number of total liquid trap-out trays are installed at several intermediate trays between these two zones. There are N_M trays in the middle of the column. No reaction occurs anywhere in the column. Reaction only occurs in the external side reactors.

The liquid trap-out trays collect all of the liquid coming down from the upper trays. The vapor from the tray below each trap-out tray flows up through the chimney to the tray above. There is no vapor–liquid contacting on the trap-out trays. Liquid from each trap-out tray is pumped to a sufficiently high pressure so that the material stays liquid at the higher temperatures in the external side reactors. The reaction occurs in adiabatic plug flow reactors. Reactor effluents are fed back to the column on the tray below the trap-out tray from which they were withdrawn. The material partially vaporizes as it flashes into the column from the higher pressure and higher temperature external reactor.

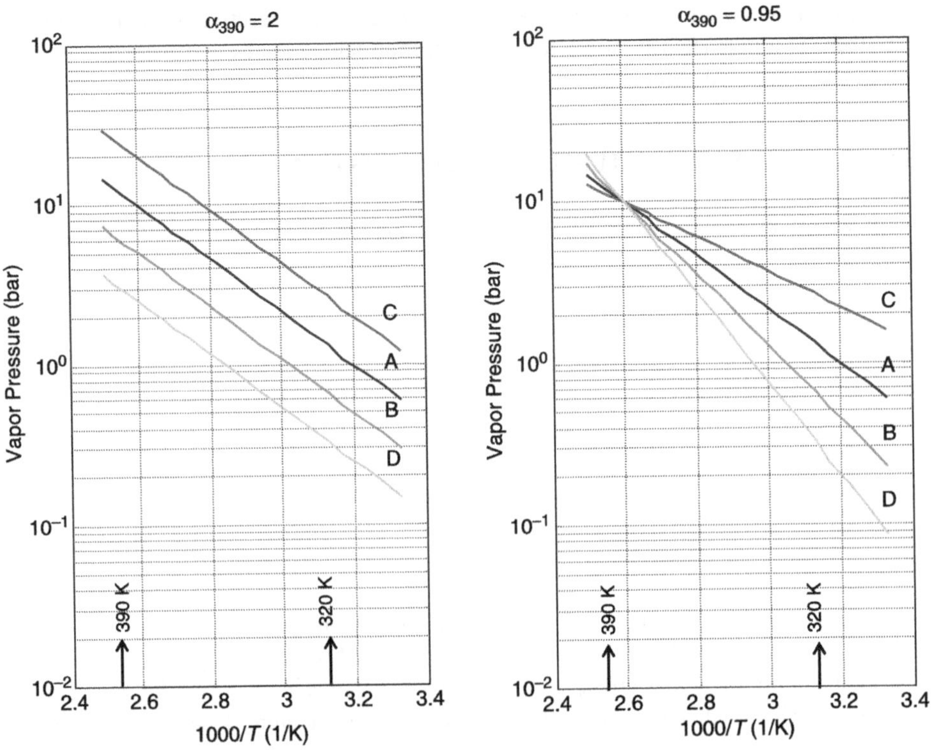

Figure 16.2 Vapor pressures for $\alpha_{390} = 2$ and 0.95 values.

Light product C leaves in the distillate, and heavy product D is removed from the bottoms.

16.2.1 Assumptions and Specifications

The column/side reactors process has a large number of design optimization variables. A number of specifications and assumptions are made to reduce this number to a manageable level. The design objective is to obtain 95% conversion for fixed fresh feed flowrates of 12.6 mol/s of each reactant. The product purities of distillate stream $x_{D,C}$ and bottoms stream $x_{B,D}$ are both 95 mol%. The two fresh feedstreams of reactants A and B are fed into the column at the trays immediately above the lower and upper trap-out trays, respectively.

The reactors are assumed to be adiabatic, and the holdups of each reactor are the same. We assume that the number of trays between each liquid trap-out tray is the same. Other assumptions are theoretical trays, equimolal overflow, saturated liquid feeds and reflux, total condenser, and partial reboiler.

Based on these specifications and simplifying assumptions, there are five optimization variables:

1. Number of external side reactors N_{EXT}
2. Holdup of each reactor V_R
3. Number of trays between each liquid trap-out N_{LT}

4. Number of stripping trays N_S
5. Number of rectifying trays N_R

The kinetic, vapor–liquid equilibrium, and economic parameters are the same as those used in Chapter 3. The steady-state design objective is the minimum TAC.

16.2.2 Reactor and Column Equations

The tray temperatures of the trap-out trays are the inlet temperatures to the external reactors. The ordinary differential equations of the plug flow reactor for steady-state adiabatic operation are integrated from 0 to the total reactor volume V_R, keeping track of how the component compositions and temperature change along the reactor. These equations are the following:

$$\begin{aligned} \frac{dx_\mathrm{A}}{dV} = \frac{dx_\mathrm{B}}{dV} &= -(k_F x_\mathrm{A} x_\mathrm{B} - k_B x_\mathrm{C} x_\mathrm{D})/F_\mathrm{EXT} \\ \frac{dx_\mathrm{C}}{dV} = \frac{dx_\mathrm{D}}{dV} &= +(k_F x_\mathrm{A} x_\mathrm{B} - k_B x_\mathrm{C} x_\mathrm{D})/F_\mathrm{EXT} \end{aligned} \tag{16.1}$$

where the kinetics vary with temperature and F_EXT is the flowrate of liquid to the external reactor (mol/s). Because the reactor is adiabatic, the temperature is directly related to composition.

$$T_\mathrm{out} = T_\mathrm{in} + \frac{(x_\mathrm{in,A} - x_\mathrm{out,A})(-\lambda)}{c_p Mw} \tag{16.2}$$

where $x_\mathrm{in,A}$ is the mole fraction of reactant A in the feed to the reactor, $x_\mathrm{out,A}$ is the mole fraction of reactant A in the effluent from the reactor, λ is the heat of the reaction (kJ/mol of A reacted), c_p is the heat capacity (kJ kg^{-1} K^{-1}), and M_w is the molecular weight (kg/kmol).

The steady-state vapor and liquid rates are constant through the stripping and rectifying sections because equimolal overflow is assumed. However, because the reaction is exothermic, some vapor is produced as the liquid from the high pressure reactor flashes into the low pressure column. This results in an increase of the vapor flowrate at each external reactor location and a corresponding decrease of the liquid rate below the external reactor location.

The quantity of this vapor is calculated from the heat generated by the reaction occurring in the external reactor, the flowrate of material into the reactor, and the thermal properties.

$$\Delta V_\mathrm{EXT} = \frac{Q_\mathrm{EXT}}{\Delta H_v} = \frac{F_\mathrm{EXT}(x_\mathrm{in,\,A} - x_\mathrm{out,\,A})(-\lambda)}{\Delta H_v} \tag{16.3}$$

where ΔH_v is the heat of vaporization (kJ/mol). Thus, there are different liquid and vapor rates in the various sections of the column.

The vapor–liquid equilibrium is assumed ideal. The column pressures are set using the vapor pressures P^S of pure components and liquid compositions in the reflux drum $x_{\mathrm{D},j}$ at 320 K (so that cooling water can be used in the condenser). Temperature T_i and vapor composition $y_{i,j}$ on tray i can be calculated with given pressure P and tray liquid composition $x_{i,j}$. These bubblepoint calculations can be done by the Newton–Raphson iterative convergence method:

$$P = \sum_{j=1}^{NC} x_{i,j} P^S_{i,j(T)} \tag{16.4}$$

$$y_{i,j} = \frac{P^S_{i,j}}{P} x_{i,j} \tag{16.5}$$

Simultaneous solution of the large set of nonlinear and algebraic equations is difficult, especially with the high degree of nonlinearity due to reaction kinetics. The steady-state solution of these equations is achieved by using the relaxation technique. This method is an efficient and robust way of solving this large set of equations. A value for the reflux flowrate R is chosen, and the dynamic equations of the system (mole fractions through the column) are integrated in time until steady-state conditions are achieved. For this study a 95% conversion is desired, so the composition of C in the distillate $x_{D,C}$ must be 95 mol% (as well as the composition of D in the bottoms). The reflux flowrate is varied until these compositions are achieved.

The dynamic component balances for the column are the conventional ordinary differential equations, except for the reactor effluent return trays:

$$\frac{d(x_{i,j}M_i)}{dt} = (F_{\text{EXT}} - \Delta V_{\text{EXT}})x_{\text{out}i,j} + V_{i-1}y_{i-1,j} - L_i x_{i,j} - V_i y_{i,j} \tag{16.6}$$

The column diameter is set by the largest vapor rate, which is V_{NT} in the top section of the column. Energy consumption is set by the vapor rate V_S at the bottom of the column.

16.2.3 Design Optimization Procedure

A five-dimensional grid search method is used to find the optimum values of the five design optimization variables. The following steps are used in the design procedure:

1. Fix the number of reactors N_{EXT} at a small value.
2. Fix the holdup for each reactor V_R at a small value.
3. Fix the number of trays N_{LT} between each liquid trap-out tray at a small value.
4. Fix the number of stripping trays N_S at a small value.
5. Fix the number of rectifying trays N_R at a small value.
6. The flowrates of the two fresh feeds are fixed at 12.60 mol/s.
7. The flowrates of the distillate and bottoms are fixed at 12.60 mol/s.
8. The vapor boilup V_S is manipulated by a proportional controller to control the level in the column base. The reflux-drum level is not controlled.
9. The reflux flowrate is manipulated by a PI controller to drive the composition of product C in the distillate to its desired value. This also sets the purity of D in the bottoms product.

10. The values for adiabatic reactors (x_{out}, T_{out}, and ΔV_{EXT}) are calculated with Eqs. (16.1)–(16.3).
11. The liquid and vapor rates through the stripping and rectifying sections are calculated using Eq. (16.3) together with the equimolal overflow assumption.
12. The vapor compositions and temperatures on each tray are computed using bubblepoint calculations with Eqs. (16.4) and (16.5).
13. The time derivatives of component material balances are evaluated on all trays in the column.
14. All ordinary differential equations are integrated using the Euler algorithm.
15. Steps 8–14 are repeated until the system achieves the convergence criterion CC, which is $CC = \max|dx_{i,j}/dt| \leq 10^{-6}$ (the largest time derivative of any component on any tray is $<10^{-6}$).
16. With the data obtained, the TAC is calculated by combining the energy cost with the annual capital cost.
17. The number of rectifying trays N_R is varied over a wide range, and steps 6–16 are repeated for each value of N_R, generating its corresponding TAC.
18. Then the number of the stripping trays N_S is varied over a range, and steps 5–17 are repeated.
19. Next, the number of trays between each liquid trap-out tray (location of liquid trap-out trays) is varied over a range, and steps 4–18 are repeated.
20. The holdup for each reactor V_R is varied over a range, and steps 3–19 are repeated.
21. The number of reactors N_{EXT} is changed over a wide range, and steps 2–20 are repeated.
22. Finally, the minimum TAC is selected as the optimum economical steady-state design for the given α_{390} value.

16.2.4 Results and Discussion

Optimum designs were developed for a range of α_{390} values. Table 16.1 gives optimum design and operating parameters. The columns have different numbers of trays. They operate at the same pressure because the reflux-drum temperature (320 K) and the distillate compositions are the same. However, the temperatures in the lower sections of the columns are quite different. Figure 16.3*a* gives the temperature profiles in the columns for each case. The lower stripping section temperatures result from the changing vapor pressures of heavier components D and B. A comparison of the two graphs shown in Figure 16.2 shows that the vapor pressures of D and B in the $\alpha_{390} = 0.95$ case at temperatures above 320 K are higher than they are for the constant relative volatility $\alpha_{390} = 2$ case. Figure 16.3*b* gives composition profiles for three values of α_{390}. All of the composition profiles are quite similar except for component A. The A profiles have two peaks in all cases, but the upper peak for the $\alpha_{390} = 2$ case is much larger than for the other α_{390} values.

Case with $\alpha_{390} = 0.95$. Figure 16.4 displays how the annual capital and energy costs are affected with the variation of two optimization variables: the holdup of reactors V_R and the number of reactors N_{EXT} for $\alpha_{390} = 0.95$. A payback period of 3 years is used to

TABLE 16.1 Optimization Results of Distillation Column/Side Reactor Design

	Previous Design	Current Designs			
α_{390}	0.95	0.95	1.25	1.50	2.00
Design variables					
N_{EXT}	3	3	3	4	4
V_R (kmol)	60	80	50	35	20
N_S	20	17	12	9	7
N_R	29	7	7	7	8
N_{LT}	5	2	2	1	1
Design parameters					
N_T	60	31	26	23	22
V_S (mol/s)	58.4	55.86	44.95	37.71	31.66
R (mol/s)	63	60.51	49.60	42.36	36.32
D_C (m)	1.34	1.32	1.22	1.14	1.08
A_r (m^2)	86.1	82.16	66.11	55.46	46.57
A_c (m^2)	186	179.48	152.69	134.93	120.09
D_R (m)	0.785	0.860	0.735	0.652	0.543
L_R (m)	7.85	8.60	7.35	6.53	5.42
Capital cost (10^3 \$)					
Column	500	289.3	230.5	195.5	177.3
Tray		10.9	8.1	6.5	5.7
Heat exchanger	350	341.0	302.9	276.1	252.6
Reactor	214	253.0	188.8	201.6	142.3
Energy cost (10^3 \$/year)	252	240.7	193.7	162.5	136.5
TAC (10^3 \$/year)	613	538.8	437.1	389.1	329.1

calculate the annual capital cost from the total capital investment. All cost graphs are plotted versus V_R for three different values of reactor number N_{EXT} with fixed $N_T = 31$, $N_S = 17$, and $N_R = 7$. Although the total number of trays in the middle zone is equal for all three N_{EXT} values, the number of trays between each liquid trap-out tray differs because of the different reactor numbers.

As the lower left graph in Figure 16.4 shows, an increase in the number of reactors results in higher reactor cost, and reactor cost increases with an increase in reactor holdup. In contrast, the costs of the column, the heat exchangers (reboiler and condenser), and energy show an exponentially decaying curve as the reactor volume is increased. This is attributable to the dependence of these costs on the vapor boilup. The increase of the reactor holdups increases the conversion of the reactants. Because less reactant is fed back to the column from the external reactors, the separation is easier for columns. The result is a decrease in reflux and vapor boilup required to achieve the desired purities of product streams. Therefore, the costs related to the separation decrease as the reactor holdup increases. When the columns with different reactor numbers are compared, these separation related costs decrease as more reactors are used.

Because of the decreasing column cost and heat exchanger cost, the capital cost initially decreases with an increase in reactor holdup. However, at some point the increasing reactor cost becomes more dominant and total capital cost begins to increase.

Although the reactor cost increases with increases in the number and holdup of the reactors, the opposite is true for separation costs. Increasing the reactor number or holdups

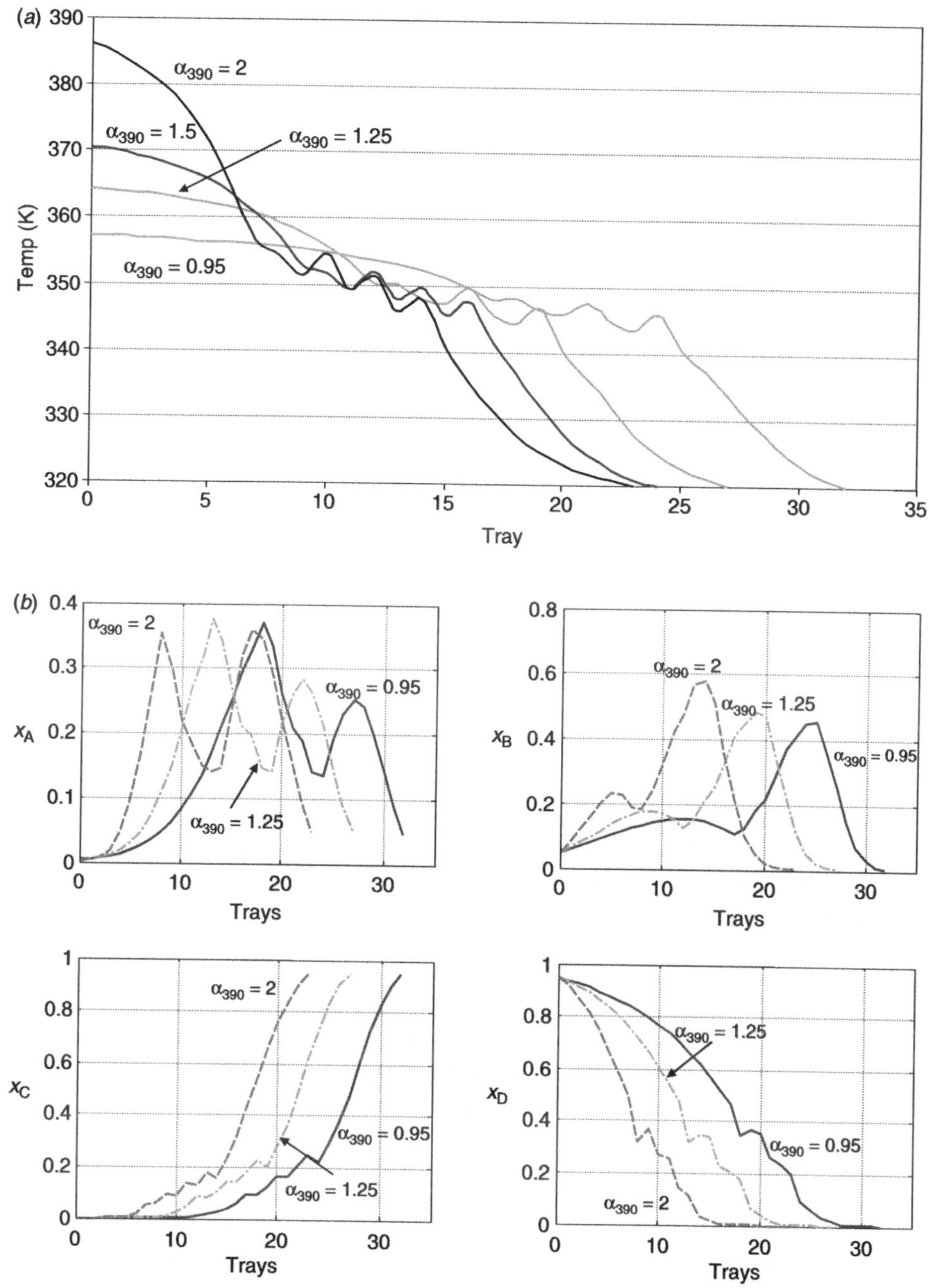

Figure 16.3 (*a*) Temperature and (*b*) composition profiles.

reduces separation costs (heat exchanger and column capital costs and energy cost). Thus, there is a minimum in the TAC curve at a certain number of reactors and at a certain value of reactor holdup because of the classical tradeoff between reactor cost and separation cost. When $\alpha_{390} = 0.95$ with constant numbers of the total, stripping, and rectifying trays, the optimum reactor number is 3, and the optimum reactor holdups is 80 kmol for each reactor.

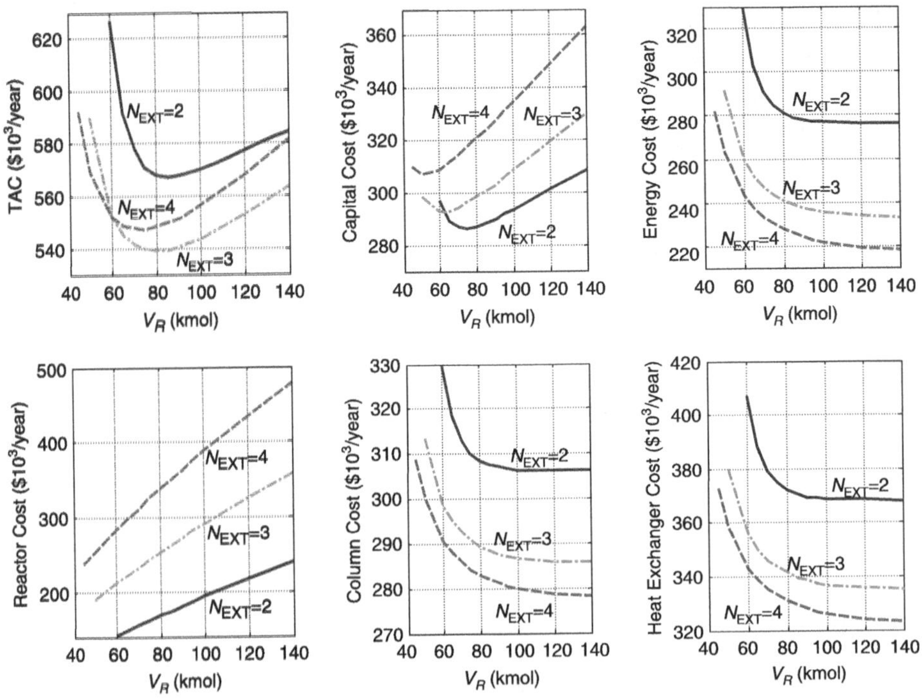

Figure 16.4 Effects of reactor numbers and holdups.

Figure 16.5 shows the effects of changes in the number of stripping and rectify stages for the $\alpha_{390} = 0.95$ case. All cost figures are plotted versus stripping stages N_S for four different values of rectifying trays N_R. The reactor holdups, number of reactors, and number of trays between each liquid trap-out tray are kept constant at $V_R = 80$, $N_{EXT} = 3$, and $N_{LT} = 2$, respectively.

Because the number and holdups of the reactors are kept constant, the reactor costs do not change. The column cost increases as the number of stripping and rectifying trays increase. However, the upper right graph in Figure 16.5 shows that there is a minimum in the energy cost, which is directly related to vapor boilup. The lower right graph shows that heat exchanger cost curves track energy costs directly.

As the number of stripping trays increases, the amount of vapor boilup required decreases initially. However, increasing the number of stripping trays above the optimum results in an increase of the required vapor boilup. This occurs because the extra separation occurring with more stripping stages concentrates the products in the middle zone and shifts the chemical equilibrium back to the reactants on the lower part of the middle zone. For example, the concentration of reactant A on the lowest tray of the middle zone increases from 0.3711 to 0.4396 as the N_S is changed from 17 (the optimum) to 22. However, the concentration of product D at the same location decreases from 0.3432 to 0.3073. Therefore, it takes more energy to keep component A from leaving in the bottoms. Thus, there is an optimum stripping tray number that minimizes vapor boilup and the resulting energy cost and heat exchanger cost.

The change in column costs over the range of separation trays shown in these graphs is relatively large (260–360 10^3 \$/year) compared to the change in heat exchanger cost

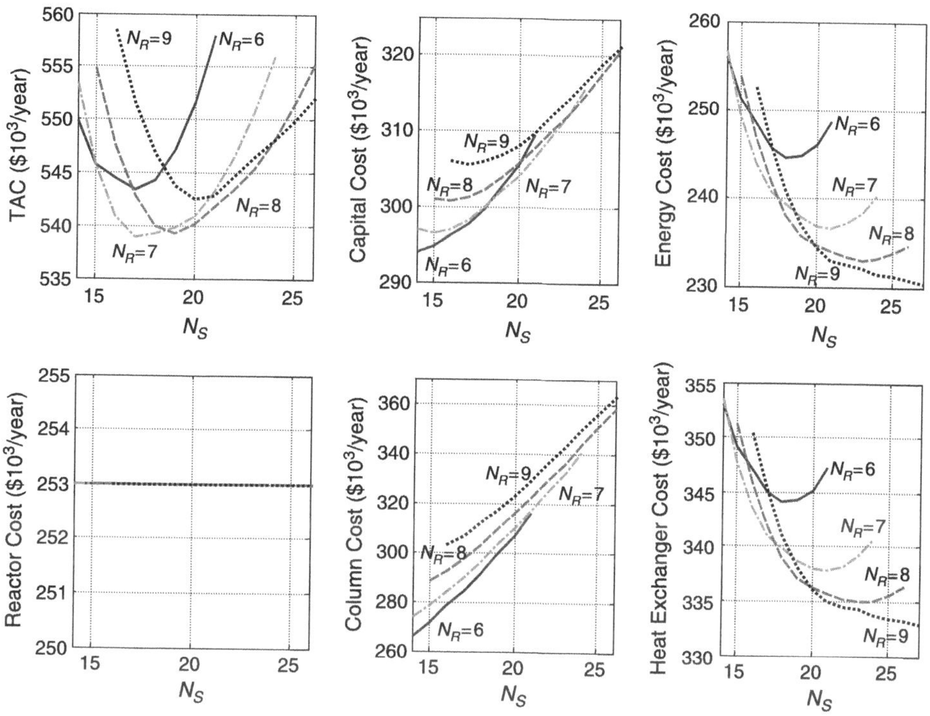

Figure 16.5 Effects of stripping and rectifying tray numbers.

(330–355 10^3 \$/year). Therefore, column cost, which decreases with decreasing numbers of separation trays, is the most dominant factor among the three capital cost factors. Thus, the optimum capital cost shifts to smaller numbers of stripping and rectifying trays, but the optimum energy cost shifts to larger numbers of stripping and rectifying trays. The result is a minimum in the TAC curve at a certain number of stripping and rectifying trays. For $\alpha_{390} = 0.95$ with constant reactor numbers $N_{EXT} = 3$ and constant reactor holdups $V_R = 80$ kmol, the optimum number of stripping and rectifying trays are $N_S = 17$ and $N_R = 7$, respectively.

It is interesting to compare these design results with the approximate heuristic design found in some initial work by Kaymak and Luyben.[1] The first two columns in Table 16.1 give a direct comparison. The improved design has fewer trays and larger reactors. The TAC is 13% lower.

Other α_{390} Cases. Figure 16.6 provides the optimum design results for the column/side reactors process for a range of temperature-dependent relative volatilities ($\alpha_{390} = 0.95$–2). These results show that there is little change in the optimum number of reactors. As the reference α_{390} decreases, the optimum number of reactors also slightly decreases from four to three. The inlet and outlet temperatures of the reactors decrease with the decrease of the reference α_{390} value.

[1]D. B. Kaymak and W. L. Luyben, Design of distillation columns with external side reactors, *Ind. Eng. Chem. Res.* **43**, 8049 (2004).

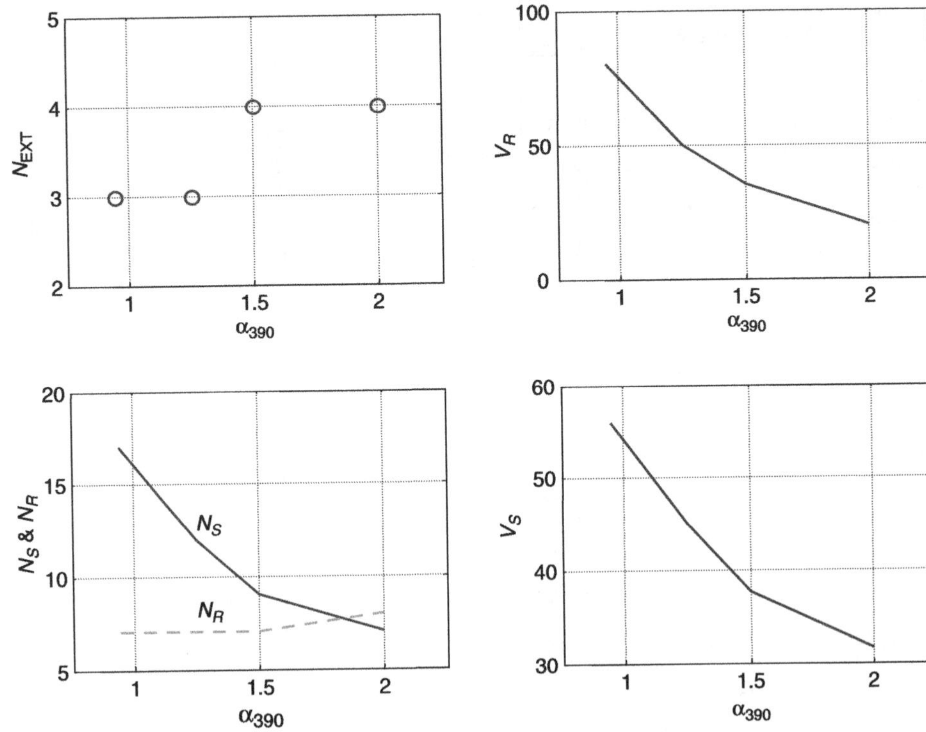

Figure 16.6 Results for column/side reactors process with different α_{390} values.

However, a lower temperature is unfavorable for reaction because the reaction rates get smaller. The result is a rapid increase in the required reactor holdups. A lower reference α_{390} value results in more difficult separation, as expected. Thus, there is an increase in the required vapor boilups and number of stripping trays. Because the operating pressure at the top is kept constant for all α_{390} cases, there is no significant change in the number of rectifying trays. Note that the optimum number of stripping trays is larger than the optimum number of rectifying trays. This is caused by the higher temperatures in the lower part of the column, which means lower relative volatilities.

Results for these four different α_{390} cases are given in Table 16.1. The values of the five design optimization variables are the number of reactors, the holdup of each reactor, the number of stripping trays, the number of rectifying trays, and the number of trays between each liquid trap-out tray. Table 16.1 and the upper graph in Figure 16.7 show that reducing the reference relative volatility increases both capital and energy costs, which results in an increase of TAC.

The lower graph in Figure 16.7 compares the economic optimum steady-state design of the column/side reactor process with those of the reactive distillation column and the multiunit conventional process. The reactive distillation column is the most economical alternative for the $\alpha_{390} = 2$, where there is no reaction/separation temperature mismatch. The column/side reactor process becomes more attractive as the mismatch of reaction/separation temperatures becomes more severe. The distillation column with a side reactor is economically superior for reference relative volatilities that are smaller than 1.5 for this case study.

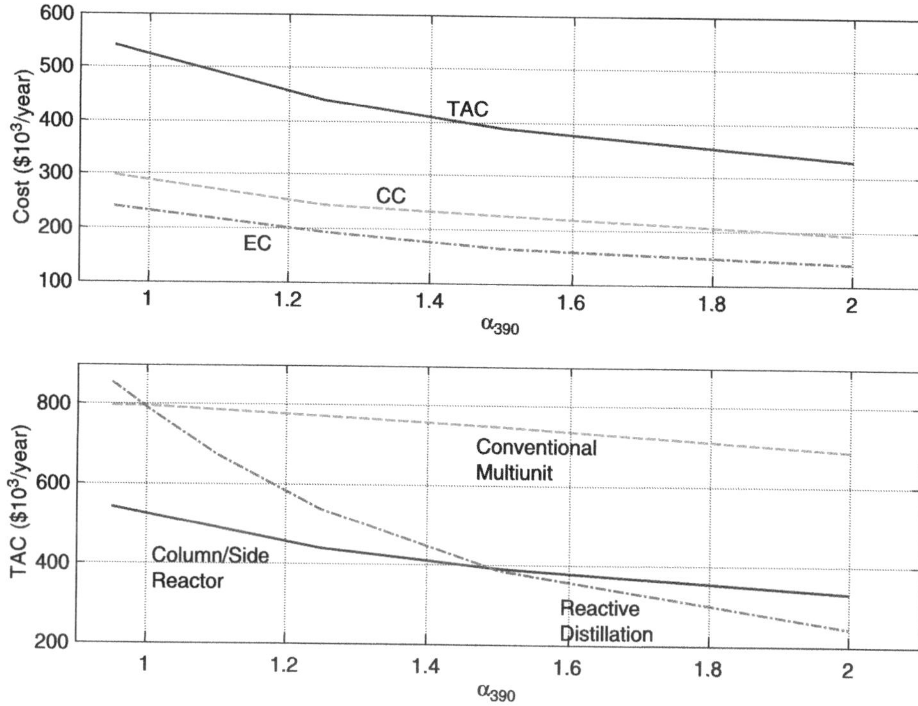

Figure 16.7 Comparison of alternative processes.

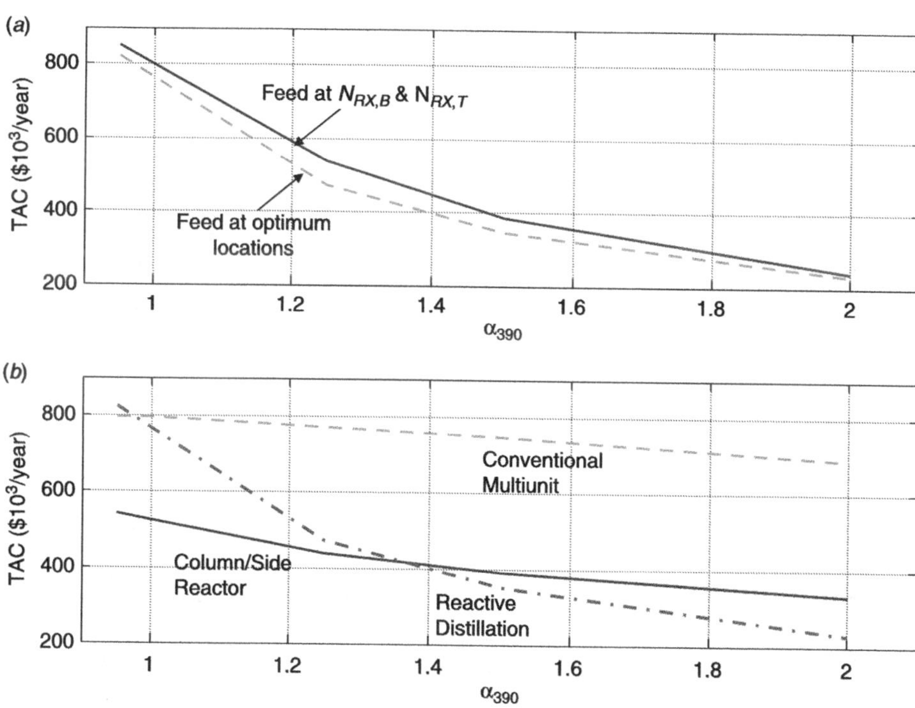

Figure 16.8 Effect of feed tray locations.

TABLE 16.2 Results for Reactive Distillation Columns with Optimized Feed Tray Location

	α_{390}							
	0.95		1.25		1.5		2	
Design variables								
N_S	14		11		9		5	
N_{RX}	68		25		13		7	
N_R	3		5		6		5	
N_{F1}/N_{F2}	15/82	29/68	12/36	20/28	10/22	13/19	6/12	7/11
P (bar)	3.75		5.50		7.00		8.50	
Design parameters								
N_T	85		41		28		17	
V_S (mol/s)	99.15	94.76	68.50	58.18	48.82	42.67	28.79	26.65
R (mol/s)	103.78	99.40	73.14	62.82	53.46	47.31	33.42	31.28
D_c (m)	1.53	1.50	1.21	1.13	1.00	0.95	0.80	0.78
A_r (m^2)	145.83	139.38	100.75	85.57	71.81	62.76	42.34	39.20
A_c (m^2)	285.71	274.95	210.48	185.17	162.18	147.09	112.98	107.73
Capital cost (10^3 \$)								
Heat exchanger	474.1	461.6	382.4	348.8	316.7	294.6	240.9	232.0
Column	763.0	747.1	329.3	307.4	198.8	188.7	104.5	101.8
Tray	37.7	36.6	12.5	11.4	6.4	6.0	2.7	2.6
Energy cost (10^3 \$/year)	427.2	408.4	295.2	250.7	210.4	183.9	124.0	114.8
TAC (10^3 \$/year)	852.2	823.5	536.6	473.2	384.4	347.0	240.1	227.0

The results in the first columns of each α_{390} case are the optimum designs with the assumption that feed locations N_{F1} and N_{F2} are at the bottom and top of the reactive zone, respectively. The results in the second column of each α_{390} case are the optimum designs where feed locations are also taken as design optimization variables while other design variables are kept at their optimum values.

16.2.5 Reactive Column with Optimum Feed Tray Locations

The results shown in Figure 16.7 for the reactive distillation column are based on designs in which the two fresh feedstreams are introduced at the two ends of the reactive zone. The lighter reactant is fed at the bottom reactive tray, and the heavier reactant is fed at the top reactive tray.

Figure 16.8 and Table 16.2 give results for when this assumption is relaxed. The optimum locations of the two fresh feeds that minimize TAC are found to be inside the reactive zone. For example, as shown in Table 16.2 for the base case $\alpha_{390} = 2$, the lower feedstream is moved from tray 6 up to tray 7 and the upper feedstream is moved from

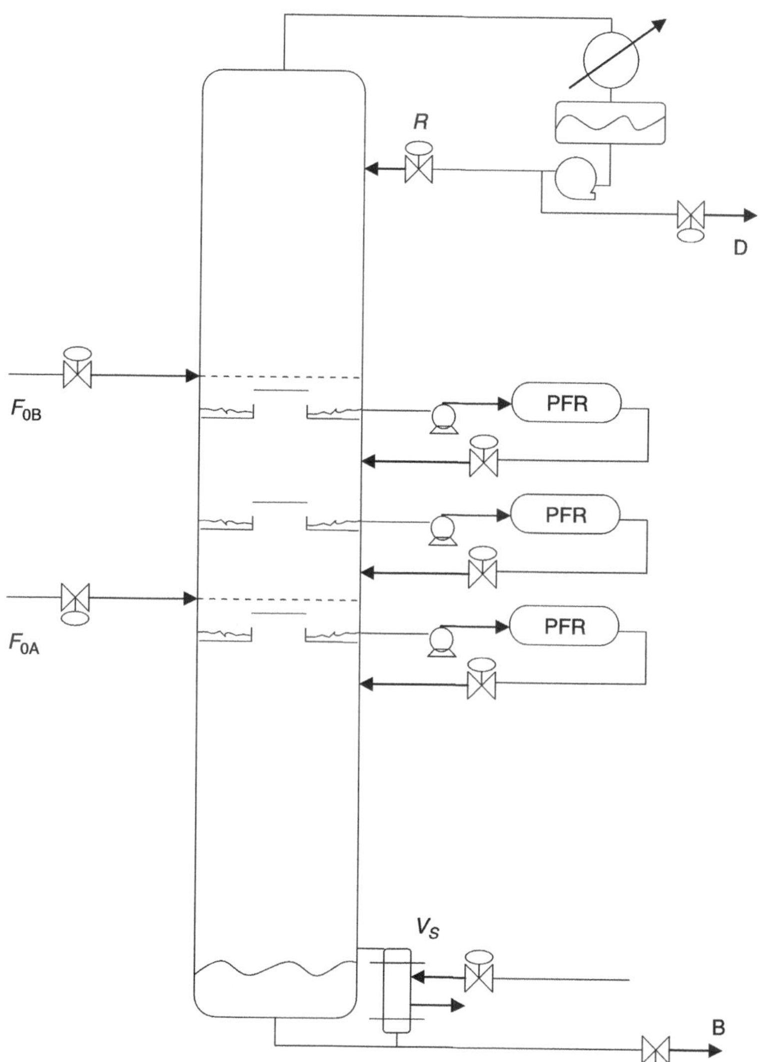

Figure 16.9 Column/side reactor process.

tray 12 down to tray 11. There is a reduction in the vapor boilup from 28.78 to 26.65 mol/s, which leads to a significant reduction in TAC from 240.1 to 227.0 10^3 \$/year.

For the case with $\alpha_{390} = 0.95$, the lower feedstream is moved from tray 15 up to tray 29 and the upper feedstream is moved from tray 82 down to tray 68. There is a reduction in the vapor boilup from 99.15 to 94.76 mol/s, which leads to a significant reduction in TAC from 852.2 to 823.5 10^3 \$/year.

Comparing the lower graphs in Figures 16.7 and 16.8 shows that the break-even point between the reactive distillation and column/side reactor process moves from an α_{390} of about 1.5 down to about 1.4.

16.3 CONTROL OF QUATERNARY IDEAL SYSTEM

16.3.1 Dynamic Tubular Reactor Model

In the steady-state design discussed earlier, the tubular reactors are modeled using ordinary differential equations that describe the variation of the temperature and compositions with axial length. Because we are interested in the dynamic behavior of the system for this control study, the tubular reactor model that is used must describe the changes in variables in both time and axial position. A rigorous approach is to use partial differential equations. A more simple method is to use a "lumped" model in which the spatial variations in temperature and composition variables are approximated with a series of CSTRs. With a reasonable number of lumps, this simple model gives a steady-state profile similar to the real steady-state profile obtained when using the rigorous steady-state equations. Each lump has its dynamic component and energy balances with variables needed to be numerically integrated in time: dT/dt, dx/dt, and so forth.

In these reactors, both the process liquid and the catalyst have thermal capacitance, so the temperatures of these two phases can be dynamically different in each lump with heat transfer between the solid and liquid phases. Of course, the liquid compositions change dynamically from lump to lump because of the convective flows in and out and because of the reaction. The kinetics are based on the volume of the liquid. This brings up the issue of specifying the amount of liquid and the amount of solid catalyst in a given total vessel volume. It is assumed that the void volume of the catalyst is 0.5, so the vessel is half filled with liquid and half filled with liquid.

The reaction occurs at the catalyst site, so that is where the heat of reaction affects the catalyst temperature. There is also heat transfer from the catalyst to the liquid. This means that there will be a temperature difference between the liquid and the solid even under steady-state conditions. The energy equation for the solid catalyst in the nth lump is

$$\frac{dT_{\text{cat},n}}{dt} = \frac{-\lambda \cdot \rho_{\text{liq}} V_{\text{liq}}/M_{\text{liq}}}{\rho_{\text{cat}} V_{\text{cat}} C_{P\text{cat}}} \left(k_F x_{\text{A},n} x_{\text{B},n} - k_B x_{\text{C},n} x_{\text{D},n}\right) - \frac{UA}{\rho_{\text{cat}} V_{\text{cat}} C_{P\text{cat}}} \left(T_{\text{cat},n} - T_{\text{liq},n}\right) \quad (16.7)$$

The energy equation for the liquid in each lump includes the convective terms associated with the flow of liquid in and out of the lump at their respective temperatures.

$$\frac{dT_{\text{liq},n}}{dt} = \frac{M_{\text{liq}} F_{\text{liq},n}}{\rho_{\text{liq}} V_{\text{liq}}} \left(T_{\text{liq},n-1} - T_{\text{liq},n}\right) + \frac{UA}{\rho_{\text{liq}} V_{\text{liq}} C_{P_{\text{liq}}}} \left(T_{\text{cat},n} - T_{\text{liq},n}\right) \quad (16.8)$$

The last term in both of these energy balances comes from the heat transfer between the catalyst and the liquid.

Two assumptions have been made to calculate the values of the heat transfer coefficient U and area A needed for these equations: the temperature difference between the liquid and solid is ~2 K at steady state, and the catalyst particles are spherical with a 0.002-m diameter.

The reaction directly affects the compositions in the liquid phase. The forward and reverse specific reaction rates follow the Arrhenius law:

$$k_F = \alpha_F e^{-E_F/\mathrm{RT}_{\mathrm{cat}}} \tag{16.9}$$

$$k_B = \alpha_B e^{-E_B/\mathrm{RT}_{\mathrm{cat}}} \tag{16.10}$$

The overall reaction rate is based on the concentrations in mole fractions and liquid holdups in moles. To avoid confusion, the specific reaction rates used in this control section (with the reactor half full of catalyst) are twice those used in the earlier design section in which catalyst was not considered. Thus, the total reactor volumes are the same as those used in the steady-state design.

$$\frac{dx_{i,n}}{dt} = \frac{F_{\mathrm{liq},n}}{\rho_{\mathrm{liq}} V_{\mathrm{liq}}/M_{\mathrm{liq}}}\left(x_{i,n-1} - x_{i,n}\right) \pm \left(k_F x_{\mathrm{A},n} x_{\mathrm{B},n} - k_R x_{\mathrm{C},n} x_{\mathrm{D},n}\right) \tag{16.11}$$

Because the reaction is exothermic, some vapor is produced as the liquid from the high pressure reactor flashes into the low pressure column. This results in an increase of the vapor flowrate at each external reactor location and a corresponding decrease of the liquid rate below the external reactor location. The quantity of this vapor is calculated from the heat generated by the reaction occurring in the external reactor, the flowrate of material into the reactor, and the thermal properties. The steady-state vapor and liquid rates are constant through the stripping and rectifying sections because equimolal overflow is assumed. However, there are different liquid and vapor rates in the middle sections of the column.

Figure 16.9 shows the system that was studied, and Figure 16.10 gives the composition and temperature profiles for the column. The left graphs are steady-state results using rigorous ordinary differential equation models for the three reactors. The right graphs are steady-state results using the lumped reactor model. The column profiles are essentially identical. Figure 16.11 gives composition and temperature profiles for each of the three reactors using either the rigorous reactor model (solid lines) or the lumped reactor model (dashed lines). These profiles are quite similar.

16.3.2 Control Structures

In practical applications it is desirable to use inferential temperature measurements whenever possible instead of direct composition measurements. Composition analyzers have higher cost, require more maintenance, and can introduce dead time into the control loop. Therefore, we first explore a control structure that does not have any composition analyzer. The performance of this two-temperature control structure will then be compared with a structure that uses a composition analyzer.

The control structures are single-input, single-output structures with PI controllers for temperature and composition and proportional-only controllers for levels with gains of 2. The Tyreus–Luyben tuning method is used to tune the temperature and composition

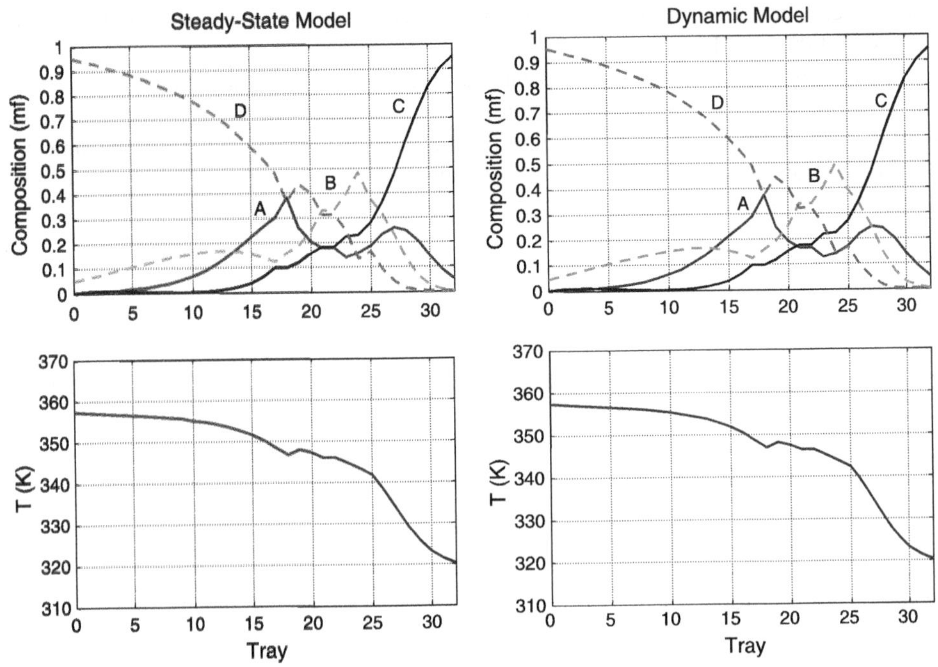

Figure 16.10 Composition and temperature profiles for column.

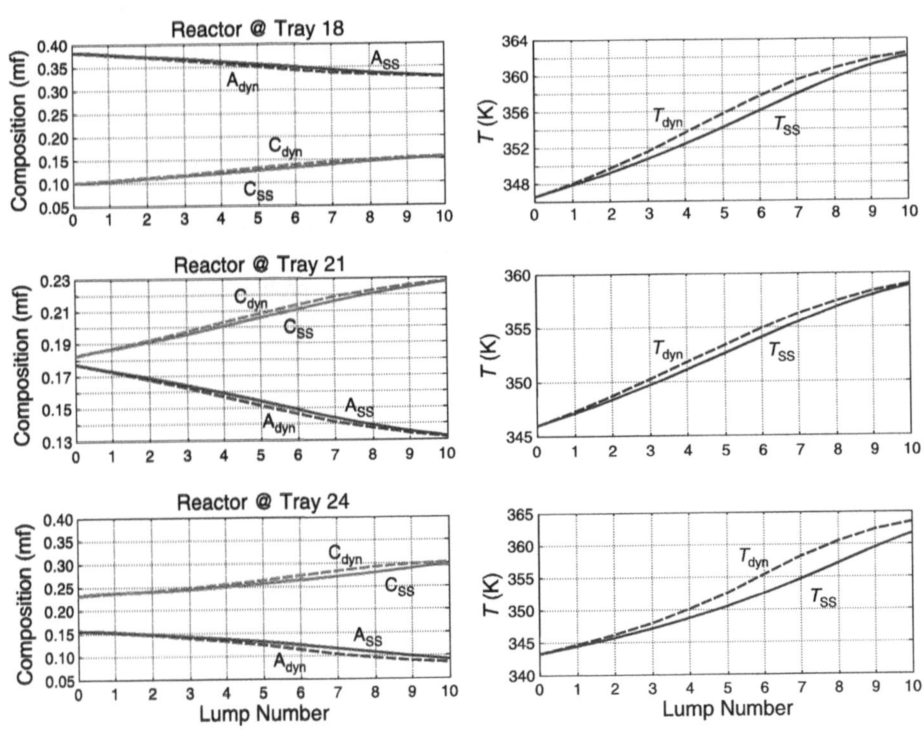

Figure 16.11 Composition and temperature profiles for reactors (length = 8.6 m).

controllers. The ultimate gain and ultimate period necessary for this method are obtained using the relay-feedback test. Two first-order measurement lags of 60 s each are used in all temperature loops, and a 3-min dead time is used for the composition analyzer. All of the control valves in the process are designed to be half open at steady state. Perfect level control of the trap-out trays is assumed, so the flowrate through each reactor is the same as that entering the trap-out tray. The compositions of the reactor feeds change dynamically because of the holdup on the trap-out tray (400 mol) and changes in the composition of the liquid stream entering the trap-out tray. There is no vapor–liquid contacting on the trap-out tray.

The effectiveness of the control structure is demonstrated by using dynamic simulations and subjecting the processes to production rate changes (ΔF_{0j} or ΔV_S) and feed composition disturbances (Δz_{0A} and Δz_{0B}).

Control Structure CS7. Control structure CS7 is based on an article by Roat and coworkers,[2] and it is shown in Figure 16.12. Two temperature controllers manipulate the two fresh feed flowrates to maintain the temperatures on two trays. Reboiler heat input is flow controlled and serves as the production rate handle. The base level is controlled by manipulating the bottoms flowrate, and the reflux-drum level is controlled by the reflux flowrate. The reflux ratio is controlled by measuring the reflux flowrate, multiplying this by the reciprocal of the desired reflux ratio, and sending this signal to a remotely set cascade flow controller on the distillate stream. The column pressure is assumed to be constant and controlled by condenser heat removal. Note that the liquid levels on the three trap-out trays are held constant by manipulating the three valves in the liquid streams from the reactors back to the column (controllers not shown in Fig. 16.12). The pressures in the liquid filled reactors are set by the pump head.

The selection of the trays to temperature control is the main issue in this structure. From the rigorous steady-state simulation, the gain matrix between the inputs (two fresh feed flowrates) and the outputs (the temperatures on all trays) is calculated numerically. Using these steady-state gains, the locations of the most sensitive trays whose temperatures are to be controlled are selected by applying SVD analysis. Some engineering judgment must also be exercised in this selection. The procedure is discussed in detail by Kaymak and Luyben.[3,4]

The steady-state gains (K_{F0A} and K_{F0B}) between tray temperatures and feed flowrates and their related SVD results for the CS7 structure are presented in Figure 16.13. The steady-state gains between tray temperatures and the F_{0A} feed are negative (an increase in feed flowrate decreases tray temperature) throughout the column except for a small region in the lower stripping section that has very small positive values. In contrast, the steady-state gains between the tray temperatures and the F_{0B} feed are mostly positive, especially in the most temperature-sensitive regions. Note that the most sensitive temperature measurements to changes in the fresh feedstreams are in the rectifying section with a negative maximum at tray 27 for F_{0A} and with a positive maximum at tray 26 for F_{0B}.

[2]S. Roat, J. Downs, E. Vogel, and J. Doss, Integration of rigorous dynamic modeling and control system synthesis for distillation columns, In J. D. Birdwell and J. R. Cockette, Editors, *Chemical Process Control—CPC III*, Elsevier, New York, 1986.

[3]D. Kaymak and W. L. Luyben, Comparison of two types of two-temperature control structures for reactive distillation columns, *Ind. Eng. Chem. Res.* **44**, 4625–4640 (2005).

[4]D. B. Kaymak and W. L. Luyben, Evaluation of a two-temperature control structure for a two-reactant/two-product type of reactive distillation column, *Chem. Eng. Sci.* **61**, 4432–4450 (2006).

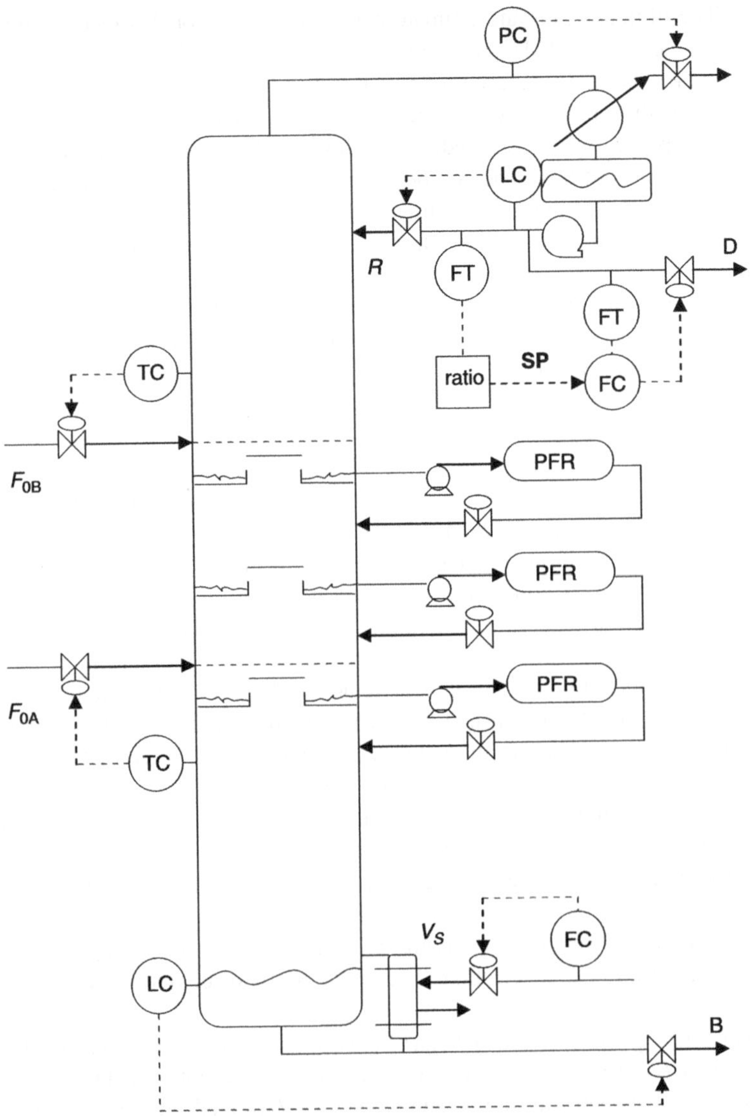

Figure 16.12 Control structure CS7.

However, because these two trays are so close to each other, it is not possible to use both of them. To overcome this problem, we look for secondary sensitive regions for both gains (second peaks that are smaller than the biggest ones). By looking at the curves in Figure 16.13, these secondary peaks are found in the upper stripping section (tray 16 for both gains).

Although this problem may be solved by selecting one of these trays to pair with one of the fresh feedstreams, a bigger problem still exists. Although both primary and secondary sensitive trays have *negative* gains for changes in F_{0A}, the gains are *positive* for changes in F_{0B}. Therefore, there will be a difference in the actions of control loops, and this may result in difficulties for this control structure. Kaymak and Luyben[3,4] pointed out that the actions

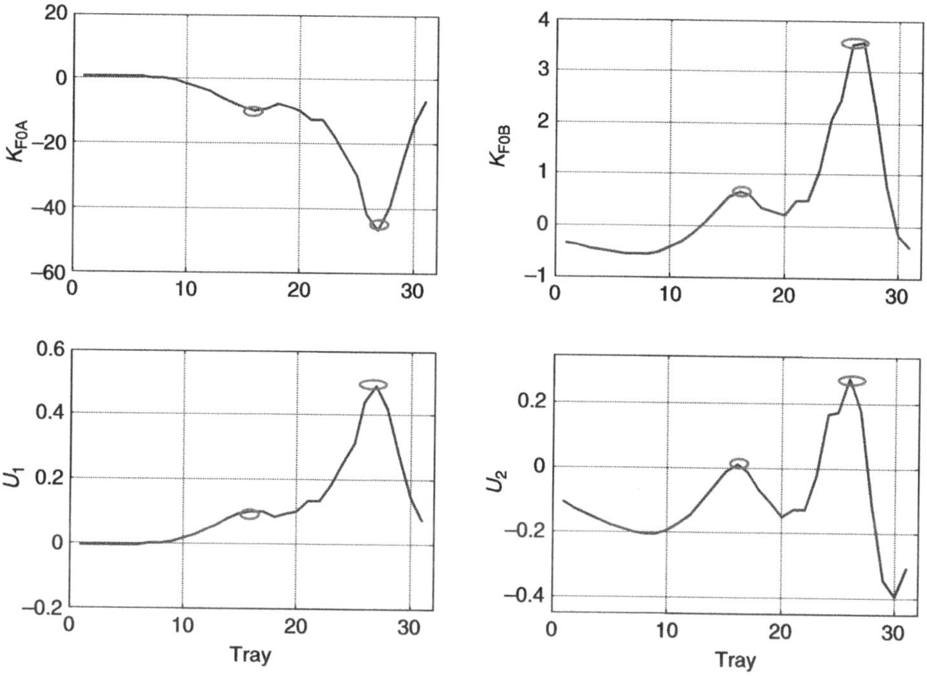

Figure 16.13 Gains and SVD for CS7.

of the two temperature controllers used in a reactive distillation column should be the same (e.g., both positive).

Controlling an upper tray in the rectifying section by manipulating the fresh feed flowrate F_{0B} (upper feed flowrate) and controlling a tray in the stripping zone by manipulating the fresh feed flowrate F_{0A} (lower feed flowrate) sounds more reasonable than the reverse. Therefore, the F_{0A}/T_{16} and F_{0B}/T_{26} pairings are first selected as the control loops instead of the alternative pairings of F_{0A}/T_{27} and F_{0B}/T_{16}.

Figure 16.14 gives relay-feedback test results for both loops used in this structure. Although the amplitude and period look reasonable for the F_{0A}/T_{16} pairing, the shark-tooth shape of the tray 26 temperature response is characteristic of a process with an inverse response. This is confirmed by the controller gain and reset time values calculated from test data. As given in Table 16.3, the values for the F_{0A}/T_{16} loop have reasonable values. However, the controller for tray 26 has an unrealistic reset time of 8600 min.

Therefore, the alternative structure with the F_{0A}/T_{27} and F_{0B}/T_{16} pairings is checked by applying the relay-feedback test. This structure also results in an inverse response that is revealed in the huge reset time of the F_{0A}/T_{27} loop. Thus, switching the pairing does not solve the problem.

Figure 16.15 shows the response of CS7 to $\pm 5\%$ step changes in the production rate (vapor boilup) using the F_{0A}/T_{16} and F_{0B}/T_{26} pairings. The response of the system is stable, but it takes a long time for the T_{26} temperature to go back to its setpoint for both positive and negative disturbances. This long settling time is a problem related to the inverse response and the huge reset time. A bigger problem is that even fairly small 5% disturbances result in large changes in the purities of product streams $x_{D,C}$ and $x_{B,D}$. While one

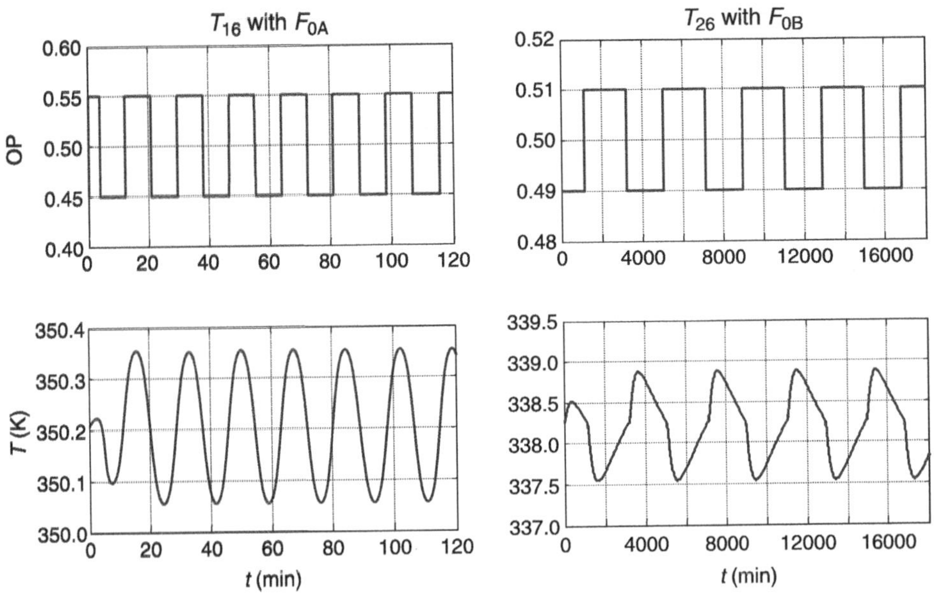

Figure 16.14 Relay-feedback test results for CS7 pairings.

of the purities settles down to a new steady state above its setpoint, the other one goes to a new steady state that is too far below its setpoint.

This problem occurs because the two temperature controllers in the CS7 structure give different initial responses as a result of their opposite actions. An increase in vapor boilup produces an increase in the temperature on both control trays (tray 16 and tray 26). The tray 16 temperature controller has direct action, so it increases the F_{0A} feed flowrate. However, the tray 26 temperature controller has reverse action, so it decreases the F_{0B} feed flowrate. Not enough reactant B is added in a timely enough manner to prevent light reactant A from going to the upper section of the column. Once this breakthrough occurs, the column moves to a different operating condition where the impurity of A in the distillate stream increases and product purities are lower. Thus, we can conclude that control structure CS7 is not a good choice for controlling this column/side reactor process.

Control Structure CS5. As mentioned in the previous section, it is desirable to use inferential temperature measurements instead of direct composition measurements whenever possible. However, the results show that control structure CS7 is not an effective control structure, at least with the optimum steady-state design studied here. It appears that some direct composition information about the reactant inventory inside the system is required for a more effective control system because the column is designed for neat

TABLE 16.3 Tuning Parameters for Control Structure CS7

Pairing	Span	K_C	τ_I
F_{0A}/T_{16}	10	1.37	38.0
F_{0B}/T_{26}	10	0.06	8600

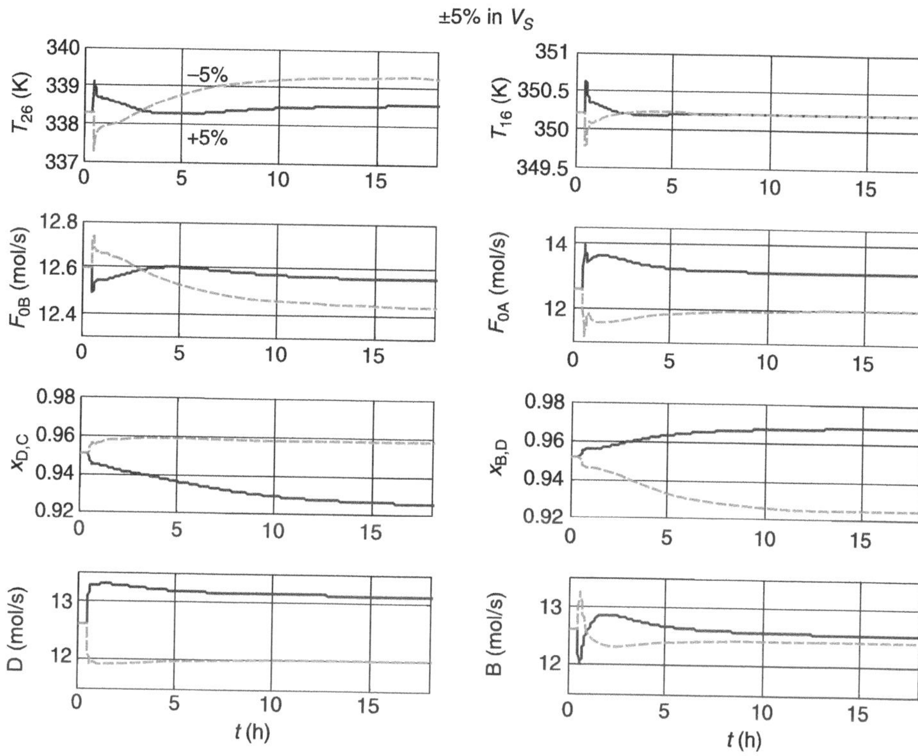

Figure 16.15 A ±5% step change in production rate handle ΔV_S.

operation. Thus, another control structure CS5 that includes one composition analyzer is explored in this section.

This control structure differs from control structure CS7 only in how the fresh feedstreams and the reboiler heat input are manipulated. Figure 16.16 shows that the temperature of a tray near the bottom of the column is used to infer bottoms product purity, and it is controlled by manipulating the vapor boilup. Tray 16, which has the biggest steady-state gain (K_{VS}) in the stripping section, is selected. Distillate purity is not controlled. Fresh feed F_{0B} is flow controlled and serves as the production rate handle.

A composition analyzer is used to detect the inventory of component A in the system. Feedback can be used to prevent the gradual buildup or depletion of reactant A. Thus, the flowrate of the other fresh feed F_{0A} is manipulated by a composition controller that maintains the concentration of component A on a selected tray.

In an attempt to see which tray is the best for detecting the component A concentration, a disturbance sensitivity approach is explored. The compositions of both products are driven to their correct values by varying the reflux ratio and heat input at different feed flowrates. The composition and temperature profiles have to change to give "perfect" composition control of the two products. The changes of the composition of component A through the column at ±20% of the original feed flowrates are shown in Figure 16.17. The figure shows that the tray with the least changing composition (tray 20) is selected as the preferred one to control by manipulating the flowrate of

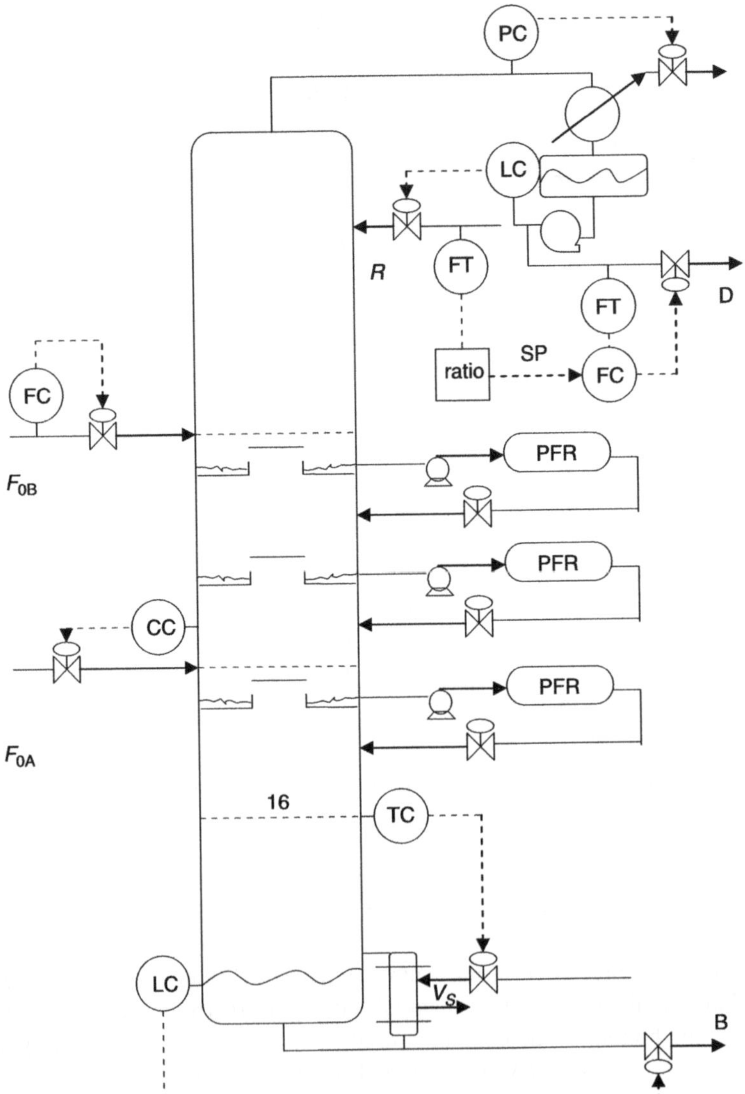

Figure 16.16 Control structure CS5.

the other fresh feed F_{0A}. This tray is also the effluent of the second (middle) reactor returned to the column.

Figure 16.18 gives relay-feedback test results for both loops: $F_{0A}/x_{A,20}$ and V_S/T_{16}. The unusual sharp spikes in the composition on tray 20 are caused by the step changes in F_{0A}, which is fed directly on tray 20. The amplitudes and periods look reasonable for both loops. These results are also confirmed by the controller gain and reset time values listed in Table 16.4. There is no inverse response (huge reset times) problem. There is also no action related problem because both loops have the same action.

To check the effectiveness of this control structure, quite large $\pm 20\%$ step changes in the production rate (fresh feedstream F_{0B}) are applied to the system. Figure 16.19

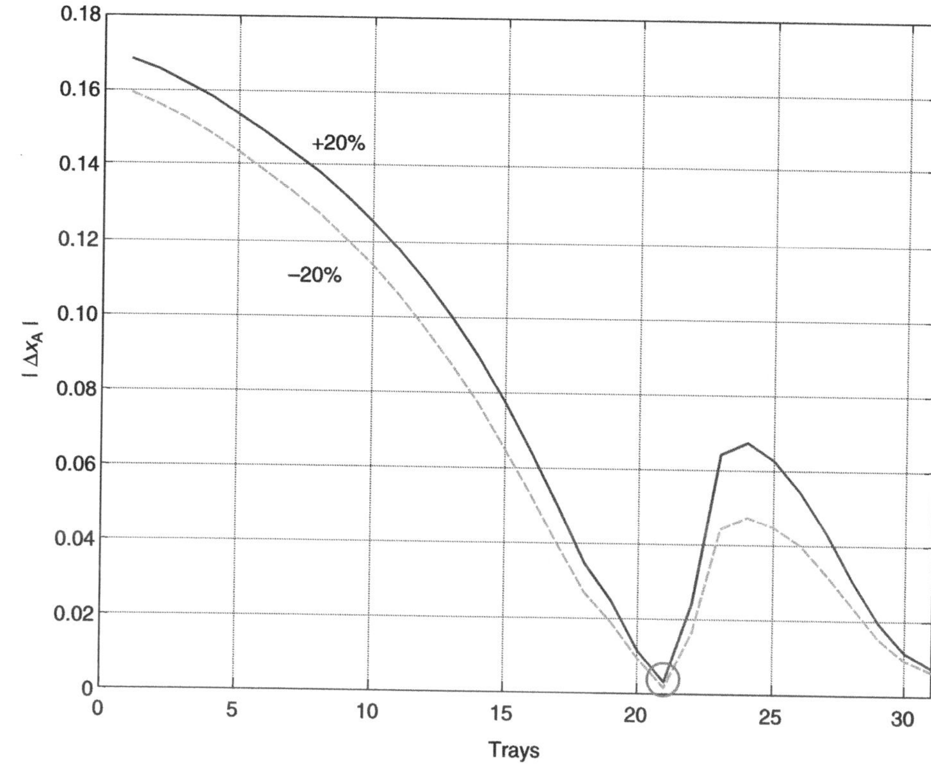

Figure 16.17 Disturbance sensitivity results.

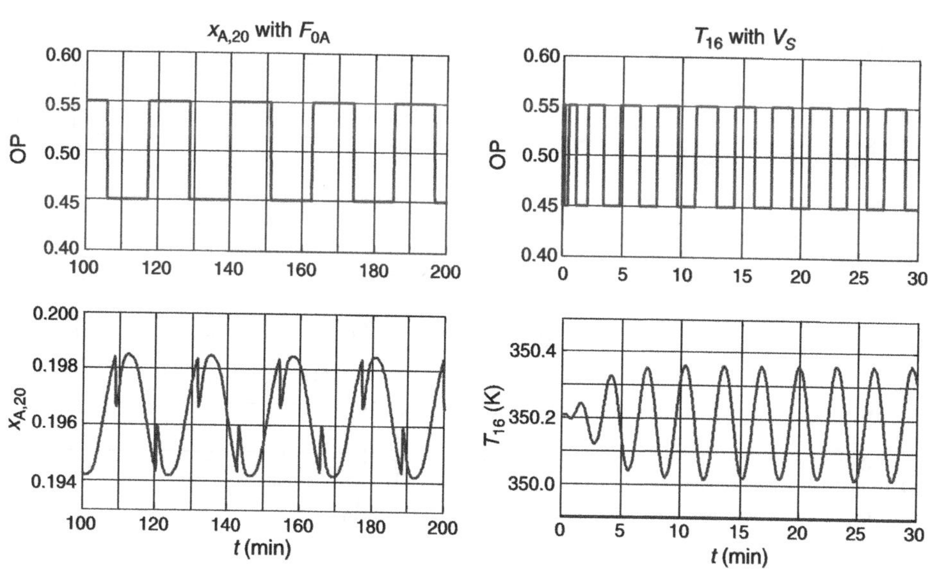

Figure 16.18 Relay-feedback test results for CS5 pairings.

TABLE 16.4 Tuning Parameters for Control Structure CS5 with Tray 20

Pairing	Span	K_C	τ_I
$F_{0A}/x_{A,20}$	0.5	4.79	49.9
V_S/T_{16}	10	1.30	7.04

shows the response of the CS5 structure to these disturbances. Both controlled variables go back to their setpoints for both positive and negative disturbances in a short time. In addition, even for these very large ±20% step changes, the purities of product streams $x_{D,C}$ and $x_{B,D}$ settle down smoothly to steady-state values that are within ±1% of the 95% specification value. Therefore, the CS5 structure is a much better choice for controlling this column/side reactor process compared to the CS7 structure.

The responses of CS5 with the tray 20 composition controlled for changes in feed compositions are given in Figures 16.20 and 16.21. In Figure 16.20 the composition of fresh feed F_{0A} is changed from pure A to either 95 mol% A with 5 mol% B (solid lines) or 90 mol% A with 10 mol% B (dashed lines). Control structure CS5 handles this type of

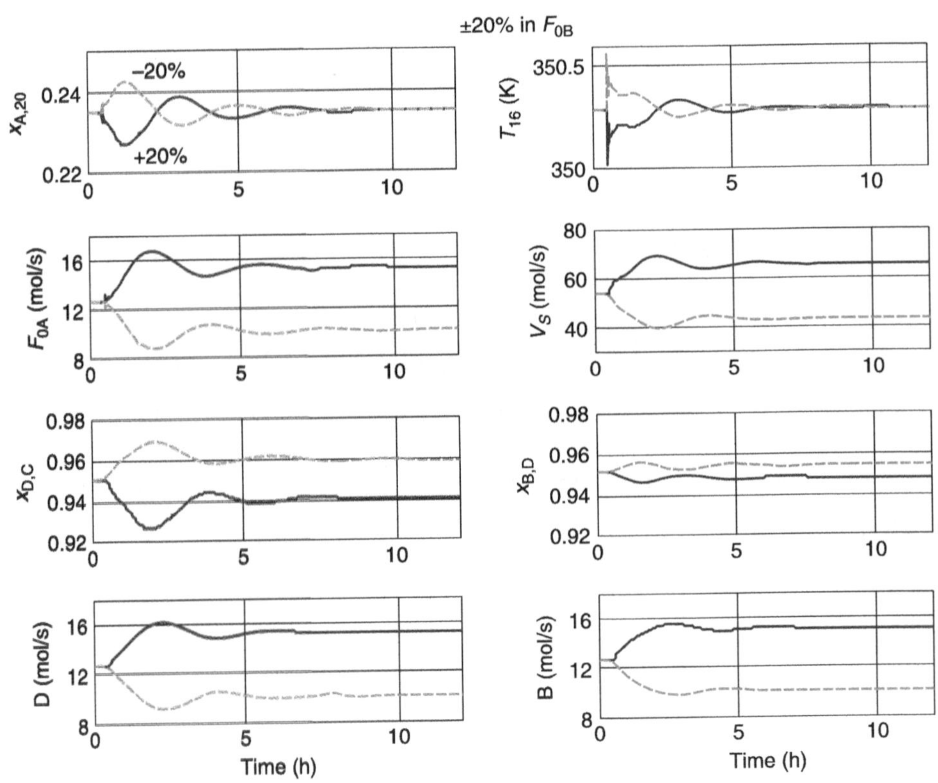

Figure 16.19 A ±20% step change in production rate handle ΔF_{0B}; control $x_{A,20}$.

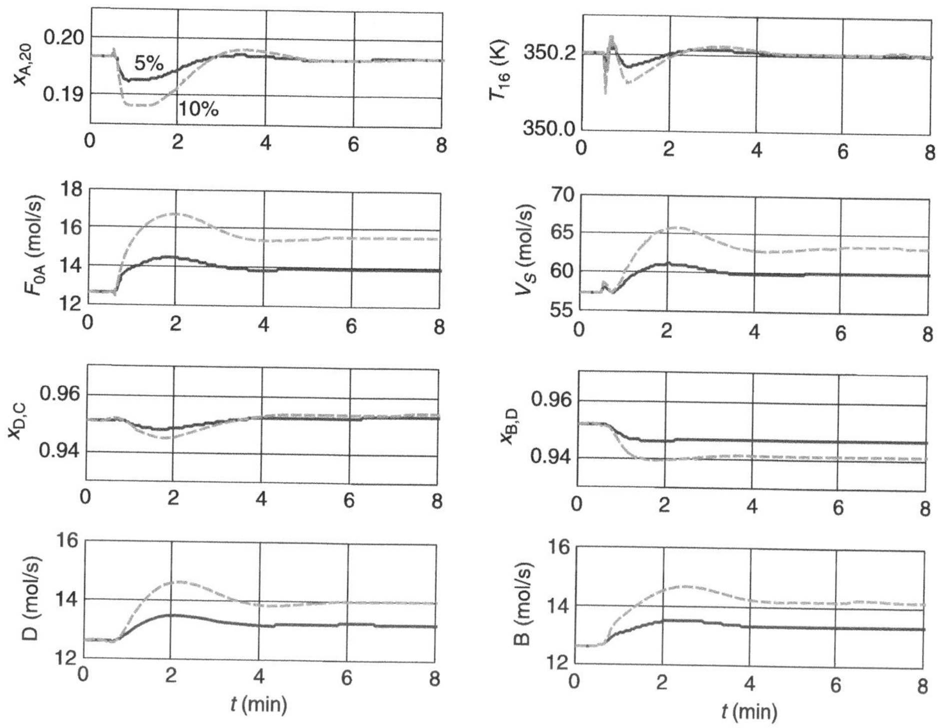

Figure 16.20 Impurity of B in fresh feed composition Δz_{0A}.

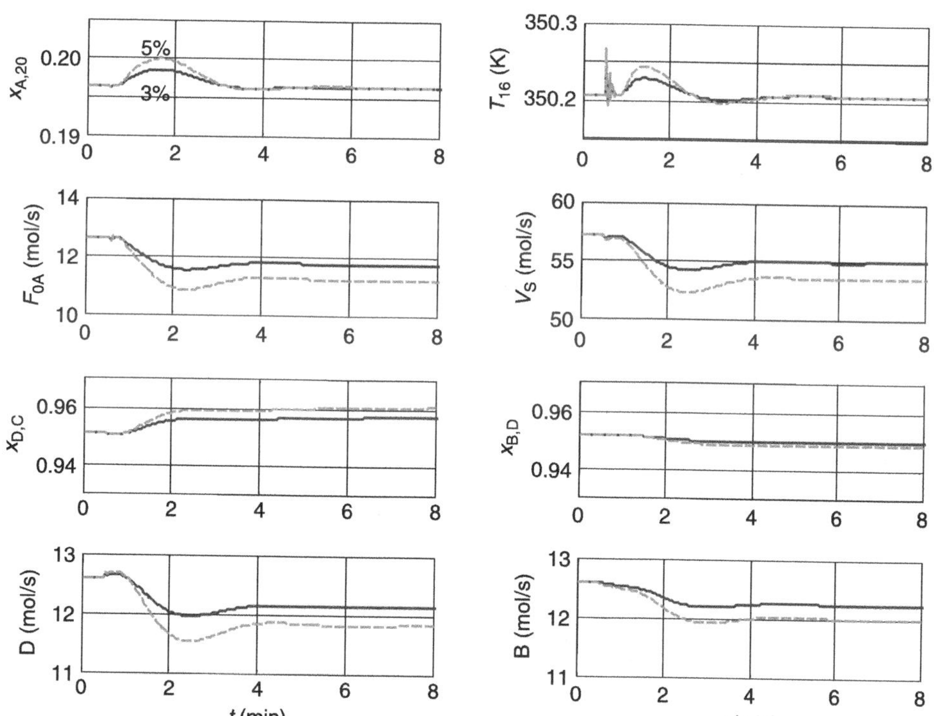

Figure 16.21 Impurity of A in fresh feed composition Δz_{0B}.

disturbance quite well. The purities of both products are maintained near the desired 95% specification. However, as the impurity amount is increased to 10 mol% B, the purity of the bottoms moves to the limit of 1% below the setpoint. Therefore, $x_{B,D}$ will deviate by more than 1% for large feed composition disturbances.

Figure 16.21 gives the responses of CS5 to 3% and 5% impurities of A in fresh feedstream F_{0B}. The control structure keeps the product purities within 1% of their specified values for the 3% feedstream impurity, but not for the 5% one.

16.4 DESIGN OF COLUMN/SIDE REACTOR PROCESS FOR ETHYL ACETATE SYSTEM

In this section we will investigate the design of the side reactor configuration for reactive distillation in which the reactive zone is placed at the lower section of the reactive column with no product removal from the column base. This is the type II configuration discussed in Chapter 7. Based on the optimal reactive distillation design, an improved side reactor configuration evolves gradually after going through a transitional step. The production of ethyl acetate[5] is used to illustrate the design procedure.

16.4.1 Process Description

The reaction for acetic acid (HAc) esterification with ethanol (EtOH) to produce ethyl acetate (EtAc) and water (H_2O) can be expressed as

$$\text{HAc} + \text{EtOH} \underset{k_B}{\overset{k_F}{\Longleftrightarrow}} \text{EtAc} + \text{H}_2\text{O} \qquad (16.12)$$

The reversible reaction is described using a pseudohomogeneous model based on the mole fraction of component i (x_i) and catalyst weight (m_{cat}). Model parameters are taken from Hangx et al.[6] for an esterification reaction catalyzed by a Purolite CT179 ion-exchange resin. The rate expression can be written as

$$\begin{aligned} R &= m_{\text{cat}} \times (k_F x_{\text{HAc}}^{1.5} x_{\text{EtOH}} - k_B x_{\text{EtAc}} x_{\text{H}_2\text{O}}) \\ k_F &= 4.24 \times 10^3 \exp(-48300/\text{RT}) \\ k_B &= 4.55 \times 10^5 \exp(-66200/\text{RT}) \\ K_{\text{EQ}} &= 3.50\ (T = 350\,K) \end{aligned} \qquad (16.13)$$

This corresponds to an exothermic reaction with an equilibrium constant slighter greater than 1. In the process simulation, a catalyst density of 770 kg/m^3 is assumed to calculate the total volume occupied by the catalyst in a reactive tray.

[5] I. K. Lai, S.-B. Hung, W.-J. Hung, C.-C. Yu, M.-J. Lee, and H.-P. Huang, Design and control of reactive distillation for ethyl and isopropyl acetates production with azeotropic feeds, *Chem. Eng. Sci.* **62**, 878 (2007).

[6] G. Hangx, G. Kwant, H. Maessen, P. Markusse, and I. Urseanu, Reaction kinetics of the esterification of ethanol and acetic acid towards ethyl acetate, Deliverable 22, Technical Report to the European Commission, Intelligent Column Internals for Reactive Separations (INTINT), 2001.

TABLE 16.5 Normal Boiling Point Ranking for Pure Components and Azeotropes

Component/Azeotrope	Temp. (°C)
EtOH–EtAc–H_2O	70.09
EtOH–EtAc	71.81
EtOH–H_2O	78.18
EtAc	77.2
EtAc–H_2O	70.37
EtOH	78.31
H_2O	100
HAc	118.01

To account for nonideal vapor–liquid equilibrium and possible VLLE for this quaternary system, the NRTL model is used to calculate the activity coefficients. Model parameters are taken from Chapter 7. Vapor-phase nonideality caused by the dimerization of acetic acid is also taken into consideration using the Hayden–O'Connell second virial coefficient model. Aspen Plus built-in parameters values are used.

With the thermodynamic models available, the phase behavior can be predicted. Table 16.5 provides the normal boiling point temperatures for pure components and azeotropic temperatures. The phase behavior exhibits a ternary minimum boiling azeotrope: ethanol, methyl acetate, and water at 70.09 °C (Fig. 16.22). From the normal boiling point ranking, this complex system can be viewed as a ternary system conceptually, where the reactants acetic acid and ethanol can be treated as HK and IK, respectively. The ternary minimum boiling azeotrope is considered as the product, which is the LK. Thus, the reactive zone should be placed in the lower section of the reactive distillation column while the pseudoproduct (liquid with an almost azeotropic composition) is removed from the top of the column. This pseudoproduct is decanted and the water product is withdrawn from the aqueous phase, and the organic phase is further purified to obtain the acetate product. This is the exact type II configuration discussed in Chapter 7.

16.4.2 Conceptual Design

Figure 16.23 shows the reactive distillation flowsheet for EtAc production. The reactive zone is placed in the lower section of the reactive distillation column. The condensed overhead vapor from the reactive distillation column is decanted. The aqueous product is withdrawn from the decanter and the organic phase is split, with a portion returned to the reactive distillation column and a portion fed to a stripping column that produces high purity EtAc. In the reactive distillation column, because the column base has much larger residence time than a tray, the catalyst holdup in the column base is assumed to be 10 times that on a reactive tray. The two fresh feeds into the reactive distillation are the alcohol feed below the azeotropic composition (87% EtOH and 13% H_2O) and industrial grade acid feed (95 mol% HAc and 5% H_2O). Following the work of Lai et al.,[5] the

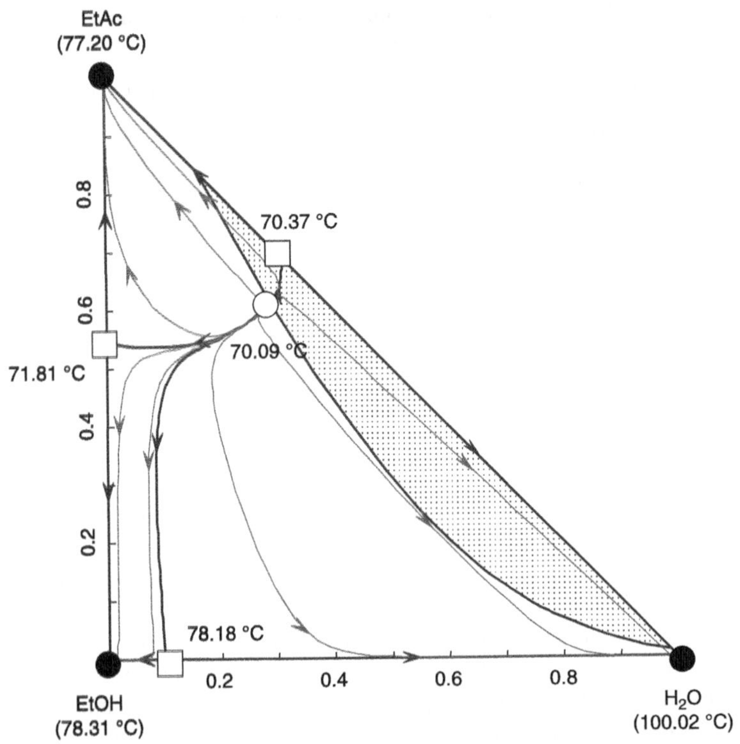

Figure 16.22 The RCM diagram for EtOH–EtAc–H_2O system.

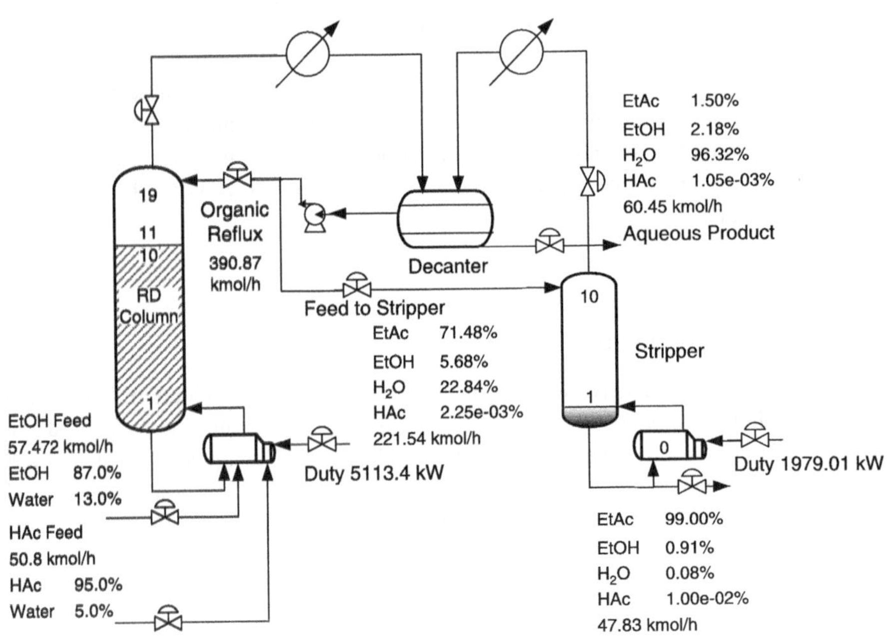

Figure 16.23 Reactive distillation flowsheet of EtAc system.

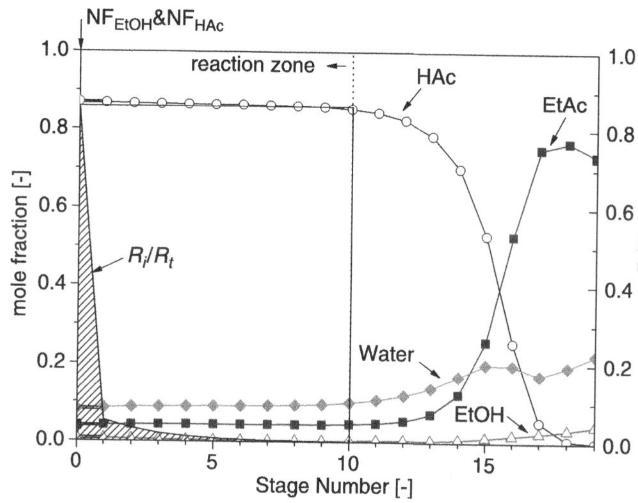

Figure 16.24 Composition profile in reactive distillation column; (shaded area) fraction of total conversion in each tray and close to 90% of total conversion occurring in column base.

optimized flowsheet consists of a reactive distillation column and a stripper as shown in Figure 16.23. The flowsheet is obtained by minimizing the TAC:

$$\text{TAC} = \text{operating cost} + \frac{\text{capital cost}}{\text{payback period}} \tag{16.14}$$

Here, a payback time of 3 periods is used. The operating cost includes the cost of steam, cooling water, and catalyst. The capital cost is the costs of the column, trays, and heat exchangers. A catalyst life of 3 months is assumed.

The reactive distillation column has 10 reactive trays plus a reactive column base, 9 rectifying trays, and a decanter. The stripper column has 9 trays. Figure 16.24 gives the composition profile of the reactive distillation. From now on, emphasis will be placed on the reactive distillation column, and design alternatives will be sought while keeping the same design and product specifications for the stripper.

Single Reactive Tray. The composition profile in the reactive distillation column (Fig. 16.24) reveals that the concentration of the heavy reactant (HAc) remains fairly constant (~90%) throughout the reactive zone, but the light reactant (EtOH) has a very low concentration in the column base (~1%) and the composition decreases toward the end of the reactive zone. Of more importance, close to 90% of the total conversion occurs in the column base, which is shown in the shaded area in Figure 16.24. This naturally suggests a coupled reactor/separator configuration.[7] The configuration in which all of the catalyst is placed in the column base is called the single reactive tray configuration. This flowsheet is

[7]C. K. Yi and W. L. Luyben, Design and control of coupled reactor/column systems. 1. A binary coupled reactor/rectifier, *Comput. Chem. Eng.* **21**, 25 (1997).

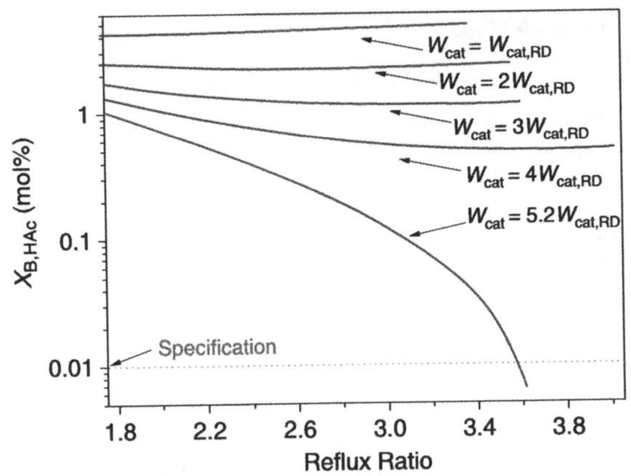

Figure 16.25 Catalyst loading in single reactive tray configuration versus achievable acetic acid conversion under different reflux ratios.

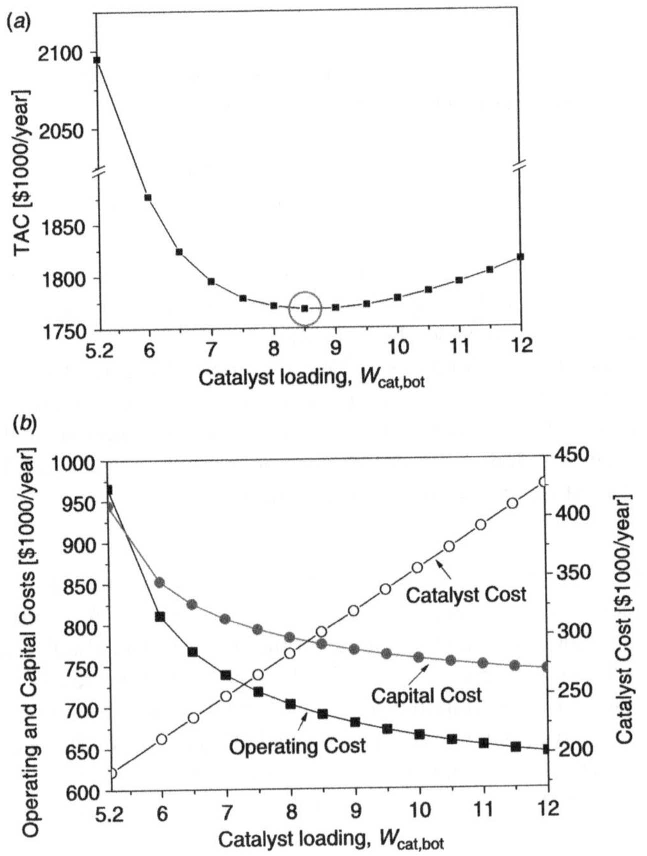

Figure 16.26 (*a*) Effect of catalyst loading in column base ($W_{cat,bot}$) on TAC for single reactive tray configuration and (*b*) tradeoff of operation cost (excluding catalyst cost), capital cost, and catalyst cost.

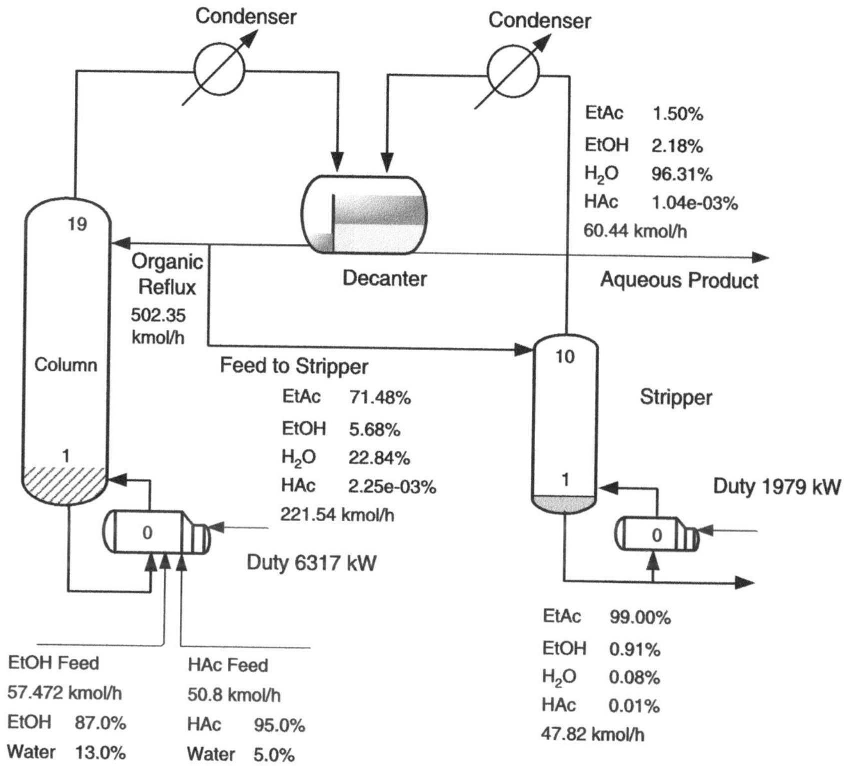

Figure 16.27 Optimized single reactive tray configuration (reactive distillation single reactive tray) for EtAc production.

intuitively appealing because it has the advantage of easy catalyst replacement. However, simulation results show that, with the same catalyst weight as that of the reactive distillation column ($W_{cat,RD} = W_{cat,bot} + W_{cat,trays}$), the single reactive tray configuration is infeasible for the given purity specification. In fact, the required catalyst loading is $5.2 \times W_{cat,RD}$. Figure 16.25 clearly shows that the acid concentration at the column overhead cannot reach the specification unless a substantial increase in the catalyst loading is made. Thus, the performance of the single reactive tray configuration is not as good as that of the reactive distillation. The conversion is only 93% with the same catalyst loading ($1 \times W_{cat,RD}$) as opposed to 99% for reactive distillation.

To obtain an improved design for the single reactive tray configuration, the TAC is computed as the amount of catalyst in the column base $W_{cat,bot}$ is varied. As the catalyst holdup $W_{cat,bot}$ is increased, the capital and energy costs decrease and the catalyst cost increases. The optimized single reactive tray design gives a catalyst loading of $8.5 \times W_{cat,RD}$, and the energy cost increases by 24% as shown in Figure 16.26. The improved single reactive tray design is demonstrated in Figure 16.27 with an organic phase reflux ratio of 2.27 as opposed to 1.76 for reactive distillation. The optimized single reactive tray configuration has a TAC that is 30% greater than that of the reactive distillation configuration (Fig. 16.28). The result reveals that, despite having almost 90% conversion in the column base for reactive distillation, additional reaction stages are

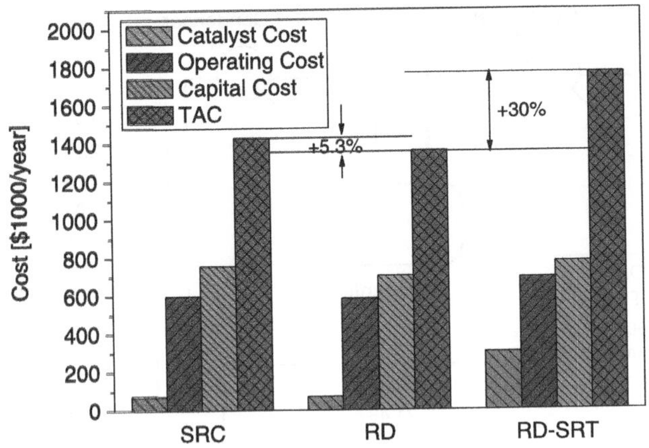

Figure 16.28 Comparison of TACs for EtAc production using reactive distillation (RD), single reactive tray reactive distillation (RD-SRT), and side reactor configuration (SRC).

essential to achieve the high purity specification (i.e., high conversion) with acceptable capital and energy costs.

Side Reactor Configuration. If a second reaction stage should be included, the questions then become, what type of reaction system should be used, and where should the

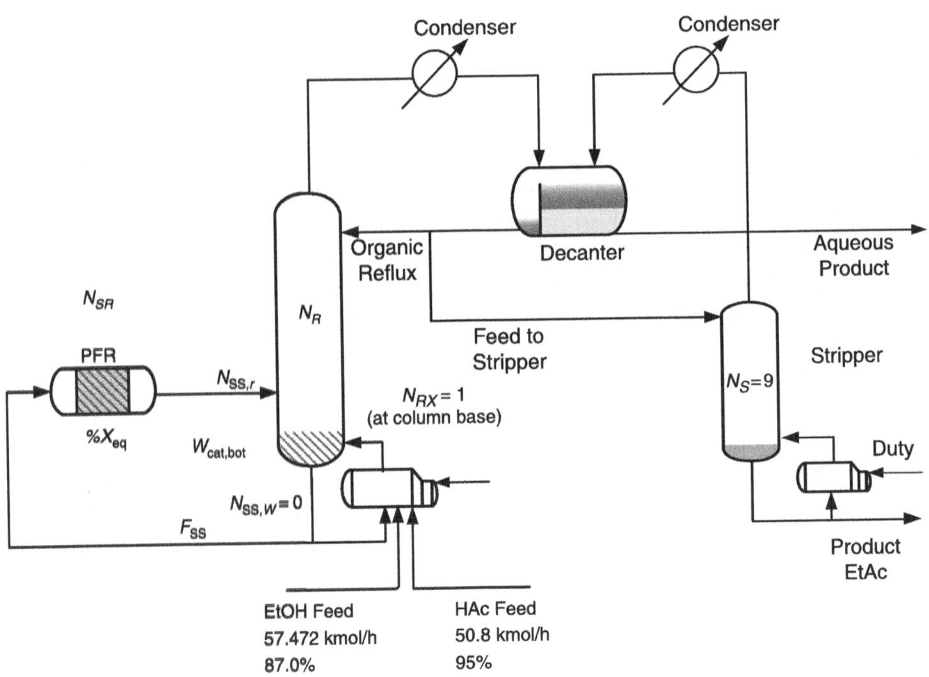

Figure 16.29 Side reactor configuration; design parameters in italic. PFR, plug flow reactor.

reactor be placed? The answer to the first question is relatively simple. From a maintenance perspective, an external side reactor is preferred (compared to an internal one). The second question actually addresses two issues: where should the side stream be withdrawn ($N_{SS,w}$, feed to the side reactor) and where should the side stream be returned to the column ($N_{SS,r}$, effluent from the side reactor)? The side stream is withdrawn from the location in the column where the reactants are most abundant. For the EtAc production system, this is from the column base. The selection of the side stream return tray $N_{SS,r}$ is less obvious, and it depends on other design parameters. In theory, finding the side stream return tray is analogous to finding the optimal feed tray location for a feedstream. However, in the side reactor configuration, this feed composition and column composition are interrelated. In other words, the side reactor effluent composition depends on the reactor size and reactor feed composition, which are functions of the side stream return tray.

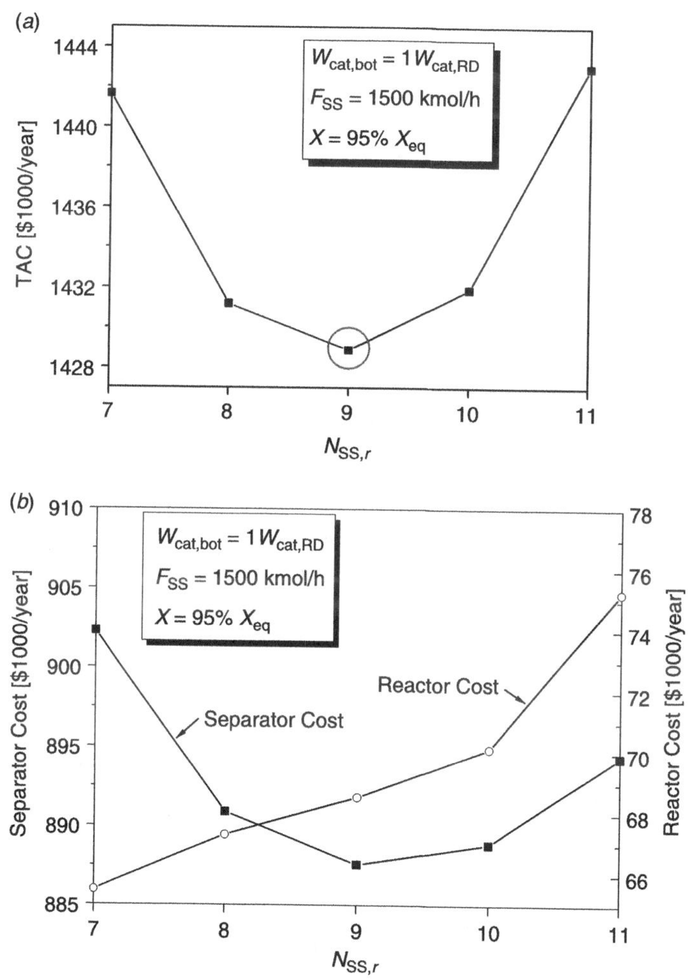

Figure 16.30 (*a*) Effect of side stream return tray ($N_{SS,r}$) on TACs and (*b*) tradeoff of separator cost (capital and operating costs) and reactor cost.

Thus, the design variables include catalyst holdup $W_{cat,bot}$ in the column base, side stream flowrate F_{SS} (side reactor feed flowrate), the percentage of equilibrium conversion ($\%X_{EQ}$) in the side reactor, and side stream (reactor effluent) return tray $N_{SS,r}$, as shown in Figure 16.29. We assume that the reactor is an adiabatic plug flow reactor with an aspect ratio (length/diameter) of 5. A systematic design procedure is proposed to find the optimal configuration. All of the simulations are carried out in Aspen Plus using the RadFrac and RPlug modules. Given the production rate and product specifications, the design steps are the following:

1. Set the total number of trays equal to that of the reactive distillation ($N_T = 19$).
2. Withdraw the side stream from the bottom of the distillation column ($N_{SS,w} = 0$).

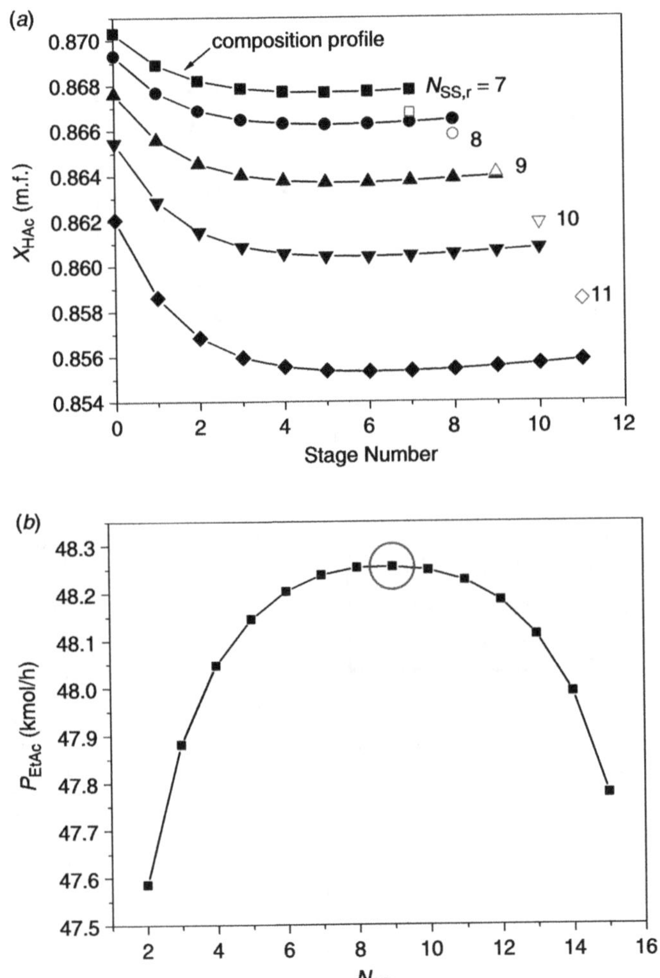

Figure 16.31 (*a*) Comparison of HAc concentration at side reactor effluent (open symbols) and in return tray (solid symbols: composition profile in column up to return tray) as side stream return tray is varied and (*b*) effect of side stream return tray ($N_{SS,r}$) on total generation of EtAc.

3. Set the number of side reactors to 1 ($N_{SR} = 1$).
4. Guess the catalyst holdup in the column base (e.g., $W_{cat,bot} = r_W \times W_{cat,RD}$).
5. Guess a side stream flowrate (e.g., $F_{SS} = r_F \times R_{org}$), which is expressed in terms of multiples (r_F) of the organic reflux flowrate (R_{org}).
6. Guess the percentage of the equilibrium conversion for the side reactor (e.g., $\%X_{eq} = 80\%$).
7. Guess a side stream return tray $N_{SS,r}$.
8. Adjust the organic reflux ratio (R_{org}/D_{org}) until the HAc specification in the overhead (<100 ppm) is reached, adjust the stripper heat duty until the EtAc specification at the stripper bottoms (>99 mol%) is achieved, and adjust the reactor size until the $\%X_{eq}$ is achieved. It is much easier to perform this in a sequential manner in order to avoid infeasible design parameters.

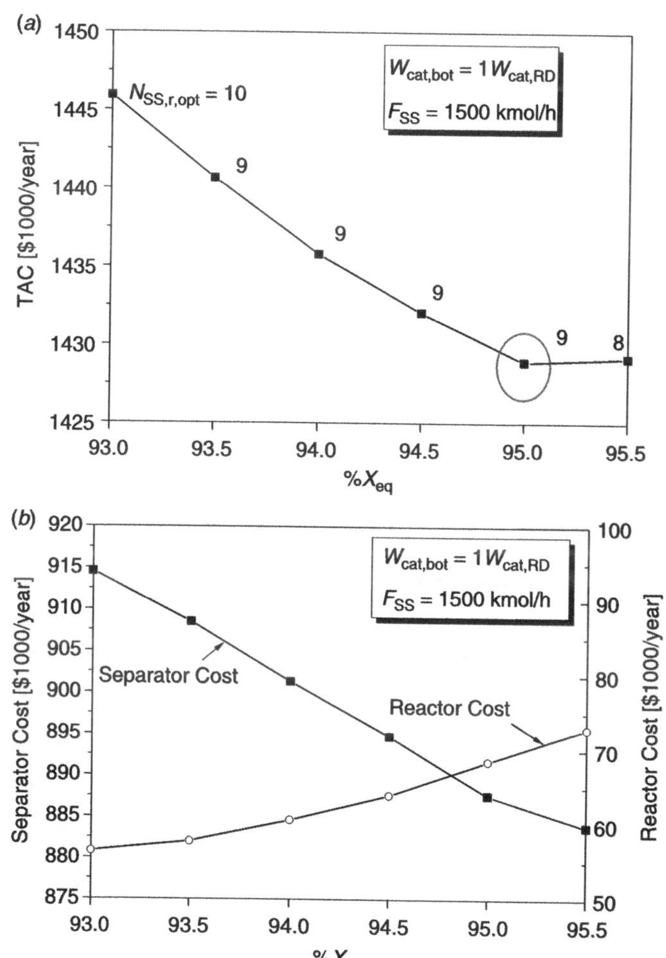

Figure 16.32 (*a*) Effects of percentage of equilibrium conversion ($\%X_{eq}$) of side reactor on TACs and (*b*) tradeoff in separator cost and reactor cost.

9. Go back to step 7 and change $N_{SS,r}$ until the TAC is minimized.
10. Go back to step 6 and change $\%X_{eq}$ until the TAC is minimized.
11. Go back to step 5 and change F_{SS} until the TAC is minimized.
12. Go back to step 4 and change $W_{cat,bot}$ until the TAC is minimized.

The effects of side stream return tray $N_{SS,r}$ are shown in Figure 16.30*a*. Returning the reactor effluent on tray 9 ($N_{SS,r} = 9$) corresponds to the minimum TAC. The tradeoff comes from the reactor cost and the separator cost, as shown in Figure 16.30*b*. The side stream is a large flow consisting of reactants and products (90 mol% HAc). From the separation cost point of view, it is optimal to feed the effluent from the side reactor onto a tray with a similar composition. In contrast, moving the return tray location

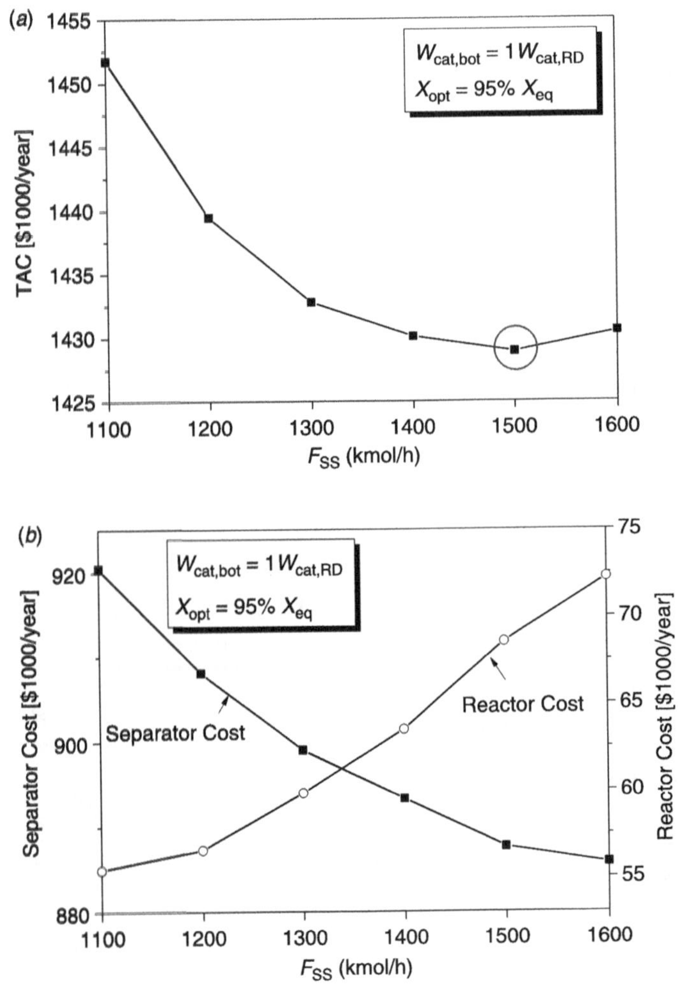

Figure 16.33 (*a*) Effect of side stream flowrate (F_{SS}) on TACs and (*b*) tradeoff in separator cost and reactor cost.

upward leads to a decrease in acetic acid concentration, as shown in Figure 16.31*a*, so the side reactor needs a larger reactor volume and more catalyst. In Figure 16.31 we observe that 1) the optimal return tray for the side reactor effluent is the tray with the liquid composition closest to that of the effluent, and 2) the production rate of the EtAc is at its maximum when $N_{SS,r} = 9$. Note that the tray number is counted from the bottom up.

Next, we study the effects of reactor conversion, which is expressed in terms of the $\%X_{eq}$ to the side reactor configuration design. Figure 16.32*a* shows that $\%X_{eq} = 95$ is the optimal conversion. Again, we look at the reactor cost and the separator cost, which are provided in Figure 16.32*b*. As $\%X_{eq}$ increases, the reactor cost increases monotonically because of a larger reactor volume and more catalyst needed for the additional reaction. At the same time, the separator cost decreases because of less reaction in the column base.

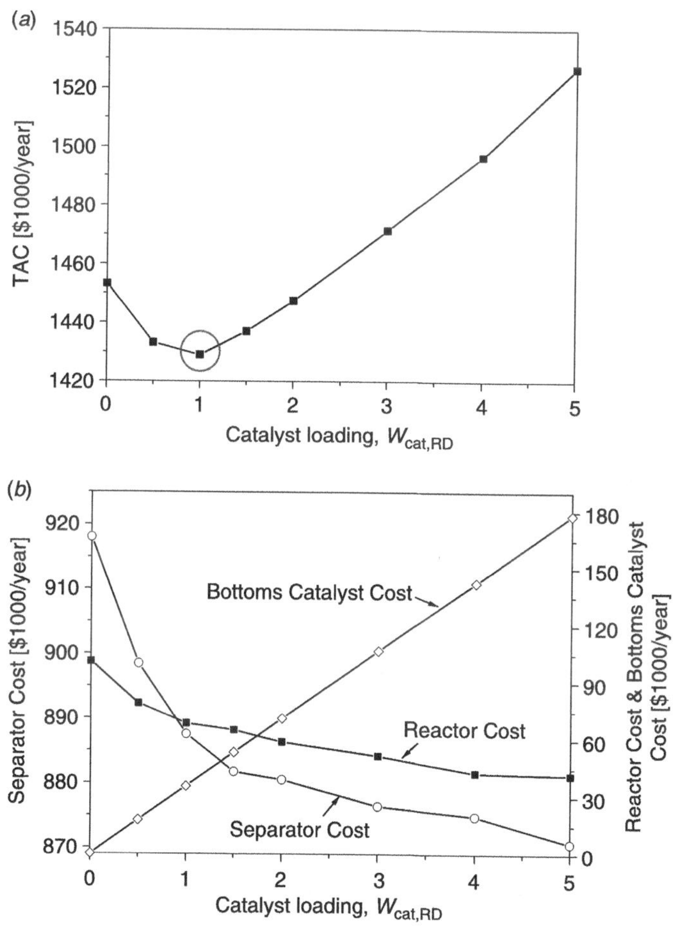

Figure 16.34 (*a*) Effect of catalyst loading in column base ($W_{cat,bot}$) on TACs and (*b*) tradeoff of separator cost (excluding catalyst cost), reactor cost, and bottoms catalyst cost.

The effects of side stream flowrate F_{SS} on the side reactor configuration design are demonstrated in Figure 16.33*a*. The optimum side stream flowrate is 1500 kmol/h. A larger side stream flowrate requires a larger reactor volume and more catalyst to achieve the same conversion, which leads to an increase in the reactor cost, as shown in Figure 16.33*b*. Because more reaction in the side reactor reduces the loading of the column, the separator cost decreases.

Finally, we address the effects of catalyst loading in the column base ($W_{cat,bot}$) on the side reactor configuration design. Figure 16.34*a* shows that $W_{cat,bot} = W_{cat,RD}$ is the optimal catalyst amount. As the bottoms catalyst increases, more reaction occurs in the column base. Therefore, the separator requires less vapor flowrate, and the volume of the side reactor is smaller. The tradeoff is revealed in Figure 16.34*b*. If there is no catalyst in the column base ($W_{cat,bot} = 0$, only a side reactor for the reaction), this configuration can still work. However, it requires a larger flowrate of the side stream and a considerable amount of catalyst to achieve the product specification.

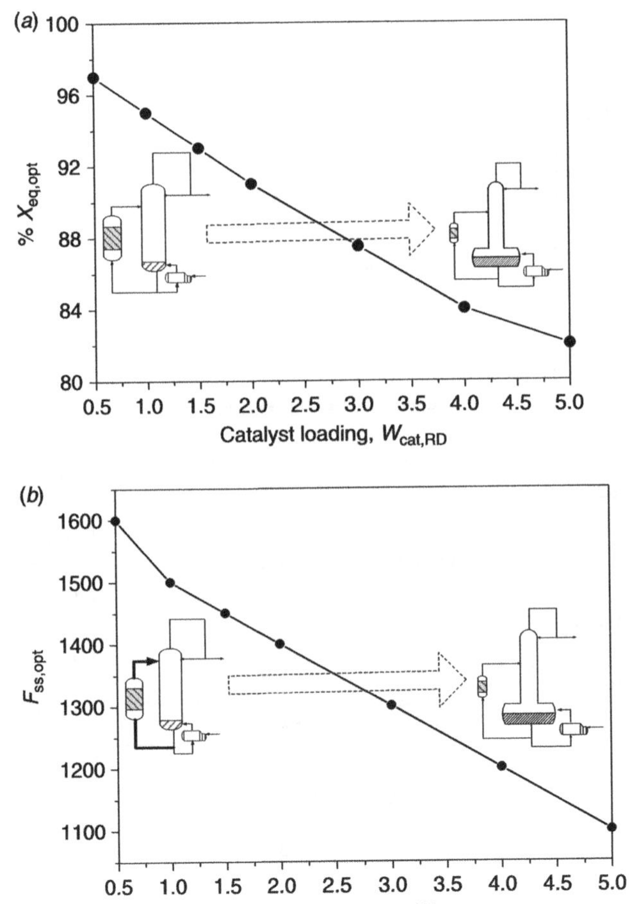

Figure 16.35 (*a*) Relationships between design variables: catalyst loading versus optimized percentage of equilibrium conversion (%X_{eq}; to achieve minimum TACs) and (*b*) catalyst loading versus the optimized side stream flowrate (to achieve minimum TACs).

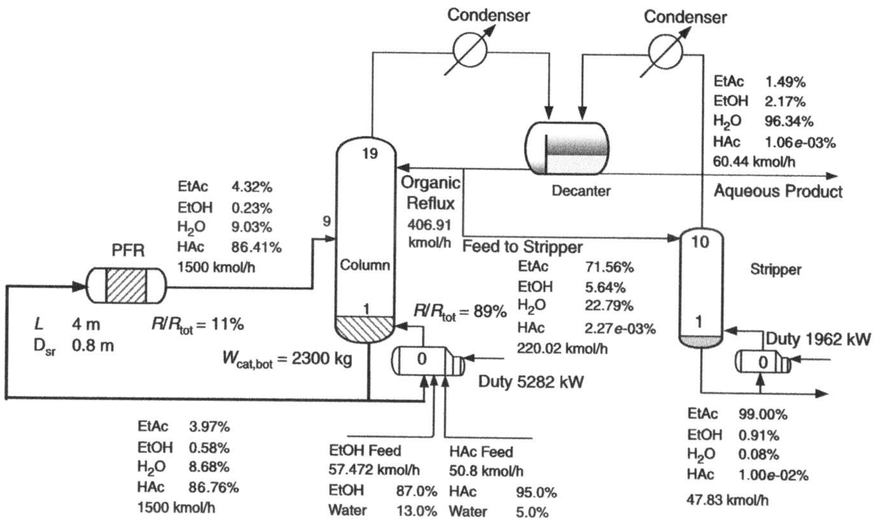

Figure 16.36 Optimized side reactor configuration flowsheet for EtAc production.

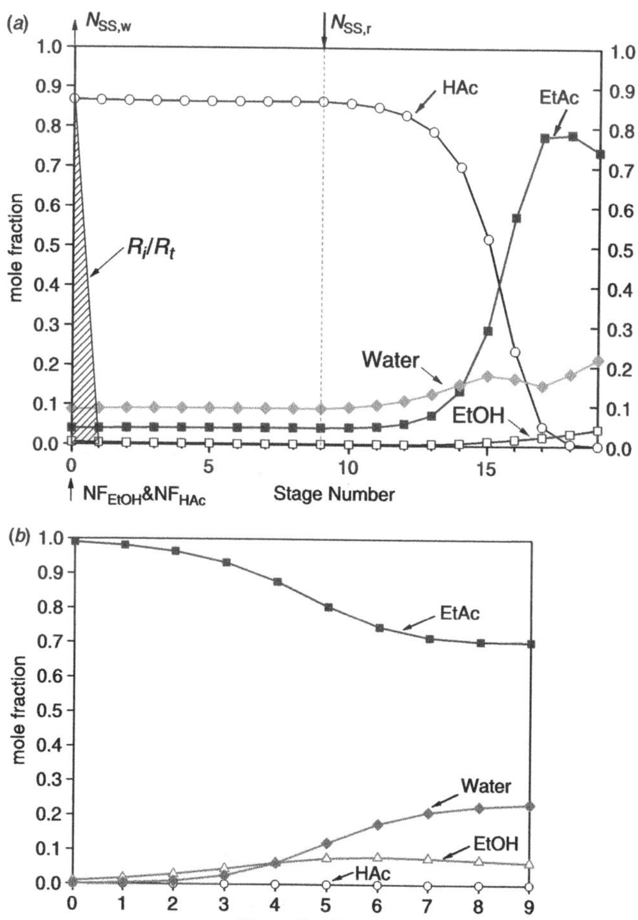

Figure 16.37 Composition profile in side reactor configuration: (*a*) reactive distillation column (shaded area, fraction of total conversion in each tray) and (*b*) stripping column.

TABLE 16.6 **Parameter Values of Steady-State Designs for Reactive Distillation (RD), Single Reactive Tray (SRT), and Side Reactor Configuration (SRC)**

Flowsheet	RD		SRT		SRC	
Column Configuration	RD Column	Stripper	Column	Stripper	Column	Stripper
Total trays including reboiler	20	10	20	10	20	10
No. of trays						
Stripping section (N_S)		9		9		9
Reactive section (N_{RX})	11		1		1	
Rectifying section (N_R)	9		19		19	
Reactive tray	0–10		0		0	
Acetic acid feed tray	0		0		0	
Alcohol feed tray	0		0		0	
Feed flowrate (kmol/h)						
Acid	50.80		50.80		0.80	
Alcohol	57.47		57.47		57.47	
Product flowrate (kmol/h)						
Top	60.45		60.44		60.443	
Bottom		47.82		47.83		47.83
$X_{D,aq}$						
Acid (mf)	0.00001		0.00001		0.00001	
Alcohol (mf)	0.02182		0.0218		0.0217	
Acetate (mf)	0.01498		0.015		0.0149	
Water (mf)	0.96391		0.9631		0.9619	
X_B						
Acid (ppm)		100		100		100
Alcohol (mf)		0.0091		0.0091		0.0091
Acetate (mf)		0.9900		0.99		0.99
Water (mf)		0.0008		0.0008		0.0008
Duty (kW)						
Condenser	−4789.92	−1668.7	−5845.6	−1669	−4941.3	−1653.9

Subcooling	−881.45		−1030.0		−901.2	
	Reboiler	5113.4	1979.1	6316.9	1979.4	5282.5
						1962.4
Column diameter (m)	2.043	1.37	2.35	1.37	2.27	1.36
Weir height (m)	0.1016	0.0508	0.1016	0.0508	0.1016	0.0508
Decanter temperature (°C)	40		40		40	
Side reactor (m)						
Diameter (m)					0.8	
Length (m)					4	
Heat transfer area (m^2)						
Condenser	177.1	61.9	215.6	61.9	182.6	61.3
Subcooling	179.96		210.3		183.99	
Reboiler	190.7	83.7	235.6	83.7	196.9	82.9
Total capital cost ($1000)	2103.0		2325.0		2269.3	
Column	521.4		583.7		566.1	
Reactor	0		0		90.1	
Column trays	91.42		109.3		104.2	
Heat exchangers	1490.2		1632.1		1508.3	
Total operating cost ($1000/year)	656.10		993.2		672.5	
Catalyst cost	71.30		302.8		73.9	
Energy cost	584.90		690.3		598.6	
TAC ($1000/year)	1357.12		1768.16		1428.89	

In summary, the tradeoff comes from the competition between the reaction in the column base and the side reactor. If we increase the reaction in the column base, it will reduce the loading of the side reactor and consequently the side reactor will require a smaller side stream flowrate and a smaller percentage of equilibrium conversion, as shown in Figure 16.35.

For the case of one side reactor, the TAC only increases by a factor of 5% compared to that of the reactive distillation design, and we conclude that an additional external reactor will not be necessary for the side reactor configuration. Specifically, the energy cost of the side reactor configuration increases by a factor of 3% while using 1.67 times the amount of the catalyst, as compared to the reactive distillation process. Figure 16.36 shows the optimized flowsheet where the side stream flowrate is 3.69 times the organic reflux flowrate and the catalyst loading in the column base and side reactor are $W_{cat,bot} = W_{cat,RD}$ and $W_{cat,SR} = 0.67 \times W_{cat,RD}$, respectively. This corresponds to 95% of the equilibrium conversion for the side reactor. Moreover, 89% of the total conversion occurs in the column base while the remaining 11% happens in the side reactor. Composition profiles for the side reactor configuration process (Fig. 16.37) show that the composition profile of the side reactor configuration is almost the same as that of the reactive distillation (see Fig. 16.24). The parameter values of the steady-state designs of all three configurations (reactive distillation, single reactive tray, and side reactor configuration) are summarized in Table 16.6.

16.5 CONTROL OF COLUMN/SIDE REACTOR PROCESS FOR ETHYL ACETATE SYSTEM

The number of studies on the control of reactive distillation columns has grown steadily in the past decade. Luyben and colleagues propose eight control structures for the neat reactive distillation of the ideal system (A + B ⇔ C + D): CS1–CS6,[8] CS6–CS7,[9] and CS7–CS8.[1,3] Chapter 13 explores control structure design for acetic acid esterification with different alcohols (ranging from C1 to C5), which are categorized into three types of flowsheets. As for the control of various coupled reactor/column systems, several articles have been published, such as Yi and Luyben[7] and Chiang et al.[10] However, the control of the side reactor configuration has not been discussed in the literature. In this section, a systematic approach to design the control structure is suggested. The design steps are the following:

1. Determine manipulated variables.
2. Use NRG (nonsquare RGA) to determine temperature control trays.

[8]M. A. Al-Arfaj and W. L. Luyben, Comparison of alternative control structures for an ideal two-product reactive distillation column, *Ind. Eng. Chem. Res.* **39**, 3298–3307 (2000).

[9]M. A. Al-Arfaj and W. L. Luyben, Comparative control study of ideal and methyl acetate reactive distillation, *Chem. Eng. Sci.* **57**, 5039–5050 (2002).

[10]S. F. Chiang, C. L. Kuo, C. C. Yu, and D. S. H. Wong, Design alternatives for amyl acetate process: Coupled reactor/column and reactive distillation, *Ind. Eng. Chem. Res.* **41**, 3233–3246 (2002).

3. Use a decentralized PI controller.
4. Use RGA for variable pairing.
5. Use relay feedback to find K_u and P_u.
6. Use Tyreus–Luyben tuning to find controller settings.

16.5.1 Determining Manipulated Variables

Compared to reactive distillation, the side reactor configuration has 1 extra manipulated variable, side stream flowrate F_{SS}, in addition to two feedflows (F_{EtOH} and F_{HAc}), two reboiler duties ($Q_{R,RD}$ and $Q_{R,STR}$), two condenser duties, one subcooling duty, aqueous and organic distillate flowrates, organic reflux flowrate R_{org}, and stripper bottoms flowrate. Similar to the control of the reactive distillation system, the alcohol feedflow is selected as the throughput manipulator. We are left with 11 manipulated variables.

In the EtAc side reactor configuration process, there are six inventory control loops required, which include the control of four liquid levels (the reactive distillation column base level, the organic phase level in the decanter, the aqueous phase holdup in the decanter, and the stripper base level) and the two column pressures of the reactive distillation and stripping columns. Basic inventory and related loops are arranged in the following ways. The top pressures of two columns are controlled by manipulating the condenser duties. The decanter temperature is controlled by changing the chilled water flow. As for liquid inventories, except for the reactive distillation column base, the remaining liquid levels are controlled by their liquid outlet flow as shown in Figure 16.38.

For the reactive distillation column base-level control, we have three candidates: side draw flow F_{SS}, organic reflux flow R_{org}, and reboiler duty $Q_{R,RD}$. Because the side draw flow has 0 gain effect on the base level and the organic reflux flow gives slow dynamics,[11] we choose the heat input of the reactive distillation column $Q_{R,RD}$ to control the base level. In doing this, we have four manipulated variables left: F_{SS}, F_{HAc}, R, and $Q_{R,STR}$.

After selecting the inventory controls, the remaining manipulated variables (HAc feed, organic reflux flow, and stripper heat duty) are used for maintaining the stoichiometric balance and for product quality control (purities of the ethyl acetate and water products). In the stripper, the acetate product purity is controlled by the stripper heat input $Q_{R,STR}$, and a temperature is used to infer the product composition. From the RCM (Fig. 16.22), the water purity is determined by the tie line of the LL equilibrium, so it is left uncontrolled. In order to maintain the stoichiometric balance, we use a temperature control to adjust the feed ratio F_{HAc}/F_{EtOH} (FR). Thus, we have a 2×2 multivariable system. From a unitwise perspective, the pairing for the manipulated variables and the controlled variables are FR$-T_{RD}$ and $Q_{R,STR}-T_{STR}$.

We have two manipulated variables left, side draw flow F_{SS} and organic reflux flowrate R_{org}. For the side stream flowrate, one can choose to ratio the side stream flowrate to the feed or simply fix F_{SS} to its setpoint. The latter is utilized here because it is a relatively large flowrate ($F_{SS} = 1500$ kmol/h) compared to the reflux flowrate ($R = 407$ kmol/h) and frequent perturbations may introduce unnecessary disturbances to the lower section of the column. Two possible approaches can be taken for the organic reflux. One is to fix the reflux ratio RR. The other is to ratio the reflux to the feed (R/F_{EtOH}). Therefore, we

[11] I. L. Chien, Y. P. Teng, H. P. Huang, and Y. T. Tang, Design and control of an ethyl acetate process: Coupled reactor/column configuration, *J. Process Control* **15**, 435–449 (2005).

(*a*)

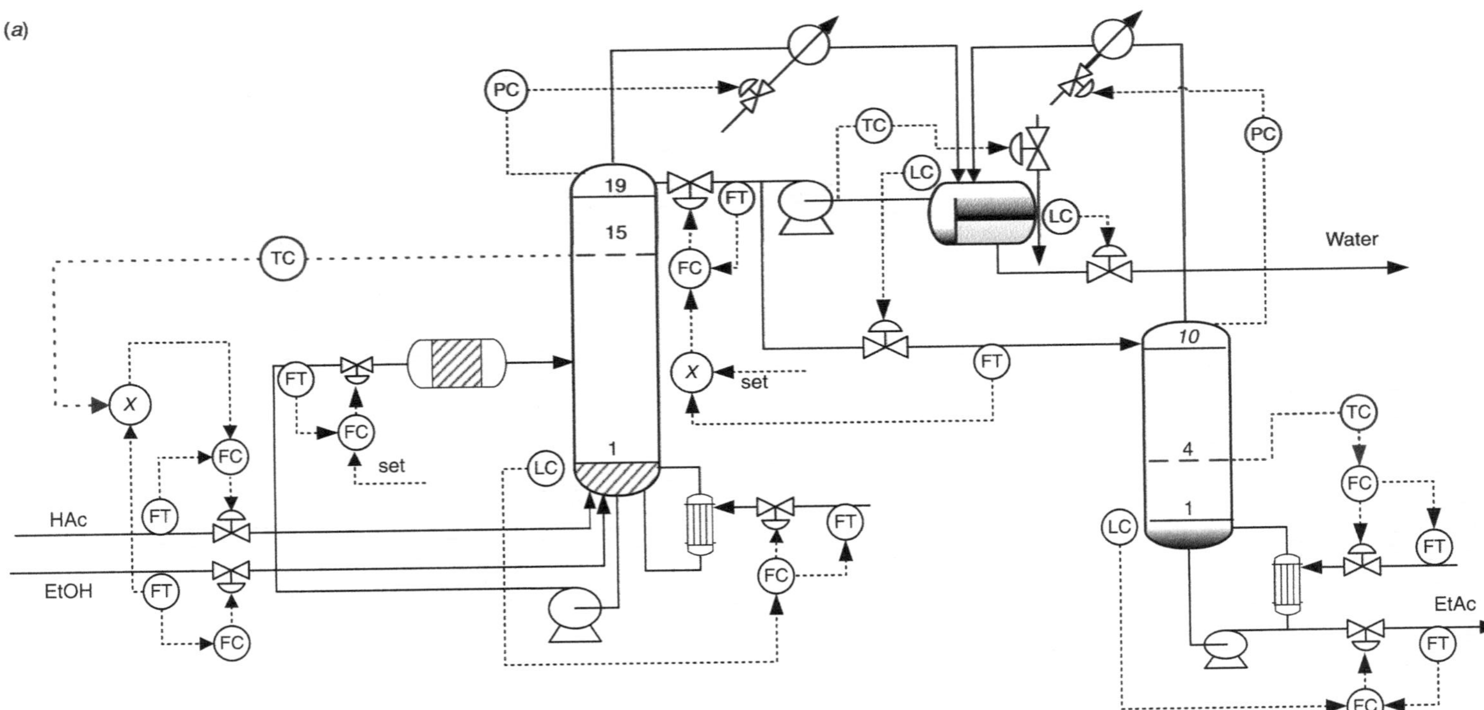

Figure 16.38 Control of side reactor configuration process: (*a*) CS1 (fixing RR), (*b*) CS2 (fixing R/F ratio), and (*c*) CS3 (CS2 with parallel cascade control on product purity).

(b)

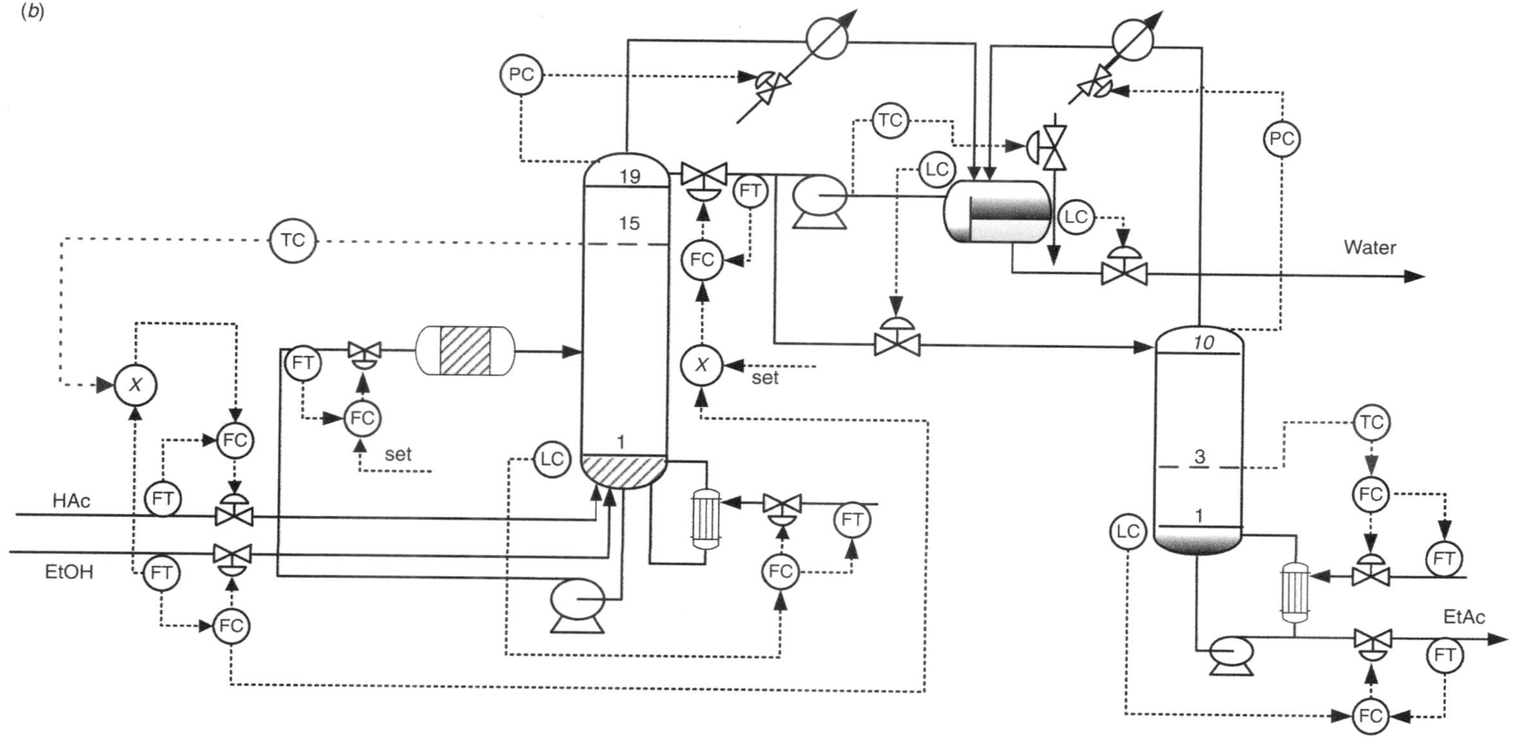

Figure 16.38 *(Continued)*.

(c)

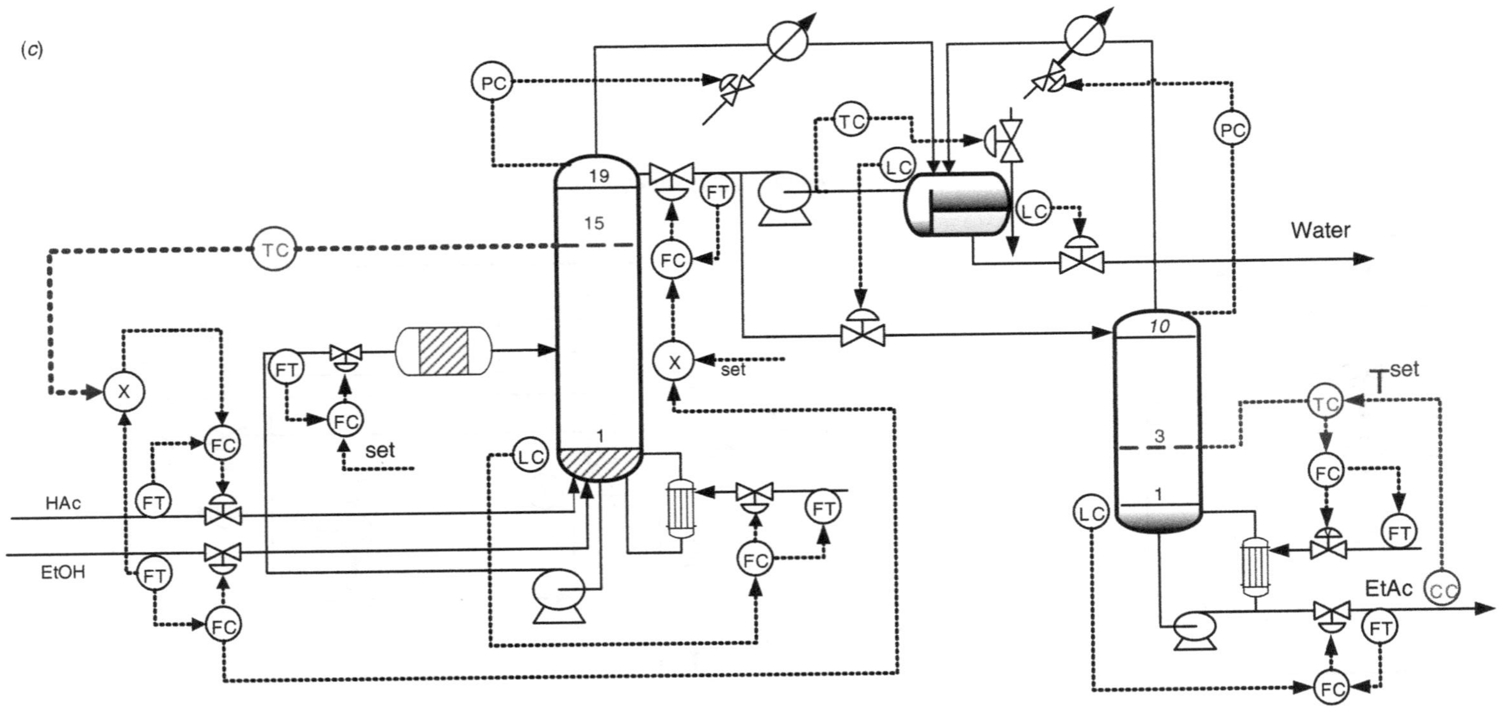

Figure 16.38 *(Continued)*.

have two possible control structures: CS1 for keeping RR constant and CS2 for fixing the reflux to feed ratio as shown in Figure 16.38.

16.5.2 Selection of Temperature Control Trays

Sensitivity analyses are performed for $\pm 0.01\%$ variations in the corresponding manipulated variables (FR and $Q_{R,\mathrm{STR}}$). Results are shown in Figure 16.39 for CS1 and Figure 16.40 for CS2. In both control structures, an *increase* in $Q_{R,\mathrm{STR}}$ leads to a *decrease* in the tray temperatures of the reactive distillation column because of a larger recycle flowrate for the reactive distillation column. As for the feed ratio (FR) changes, an increase in the heavy reactant results in the main column (reactive distillation column) tray temperatures increasing while the stripper temperatures show little variation. At the completion of the sensitivity analyses, NRG[12] is used to find the temperature control trays in each column. The largest row sum of the NRG in each column is selected as the temperature control trays (Fig. 16.41). Note that the temperature control trays selected by the NRG are consistent with the most sensitive trays in Figures 16.39 and 16.40. The analyses select temperature control trays $T_{\mathrm{RD},15}$ and $T_{\mathrm{STR},4}$ for CS1 and $T_{\mathrm{RD},15}$ and $T_{\mathrm{STR},3}$

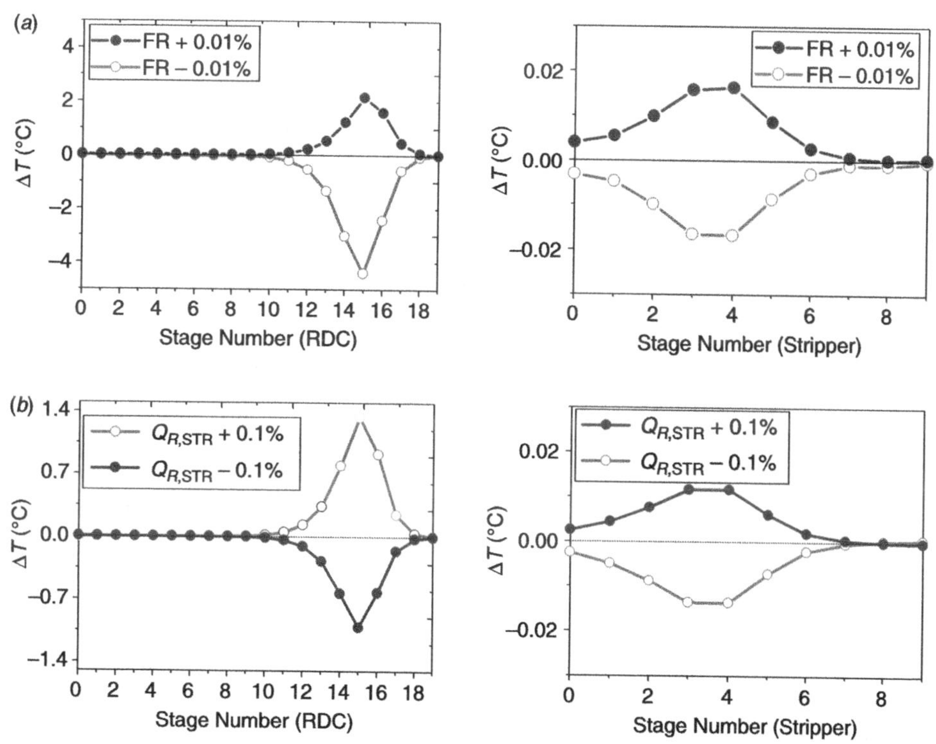

Figure 16.39 Sensitivities of tray temperatures for $\pm 0.1\%$ manipulated variables changes in CS1: (*a*) feed ratio changes and (*b*) stripper reboiler duty changes.

[12] J. W. Chang and C. C. Yu, The relative gain for non-square multivariable systems, *Chem. Eng. Sci.* **45**, 1309–1323 (1990).

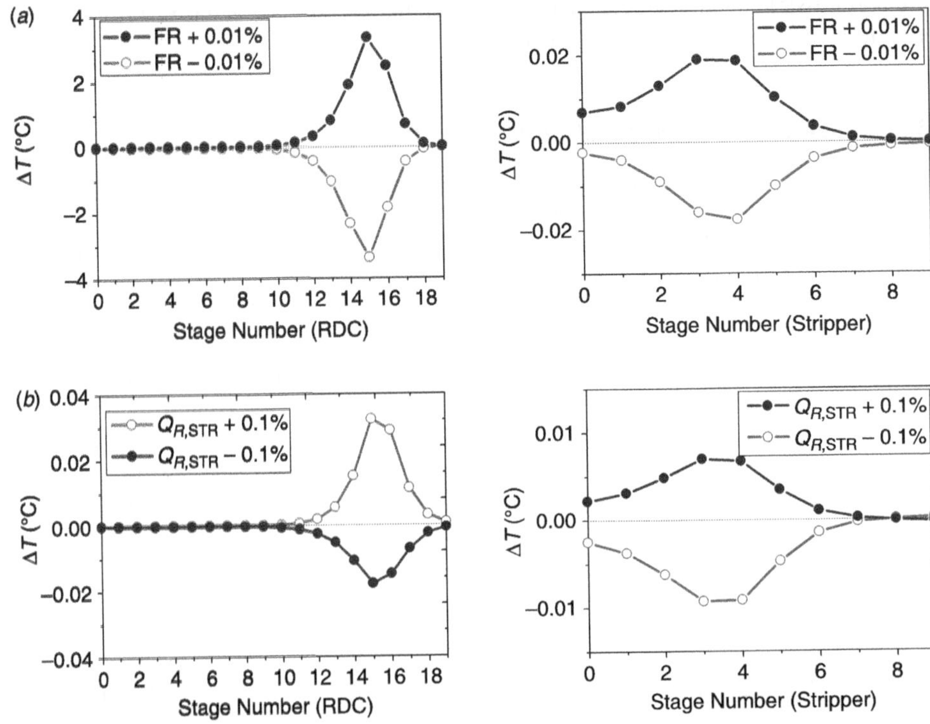

Figure 16.40 Sensitivities of tray temperatures for $\pm 0.1\%$ manipulated variables changes in CS2: (*a*) feed ratio changes and (*b*) stripper reboiler duty changes.

for CS2. Next, the RGAs are computed for variable pairing (Table 16.7). They give the following controller pairing:

$$
\begin{aligned}
\text{CS1:}\quad & \text{FR} - T_{\text{RD},15} \\
& Q_{R,\text{STR}} - T_{\text{STR},4} \\
\text{CS2:}\quad & \text{FR} - T_{\text{RD},15} \\
& Q_{R,\text{STR}} - T_{\text{STR},3}
\end{aligned}
$$

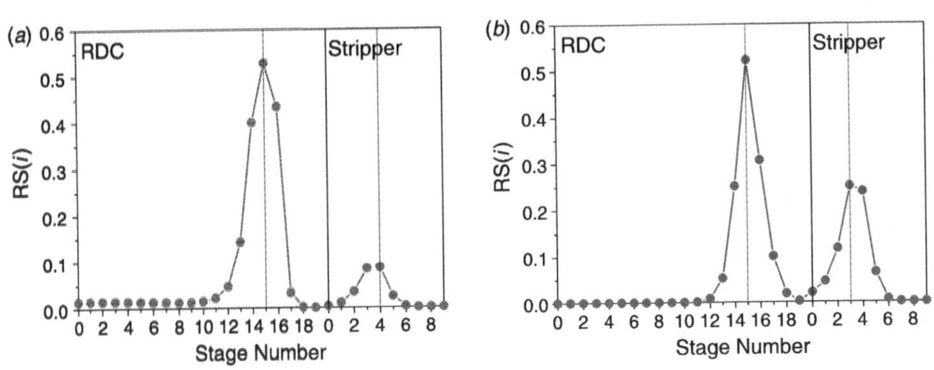

Figure 16.41 Row sums of NRG for reactive distillation column and stripper for (*a*) CS1 and (*b*) CS2.

TABLE 16.7 Variable Pairing, Results from Relay Feedback Tests, and Final Controller Settings

	Pairing	RGA	Results of Relay		Control Parameters	
			K_u	P_u (min)	K_c	τ_I (min)
CS1	FR–T_{RD}	$T_{RD,15}$ $T_{STR,4}$	144.58	7.14	1.93[a]	357[a]
	$Q_{R,STR}$–T_{STR}	$\Lambda = \begin{bmatrix} 0.67 & 0.33 \\ 0.33 & 0.67 \end{bmatrix}$ FR, $Q_{R,STR}$	60.71	1.56	6.75[a]	9.36[a]
CS2	FR–T_{RD}	$T_{RD,15}$ $T_{STR,3}$	25.06	12.96	1.67[a]	129.6[a]
	$Q_{R,STR}$–T_{STR}	$\Lambda = \begin{bmatrix} 0.98 & 0.02 \\ 0.02 & 0.98 \end{bmatrix}$ FR, $Q_{R,STR}$	57.93	1.56	19.31	3.12
CS3	FR–T_{RD}	—	27.22	12.24	0.91[a]	244.8[a]
	$x_{B,EtAc}$–T_{STR}^{SET}		16.92	19.32	5.64	38.64
	$Q_{R,STR}$–T_{STR}		57.93	1.56	9.65[a]	6.24[a]

[a]Further detuning from the Tyreus–Luyben settings.

16.5.3 Controller Design

Once the decentralized control structure is in place, dynamic simulations are performed using Aspen Dynamics. A third-order 0.5-min time lag is assumed for temperature measurement[13] and a dead time of 5 min is assumed for the composition analyzer. The liquid level is controlled using a proportional-only controller. PI controllers are used for flow, pressure, composition, and temperature controls. Relay-feedback tests are performed on the temperature loops to find the ultimate gains (K_u) and ultimate period (P_u) of each temperature control loop, and initial controller parameters are calculated according to $K_c = K_u/3$ and $\tau_I = 2P_u$. In order to obtain an acceptable damping, further detuning from the initial settings are required for some of the loops. Controller settings for all loops in CS1 and CS2 are summarized in Table 16.7.

16.5.4 Performance

The control performance is tested for feedflow and feed composition disturbances (ethanol and acetic acid). Figure 16.42 presents the dynamic responses of CS1 and CS2 for ±20% feedflow changes. The results show that CS2 has fast (<5-h settling time) and symmetric responses but CS1 exhibits much slower (~10 h) and oscillatory responses. The reason for that is that we have a severe interaction between the two process units (reactive distillation and stripper) for CS1 compared to that of CS2. This can be seen in Figures 16.39 and 16.40 when one compares the reactive distillation temperature sensitivity for the stripper reboiler duty changes or in the RGA values in Table 16.7. It can also be seen that with temperature control (CS1 and CS2) steady-state errors exist for both product compositions (0.05% error for 20% production rate changes). For HAc feed composition disturbances (±5 mol%) and EtOH feed composition disturbances (−5 and −10 mol%), the CS2 structure

[13]W. L. Luyben, B. D. Tyreus, and M. L. Luyben, *Plantwide Process Control*, McGraw-Hill, New York, 1998.

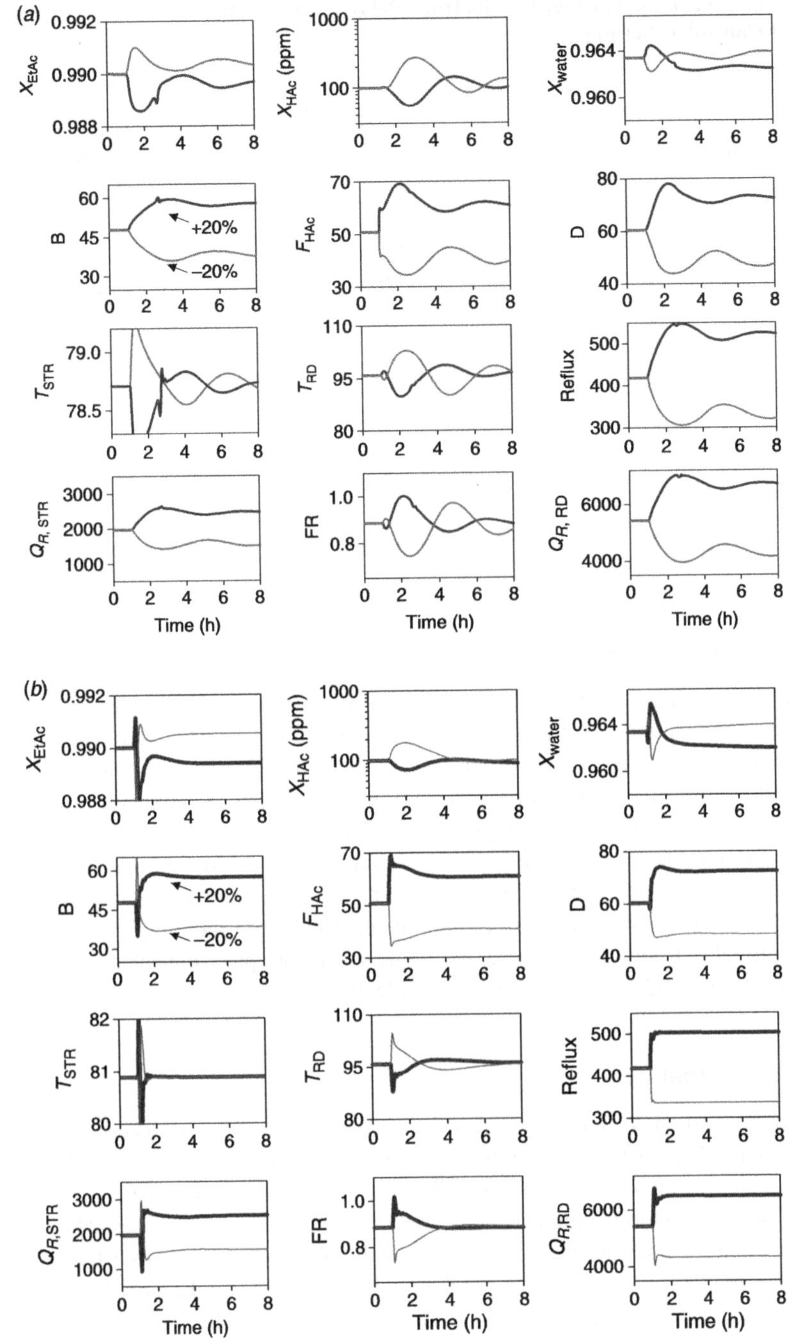

Figure 16.42 Load responses for ±20% production rate changes using (*a*) CS1 and (*b*) CS2.

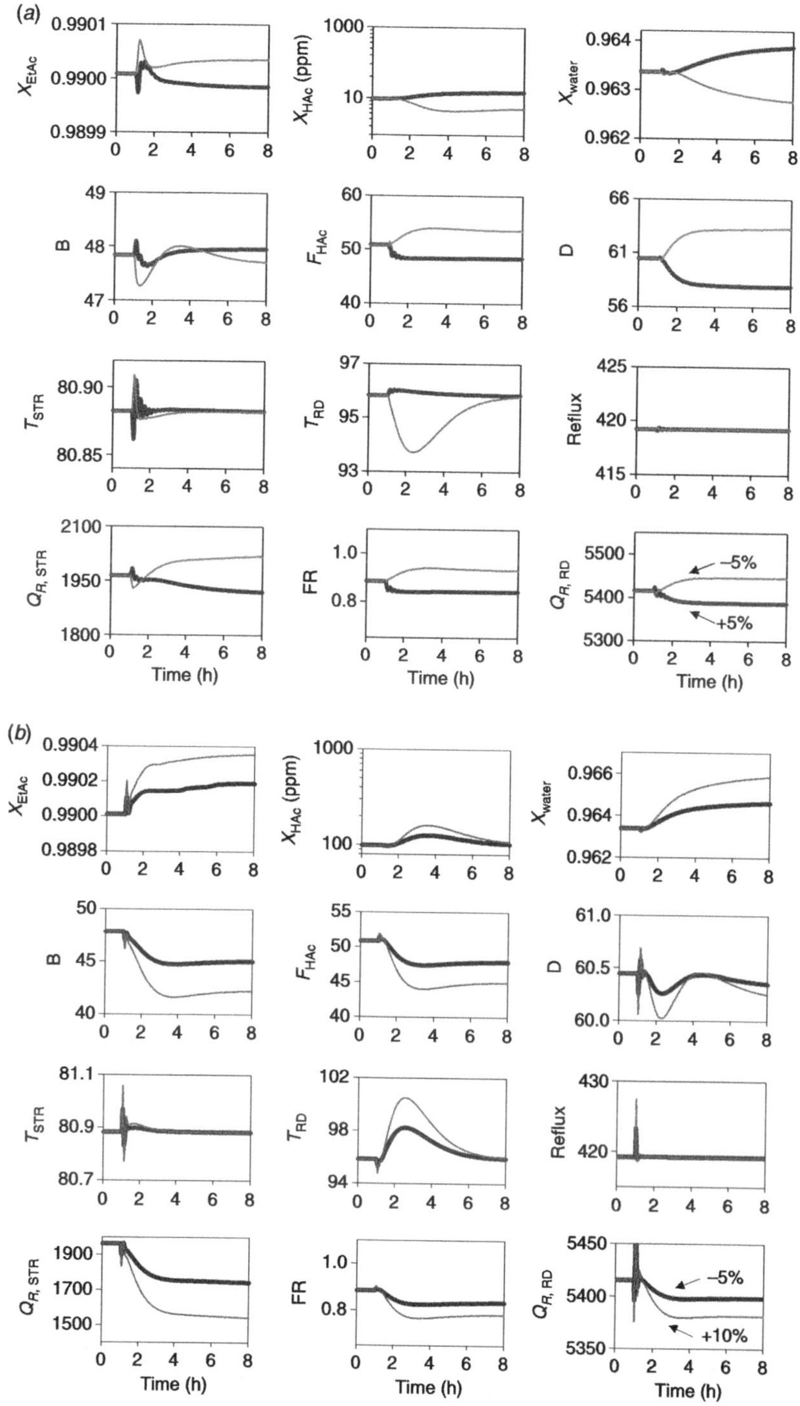

Figure 16.43 Load responses for feed composition disturbances using CS2 (*a*) ± 5 mol% HAc feed composition changes and (*b*) -5 and -10 mol% EtOH feed composition changes.

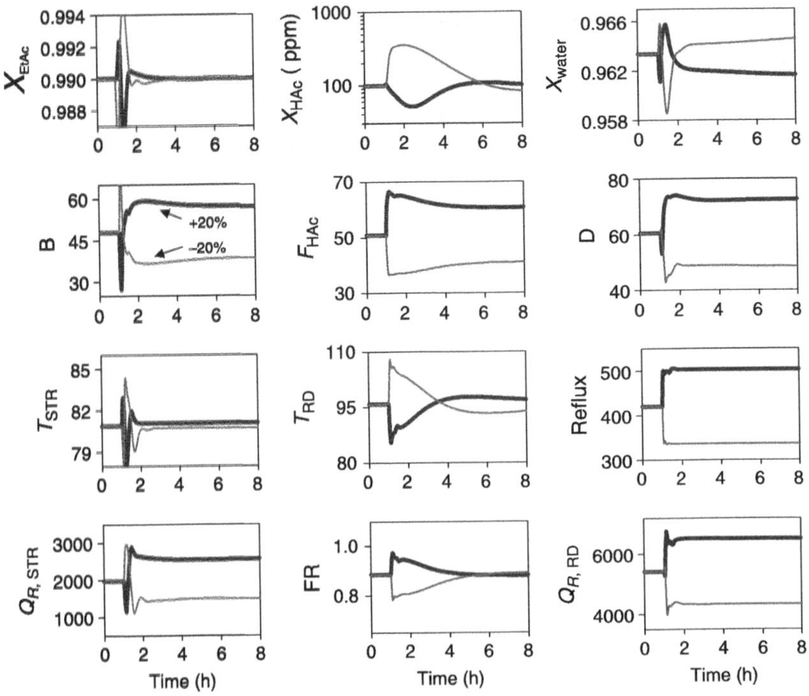

Figure 16.44 Load responses for ±20% production rate changes using temperature/composition cascade control (CS3) in product purity loop.

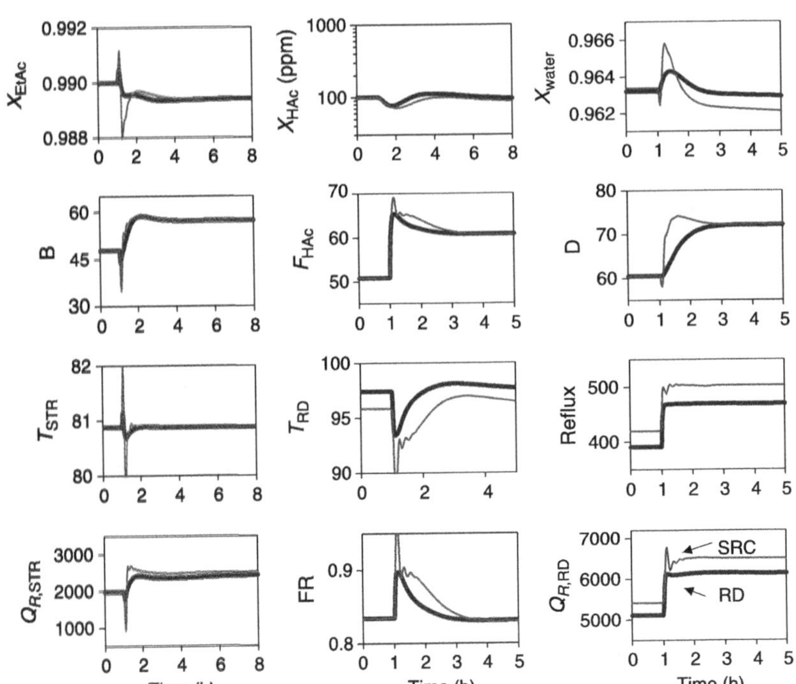

Figure 16.45 Load responses for 20% production rate increase using CS2 for side reactor configuration and reactive distillation.

still works very well with fast responses (Fig. 16.43). The results clearly indicate that, instead of fixing the reflux ratio, it is preferable to fix the reflux to feed ratio for this dual-temperature control of the side reactor configuration. Of more importance, reasonable control performance can be obtained using simple temperature control to handle production rate and composition variations.

16.5.5 Extension to Composition Control

To eliminate the steady-state offset in acetate purity, we also consider composition control. A parallel cascade control structure is used in which the setpoint of the stripper temperature controller is reset by the output of the composition controller. Thus, we have one composition control loop $x_{\mathrm{B,EtAc}}{-}T_{\mathrm{STR}}^{\mathrm{SET}}$ and two temperature control loops $Q_{R,\mathrm{STR}}{-}T_{\mathrm{STR}}$ and $Q_{R,\mathrm{RD}}{-}T_{\mathrm{RD}}$, with constant R/F_{EtOH} (CS3, as shown in Fig. 16.38). Relay-feedback tests are performed to find the ultimate gains K_u and ultimate period P_u of each control loop and initial controller parameters are calculated according to $K_c = K_u/3$ and $\tau_I = 2P_u$. Controller settings for all loops are summarized in Table 16.7. The control performance is tested for feedflow disturbances. Figure 16.44 shows the dynamic responses of CS3 for $\pm 20\%$ feedflow changes. The control performance of CS3 is quite similar to that of the temperature control (CS2), except that steady-state error in product purity is eliminated.

16.5.6 Comparison with Reactive Distillation Temperature Control

The steady-state economics of the side reactor configuration are comparable to that of the reactive distillation, while giving advantages in catalyst replacement and ease of maintenance. It is interesting to see how the side reactor configuration dynamic behavior compares to that of reactive distillation. Using the same control structure for both process configurations, we compare the dynamic performance of the side reactor configuration with that of reactive distillation. The CS2 structure has two temperature loops ($Q_{R,\mathrm{STR}}{-}T_{\mathrm{STR}}$ and $Q_{R,\mathrm{RD}}{-}T_{\mathrm{RD}}$) with a constant reflux to feed ratio R/F_{EtOH}. The results given in Figure 16.45 demonstrate that the speeds of responses for the compositions are about the same for both configurations. Unitwise, the reactive distillation shows a slightly smaller settling time in the main column and the stripper exhibits a similar settling time.

16.6 CONCLUSION

Reaction and separation temperatures cannot be set independently in a conventional reactive distillation column. If a temperature mismatch exists between the temperature favorable for reaction and those favorable for separation, other alternative flowsheets should be considered. The results shown in this chapter demonstrate that the use of a distillation column with external side reactors can provide a more economical process because the distillation column and the side reactors can be operated at their optimum temperatures. Both steady-state design and dynamic control of these systems were explored.

The use of side reactors was illustrated in an ideal quaternary system and in the real chemical system of acetic acid esterification using ethanol (side reactor configuration). A gradual transition from the reactive distillation design to a single reactive tray column design and then to the side reactor configuration was demonstrated. The results demonstrated that the side reactor configuration is economically competitive, having only a 5% increase in TAC, and has a simple flowsheet structure, with one external side reactor plus a reactive column base. Finally, the operability of the side reactor configuration process was tested for feedflow and feed composition disturbances. The results showed that reasonable control performance could be obtained using simple temperature control schemes and the dynamic behavior was comparable to that of reactive distillation.

CHAPTER 17

EFFECTS OF BOILING POINT RANKINGS ON THE DESIGN OF REACTIVE DISTILLATION

In this chapter, we attempt to provide a systematic design procedure to determine the process configuration, relative position of the reactive zone, and separation sections for reaction systems with different boiling point rankings. Instead of investigating real chemical systems, ideal chemical reaction systems with different relative volatility rankings are studied. This provides a gradual transition as the reaction and separation properties change. The reaction under consideration is a reversible reaction, $A + B \Leftrightarrow C + D$, and this constitutes a quaternary system with 24 (4!) possible relative volatility arrangements. These 24 systems can be further grouped into six categories $4!/(2!{\cdot}2!)$ according to the ranking of the relative volatilities of the reactants and products. The likely process configurations are explored and the design optimized based on TAC. The results clearly indicate that the relative volatility rankings play a dominant role in the reactive distillation configuration and the TAC varies by a factor of $\sim$7 as we move from the most favorable case (reactants are IKs) to the least favorable relative volatility ranking (products are IKs). Finally, heuristics are given to correlate the relative volatility ranking to the TAC.

17.1 PROCESS AND CLASSIFICATION

17.1.1 Process

We use quaternary ideal reactive distillation systems to illustrate the effects of boiling point ranking on conceptual design. Consider the following second-order liquid-phase reversible reaction:

$$A + B \Longleftrightarrow C + D \tag{17.1}$$

Reactive Distillation Design and Control. By William L. Luyben and Cheng-Ching Yu

The forward and backward specific rates following the Arrhenius law on tray j are given by

$$k_{Fj} = a_F e^{-E_F/RT_j} \tag{17.2}$$

$$k_{Bj} = a_B e^{-E_B/RT_j} \tag{17.3}$$

where a_F and a_B are the preexponential factors for the forward and backward reactions, respectively; E_F and E_B are the activation energies; and T_j is the absolute temperature on tray j. The reaction rate on tray j can be expressed in terms of mole fractions ($x_{j,i}$) and the liquid holdups (M_j).

$$R_{j,i} = \nu_i M_j (k_{Fj} x_{j,\mathrm{A}} x_{j,\mathrm{B}} - k_{Bj} x_{j,\mathrm{C}} x_{j,\mathrm{D}}) \tag{17.4}$$

where $R_{j,i}$ is the reaction rate of component i on the jth tray (mol/s), ν_i is the stoichiometric coefficient that takes a negative value for the reactants, and M_j is the reactive holdup on tray j (mol). Note that the kinetic holdup comes from the tray sizing by considering a weir height of 10 cm and the column diameter is determined from the maximum vapor rate by assuming an F factor of 1. For the case when the reaction occurs in the reboiler and/or the condenser, the maximum kinetic holdup is assumed to be 20 times the tray holdup. In other words, the reactive holdup depends on column geometry and it cannot be set arbitrarily. This is typically true for ion-exchange resin catalyzed reactions. The overall reactive holdup is transformed into an equivalent catalyst weight, and the unit price of Amberlyst 15 is used for the cost estimation. Note that the amount of reactive holdup, consequently the Damköhler number, has an important impact on the feasibility and design of reactive distillation columns.[1,2]

Assumptions made in this work include the following:

1. Equimolar reactant feed is assumed, that is, $F_{0\mathrm{A}} = F_{0\mathrm{B}}$, where F_{0i} is the feed flowrate of reactant i. This implies a neat reactive distillation design, as opposed to an excess reactant design (cf. Chap. 4).
2. Forward reaction rate k_F is specified as 8 mol/s at 366 K, and backward rate k_B is set to 4 mol/s at the same temperature (i.e., $K_{\mathrm{EQ}} = k_F/k_B = 2$ at 366 K). Kinetic and physical property data are given in Table 17.1.
3. The reactive holdup on tray j (M_j) is computed from the column geometry by assuming a weir height of 10 cm, and the reactive holdups in the reboiler and condenser are taken to be 20 times the tray holdup.
4. Ideal vapor–liquid equilibrium is assumed, in which constant relative volatilities are used. The tray temperature is computed from the bubblepoint temperature calculation provided with the Antoine vapor pressure coefficients (Table 17.1). That is,

$$P = x_{j,\mathrm{A}} P^S_{\mathrm{A}(T_j)} + x_{j,\mathrm{B}} P^S_{\mathrm{B}(T_j)} + x_{j,\mathrm{C}} P^S_{\mathrm{C}(T_j)} + x_{j,\mathrm{D}} P^S_{\mathrm{D}(T_j)} \tag{17.5}$$

[1]F. Chen, R. S. Huss, M. F. Malone, and M. F. Doherty, Multiple steady states in reactive distillation: Kinetic effects, *Comput. Chem. Eng.* **26**, 81–93 (2002).

[2]F. Chen, R. S. Huss, M. F. Malone, and M. F. Doherty, Simulation of kinetic effects in reactive distillation, *Comput. Chem. Eng.* **24**, 2457–2473 (2000).

TABLE 17.1 Physical Properties for Process Studied

Activation energy (cal/mol)	Forward (E_F)	12,000			
	Backward (E_B)	17,000			
Specific reaction rate at 366 K (kmol s^{-1} kmol^{-1})	Forward (k_F)	0.008			
	Backward (k_B)	0.004			
Heat of reaction (cal/mol)	λ	−5000			
Heat of vaporization (cal/mol)	ΔH_v	6944			
Relative volatilities (LLK/LK/HK/HHK)	$\alpha_{LLK} > \alpha_{LK} > \alpha_{HK} > \alpha_{HHK}$	8/4/2/1			
		LLK	LK	HK	HHK
Vapor pressure constants[a]	A_{VP}	13.04	12.34	11.65	10.96
$\ln P_i^S = A_{VP,i} - B_{VP,i}/T$	B_{VP}	3862	3862	3862	3862

[a]Pressure is in bar, and temperature is in Kelvin.

where P is the total pressure and P^S is the vapor pressure. Because we will change the boiling point ranking of the reactants (A and B) and products (C and D), the Antoine coefficients are given for lighter than light key (LLK), light key (LK), heavy key (HK), and heavier than heavy key (HHK). Note that, as the result of constant relative volatility, the Antoine coefficients B_{VP} are the same for all four components.

5. The heat of reaction vaporizes some liquid on the reactive trays. Therefore, the vapor flowrate increases up through the reactive zone, and the liquid flowrate decreases down through the reactive zone.[3] Counting trays from the bottom up, one obtains

$$V_j = V_{j-1} - (\lambda/\Delta H_V)R_{j,i} \tag{17.6}$$

$$L_j = L_{j+1} + (\lambda/\Delta H_V)R_{j,i} \tag{17.7}$$

where λ is the heat of reaction (−5000 cal/mol) and ΔH_v is the heat of vaporization (6944 cal/mol).

6. The liquid hydraulic time constant (β) is included by using a linearized form of the Francis weir formulation, and β is set to 6 s.
7. Vapor holdup and pressure drop are neglected. The dynamic reactive distillation model describing material and energy balances is described by Al-Arfaj and Luyben.[3] The computer code of Kaymak et al.[4] is used here with some modification to accommodate process configuration changes. The computer code is programmed in Matlab M-files that can perform economical analysis as the flowsheet converges.

[3]M. A. Al-Arfaj and W. L. Luyben, Comparison of alternative control structures for an ideal two-product reactive distillation column, *Ind. Eng. Chem. Res.* **39**, 3298–3307 (2000).

[4]D. B. Kaymak, W. L. Luyben, and O. J. Smith IV, Effect of relative volatility on the quantitative comparison of reactive distillation and conventional multi-unit systems, *Ind. Eng. Chem. Res.* **43**, 3151–3162 (2004).

17.1.2 Classification

For a quaternary system (A, B, C, and D), there are 24 (4!) possible boiling point rankings (Fig. 17.1). Here, we use LLK, LK, HK, and HHK to denote the component ranging from the lightest boiler to the heaviest boiler. In terms of relative volatilities, these four boilers can be expressed in the following order:

$$\alpha_{LLK} > \alpha_{LK} > \alpha_{HK} > \alpha_{HHK} \tag{17.8}$$

Because these two reactants (A and B) are interchangeable and the same applies to the two products (C and D), this leaves us with 6 (24/2/2) possible configurations. Figure 17.1 shows all 24 possible boiling point rankings, and they can be further reduced to 6 distinct

LLK	LK	HK	HHK	No.	Configuration*	Type**
A	B	C	D	1	(1)	$\mathrm{II_R}$
		D	C	2	(1)	$\mathrm{II_R}$
	C	B	D	3	(2)	$\mathrm{III_R}$
		D	B	4	(3)	$\mathrm{I_R}$
	D	B	C	5	(2)	$\mathrm{III_R}$
		C	B	6	(3)	$\mathrm{I_R}$
B	A	C	D	7	(1)	$\mathrm{II_R}$
		D	C	8	(1)	$\mathrm{II_R}$
	C	A	D	9	(2)	$\mathrm{III_R}$
		D	A	10	(3)	$\mathrm{I_R}$
	D	A	C	11	(2)	$\mathrm{III_R}$
		C	A	12	(3)	$\mathrm{I_R}$
C	A	B	D	13	(4)	$\mathrm{I_P}$
		D	B	14	(5)	$\mathrm{III_P}$
	B	A	D	15	(4)	$\mathrm{I_P}$
		D	A	16	(5)	$\mathrm{III_P}$
	D	A	B	17	(6)	$\mathrm{II_P}$
		B	A	18	(6)	$\mathrm{II_P}$
D	A	B	C	19	(4)	$\mathrm{I_P}$
		C	B	20	(5)	$\mathrm{III_P}$
	B	A	C	21	(4)	$\mathrm{I_P}$
		C	A	22	(5)	$\mathrm{III_P}$
	C	A	B	23	(6)	$\mathrm{II_P}$
		B	A	24	(6)	$\mathrm{II_P}$

*6 distinct possible rankings

**3 process types (I-III)with the reactant or product as the lightest one

Figure 17.1 All 24 possible boiling point rankings, six distinct configurations, and three process types for possible reactant/product relative volatilities distribution.

TABLE 17.2 Classification for Process Types

Reaction $R1 + R2 \Leftrightarrow P1 + P2$		
Type I (one group)	I_p	$\alpha_{P1} > \alpha_{R1} > \alpha_{R2} > \alpha_{P2}$
	I_R	$\alpha_{R1} > \alpha_{P1} > \alpha_{P2} > \alpha_{R2}$
Type II (two groups)	II_p	$\alpha_{P1} > \alpha_{P2} > \alpha_{R1} > \alpha_{R2}$
	II_R	$\alpha_{R1} > \alpha_{R2} > \alpha_{P1} > \alpha_{P2}$
Type III (alternating)	III_p	$\alpha_{P1} > \alpha_{R1} > \alpha_{P2} > \alpha_{R2}$
	III_R	$\alpha_{R1} > \alpha_{P1} > \alpha_{R2} > \alpha_{P2}$

configurations for a quaternary system. According to the *distribution* of reactants and products in relative volatility ranking, we can further classify these 6 configurations into three types as shown in Table 17.2.

Type I: One Group. In the first case the two reactants are LK and HK, and the two products are LLK and HHK. In other words, we have $\alpha_{P1} > \alpha_{R1} > \alpha_{R2} > \alpha_{P2}$, where P_i is the ith product and R_i is the ith reactant. This forms a distinct group with the reactants in the middle (Table 17.2). In this case, one of the *products* is the lightest boiler, and this case is denoted as type I_p. Similarly, if two products form one distinct group (as LK and HK) while having the reactants as LLK and HHK, this is also classified as type I. However, if one *reactant* is the lightest boiler, this case is called type I_R (Table 17.2). Figure 17.1 shows four possible scenarios for type I_p and another four possible cases for type I_R. In the subsequent development we use $\alpha_C > \alpha_A > \alpha_B > \alpha_D$ to describe type I_p and $\alpha_A > \alpha_C > \alpha_D > \alpha_B$ to represent type I_R.

Type II: Two Groups. The second situation is when the relative volatilities of the two reactants and the two products are adjacent to each other. This forms two distinct groups for the reactants and the products in the relative volatility ranking as shown in Table 17.2. The first example is that the *products* are LLK and LK and the two reactants are HK and HHK. That implies $\alpha_{P1}>\alpha_{P2}>\alpha_{R1}>\alpha_{R2}$. Because one of the products is the lightest component, it is denoted as type II_p. Figure 17.1 shows four possible scenarios for type II_p, and we use $\alpha_C > \alpha_D > \alpha_A > \alpha_B$ to represent this type. In contrast, if the *reactants* are LLK and LK and the two products are HK and HHK, we have type II_R. Table 17.2 shows type II_R with two distinct groups in the relative volatility ranking. Here we use $\alpha_A > \alpha_B > \alpha_C > \alpha_D$ to describe type II_R.

Type III: Alternating. The third case corresponds to the situation when the relative volatility ranking is arranged in an alternating manner for the reactants and the products as demonstrated in Table 17.2. For example, if the sequence of the relative volatility in a descending order is product 1, reactant 1, product 2, and reactant 2 (i.e., $\alpha_{P1}>\alpha_{R1}>\alpha_{P2}>\alpha_{R2}$), we have the case of type III_p. Figure 17.1 shows four possible cases for type III_p, and we use $\alpha_C > \alpha_A > \alpha_D > \alpha_B$ to represent it. In contrast, if the reactants are LLK and HK and the two products are LK and HHK, we have type III_R. Figure 17.1 shows four possible combinations for type III_R. In this study we use $\alpha_A > \alpha_C > \alpha_B > \alpha_D$ to represent type III_R.

17.2 RELAXATION AND CONVERGENCE

We now address some simulation and convergence issues in the design of reactive distillation as the process configuration changes. The relaxation approach[5] is taken here. For a given reactive distillation design (e.g., number of reactive, rectifying, and stripping trays; feed flowrates; and feed tray locations), a control system is set up to drive the process toward the design specifications. Given the initial guesses for tray compositions, the set of ordinary differential equations is integrated from the initial conditions until a steady state is reached. Note that the control structure for relaxation is very different from the one in control practice. The reason is that perfect flow control can be achieved in a computer simulation so the problem of stoichiometric imbalance does not occur. For example, for the type I_p system shown in Figure 17.2*a* with the product as the lightest component, all four external flows (inlet and outlet streams) are under flow control. The composition of one of the products is controlled by adjusting the reflux flow, and the bottoms level is maintained by changing the vapor boilup. In other words, given the feeds, the 2 control degrees of freedom can be considered to be the distillate composition and distillate flowrate. These fix the bottoms flowrate and the bottoms composition under steady-state conditions.

Another example is also a type I system, but one of the reactants is the lightest component, type I_R. This leads to a rather unconventional reactive distillation column, as will be explained later. The product is withdrawn from the middle of the column as a liquid side stream. Figure 17.2*b* shows the control structure used to obtain steady-state conditions. Again, the two feeds are under flow control, the product composition is controlled by adjusting vapor boilup, the reflux-drum level is maintained by changing the reflux flow, and the base level is controlled by the side draw flowrate. This type of control structure will never be implemented in practice because in a real system the assumption of perfect flow control can never be realized. Given two feeds, this system has only 1 control degree of freedom, which is side stream composition. A convergence criterion is used to verify the steady-state condition. In this work, the following criterion is used:

$$\varepsilon = \max\left(\left|x_{B,i}^{k} - x_{B,i}^{k-1}\right|, \left|x_{j,i}^{k} - x_{j,i}^{k-1}\right|, \left|x_{D,i}^{k} - x_{D,i}^{k-1}\right|\right)$$
$$< 10^{-4}\ \forall i = 1, \mathrm{NC}\ \forall j = 1, \mathrm{NT} \tag{17.9}$$

where $x_{j,i}$ denotes the *i*th component on the *j*th tray, $x_{B,i}$ is the bottoms composition for the *i*th component, $x_{D,i}$ is the distillate composition of the *i*th component, NC is the number of components, NT is the total number of trays, and superscript k is the discrete time index.

For the given convergence criterion with an integration step size of 1 s, it takes 7 h (process time, Fig. 17.3*a*) to converge a type I_p flowsheet (Fig. 17.2*a*), which corresponds to a 69-s CPU time on a Pentium 4 PC. For the difficult reactive distillation process type I_R (Fig. 17.2*b*), it takes 160 h of process time to meet the convergence criterion. Figure 17.3*b* shows extremely slow composition dynamics in the side stream composition. This

[5] W. L. Luyben, Use of dynamic simulation to converge complex process flowsheets, *Chem. Eng. Edu.* Spring issue, 142–149 (2004).

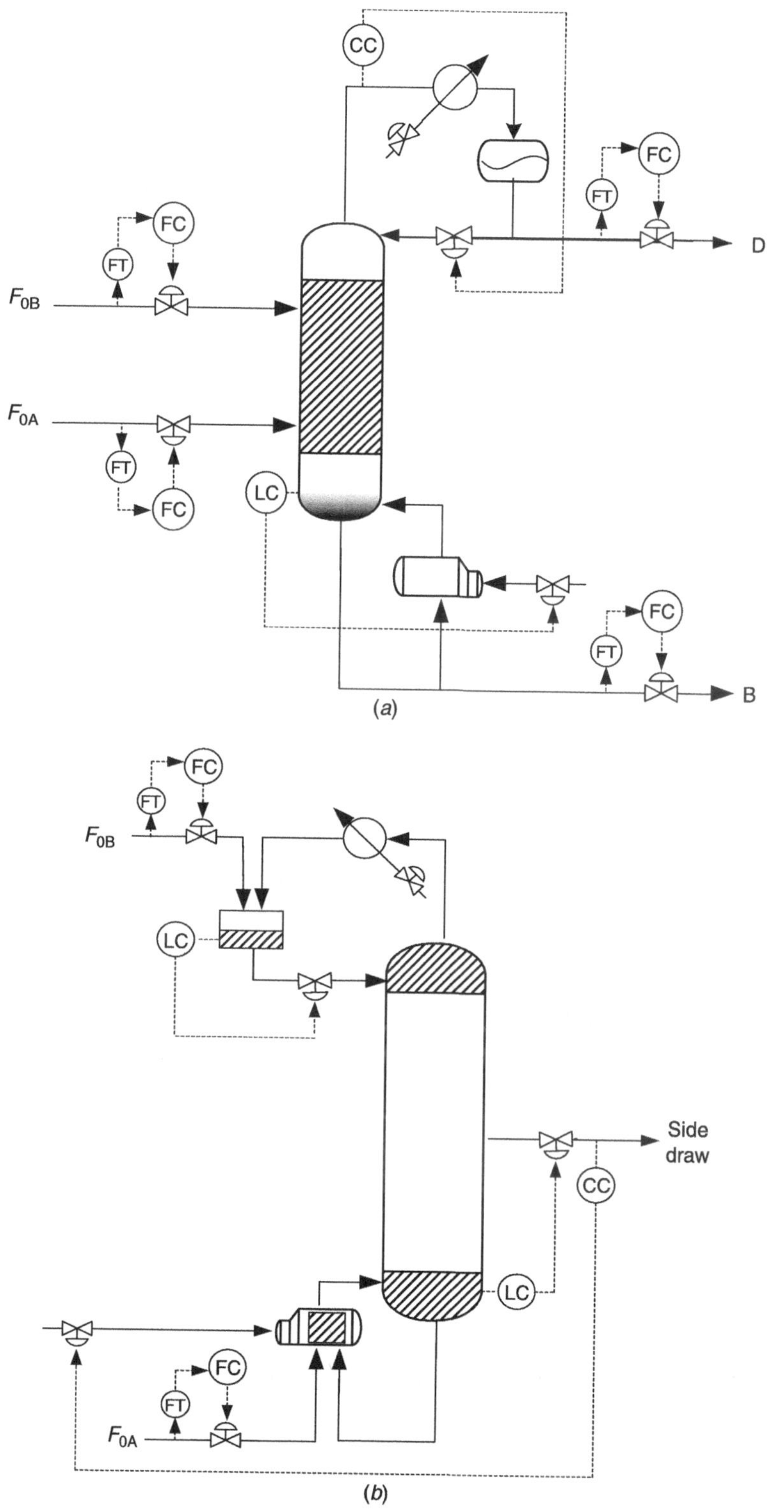

Figure 17.2 Control structure used for dynamic simulation using relaxation for (*a*) type I_p: LK + HK ⇔ LLK + HHK and (*b*) type I_R: LLK + HHK ⇔ LK + HK.

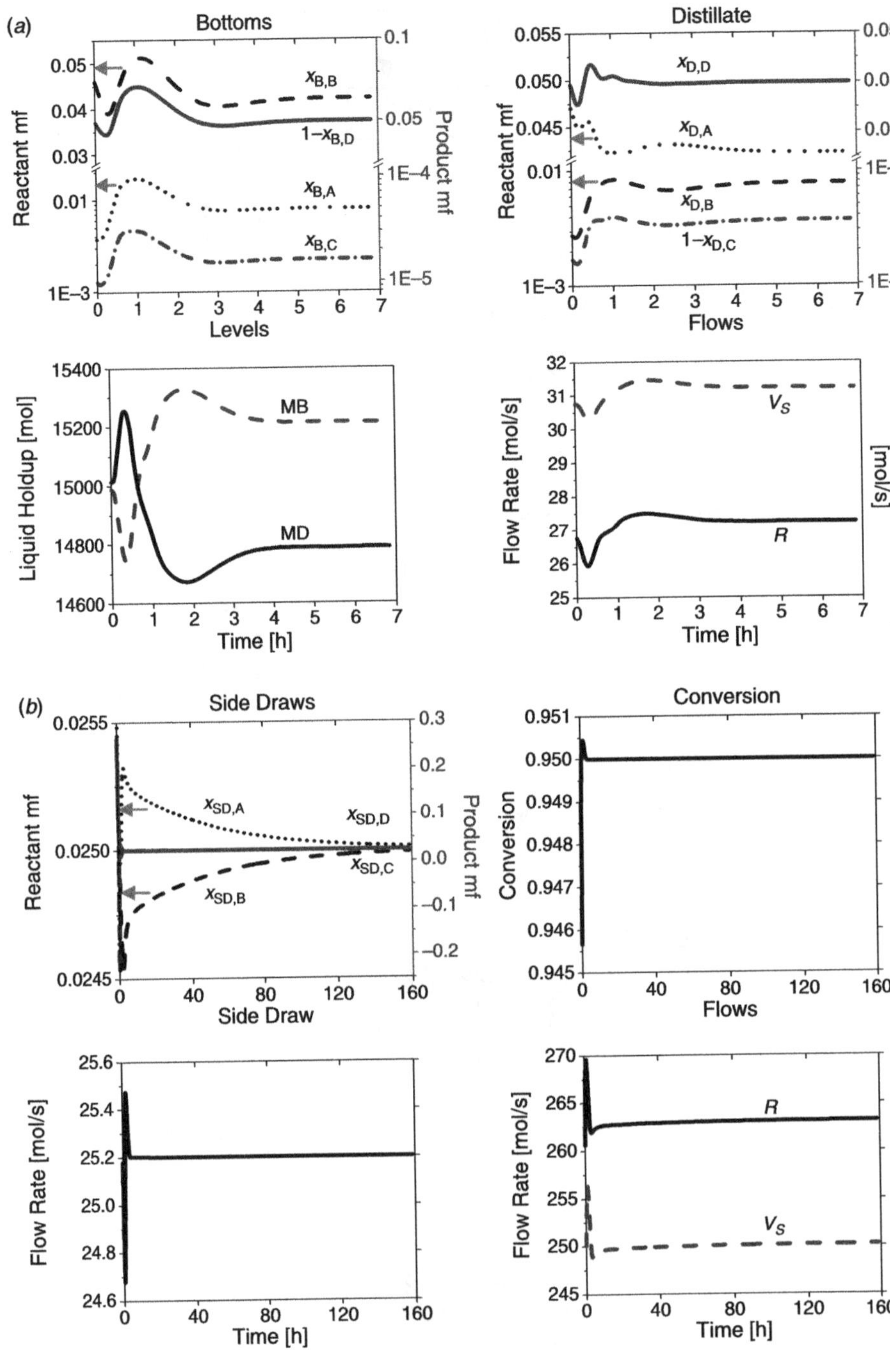

Figure 17.3 Converging dynamic simulation using relaxation for (*a*) type I_p: LK + HK ⇔ LLK + HHK and (*b*) type I_R: LLK + HHK ⇔ LK + HK.

corresponds to a 1800-s CPU time on a Pentium 4 PC. Moreover, in many cases, the flowsheet may not converge after 200 h if the initial guesses are not appropriate. For the control part, little improvement in the speed of convergence is achieved by using variable gain controllers, for example, using larger controller gains as the error becomes smaller. The results presented here clearly indicate that the simulation of reactive distillation systems remains a very difficult problem, and simulations frequently fail to converge. Unless further improvement in numerical simulation is made, automated design via mathematical programming is difficult, if not impossible.

17.3 PROCESS CONFIGURATIONS

Before getting into exploring process configurations, the design procedure is outlined. The design variables (e.g., location of reactive zone, tray numbers, feed locations etc.) are identified first, and then all combinations are exhaustively searched to find the finalized design. The objective function for design is the TAC, and the specifications are 95% purity levels for the products. The design problem can be formulated as

$$\begin{aligned} &\underset{X}{\text{minimize}}\,\text{TAC} \\ &\text{subject to: } x_{\text{product,C}} \geq 0.95 \\ &\qquad\qquad\quad x_{\text{product,D}} \geq 0.95 \end{aligned} \tag{17.10}$$

where X is the vector of design variables and TAC is defined as

$$\text{TAC} = \text{operating cost} + \frac{\text{capital cost}}{\text{payback period}} \tag{17.11}$$

A payback period of 3 is used here. The formula for the TAC computation is taken from Kaymak and colleagues.[4] As a result of different relative volatility rankings, the process may vary from a one-column configuration to a two-column configuration, with different product withdraw locations. We use a type I_p example to illustrate the design procedure.

$$\begin{array}{ccccccc} \text{A} & + & \text{B} & \Longleftrightarrow & \text{C} & + & \text{D} \\ \text{LK} & & \text{HK} & & \text{LLK} & & \text{HHK} \end{array} \tag{17.12}$$

Given the product specification, the design optimization variables are number of reactive trays N_{RX}, numbers of rectifying and stripping trays N_R and N_S, and feed tray locations for heavy and light reactants NF_{heavy} and NF_{light}. Instead of blindly exhausting all possible combinations of design variables, a systematic design procedure is used. Note that process knowledge is needed to place the reactive zone(s). A simple rule is to place the reactive zone(s) at locations where the reactants are most *abundant*. For the type I_p process, the reactive zone is located at the midsection of the column.

1. Set the reactants feed ratio to 1 (i.e., $\text{FR} = F_{\text{light}}/F_{\text{heavy}} = 1$).
2. Place the reactive zone in the *midsection* of the column, and fix the number of reactive trays N_{RX}.

3. Place heavy reactant feed NF_{heavy} on the top of the reactive zone, and introduce light reactant feed NF_{light} on the lowest tray of the reactive zone.
4. Guess the number of trays in rectifying section N_R and stripping section N_S.
5. Run the simulation using the dynamic model with feedback control to meet the product specification.
6. Return to step 4 and change N_R and N_S until the TAC is minimized.
7. Return to step 3 and find the feed locations (NF_{heavy} and NF_{light}) until the TAC is minimized.
8. Return to step 2 and vary N_{RX} until the TAC is minimized.

17.3.1 Type I: One Group

Type I_p: LK + HK = LLK + HHK. This is the most popular reactive distillation configuration, which has been studied by a number of authors.[3,4] In real chemical systems, the production of diphenyl carbonate belongs to this class.[6] The *net* reaction can be expressed as

dimethyl carbonate	+	2 phenol	⟺	2 methanol	+	diphenyl carbonate
LK		HK		LLK		HHK

Note that many real chemical systems have azeotropes associated with quaternary systems, and it may make the neat reactive distillation design infeasible. However, for the real chemical systems illustrated here, the placement of the reactive zone is the same as that of the ideal systems, despite having azeotropes. The boiling point ranking leads to an easy separation between the reactants and products. The two products leave the reactive section from opposite sides of the reactive zone while the two intermediate boilers (the reactants) are kept in the reactive zone. This is the most favorable boiling point arrangement possible for a quaternary reactive distillation system.

Figure 17.4*a* shows the effects of separation trays and reactive trays on the TAC. Because of the symmetry in the relative volatilities between the two products and the two reactants in the separation sections of the column, the numbers of trays in the rectifying section and the stripping section are assumed to be the same, that is, $N_R = N_S$. Figure 17.4*a* indicates that the optimum number of reactive trays N_{RX} is 16 and the optimum number of separation trays $N_R = N_S$ is 4. The tradeoff comes from the fact that in the vicinity of the optimum the operating cost goes down and the capital cost goes up as we increase the number of separation trays. Figure 17.4*b* shows that the TAC is quite sensitive to changes in the feed tray locations, which is consistent with the findings in Chapters 7 and 18. When the heavy reactant feed and the light reactant feed are five trays apart, the design gives the lowest TAC, and the optimum corresponds to $NF_{\mathrm{heavy}} = 16$ and $NF_{\mathrm{light}} = 11$. This results in a TAC of \$254,170. Detailed process parameters are provided in Table 17.3.

[6]S. Fukuoka, T. Watanabe, and T. Dozono, Methods for producing a crystallized aromatic carbonate and a crystallized aromatic polycarbonate obtained thereby, US Patent 4,948,871, 1990.

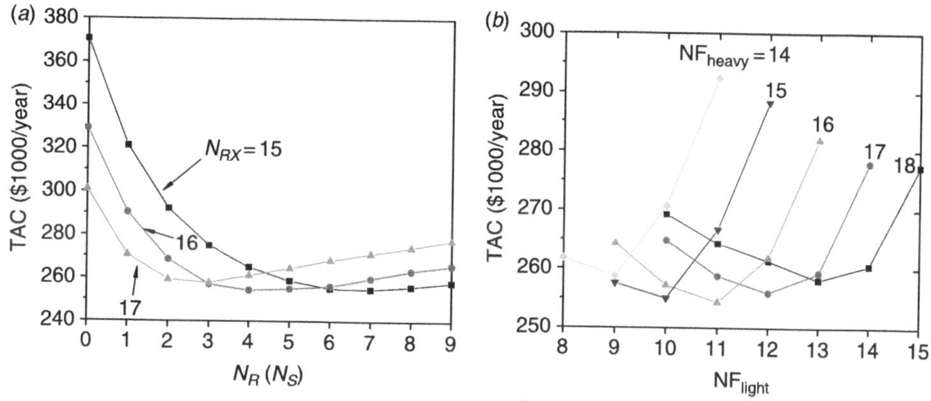

Figure 17.4 Relationship between TAC and design variables for type I_p process: (*a*) number of separation trays (N_R and N_S) versus TAC for different N_{RX} and (*b*) feed tray locations (NF_{heavy} and NF_{light}) with $N_{RX} = 22$ and N_R (N_S) = 4.

Figure 17.5 demonstrates that, for type I_p reactive distillation with products as LLK and HHK, the reactive zone is located at the middle of the column and the light product (LLK) and the heavy product (HHK) are removed from the top and bottoms of the column, respectively. This is intuitively correct because we have higher reactant (intermediate boilers) concentrations in the midsection of the column. As will become clear later, one of the most important elements in the conceptual design of reactive distillation is placing the reactive zone at the right location. The composition profile of the final design is given in Figure 17.6 and, as expected, significant amounts of reactants are present in the reactive zone (between two dashed lines) and product compositions increase gradually toward the top and bottoms of the column. Figure 17.6 also shows the fraction of total conversion (R_i/R_t) in each reactive tray. These results indicate that all reactive trays are well utilized in the reactive zone.

Before leaving this section, we explore the effects of reactive holdups on the designs (assumption 3 in design). The weir height is varied from 5 to 40 cm (likely the lower and upper limits for real columns). Figure 17.7 shows that the TAC changes −10% (5 cm) to +10% (40 cm) from the base case design. Instead of using two extreme values, in this work a modest weir height of 10 cm is used in all designs.

Another issue is that the effect of pressure on the design is not explored here because we are assuming constant relative volatility systems. The column pressure is fixed at 8 bar in this work. Pressure is very important in reactive distillation because of the effect of temperature on both vapor–liquid equilibrium and reaction kinetics. For exothermic reactions, the optimum column pressure is affected by the competing effects of temperature on the specific reaction rates and the chemical equilibrium constant.

Type I_R: LLK + HHK = LK + HK. This corresponds to the following reaction:

$$\underset{\text{LLK}}{\text{A}} + \underset{\text{HHK}}{\text{B}} \Longleftrightarrow \underset{\text{LK}}{\text{C}} + \underset{\text{HK}}{\text{D}} \tag{17.13}$$

TABLE 17.3 **Final Design for All Six Process Flowsheets**

	I_p RD	I_R RD	SD	II_p RD	SD	II_R RD	SD	III_p RD	III_R RD
Total trays	24	32	48	11	54	28	62	59	96
Trays Stripping section (N_S)	4					3		2	35
Reactive section (N_{RX})	16	$2^a + 4^b$		6^a		25^b		49	57
Rectifying section (N_R)	4			5				8	4
Reactive trays	5–20	0–1, 27–32		0–5		4–28		9–57	5–61
Feed tray									
Heavy reactant (NF_{heavy})	11	32		0		28		57	55
Light reactant (NF_{light})	16	0		0		4		46	5
Feed flowrate (kmol/h)									
Heavy reactant	45.36	45.36		45.36		45.36		45.36	45.36
Light reactant	45.36	45.36		45.36		45.36		45.36	45.36
Flowrate (kmol/h)									
Top product	45.36		45.36	90.72	45.36		45.36	45.36	45.36
Bottom product	45.36		45.36		45.36	90.72	45.36	45.36	45.36
Reflux	95.15	931.03	94.32	93.71	97.884	446.18	102.38	99.40	270.68
Vapor boilup rate (kmol/h)	109.45	881.28	139.68	152.57	143.25	368.53	147.71	113.72	285.05
Column diameter (m)	0.692	1.740	0.700	0.801	0.698	1.136	0.733	0.702	1.053
Weir height (m)	0.100	0.100	0.100	0.100	0.100	0.100	0.100	0.100	0.100
Heat transfer area (m^2)									
Condenser	96.75	603.95	95.26	125.76	97.69	256.33	100.74	98.74	215.53
Reboiler	45.27	360.06	57.07	62.34	58.53	150.57	60.36	46.47	116.46
Total capital cost ($1000)	349.45	1198.52	457.83	346.30	482.52	686.91	529.76	488.54	988.36
Column	119.13	379.03	209.34	68.64	229.30	221.02	269.87	247.90	564.00
Column trays	3.12	16.21	6.32	1.62	7.08	7.53	8.77	7.81	23.82
Heat exchangers	227.83	803.29	242.16	276.03	246.15	458.35	251.12	232.83	400.54
Total operating cost ($1000/year)	137.68	1176.4	167.22	207.80	171.47	513.16	176.83	157.24	396.35
Catalyst	6.68	121.53		25.16		72.03		21.10	55.15
Energy	131.01	1054.9	167.22	182.63	171.47	441.14	176.83	136.14	341.20
TAC ($1000/year)	254.17	1575.9	319.90	323.23	332.31	742.13	353.41	320.08	725.80
TAC (entire process, $1000/year)	254.17	1895.8		655.54		1095.5		320.08	725.80

Column configurations: RD, reactive distillation column; SD, simple distillation column.

[a]Including a reactive reboiler.

[b]Including a reactive condenser.

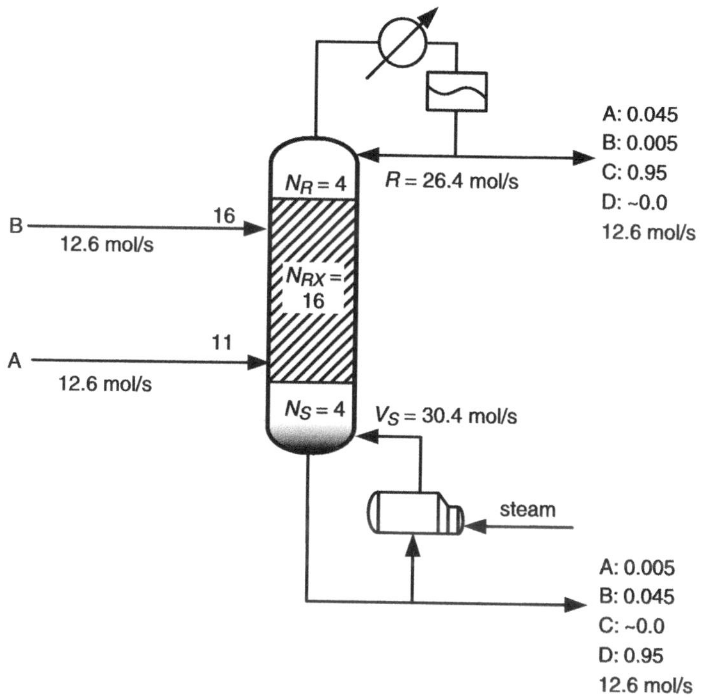

Figure 17.5 Final design for type I_p process (LK + HK ⇔ LLK + HHK).

The boiling point ranking clearly indicates that this is a very difficult reactive distillation system because large amounts of reactants A and B can hardly coexist in the liquid phase (one is the LLK and the other is the HHK). This is the worst case scenario for the forward reaction (extremely favorable for the backward reaction). In the production of

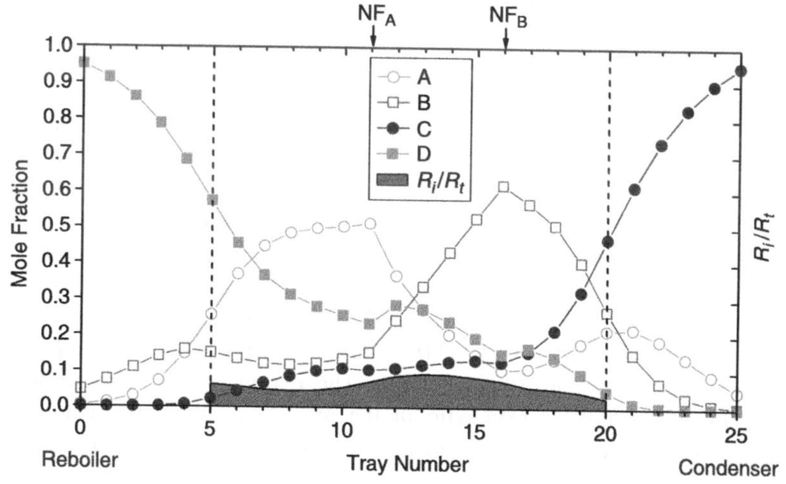

Figure 17.6 Composition profile, feed locations, and fraction of total conversion (R_i/R_t) for type I_p process (LK + HK ⇔ LLK + HHK).

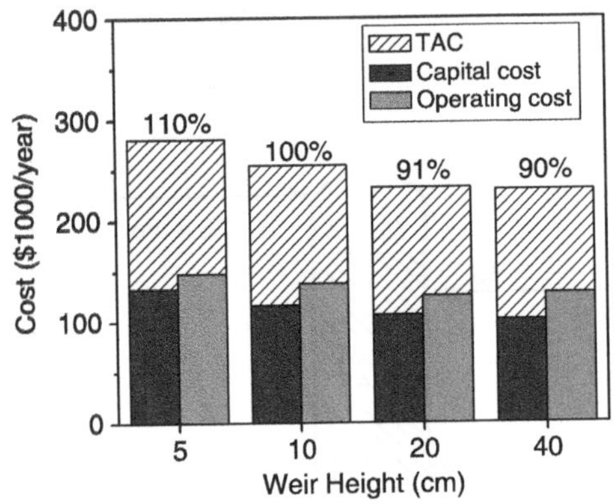

Figure 17.7 Effects of weir height (reactive holdup) on the TAC with 10-cm weir height as base case.

biobased esters and biodiesel, many processes fall into this class. A typical example is the recovery of lactic acid.[7] The reaction can be expressed as

$$\underset{\text{LLK}}{\text{methanol}} + \underset{\text{HHK}}{\text{lactic acid}} \Longleftrightarrow \underset{\text{LK}}{\text{water}} + \underset{\text{HK}}{\text{methyl lactate}}$$

Unless we can consume all of the HHK before it reaches the bottom of the reactive zone and react away all of the LLK before it reaches the top of the reactive distillation, a single column configuration (e.g., Fig. 17.5) is not possible. This is typically true for the neat design.

A more likely configuration is to place the reactive zones at the opposite ends of the column where significant amounts of LLK and HHK (two reactants) are present. Then the heavy reactant (HHK) is introduced into the top of the column where a significant amount of the other reactant (LLK) is present. Similarly, the light reactant (LLK) is fed to the bottom of the column where a significant amount of the heavy reactant (HHK) is present (Fig. 17.8). The next question then becomes, how do we withdraw the product? The answer is actually quite simple. A side stream from the reactive distillation column will contain a mixture of two products. Thus, we need an additional column to separate these two products. Once the correct process configuration is set, design variables are easily identified. They are the numbers of reactive stages at two different zones, number of separation trays, and side draw location. Note that the feed locations shown in Figure 17.8 are actually the optimal ones, and we will not show the effect of feed tray locations in design for type I_R. In addition, note here that the separation column is designed by setting the total number of trays to 2 times the minimum number of trays, that is, $N_T = 2N_{\min}$.

Figure 17.8 gives the process flowsheet, and Table 17.3 gives parameter values. This flowsheet results in a TAC of $1,576,000 for the reactive distillation column alone, which is 620% of that of the type I_p system. The reason is that, in order to achieve high conversion, the reflux to feed ratio in the reactive distillation is extremely high (20.5) and the vapor boilup to feed ratio is 19.1. This corresponds to a column of 30

[7] J. I. Choi and W. H. Hong, Recovery of lactic acid by batch distillation with chemical reactions using ion exchange resin, *J. Chem. Eng. Japan.* **32**, 184–189 (1999).

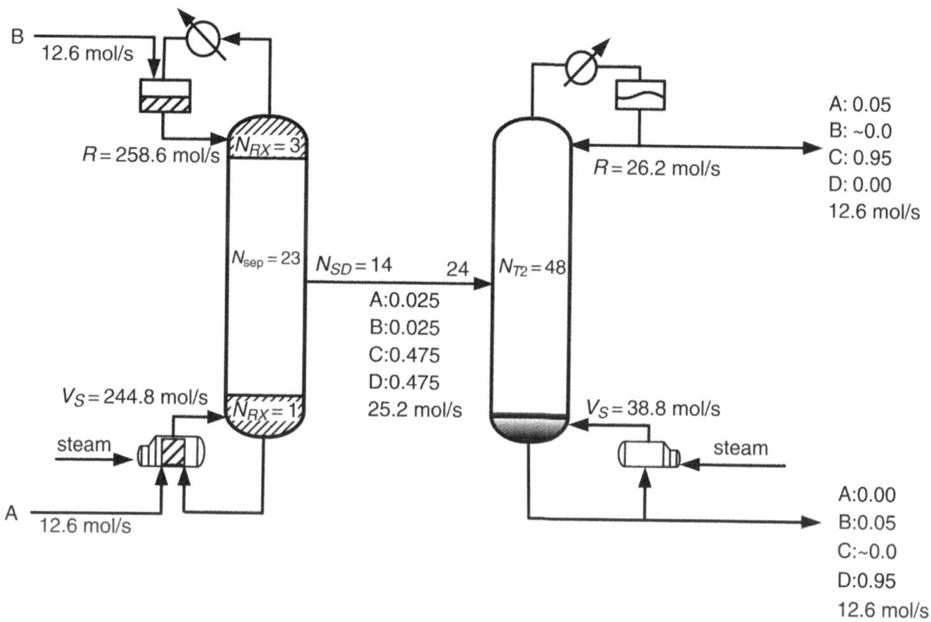

Figure 17.8 Final design for type I_R process (LLK + HHK ⇔ LK + HK).

trays where the reactive zones are located at the top and bottoms of the column with reactive holdups of 20 and 1 times the tray reactive holdup in the condenser and reboiler, respectively. Because the catalyst cost is also included in the TAC calculation, the reactive holdups in the reboiler and condenser are also design variables theoretically (not exceeding 20 times the tray holdup). In this case, the reboiler reactive holdup is varied to minimize the TAC.

A mixture of two products (LK and HK) is withdrawn from the middle of the separation section (tray 14), and this side stream is fed into a simple distillation column. The second column has a total of 48 trays with a moderate reflux ratio (2.08) and boilup ratio (3.08). Figure 17.9*a* displays the composition profile for the reactants (A and B) and products (C and D). As expected, the concentrations of the two reactants are drastically different in the reactive zones. For example, the LLK (reactant A) composition exceeds 60% toward the top reactive zone while the concentration of the heavy reactant (HHK, reactant B) remains below 2%, despite the fresh feed of HHK being introduced in the condenser. Similar behavior is observed in the lower reactive zone. At the side draw location, the compositions of both products are the same (stoichiometric balance), as shown in Figure 17.9*a*.

The product stream is fed to a distillation column for further separation to obtain 95% C and D. Figure 17.9*b* shows the composition profile of the distillation column with rather symmetric profiles.

17.3.2 Type II: Two Groups

Type II_p: HK + HHK = LLK + LK. This corresponds to the following reaction.

$$\begin{array}{ccccccc} A & + & B & \Longleftrightarrow & C & + & D \\ HK & & HHK & & LLK & & LK \end{array} \tag{17.14}$$

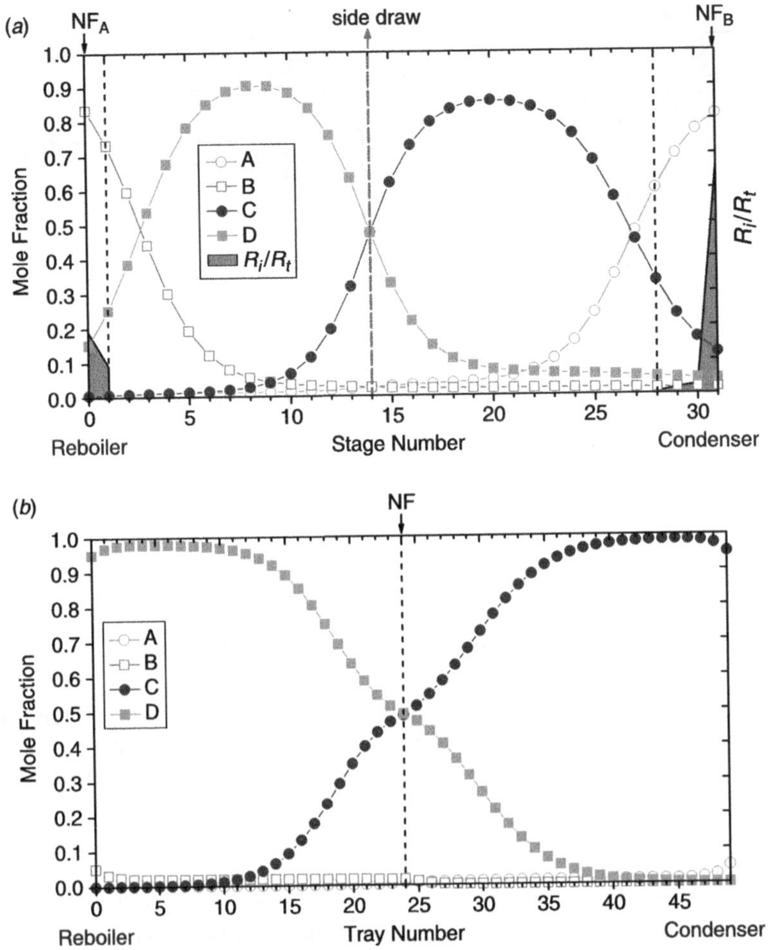

Figure 17.9 Composition profile, feed locations, and fraction of total conversion (R_i/R_t) for type I_R process (LLK + HHK ⇔ LK + HK) in (*a*) reactive distillation column and (*b*) simple distillation column.

The boiling point ranking indicates that both reactants (A and B) are heavier than the products (C and D). These two reactants are more concentrated toward the lower part of the reactive distillation column, and the two products can be separated easily from the top of the column. Following the rule of "placing the reactive zone at a location with high reactant concentration," the lower section of the column should be made reactive while the separation between reactants and products is carried out in the upper section of the column. This leads to the two-column distillation configuration shown in Figure 17.10. This configuration is actually quite similar to that of ethyl acetate and isopropyl acetate reactive distillation columns (Type II in Chapter 7). The overhead product of the reactive distillation column is fed to a simple distillation column to separate the two products.

Because the conversion is not 100% in the reactant distillation, the unconverted heavy reactants (A and B) need to be separated from the light products (C and D) in the distillation column. The boiling point ranking C(LLK), D(LK), A(HK), and B(HHK)

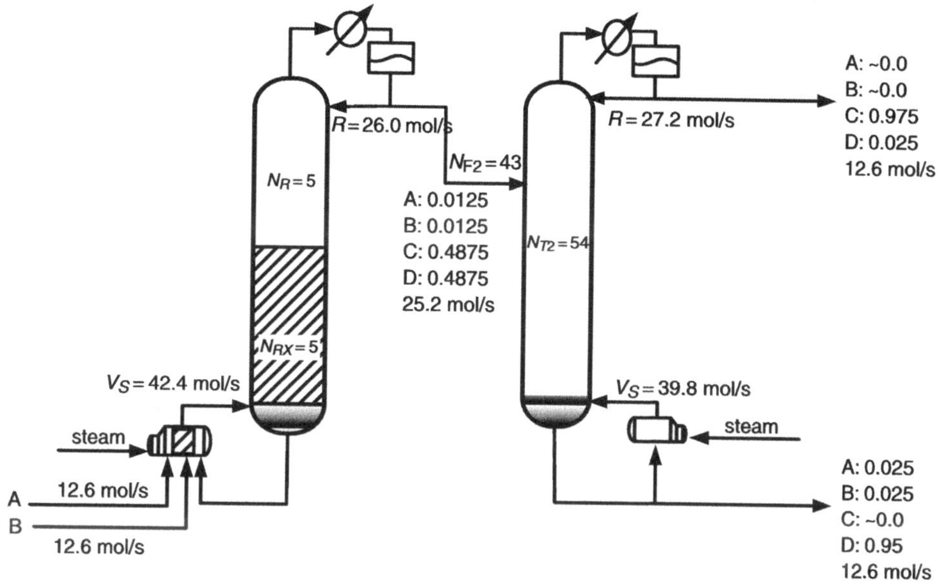

Figure 17.10 Final design for type II$_p$ process (HK + HHK ⇔ LLK + LK).

indicates that both reactants will end up in the bottoms of the product column. We have D as the impurity in the column overhead and all of the unconverted reactants A and B will end up in the column base with no C in the bottoms. In order to keep product compositions at no less than 95%, the conversion in the reactive distillation column is set to a value greater than 95% (97.5% in this case). This results in the following product composition: 95% D (LK) and 97.5% C (LLK). Figure 17.10 gives the final design that minimizes the TAC. Table 17.3 gives the parameter values of this design.

The flowsheet gives a TAC of $665,000, which is 158% higher than that of type I$_p$. The TAC of the reactive distillation column is $323,000, which is slightly higher than that of the type I$_p$ reactive distillation column (TAC = $254,000). The reactive column has five reactive trays plus a reactive reboiler and five rectifying trays. The vapor boilup to feeds ratio is around 2, and the reflux ratio is close to 1. Actually, this is a relatively simple reactive distillation column with reasonable energy consumption. The reaction takes place in the *high* temperature zone of the column, and we have a large reactive holdup (the holdup in the column base is 20 times the tray reactive holdup) in the column base when the temperature is the highest. In other words, the location of the reactive zone (Fig. 17.10) facilitates the reactive separation.

Figure 17.11*a* confirms that most of the conversion takes place in the reboiler, and the remaining reactive trays carry the reaction further to the desired conversion. The reactant composition of HHK (B) is kept high in the reactive zone and the other reactant A (HK) is maintained at an almost constant level to ensure that the forward reaction dominates. Toward the top of the reactive distillation column, two light products are separated from the two heavier reactants (Fig. 17.11*a*).

The TAC of the second column ($332,000) is about the same as that of the reactive distillation column. This is a simple distillation with heavy impurities. Thus, the bottoms composition is kept to the specification of 95% D (LK) with the impurity being a mixture of both A and B. The top product has a composition of 97.5% C with 2.5% D as the impurity. This column has 54 trays, and the feed is introduced into tray 43.

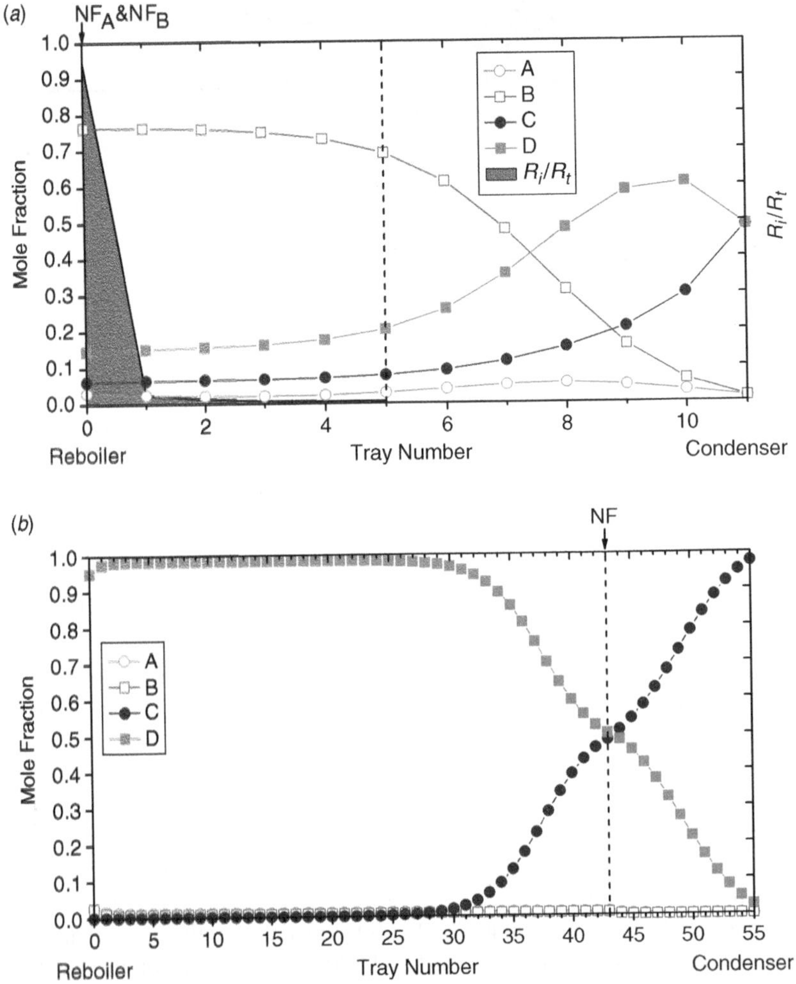

Figure 17.11 Composition profile, feed locations, and fraction of total conversion (R_i/R_t) for type II_P process (HK + HHK ⇔ LLK + LK) in (*a*) reactive distillation column and (*b*) simple distillation column.

The distillation column has a moderate energy consumption with a boilup ratio of 3.36 and a reflux ratio of 2.07. Figure 17.11*b* shows the composition profile in the distillation column where the two unreacted (heavy) reactants go out the bottom of the column.

Type II_R: LLK + LK = HK + HHK. This case corresponds to the following reaction:

$$\begin{matrix} \text{A} & + & \text{B} & \Longleftrightarrow & \text{C} & + & \text{D} \\ \text{LLK} & & \text{LK} & & \text{HK} & & \text{HHK} \end{matrix} \tag{17.15}$$

In terms of reactants and products, the boiling point ranking is just the opposite of type II_p [Eq. (17.15) vs. Eq. (17.14)]. Now the two reactants are lighter than the two products. Thus, these two reactants are more concentrated toward the upper part of the reactive distillation column, and the two products can be withdrawn from the bottom of the column. A typical example is the acetalization reaction for the synthesis of ethylal.[8] The reaction can be expressed as

$$\underset{\text{LLK}}{\text{formaldehyde}} + \underset{\text{LK}}{\text{ethanol}} \Longleftrightarrow \underset{\text{HK}}{\text{ethylal}} + \underset{\text{HHK}}{\text{water}}$$

Thus, the upper section of the column should be made reactive while the separation is carried out in the stripping section. This gives the reactive distillation configuration shown in Figure 17.12. The reaction takes place in the reactive distillation column, and the bottoms product is then fed to a simple distillation column to separate the two products. Because the two unreacted reactants (LLK and LK) are lighter than the two products (HK and HHK), the conversion in the reactive distillation column is set to 97.5% such that both products, HK in particular, can meet the 95% specification. The product compositions are 95% C (HK), along with two light reactants, and 97.5% D (HHK), along with 2.5% HK (C). Following the design procedure, the final design is shown in Figure 17.12. Table 17.3 gives the parameter values of this design.

In terms of topology, this process looks like the mirror image of type II_p in which the reactants and two products switch ends of the columns. However, the flowsheet gives a TAC of $1,096,000, which is 150% of that of type II_p. Moreover, the TAC of the reactive distillation column is $742,000, which is twice the cost of the reactive distillation for type II_p (TAC = $323,000). The reason for that is actually quite simple. The reaction takes place in the *low* temperature zone (upper section of the column). This leads to a reactive distillation with 24 reactive trays plus a reactive condenser and 3 stripping trays. As

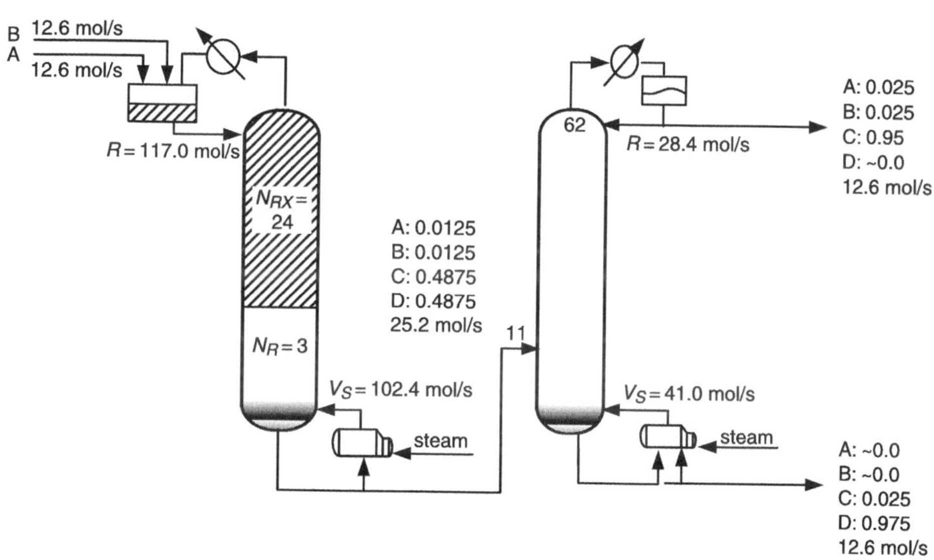

Figure 17.12 Final design for type II_R process (LLK + LK ⇔ HK + HHK).

[8]S. P. Chopade and M. M. Sharma, Reaction of ethanol and formaldehyde: Use of versatile cation-exchange resins as catalyst in batch reactors and reactive distillation columns, *Reactive and Functional Polymers*. **32**, 53–64 (1997).

expected, the LK (component D) is introduced into the condenser, and the LLK (component C) is fed into the reactive zone that is 9 trays below (Fig. 17.12). The reflux/feeds ratio is around 4, and the boilup ratio is also close to 4. These values are much higher than that of type II_p, which lead to higher energy consumption in the reactive distillation column.

This example illustrates the conflict between reactant concentration and reaction temperature for a chemical reaction, and this is a particularly important feature of reactive distillation. Because of the reversible reaction, the reactive zone should be placed where the reactants are most abundant. In the type II_p example considered in this section, this occurs in the upper section of the reactive distillation column, which is the *low* temperature zone. It is interesting to note that the reaction temperature is the lowest (reflux drum) when the product of the two-reactant concentration is the highest. Figure 17.13*a* indicates that

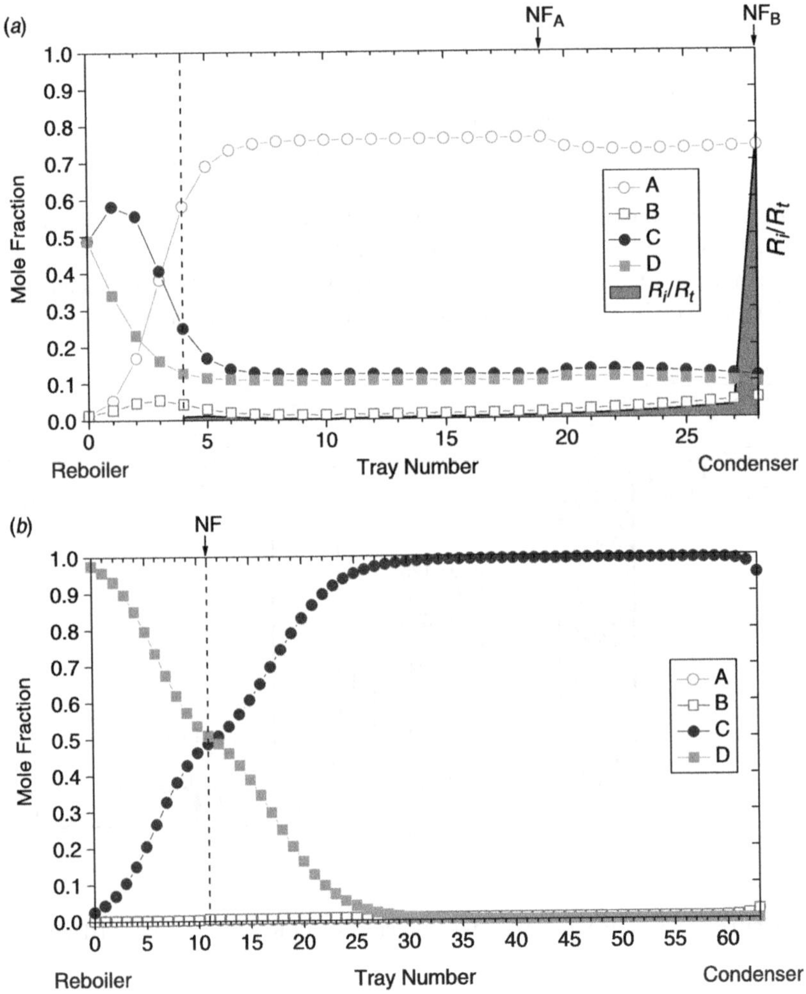

Figure 17.13 Composition profile, feed locations, and fraction of total conversion (R_i/R_t) for type II_R process (LLK + LK ⇔ HK + HHK) in (*a*) reactive distillation column and (*b*) simple distillation column.

most of a significant portion of the reaction takes place in the condenser, and the remaining reactive trays carry the reaction further to the desired specification. The composition profile is qualitatively similar to that of II_p, except the reactive zone is much longer. Toward the bottoms of the reactive distillation column, the two heavy products reach 48.75% with 2.5% of unreacted reactants.

The TAC of the second column ($353,000) is only half of that of the reactive distillation column. This is a simple distillation with light impurities. The reflux ratio is around 2, and the boilup ratio is a little higher than 3. Thus, the top composition achieves the specification (95% D, LK component) with the two reactants as the impurities. The bottoms composition is 97.5% D with 2.5% C as the only impurity. This column has 62 trays, and the feed is introduced into tray 11. Figure 17.13*b* provides the composition profiles in the distillation column. The two unreacted (light) reactants end up on the top of the column.

Remember that a fixed pressure of 8 bar is being used in all of these cases for purposes of comparison. The optimum pressure can be found in each case, and this will affect the designs and economics. For example, the pressure in the case with light reactants could be increased so that higher temperatures occur in the top of the reactive distillation column. If we limit the temperature in the reactive zone to no greater than 440 K and include the column pressure as a design variable, this leads to a pressure of 12 bar for type II_p (light reactants) and 6 bar for type II_R (heavy reactants). However, the ratio of TACs for these two types (TAC of type II_p/TAC of type II_R) is about the same as compared to that in Table 17.3.

17.3.3 Type III: Alternating

Type III_p: LK + HHK = LLK + HK. When one of the products (C) is the lightest component, we have the following reaction:

$$\underset{\text{LK}}{\text{A}} + \underset{\text{HHK}}{\text{B}} \Longleftrightarrow \underset{\text{LLK}}{\text{C}} + \underset{\text{HK}}{\text{D}} \tag{17.16}$$

The boiling point ranking indicates that if we consume the heavy reactant (HHK component B) before it reaches the bottom of the reactive zone, the heavy product (HK component D) can be obtained from the column base. The scenario is simpler toward the top of the column because the light product (LLK component C) can be withdrawn from the top while the light reactant (LK component A) can be prevented from leaving the reactive zone. The popular methyl acetate production via acetic acid esterification (Chap. 7) example falls into this category.

$$\underset{\text{LK}}{\text{methanol}} + \underset{\text{HHK}}{\text{acetic acid}} \Longleftrightarrow \underset{\text{LLK}}{\text{methyl acetate}} + \underset{\text{HK}}{\text{water}} \tag{17.17}$$

Because two products are withdrawn from the opposite ends of the column, the reactive zone is placed in the middle. This column configuration is shown in Figure 17.14. Table 17.3 gives the parameter values of this design.

This flowsheet has a TAC of $321,000, which is the second lowest in all six cases (126% of that of type I_p). The reactive distillation column has a total of 59 trays with 49 reactive trays, 8 stripping trays, and 2 rectifying trays. Two feeds (11 trays apart) are introduced into

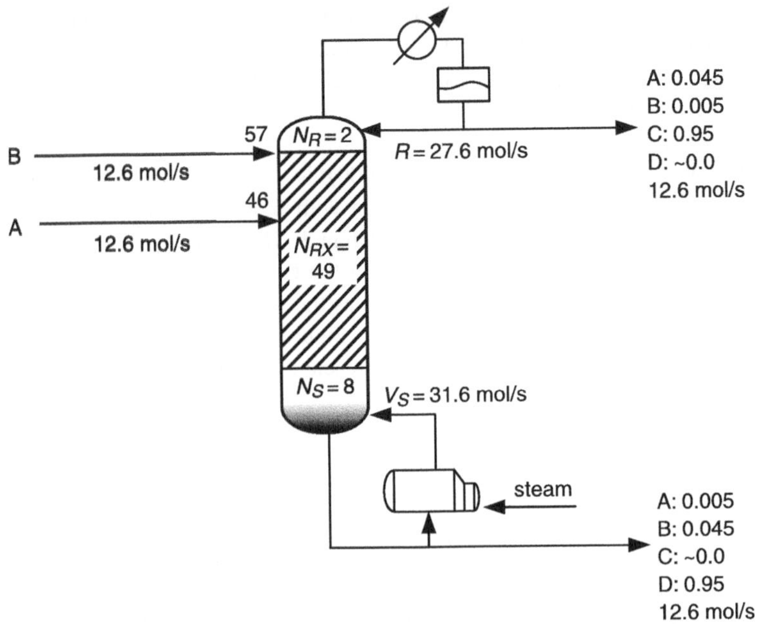

Figure 17.14 Final design for type III_p process (LK + HHK ⇔ LLK + HK).

the upper section of the reactive trays as shown in Figure 17.14. Both the reflux ratio and boilup ratio are slightly greater than 2.

Actually, this is a relatively simple reactive distillation column with moderate energy consumption, despite having a relatively large number of reactive trays. Having one reactant that is the HHK (reactant B) has its advantages and disadvantage. The HHK increases tray temperatures when we have significant amounts of this heavy reactant, which is advantageous for the reaction (see the temperature profile in Fig. 17.15*b*). The down side is that we have to react away almost all of the HHK in the reactive zone (otherwise it will end up in the column base), and this leads to a large number of reactive trays. This is clearly illustrated in Figure 17.15*a*, where we have a very small amount of conversion between tray 9 and tray 40. The purpose for this portion of the reactive trays is to consume the remaining heavy reactant (HHK component B). The composition of the LK reactant (component A) is kept fairly constant below the feed point to ensure the dominance of the forward reaction (Fig. 17.15*a*). The two products are further purified in the top and bottom rectifying and stripping sections of the reactive distillation column to meet the specifications.

Type III_R: LLK + HK = LK + HHK. For this alternating type, when one of the reactants (A) is the lightest component, we have the following reaction:

$$\begin{array}{ccccccc} A & + & B & \Longleftrightarrow & C & + & D \\ \text{LLK} & & \text{HK} & & \text{LK} & & \text{HHK} \end{array} \tag{17.18}$$

The boiling point ranking suggests that if we consume the light reactant (LLK component A) toward the top of the column, the light product (LK component C) can be obtained from

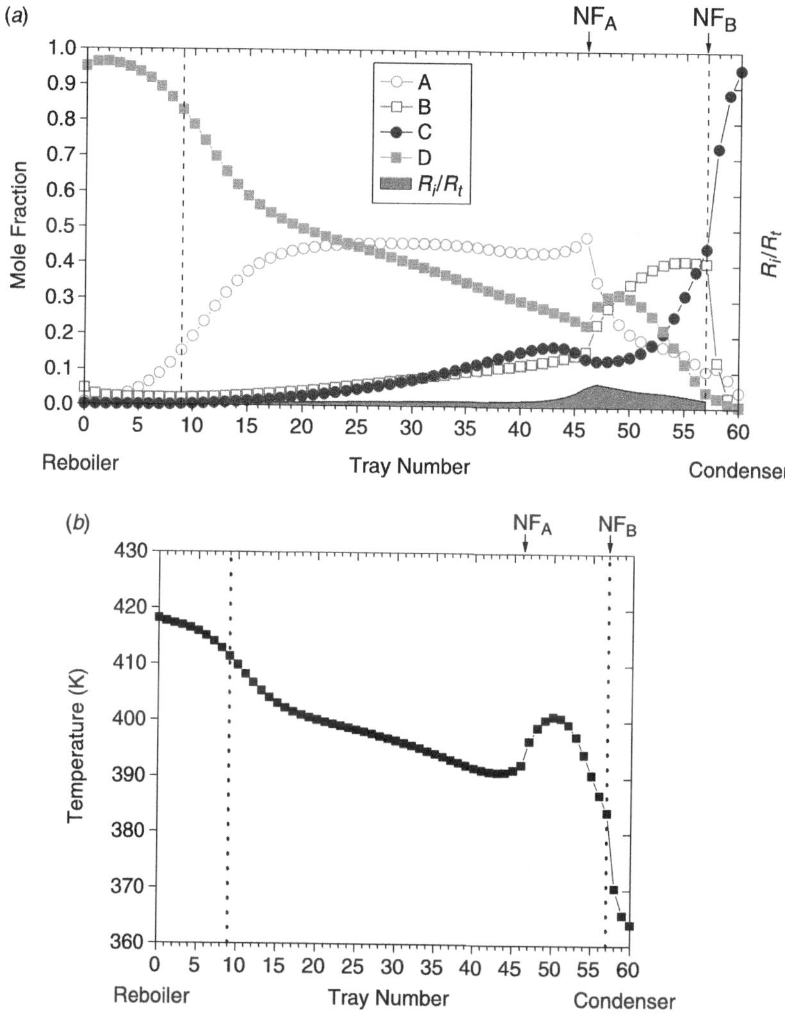

Figure 17.15 (*a*) Composition profile, feed locations, and fraction of total conversion (R_i/R_t) and (*b*) temperature profile for type III$_p$ process (LK + HHK $\Leftrightarrow$ LLK + HK).

the column overhead. The scenario is simpler toward the bottoms of the column because the heavy reactant (HHK component D) can be separated easily from the heavy reactant (HK component B). Methyl acetate hydrolysis[9] is a typical example.

$$\underset{\text{LLK}}{\text{methyl acetate}} + \underset{\text{HK}}{\text{water}} \Longleftrightarrow \underset{\text{LK}}{\text{methanol}} + \underset{\text{HHK}}{\text{acetic acid}} \tag{17.19}$$

Note that the neat design is considered here instead of the excess reactant design often seen in methyl acetate hydrolysis. Because the light product (LK component C) and heavy

[9] M. F. Malone, R. S. Huss, and M. F. Doherty, Green chemical engineering aspects of reactive distillation, *Environ. Sci. Techno.* **37**, 5325–5329 (2003).

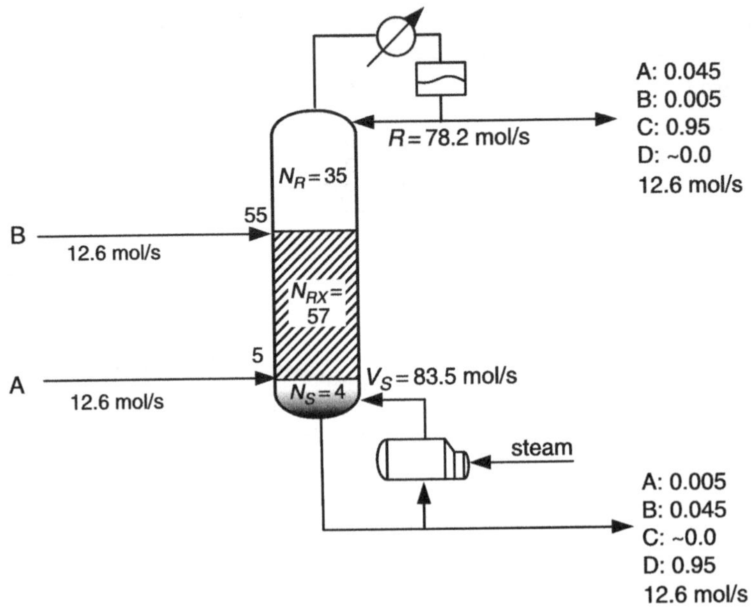

Figure 17.16 Final design for process type III_R (LLK + HK ⇔ LK + HHK).

product (HHK component D) are withdrawn from the top and bottoms of the column, the reactive zone is placed in the middle. Figure 17.16 gives the column configuration. Following the proposed design procedure, the final design is shown in Table 17.3 with detailed parameter values.

We might think that this is simply the mirror image of type III_p. However, this flowsheet has a TAC of $726,000, which is more than twice that of type III_p. The reactive distillation column has a total of 96 trays comprising 57 reactive trays, 4 stripping trays, and 35 rectifying trays. Two feeds are introduced into the top and bottoms of the reactive zone as shown in Figure 17.16. The reflux ratios and boilup ratios are much higher than those of type III_p with values greater than 6. Actually, this is a capital intensive reactive distillation column with relatively high energy consumption.

Liquid-phase reactions with one reactant being the LLK (reactant A) pose a difficult design, especially for reactive distillation. The reasons are 1) we have to consume all of the LLK (reactant A) toward the top of the column and 2) we have to maintain a high concentration for the heavy reactant (HK component B) to ensure that the forward reaction is dominant. Thus, 57 reactive trays are used to react away most of the LLK (reactant A) as is clearly shown in Figure 17.17*a*, where we have a very small amount of conversion between tray 30 and tray 61. Leaving the top of the reactive zone of the column, we have almost equal molar amounts of LK (product C) and HK (reactant B) with a trace amount of LLK (reactant A). A large number of rectifying trays (35 trays) are employed to return HK back to the reactive zone while having LLK and LK as the top product. The scenario toward the bottoms is much simpler because we are performing separation between HK (reactant B) and HHK (product C). Thus, only 4 stripping trays are required. The composition profiles produce a very interesting temperature profile (Fig. 17.17*b*) with two distinct plateaus.

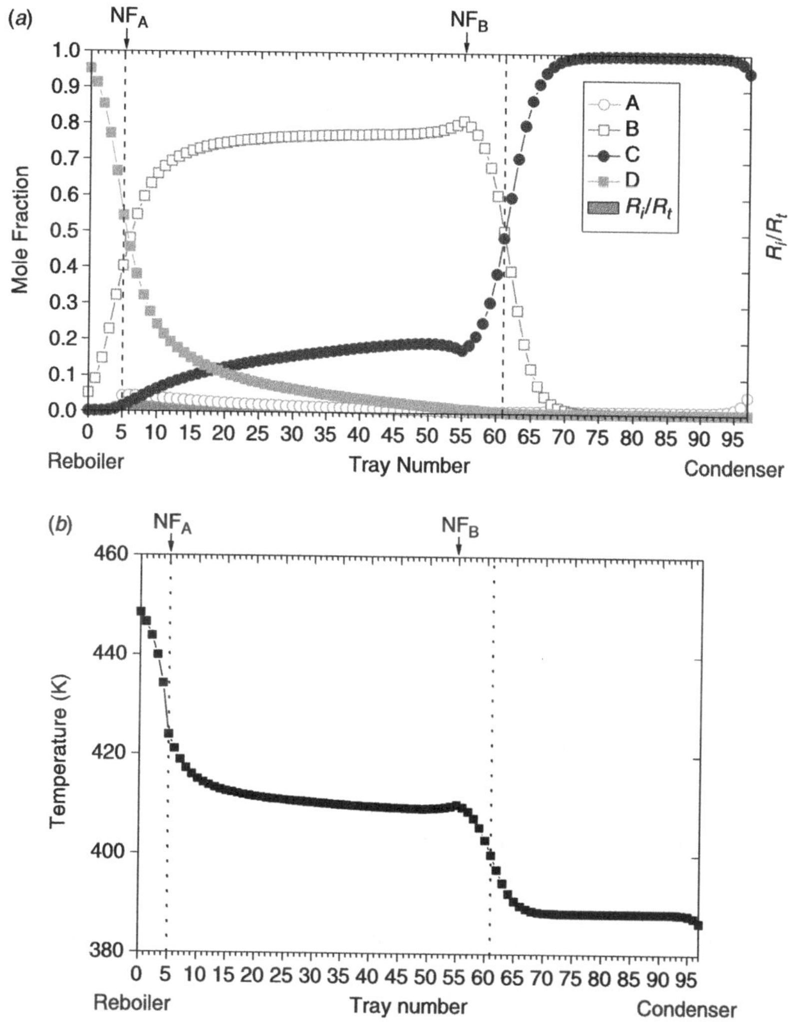

Figure 17.17 (*a*) Composition profile, feed locations, and fraction of total conversion (R_i/R_t) and (*b*) temperature profile for process type III_R (LLK + HK ⇔ LK + HHK).

17.4 RESULTS AND DISCUSSION

17.4.1 Summary

For the reversible reaction A + B ⇔ C + D, we have explored the design of reactive distillation systems for all possible relative volatility rankings. Six distinct configurations are further classified into three different types (type I: one group, type II: two groups, and type III: alternating) according to the relative position of the reactants and the products in the relative volatility sequence (Fig. 17.1). The subscripts p and R are used to denote the case when the product (type I_p, type II_p, and type III_p) or the reactant (type I_R, type II_R, and type III_R) is the lightest component.

Following a systematic design procedure, process flowsheets are developed and corresponding TACs (operating cost + discounted capital cost) are computed. The TACs range from \$254,000 to \$1,896,000, which is a factor of 7.5. The flowsheets use either one-column or two-column schemes. The number of trays in the reactive distillation column varies from 11 to 96. The energy cost in the reactive distillation varies from \$131,000 to \$1,055,000, which is a factor of 8. The reactive zone can be placed in the *middle*, in the *upper* section, in the *lower* section, or at *both* ends. All of these differences come from shuffling the relative volatilities of the reactants and products in a range of 8, 4, 2, and 1.

It may appear that the outcomes are case based. However, when we carefully examine the TAC and the process, some observations can be made. Figure 17.18 ranks the TACs for corresponding process flowsheets. In terms of the relative volatility sequence, we have the following order for reaction $A + B \Leftrightarrow C + D$:

$$\begin{array}{ll} \text{lowest TAC} & \alpha_C > \alpha_A > \alpha_B > \alpha_D \\ & \alpha_C > \alpha_A > \alpha_D > \alpha_B \\ \downarrow & \alpha_C > \alpha_D > \alpha_A > \alpha_B \\ & \alpha_A > \alpha_C > \alpha_B > \alpha_D \\ & \alpha_A > \alpha_B > \alpha_C > \alpha_D \\ \text{highest TAC} & \alpha_A > \alpha_C > \alpha_D > \alpha_B \end{array} \tag{17.20}$$

The TAC ranking clearly indicates that the group (the three in the lower part) with higher TACs has one of the reactants (A) being the lightest component. This leads to the following heuristic:

Heuristic H1: It is not favorable to have one of the *reactants* as the *lightest* component.

The explanation is quite straightforward: *light* component implies *low* reaction temperature. Certainly, we would prefer that the reaction take place at the high temperature zone instead of the lower one. Even in the same group [lower or upper three in Eq. (17.20)], we can observe that when the relative volatility between two products is further apart, the TAC will be lower. Thus, the next heuristic addresses the relative volatilities of products.

Heuristic H2: Prefer the case when the *relative volatility between two products is large* (preferably separated by reactants).

The reason is that it facilitates the separation for the products. With everything being equal (from heuristics H1 and H2), the relative volatility between reactants is also a useful measure.

Heuristic H3: Prefer the case when the *relative volatility between two reactants is small* (preferably not separated by products).

The explanation is that we can have higher reactant concentrations, which is favorable for the forward reaction. The three heuristics presented here are somewhat straightforward and can be combined into one:

Prefer the case that reactants are intermediate with a small difference of relative volatility and products are the most and the least volatile with a large difference of relative volatility.

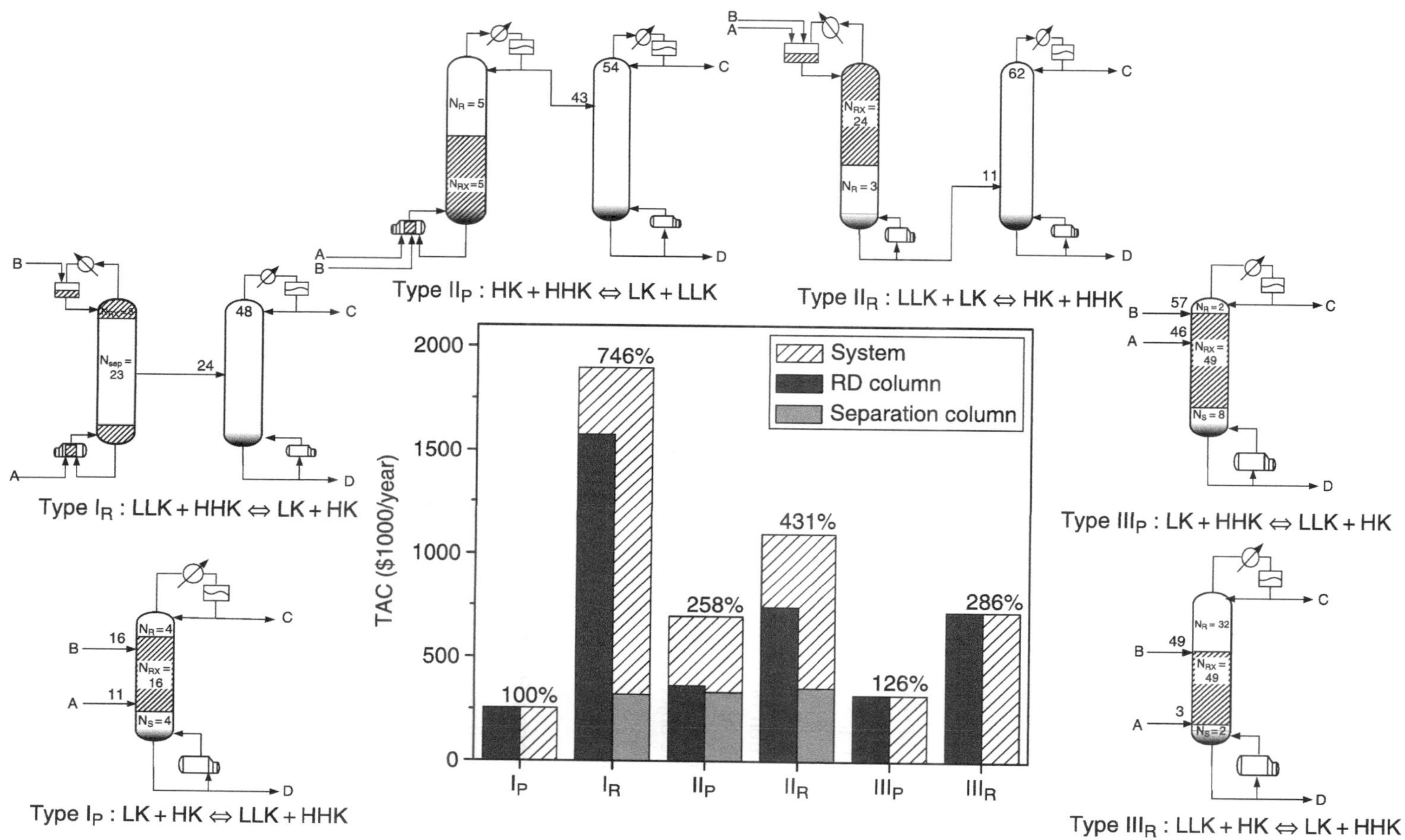

Figure 17.18 TAC and corresponding column costs for all six process flowsheets.

The flowsheet with the lowest TAC (type I_p) is favored by all three heuristics (reactant is not the lightest one, relative volatility between two products is the largest, and relative volatility between two products is the smallest). The flowsheet with the highest TAC (type I_R) is a difficult reactive distillation as suggested by the heuristics (reactant is the lightest one, relative volatility between two products is the smallest, and relative volatility between two products is the largest).

Before leaving this section, we will comment on the scenario in which the forward reaction A + B ⇔ C + D, obtaining products C and D, and backward reaction C + D ⇔ A + B, converting back to A and B, are performed sequentially in two reactive distillation columns for the purpose of separation. A notable example is the recovery of dilute polyol (e.g., propylene glycol and/or ethylene glycol) from aqueous solution via acetalization.[10,11] For the separation of a 5 wt% polyol (e.g., propylene glycol) from aqueous solution, the cost of conventional distillation will be expensive because water is the low boiler. An alternative is to use a reactive distillation to convert the polyol into acetal, which is far more volatile than the polyol, making them easily separate from water and able to be withdrawn from the column overhead. The general reaction can be expressed as

$$\text{polyol (HHK)} + \text{aldehyde (LLK)} \Longleftrightarrow \text{cyclic acetal (LK)} + \text{water (HK)}$$

In the second step, the reverse reaction is carried out to convert the acetal back to the polyol and aldehyde. Figure 17.18 clearly shows that for the mentioned forward and backward reactions, different reactive distillation configurations should be used. For example, if the forward reaction (to form C and D) belongs to type I_p, the type I_R configuration should be used to convert C and D back to A and B. For types I and II, the placement of the reactive zones is quite different (Fig. 17.18) and we generally need two reactive distillation columns to achieve the objective. The type III configurations, by contrast, possibility exist for using the same reactive distillation column to perform reaction and separation in the same equipment.

17.4.2 Excess Reactant Design

Thus far, we have considered only the neat design. The excess reactant design may be preferable for systems with high TACs, especially for types I_R, II_R, and possibly III_R. The motivation comes from the fact that some of the reactant concentrations are so low that a large reflux ratio and/or boilup ratio are required to achieved 95% conversion. We can refer to the composition profile of reactants A toward the column base and B toward the top in Figure 17.9 for type I_R. Other examples are reactant B in the reactive zone (Fig. 17.13) for type II_R and reactant A in the reactive zone (Fig. 17.17) for type III_R. The excess reactant design is a simple means to achieve an improved reactant composition profile.

Type I_R is used to illustrate the design. The design questions then become, 1) which reactant should be in excess (e.g., light reactant or heavy reactant) and (2) what will be the separation sequence for this recycle plant? Recall that in the reaction LLK + HHK ⇔ LK+

[10] J. Hao, H. Liu, and D. Liu, Novel route of reactive extraction to recover 1,3-propanediol from a dilute aqueous solution, *Ind. Eng. Chem. Res.* **44**, 4380–4385 (2005).

[11] A. D. Dhale, L. K. Myrant, S. P. Chopade, J. E. Jackson, and D. J. Miller, Propylene glycol and ethylene glycol recovery from aqueous solution via reactive distillation, *Chem. Eng. Sci.* **59**, 2881–2890 (2004).

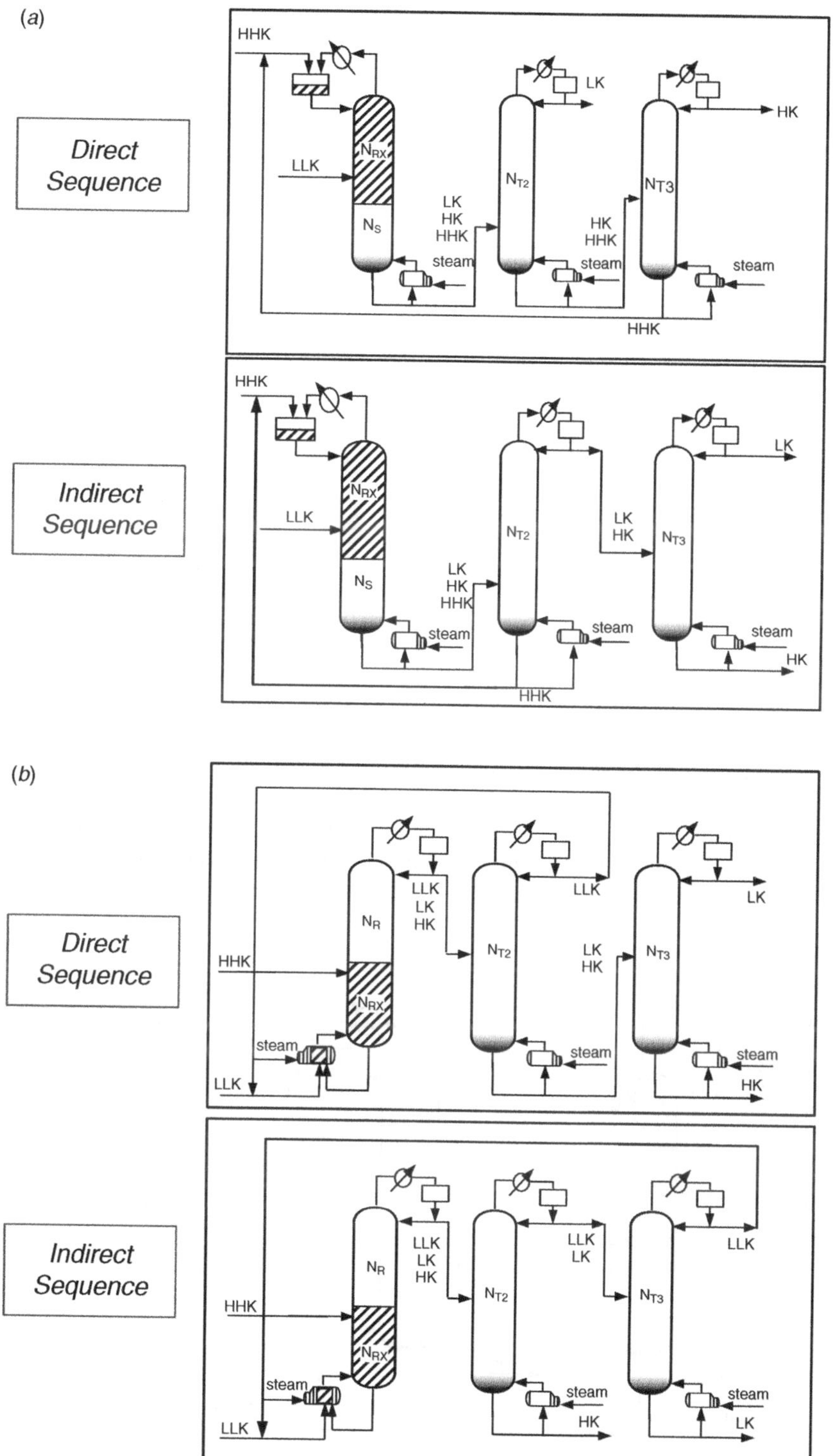

Figure 17.19 Possible process configurations for type I_R (LLK + HHK $\Leftrightarrow$ LK + HK) with excess reactant design: (*a*) heavy reactant (HHK) excess and (*b*) light reactant (LLK) excess.

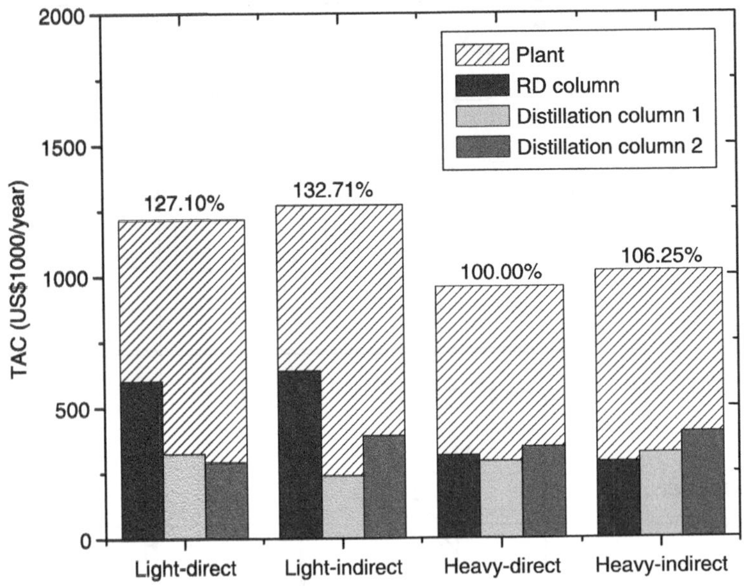

Figure 17.20 Total annual costs for type I_R (LLK + HHK ⇔ LK + HK) with four possible excess reactant designs: light reactant (LLK) excess with direct and indirect separation sequences and heavy reactant (HHK) excess with direct and indirect separation sequences.

HK for a type I_R system, we have the following two options for each choice: 1) heavy reactant excess versus light reactant excess and 2) direct versus indirect separation sequence. This leads to four design alternatives as shown in Figure 17.19. For the case in which the heavy reactant (HHK) is in excess, we have three components (LLK, LK, and HK) leaving the reactive distillation column for further separation. Essentially, all of the LLK

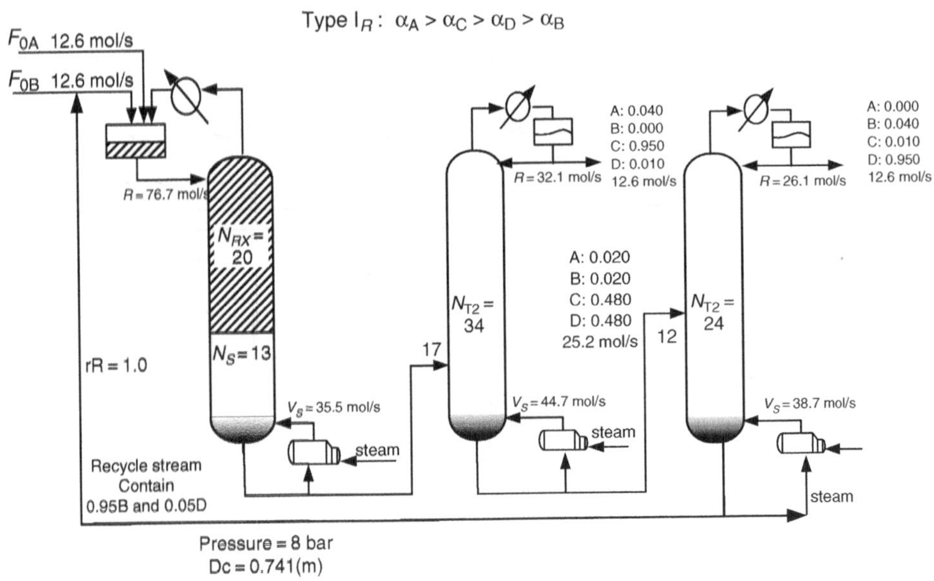

Figure 17.21 Optimized flowsheet for type I_R process (LLK + HHK ⇔ LK + HK) with excess reactant design.

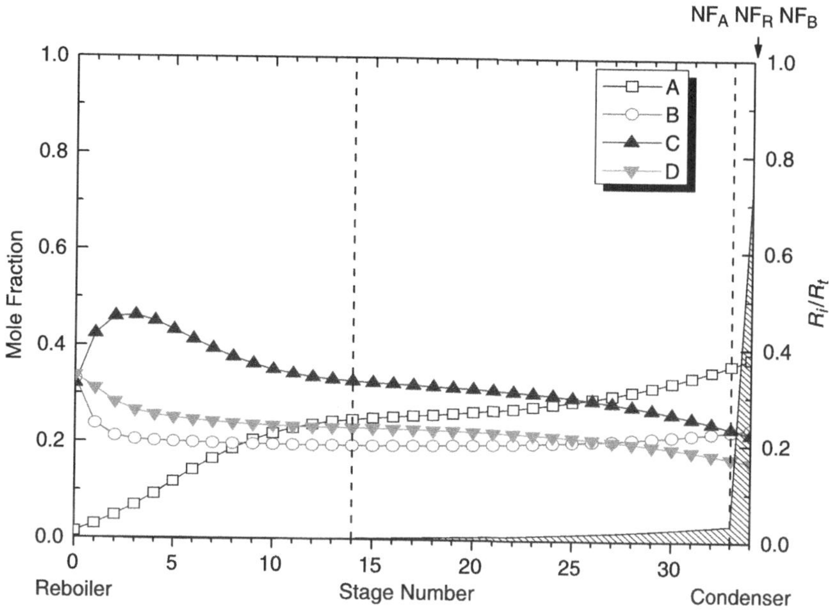

Figure 17.22 Composition profile, feed locations, and fraction of total conversion (R_i/R_t) for type I_R process (LLK + HHK ⇔ LK + HK) with excess reactant design.

reactant has been consumed, except for a small amount that appears as an impurity in the distillate stream that contains the LK product.

Two column sequences are possible for this ternary separation: direct and indirect (Fig. 17.19*a*). A similar scenario applies when the light reactant is in excess (Fig. 17.19*b*). The two conventional columns must separate the ternary mixture of LLK, LK, and HK.

The design procedure is similar to that of Section 17.3 except for an additional design variable: the recycle flowrate. A large recycle flowrate implies a higher reactant concentration for the excess reactant while having a greater recycle cost. Thus, it becomes a dominant design variable for excess reactant design.

Figure 17.20 shows that the heavy reactant excess design is favored over the light reactant excess design by a factor of almost 30%. As for the column sequencing, the "direct" separation sequence is preferred over the "indirect" sequence. For the separation sequence, the LLK is boiled up twice for the indirect sequence as compared to that of the direct sequence where the LLK is only boiled up once. Therefore, we expect a lower separation cost using the direct sequence.

As for heavy reactant or light reactant excess, Figure 17.9 reveals that we have a higher heavy reactant (B) concentration toward the column top as compared to the concentration of light reactant (A) toward the column base. In other words, relatively speaking, it is easier to increase the heavy reactant concentration compared to that of the light reactant. Figure 17.21 gives the TAC optimal design for type I_R for the excess reactant design. The composition profile in Figure 17.22 indicates that the composition of the heavy reactant (B) is maintained at ~20% throughout the reactive zone, and this is a significant increase from the neat design (cf. Fig. 17.9). Of more importance, the TAC of the excess reactant design is only 50% of that of the neat design. This means that, for cases with poor reactant distribution, excess reactant is an attractive design alternative.

17.5 CONCLUSION

This chapter explored the effects of relative volatility rankings on the design of reactive distillation systems. Ideal reactive distillation systems were used to illustrate the process flowsheet generation as the ranking varied. For a quaternary system with a second-order reversible reaction (A + B $\Leftrightarrow$ C + D), the 24 possible relative volatility rankings were classified into three process types based on the distribution of reactants and products in the relative volatility sequence (types I, II, and III for one group, two groups, and alternating, respectively). Each type was further denoted by the nature (product p or reactant R) of the lightest component. A systematic design procedure was used to generate a process flowsheet as the relative volatility ranking changed. The TAC was used to evaluate the appropriateness of different designs.

The assumptions made in this work include 1) ideal vapor–liquid equilibrium, 2) equal molar feed (neat process), 3) the reactive holdup set by the column diameter, and 4) a sequential approach for optimization. It is interesting to note that the reactive zone can be placed at the upper section, lower section, middle, or both ends of the reactive distillation column, depending on the sequences of the relative volatilities. The principle is actually quite simple: place the reactive zone where the reactants are most abundant and introduce the feeds to facilitate the reaction (considering the composition effect).

Among the various types of systems, the flowsheet could have either one or two columns. The TACs varied by a factor of 7.5, the energy costs changed by a factor of 8, and the number of trays in the reactive distillation column ranged from 11 to 96. Physical explanations for the effects of combined reaction/separation were given. Heuristics were provided to evaluate potential difficulties in reactive distillation design.

As discussed in the final section of this chapter, an excess reactant design is more attractive than the neat design for some of the systems with difficult relative volatility rankings.

CHAPTER 18

EFFECTS OF FEED TRAY LOCATIONS ON DESIGN AND CONTROL OF REACTIVE DISTILLATION

The effects of feed locations on the design of reactive distillation systems are explored in this chapter. Ideal reactive distillation systems are used to illustrate the advantage of feed tray optimization in design and control. The process system studied is type I_p as defined in Chapter 17, where we have the following reaction: A + B $\Leftrightarrow$ C + D. Reactants A and B are LK and HK, respectively, and products C and D are LLK and HHK, respectively. Process parameters such as relative volatilities between reactants, relative volatilities between products, column pressure, activation energies, and preexponential factors are varied to seek possible generalization. For all of the systems that were studied, the percentage of energy savings ranges from 6% to 47%, which is obtained by simply rearranging the feed locations. Finally, the idea of dynamically manipulating the feed tray locations is explored.

18.1 PROCESS CHARACTERISTICS

Consider an ideal reactive distillation (Fig. 18.1) with a reversible liquid-phase reaction in the reactive section:

$$\mathrm{A + B} \Longleftrightarrow \mathrm{C + D}$$
$$\mathrm{LK + HK} \Longleftrightarrow \mathrm{LLK + HHK}$$

The forward and backward specific rates following the Arrhenius law on tray j are given by

$$k_{Fj} = a_F e^{-E_F/RT_j} \tag{18.1}$$

$$k_{Bj} = a_B e^{-E_B/RT_j} \tag{18.2}$$

where a_F and a_B are the preexponential factors for the forward and backward reactions, respectively; E_F and E_B are the activation energies; and T_j is the absolute temperature on

Reactive Distillation Design and Control. By William L. Luyben and Cheng-Ching Yu

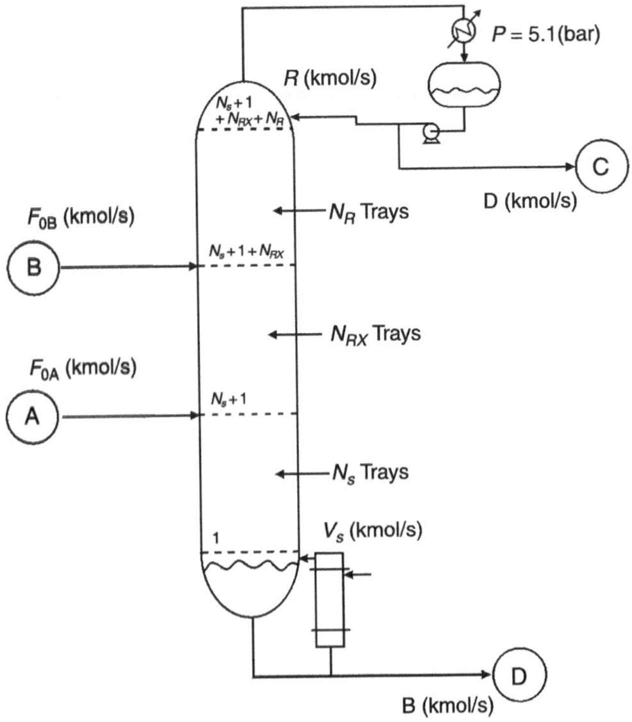

Figure 18.1 Reactive distillation with N_R rectifying trays, N_{RX} reactive trays, and N_S stripping trays with conventional feed arrangement ($N_{F,\mathrm{B}} = N_{RX,\mathrm{top}}$ and $N_{F,\mathrm{A}} = N_{RX,\mathrm{bot}}$).

tray j. The reaction rate on tray j can be expressed in terms of mole fractions ($x_{j,i}$) and the liquid holdups (M_j).

$$\aleph_{j,i} = \nu_i M_j (k_{Fj} x_{j,A} x_{j,B} - k_{Bj} x_{j,C} x_{j,D}) \qquad (18.3)$$

where $\aleph_{j,i}$ is the reaction rate of component i on the jth tray (kmol/s), ν is the stoichiometric coefficient that takes a negative value for the reactants, and M_j is the molar holdup on tray j (kmol) that is assumed to be constant throughout the simulation. This is typically true for catalyst weight based kinetics.

The assumptions made in this work include the following:

1. Forward reaction rate k_F is specified as 0.008/s at 366 K, and backward reaction rate k_B is set to 0.004/s at the same temperature. Kinetic and physical property data for the system are given in Table 18.1.[1]
2. A molar holdup (M_j) of 1000 mol is assumed.[1]
3. Ideal vapor–liquid equilibrium is assumed, in which constant relative volatilities are used. The tray temperature is computed using the Antoine vapor pressure equation (Table 18.1). Note that as the result of constant relative volatility, the Antoine coefficients (B_{VP}) are the same for all four components.
4. Vapor holdup and pressure drop are neglected.

[1]M. A. Al-Arfaj and W. L. Luyben, Comparison of alternative control structures for an ideal two-product reactive distillation column, *Ind. Eng. Chem. Res.* **39**, 3298–3307 (2000).

TABLE 18.1 Physical Properties for High Activation Energies Case

Activation energy (cal/mol)	Forward (E_F)			30,000	
	Backward (E_B)			40,000	
Specific reaction rate at 366 K (kmol s^{-1} kmol^{-1})	Forward (k_F)			0.008	
	Backward (k_B)			0.004	
Heat of reaction (cal/mol)				−10,000	
Heat of vaporization (cal/mol)				6944	
Relative volatilities ($\alpha_C/\alpha_A/\alpha_B/\alpha_D$)				8/4/2/1	
		C	A	B	D
Vapor pressure constants[a]	A_{VP}	13.04	12.34	11.65	10.96
	B_{VP}	3862	3862	3862	3862

[a] $\ln P_i^S = A_{VP,i} - B_{VP,i}/T$, where T is the temperature (K) and P_i^S is the vapor pressure of pure component i (bar).

As shown in the Figure 18.1, the column is divided into three sections. The middle section is the reactive zone containing N_{RX} trays. The rectifying section has N_R trays, and the stripping section has N_S trays. Thus, we are considering a reactive distillation column in which the reaction only occurs in the reactive section, which implies a reaction catalyzed by a solid catalyst.

The relative volatilities of the components are in the following order:

$$\alpha_C > \alpha_A > \alpha_B > \alpha_D$$

Products C and D are the lightest and heaviest components, respectively, with reactants A and B being the intermediate boilers. The thermodynamic behavior indicates that we should remove light product C from the distillate and obtain heavy product D from the bottoms. Figure 18.1 also shows that fresh feedstream F_{0A} containing reactant A is fed to the bottom of the reactive zone, and heavy reactant B is fed to the top of the reactive zone. With this feed arrangement, the light reactant travels up the reactive zone and the heavy reactant comes down from the top of the reactive zone. This provides a countercurrent mixing for the reactants and, in theory, it is favorable for the reaction to proceed forward. Light product C moves upward toward the rectifying section and is further purified to meet the product specification. Similarly, heavy product D travels downward and is further purified in the stripping section.

18.1.1 Modeling

In the Figure 18.1, the component balances for the column can be expressed as rectifying and stripping trays:

$$\frac{d(x_{j,i}M_j)}{dt} = L_{j+1}x_{j+1,i} + V_{j-1}y_{j-1,i} - L_jx_{j,i} - V_jy_{j,i} \tag{18.4}$$

reactive trays:

$$\frac{d(x_{j,i}M_j)}{dt} = L_{j+1}x_{j+1,i} + V_{j-1}y_{j-1,i} - L_jx_{j,i} - V_jy_{j,i} + \Re_{j,i} \tag{18.5}$$

and feed trays:

$$\frac{d(x_{j,i}M_j)}{dt} = L_{j+1}x_{j+1,i} + V_{j-1}y_{j-1,i} - L_jx_{j,i} - V_jy_{j,i} + \Re_{j,i} + F_jz_{j,i} \qquad (18.6)$$

Here, $x_{j,i}$ and $y_{j,i}$ denote the liquid and vapor mole fraction of component i on tray j, with L_j and V_j corresponding to the liquid and vapor flowrates for the jth tray. A liquid hydraulic time constant (β) is included by using a linearized form of the Francis weir formulation, and β is set to 6 s in this work. Because equimolar overflow is assumed, the vapor and liquid flowrates are constant throughout the stripping and rectifying sections but change in the reactive zone as a result of the exothermic reaction.[1] The heat of reaction vaporizes some liquid on each tray in this section. Therefore, the vapor flowrate increases up through the reactive zone, and the liquid flowrate decreases down through the reactive zone. The expressions are the same as those of Eqs. (17.6) and (17.7).

The vapor–liquid equilibrium is assumed to be ideal and the bubblepoint temperature calculation is used to find the tray temperature (see Table 18.1 for the vapor pressure data of pure components). The column pressure is fixed at 5.1 bar unless otherwise noted.

18.1.2 Steady-State Design

Typical design variables of a reactive distillation column include column pressure P, number of reactive trays N_{RX}, number of trays in the stripping N_S and rectifying N_R sections, and the locations of the feed trays ($N_{F,A}$ and $N_{F,B}$, respectively) as shown in Figure 18.1.

In this work, P is fixed at a constant value and N_{RX} is selected to ensure the desired conversion with the conventional feed arrangement (heavy reactant to the top of the reactive zone and light reactant to the bottom of the reactive zone). The number of trays in the stripping and rectifying sections is set to twice of the minimum number of trays according to the Fenske equation,[2] which is

$$N = 2 \cdot N_{\min} = \frac{2 \cdot \ln\left(\frac{x_{D,LK}}{x_{D,HK}} \frac{x_{B,HK}}{x_{B,LK}}\right)}{\ln\left(\frac{\alpha_{LK}}{\alpha_{HK}}\right)} \qquad (18.7)$$

For the rectifying tray number N_R we use the liquid composition right above the reactive tray for x_B, and for N_S we use the vapor phase composition right below the reactive zone as x_D. This leaves us with the feed tray locations as the design variables. Because all tray numbers are determined, the effects of feed tray locations can be compared by simply looking at the energy consumption (i.e., vapor boilup rate).

18.1.3 Base Case

Equations describing the material balances were programmed in Fortran, and all simulations were carried out on a Pentium PC. Note that the convergence of the reactive distillation is far more difficult than conventional distillation. Typically, a steady-state simulation is

[2]J. M. Douglas, *Conceptual Design of Chemical Process*, McGraw-Hill, New York, 1988.

carried out in a two-step procedure. First, the Wang–Henke method is used to converge the flowsheet (MESH equations) to a certain degree (actually to the point at which the objective function fluctuates). Second, the temperature and composition profiles are fed to a dynamic program that is integrated until temperatures and compositions converge (relaxation approach).

Saturated liquid feeds are assumed, and the two feed flowrates are 0.0126 kmol/s each of pure A and pure B. The two feeds are introduced on trays $N_{F,\mathrm{A}}$ and $N_{F,\mathrm{B}}$, respectively. In the base case, the feeds are introduced to the bottom ($N_{RX,\mathrm{bot}}$) or the top ($N_{RX,\mathrm{top}}$) of the reactive section (see Fig. 18.1 and base case in Table 18.2). Note that this is the typical feed arrangement for reactive distillation systems, which is termed the *conventional* feed arrangement. The conversion is specified to be 95% in this work, which corresponds to purities of 95% C in the distillate and 95% D in the bottoms.

Figure 18.2 (thickest line) shows the composition profiles of all four components at the nominal design. Reactant A has the highest concentration (x_A) on the feed tray ($N_{F,\mathrm{A}} = 9$). The profile shows that x_A decreases toward the upper reactive zone as a result of the reaction and also decreases toward the bottoms of the column as a result of separation. Similar behavior is observed for heavy reactant B. Both light product C and heavy product D meet the specification toward the ends of the column.

Al-Arfaj and Luyben[1] show the steady-state temperature profile in which a nonmonotonic temperature profile is observed. A local temperature minimum occurs on the lower feed tray because of the presence of a significant amount of light reactant A. This behavior is not uncommon for reactive distillation columns but is rarely seen in conventional distillation systems.

18.1.4 Feed Locations Versus Reactants Distribution

We emphasize that the reactive section of a reactive distillation column can be viewed as a cascade-type two-phase reactor with the reactor temperature determined by the bubblepoint temperature of the tray liquid-phase composition. It is clear that the composition and temperature profiles will certainly affect the performance of the reactive zone, and the feed tray locations appear to be one of the most effective variables for these profiles' redistribution. In this section, we are interested in how the composition profile will be affected by changing the feed tray location. Each individual feed tray location is changed while holding the other constant.

Initially, we fix the feed location of the A component at the bottom of the reactive zone ($N_{F,\mathrm{A}} = 9$) and change the feed location of the B component from the top of the reactive zone to the bottom. Figure 18.2*a* gives composition profiles for three values of the feed location of B ($N_{F,\mathrm{B}} = 19$, 14, or 11). As the feed location of component B moves down the column, the mole fraction of heavy reactant B (x_B) increases toward the lower section of reactive zone. That means we have a wider and less variation in the distribution of component B throughout the reactive zone as the feed tray is lowered. Consequently, the mole fraction of light reactant A (x_A) becomes smaller in the lower reactive zone while the profiles of the two products (x_C and x_D) remain qualitatively similar.

This rearrangement of the reactant composition certainly alters the "fraction of total conversion" (i.e., conversion in each tray divided by the total conversion) as well as the temperature profile in the reactive zone. Figure 18.3 provides the profiles of the fraction of total conversion, the compositions of the reactants, and the temperature in the reactive section as $N_{F,\mathrm{B}}$ varies. When we move $N_{F,\mathrm{B}}$ down, both reactant B and conversion increase

TABLE 18.2 Effects of Feed Locations on Design for Systems with Different Relative Volatilities and Rate Constants[a]

	Base Case	$\frac{\alpha_A}{\alpha_B} = 2$	$\frac{\alpha_A}{\alpha_B} = 1.5$	$\frac{\alpha_A}{\alpha_B} = 3.0$	$\frac{\alpha_C}{\alpha_A} = 4.0$	$\frac{\alpha_B}{\alpha_D} = 4.0$	Optimal P (9 bar)	Low k	High k
$\alpha_C/\alpha_A/\alpha_B/\alpha_D$	8/4/2/1	8/4/2/1	6/3/2/1	12/6/2/1	16/4/2/1	16/8/4/1	8/4/2/1	8/4/2/1	8/4/2/1
$K_{F,366}$ (s^{-1})	0.008	0.008	0.008	0.008	0.008	0.008	0.008	0.0048	0.016
$K_{B,366}$ (s^{-1})	0.004	0.004	0.004	0.004	0.004	0.004	0.004	0.0024	0.008
$N_S/N_{RX}/N_R$	8/11/9	8/11/9	9/11/9	8/11/8	5/11/5	6/11/6	8/11/9	8/11/9	6/11/6
$N_{RX,bot}/N_{RX,top}$	9/19	9/19	10/20	9/19	6/16	7/17	9/19	9/19	7/17
$N_{F,A}/N_{F,B}$	9/19	11/15	11/13	10/17	12/14	8/10	14/17	9/17	10/13
$X_{D,C}/X_{B,D}$	0.95/0.95	0.95/0.95	0.95/0.95	0.95/0.95	0.95/0.95	0.95/0.95	0.95/0.95	0.95/0.95	0.95/0.95
F_{0A}, F_{0B}, D, B (kmol/s)	0.0126	0.0126	0.0126	0.0126	0.0126	0.0126	0.0126	0.0126	0.0126
R (kmol/s)	0.0366	0.0332	0.0409	0.0274	0.0198	0.0284	0.0263	0.0445	0.0277
V_S (kmol/s)	0.0320	0.0285	0.0362	0.0227	0.0152	0.0237	0.0217	0.0399	0.0230
Energy savings[b] (%)	0	−10.9	−15.2	−6.9	−46.8	−15.6	−27.1	−5.5	−21.9

[a]Temperature-sensitive kinetics; $E_F = 30{,}000$ cal/mol and $E_B = 40{,}000$ cal/mol.

[b]Compared to the conventional feed arrangement (i.e., $N_{RX,bot} = N_{F,A}$ and $N_{RX,top} = N_{F,B}$).

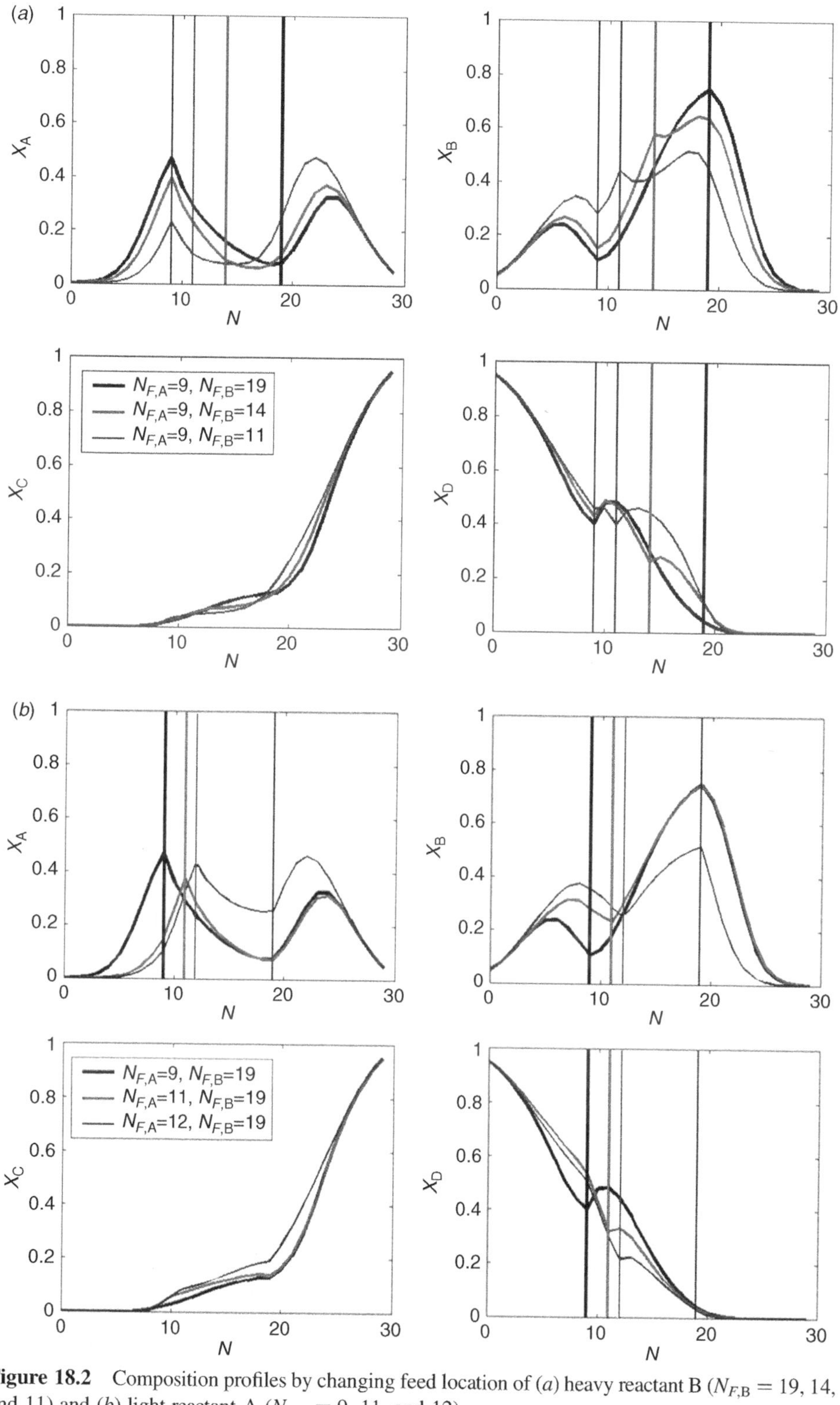

Figure 18.2 Composition profiles by changing feed location of (*a*) heavy reactant B ($N_{F,B} = 19$, 14, and 11) and (*b*) light reactant A ($N_{F,A} = 9$, 11, and 12).

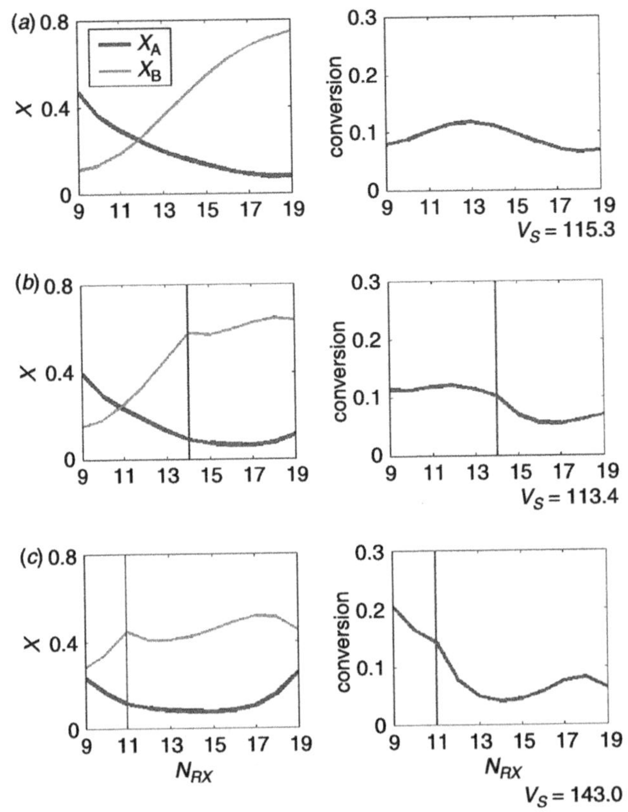

Figure 18.3 Reactant compositions, fraction of total conversion, and temperature profiles in reactive zone for different feed locations of heavy reactant B: $N_{F,\mathrm{B}} =$ (*a*) 19, (*b*) 14, and (*c*) 11.

in the lower reactive section. This implies that the lower reactive trays are better utilized but at the cost of smaller conversion in the upper reactive trays (i.e., upper reactive trays are underutilized; Figs. 18.3*a*–18.3*c*). It seems a balanced usage of the reactive trays is necessary to achieve optimality, and this means an optimal feed location exists for component B. The energy consumption (vapor rate V_S to be exact) is a good measure of column performance. In this case, the vapor rate changes from 0.0320 to 0.0315 kmol/s and then to 0.0397 kmol/s as $N_{F,\mathrm{B}}$ changes from 19 to 14 and then to 11, respectively. These results clearly indicate that the energy penalty can be significant if we place the feed at an inappropriate location, and the conventional design seems to be a pretty good choice.

The same analysis can be applied to the feed location of light component A. Now we fix the $N_{F,\mathrm{B}}$ at the top reactive zone (i.e., $N_{F,\mathrm{B}} = N_{RX,\mathrm{top}}$) and vary $N_{F,\mathrm{A}}$. Figure 18.2*b* gives the results for three feed locations ($N_{F,\mathrm{A}} = 9$, 11, and 12). The mole fraction of A increases toward the top of the reactive section while the mole fraction of heavy reactant B decreases as $N_{F,\mathrm{A}}$ is increased. However, heavy reactant B increases toward the bottom of the reactive zone. This reactant redistribution leads to significantly different energy consumption. The vapor rate changes from 0.0320 to 0.0292 kmol/s and then to 0.0386 kmol/s as $N_{F,\mathrm{A}}$ moves from 9 to 11 to 12, respectively.

These results clearly indicate that feed tray locations are important design/operation parameters. Improved process design can be achieved by simply adjusting the feed locations. One question then arises: how much energy can be saved if we adjust the feed locations simultaneously?

18.1.5 Optimal Feed Locations

Finding the optimal feed locations can be formulated as an optimization problem in which the vapor rate is minimized by varying the feed tray locations.

$$\begin{aligned} &\underset{N_{F,\mathrm{B}},\,N_{F,\mathrm{A}}}{\text{minimize}}\ V_S \\ &\text{subject to: } X_{\mathrm{D,C}} = X_{\mathrm{B,D}} = 0.95 \end{aligned} \tag{18.8}$$

Because the total tray number N_T is finite, we can find the optima by exhausting all N_T^2 possibilities. It is reasonable to restrict the search space to the reactive zone so that the possible choices are further reduced to N_{RX}^2. Here, we take a brute force approach by fixing $N_{F,\mathrm{A}}$ first while varying $N_{F,\mathrm{B}}$ until a minimum V_S is found. Next, $N_{F,\mathrm{A}}$ is changed, and the procedure repeats itself until a global minimum is located. Figure 18.4 shows the variation of the vapor rate for various values of the two feed locations. The results show that we should move the feed location of heavy reactant B down to $N_{F,\mathrm{B}} = 15$ (from 19) and move the feed tray of light reactant A up to $N_{F,\mathrm{A}} = 11$ (from 9). This corresponds to a 10.9% energy

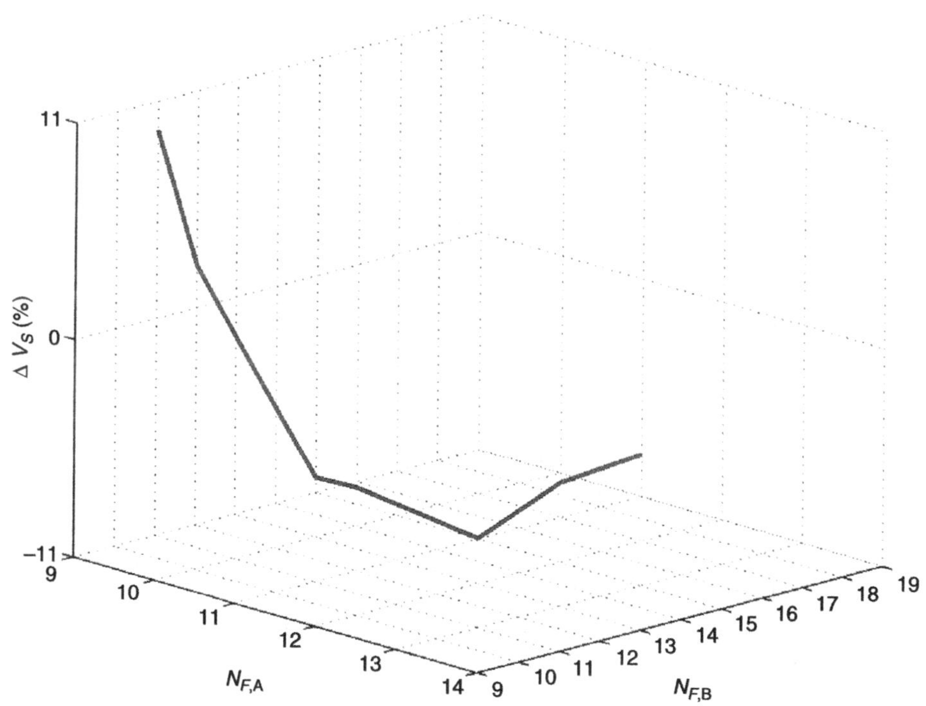

Figure 18.4 Feed tray locations and corresponding energy consumption (compared to base case) from optimization results.

savings compared to the conventional feed arrangement (see Table 18.2). Furthermore, simulation results from this and many other examples reveal that the feed location of the heavy reactant should *not* be placed *lower* then the feed tray of the light reactant. This reduces the search space to $(N_{RX}+1)N_{RX}/2$.

In addition to the percentage of energy savings, comparisons are also made in terms of the profiles of temperature, compositions, fraction of total conversion, and forward and reverse specific reaction rates. Figure 18.5 shows that the optimal feed arrangement (Fig. 18.5*b*) has a much *sharper* temperature profile in the reactive zone than the conventional feed locations (Fig. 18.5*a*). The tray temperature almost reaches 390 K in the optimal case, but barely reaches 380 K in the conventional case. Further, the profiles of the tray conversion and rate constant also take a qualitative shape similar to that of the temperature.

The composition profiles in Figure 18.5 explain what is happening. First, as a result of moving $N_{F,\mathrm{B}}$ downward and $N_{F,\mathrm{A}}$ upward, we have nonmonotonic *reactant* distributions for the optimal case as opposed to the monotonic *reactant* distribution for the conventional one. This is advantageous for the forward reaction. Second, we obtain an almost monotonic *product* distribution for the optimal case, especially for heavy product D. The x_D (tray composition of product D) almost reaches 60% at the bottom of the reactive section, which has a profound effect on the temperature profile. In contrast, the mole fraction of D gives a nonmonotonic profile for the conventional case, and x_D takes a downturn toward the bottom of the reactive tray as a result of dilution from the excess light reactant A that is introduced in the bottom of the reactive zone. The results presented in Figure 18.5 reveal the complicated interaction between temperature and composition in the reactive zone,

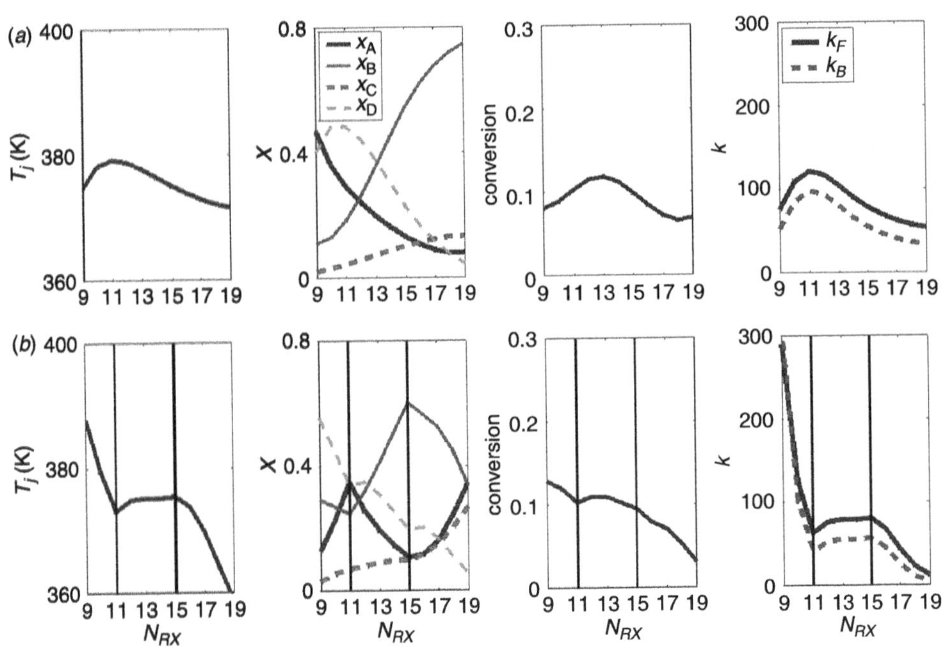

Figure 18.5 Profiles of temperature, composition, fraction of total conversion, and reaction rate constants in reactive zone for base case with (*a*) conventional feed arrangement ($N_{F,\mathrm{A}}=9$ and $N_{F,\mathrm{B}}=19$) and (*b*) optimal feed arrangement ($N_{F,\mathrm{A}}=11$ and $N_{F,\mathrm{B}}=15$) with 11% energy savings.

and it is almost certain that a better profile can always be achieved by varying the feed locations.

In summary, for the system with relative volatilities of $\alpha_C/\alpha_A/\alpha_B/\alpha_D = 8/4/2/1$, we should move the feed locations of the heavy reactant downward and the light reactant upward. In terms of the search space for the optimal feed trays, we have the following heuristic:

> *Heuristic H1:* Never place the heavy reactant feed below the feed tray of the light reactant (similarly, do not place the light reactant feed above the feed tray of the heavy reactant).

18.2 EFFECTS OF RELATIVE VOLATILITIES

Only one specific example has been explored thus far. It is interesting to see whether the results can be extended to different cases (e.g., different relative volatilities) and how this vapor–liquid equilibrium change will impact the location of optimal feed trays and the percentage of energy savings. Note that for every case studied, the column is *redesigned* using the procedure in Section 18.1. This means the columns may have different N_R, N_S, and N_{RX} and the location of the feed trays are described in terms of their relative position in the reactive zone.

18.2.1 Changing Relative Volatilities of Reactants

In this section we will explore the effects of the relative volatilities of the reactants on the optimum feed tray locations. By relative volatilities of reactants, we mean that the separation between the two reactants (A and B) becomes easier or more difficult while keeping the relative volatilities between adjacent reactants and products constant at 2. Two cases are studied: one is a more difficult separation (i.e., $\alpha_A/\alpha_B = 3/2$) and the other is an easier one (i.e., $\alpha_A/\alpha_B = 6/2$) compared to the base case (i.e., $\alpha_A/\alpha_B = 4/2$).

In the first case, the relative volatilities are $\alpha_C = 6$, $\alpha_A = 3$, $\alpha_B = 2$, and $\alpha_D = 1$. With the conventional feed arrangement (reactants fed at the two ends of the reactive zone), we have 33% more energy consumption (0.0428 kmol/s) compared to that of the base case. This shows that, similar to conventional distillation, difficult separation, even between reactants A and B, requires more energy. Moreover, Figure 18.6 shows that the composition of A is higher toward the lower reactive zone compared to the base case, and this leads to a decrease in product D composition, which subsequently requires a larger vapor rate to meet the specification. The optimization procedure predicts that the optimum feed trays are $N_{F,A} = 11$ and $N_{F,B} = 13$ (Fig. 18.6*b*), and this configuration produces a 15.2% energy savings (from 0.0428 to 0.0363 kmol/s) over the conventional feed arrangement (Table 18.2). Note that the percentage of energy savings is computed with respect to the conventional feed arrangement in each case. We immediately observe that the two feeds move *closer* to each other (only two trays apart), and a nonmonotonic reactant composition distribution can be seen (Fig. 18.6*b*). Similar to the base case (e.g., Fig. 18.5*b*), we also have an almost monotonic composition distribution in D. This results in a higher temperature in the lower section of the reactive trays and leads to a higher reaction rate and consequently higher conversion, as shown in Figure 18.6*b*.

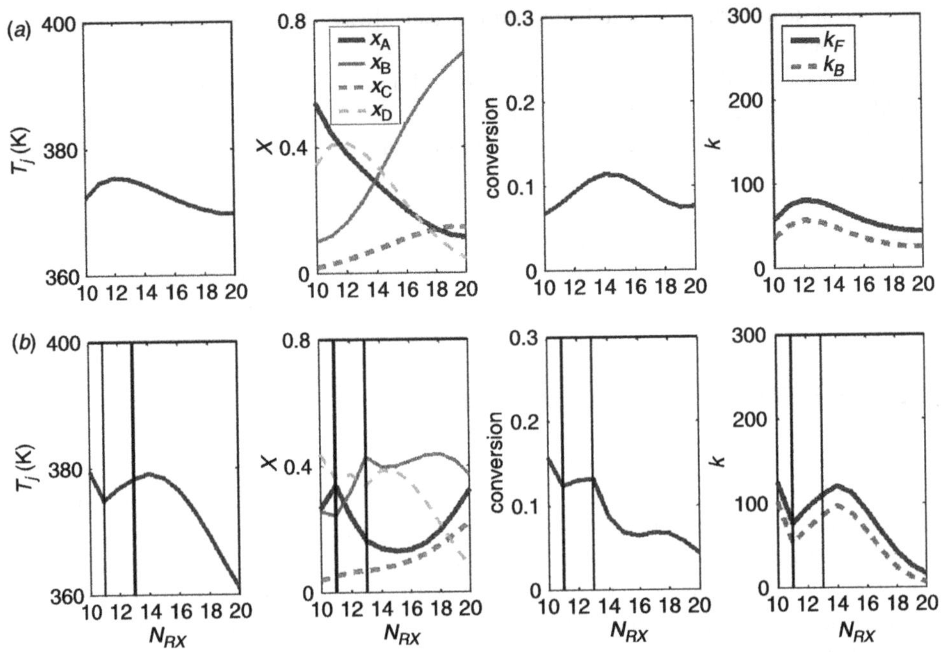

Figure 18.6 Profiles of temperature, composition, fraction of total conversion, and reaction rate constants in reactive zone for $\alpha_C/\alpha_A/\alpha_B/\alpha_D = 6/3/2/1$ system with (*a*) conventional feed arrangement ($N_{F,A} = 10$ and $N_{F,B} = 20$) and (*b*) optimal feed arrangement ($N_{F,A} = 11$ and $N_{F,B} = 13$) with 15% energy savings.

The second case is just the opposite, where we have easy separation between the two reactants. In this example, the relative volatilities are $\alpha_C = 12$, $\alpha_A = 6$, $\alpha_B = 2$, and $\alpha_D = 1$. Unlike in the previous example, the energy consumption (0.0227 kmol/s) is only 84.4% (0.0227/0.0320 * 100%) of the base case. This again reconfirms that what is true in conventional distillation is also true in reactive distillation: easier separation requires less energy. We also observed that the composition of D is higher in the lower reactive zone compared to the other two cases (cf. Figs. 18.5*a* and 18.6*a*), and this implies a smaller vapor rate to meet the specification. The optimization calculations show that the optimum feed trays are $N_{F,A} = 10$ and $N_{F,B} = 17$, and this corresponds to a 6.9% energy savings (from 0.0244 to 0.0227 kmol/s) over the conventional feed arrangement. Table 18.2 reveals that the two feeds move *away* from each other, and a little improved reactant composition distribution can be observed. In fact, the optimal feed trays are located quite close to the conventional feed trays (e.g., only one and two trays away from the ends of the reactive zone). An almost monotonic composition distribution in D is also observed that leads to a slightly higher temperature in the lower section of the reactive trays but not by much. This explains why the improvement is not as significant as in the previous case.

18.2.2 Changing Relative Volatilities of Products

Now we consider the cases in which the relative volatilities between the two products and their adjacent reactants are different from the base case value of 2. Two cases are explored: one is when light product C is easier to separate from light reactant A (i.e., $\alpha_C/\alpha_A = 4$), and the other is when the relative volatility between heavy reactant B and heavy product D (i.e., $\alpha_B/\alpha_D = 4$) is larger than the base case value of 2.

In the first case, we have $\alpha_C = 16$, $\alpha_A = 4$, $\alpha_B = 2$, and $\alpha_D = 1$. With the conventional feed arrangement, the energy consumption (0.0285 kmol/s) is 10.9% less than the base case because of the large relative volatility between C and A. The optimization calculations show that the optimum feed trays become $N_{F,A} = 12$ and $N_{F,B} = 14$ (Fig. 18.7*b*). Compared to the conventional feed arrangement, this corresponds to a 46.8% energy savings (from 0.0285 to 0.0152 kmol/s)! This is a very significant energy savings by very simple means (feed rearrangement).

Two observations can be made immediately. First, the two feeds are quite *close* to each other and the feed locations move to the upper section of the reactive zone. Second, the fraction of total conversion is distributed relatively uniformly throughout the reactive zone (at least compared to other cases) as shown in Figure 18.7*b*. This implies that none of the reactive trays are underutilized, and this is achieved because of the interplay between the composition and temperature distributions (e.g., showing temperature increases whenever necessary). An almost monotonic composition distribution in D is also seen; a significantly smaller amount of product C is also observed in the upper reactive zone, which allows for higher reactant concentration. All of these factors result in a much smaller vapor rate compared to the conventional feed arrangement.

The second example is just the opposite. We have an easy separation between the heavy reactant and the heavy product. In this case, the relative volatilities are $\alpha_C = 16$,

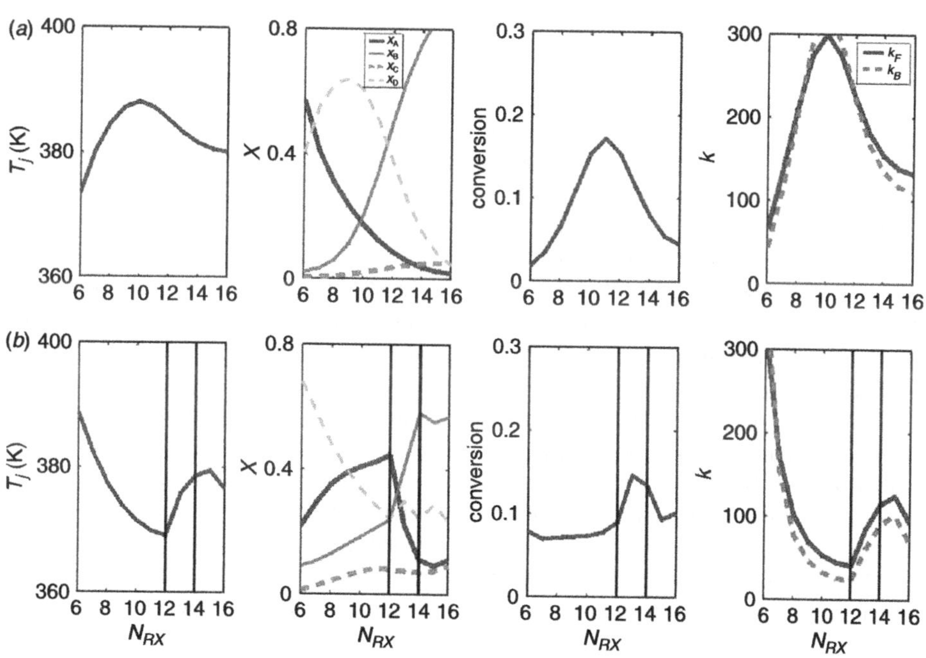

Figure 18.7 Profiles of temperature, composition, fraction of total conversion, and reaction rate constants in the reactive zone for $\alpha_C/\alpha_A/\alpha_B/\alpha_D = 16/4/2/1$ system with (*a*) conventional feed arrangement ($N_{F,A} = 6$ and $N_{F,B} = 16$) and (*b*) optimal feed arrangement ($N_{F,A} = 12$ and $N_{F,B} = 14$) with 46.8% energy savings.

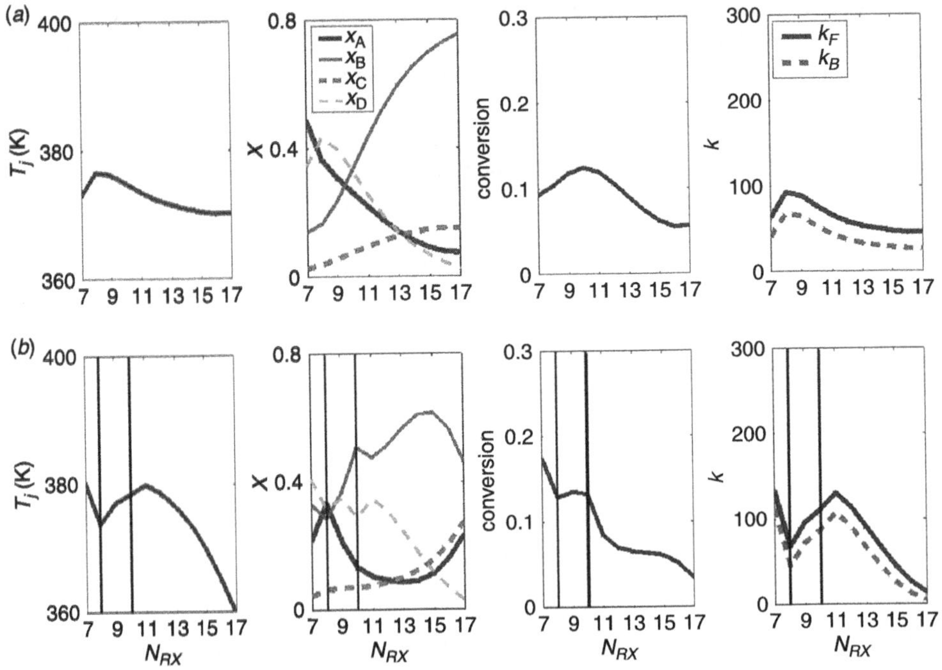

Figure 18.8 Profiles of temperature, composition, fraction of total conversion, and reaction rate constants in reactive zone for $\alpha_C/\alpha_A/\alpha_B/\alpha_D = 16/8/4/1$ system with (*a*) conventional feed arrangement ($N_{F,A} = 7$ and $N_{F,B} = 17$) and (*b*) optimal feed arrangement ($N_{F,A} = 8$ and $N_{F,B} = 10$) with 15.6% energy savings.

$\alpha_A = 8$, $\alpha_B = 4$, and $\alpha_D = 1$. For the conventional feed arrangement, the energy consumption (0.0281 kmol/s) is only 87.8% of the base case. The concentration of D is smaller toward the lower reactive zone compared to the other two cases (cf. Figs. 18.5*a* and 18.8*a*), and this allows a higher reactant concentration in the same section. The optimization calculations show that the optimum feed trays are $N_{F,A} = 8$ and $N_{F,B} = 10$ (Fig. 18.8*b*), which corresponds to a 15.6% energy savings (from 0.0281 to 0.0237 kmol/s) over the conventional feed arrangement (Table 18.2). Note also that these two feeds are quite *close* to each other and they are located in the lower section of the reactive zone, as can be seen in Figure 18.8*b*. In addition to an almost monotonic composition distribution in D, there is also a high concentration of B throughout the reactive zone in Figure 18.8*b*, and this improves the effectiveness of the reactive trays. This is allowed because B can be separated easily from heavy product D. However, unlike in the previous case, the decreasing trend of the temperature toward the upper reactive zone leads to a monotonically decreasing fraction of total conversion in the same direction (Fig. 18.8*b*). The underutilized reactive trays in the upper reactive zone explain why the margin of improvement is not quite as significant as in the previous case.

18.2.3 Summary

These results clearly indicate that the feed locations are important design parameters, and significant energy savings (ranging from 7% to 47%) will result if we place the feed trays optimally (Table 18.2). As for the specific feed locations, the following heuristics are useful:

> *Heuristic H2:* Place the light and heavy reactants' feed location *close* to each other when the relative volatility between the reactants is *small* (e.g., Fig. 18.6*b*). Similarly, move the feed tray locations *away* from each other when the relative volatility between the reactants is *large*.
>
> *Heuristic H3:* When the relative volatility between the *light reactant and the light product is large*, move the feed locations *upward* (i.e., to the upper reactive zone, e.g., Fig. 18.7*b*). Similarly, when the relative volatility between the *heavy reactant and the heavy product is large*, move the feed locations *downward* (i.e., to the lower reactive zone, e.g., Fig. 18.8*b*).

These were observed not only for systems with base case kinetics (Table 18.2) but also for processes with reaction kinetics that are less temperature sensitive (smaller activation energies), as discussed in the following section.

18.3 EFFECTS OF REACTION KINETICS

In this section we explore the effects of reaction kinetics on the optimal feed tray locations and the resulting energy savings. Two scenarios are studied. In the first case, the activation energies of the forward and backward reactions are both reduced by an order of magnitude, which implies a less temperature-sensitive reaction rate. In the second case, the specific reaction rates of both the forward and backward reactions are varied. We are interested in how these changes will impact the optimal feed locations and the corresponding energy savings.

18.3.1 Reducing Activation Energies

Consider the case in which the forward and backward activation energies are both reduced to 3000 cal/mol, which is an order of magnitude smaller than used in the base case. The rate constants at 366 K are kept the same as in the base case (0.008 and 0.004 kmol s^{-1} $kmol^{-1}$ for the forward and backward reactions, respectively) by changing the preexponential factors. These changes impact the reactive distillation column in two ways. First, the reactions are not quite as temperature sensitive as in the previous case. Second, the heat of reaction is 0, as opposed to the previous case where 10,000 cal of energy are released for every mole of reactant converted.

Base Case with Low Activation Energies. Figure 18.9*a* shows that, with the conventional feed arrangement, the profiles for the temperature, composition, and fraction of total conversion are qualitatively similar to those of the higher activation example (Fig. 18.5*a*). However, the profile of the backward reaction rate is smaller. The energy consumption is higher in the present example compared to one with a higher activation energy (0.0393 vs. 0.0320 kmol/s). The reason for this is that energy is no longer released from the reactions, and the effect of direct heat integration disappears. The optimization calculations show that the optimal feed trays becomes $N_{F,A} = 11$ and $N_{F,B} = 15$, and this results in a 10.7% energy savings when compared to the conventional feed arrangement (see

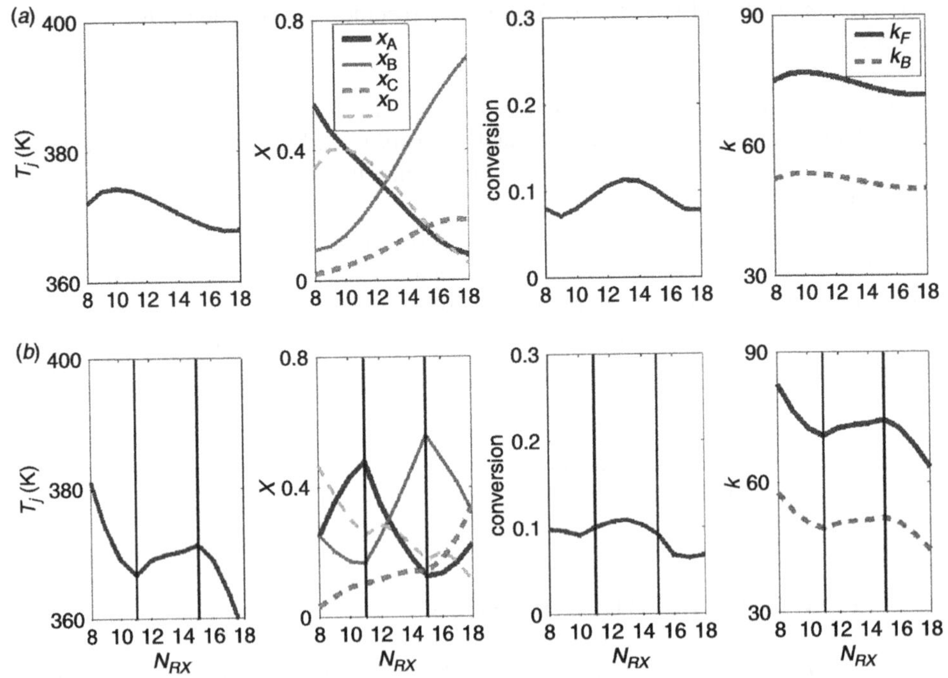

Figure 18.9 Profiles of temperature, composition, fraction of total conversion, and reaction rate constants in reactive zone for base case with *low activation energy* using (*a*) conventional feed arrangement ($N_{F,A} = 8$ and $N_{F,B} = 18$) and (*b*) optimal feed arrangement ($N_{F,A} = 11$ and $N_{F,B} = 15$) with 11% energy savings.

Table 18.3). Figure 18.9*b* demonstrates that, similar to the example of high activation energy, these two fresh feedstreams are separated by four trays. However, unlike the previous case, the relative position moves up slightly. This seems reasonable because we have a less temperature-sensitive reaction, and the effect of the temperature profile is not as important as in the previous example. This indicates that the composition profile is much more important in this example. The fraction of total conversion on the reactive trays clearly indicates a nonmonotonic profile throughout the reactive zone, compared to the high activation energy case (Fig. 18.9*b* vs. 18.5*b*). The temperature insensitivity is also illustrated in the profile of the rate constants where the ratio of the maximum over the minimum rate constants is 1.3 in this case, while the high activation energy example gives a value of 22! In terms of energy savings via feed tray optimization, both cases show quite similar results (10.9% vs. 10.7% as shown in Tables 18.2 and 18.3, respectively).

Changing Relative Volatilities of Reactants. Similar to the previous example, we first explore the case of difficult separation between the two reactants. The relative volatilities are $\alpha_C = 6$, $\alpha_A = 3$, $\alpha_B = 2$, and $\alpha_D = 1$. Optimization calculations show that the optimal feed trays are $N_{F,A} = 13$ and $N_{F,B} = 15$, and a 19.1% energy savings (from 0.0482 to 0.0390 kmol/s) can be achieved by the feed rearrangement (Table 18.3). It is

TABLE 18.3 Effects of Feed Locations on Design for Systems with Different Relative Volatilities[a]

	Base Case	$\frac{\alpha_A}{\alpha_B}=2$	$\frac{\alpha_A}{\alpha_B}=1.5$	$\frac{\alpha_A}{\alpha_B}=3.0$	$\frac{\alpha_C}{\alpha_A}=4.0$	$\frac{\alpha_B}{\alpha_D}=4.0$
$\alpha_C/\alpha_A/\alpha_B/\alpha_D$	8/4/2/1	8/4/2/1	6/3/2/1	12/6/2/1	16/4/2/1	16/8/4/1
$k_{F,375}$ (s^{-1})	0.0215	0.0215	0.0215	0.0215	0.0215	0.008
$k_{B,375}$ (s^{-1})	0.0150	0.0150	0.0150	0.0150	0.0150	0.004
$N_S/N_{RX}/N_R$	7/11/8	7/11/8	8/11/8	8/11/8	5/11/6	6/11/6
$N_{RX,bot}/N_{RX,top}$	8/18	8/18	9/19	9/19	6/16	9/19
$N_{F,A}/N_{F,B}$	8/18	11/15	13/15	10/17	14/15	9/12
$X_{D,C}/X_{B,D}$	0.95/0.95	0.95/0.95	0.95/0.95	0.95/0.95	0.95/0.95	0.95/0.95
F_{OA}, F_{OB}, D, B (kmol/s)	0.0126	0.0126	0.0126	0.0126	0.0126	0.0126
R (kmol/s)	0.0267	0.0225	0.0264	0.0185	0.0105	0.0185
V_S (kmol/s)	0.0393	0.0351	0.0390	0.0311	0.0231	0.0311
Energy savings[b] (%)	0	−10.7	−19.1	−5.4	−33.6	−14.7

[a]Temperature-insensitive kinetics; $E_F = 3000$ cal/mol and $E_B = 3000$ cal/mol.
[b]Compared to conventional feed arrangement (i.e., $N_{RX,bot} = N_{F,A}$ and $N_{RX,top} = N_{F,B}$).

clear that heuristic H2 applies here, and the percentage of energy saved is quite similar to the high activation energy counterpart (Table 18.2).

Next, we examine the case of easier separation between reactants. The relative volatilities are $\alpha_C = 12$, $\alpha_A = 6$, $\alpha_B = 2$, and $\alpha_D = 1$. Feed tray location optimization gives $N_{F,A} = 10$ and $N_{F,B} = 17$. In this case, only 5.4% energy can be saved (from 0.0328 to 0.0310 kmol/s) because the conventional feed arrangement is a good design to begin with (Table 18.3). Heuristic H2 again applies here, and the percentage of energy saved is also similar to the high activation energy counterpart (Table 18.2).

Changing Relative Volatilities of Products. Easy separation between the light reactant and the light product is studied first. The relative volatilities in this case are $\alpha_C = 16$, $\alpha_A = 4$, $\alpha_B = 2$, and $\alpha_D = 1$. Recall that this scenario results in the largest energy savings for the high activation energy case (Fig. 18.7*b*). Optimization calculations demonstrate that the optimal feed trays are $N_{F,A} = 14$ and $N_{F,B} = 15$, and a 33.6% energy savings (from 0.0348 to 0.0231 kmol/s) can be achieved by the feed rearrangement (Table 18.3). Heuristic H2 applies here, and the percentage of energy saved is quite significant but not quite as large as that of the high activation energy counterpart (33.6% vs. 46.8%).

Finally, the case of easy separation between the heavy reactant and the heavy product is explored. The relative volatilities are $\alpha_C = 16$, $\alpha_A = 8$, $\alpha_B = 4$, and $\alpha_D = 1$. Feed tray location optimization gives $N_{F,A} = 9$ and $N_{F,B} = 12$. In this case, 14.7% energy can be saved (from 0.0365 to 0.0311 kmol/s; Table 18.3). Once again heuristic H2 applies, and the percentage of energy saved is also similar to the high activation energy counterpart (Table 18.2).

Summary. For the low activation energy kinetics, the reaction rate constants are relatively insensitive to temperature. Thus, the composition profile plays a more important role than the temperature profile. Feed tray locations are useful handles for composition redistribution in the reactive zone, an important design variable. Generally, the average percentage of energy savings is a little less than the high activation energy counterpart (cf. Tables 18.2 and 18.3). This example clearly illustrates that heuristics H2 and H3 offer good guidelines for placement of feeds.

18.3.2 Effects of Preexponential Factor

In this section the rate constants are reduced or increased at 366 K by adjusting the preexponential factors while fixing the activation energies at 30,000 and 40,000 cal/mol (high activation energy case). These can be viewed as two different systems or the scenarios of changing catalyst activity. The optimal feed tray location and percentage of energy savings are of interest.

Low Reaction Rate Constant. Here we consider a lower reaction rate constant with relative volatilities $\alpha_C/\alpha_A/\alpha_B/\alpha_D = 8/4/2/1$. The specific reaction rates at 366 K are changed to *60% of the nominal values* ($k_F = 0.0048$ and $k_B = 0.0024$).

Because the reaction rates are almost 60% smaller, under the conventional feed arrangement the vapor rate increases by a factor of 30% to meet product specifications compared to the nominal case (Table 18.2). Product D takes a slightly lower value than the base case (Fig. 18.5*a*), and more energy is required to separate the product from the reactants.

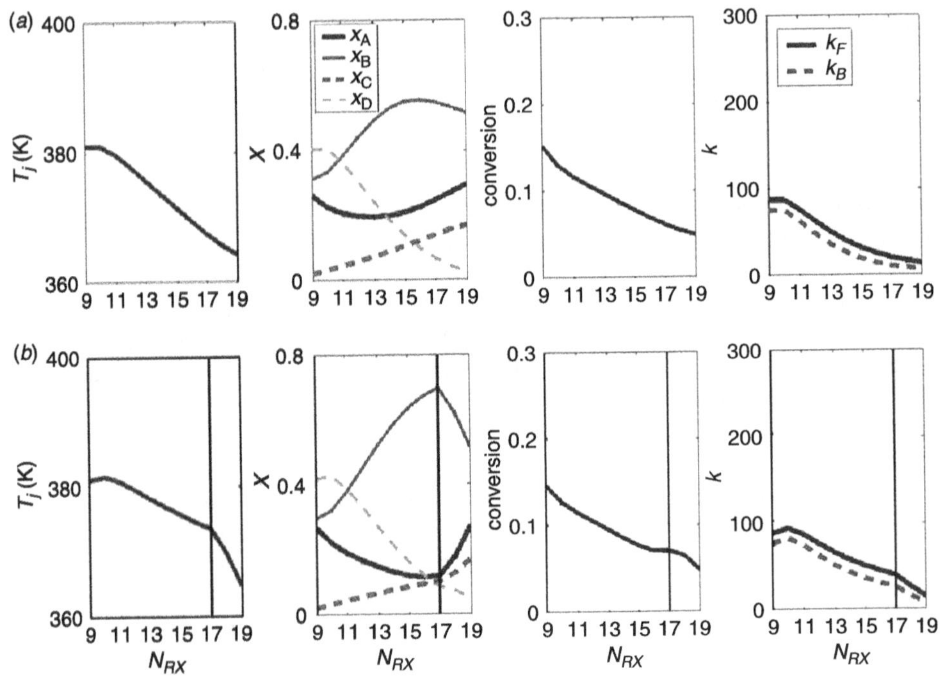

Figure 18.10 Profiles of temperature, composition, fraction of total conversion, and reaction rate constants in reactive zone for $\alpha_C/\alpha_A/\alpha_B/\alpha_D = 8/4/2/1$ system when *rate constants reduced to 60%* with (*a*) conventional feed arrangement ($N_{F,A} = 9$ and $N_{F,B} = 19$) and (*b*) optimal feed arrangement ($N_{F,A} = 9$ and $N_{F,B} = 17$) with 5% energy savings.

Following the optimization procedure, the optimal feed locations are $N_{F,A} = 9$ and $N_{F,B} = 17$. This results in only a 5.5% energy savings (Table 18.2).

Figure 18.10 shows the profiles for the temperature, composition, conversion, and rate constants in the reactive zone. Because of smaller rate constants, the reactants do not decrease as fast as the case of high rate constants (e.g., Fig. 18.5*a*). In fact, there is a favorable reactant composition profile, and this is essential for good performance. In other words, the conventional feed arrangement is already a good choice, and the optimized feed trays provide a small improvement in the reactant distribution.

High Reaction Rate Constant. In this case, the reaction rate constants are *doubled* at 366 K (i.e., $k_F = 0.016$ and $k_B = 0.008$). Because of higher reaction rates, the numbers of trays in the stripping and rectifying sections decrease, and the vapor rate is also reduced by a 8%. With the conventional feed arrangement, the reactant compositions decrease drastically toward the opposite ends of the reactive zone as shown in Figure 18.11*a*. These are not favorable profiles from the reaction standpoint. We expect that optimizing the feed tray location can reduce operating cost. Feed tray optimization shows that the optimal feed trays are $N_{F,A} = 10$ and $N_{F,B} = 13$ (Fig. 18.11*b*). This optimized feed arrangement results in a 21.9% savings in the vapor rate, which is quite significant compared to that of the base case (Table 18.2). The fractional conversion is also distributed relatively uniformly throughout the reactive section compared to the conventional feed arrangement.

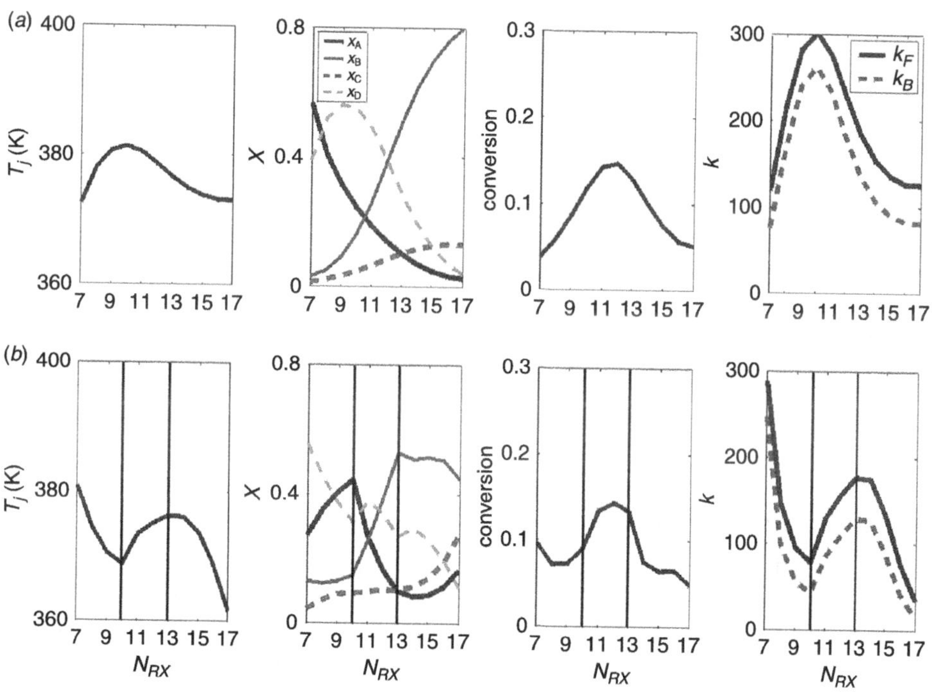

Figure 18.11 Profiles of temperature, composition, fraction of total conversion, and reaction rate constants in reactive zone for $\alpha_C/\alpha_A/\alpha_B/\alpha_D = 8/4/2/1$ system when *rate constants increased to 200%* with (*a*) conventional feed arrangement ($N_{F,A} = 7$ and $N_{F,B} = 17$) and (*b*) optimal feed arrangement ($N_{F,A} = 10$ and $N_{F,B} = 13$) with 22% energy savings.

Summary. We gained some insight by studying these three systems with low (60%), nominal (100%), and high (200%) rate constants. First, as expected, the energy consumption increases as the rate constant becomes smaller (Table 18.2), which is true for either the conventional feed arrangement or the optimized feed locations. Second, the percentage of energy saved by optimizing the feed locations increases as the rate constant becomes larger. The reason for this is that the shapes of the reactant profiles are not favorable for reaction when the rate constants are large under the conventional feed arrangement. Third, the optimal feed trays move closer to each other when the rate constants go up. This results in the following heuristic:

> *Heuristic H4:* Place the feed trays *away from each other* when the rate constants become *smaller* (e.g., Fig. 18.10*b*). Similarly, move the feed tray locations *closer to each other* when the rate constants become *larger* (e.g., Fig. 18.11*b*).

This heuristic is useful in design as well as for operation when the reaction rate constants vary, for example, in catalyst deactivation.

18.4 OPERATION AND CONTROL

These analyses clearly show that an improved design can be achieved by treating the feed tray location as an optimization variable, and the results indicate that significant energy savings can be obtained by simply rearranging the feeds. However, these analyses are limited to the design aspects with different thermodynamics parameters (i.e., relative volatilities) as well as kinetics parameters (i.e., activation energy and preexponential factor). In this section we are more interested in how this finding will affect the operation and control of reactive distillation columns.

Despite clear economic incentives for reactive distillation systems, there are relatively few articles that study the dynamics and control of reactive distillation columns. Al-Arfaj and Luyben[1] give a review of the literature dealing with the closed-loop control of reactive distillation systems. Several control structures for an ideal two-product reactive distillation system and real chemical systems[3,4] have been proposed. One important principle in the control of reactive distillation is that we need to control one intermediate composition (or tray temperature) in order to maintain the stoichiometric balance between the two reactant components.[1]

18.4.1 Optimal Feed Location for Production Rate Variation

In general, production rate variation is one of the most important load disturbances in plant-wide control and operation, and it can be measured, which is more important. Here we are interested in whether significant energy savings can be obtained by adjusting the feed tray locations as the production rate changes. If the operating cost can be appreciably reduced by

[3]M. A. Al-Arfaj and W. L. Luyben, Comparative control study of ideal and methyl acetate reactive distillation, *Chem. Eng. Sci.* **57**, 5039–5050 (2002).

[4]S. G. Huang, C. L. Kuo, S. B. Hung, Y. W. Chen, and C. C. Yu, Temperature control of heterogeneous reactive distillation: Butyl propionate and butyl acetate esterification, *AIChE J.* **50**, 2203–2216 (2004).

dynamically adjusting the feed locations, this implies that the feed tray location is not only a dominant design variable but also a useful manipulated variable for control.

We use the base case (Table 18.2) as an example in which the optimal feed trays are $N_{F,\mathrm{A}} = 11$ and $N_{F,\mathrm{B}} = 15$ with the design conditions shown in Figure 18.5*b*. The control objective is to maintain the product compositions (C and D) at 95%. Both positive and negative production rate variations are explored.

First, consider the case with a +40% feed flowrate increase. The optimization is performed to find the optimal feed locations by minimizing the vapor rate. We obtain $N_{F,\mathrm{A}} = 10$ and $N_{F,\mathrm{B}} = 16$, which corresponds to a 28% energy savings! This is not totally unexpected because an increase in the production can be viewed, in a sense, as a reduction of reaction capability (a smaller residence time or a decrease in the reaction rate constant). Therefore, we should move the feed trays away from each other as suggested by heuristic H4. However, the percentage of energy savings is larger than our expectation because we did not explore the nonoptimal cases in Section 18.3 (what if the feed locations are placed incorrectly?). The findings here clearly show that the optimal feed trays can change as the feed rate varies.

Second, the optimization is carried out for a 40% decrease in the feed flowrate. The optimal feed trays become $N_{F,\mathrm{A}} = 12$ and $N_{F,\mathrm{B}} = 15$, and a 9% savings in the vapor rate is observed. As pointed out earlier, this has the same effect as that from reaction rate increases, and we should move the feed trays closer to each other. The results clearly indicate that we should change the feed tray locations as the production rate changes because 9% or 28% energy can be saved by simply moving the feed trays. The next question then becomes, how can we implement such a control strategy? The coordinated control of Doukas and Luyben[5] shed some light in this direction.[6]

18.4.2 Control Structure

Before delving into the feed rearranging control structure, we first construct the fundamental control configuration for the reactive distillation with two feeds. Recall that, unlike the control of conventional distillation systems, we need to control the internal composition (or temperature) to maintain stoichiometric amounts of the two fresh feeds.[1] For the purpose of illustration in this work, we choose to control the composition of reactant A on tray 13 where a large change in the composition of A is observed (Fig. 18.5*b*). Thus, we have three compositions to be controlled: top composition of C, bottoms composition of D, and composition A on tray 13. For the manipulated variables, the ratio scheme is used; these three ratios are reflux ratio, boilup ratio, and feed ratio. Figure 18.12 shows the control structure.

The steady-state sensitivity analysis gives the following steady-state gain matrix ($\mathbf{K_p}$) between the controlled and manipulated variables:

$$\begin{bmatrix} x_{\mathrm{A}} \\ x_{\mathrm{D,C}} \\ x_{\mathrm{B,D}} \end{bmatrix} = \begin{bmatrix} 20.61 & 7.53 & -8.94 \\ 0.2 & 0.18 & -0.1 \\ 0.32 & -0.13 & 0.26 \end{bmatrix} \begin{bmatrix} F_{0\mathrm{A}}/F_{0\mathrm{B}} \\ R/D \\ V_S/B \end{bmatrix}$$

[5]N. Doukas and W. L. Luyben, Control of sidestream columns separating ternary mixtures, *InTech* **25**(6), 43–48 (1976).

[6]D. M. Chang, C. C. Yu, and I. L. Chien, Coordinated control of blending systems, *IEEE Trans. Control Syst. Technol.* **6**, 495–506 (1998).

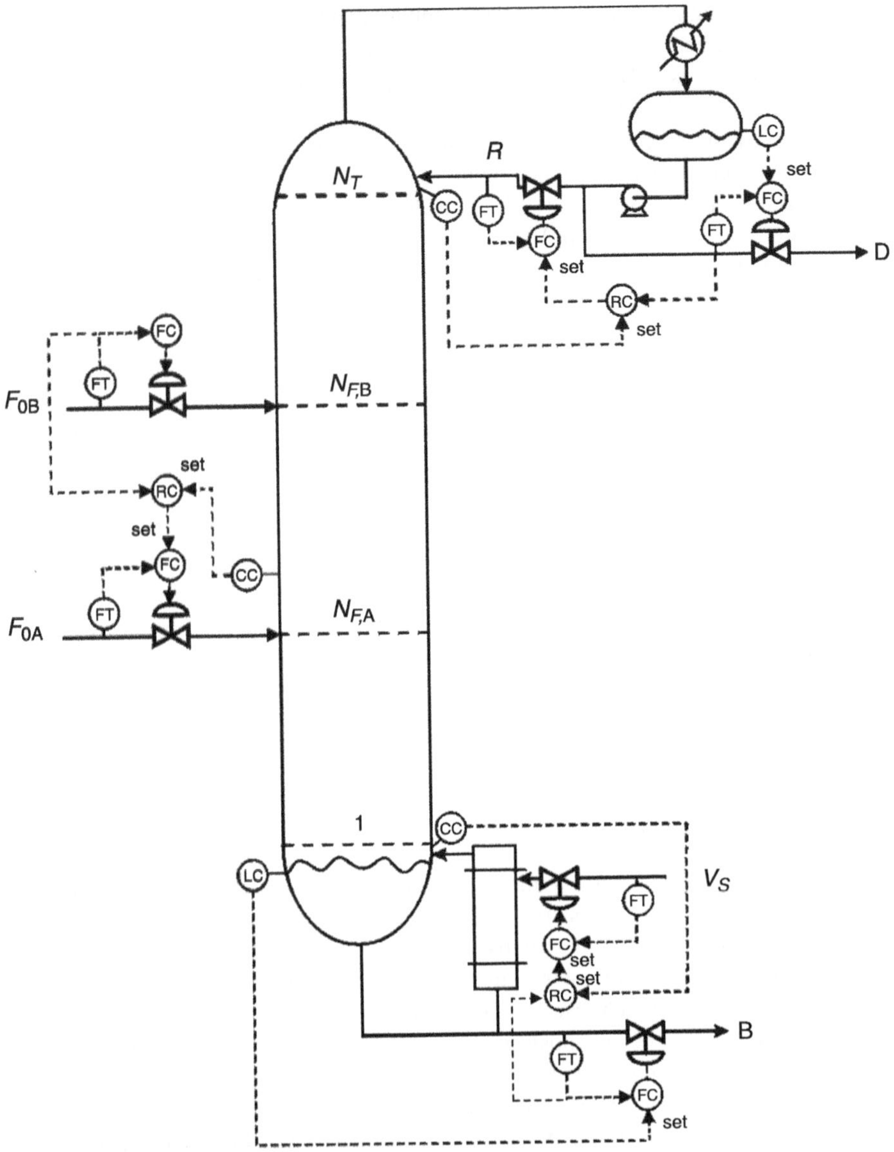

Figure 18.12 Control structure of reactive distillation using feed ratio, reflux ratio, and boilup ratio as manipulated variables with *fixed* feed locations.

The RGA (Λ) can be computed from the gain matrix from the relationship $\boldsymbol{\Lambda} = \mathbf{K_p} \otimes \mathbf{K_p^T}$. Thus, we have

$$
\begin{array}{c} \\ \Lambda = \end{array}
\begin{array}{c}
\begin{array}{ccc} F_{0A}/F_{0B} & R/D & V_S/B \end{array} \\
\left[\begin{array}{ccc} 0.85 & -0.78 & 0.93 \\ -0.20 & 1.83 & -0.63 \\ 0.35 & -0.05 & 0.7 \end{array}\right]
\end{array}
\begin{array}{c} x_A \\ x_{D,C} \\ x_{B,D} \end{array}
$$

The result of the RGA indicates that we should pair the internal composition with the feed ratio, pair the top composition with the reflux ratio, and pair the bottoms composition with the boilup ratio ($x_A - F_{0A}/F_{0B}$, $x_{D,C} - R/D$, and $x_{B,D} - V_S/B$). After the variable pairing, the basic control loops are in place. Figure 18.12 shows the control configuration for the reactive distillation without feed rearrangement.

1. Fresh feed F_{0B} is the throughput manipulator that is flow controlled.
2. Fresh feed F_{0A} is ratioed to F_{0B}, and the ratio is set by the tray 13 composition (x_A) controller.
3. The top composition of C is maintained by changing the reflux ratio.
4. The bottoms composition of D is controlled by changing the boilup ratio.
5. The base level is controlled by manipulating the bottoms flowrate.
6. The reflux-drum level is maintained by adjusting the distillate flowrate.

This structure consists of three composition loops and two level loops. We employ a decentralized control structure with PI controllers for the composition loops, and perfect level control is assumed for the level loops. In the identification phase, the relay feedback method is used to obtain the ultimate gain and ultimate period, and the controllers are tuned using the Tyreus–Luyben tuning method. Note that 5 min of analyzer dead time was assumed for these composition measurements.

Because both fresh feedflows are measured, we can coordinate the feed location as the production rate changes. We take the upper feedflow as an example to illustrate the feed rearrangement. Nominally, the feed tray for the heavy reactant is tray 15; as the flowrate (F_{0B}) increases by 40%, the feed location should be switched to tray 16. Thus, the idea is to use the feed tray location as a manipulated variable as the feedflow changes.[5,7] Instead of making a discontinuous switch, a linear combination of valve openings between trays 15 and 16 is employed (split-ranged valves) that provides a gradual transition as the feed flowrate increases.

We use the fresh feed of B (F_{0B}) to illustrate the coordination between the valve openings into trays 15 and 16. At the nominal design flowrate ($F_{0B}/\bar{F}_{0B} = 0$ in the box in the upper corner of Fig. 18.13), only the valve to tray 15 is open, and the valve to tray 16 is shut ($OP_{15} = 1$ and $OP_{16} = 0$ in Fig. 18.13, where OP_i is the operating variable or fraction of valve opening to feed tray i). As the feed flowrate increases (i.e., $F_{0B}/\bar{F}_{0B} > 0$), the valve to tray 16 is gradually opened ($OP_{16} > 0$) while closing the valve to tray 15 ($OP_{15} < 1$). When the feed flowrate reaches a 40% increase ($F_{0B}/\bar{F}_{0B} = 0.4$), only the valve to tray 16 remains open, and the valve to tray 15 is totally closed ($OP_{15} = 0$ and $OP_{16} = 1$). This provides a mechanism to coordinate the openings of these two valves. The same idea can be extended to the feedflow of A, switching between trays 10, 11, and 12, which is shown in the lower box in Figure 18.13. This structure can be implemented in either conventional control hardware or distributed control systems with little difficulty.

18.4.3 Closed-Loop Performance

Next, the closed-loop performance of both control structures (with and without feed rearrangement) is evaluated. We first consider the case of a 40% production rate increase.

[7] W. L. Luyben, *Process Modeling, Simulation and Control for Chemical Engineers*, Second Edition, McGraw-Hill, New York, 1989.

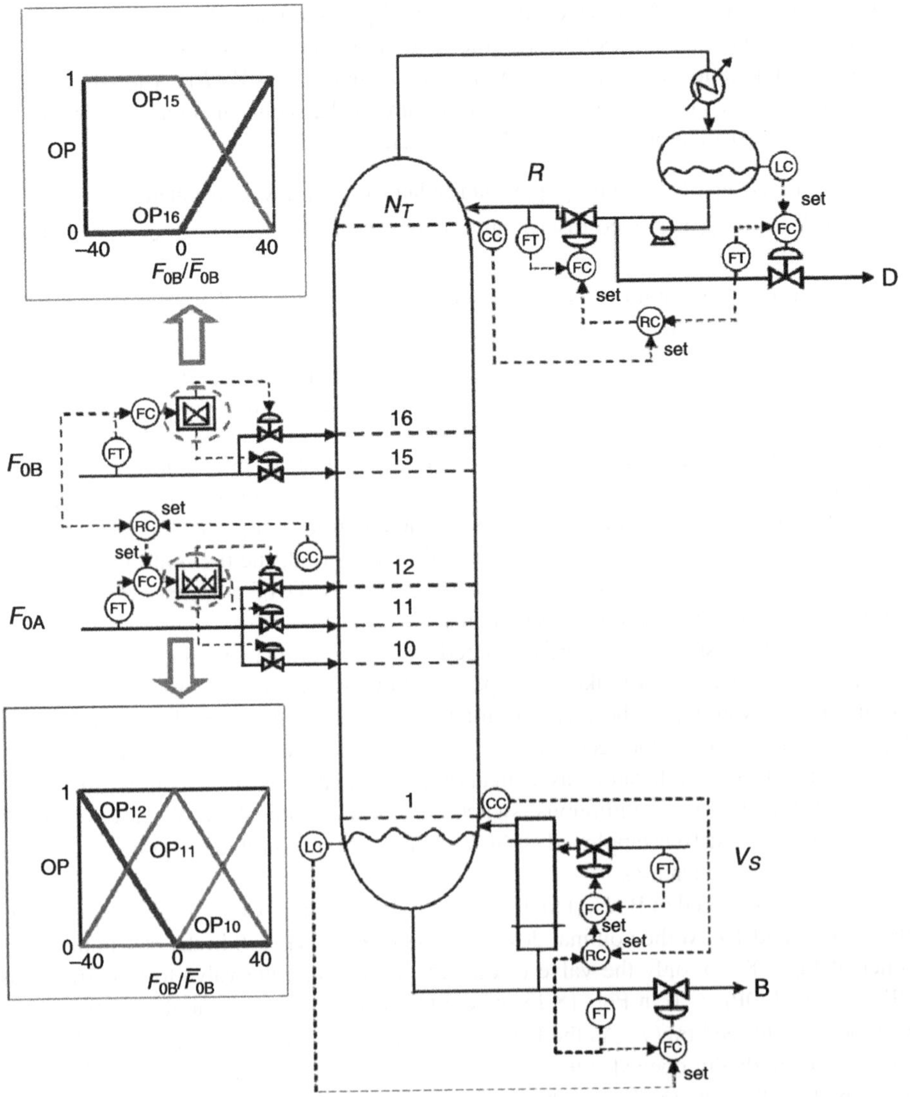

Figure 18.13 Control structure of reactive distillation with *variable* feed locations as production rate changes; OP_i, operating variable or fraction of valve opening to feed tray *i*.

The control structure with feed tray manipulation ("coordinated control") gives fast dynamics in the product composition, as can be seen in Figure 18.14 (solid lines) where the top and bottoms compositions return to setpoint in less than 10 h. In contrast, the conventional control structure shows a little slower dynamic response with the product compositions taking more than 10 h to return to their setpoints. Of more importance, the coordinated control structure results in a 21% energy savings compared to the conventional control structure, which can be seen from the smaller vapor rate in Figure 18.14.

Note that a 21% energy savings is smaller than the 28% from steady-state analysis, and the reason is that we fix the tray 13 composition of A to the nominal value. Nonetheless, the amount of energy saved is still quite significant. Figure 18.15 shows the responses for a

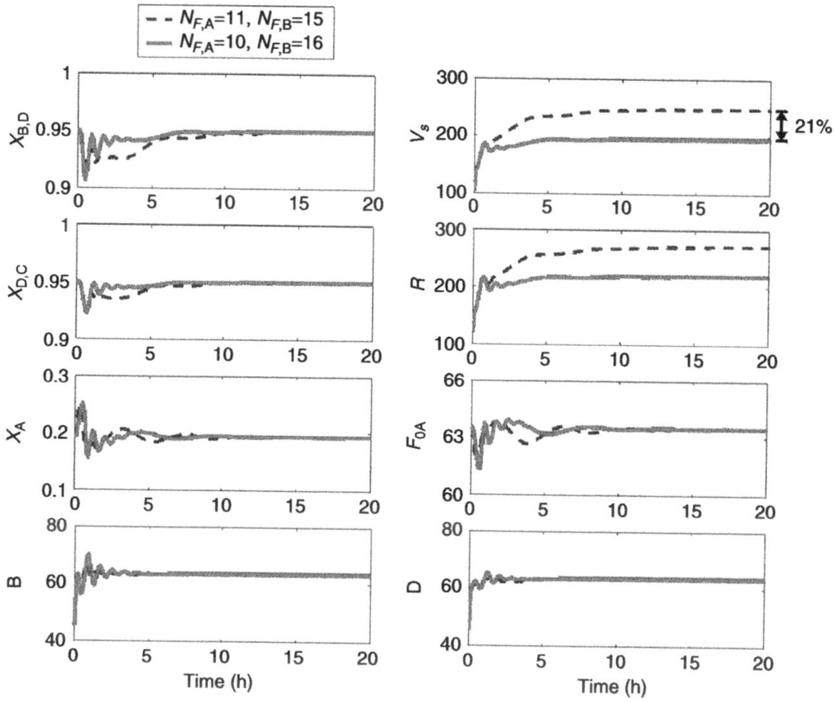

Figure 18.14 Closed-loop responses for +40% production rate increase with fixed feed locations (dashed) and coordinated feed trays (solid; with 21% energy savings).

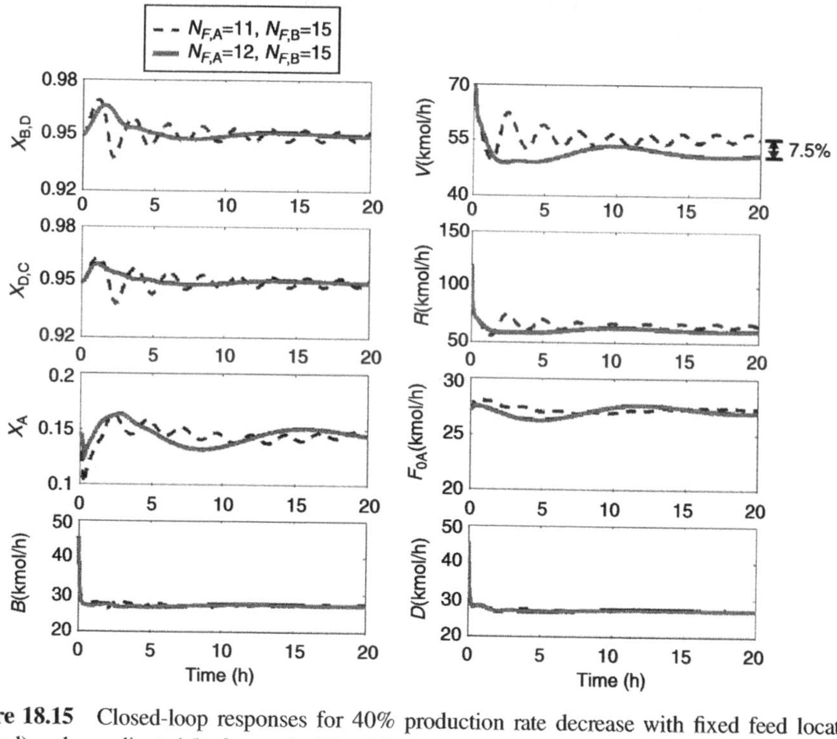

Figure 18.15 Closed-loop responses for 40% production rate decrease with fixed feed locations (dashed) and coordinated feed trays (solid; with 7.5% energy savings).

40% step decrease in the production rate. Faster dynamics for top and bottoms product compositions are again observed for the coordinated control structure. Moreover, a 7.5% energy savings can be achieved with this improved dynamics.

The results presented in this section clearly show that the concept of optimal feed tray location can be carried over to process operation and control. With a simple modification in the control structure, improved closed-loop performance can be achieved while realizing substantial energy savings at the same time.

18.5 CONCLUSION

This chapter explored the effects of feed locations on the design of reactive distillation, and ideal reactive distillation systems were used to illustrate the advantage of feed tray optimization. Reactive distillation columns with various process parameters were studied. They included the relative volatilities between reactants, relative volatilities between products, activation energies, and preexponential factors. The results from all systems that were studied indicated a 6% to 47% energy savings, which could be achieved by simply rearranging the feed locations. Because the temperature and composition profiles play a vital role in the effective utilization of the reactive section, the optimal feed locations are essential to obtain improved performance. Qualitatively, heuristics were given to place the feeds at the vicinity of optimal locations. Quantitatively, a systematic procedure was proposed to find the right feed trays. Finally, the idea of optimal feed trays could be carried over to the control of the reactive distillation system. First, steady-state analysis was performed to find the optimal feed trays as the measurable load variable changed. Second, a coordinated control structure was proposed to rearrange the feeds as the disturbance came into the system. The results indicated that substantial energy could be saved during process operation by feed rearrangement while showing improved closed-loop dynamics.

APPENDIX

CATALOG OF TYPES OF REAL REACTIVE DISTILLATION SYSTEMS

A ⇔ B		
A	B	Reference
α-isophorone	β-isophorone	Suyama et al., 2002
n-butane	isobutane	Lebas et al., 1999
1,4-dichloro-2-butene	1,2-dichloro-3-butene	Lee et al., 2000
isobutane	*n*-butane	Gartside and Greene, 2005
2-methyl-1-butene	3-methyl-1-butene	Powers, 2004
2-methyl-2-butene	3-methyl-1-butene	Powers, 2004
cyclohexanone oxime	ε-caprolactam	Lee et al., 2000
cis-2-butene	1-butene	Powers, 2004
isoolefins (c4)	octane boosting oligomers	Vora and Hammershaimb, 2000
n-paraffin	isoparaffins	Lebas et al., 1999
trans-2-butene	1-butene	Powers, 2004

2A ⇔ B		
A	B	Reference
acetone	diacetone alcohol	Podrebarac et al., 1998
cyclopentadiene	dicyclopentadiene	Rosinson and Dilliland, 1950
isobutyraldehyde	isobutylisobutyrate	Lee et al., 2000
isobutene	diisobutene	Kamath et al., 2006
n-butenes	octenes	Loescher et al., 2006

3A ⇔ B		
A	B	Reference
formaldehyde	trioxane	Hu et al., 1999

$A + B \Leftrightarrow C$			
A	B	C	Reference
1-butanol	acetylene	butyl vinyl ether	Pisarenko et al., 2002
acetaldehyde	acetic anhydride	vinyl acetate	Zoeller et al., 1998
acetic acid	cyclohexene	cyclohexyl acetate	Saha and Sharma, 1996
acetone	hydrogen	isopropanol	Gelbein, 2002
acrolein (2-propenal)	substituted anilines	Schiff base	Patel et al., 2000
acrylic acid	cyclohexene	cyclohexyl acrylate	Saha and Sharma, 1996
acytelene 1,3-butadiene	hydrogen	1-butene	Imai, 2000
alkyl anthraquinone	hydrogen	hydrogen peroxide	Gotze and Bailer, 1999
alkyl phenyl oxalate	diphenyl oxalate	dialkyl oxalate	Nishihira et al., 2000
aniline	hydrogen	cyclohexylamine	Garg et al., 2005
α-methyl styrene	hydrogen	cumene	Hildreth and Wyckoff, 1999
benzene	ethylene	ethyl benzene	Pohl, 2006
benzene	propylene	cumene	Pohl, 2006
benzene	α-olefins (e.g., $C_{12}H_{24}$)	linear alkyl benzene	Smith et al., 2006
benzene	1-dodecene	C-12 alkylbenzene	Lei et al., 2003
benzene	hydrogen	cyclohexane	Gildert, 2001
butadiene	sulfur dioxide	butadiene sulfone	AIChE, 1970
caprolactam	water	nylon 6	Stuart et al., 2002
chlorine	acetic acid	monochloroacetic acid	Crouzen et al., 2006
chlorobenzene	nitric acid	nitrochlorobenzene	Belson, 1990
cyclohexene	water	cyclohexanol	Khaledi and Bishnoi, 2006
diisobutylene	ethylene	neohexene	Podrebarac, 2003
dimethylamine	ethylene oxide	dimethylethanolamine	Garg et al., 2005
ethanol	propylene oxide	propylene glycol monoethyl ether	Wen et al., 2001
ethanol	ethylene oxide	2-ethoxy ethanol	Zhang and Wan, 1991
ethanol	ethylene oxide	ethyl cellosolve	Zheng et al., 2004
ethylene oxide	water	ethylene glycol	Liu, Qu, and Wang, 2002
formic acid	cyclohexene	cyclohexyl formate	Saha and Sharma, 1996
hexamethylene diamine	adipic acid	nylon 6,6 prepolymer	Doherty and Buzad, 1992
higher molecular olefins (C_{15+})	C_3–C_8 olefins	detergent range olefins (C_{10}–C_{14})	Podrebarac, 2006
hydrocracked oil fraction	hydrogen	hydrotreated fraction	Maraschino, 2002
isoamylene	water	*tert*-amyl alcohol	Gonzalez et al., 1997

A + B ⇔ C			
A	B	C	Reference
isobutanol	isobutene	isobutyl *tert*-butyl ether	Bessling et al., 2002
isobutene	water	*tert*-butanol	Reusch et al., 2006
isobutane	propylene/ butylene	highly branched paraffins	Huss and Kennedy, 1990
isophorone	hydrogen	trimethylcyclohexanol	Schmitt, 1996
isobutylene	isobutylene	tetramethyl ethylene	Podrebarac, 2003
isopropanol	propylene	diisopropyl ether	Marker et al., 1996
maleic anhydride	methanol	dimethyl maleate	Hao et al., 2003
mesityl oxide	hydrogen	methyl isobutyl ketone (MIBK)	Vandersall and Weinand, 2005
methanol	isobutene	methyl *tert* butyl ether	Ryu and Gelbein, 2002
methanol	isoamylene	*tert*-amyl methyl ether	Kolodziej et al., 2004
methanol	2-ethyl-1-butene	3-methoxy-3-methylpentane	Hendriksen et al., 1998
methanol	2,4,4-trimethyl pentenes	2-methoxy-2,4,4-trimethyl pentane	Rihko-Struckmann et al., 2004
mixed anhydride	triflic anhydride	carboxylic anhydride	Hembre and Lin, 2002
methyl acetylene	hydrogen	propylene	Stanley and Gildert, 2002
olefins (C_3–C_6)	hydrogen	propylene	Gartide and Gildert, 2002
propylene	water	isopropanol	Sakuth and Peters, 1998
propylene	isopropanol	diisopropyl ether	Marker et al., 1998
propadiene	hydrogen	propylene	Stanley and Gildert, 2002
pyridines	carboxylic acids	salt	Grosser et al., 1987
silylated compounds	sulfuric acid	silylated sulfuric acid ester	Lopez, 2000
toluene	propylene	isopropyltoluene	Hendriksen et al., 1999

A + B ⇔ 2C			
A	B	C	Reference
acetic anhydride	water	acetic acid	Costa and Canepa, 1969
benzene	xylene	toluene	Okawa et al., 1993
2-butene	ethylene	propylene	Podrebarac, 2006
diisopropyl ether	water	isopropyl alcohol	Burton, 2005

A + 2B ⇔ C			
A	B	C	Reference
acetylene	hydrogen	ethane	Ryu, 2006
glyoxal	acetaldehyde	bis(cyclic acetal)	Sharma, 1995

A ⇔ B + C			
A	B	C	Reference
2-cyclohexene-1-ol	1,3-cyclohexadiene	water	Ishida et al., 1997
2-propanol	acetone	hydrogen	Gaspillo et al., 1998
3,3-dimethyl butanol	3,3-dimethyl butanal	hydrogen	Jerry et al., 2003
butanediol	tetrahydrofuran	water	Liu, Zhang, and Gao, 2002
cumene hydroperoxide	phenol	acetone	Doron and Jose, 2002
decalin	naphthalene	hydrogen	Shinohara et al., 2004
dimethyl xylylene dicarbamete	xylylene diisocyanate	methanol	Okawa et al., 1993
glycerol	acetol	water	Chin et al., 2006
isopropanol	acetone	hydrogen	Chung et al., 1997
methanol	dimethyl ether	water	An et al., 2004
phenyl-ethanol	styrene	water	Lange and Otten, 2006
tert-butanol	isobutylene	water	Zhao et al., 2004
tert-butanol	isobutene	water	Qi and Sundmacher, 2006
trichlorosilane	silane	silicon tetrachloride	Müller et al., 2005

2A ⇔ B + C			
A	B	C	Reference
1-butene	propylene	pentene	Jung et al., 1987
1-butene	ethylene	2-*trans* hexene	Jung et al., 1987
2-pentene	2-butene	3-hexene	Hoffmaster and Hauan, 2006
toluene	benzene	xylene	Stitt, 2002

3A ⇔ B + 2C			
A	B	C	Reference
acetone	isophorone	water	Schmitt, 1996

A + B ⇔ C + D				
A	B	C	D	Reference
1-butene	2-butene	propene	2-pentene	Gartside, 2004
2-keto-L-gulonic acid	butanol	2-keto-L-gulonic acid butyl ester	water	Domschke et al., 2006
3-pentenoic acid	methanol	methyl-3-pentenoate	water	Sherman et al., 1996
3,3-dimethyl butyl ester	water	3,3-dimethyl butanol	mineral acid	Ebner et al., 2004
acetic acid	methanol	methyl acetate	water	Tang et al., 2005
acetic acid	ethanol	ethyl acetate	water	Tang et al., 2005
acetic acid	isopropanol	isopropyl acetate	water	Tang et al., 2005
acetic acid	*n*-propanol	*n*-propyl acetate	water	Huang et al., 2005
acetic acid	isobutanol	isobutyl acetate	water	Lederer et al., 2002
acetic acid	*n*-butanol	*n*-butyl acetate	water	Tang et al., 2005
acetic acid	isoamyl alcohol	isoamyl acetate	water	Saha et al., 2005
acetic acid	amyl alcohol	amyl acetate	water	Tang et al., 2005
acetic acid	*n*-hexanol	*n*-hexyl acetate	water	Schmitt and Hasse, 2006
acetaldehyde	ethylene glycol	2-methyl-1,3-dioxolane	water	Dhale et al., 2004
acetaldehyde	propylene glycol	2,4-dimethyl-1,3-dioxolane	water	Adams and Seider, 2006
acetone	phenol	bisphenol A	water	Smith and Gelbein, 2005
acetone	citral	pseudoionone	water	Smith and Lawrence, 2006c
acrylic acid	*n*-butanol	*n*-butyl acrylate	water	Zeng et al., 2006
acrylic aicd	ethanol	ethyl acrylate	water	Jelinek and Hlavacek, 1976
alkylate	hydrogen	hydrogen sulfide	sulfur-free alkylate	Smith and Lawrence, 2006b
ammonia	alcohols	amine	water	Garg et al., 2005
amino nitrile	water	polyamide	ammonia	Leemann et al., 2002
2,3-butanediol	formaldehyde	4,5-dimethyl-1,3-dioxolane	water	Broekhuis et al., 1994
benzene	nitrous oxide	phenol	nitrogen	Hamilton, 2005
benzene	nitric acid	nitrobenzene	water	Thelen et al., 1977
butyl acetate	ethanol	ethyl acetate	*n*-butanol	Dil'man et al., 2005
butyl acrylate	dimethyl amino ethanol	dimethylaminoethyl acrylate	butanol	Geisendoerfer et al., 2007
carbon monoxide	hydrogen	hydrocarbons	oxygenates	Prakasa and George, 2003
diacetate ester of poly(tetramethylene ether)	alkanol	poly(tetramethylene ether glycol)	alkanol acetate ester	Dorai, 1971
dialkyl carbonate	phenol	diaryl carbonate	alcohol	Bruin et al., 2001

A + B ⇔ C + D				
A	B	C	D	Reference
dialkyl carbonate	ethanol	diaryl carbonate	alcohol	Luo and Xiao, 2001
dialkyl oxalate	phenol	alkyl phenyl oxalate	alcohol	Nishihira et al., 2000
diaryl carbonate	bisphenols	polycarbonate	phenol	John et al., 2006
dibenzothiophene	hydrogen	hydrogen sulfide	biphenyl (bPh)	Vargas-Villamil et al., 2004
dichlorodimethyl silane	chlorine	dichlorosilane	overchlorinated products	Barnum and Blaisdell, 1995
dichlorobenzene	chlorine	trichlorobenzene	hydrogen chloride	Feathers and Mansell, 1992
dodecanoic acid	methanol	methyl dodecanoate	water	Kiss et al., 2006
dodecanoic acid	2-ethyl hexanol	2-ethylhexyl dodecanoate	water	Ometa et al., 2003
dodecanoic acid	1-propanol	1-propyl dodecanoate	water	Kiss et al., 2006
ethanol	*tert*-butyl alcohol	ethyl *tert*-butyl ether	water	Yu et al., 2002
fatty methylester	isopropanol	isopropyl ester	methanol	Reepmeyer et al., 2004
formaldehyde	methanol	methylal	water	Hagen and Spangler, 2002
formaldehyde	methanol	methylal	water	Ott et al., 2005
formaldehyde	ethanol	ethylal	water	Chopade and Sharma, 1997
formaldehyde	ethylene glycol	1,3-dioxolane	water	Chopade et al., 2003
formaldehyde	propylene glycol	4-methyl-1,3-dioxolane	water	Broekhuis et al., 1994
formic acid	ethanol	ethyl formate	water	Rhim et al., 1985
glycinonitrile	water	glycine	ammonia	Toshiya et al., 2005
isobutylene	2-butene	2-methyl-2-butene	propylene	Podrebarac, 2006
ketone	ethanol	acetone	chiral alcohol	Okasinski and Doherty, 2003
lactic acid	methanol	Methyl lactate	water	Asthana et al., 2006
lactic acid	ethanol	ethyl lactate	water	Asthana et al., 2005b
lauric acid	propanol	propyl laurate	water	Dimian et al., 2004
methyl acetate	water	methanol	acetic acid	Lin et al., 2008
methyl acetate	*n*-butanol	butyl acetate	methanol	Luyben et al., 2004
methyl benzoate	benzyl alcohol	benzyl benzoate	methanol	Tang et al., 1984
Methyl lactate	water	lactic acid	methanol	Ma et al., 2005

(*Continued*)

A + B ⇔ C + D				
A	B	C	D	Reference
	methanol	ethyl lactate	ethanol	Miller, Ashanta, Kolah, and Lira, 2006
methyl lactate	methanol	ethyl lactate	ethanol	Miller, Ashanta, Kolah, and Lira, 2006
methyl glyoxal	methanol	methyl glyoxal dimethyl acetal	water	Groening Carsten et al., 1996
methyl formate	water	methanol	formic acid	Saari et al., 2002
methyl phenyl ketones (*r*, *s*)	ethanol	acetone	phenyl ethyl alcohol (*r*)	Okasinski and Doherty, 2003
monobutyl phthalate	butanol	dibutyl phthalate	water	Berman et al., 1948
myristic acid	isopropanol	isopropyl myristate	water	Bock et al., 1997
mono-*para*-nitrobenzoic acid	glycerol	mono-*para*-nitrobenzoate	water	Baker et al., 1976
m-xylene	sodium *p*-xylene	sodium *m*-xylene	*p*-xylene	Terrill et al., 1985
naphtha	hydrogen	hydrogen sulfide	sulfur-free naphtha	Smith and Lawrence, 2006a
n-hexanoic acid	*n*-octanol	octyl hexanoate	water	Klöker et al., 2005
palmitic acid	isopropanol	isopropyl palmitate	water	Bhatia et al., 2006
pentenoyl chloride	water	pentenoic acid	hydrogen chloride	Murphree and Ozer, 1996
propylene oxide (*r*, *s*)	water	propylene oxide (*s*)	propylene glycol (*r*)	Okasinski and Doherty, 2003
primary amine	mesityl oxide	acetone	imine	Smith and Lawrence, 1982
propionic acid	ethanol	*n*-butyl propionate	water	Bhanvase et al., 2006
propionic acid	*n*-butanol	*n*-butyl propionate	water	Wu et al., 2004
tert-amyl alcohol	ethanol	*tert*-amyl ethyl ether	water	Aiouache and Goto, 2003
tert-butyl bromide	ethylene	1-neohexyl bromide	neohexene	Lin et al., 2003
toluene	chlorine	benzyl chloride	hydrochloric	Xu and Dudukovic, 1999
vinyl stearate	acetic acid	vinyl acetate	stearic acid	Geelen and Wijffels, 1965
glycerides (vegetable oil)	methanol	alkyl esters (biodiesel)	glycerol	He et al., 2006
vacuum gas oil, a diesel boiling range petroleum fraction	hydrogen	hydrogen sulfide	sulfur-free petroleum fraction	Dean, 2006

$A + 2B \Leftrightarrow C + 2D$

A	B	C	D	Reference
2,3-butylene glycol	acetic acid	2,3-butylene glycol diacetate	water	Schniepp et al., 1945
dimethyl carbonate	ethanol	diethyl carbonate	methanol	Dragomir and Jobson, 2005
methanol	urea	dimethyl carbonate	ammonia	Ryu and Gelbein, 2006
methyl acetate	water	dimethyl ether	acetic acid	Hoyme, 2003
poly(tetramethylene glycol diacetate)	methanol	poly(tetramethylene glycol)	methyl acetate	Anonymous, 1999

$A \Leftrightarrow 2C,\ B + C \Leftrightarrow D$

A	B	C	D	Reference
dicyclopentadiene	hydrogen	cyclopentadiene	cyclopentane	Silverberg et al., 2000

$2A \Leftrightarrow C,\ B + C \Leftrightarrow D$

A	B	C	D	Reference
isobutene	hydrogen	diisobutene	isooctane	Gildert and Loescher, 2001

$2A \Leftrightarrow B,\ B \Leftrightarrow C + D$

A	B	C	D	Reference
acetone	diacetone alcohol	mesityl oxide	water	Nicol, 2003
butanal	intermediate of aldol condensation	2-ethyl hexenal	water	Mori et al., 1997

$2A \Leftrightarrow B + C,\ 2A \Leftrightarrow D$

A	B	C	D	Reference
isobutylene	tetramethylethylene	ethylene	diisobutylene	Podrebarac, 2006

$A \Leftrightarrow C + D,\ B + D \Leftrightarrow E$

A	B	C	D	E	Reference
monomethyl succinate	water	methanol	succinic anhydride	succinic acid	Shekhawat et al., 2006

2A ⇔ B + C, B + D ⇔ E					
A	B	C	D	E	Reference
acetone	mesityl oxide	water	hydrogen	methyl isobutyl ketone	Lawson and Nkosi, 1999

2A ⇔ C + 2D, C + B ⇔ A + D					
A	B	C	D	E	Reference
methanol	dimethylamine	methyl formate	hydrogen	dimethylformamide	Maliszewskyj et al., 2004

A + B ⇔ C, A + C ⇔ D				
A	B	C	D	Reference
ethanol	formaldehyde	hemiacetal	acetal	Sharma, 1995
benzene	propylene	cumene	diisopropylbenzene	Shoemaker and Jones, 1987

A + B ⇔ C, 2C ⇔ B + D				
A	B	C	D	Reference
ethylene	water	ethanol	diethylether	Eguchi et al., 1987
formaldehyde	methanol	hemiformal	higher hemiformal	Hasse et al., 1990
formaldehyde	water	methylene gloycol	polyoxymethylene	Hasse et al., 1990

A + B ⇔ 2C, C + D ⇔ E					
A	B	C	D	E	Reference
dimethyl ether	water	methanol	carbon monoxide	acetic acid	Voss, 2001

A + B ⇔ D, C + D ⇔ A + E					
A	B	C	D	E	Reference
methanol	carbon monoxide	water	methyl formate	formic acid	Lee et al., 2000
cyclohexene	formic acid	water	cyclohexyl formate	cyclohexanol	Sharma, 1995
carbon dioxide	propylene oxide	water	propylene carbonate	1,2-propanediol	Von Hebel and Lange, 2005

$A + B \Leftrightarrow C, C \Leftrightarrow D + E$					
A	B	C	D	E	Reference
triacetone amine	hydrazine	4-oxo-2,2,6,6-tetramethylhydrazone	norpempidine	nitrogen	Kampmann and Stuhlmuller, 1997
acetic anhydride	acetaldehyde	ethylene diacetate	acetic acid	vinyl acetate	Zoeller et al., 1998

$A + B \Leftrightarrow C, C + B \Leftrightarrow D + E$					
A	B	C	D	E	Reference
ethylene carbonate	methanol	2-hydroxyethyl methyl carbonate	dimethyl carbonate	ethylene glycol	Scott et al., 2003 Kanamaru, 2003

$A + B \Leftrightarrow D, A + C \Leftrightarrow E$					
A	B	C	D	E	Reference
2-methyl-2-butene	ethanol	water	*tert*-amyl ethyl ether	*tert*-amyl alcohol	Sahapatsombud et al., 2005
formaldehyde	methanol	water	hemiformal	methylene glycol	Hasse et al., 1990

$A + B \Leftrightarrow C + D, 2B \Leftrightarrow E$					
A	B	C	D	E	Reference
methanol	acetic acid	methyl acetate	water	acetic dimer	Agreda et al., 1990

$A + B \Leftrightarrow C + D, 2A \Leftrightarrow E + D$					
A	B	C	D	E	Reference
ethanol	acetic acid	ethyl acetate	water	diethyl ether	Castier et al., 1989

$A + B \Leftrightarrow C + D, B + C \Leftrightarrow D + E$

A	B	C	D	E	Reference
adipic acid	methanol	adipic acid monomethyl ester	water	dimethyl adipate	Hung et al., 2008
dimethyl carbonate	ethanol	methyl and ethyl carbonate	methanol	diethyl carbonate	Luo and Xiao, 2001
dichlorosilane	hydrogen chloride	trichlorosilane	hydrogen	silicon tetrachloride	Ung and Doherty, 1995b
di-*tert*-butylbenzene	*m*-xylene	*tert*-butylbenzene	*tert*-butyl-*m*-xylene	benzene	Ung and Doherty, 1995a
succinic acid	methanol	adipic acid monomethyl ester	water	dimethyl adipate	Asthana et al., 2005a

$A + B \Leftrightarrow C + D, 2C \Leftrightarrow A + E$					
A	B	C	D	E	Reference
dimethyl carbonate	phenol	phenylmethyl carbonate	methanol	diphenyl carbonate	de Bruin et al., 2004
water	methyl acetate	methanol	acetic acid	dimethyl ether	Hoyme, 2003

$A + B \Leftrightarrow C + D, A + B \Leftrightarrow C + E$					
A	B	C	D	E	Reference
m-xylene	di-*tert*-butyl benzene	*tert*-butyl xylene	*tert*-butyl benzene	benzene	Venkataraman et al., 1990

$2A + B \Leftrightarrow 2C + 4D, C + 2D \Leftrightarrow E$					
A	B	C	D	E	Reference
methane	oxygen	carbon monoxide	hydrogen	methanol	Allison et al., 2004

$A \Leftrightarrow B, B \Leftrightarrow C, B \Leftrightarrow D$				
A	B	C	D	Reference
m-xylene	*o*-xylene	*p*-xylene	ethylene benzene	Castier et al., 1989

$A + B \Leftrightarrow C, C + B \Leftrightarrow D, 2B \Leftrightarrow E$					
A	B	C	D	E	Reference
isobutane	isobutene	trimethyl pentane	dodecane	dimethyl hexane	Albright et al., 1988

A + B $\Leftrightarrow$ C, C + B $\Leftrightarrow$ D, D + B $\Leftrightarrow$ E					
A	B	C	D	E	Reference
ethylene oxide	ammonia	monoethanolamine	diethanolamine	triethanolamine	Garg et al., 2005
monomethyl amine	ethylene oxide	methyl ethanolamine	methyl diethanolamine	methyl triethanolamine	Garg et al., 2005
ammonia	propylene oxide	monoisopropanolamine	diisopropanolamine	triisopropanolamine	Garg et al., 2005

$A + B \Leftrightarrow C + D, B + D \Leftrightarrow C + E, B + E \Leftrightarrow C + F$						
A	B	C	D	E	F	Reference
citral acid	ethanol	water	monoethyl citrate	diethyl citrate	triethyl citrate	Miller, Ashanta, Kolah, Vu, and Lira, 2006

$A + B \Leftrightarrow C + D + E$					
A	B	C	D	E	Reference
ethyl acetate	*R*-OH (*r*, *s*)butyl or isoamyl	*R*-acetate (*r*)	ethanol	*R*-OH (*s*)	Okasinski and Doherty, 2003
propylene chlorohydrin	calcium dichloride	propylene oxide	calcium dihydroxide	water	Bezzo et al., 1999

$A + B + C \Leftrightarrow D + E$					
A	B	C	D	E	Reference
acetic acid	ethylene	oxygen	vinyl acetate	water	Adams et al., 2003

$A + B \Leftrightarrow D, C + D \Leftrightarrow E + F$						
A	B	C	D	E	F	Reference
propylene oxide	carbon dioxide	methanol	propylene carbonate	1,2-propanediol	dimethyl carbonate	Von Hebel and Lange, 2005

A + B ⇔ D, C + D ⇔ E + F								
A	B	C	D	E	F	G	H	Reference
carbon dioxide	water	calcium acetate	ethanol	carbonic acid	calcium carbonate	acetic acid	ethyl acetate	Verser and Eggeman, 2006

REFERENCES

Adams T. A. II and W. D. Seider, Semicontinuous distillation with chemical reaction in a middle vessel, *Ind. Eng. Chem. Res.* **46**, 5548–4460 (2006).

Adams J., W. Groten, and S. Nemphos, Process for vinyl acetate, U.S. Patent 6,620,465 (2003).

Agreda V. H., L. R. Partin, and W. H. Heise, High-purity methyl acetate via reactive distillation, *Chem. Eng. Prog.* **86**, 40–46 (1990).

AIChE, *Student Contest Problem*, AIChE, New York, 1970.

Aiouache F. and S. Goto, Reactive distillation–pervaporation hybrid column for *tert*-amyl alcohol etherification with ethanol, *Chem. Eng. Sci.* **58**, 2465–2477 (2003).

Albright L. F., M. A. Spalding, J. Faunce, and R. E. Eckert, Alkylation of isobutane with C4 olefins. 3. Two-step process using sulfuric acid as catalyst, *Ind. Eng. Chem. Res.* **27**, 391–397 (1988).

Allison J. D., H. A. Wright, T. H. Harkins, and D. S. Jack, Use of catalytic distillation reactor for methanol synthesis, U.S. Patent 6,723,886 (2004).

An W. Z., K. T. Chuang, and A. R. Sanger, Dehydration of methanol to dimethyl ether by catalytic distillation, *Can. J. Chem. Eng.* **82**, 948–955 (2004).

Anonymous, *European Chemical News* **29**, March (1999).

Asthana N., A. Kolah, D. Vu, C. T. Lira, and D. J. Miller, Reactive distillation for the biorefinery: Production of organic acid esters, In *ACS National Meeting Book of Abstracts*, ACS, Washington, D.C., 2005a.

Asthana N., A. Kolah, D. T. Vu, C. T. Lira, and D. J. Miller, A continuous reactive separation process for ethyl lactate formation, *Org. Process Res. Dev.* **9**, 599–607 (2005b).

Asthana N. S., A. K. Kolah, D. T. Vu, C. T. Lira, and D. J. Miller, A kinetic model for the esterification of lactic acid and its oligomers, *Ind. Eng. Chem. Res.* **45**, 5251–5257 (2006).

Baker J. A., A. Gray, and J. F. Benford, Esterification of nitrobenzoic acids, U.S. Patent 3,948,972 (1976).

Barnum C. S. and C. T. Blaisdell, Reactive distillation process for free radical halogenation, U.S. Patent 5,449,801 (1995).

Reactive Distillation Design and Control. By William L. Luyben and Cheng-Ching Yu

Belson D. J., A distillation method of aromatic nitration using azeotropic nitric acid, *Ind. Eng. Chem. Res.* **29**, 1562–1565 (1990).

Berman S. and A. Isbenjian, Esterification—Continuous production of dibutyl phthalate in a distillation column, *Ind. Eng. Chem.* **40**, 2139–2148 (1948).

Bessling B., J. Knab, W. Brox, and B. Lohe, Method and device for obtaining isobutenes from conjugated hydrocarbons, U.S. Patent 6,362,386 (2002).

Bezzo F., A. Bertucco, A. Forlin, and M. Barolo, Steady-state analysis of an industrial reactive distillation column, *Sep. Purif. Technol.* **16**, 251–260 (1999).

Bhanvase B. A., Y. R. Soman, and R. V. Naik, Reactive distillation studies, *Chem. Eng. World* **41**, 43–47 (2006).

Bhatia S., A. L. Ahmad, A. R. Mohamed, and S. Y. Chin, Production of isopropyl palmitate in a catalytic distillation column: Experimental studies, *Chem. Eng. Sci.* **61**, 7436–7447 (2006).

Bock H., G. Wozny, and B. Gutsche, Design and control of a reaction distillation column including the recovery system, *Chem. Eng. Process* **36**, 101–109 (1997).

Broekhuis R. R., S. Lynn, and C. J. King, Recovery of propylene glycol from dilute aqueous solutions via reversible reaction with aldehydes, *Ind. Eng. Chem. Res.* **33**, 3230–3237 (1994).

Bruin P. R., J. S. Law, and V. A. Vriens, Method and apparatus for the continuous production of diaryl carbonate, U.S. Patent Application 20010021786 (2001).

Burton P. E., Process for hydrolyzing di-isopropyl ether to isopropyl alcohol by catalytic distillation using a solid acid catalyst, U.S. Patent 6,906,229 (2005).

Castier M., P. Rasmussen, and A. Fredenslund, Calculation of simultaneous chemical and phase equilibria in nonideal systems, *Chem. Eng. Sci.* **44**, 237–248 (1989).

Chin C. W., M. A. Dasari, G. J. Suppes, and W. R. Sutterlin, Dehydration of glycerol to acetol via catalytic reactive distillation, *AIChE J.* **52**, 3543–3548 (2006).

Chopade S. P., A. D. Dhale, C. W. Kiesling, A. M. Clark, J. E. Jackson, and D. J. Miller, Process for the recovery of a polyol from an aqueous solution, U.S. Patent 6,548,681 (2003).

Chopade S. P. and M. M. Sharma, Acetalization of ethylene glycol with formaldehyde using cation-exchange resins as catalysts: Batch versus reactive distillation, *React. Funct. Polym.* **34**, 37–45 (1997).

Chung Y., H. K. Jeong, H. K. Song, and W. H. Park, Modelling and simulation of the chemical reaction heat pump system adopting the reactive distillation process, *Comput. Chem. Eng.* **21**, S1007–S1012 (1997).

Costa P. and B. Canepa, Sul calcolo dolonna per la distikkazione di miscele reagenti, *Quad. Ing. Chim. Ital.* **5**, 113–121 (1969).

Crouzen J., J. J. Gorissen, and C. J. G. Van Strien, Process for the preparation of monochloroacetic acid, U.S. Patent 7,135,597 (2006).

de Bruin P. R., J. S. Law, and V. A. Vriens, Method and apparatus for the continuous production of diaryl carbonates, U.S. Patent 6,767,517 (2004).

Dean C., Desulfurization of a naphtha gasoline stream derived from a fluid catalytic cracking unit, U.S. Patent 7,122,114 (2006).

Dhale A. D., L. K. Myrant, S. P. Chopade, J. E. Jackson, and D. J. Miller, Propylene glycol and ethylene glycol recovery from aqueous solution via reactive distillation, *Chem. Eng. Sci.* **59**, 2881–2890 (2004).

Dil'man V. V., V. A. Lotkhov, A. A. Lipatova, S. Y. Kvashnin, and N. N. Kulov, Estimation of the mass-transfer efficiency in catalytic distillation in a packed column, *Theor. Found. Chem. Eng.* **39**, 455–462 (2005).

Dimian A. C., F. Omota, and A. Bliek, Entrainer-enhanced reactive distillation, *Chem. Eng. Process* **43**, 411–420 (2004).

Doherty M. F. and G. Buzad, Reactive distillation by design, *Chem. Eng. Res. Des.* **70**, 448–458 (1992).

Domschke T., M. Merger, G. Grossmann, and T. Faust, Method for the production of 2-keto-L-gulonic acid C_4C_{10} alkyl esters, U.S. Patent 7,091,375 (2006).

Dorai S., Alkanolysis of polyether polyol esters by reactive distillation, U.S. Patent 5,852,218 (1971).

Doron L. and S. Jose, Apparatus for producing phenol using reactive distillation, U.S. Patent 6,441,252 (2002).

Dragomir R. M. and M. Jobson, Conceptual design of single-feed hybrid reactive distillation columns, *Chem. Eng. Sci.* **60**, 4377–4395 (2005).

Ebner J. R., Z. Guo, A. Hershman, L. M. Klein, W. D. McGhee, M. Paster, and I. Prakash, Process for the preparation of 3,3-dimethylbutanal, U.S. Patent 6,803,487 (2004).

Eguchi K., T. Tokiai, and H. Arai, High pressure catalytic hydration of olefins over various proton-exchanged zeolites, *Appl. Catal.* **34**, 275–287 (1987).

Feathers R. E. and J. D. Mansell, Chlorinated benzenes, U.S. Patent 5,149,892 (1992).

Garg D., S. N. Shah, M. J. Okasinski, and A. S. Drayton-Elder, Process for producing alkanolamines, U.S. Patent 6,846,959 (2005).

Gartide R. and G. Gildert, Cracked gas processing and conversion for propylene production, U.S. Patent 6,420,619 (2002).

Gartside R. J., Process for the production of linear alpha olefins and ethylene, U.S. Patent 6,727,396 (2004).

Gartside R. J. and M. I. Greene, Olefin isomerization process, U.S. Patent 6,875,901 (2005).

Gaspillo P. A. D., L. C. Abella, and S. Goto, Dehydrogenation of 2-propanol in reactive distillation column for chemical heat pump, *J. Chem. Eng. Jpn.* **31**, 440–444 (1998).

Geelen H. and J. B. Wijffels, Use of a distillation as a chemical reactor, *Chimica & l Industria*, **47**, 236 (1965).

Geisendoerfer M., G. Nestler, J. Schroeder, and H. Vandenmersch, Method for producing (meth)acrylic acid esters, US Patent 7,294,240 (2007).

Gelbein A. P., Three stage propylene oxide process, U.S. Patent 6,900,360 (2002).

Gildert G. R., Hydrogenation of benzene to cyclohexane, U.S. Patent 6,187,980 (2001).

Gildert G. R. and M. E. Loescher, Catalytic distillation process for the production of C8 alkanes, U.S. Patent 6,274,783 (2001).

Gonzalez J. C., H. Subawalla, and J. R. Fair, Preparation of *tert*-amyl alcohol in a reactive distillation column. 2. Experimental demonstration and simulation of column characteristics, *Ind. Eng. Chem. Res.* **36**, 3845–3853 (1997).

Gotze L. and O. Bailer, Reactive distillation with Katapak-S, *Sulzer Tech. Rev.* **81**, 29–31 (1999).

Groening C., K. Ebel, G. Kaibel, J. Therre, J. Koopmann, H. Menig, G. Fritz, and R. Dietz, Production of methylglyoxal dimethyl acetal, Jpn. Patent 08176056 (1996).

Grosser J. H., M. F. Doherty, and M. F. Malone, Modeling of reactive distillation systems, *Ind. Eng. Chem. Res.* **26**, 983–989 (1987).

Hagen G. P. and M. J. Spangler, Preparation of polyoxymethylene dialkane ethers by catalytic conversion of formaldehyde formed by dehydrogenation of methanol or dimethyl ether, U.S. Patent 6,350,919 (2002).

Hamilton D. M., Process for catalytic hydroxylation of benzene, U.S. Patent 6,900,358 (2005).

Hao X. R., L. Y. Dong, Z. Q. Yu, W. Wang, and W. C. Wang, Synthesis of dimethyl maleate by catalytic distillation [in Chinese], *Petrochem. Technol.* **32**, 187 (2003).

Hasse H., I. Hahnenstein, and G. Maurer, Revised vapor–liquid equilibrium model for multicomponent formaldehyde mixtures, *AIChE J.* **36**, 1807–1814 (1990).

He B. B., A. P. Singh, and J. C. Thompson, A novel continuous-flow reactor using reactive distillation for biodiesel production, *Trans. Am. Soc. Agri. Eng.* **49**, 107–112 (2006).

Hembre R. and R. Lin, Process for the preparation of triflic anhydride, U.S. Patent 6,469,206 (2002).

Hendriksen D. E., G. G. McGlamery, M. J. Keenan, and D. D. Pete, Process for preparing 3-methyl-2-pentene, U.S. Patent 5,731,486 (1998).

Hendriksen D. E., J. R. Lattner, and M. J. G. Janssen, Alkylation process using zeolite beta, U.S. Patent 6,002,057 (1999).

Hildreth J. M. and N. J. Wyckoff, Removal of α-methyl styrene from cumene, U.S. Patent 5,905,178 (1999).

Hoffmaster W. R. and S. Hauan, Using feasible regions to design and optimize reactive distillation columns with ideal VLE, *AIChE J.* **52**, 1744–1753 (2006).

Hoyme C. A., Reactive distillation process for hydrolysis of esters, U.S. Patent 6,518,465 (2003).

Hu M., X. G. Zhou, and W. K. Yuan, Simulation and optimization of a coupled reactor/column system for trioxane synthesis, *Chem. Eng. Sci.* **54**, 1353–1358 (1999).

Huang Y. S., E. U. Schlunder, and K. Sundmacher, Feasibility analysis of membrane reactors—Discovery of reactive azeotropes, *Catal. Today* **104**, 360–371 (2005).

Hung S. B., I. K. Lai, H. P. Huang, M. J. Lee, and C. C. Yu, Reactive distillation for two-stage reaction systems: Adipic and glutraic acid esterification, *Ind. Eng. Chem. Res.* **47**, 3076–3087 (2008).

Huss A., Jr. and C. R. Kennedy, Hydrocarbon processes comprised of catalytic distillation using Lewis acid promoted inorganic oxide catalyst systems, U.S. Patent 4,935,577 (1990).

Imai T., 1,3-Butadiene separation from a crude C_4 stream using catalytic extractive distillation, U.S. Patent 6,040,489 (2000).

Ishida H., M. Ono, S. Kaji, and A. Watanabe, Synthesis of 1,3-cyclohexadiene through liquid phase dehydration of 2-cyclohexen-1-ol in aqueous solution using zeolite catalyst, *Nippon Kagaku Kaishi* **4**, 267–275 (1997).

Jelinek J. and V. Hlavacek, Steady state countercurrent equilibrium stage separation with chemical reaction by relaxation method, *Chem. Eng. Commun.* **2**, 79–85 (1976).

Jerry E., G. Zhi, H. Arnold, K. Loraine, M. William, P. Mark, and P. Indra, Process for the preparation of 3,3-dimethylbutanal, U.S. Patent 6,573,409 (2003).

John G., E. Dikow, C. Mendoza-Frohn, G. Ronge, R. Bachmann, L. Gottschalk, and M. Prein, Reactive rectification, U.S. Patent 7,071,283 (2006).

Jung C. W., P. E. Garrou, and G. R. Strickler, Disproportionation of alkenes, U.S. Patent 4,709,115 (1987).

Kamath R. S., Z. W. Qi, K. Sundmacher, P. Aghalayam, and S. M. Mahajani, Comparison of reactive distillation with process alternatives for the isobutene dimerization reaction, *Ind. Eng. Chem. Res.* **45**, 2707–2714 (2006).

Kampmann D. and G. Stuhlmuller, Process for the continuous preparation of 2,2,6,6-tetramethylpiperidine, U.S. Patent 5,663,351 (1997).

Kanamaru T., Method for producing dimethyl carbonate and ethylene glycol, Jpn. Patent 2003342209 (2003).

Khaledi R. and P. R. Bishnoi, A method for modeling two- and three-phase reactive distillation columns, *Ind. Eng. Chem. Res.* **45**, 6007–6020 (2006).

Kiss A. A., A. C. Dimian, and G. Rothenberg, Solid acid catalysts for biodiesel production—Towards sustainable energy, *Adv. Synth. Catal.* **348**, 75–81 (2006).

Kolodziej A., M. Jaroszynski, W. Salacki, W. Orlikowski, K. Fraczek, M. Klöker, E. Y. Kenig, and A. Gorak, Catalytic distillation for TAME synthesis with structured catalytic packings, *Chem. Eng. Res. Des.*, **82**, 175–184 (2004).

Klöker M., E. Y. Kenig, A. Hoffmann, P. Kreis, and A. Górak, Rate-based modelling and simulation of reactive separations in gas/vapour–liquid systems, *Chem. Eng. Process* **44**, 617–629 (2005).

Lange J.-P. and V. Otten, Dehydration of phenyl-ethanol to styrene: Zeolite catalysis under reactive distillation, *J. Catal.* **238**, 6–12 (2006).

Lawson K. H. and B. Nkosi, Production of MIBK using catalytic distillation technology, U.S. Patent 6,008,416 (1999).

Lebas E., S. Jullian, C. Travers, P. Capron, J.-F. Joly, and M. Thery, Paraffin isomerisation process using reactive distillation, U.S. Patent 5,948,948 (1999).

Lederer J., J. Kolena, J. Hanika, W. W. Levering, O. Bailer, P. Moravek, Q. Smejkal, and V. Macek, Process and apparatus for the production of butylacetate and isobutylacetate, U.S. Patent 6,458,992 (2002).

Lee J. W., S. Hauan, and A. W. Westerberg, Graphical methods for reaction distribution in a reactive distillation column, *AIChE J.* **46**, 1218–1233 (2000).

Leemann M., V. Hildebrandt, H. Thiele, and S. Espig, Production of polyamides by reactive distillation, U.S. Patent 6,358,373 (2002).

Lei Z. G., C. Y. Li, B. H. Chen, W. Erqiang, and J. C. Zhang, Study on the alkylation of benzene and 1-dodecene, *Chem. Eng. J.* **93**, 191–200 (2003).

Lin F., C. Li, and X. Chen, Synthesis of neohexene, *Huaxue Fanying Gongcheng Yu Gongyi* **19**, 40–44 (2003).

Lin Y. D., J. H. Chen, J. K. Cheng, H. P. Huang, and C. C. Yu, Process alternative for methyl acetate conversion using reactive distillation: 1. Hydrolysis, *Chem. Eng. Sci.*, **63**, 1668–1682 (2008).

Liu H.-W., Y.-X. Qu, and W.-C. Wang, Simulation of hydration process of water and ethylene oxide by reactive distillation to produce glycol [in Chinese], *J. Beijing Univ. Chem. Technol.* **29**, 18 (2002).

Liu Q.-L., F. Zhang, and H.-Q. Gao, Synthesis of tetrahydrofuran from butanediol by reactive distillation [in Chinese], *Chem. Eng. (China)* **30**, 75 (2002).

Loescher M. E., D. G. Woods, M. J. Keenan, S. E. Silverberg, and P. W. Allen, Oligomerization process, U.S. Patent 7,145,049 (2006).

Lopez R., Removal of silylated compounds from solvent and gas waste streams, U.S. Patent 6,013,821 (2000).

Luo H. P. and W. D. Xiao, A reactive distillation process for a cascade and azeotropic reaction system: Carbonylation of ethanol with dimethyl carbonate, *Chem. Eng. Sci.* **56**, 403–410 (2001).

Luyben W. L., K. M. Pszalgoeski, M. R. Schaefer, and C. Siddons, Design and control of conventional and reactive distillation processes for the production of butyl acetate, *Ind. Eng. Chem. Res.* **43**, 8014–8025 (2004).

Ma L., Y. Zhang, and J. Yang, Purification of lactic acid by heterogeneous catalytic distillation using ion-exchange resins, *Chin. J. Chem. Eng.* **13**, 24–31 (2005).

Maliszewskyj R. J., M. G. Turcotte, and J. W. Mitchell, Dimethylformamide synthesis via reactive distillation of methyl formate and dimethylamine, U.S. Patent 6,723,877 (2004).

Maraschino M. J., Apparatus and process for hydrogenations, U.S. Patent 6,407,300 (2002).

Marker T. L., G. A. Funk, P. T. Barger, and H. U. Hammershaimb, Two-stage process for producing diisopropyl ether using catalytic distillation, U.S. Patent 5,504,258 (1996).

Marker T. L., G. A. Funk, P. T. Barger, and H. U. Hammershaimb, Two-stage process for producing diisopropyl ether using catalytic distillation, U.S. Patent 5,744,645 (1998).

Miller D. J., N. Ashanta, A. Kolah, and C. T. Lira, Improved process for production of organic acid esters, U.S. Patent Application 200614977 (2006).

Miller D. J., N. Ashanta, A. Kolah, D. T. Vu, and C. T. Lira, Process for reactive esterification distillation, U.S. Patent Application 20060252956 (2006).

Mori T., K. Fujita, Y. Kajita, and M. Takai, Method for producing a dimerized aldehyde, U.S. Patent 5,667,644 (1997).

Müller D., G. Ronge, J.-P. Schäfer, H.-J. Leimkühler, U. Strauss, and H.-D. Block, Method and facility for producing silane, U.S. Patent 6,942,844 (2005).

Murphree B. and R. Ozer, Preparation of pentenoic acid, U.S. Patent 5,536,873 (1996).

Nicol W., Comparing catalytic distillation to separate reaction and distillation for the production of diacetone alcohol, *Chem. Eng. Res. Des.* **81**, 1026–1032 (2003).

Nishihira K., S. Tanaka, Y. Nishihida, and S. Fujitsu, Process for producing a diaryl oxalate, U.S. Patent 6,018,072 (2000).

Okasinski M. J. and M. F. Doherty, Simultaneous kinetic resolution of chiral propylene oxide and propylene glycol in a continuous reactive distillation column, *Chem. Eng. Sci.* **58**, 1289–1300 (2003).

Okawa T., Y. Sato, H. Igarashi, and S. Suzuki, Process for producing xylylene diisocyanate, U.S. Patent 5,196,512 (1993).

Omota F., A. C. Dimian, and A. Bliek, Fatty acid esterification by reactive distillation. Part 1: Equilibrium-based design, *Chem. Eng. Sci.* **58**, 3159–3174 (2003).

Ott M., H. Schoenmakers, and H. Hasse, Distillation of formaldehyde containing mixtures: Laboratory experiments, equilibrium stage modeling and simulation, *Chem. Eng. Process* **44**, 687–694 (2005).

Patel N. R., V. E. Lewis, and M. D. Enderson, Method for removal of aldehydes from chemical manufacturing production streams during distillative purification, U.S. Patent 6,074,532 (2000).

Pisarenko O., A. Yu, L. Serafimov, and V. Timofeev, Method for the manufacture of butyl vinyl ether from 1-butanol and acetylene, Russ. Patent 304,678 (2002).

Podrebarac G. G., Olefin metathesis in a distillation column reactor, U.S. Patent 6,583,329 (2003).

Podrebarac G. G., Olefin metathesis, U.S. Patent 7,041,861 (2006).

Podrebarac G. G., F. T. T. Ng, and L. Rempel, Production of diacetone alcohol with catalytic distillation. Part I: Catalytic distillation experiments, *Chem. Eng. Sci.* **53**, 1067–1075 (1998).

Pohl S. L., Process for the production of alkylbenzene with ethane stripping, U.S. Patent 7,071,369 (2006).

Powers D. H., Alpha olefin production, U.S. Patent 6,768,038 (2004).

Prakasa A. and S. George, Fuel and oil detergents, U.S. Patent 6,630,430 (2003).

Qi Z. W. and K. Sundmacher, Multiple product solutions of *tert*-butyl alcohol dehydration in reactive distillation, *Ind. Eng. Chem. Res.* **45**, 1613–1621 (2006).

Reepmeyer F., J. U. Repke, and G. Wozny, Time optimal start-up strategies for reactive distillation columns, *Chem. Eng. Sci.* **59**, 4339–4347 (2004).

Reusch D., A. Beckmann, F. Nierlich, and A. Tuchlenski, Method for producing *tert*-butanol by means of reactive rectification, U.S. Patent 7,115,787 (2006).

Rhim J. K., S. Y. Bae, and H. T. Lee, Isothermal vapor–liquid equilibrium accompanied by esterification. Ethanol–formic acid system, *Int. Chem. Eng.* **25**, 551–557 (1985).

Rihko-Struckmann L. K., R. S. Karinen, A. O. I. Krause, K. Jakobsson, and J. R. Aittamaa, Process configurations for the production of the 2-methoxy-2,4,4-trimethylpentane—A novel gasoline oxygenate, *Chem. Eng. Process* **43**, 57–65 (2004).

Rosinson C. S. and E. R. Dilliland, *Elements of Fractional Distillation*, Fourth Edition, McGraw-Hill, New York, 1950.

Ryu J. Y., Process for the selective hydrogenation of alkynes, U.S. Patent 7,041,860 (2006).

Ryu J. Y. and A. P. Gelbein, Process and catalyst for making dialkyl carbonates, U.S. Patent 6,392,078 (2002).

Ryu J. Y. and A. P. Gelbein, Process for making dialkyl carbonates, U.S. Patent 7,074,951 (2006).

Saari K., E. Tirronen, A. Vuori, and M. Lahtinen, Method for preparing formic acid, U.S. Patent 6,429,333 (2002).

Saha B., A. Alqahtani, and H. T. R. Teo, Production of iso-amyl acetate: Heterogeneous kinetics and techno-feasibility evaluation for catalytic distillation, *Int. J. Chem. React. Eng.* **3**, A30 (2005).

Saha B. and M. M. Sharma, Esterification of formic acid, acrylic acid and methacrylic acid with cyclohexene in batch and distillation column reactors: Ion-exchange resins as catalysts, *React. Funct. Polym.* **28**, 263–278 (1996).

Sahapatsombud U., A. Arpornwichanop, S. Assabumrungrat, P. Praserthdam, and S. Goto, Simulation studies on reactive distillation for synthesis of *tert*-amyl ethyl ether, *Kor. J. Chem. Eng.* **22**, 387–392 (2005).

Sakuth M. and U. Peters, Process for cracking tertiary butyl alcohol in a reaction distillation column, U.S. Patent 5,849,971 (1998).

Schmitt K., New acetone chemistry in Germany, Scholven report [Reprint], *Chem. Ind. Int.* (1996).

Schmitt M. and H. Hasse, Chemical equilibrium and reaction kinetics of heterogeneously catalyzed *N*-hexyl acetate esterification, *Ind. Eng. Chem. Res.* **45**, 4123–4132 (2006).

Schniepp L. E., J. W. Dunning, and E. C. Lathrop, Continuous process for acetylation, *Ind. Eng. Chem.* **37**, 872–877 (1945).

Scott B., S. Jose, and J. Zhaozhong, Co-production of dialkyl carbonates and diols with treatment of ydroxyl alkyl carbonates, U.S. Patent 6,573,396 (2003).

Sharma M. M., Some novel aspects of cationic ion-exchange resins as catalysts, *React. Funct. Polym.* **26**, 3–23 (1995).

Shekhawat D., J. E. Jackson, and D. J. Miller, Process model and economic analysis of itaconic acid production from dimethyl succinate and formaldehyde, *Bioresour. Technol.* **97**, 342–347 (2006).

Sherman S. R., C. A. Eckert, and L. S. Scott, Modeling multicomponent equilibria from binary equilibrium data for reacting systems, *Chem. Eng. Process* **35**, 363–371 (1996).

Shinohara C., S. Kawakami, T. Moriga, H. Hayashi, S. Hodoshima, Y. Saito, and S. Sugiyama, Local structure around platinum in Pt/C catalysts employed for liquid-phase dehydrogenation of decalin in the liquid-film state under reactive distillation conditions, *Appl. Catal. A: Gen* **266**, 251–255 (2004).

Shoemaker J. D. and E. M. Jones, Jr., Cumene by catalytic distillation, *Hydrocarbon Process* **66**, 57–58 (1987).

Silverberg S., L. Sanchez, and J. Lattner, Use of catalytic distillation to produce cyclopentane or cyclopentene, U.S. Patent 6,100,435 (2000).

Smith G. A., P. R. Anantaneni, S. S. Ashrawi, R. M. Smadi, J. F. Knifton, and M. Stockton, Alkylbenzene detergents with high 2-isomer content, U.S. Patent 7,008,914 (2006).

Smith J. and A. Lawrence, Imines from mesityl oxide, U.S. Patent 4,332,968 (1982).

Smith J. and A. Lawrence, Downflow process for hydrotreating naphtha, U.S. Patent 7,125,484 (2006a).

Smith J. and A. Lawrence, Method of removing entrained sulfuric acid from alkylate, U.S. Patent 7,126,038 (2006b).

Smith J. and A. Lawrence, Supported metal oxides as catalysts for aldol condensations, U.S. Patent 7,098,366 (2006c).

Smith L. A. and A. P. Gelbein, Process for the production of bisphenol-A, U.S. Patent 6,939,994 (2005).

Stanley S. J. and G. R. Gildert, Process for the removal of MAPD from hydrocarbon streams, U.S. Patent 6,414,205 (2002).

Stitt E. H., Reactive distillation for toluene disproportionation: A technical and economic evaluation, *Chem. Eng. Sci.* **57**, 1537–1543 (2002).

Stuart F., M. Ernest, and M. Ahmet, Process for converting caprolactam to nylon 6, U.S. Patent 6,479,620 (2002).

Suyama K., N. Kiyota, T. Konishi, and Y. Matsumura, Method for producing trimethylhydroquinone, U.S. Patent 6,441,249 (2002).

Tang S., S. Li, J. Han, and B. Chen, Study of the process of reactive distillation for preparing benzyl benzoate, *Chem. React. Eng. Technol.* **39**, 11 (1984).

Tang Y. T., Y. W. Chen, H. P. Huang, C. C. Yu, S. B. Hung, and M. J. Lee, Design of reactive distillations for acetic acid esterification, *AIChE J.* **51**, 1683–1699 (2005).

Terrill D. L., L. F. Sylvestre, and M. F. Doherty, Separation of closely boiling mixtures by reactive distillation. 1. Theory, *Ind. Eng. Chem. Process Des. Dev.* **24**, 1062–1071 (1985).

Thelen B., W. Auge, and K.-W. Thiem, Process for the production of aromatic mononitro compounds, U.S. Patent 4,064,147 (1977).

Toshiya A., K. Kawakami, and K. Otsubo, Process for producing glycine, U.S. Patent 6,916,638 (2005).

Ung S. and M. F. Doherty, Synthesis of reactive distillation systems with multiple equilibrium chemical reactions, *Ind. Eng. Chem. Res.* **34**, 2555–2565 (1995a).

Ung S. and M. F. Doherty, Vapor–liquid phase equilibrium in systems with multiple chemical reactions, *Chem. Eng. Sci.* **50**, 23–48 (1995b).

Vandersall M. T. and R. A. Weinand, Metal-doped sulfonated ion exchange resin catalysts, U.S. Patent 6,977,314 (2005).

Vargas-Villamil F. D., J. O. Marroquin, C. de la Paz, and E. Rodriguez, A catalytic distillation process for light gas oil hydrodesulfurization, *Chem. Eng. Process* **43**, 1309–1316 (2004).

Venkataraman S., W. K. Chan, and J. F. Boston, Reactive distillation using Aspen Plus, *Chem. Eng. Prog.* **86**, 45–54 (1990).

Verser D. W. and T. J. Eggeman, Process for producing ethanol from corn dry milling, U.S. Patent 7,074,603 (2006).

Von Hebel K. L. and J.-P. Lange, Process for the preparation of propanediol, U.S. Patent 6,897,343 (2005).

Vora B. V. and H. U. Hammershaimb, Oligomer production with catalytic distillation, U.S. Patent 6,025,533 (2000).

Voss B., Acetic acid reactive distillation process based on DME/methanol carbonylation, U.S. Patent 6,175,039 (2001).

Wen L.-Y., X.-L. Liu, J. Li, F.-X. Cui, and M.-Y. Li, Zeolite and beta catalyst for synthesis of propylene glycol monoethyl ether by catalytic distillation [in Chinese], *Petrochem. Technol.* **30**, 749 (2001).

Wu K.-C., W.-T. Liu, R.-Y. Chang, and C.-S. Tsai, Process for preparing a carboxylic ester, U.S. Patent 6,730,806 (2004).

Xu Z. and M. P. Dudukovic, Modeling and simulation of semi-batch photo reactive distillation, *Chem. Eng. Sci.* **54**, 1397–1403 (1999).

Yu Z.-Q., B.-L. Gao, and Y.-H. Jia, Mathematical model for selective hydrogenation of C3 streams in a catalytic distillation column [in Chinese], *Chem. React. Eng. Technol.* **18**, 12 (2002).

Zeng K. L., C. L. Kuo, and I. L. Chien, Design and control of butyl acrylate reactive distillation column system, *Chem. Eng. Sci.* **61**, 4417–4431 (2006).

Zhang W. and W. L. Wan, Synthesis of 2-ethoxyethanol using catalytic distillation, *Appl. Catal.* **74**, 20 (1991).

Zhao G.-S., B.-L. Yang, and X.-H. Yang, Catalytic dehydration of *tert*-butyl alcohol using cation exchange resin [in Chinese], *J. Chem. Eng. Chin. Univ.* **18**, 719–723 (2004).

Zheng Y. X., F. T. T. Ng, and G. L. Rempel, A comparison of a pseudo-homogeneous non-equilibrium model and a three-phase non-equilibrium model for catalytic distillation, *Chem. Eng. J.* **100**, 119–127 (2004).

Zoeller J. R., D. W. Lane, E. H. Cwirko, J. D. W. Fuller, and S. D. Barnicki, Process for generating vinyl carboxylate esters, U.S. Patent 5,821,384 (1998).

INDEX

Acetic acid esterification, 147
Acetic acid dimerization, 149
Activation energy, 533
Activity coefficients, 182
Allgower's nonlinearity measure, 355, 359
Antoine equation, 62, 433, 488
Aspen Split, 186
Average temperature control, 399

Biodiesel, 500
Boiling-point rankings, 487
Bubblepoint calculation, 46, 436, 488, 522, 523

Cascade temperature/composition, 285, 334, 345, 485
Coordinated control, 539
Chemical equilibrium constant, 27, 49
Closedloop performance, 247
Column/side reactor, 431
Conceptual design, 181, 186
Conflict between design and control, 24, 241, 249
Control degrees of freedom, 261
Controllability, 244
Controller action, 241, 247, 253
Controller gain and reset, 256
Convergence in RADFRAC, 218
Convergence of recycles, 199, 209
Conversion, 54
Counterintuitive results, 23, 206, 207, 259

Damkohler number, 488
Decanter, 150
Decomposition reaction, 324
Design procedure, 122, 128, 139, 158, 436, 466, 495, 517
Design spec/vary, 199, 218
Difference point method, 134
Direct sequence, 516
Distillation boundary, 186
Disturbance sensitivity, 453
Dual composition control, 329
DuPont, 2

Eastman Chemical, 2
Eastman control structure, 241, 245
Equilibrium RD model, 225, 234
Estimates, 199
Excess reactant operation, 7, 71, 242, 514
Exporting to Aspen Dynamics, 397, 407
External reactor, 433
Extractive distillation, 209
Euler integration, 47, 437

Reactive Distillation Design and Control. By William L. Luyben and Cheng-Ching Yu

Feasibility analysis, 134
Feed ratio, 159, 356, 475, 495
Feed tray location, 33, 445, 519, 523
Fenske equation, 40, 42, 522
Fragility of simulation, 177
Francis weir formula, 522

Hayden–O'Connell, 149
Holdup on reactive trays, 20, 254

Indirect sequence, 516
Inerts in ternary system, 99, 310
Input multiplicity, 360, 381
Internal composition control, 272, 539
Inverse response, 250, 256, 451

LHHW, 214, 230
Literature survey, 11
Liquid hydraulic time constant, 489, 522
Lumped tubular reactor model, 446

Minimum number of trays, 500
Model inversion, 363
Multiunit process, 38, 432

Neat operation, 7, 71, 243
Newton–Raphson, 436
Nonsquare relative gain, 363, 366, 474, 479
NRTL, 149
Number reactive trays, 22

On-aim control, 388
On-demand control, 261

Packing sizing, 408
Payback period, 437
Plumbing, first law, 399
Prereactor, 195, 390
Pressure-driven dynamic simulation, 407
Pressure effects, 5, 24, 94, 497
Pressure-swing distillation, 206
Pseudoneat operation, 193

Rangeability, 241, 256
Reactant losses in ternary system, 102
Reactive azeotrope, 136
Reactive tray holdup, 94
Recovery column, 75
Rectifying trays, 32
Recycle impurity, 54
Relative volatility, 26, 30, 487, 491, 531
  Temperature dependent, 30, 61, 433
Relaxation method, 436, 492, 523
Relay-feedback testing, 246, 266, 296, 313, 335, 344, 449, 451, 454
  Sequential, 363, 381, 413, 421, 424
Residue curve, 186, 193
Residue curve maps, 150, 475
RGA, 363, 366, 369, 475, 540
Robustness, 255, 363

Sensitivity analysis, 363, 479, 539
Shark-tooth shape, 451
Side reactor, 464
Sign reversal, 358
Split-range valves, 541
Steady-state gains, 246, 265, 303, 320, 326, 449
Step test, 252
Stripping trays, 32, 94
Suboptimal design, 249
SVD, 246, 250, 307, 320, 326, 344, 449

Temperature control tray location, 246
Temperature mismatch, 8
Ternary map analysis, 181, 459
Ternary mixing rule, 136, 185
Ternary system, 89
Ternary decomposition reaction, 119
Tradeoff between design and control, 249
Tradeoff between reactor and separation, 50, 439
Transformed variables, 136
Trap-out tray, 432
Tray holdup, 6
Tray sizing, 397, 408
Tubular reactor dynamic model, 446
Two-temperature control, 376, 447
Txy diagram, 182
Tyreus–Luyben tuning, 246, 266, 296, 313, 326, 335, 345

Ultimate gain, 246, 252, 256, 285, 296
Ultimate period, 246, 252, 256, 285, 296
UNIQUAC, 149
Underwood equation, 40, 42

Wang–Henke method, 523

Printed and bound by CPI Group (UK) Ltd, Croydon, CR0 4YY
18/09/2026
14965012-0002